T0093882

Environmental Contamination Remediation and Management

There are many global environmental issues that are directly related to varying levels of contamination from both inorganic and organic contaminants. These affect the quality of drinking water, food, soil, aquatic ecosystems, urban systems, agricultural systems and natural habitats. This has led to the development of assessment methods and remediation strategies to identify, reduce, remove or contain contaminant loadings from these systems using various natural or engineered technologies. In most cases, these strategies utilize interdisciplinary approaches that rely on chemistry, ecology, toxicology, hydrology, modeling and engineering.

This book series provides an outlet to summarize environmental contamination related topics that provide a path forward in understanding the current state and mitigation, both regionally and globally.

Topic areas may include, but are not limited to, Environmental Fate and Effects, Environmental Effects Monitoring, Water Re-use, Waste Management, Food Safety, Ecological Restoration, Remediation of Contaminated Sites, Analytical Methodology, and Climate Change.

Jonás García-Rincón · Evangelos Gatsios ·
Robert J. Lenhard · Estella A. Atekwana ·
Ravi Naidu
Editors

Advances in the Characterisation and Remediation of Sites Contaminated with Petroleum Hydrocarbons

 Springer

Editors
Jonás García-Rincón
Legion Drilling & Numac
Wetherill Park, NSW, Australia

Evangelos Gatsios
INTERGEO
Aspropyrgos, Greece

Robert J. Lenhard
San Antonio, TX, USA

Estella A. Atekwana
Department of Earth and Planetary Sciences
University of California Davis
Davis, CA, USA

Ravi Naidu
Advanced Technology Centre
The University of Newcastle
Callaghan, NSW, Australia

ISSN 2522-5847 ISSN 2522-5855 (electronic)
Environmental Contamination Remediation and Management
ISBN 978-3-031-34449-7 ISBN 978-3-031-34447-3 (eBook)
https://doi.org/10.1007/978-3-031-34447-3

This Springer imprint is published by the registered company Springer Nature Switzerland AG
The registered company address is: Gewerbestrasse 11, 6330 Cham, Switzerland

Preface

Petroleum hydrocarbons (PHCs) are among the most common and widespread contaminants in urban and industrial environments. Following a release from containment facilities, these complex mixtures of organic compounds can be found as non-aqueous phase liquids (NAPLs) in the subsurface, where they may act as a persistent source of contamination for soil and groundwater systems. Common petroleum fuels such as petrol (gasoline), diesel, and kerosene-type jet fuel are lighter-than-water NAPLs (LNAPLs).

Subsurface characterisation and remediation of sites contaminated with PHCs is often challenging due to the multi-fluid, multi-component nature of the problem and the various physical, chemical, and biological processes involved in a dynamic and naturally heterogeneous hydrogeological setting. Establishing an effective and sustainable strategy to manage risks posed by PHCs must be done considering all stakeholders and existing regulatory framework.

The objective of this book is to give visibility to advances in the characterisation and remediation of PHC-impacted sites, as well as pertinent concepts and methods that may still be underutilised by the remediation industry internationally. Potential future lines of research and development as well as best practices that provide guidance to contaminated land practitioners are presented throughout the book.

The book contains three sections: Chaps. 1–6 mainly address understanding and analysing the fate and transport of PHCs in the subsurface; Chaps. 7–11 focus on site characterisation tools; and Chaps. 12–18 discuss in situ remediation technologies and site management strategies.

In the first section, Chap. 1 serves as an introduction to the rest of the book and presents an overview of the main challenges found in PHC-contaminated sites. Chapter 2 critically underpins our assessment of LNAPL distribution by examining the historical development of constitutive relationships between water, LNAPL, and air governed by fluid saturations and capillary pressures. Chapter 3 shows in more detail how these constitutive relationships enable the estimation of subsurface LNAPL volumes, in particular in unconsolidated media with multi-modal pore distributions. Chapter 4 further explores the effects of subsurface heterogeneity on LNAPL distribution by discussing the application of geological-based sequence stratigraphy to both site investigation and remediation. Chapter 5 focuses on natural source zone depletion (NSZD) mechanisms, which are essential to understand the fate and transport of petroleum NAPLs. Lastly, Chap. 6 addresses petroleum vapour intrusion and the risks linked to it.

In the second section, Chap. 7 discusses high-resolution site characterisation (HRSC) approaches for the investigation of subsurface physical properties by using direct push, nuclear magnetic resonance (NMR), and groundwater tracing technologies, and Chap. 8 focuses on high-resolution targeted NAPL delineation methods. Chapter 9 explains the influence of PHC biodegradation on geophysical measurements and how biogeophysics can contribute to optimise site investigation outcomes. Chapters 10 and 11 consider, respectively, the use of molecular biological tools (MBTs) and compound-specific isotope analysis (CSIA) to determine the presence and activity of PHC-degrading microorganisms and assess remediation performance.

In the third section, each chapter examines the application of different in situ remediation technologies. Chapter 12 evaluates the utility and challenges of estimating LNAPL transmissivity and implementing hydraulic recovery systems. Chapter 13 deliberates how NSZD can be incorporated into a site management strategy. Chapter 14 stresses the continued central role of bioremediation. Chapter 15 examines the range of in situ chemical oxidation (ISCO) possibilities. Chapter 16 explores the use of activated carbon injectates as a sorptive and treatment substrate. Chapter 17 discusses the novel application of foams as a remediation and blocking agent. Chapter 18 presents advances in low-temperature thermal remediation.

Not all topics could be covered and more could have been included regarding certain subjects (e.g. sustainability and resiliency considerations for site management, cutting-edge laboratory analytical techniques, or assessment of risks posed by PHC biodegradation by-products), but we hope this book will contribute to the increased adoption of improved theories and methods within our industry and promote further research in the fields of contaminated site characterisation and in situ remediation.

Sydney, Australia   Jonás García-Rincón
Athens, Greece   Evangelos Gatsios
San Antonio, USA   Robert J. Lenhard
Davis, USA   Estella A. Atekwana
Newcastle, Australia   Ravi Naidu

Reviewers

Note: Chapter reviews represented the reviewers' experience and opinion and were not necessarily made on behalf of their organisations

Michael Annable, University of Florida
Tanya Astbury, formerly at Shell and VIVA Energy
Dawit Bekele, Douglas Partners/University of Newcastle
Barbara Bekins, USGS
Francis H. Chapelle, Chapelle and Associates
Rick Cramer, Burns and McDonnell
Susan De Long, Colorado State University
Eduardo de Miguel, Technical University of Madrid
Everton de Oliveira, Hydroplan/Groundwater Project
Kenneth D. Ehman, Chevron
Mateus Knabach Evald, Finkler Sustainable Technologies
Marco Falconi, ISPRA
Juliana Gardenalli de Freitas, São Paulo Federal University
Erin Hauber, US Army Corps of Engineers
Ian Hers, Hers Environmental
William Herkelrath, USGS
Kelly Johanna Hidalgo Martínez, University of Campinas
Parisa Jourabchi, ARIS
Tissa Illangasekare, Colorado School of Mines
Seung-Woo Jeong, Kunsan National University
Thomas Kady, USEPA
Jimmy C. M. Kao, National Sun Yat-Sen University
Saeed Kiaalhosseini, Golder-WSP
Steve K. Kalule, USK Consulting
Ravi Kolhatkar, Chevron
Mark Kram, Groundswell Technologies
Poonam Kulkarni, GSI
William W. Little, W. W. Little Geological Consulting

Don Lundy, GES
Mark Lyverse, formerly at Chevron
Deqiang Mao, Shandong University
Wesley McCall, Geoprobe Systems
Kevin G. Mumford, Queen's University
Charles Newell, GSI
Dimitrios Ntarlagiannis, Rutgers University
Dan Pipp, Geoprobe Systems
Colin P. Plank Burns and McDonnell
Kavitha Ramadass, University of Newcastle
José Luis Rodríguez Gallego, University of Oviedo
Mònica Rosell, University of Barcelona
Matthew Rosseau, GHD
Michael R. Shultz, Burns and McDonnell
Natasha Sihota, Chevron
Jonathan W. Smith, Shell
Kristen Thoreson, Regenesis
Courtney Toth, University of Toronto
Eleni Vaiopoulou, Concawe
Remke van Dam, Southern Geoscience Consultants
Paul Van Geel, Carleton University
Carsten Vogt, Helmholtz-UFZ
Dave Walsh, Vista Clara
Anne Wozney, Golder-WSP
Yijun Yao, Chinese Academy of Sciences
William Zavora, Calgon Carbon Corporation
Julio Zimbron, E-Flux

Contents

Contributors

David F. Alden Tersus Environmental, Wake Forest, NC, USA

Eliot A. Atekwana Department of Earth and Planetary Sciences, University of California Davis, Davis, CA, USA

Estella A. Atekwana Department of Earth and Planetary Sciences, University of California Davis, Davis, CA, USA

Olivier Atteia UMR EPOC, Bordeaux-INP, Talence, France

G. D. Beckett AQUI-VER, Inc, Park City, UT, USA

Henri Bertin I2M-CNRS, University of Bordeaux, Nouvelle-Aquitaine, France

Daniel Bouchard GHD, Montreal, Canada

James J. Butler Jr. Kansas Geological Survey, University of Kansas, Lawrence, KS, USA

Hung K. Chang Department of Applied Geology and Centro de Estudos Ambientais, UNESP—São Paulo State University, Rio Claro, SP, Brazil

John F. Devlin Geology Department, University of Kansas, Lawrence, KS, USA

Peter Dietrich Department Monitoring and Exploration Technologies, Helmholtz Centre for Environmental Research-UFZ, Leipzig, Germany

Craig Divine Arcadis U.S. Inc, Highlands Ranch, CO, USA

Sandra M. Dworatzek SiREM, Guelph, ON, Canada

Nicolas Fatin-Rouge IC2MP, CNRS, University of Poitiers, Poitiers, France

Emily Fitzhenry INRS-ETE, Quebec, QC, Canada

Jonás García-Rincón Legion Drilling & Numac, Wetherill Park, NSW, Australia

Randy St. Germain Dakota Technologies, Inc., Fargo, ND, USA

Ron Gestler Geosyntec Consultants International, Inc., Guelph, ON, Canada

Patrick Höhener Aix-Marseille University, Marseille, France

Kayvan Karimi Askarani Civil and Environmental Engineering Department, Colorado State University, Fort Collins, CO, USA; GSI Environmental Inc., Houston, TX, USA

Trent A. Key ExxonMobil Environmental and Property Solutions Company, Spring, TX, USA

Robert J. Lenhard San Antonio, TX, USA

Gaisheng Liu Kansas Geological Survey, University of Kansas, Lawrence, KS, USA

Richard Martel INRS-ETE, Quebec, QC, Canada

Sarah M. Miles Geosyntec Consultants International, Inc., Guelph, ON, Canada

Jonah Munholland Arcadis U.S. Inc, Highlands Ranch, CO, USA

Scott Noland Remediation Products Inc., Golden, CO, USA

Leonard O. Ohenhen Department of Geosciences, Virginia Tech, Blacksburg, VA, USA

Tom Palaia Jacobs Solutions, Greenwood Village, CO, USA

Sid Park Jacobs Solutions, Greenwood Village, CO, USA

Andy Pennington Arcadis U.S. Inc, Highlands Ranch, CO, USA

Clément Portois Colas Environnement, Bordeaux, France

Davinder Randhawa Arcadis U.S. Inc, Highlands Ranch, CO, USA

Michael O. Rivett GroundH2O Plus, Birmingham, United Kingdom

Thomas Robert INRS-ETE, Quebec, QC, Canada

Silvia Rossbach Department of Biological Sciences, Western Michigan University, Kalamazoo, MI, USA

Derek Rosso Arcadis U.S. Inc, Highlands Ranch, CO, USA

Junaid Sadeque AECOM, Los Angeles, CA, USA

Tom Sale Civil and Environmental Engineering Department, Colorado State University, Fort Collins, CO, USA

Ryan C. Samuels AECOM, Los Angeles, CA, USA

Miguel A. Alfaro Soto Department of Applied Geology and Centro de Estudos Ambientais, UNESP—São Paulo State University, Rio Claro, SP, Brazil

Julie Sueker Arcadis U.S., Inc., Broomfield, USA

Dora M. Taggart Microbial Insights, Inc, Knoxville, TN, USA

Neil R. Thomson Department of Civil and Environmental Engineering, University of Waterloo, Waterloo, ON, Canada

Martinus Th. van Genuchten Department of Nuclear Engineering, Federal University of Rio de Janeiro, UFRJ, Rio de Janeiro, RJ, Brazil

Iason Verginelli Laboratory of Environmental Engineering, Department of Civil Engineering and Computer Science Engineering, University of Rome Tor Vergata, Rome, Italy

Alexandre Vicard UMR EPOC, Bordeaux-INP, Talence, France

Gary P. Wealthall Geosyntec Consultants, Atlanta, USA

Edward Winner Remediation Products Inc., Golden, CO, USA

Acronyms and Symbols

a	Empirical coefficient
A	Pre-exponential factor relating to the collision of molecules
AAPG	American Association of Petroleum Geologists
ABC	Anaerobic benzene carboxylase
AC	Activated carbon
AER	Air exchanges per hour
AF_{bio}	Attenuation factor due to biodegradation
AF_{cap}	Attenuation factor in the capillary fringe
AF_{ss}	Air attenuation factor
AGU	American Geophysical Union
API	American Petroleum Institute
AS	Air sparge
AST	Above storage tank
AT	Averaging time
ATSDR	Agency for Toxic Substances and Disease Registry
b	Empirical coefficient
B_0	Static magnetic field
B_1	Alternating magnetic perturbation
BCR	Benzoyl coenzyme A reductase
BHE	Borehole heat exchanger
b_n	In-well LNAPL thickness
BOD	Biochemical oxygen demand
BPC	Building pressure cycling
BSS	Benzylsuccinate synthase
BTEX	Benzene, toluene, ethylbenzene, and xylenes
BW	Body weight
C	Heat capacity
c	Velocity of light in a vacuum
C_a	VOC concentration at the aerobic to anaerobic interface
C_i^s	Petroleum vapour source concentration
C_{ox}^{atm}	Atmospheric oxygen concentration

C_{ox}^{min}	Minimum oxygen concentration
CBI	Carbon-based injectate
C_{cap}	VOC concentration at the top of the capillary fringe
CDC	Capillary desaturation curve
Ce	Dissolved concentration in equilibrium
$C_{w,i}^{eff}$	Effective aqueous solubility
CHP	Catalysed hydrogen peroxide
C_{indoor}	VOC concentration in indoor air
$C_{indoor, acc}$	Acceptable indoor risk-based level
C-IRMS	Combustion-isotope ratio mass spectroscopy
CIS	Cross-injection system
CMC	Critical micelle concentration
COC	Contaminant of concern
C_{ox}	Molar concentration of the oxidant
C_{PHC}	Molar concentration of the PHC compound
CPT	Cone penetration testing
C_{sg}	Soil gas concentration
CSIA	Compound-specific isotope analysis
CSM	Conceptual site model
C_{soil}	Total soil concentration
C_{source}	VOC concentration in the soil gas in correspondence of the source in groundwater
C_{ss}	VOC concentration at the top of the aerobic zone
CVOC	Chlorinated volatile organic compound
$C_{w,i}$	Aqueous phase concentration
D	Effective porous medium diffusion coefficient
D_{tot}^{eff}	Overall diffusion coefficient in the vadose zone
D^{eff}	Effective diffusion coefficient
D_{air}	Diffusion coefficient in air
dC/dz	Concentration gradient
D_{cap}	Effective diffusion coefficient in the capillary fringe
DCC	Dynamic closed chamber
d_f	Slab depth from open ground
d_i	Thickness of the i-th layer
D_{LNAPL}	Specific volume of in-situ LNAPL
dm_{COC}	Change in chemical content in the LNAPL normalised for the conservative marker
DNAPL	Dense non-aqueous phase liquid
DO	Dissolved oxygen
DP	Direct push
DPE	Dual-phase extraction
DPIL	DP injection logger
DPP	DP permeameter
DPT	Direct Push Technology
d_s	Vertical source distance from open ground

D_{soil}	Effective diffusion coefficient in the vadose zone
DTS	Distributed temperature sensing
D_v^{eff}	Effective vapour diffusion coefficient
D_{wat}	Diffusion coefficient in water
DyeLIF	Dye-enhanced laser induced fluorescence
E&P	Exploration and production
EA	Activation energy
E_{Bottom}	Energy flux at the bottom of the reference volume
EC	Electric conductivity (Chap. 4), exposure factor (Chap. 6)
ED	Exposure duration
EDB	Ethylene dibromide
EDO	Ethylbenzene/isopropylbenzene dioxygenase
EF	Annual exposure frequency
EF_{gi}	Daily exposure frequency
EGLE-RRD	Great Lakes and Energy—Remediation and Redevelopment Division
EK	Electrokinetic
EM	Electromagnetic
EMI	Electromagnetic induction
E_{NSZD}	Energy produced through NSZD over the height of energy balance element
E^o	Standard reduction potential
EOR	Enhanced oil recovery
epcap	Parameter for foam shear thinning effect
EPA	Environmental Protection Agency
epdry	Parameter controlling the abruptness of the foam collapse
EPH	Extractable petroleum hydrocarbons
ER	Electrical resistivity
ERH	Electrical resistance heating
ERI	Electrical resistivity imaging
ERT	Electrical resistivity tomography
ESS	Environmental sequence stratigraphy
E_{Sto}	Stored energy over the height of energy balance element
ETBE	Ethyl tert-butyl ether
E_{Top}	Energy flux at the top of the reference volume
f	Frequency
FAMF	Fatty acid methyl ester
fmcap	Critical capillary number
f	Remaining contaminant fraction
fcap	Smallest capillary number expected to be encountered
F_i	Foam factors
F_{dry}	Dry-out factor
FID	Flame ionisation detector
fmdry	Critical water saturation under which foam collapses
f mmob	Reference mobility reduction factor for wet foams

f_{oc}	Organic carbon fraction of the soil
f_q	Foam quality
g	Scalar gravitational constant
GCxGC-TOFMS	Two-dimensional gas chromatography with time-of-flight mass spectrometry
GC–MS	Gas chomatography–mass spectrometry
GEO	Geochemistry
GFC	Groundwater flux characterisation
GIS	Geographic information system
GPR	Ground penetrating radar
GRT	Global Remediation Technologies
GSA	Geological Society of America
GWSWI	Groundwater-surface water interface
H	Henry's law constant of the contaminant of concern (Chap. 6), applied magnetic field (Chap. 9)
h	Hydraulic head
h_{ao}	Air-LNAPL capillary head
h_{aw}	Air-water capillary head
h_c	Capillary pressure head
h_{cap}	Thickness of the capillary fringe
h_d	Displacement head
h_{ow}	LNAPL-water capillary head
HPA	High plains aquifer
HPT	Hydraulic profiling tool
HRK	High-Resolution K
HRS	Hazard Ranking System
HRSC	High-resolution site characterisation
HST	Highstand systems tract
HVAC	Heating, ventilation, and air conditioning
i	Gradient
I	Induced current
i	Fixed time
IAS	In situ air sparging
IFT	Interfacial fluid tension
IoT	Internet of Things
IP	Induced polarisation
IRGA	Infrared gas analyzer
IRZ	In situ reactive zone
ISCO	In situ chemical oxidation
ISM	In situ microcosm
ITRC	Interstate Technology and Regulatory Council
IUPAC	International Union of Pure and Applied Chemistry
IVF	Incised valley fill
IW	Injection well
J_{adv}	Advective flux

J_{diff}	Diffusive mass flux of a vapour in soil
$J_{Background}$	Gas flux associated with natural soil respiration
J_{NSZD}	Gas flux associated with NSZD
K	Thermal conductivity (Chap. 5)
k	Permeability or intrinsic permeability (Chaps. 2, 17), degradation rate (Chap. 11)
K	Hydraulic conductivity (Chaps. 7, 17), geometric factor (Chap. 9)
k'_{obs}	Observed first-order rate coefficient
K_d	The lumped mass transfer or dissolution rate coefficient
KGS	Kansas geological survey
K_n	Effective LNAPL hydraulic conductivity
K_{oc}	Organic carbon to water partition coefficient
K_{ow}	Octanol-water partition constant
k_{point}	Point attenuation rate
k_{rg}^{f}	Relative permeability modified to account for foam
k_r	Relative permeability
k_{rg}	Relative permeability for the gas phase
k_{ra}	Air relative permeability
k_{rn} or k_{ro}	LNAPL relative permeability
k_{rw}	Water relative permeability
k_v	Soil permeability to vapour flow
L	Distance
L_a	Aerobic zone thickness
L_b	Anaerobic zone thickness
LCSM	LNAPL conceptual site model
LE	Local equilibrium
LIF	Laser-induced fluorescence
LNAPL	Light non-aqueous phase liquid
LoD	Limit of detection
L_R	Diffusive reaction length
LRS	Larned Research Site
$L_{slab,c}$	Critical slab size
LST	Lowstand systems tract
m	van Genuchten parameter (Chaps. 2, 3), empirically determined constant (Chap. 7)
M	Applied magnetization
M_0	Amplitude of initial magnetization
MBT	Molecular biological tool
MCL	Maximum contaminant limit
MEP	Maximum extent practicable
MFS	Maximum flooding surface
MGPs	Manufactured gas plants
MI	Microbial Insights

MIP	Membrane Interface Probe
MLOE	Multiple lines of evidence
MNA	Monitored natural attenuation
MNSS	Methylnaphthalene succinate synthase
$M_{ox/s}$	In situ oxidant to solids mass
MPE	Multiphase extraction
MRF	Mobility reduction factor
MS	Magnetic susceptibility
MSL	Mean sea level
MT	Magnetotellurics
MTBE	Methyl tert-butyl ether
MW	Monitoring Well
n	van Genuchten curve shape parameter (Chaps. 2, 3, 12), empirically determined constant (Chap. 7)
NAIP	Natural attenuation indicating parameter
NAPL	Non-aqueous phase liquid
N_B	Bond number
N_{Ca}	Capillary number
ND	Not detected
n_e	Effective porosity
NEPA	National Environmental Policy
n_f	Density of flowing lamellae
NIF	Newport–Inglewood Fault
NMR	Nuclear magnetic resonance
NNLS	Non-negative least squares
NOD	Natural oxidant demand
NOI	Natural oxidant interaction
NOM	Natural organic matter
NPL	National Priority List
NSZD	Natural source zone depletion
OIP	Optical Image Profiler
OMA	Optical multi-channel analyzer
OoM	Order of magnitude
ORC	Oxygen release compound
ORP	Oxidation reduction potential
P	Pressure
P_a	Air pressure
PAC	Powdered activated carbon
PAH	Polycyclic aromatic hydrocarbon
P_c	Capillary pressure
P_{ce}	Entry capillary pressure
PCE	Tetrachloroethene
PCR	Polymerase chain reaction
PHC	Petroleum hydrocarbon
PHE	Phenol hydroxylase

PIANO	Paraffins, isoparaffins, aromatics, naphthenes, and olefins
PLFA	Phospholipid Fatty Acid Analysis
PMT	Photomultiplier Tube
ρ_o	LNAPL density
P_o	Pressure of the organic phase
PRB	Permeable reactive barrier
PRP	Potentially responsible party
PT	Pressure transducer
PV	Pore volume
PVC	Polyvinyl chloride
PVF	Palos Verdes fault
PVI	Petroleum vapour intrusion
PVP	Point velocity probe
ρ_w	Water density
P_w	Pressure of the aqueous phase
q	Heat flux due to conduction (Chap. 5), Darcy flux (Chaps. 7, 17)
Q	Flow rate (Chap. 7), injection rate (Chap. 17)
$Q_{building}$	Building ventilation rate
q_g	Darcy flux of gas
qPCR	Quantitative polymerase chain reaction
q_s	Surface heating and cooling
QS	Quality scan
Q_{soil}	Soil gas entry rate
q_{ss}	Subsurface heat source associated with NSZD
Q_t	Cone resistance
R	Radius of curvature of the fluid-fluid interface (Chap. 3), carcinogenic risk (Chap. 6), resistance (Chap. 9)
r	Pore radius (Chaps. 3, 17), probe radius (Chap. 7)
R_{aq}	Aqueous reaction rate
RBCA	Risk-Based Corrective Action
$R_{COC-NAPL}$	COC-specific NSZD rate in NAPL
RCRA	Resource Conservation and Recovery Act
RDEG	Ring-hydroxylating toluene monooxygenase
RF	Resistance factor
RME	Reasonable maximum exposure
RMO	Toluene monooxygenase
ROI	Radius of influence
ROST	Rapid Optical Screening Tool
RPM	Rotated potential mixing
R_t	Stable isotope ratio of the compound at time t
S	Saturation
S_a	Air saturation
SAG	Surfactant-alternating gas
SDR	Schlumberger Doll Research

SDWA	Safe Drinking Water Act
$S_{e,i}$	Effective contaminant solubility in NAPL mixture
SEM	Scanning electron microscope
SEPM	Society for Sedimentary Geology
S_i	Aqueous solubility of the pure-phase compound
SIP	Spectral induced polarisation (Chap. 9), stable isotope probing (Chap. 11)
S_m	Minimum or irreducible wetting fluid saturation
S_n or S_o	LNAPL saturation
S_{oe}	Entrapped NAPL saturation
$\overline{S}_o$	Effective LNAPL saturation
S_{or}	Residual NAPL saturation
SP	Self-potential
SPCC	Spill Prevention, Control, and Countermeasure Plan
S_r	Residual saturation
S_t	Total-liquid saturation
$\overline{S}_t$	Effective total-liquid saturation
SVE	Soil vapour extraction
S_w	Water saturation
$\overline{S}_w$	Effective water saturation
S_{wr}	Residual water saturation
t	Time
T	Absolute temperature
TAME	Tert-amyl methyl ether
TarGOST	Tar-specific Green Optical Screening Tool
TBA	Ter-butyl alcohol
T_{Back}	Measured temperatures at the background location
TCE	Trichloroethene
TDIP	Time domain induced polarisation
TEAP	Terminal electron acceptor process
TEA	Terminal electron acceptor
TEXX	Toluene, ethylbenzene, m,p-xylene, and o-xylene
THQ	Hazard quotient
T_{Imp}	Measured temperatures at an impacted location
TISR	Thermal In Situ Sustainable Remediation
T_n	LNAPL transmissivity
T_{NSZD}	Component of temperatures associated with NSZD
TPH	Total petroleum hydrocarbons
TR	Target risk
TS	Transgressive surface
TST	Transgressive systems tract
TTZ	Target treatment zone
U	Heat source/sink term
USCS	United soil classifications system
USEPA	United States Environmental Protection Agency

USGS	United States Geological Survey
UST	Underground storage tank
u_t	Superficial velocity
UVOST	Ultra-violet Optical Screening Tool
v	Velocity
$V_{building}$	Building volume
V_D	Darcy velocity of groundwater flow
v_w	Ambient groundwater seepage velocity
v_g	Local gas velocity
VOC	Volatile organic compound
VPDB	Vienna-Pee Dee Belemnite
VSMOW	Vienna Standard Mean Ocean Water
W	Mass of AC
w_i	Subcurve weight factors varying between 0 and 1
WTFZ	Water table fluctuation zone
x	Distance
X_i	Mole fraction of compound i in LNAPL mixture
XSD	Halogen-specific detector
YZ	Area perpendicular to groundwater flow
z	Elevation
z_{ao}	Height of the air-LNAPL interface in the well
z_{aw}	Height of the potentiometric surface
z_{ow}	Height of the LNAPL-water interface in the well
Z_s	Skin depth
α	Air-water van Genuchten parameter (Chaps. 2, 3, 6, 12), angle (Chap. 7), attenuation factor (Chap. 9), fractionation factor (Chap. 10), empirical parameter that depends on the porous medium, surfactant, and flow conditions (Chap. 17)
β_{ao}	Air-LNAPL scaling factor
β_{ow}	LNAPL-water scaling factor
γ	Contact angle
γ_i	Stoichiometric mass of oxygen
Δ	Isotopic shift
Δ_{Hr}	Enthalpy released during oxidation of the NAPL
$\Delta \delta^{13}C$	Carbon isotope shift
$\delta^{13}C$	$^{13}C/^{12}C$ isotopic ratio
δ^2H	$^2H/^1H$ isotopic ratio
Δh	The injection-induced pressure head changes
ΔP	Pressure difference
δV	Potential difference
ε	Dielectric permittivity (Chap. 9), enrichment factor (Chap. 10)
E	Electric field
ε_0	Permittivity of a vacuum
H	Magnetic field
η_a	Air viscosity

η_o	LNAPL viscosity
η_w	Water viscosity
θ	Contact angle
θ_a	Air-filled porosity
θ_r	Residual soil water content
θ_T	Total porosity
θ_w	Water-filled porosity
κ	Thermal diffusivity (Chap. 5), dielectric constant (Chap. 9)
λ	Pore-size index (Chap. 2), biodegradation rate (Chap. 5), first-order reaction rate constant (Chap. 6), wavelength (Chap. 8), bulk thermal conductivity (Chap. 18)
Λ	CSIA lambda parameter
μ	Magnetic permeability
μ	Kinematic viscosity
μ_w	Water viscosity
μ_f	Foam viscosity
μ_g	Gas viscosity
ρ	Density
ρ_α	Apparent resistivity
ρC	Bulk volumetric heat capacity
ρC_f	Volumetric heat capacity of the fluid
ρ_a	Air density
ρ_{LNAPL}	LNAPL density
ρ_{ro}	LNAPL-specific gravity
ρ_s	Soil bulk density
$|\sigma|$	Impedance magnitude
σ	Interfacial tension (Chap. 3), electrical conductivity (Chap. 9)
σ_{ao}	Air-LNAPL interfacial tension
σ_{aw}	Air-water interfacial tension
σ_{ow}	LNAPL-water interfacial tension
τ	Lifetime
ϕ	Porosity
ϕ_e	Effective porosity
φ	Phase shift
χ	Magnetic susceptibility
$\partial T/\partial z$	Change in temperature with respect to distance in the vertical direction

Chapter 1
Complexities of Petroleum Hydrocarbon Contaminated Sites

David F. Alden⊙, **Jonás García-Rincón**⊙, **Michael O. Rivett**⊙,
Gary P. Wealthall, and Neil R. Thomson⊙

Abstract This introductory chapter provides an overview of some challenges encountered during assessment and remediation of sites contaminated by petroleum hydrocarbons (PHCs), especially due to the presence of petroleum non-aqueous phase liquid (NAPL). PHC contamination of soil and groundwater has become recognized globally as an environmental concern. NAPL releases may go unnoticed for years and can be costly, time-consuming, and challenging to cleanup. Environmental investigations are required to assess the subsurface conditions, and, when necessary, appropriate remedial strategies are implemented. Remediation success depends on the development of a living conceptual site model, evaluation, and selection of potential treatment technologies followed by technology implementation and performance monitoring. This chapter outlines various challenges commonly found at PHC-contaminated sites associated to regulatory issues, multiphase flow mechanics, NAPL partitioning and biodegradation, and hydrogeological controls.

Supplementary Information The online version contains supplementary material available at https://doi.org/10.1007/978-3-031-34447-3_1.

D. F. Alden (✉)
Tersus Environmental, 1116 Colonial Club Rd., Wake Forest, NC 27587, United States
e-mail: david.alden@tersusenv.com

J. García-Rincón
Legion Drilling & Numac, 11–13 Metters Place, Wetherill Park, NSW 2164, Australia
e-mail: jonas.garciarincon@legiondrilling.com.au

M. O. Rivett
GroundH2O Plus, 41 Wilmington Road, Quinton, Birmingham B32 1DY, United Kingdom
e-mail: rivett@groundh2oplus.co.uk

G. P. Wealthall
Geosyntec Consultants, Atlanta, USA

N. R. Thomson
Department of Civil and Environmental Engineering, University of Waterloo, West Waterloo, Ontario N2L 3G1, Canada
e-mail: neil.thomson@uwaterloo.ca

Keywords Contaminated land management · Fuel hydrocarbons · LNAPL contamination · Source zone remediation · Sustainability

1.1 Introduction

Petroleum hydrocarbons (PHCs) are among the most common and widespread subsurface contaminants in urban and industrial environments. PHCs are primary compounds found in crude oil and its refined products. These organic liquids contain a large number of different hydrocarbons with the most common being alkanes, cycloalkanes (naphthenes), aromatic hydrocarbons, and more complex substances like asphaltenes. Some common products and liquid wastes containing PHCs include:

- Gasoline/petrol;
- Jet fuels (Jet-A, JP-4, JP-5);
- Kerosene/heating oil no. 2;
- Diesel fuel;
- Heating oil no. 4 (Bunker A) and no. 6 (Bunker C);
- Hydraulic oil, cutting oil, lubricating oil, mineral oil, dielectric fluid, dielectric mineral oil, and transformer oil;
- Crude oil and other PHC mixtures;
- Biofuels;
- Waste oil, waste vehicular crankcase oil; and
- Creosote, coal tar, and bitumen.

Residential, industrial, or transport activities that use these petroleum products may be sources of contamination through accidental or intentional spills, releases from above ground and underground storage tanks, and leaks from associated pipelines. Petroleum releases can occur at gas stations, refineries, bulk-product terminals, manufactured gas plant facilities (gasworks), airports, homes heated with oil-burning devices, pipelines, chemical manufacturing facilities, landfills, clandestine disposal sites, metal cutting facilities, automotive repair shops, oil drilling pads, and farms.

PHC products typically comprise complex mixtures of predominantly non-polar organic compounds that together form organic liquid fuels or oils. These liquids are commonly referred to as "non-aqueous phase liquids" (NAPLs), are relatively immiscible (sparingly soluble) with water, and can persist in the subsurface as sources of soil and groundwater pollution for decades. A defining property of the NAPL is its density relative to water. NAPLs are classified as either lighter than water (LNAPL) or denser than water (DNAPL). Most petroleum NAPLs are LNAPLs. As such, upon reaching the capillary fringe and the water table, LNAPL will laterally spread with limited invasion below the water table due to its buoyancy. A DNAPL such as some coal tars or creosotes, on the other hand, may penetrate vertically downward well beyond the water table through the saturated zone.

Mass transfer, natural attenuation, and natural source zone depletion (NSZD) phenomena are also critical to understanding the fate and transport of PHCs in the subsurface. Partitioning of LNAPL constituents into other phases (water, gas) result in aqueous and gas phase plumes of organic compounds that could pose risks to human health and the environment. Important classes of PHCs that are of environmental concern include the more soluble mono-aromatic hydrocarbons (i.e., benzene, ethylbenzene, toluene, and xylenes (BTEX)), polycyclic aromatic hydrocarbons (PAHs) (e.g., naphthalene, fluorene, and anthracene), and gasoline additives such as the fuel oxygenate methyl tert-butyl ether (MTBE). Microbial degradation (biodegradation) processes have been frequently found to play a pivotal role in mitigating risks derived from PHC contamination.

As conceptualized in Fig. 1.1, when LNAPL is released into an unconsolidated porous medium, it will tend to migrate downwards through the unsaturated zone under the influence of gravity with its flow path affected by variations in water content and the porous medium's properties. For instance, the presence of a finer-grained layer acting as a capillary barrier will cause lateral spreading of the LNAPL at the interface between layers. If the LNAPL pressure builds and is able to overcome the non-wetting fluid capillary entry pressure of the finer-grained layer, then it will break through the finer-grained layer. Otherwise, it will flow along the top and eventually around the finer-grained layer.

If sufficient LNAPL volume is present, the LNAPL will continue to flow downwards to the water table. Once the LNAPL reaches the water-saturated capillary fringe, downward movement of LNAPL and buoyancy forces can displace some water in the capillary fringe and in the uppermost water-saturated zone below the water table. Lateral LNAPL spreading will occur due to density contrasts with some bias of the main LNAPL body down the hydraulic gradient. This would not preclude migration in other directions in response to geological heterogeneities and preferential migration pathways in the porous or fractured geologic media and influence of local gradients induced by LNAPL pressures. The LNAPL body will eventually stabilize as the LNAPL pressure heads and gradients weaken due to LNAPL redistribution and NSZD processes.

Water-table fluctuations due to daily and seasonal variations or local groundwater pumping induce vertical migration of mobile LNAPL and the creation of an LNAPL "smear zone". As the water table falls, LNAPL mass will be transferred downward while leaving behind a trail of residual LNAPL (not occluded by water) retained in the pores. As the water table rises, part of the LNAPL will move upward, while other LNAPL mass becomes entrapped (occluded by water) due to capillary forces. Depending on the amplitude of the groundwater fluctuations, LNAPL may be trapped meters below the elevation of the mean water table in the saturated zone. In addition, residual LNAPL will be retained in pores that the LNAPL encountered as it migrated through the unsaturated zone. The amount of residual and entrapped LNAPL is a function of porous medium properties, LNAPL fluid properties, and fluid saturation history. The cross-sectional illustration of an LNAPL body in Fig. 1.1 has been shaped by one or more water-table fluctuations.

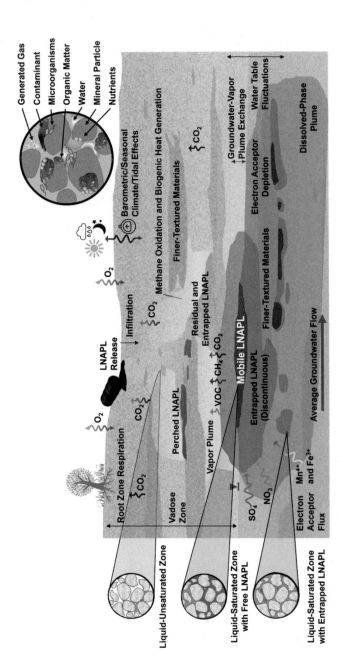

Fig. 1.1 Conceptual LNAPL architecture in the subsurface following a point release, along with petroleum hydrocarbons fate and transport mechanisms (central figure); LNAPL mobility as a function of relative saturations of air, water, and LNAPL (left circles); factors modulating *in-situ* petroleum hydrocarbon biodegradation (top right circle) (adapted from Lundy and Gogel 1988; Rivett et al. 2011; CL:AIRE 2014; Vila et al. 2019; García-Rincón 2020)

In general, water preferentially wets the geological formation with respect to LNAPL and air (which commonly are the intermediate-wetting fluid and non-wetting fluid, respectively) except in very dry soils, where water may be the non-wetting fluid. Released LNAPL to the subsurface is hence usually non-wetting preferring to occupy the more spacious pore body rather than pore throats, thereby minimizing contact with the solid grains that may remain water coated. With repeated spills, the LNAPL may transition from a non-wetting to a wetting fluid in time. Changes in LNAPL chemical composition may also change its wettability. As a result of all these factors, LNAPL architecture varies dramatically between sites and across geological settings. This is generally illustrated in the conceptual models of CL:AIRE (2014) depicting LNAPL behavior for a number of geologic settings, including mildly to highly heterogeneous unconsolidated porous media (e.g., sands to glacio-fluvial sands and gravel) and low and high matrix porosity fractured bedrock (e.g., granite to sandstone).

As shown in Fig. 1.1, a dissolved or aqueous phase plume will develop in the capillary fringe and in the saturated groundwater zone, and a vapor or gas phase plume will develop in the unsaturated zone. Since most LNAPLs are "multicomponent" comprising complex mixtures of PHC compounds, both the aqueous and gas phase plumes will likewise be multicomponent in nature. The composition of the aqueous phase plume is controlled by the dissolution rate and the effective solubility of each LNAPL component that is LNAPL composition-dependent. The composition of the gas phase plume is controlled by the volatilization rate and the effective partial pressure of each LNAPL component which is likewise LNAPL composition dependent. As the LNAPL weathers, the more soluble and volatile components in the LNAPL will become depleted, and the remaining LNAPL will become enriched with progressively less soluble and volatile compound constituents. The change in composition may influence physical properties such as LNAPL density and viscosity.

The significance of microbial degradation (biodegradation) to PHC sites cannot be overstated. While biodegradation of individual PHC compounds varies, most hydrocarbons are relatively biodegradable across a range of aerobic and anaerobic conditions ranging from nitrate reducing to methanogenic. The contribution of biodegradation to the natural attenuation of dissolved-phase hydrocarbon plumes has long been recognized and may cause sufficient attenuation of plumes prior to their reaching receptors at risk to justify the use of monitored natural attenuation (MNA) as a viable site management strategy (Wiedemeier et al. 1999; USEPA 1999b; ASTM 2004; CL:AIRE 2014; ITRC 2018). It has only been in recent years, however, that the significance of biodegradation in LNAPL source zone areas has been recognized leading to the emergent adoption of NSZD monitoring as a management option to complement or even replace active remedial interventions (Rivett and Sweeney 2019). In particular, the role of methanogenesis and direct outgassing from the LNAPL source circumventing the assumed requirement for hydrocarbons to dissolve to be biodegraded was overlooked for many years and may lead to significant vapor-based NSZD far exceeding the aqueous-based depletion (Smith et al. 2022).

Effective LNAPL problem conceptualization within specific PHC-contaminated site scenarios underpins the management of risks and appropriate intervention remedy selection. Recognition needs to be made of the breadth of risks or drivers at sites, including nuisance impairment of beneficial use of resources or aesthetic values, societal community concerns, and business factors including reputational risk and problem persistence leading to intergenerational equity issues (CL:AIRE 2014). One possible approach to risk evaluation discussed in CL:AIRE (2014) is to categorize risks by "LNAPL saturation" (largely controlling LNAPL mobility) or composition. "Saturation" risks occur when pore-space LNAPL saturations exceed residual or entrapped saturation thresholds enabling LNAPL mobility and risks of migration of the LNAPL body to receptors resulting in, for example, seepage of LNAPL oily films into surface waters, or LNAPL contamination of supply wells. Addressing saturation risks are often an initial site priority. Composition risks, on the other hand, derive from the LNAPL chemical composition, which may include constituents that have toxicity to human health or the environment and may pose direct or indirect, chronic, or acute risks. Certain constituents may also be highly volatile and flammable and therefore drive explosive risks. Composition risks hence include the migration of toxic constituents such as BTEX or PAHs within dissolved-phase plumes to supply wells or within vapor plumes migrating toward and accumulating with PHC degradation products such as methane in buildings and posing explosive risks. It is recognized within this simple bi-classification of risks that the mobility of the LNAPL will still be influenced by its composition as composition determines physical properties such as viscosity and interfacial tensions; however, such changes are usually very gradual.

The remainder of this chapter will provide a more detailed overview of challenges associated with:

- Problem recognition and regulatory issues (Sect. 1.2);
- Multiphase flow mechanics (Sect. 1.3);
- PHC chemical makeup and interactions (Sect. 1.4); and
- Geological and hydrogeological controls (Sect. 1.5).

1.2 Problem Recognition and Regulatory Environment

1.2.1 Problem recognition—The Case of Large Oil Spills

Contaminated site assessment and management has evolved to being a multidisciplinary field dealing with hydrogeological, chemical, biological, fluid mechanics, and management topics, often requiring subject-matter experts from a variety of backgrounds including environmental, chemical, and civil engineering, geology, microbiology, toxicology, statistics, legal, and project management.

Subsurface contamination by PHCs can be a result of leaking storage tanks and pipelines, accidental releases, or intentional disposals. Without accurate wet-stock reconciliation or leak-detection systems, subsurface releases can go unnoticed for years, an "invisible problem" that can be costly, time-consuming, and difficult to treat after it is discovered. Often, PHC cleanup reminds the general public's mind of large oil spills, nevertheless, cleanup of local gasoline stations is the predominant type of events managed by the soil and groundwater PHC remediation industry.

Large oil spills galvanize significant media and public attention because they have catastrophic local environmental and human health impacts and, when on land, may create complex surface water and groundwater remediation problems that can take years of effort to manage the environmental impacts. Examples include the 1989 *Exxon Valdez* oil tanker collision into a reef offshore of Alaska, USA, that drew much media coverage and attention to the detrimental environmental impacts that occurred to 2000 km of the intertidal beaches and marshes along the shoreline of Prince William Sound (Xia and Boufadel 2010). The 2010 *Deepwater Horizon* explosion and oil rig blowout illustrates the potential consequences of not following established safety protocols. The heavy oil spill from the tanker *Prestige* in 2002 affected 1900 km of the coast in the northwest of Spain, devastating the local economy and demonstrating the gargantuan effort required to restore conditions to a pre-spill state. One of the earliest and largest oil spills on land is the seemingly forgotten 18-month long hydrocarbon eruption in 1910 at Lakeview Gusher #1 in California, USA, where oil-soaked soils are still present. The United Nations Environment Program (UNEP) released a report in 2011 documenting the environmental degradation of the land and waters of the Niger River Delta in Ogoniland, Nigeria, associated with operations of the multinational petroleum industry in the area since the 1970s. This document set out urgent recommendations for cleanup and protection of drinking water and health of exposed populations. The cleanup effort is nevertheless far from being completed. Other examples of oil spills associated with unstable sociopolitical environments include the Kuwaiti oil fires caused by Iraqi military forces during the Gulf War in 1991 and the multiple attacks to the Caño Limón–Coveñas crude oil pipeline in the Arauca Department, Colombia.

Some large oil spills have yielded significant research opportunity, a foremost example of relevance to the soil and groundwater remediation community being the terrestrial oil pipeline rupture near Bemidji, Minnesota, in 1979 (Fig. 1.2). The release consisted of 1,700,000 L of crude oil, of which 400,000 L remained in the subsurface after the initial cleanup works in 1980. In 1983, the United States Geological Survey (USGS) established the National Crude Oil Spill Fate and Natural Attenuation Research Site, which has produced decades of data and has been paramount for the current understanding of natural processes at PHC-contaminated sites. The ongoing USGS research project has four primary objectives: (I) characterizing the nature, toxicity, and prevalence of partial transformation products emanating from the crude oil source; (ii) evaluating the secondary impacts (such as arsenic cycling) of biodegradation; (iii) understanding the timeframe of natural attenuation of PHC source zones; and (iv) developing field tools, methods, and data that support evaluations of environmental health effects of natural attenuation of crude oil (USGS 2021). The numerous research outcomes produced at Bemidji illustrate the importance of establishing experimental sites for the study of subsurface contamination phenomena.

The great proportion of PHC-contaminated sites requiring cleanup are relatively small sites in industrialized countries. In the United States alone, there are 135

Fig. 1.2 Crude oil reached a wetland near Bemidji after the rupture of an oil pipeline. Initial efforts to remove the floating oil were conducted by the pipeline company in 1979 using pumps and other machinery (photo courtesy of Jared Trost, USGS)

reported operating refineries and over 133,000 gas stations. Between 1998 and 2021, 564,767 petroleum releases from underground storage tanks (USTs) were reported, and 85% have been addressed to meet regulatory requirements (i.e., "closed"). (UST FAQ www.epa.gov/ust, accessed on May 28, 2022 https://www.epa.gov/ust/frequent-questions-about-underground-storage-tanks#gen11).

1.2.2 Regulatory Frameworks

In the context of a country's regulatory framework, oil spills on land are sources of soil and groundwater contamination often brought to light during environmental assessments driven by real estate transactions, suspicion of chemical leakage from existing tanks or historical land use, or as a precautionary action prior to future land use. The need to meet regulatory compliance and achieve cleanup standards agreed to mitigate risks is often a key remediation driver. Risks do not just relate to human health but also the environment, assets, reputation, and financial aspects.

Environmental law is often created in response to high-profile events in which the press generates public awareness (Wargo 2009). Once an issue gains societal recognition as a problem, it may get on the docket of government institutions through organized interest group pressure and eventually may receive enough attention by governing policymakers for decisional action (Kraft and Vig 2010). For example, the wide distribution of Rachel Carson's famous book *Silent Spring* in the early 1960s, warning about the compounding effect pesticides have on the environment focused public opinion sharply and likely incentivized the creation of the United States Environmental Protection Agency (USEPA) that now sets corrective action guidelines with respect to contaminated soils and groundwater in the United States. Extraordinary events, such as the Cuyahoga River in Ohio catching fire thirteen times between 1868 and 1969 due to petroleum NAPL on its surface, prompted establishment of the National Environmental Policy (NEPA), which mandates that federal agencies assess the potential environmental impacts of their actions and the Clean Water Act to reduce emissions from industrial facilities (Wargo 1998). Arguably the most pivotal has been the infamous New York State Love Canal (landfill) Incident in the 1970s soon after which groundwater contamination problems emerged throughout the United States where public interest and perception changed dramatically (Hadley and Newell 2012). The incident was an important catalyst for the cleanup of contaminated aquifers in the United States as it prompted the passing of the Superfund Act, technically known as the Comprehensive Environmental Response, Compensation, and Liability Act (CERCLA), which was passed in 1980. Its notoriety raised the profile of groundwater contamination internationally.

Parallels in contaminated land and groundwater legislations to manage PHC and most contaminants are evident amongst countries in North America, Europe, and in Australia. Although PHCs and LNAPL contamination are rarely directly specified within such legislations, oil/fuel storage and transport-specific regulations enable groundwater protection and are indirectly concerned with PHCs and LNAPL. Similarly, many countries have agency–government (or related national organizations) published non-statutory guideline documents specific to PHC- and LNAPL-contaminated sites that guide the assessment, characterization, and remediation of these sites specifically. By way of example, specific regulatory acts that influence PHC remediation practices in the United States are summarized below:

- *The Safe Drinking Water Act (SDWA)* sets standards for drinking water quality which consider cost, technological feasibility, and health goals (USEPA 1999a). Despite the intention that maximum contaminant limits (MCLs) had been developed to evaluate public water supply systems, the urgency of addressing the ever-increasing scope of groundwater contamination soon led to widespread use of MCLs as cleanup goals for all aqueous groundwater plumes, even those not threatening water supply systems (Hadley and Newell 2012). These stringent goals are often hard to achieve with the technologies available and our current understanding of subsurface processes.
- *The Resource Conservation and Recovery Act (RCRA)* determines corrective actions that individual states adopt, adapt, and enforce through their own agencies to address the release of hazardous chemicals. USEPA guidelines under RCRA state that cleanup should "limit the risk to human health first and then (…) ensure protection [to humans and aquatic receptors] based on maximum beneficial use

of the groundwater at a particular facility [such as drinking, agriculture, discharge to adjacent groundwater or surface water bodies]" (USEPA 2004).

- Directly relevant to NAPL risk management is regulation regarding the condition of underground storage tanks (USTs). This relies on local enforcement to find and clean up current and past leaks. For example, the code of Federal Regulations (CFR) pertaining to UST sites requires the removal of NAPL, which is referred to as "free product," to the maximum extent practicable (MEP), a definition actually left to the discretion of the "implementing agency (40 CFR §280.64)."

- The oil spill prevention program is set up to assist facilities to prevent spills and to implement countermeasures to control spills should they occur through the *Spill Prevention, Control and Countermeasure Plan (SPCC)*.

- The *Small Business Liability Relief and Brownfields Revitalization Act* stems from the USEPA's 1995 Brownfields Program that, through the use of grants, helps states, tribes, communities, and other stakeholders work together to prevent, assess, clean up, and sustainably reuse properties that may be impacted by the presence or potential presence of hazardous substances (i.e., Brownfield sites).

- The *Superfund Act* is a federal program to investigate and remediate contaminated sites that have been placed on a National Priority List (NPL) based on complexity, a preliminary Hazard Ranking System (HRS), or a health advisory from the Agency for Toxic Substances and Disease Registry (ATSDR).

Similar types of legislation are mimicked internationally to varying degrees. In some countries, contaminated land regulations are based on laws approved at national level and their implementation and the manner in which they are interpreted are often left to federal agencies. In contrast, in other countries jurisdiction for legislation may be at a state or regional level rather than national. Either way, both often create differences between regions and pose much of the burden of implementing environmental protection work on state and local governments. For example, the Brownfield Act in the United States relies on state voluntary response programs, through which states create incentives to assist in the cleanup and reuse of brownfield properties and ensure resources for long-term monitoring and other needs.

Globally, numerous published (often non-statutory) guidance developed nationally or at state/regional level exist for investigating, assessing, understanding, and addressing the presence and migration of NAPL releases. Plans typically require that sites achieve a permanent mitigation solution in terms of eliminating significant risk or removing oil and hazardous material to the most feasible extent. Some jurisdictions may set LNAPL recovery goals to the highest potential or "ceiling" values as determined from site-specific estimates of LNAPL and hydrogeological properties, while others determine that NAPL is of concern when the apparent LNAPL thickness measured in a monitoring well exceeds a certain thickness. Such approaches may not always be supported by the technical literature and current multiphase flow understanding.

1.2.3 Toward Improved Management and Regulation of PHC-Contaminated Sites

LNAPL treatment remedies focus on removing, destroying, or stabilizing LNAPL, particularly when this LNAPL poses risks if it is mobilized or can form dissolved-phase and vapor plumes of toxic constituents. Since removal is often technically and/or economically unfeasible, the industry seems to increasingly favor in-situ treatments addressing the LNAPL source zone or focusing on the aqueous or vapor-phase PHC plumes when the only objective is to create a barrier to protect sensitive receptors against contaminant exposure. In many cases, actively monitoring a contaminant plume suffices as a contamination management strategy, in which cases supporting evidence is typically gathered to show that risks are controlled.

Creating a regulatory framework is usually a political, long, arduous, and constantly evolving process. One of the factors influencing this evolution is the technical and economic challenges encountered by remediation practitioners. For example, the presence of NAPL in combination with unfavorable complexities has encouraged stakeholders to advocate to modify cleanup criteria or create technical impracticability waivers. In the United States, for example, there are various alternative cleanup end points, where instead of meeting a drinking water maximum contaminant levels, site-specific risks to potential receptors are evaluated and mitigated using both remediation techniques and engineering controls. Risk-based goals are sometimes a response to the economically prohibitive resources required to eliminate contamination. In other words, remediation objectives have evolved from practically complete removal to focus on interim stages that address site-specific needs that are evaluated in terms of intended land use and stakeholder requirements, which in turn requires effective stakeholder engagement. Hadley and Newell (2012) proposed *functional* remedial goals that incorporate contaminant mass discharge as a remediation metric.

The adoption of sustainability practices, including prioritizing smaller environmental footprints and integrating concepts such as environmental health, social equity, and economic vitality in the assessment of cleanup methods, has been gaining popularity in the remediation industry. In countries like Australia, Brazil, South Africa, the United Kingdom, and the United States, among others, the management and remediation of contaminated soil and groundwater is sometimes incorporating sustainability concepts into their guidelines and/or risk-based regulatory frameworks (Defra 2012; Scott and McInerney 2012 as cited in CL:AIRE 2014). The design of resilient remediation systems is also gaining traction as there is greater awareness on how climate change, extreme weather events (ITRC 2021), and unstable sociopolitical environments may negatively impact existing systems.

Along with sustainability and resiliency, environmental justice and indigenous and nature rights are philosophies that may influence contaminated site management. Values brought forward by the environmental justice movement have gained momentum since it first emerged in the United States in the late 1980s after the publication of *Toxic Waste and Race*, a study exposing disparities in the burden of

environmental degradation and pollution facing minority and low-income communities (Commission for Racial Justice of the United Church of Christ 1987). The
United States government begun addressing this issue by creating environmental
justice offices in the EPA and the Department of Justice. Similarly, values adopted
by indigenous movements that may drive changes in regulations and potentially
groundwater cleanup include nature's rights and concepts such as *Sumak Kawsay*, a
Quechua value embracing ancestral, communitarian knowledge and lifestyle. Based
on this philosophy, the 2008 Constitution of Ecuador incorporated the concept of *the
rights of nature*, as did the 2010 Law of the Rights of Mother Earth in Bolivia.

1.3 Multiphase Flow Mechanics

An understanding of multiphase flow processes in water-NAPL-gas systems (see
Chaps. 2, 3, and 17) is paramount for the development of sound conceptual site
models (CSMs) as well as the design of effective remediation strategies. As depicted
in Fig. 1.3, NAPL, water, and gas coexist in the pore space. NAPL behavior is understood through the interaction of forces in the system that are a function of each fluid's
properties such as viscosity, density, surface and interfacial tension, wettability, and
physical properties associated to a chemical composition. For a comprehensive
introduction to multiphase flow mechanics in porous media, the reader is referred to
Corey (1994).

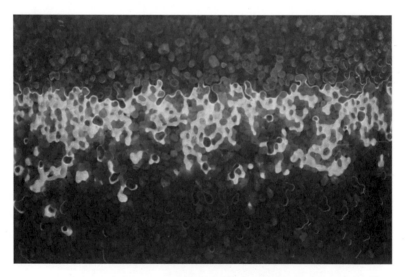

Fig. 1.3 Water-LNAPL-gas system in a sand tank where a fluorescent tracer was added to the
LNAPL (photo and video courtesy of Julio Zimbron, E-Flux). See Electronic Supplementary Material for a video illustrating how some LNAPL mass became entrapped (occluded by water) after
rising the water level in the tank (ESM_1)

LNAPL behavior may sometimes be considered counterintuitive, and its oversimplification has historically led to costly consequences. Examples include the misconception that the LNAPL thickness in a monitoring well corresponds to a similar thickness of LNAPL uniformly saturating the vicinity of the monitoring well (as opposed to a more accurate elevation-dependent LNAPL saturation profile that penetrates below the water table). Similarly, in-well LNAPL thickness has often been assumed proportional to the NAPL mass in the subsurface without considering site-specific LNAPL-retention properties or the LNAPL distribution architecture.

The fate, migration, and stability of LNAPLs markedly vary by type of LNAPL, reflecting differences in their intrinsic fluid properties. Although a low-viscosity NAPL such as gasoline moves faster and perhaps further, it may still reach hydrostatic equilibrium relatively quickly. Viscous LNAPLs like a heavy oil or coal tar creosote, on the other hand, may not equilibrate within practical time frames as the LNAPL body continues to migrate very slowly. Weathering of the LNAPL may cause properties to change; for instance, viscosity and density can increase and borderline LNAPLs such as coal tar creosote may render them becoming borderline DNAPLs with time.

Rationalization of seasonal variations of observed LNAPL thickness in monitoring wells can be better appreciated when multiphase flow concepts are considered together with geological complexities. For instance, within unconfined aquifers the thickness of LNAPL in wells is usually observed to decrease as the water table rises. In contrast, within a confined aquifer, the thickness of LNAPL in wells increases as the potentiometric surface rises since the monitoring well provides a "pressure release" for LNAPL confined by the overlying aquitard. Predicting the volume of scattered LNAPL blobs and ganglia entrapped below the water table (discontinuous, water occluded LNAPL) represents a significant challenge, but is important as these represent long-term sources of contamination. The profile of LNAPL entrapped varies with water-table elevation and may lead to seasonal changes in the mass transfer of PHCs into the dissolved-phase and vapor-phase plumes.

It is possible to encounter LNAPL at significant depths below modern water tables. Such a scenario found in the LA Basin was ascribed to sites with older, historical releases of LNAPL and subsequent rising water tables since the 1950s (as also shown in the case study in Chap. 4). It is not unusual for many cities in developed nations to show rising water tables over recent decades, especially where groundwater was long used by industries that have since demised, for instance both London and Birmingham in the UK.

Quantifying subsurface LNAPL mass requires estimating LNAPL volumes at multiple locations. Site-specific uncertainties complicate this task, such as lack of knowledge of the LNAPL source and type, formation heterogeneity at the scale of centimeters, unknown historical water-table fluctuations, and delineating the depth intervals impacted by LNAPL. Such investigations when at depth may be expensive and challenging especially in consolidated aquifer systems (e.g., sandstone or fractured rock). Often three-dimensional spatial interpolation of data is required as characterization resources are usually limited to discern the detail of LNAPL architecture ideally sought.

A NAPL mass estimate represents a critical baseline parameter used to evaluate the efficiency of remediation technologies or, combining it with estimates of NSZD rates, the longevity of the LNAPL mass (see Chaps. 5 and 13). The scarcity of monitoring points, misinterpretation of in-well LNAPL thickness values and the sensitivity of multiphase LNAPL flow to permeability within intrinsically heterogeneous geological environments means that constraining LNAPL mass estimates can often be a rather "holy grail" challenge but nevertheless important to make efforts to constrain these estimates as far as possible and recognize associated uncertainties present.

1.4 Complexities Associated with PHC NAPL Composition

Refined fuel NAPLs consist of complex mixtures of potentially hundreds of compounds blended with biofuel components such as ethanol in gasoline or fatty acid methyl ester (FAME) in diesel, octane enhancers (such as MTBE), and low-volume additives (such as detergents and anti-foaming agents). Microbial products, metabolites, and other non-PHCs such as natural organic matter or additional organic contaminants like halogenated compounds can also be found at sites impacted by petroleum NAPL releases. This complicates the definition, analysis, and quantitation of bulk parameters such as total petroleum hydrocarbons (TPH) in environmental media and has led to the reliance on a few specific compounds like BTEX or certain PAHs and the adoption of fractionation approaches for risk assessment purposes (ITRC 2018). Analytical techniques such as silica gel cleanup, non-targeted two-dimensional gas chromatography with time-of-flight mass spectrometry (GCxGC-TOFMS), or high-resolution mass spectrometry, to name a few, are being increasingly employed to investigate the fate of PHCs and better inform risk assessments (Bekins et al. 2020; Bojan et al. 2021; O'Reilly et al. 2021). As discussed in Chaps. 10 and 11, molecular biological tools (MBTs) and compound-specific isotope analysis (CSIA) are also gaining popularity to forensically investigate contaminated sites, evaluate remediation performance, and/or identify the responsible party when multiple possible contamination sources exist.

The overall LNAPL composition, along with the age of the release and weathering processes (dissolution, volatilization, and biodegradation), dictates how the bulk properties of an LNAPL, such as density, viscosity, and toxicity will vary with time. Although the majority of petroleum LNAPLs or refined individual constituents (e.g., benzene solvent) are less dense than water and, in general, cease their downward migration upon reaching the vicinity of the water table the densities of coal tar and creosotes may evolve from being marginally less dense than water to more dense than water as discussed above.

Partitioning or mass transfer of individual constituent compounds from the NAPL to adjacent gaseous (air) or aqueous phases are critical processes leading to the continual formation of vapor phase plumes in the vadose zone and dissolved-phase plumes in groundwater. Mass transfer rates from a NAPL are highly dependent upon the chemical composition of the NAPL which varies substantially across the wide range of NAPL types. Both volatilization and dissolution of the individual compounds in a multicomponent NAPL will vary with time as the composition of the NAPL changes as the effective solubility and volatility of a compound depends on its mole fraction within the NAPL. Preferential dissolution or vaporization of more soluble and volatile compounds may leave the remaining NAPL progressively enriched in less soluble and less volatile compound constituents and result in an overall declining mass transfer of hydrocarbon mass with time. The most common volatile organic compounds found in PHC NAPLs that are monitored due to their known toxicity are BTEX, some PAHs, and fuel oxygenates (e.g., MTBE). The relatively high volatility and solubility of these compounds allows them to deplete from the NAPL mass fairly rapidly, eventually reducing the composition-based risk NAPL poses with respect to these compounds.

The spatial extent of dissolved-phase PHC plumes is limited by various depletion mechanisms such as dilution by dispersion and diffusion, volatilization, biodegradation, and sorption and retention. It is thus valuable to identify site-specific limitations to natural attenuation of PHCs. Because biodegradation of dissolved-phase constituents can be the dominant contaminant mass loss process under various groundwater conditions, it is critical to review potential limitations to biodegradation. These include extreme pH levels, low temperatures, insufficient electron acceptors, and the abundance of labile carbon or the presence of inhibitory compounds. Biodegradation mechanisms and rates are a function of biogeochemical conditions which can vary in time and space. For instance, conditions can relatively quickly become anaerobic or even methanogenic at the core of a contaminant plume where PHC are used by microorganisms as a carbon and energy (electron donor) source. Aerobic respiration processes thus deplete dissolved oxygen prior to degrading all the PHC and thus need to turn to nitrates, sulfates, or other reduced compounds in groundwater for terminal electron acceptors. Between the anaerobic core and a typically aerobic fringe of the plume intermediate redox zones may be dominated by sulfate-, manganese-, iron-, or nitrate-reducing metabolic processes. Nevertheless, not all compounds are readily biodegradable under certain redox conditions and biodegradation rates can be easily inhibited by lack of electron acceptors or micronutrients, pH outside of circumneutral conditions, extreme temperatures, or in the presence of inhibitory chemicals. The natural attenuation process knowledge base will also underpin effective selection of treatment technologies. For example, biodegradability of a specific PHC can determine the suitability of bioremediation approaches such as biosparging, bioventing, bioaugmentation, or anaerobic biostimulation.

With practitioner focus on quantifying natural attenuation of dissolved plumes, it has only been recently appreciated that natural depletion of LNAPL mass can be significant (Rivett and Sweeney 2019). Of key significance, has been the comparatively recent recognition and emerging acceptance that the vapor-based mass transfer from LNAPL may have been underappreciated and may be much more relevant than aqueous phase mass transfer limited by the low solubility of most hydrocarbons (Smith et al. 2022). In particular, the importance of LNAPL losses via volatilization and methanogenesis and subsequent aerobic biodegradation of methane in the overlying unsaturated zone that mitigate contaminant breakthrough at ground surface (Lundegard and Johnson 2006). Such direct losses imply that less soluble PHC components may be degraded circumventing the need for (and commonly held previous belief) these constituents to be dissolved first in order to biodegrade in the aqueous phase. Monitoring NSZD processes may represent a source zone management strategy or be used to evaluate the effectiveness of active remediation techniques against a baseline NSZD rate. In addition, the presence of microorganisms, extracellular polymeric substances, and bubbles formed during ebullition may influence LNAPL transport. For instance, bioclogging of the pore space limits the aquifer's permeability. The evaluation of the dynamic influence on NAPL's fate due to water-table fluctuations or shallow subsurface temperature variations across climatic zones represents research opportunities.

1.5 Geological and Hydrogeological Concepts that Help Tackle LNAPL Management Challenges

The heterogeneity of hydrogeological systems inevitably renders complex LNAPL distribution, and the likelihood of complex dissolved and vapor plume generation and fate. It is unsurprising then that the remediation of such NAPL architectures and the associated plumes is likewise challenging. CL:AIRE (2014) provides generic conceptual models for a variety of geological systems, ranging from porous unconsolidated sedimentary units to karst limestone and fractured bedrock of all types. While such conceptual models are illustrative of a specific (hydro)geology setting, site-specific characterization is required to develop a CSM to underpin effective remedy development. The most challenging geological settings are those with a high degree of heterogeneity, which can range in scale from centimeters (pore properties) to kilometers (type of sediment or rock body, tectonic setting, and climate).

The sensitivity of LNAPL migration to subsurface heterogeneity inevitably poses challenges to elucidating the detail of LNAPL migration and its retention. However, some general features may be predictable with preferential migration through coarser-textured materials and wider aperture fractures in fractured or dual porosity systems as shown by the various hydrogeological scenarios in CL:AIREe (2014). Pooling or confining of LNAPL by a system's more contiguous, finer-textured units is also reasonably predictable.

In unconsolidated sediment, sedimentary rocks, and some extrusive igneous flows, primary physical factors are those developed at the time of initial deposition such as grain size, grain sorting, and grain shape may be determined from core logs. Secondary physical characteristics developed post deposition or eruption in response to surface weathering, burial, and movement of fluids are important to consider. Example processes affecting porosity and permeability include compaction (intergranular space reduction), cementation (pore filling by precipitation of minerals from groundwater), burrowing (mixing by organisms), dissolution (enlargement of pores or development of new pores through groundwater dissolution and removal), and fracturing (cracks or fissures from a variety of processes).

A particular challenge for remediation practitioners to overcome is the heterogeneous nature of porous media, not just in terms of porosity and permeability but also mineralogy, geochemistry, and biological microenvironments. The depositional environment of sedimentary units frequently results in sediments that vary greatly and abruptly both in the direction of sediment transport and perpendicular to it. For instance, fluvial systems typically consist of a mixture of channel bar deposits composed of sand and/or gravel interbedded with overbank mudstones, or perhaps silt or clay-rich units that may serve as aquitards restricting groundwater flows or act as capillary barriers impeding NAPL migration. This results in geological units that can be highly partitioned with interleaved aquifer–aquitard sequences with create a complex groundwater flow behavior (see Chap. 4 for further details on depositional environments).

Of concern to both contaminant fate and remediation efforts is the identification of preferential migration pathways as well as relatively stagnant or low-flow regions in which groundwater flow and contaminant transport are limited.

1.6   Summary

The ever-increasing value of groundwater as a freshwater resource, along with a better understanding of risks and sustainability issues associated to the release of PHC compounds will further justify the adoption of improved characterization and remediation approaches. By highlighting complexities encountered in this process, this chapter is not only a reminder to practitioners of the diligence required to carry out projects successfully but also invites the wider research community to continue to pursue underpinning science that helps address complexities encountered at the field scale.

Environmental policy is evolving by incorporating collaborative decision-making that involves multiple stakeholders, public–private partnerships, market-based incentives, and enhanced flexibility in rulemaking and enforcement (Kraft and Vig 2010). This complements the management and remediation of contaminated soil and groundwater increasingly being undertaken in the context of sustainable, risk-based regulatory frameworks.

Thousands of petroleum NAPL projects have successfully mitigated risk of exposure to hazardous chemicals, many sites classified as toxic have been restored to beneficial uses, and remediation has served as a profitable practice in the short and long term. The understanding of PHC contamination in the subsurface, the associated regulations protecting human health and the environment, and the associated cleanup methods and technologies have become more efficient over the years to overcome technical and economic challenges that have often delayed cleanups for significant periods of time.

Ultimately, CSMs take into account available information and the latest research about conditions and risks of contaminated sites. Because PHC remediation is complex, specialized characterization techniques are developed to locate and quantify contaminants, demonstrate degradation rates and mechanisms, forecast their persistence, and evaluate performance of remediation techniques. Management of PHC-impacted sites involves continued CSM updating, with the goal of mitigating risks by implementing site management strategies.

Still, much remains to be done. Site investigation and characterization can always be improved, risks better understood, implemented remedies more efficient, targeted, cost effective and sustainable, community awareness and education increased, and underpinning science advanced. These provide rationale for this volume and the subsequent chapters that seek to draw together a substantial part of our learning to date and identify future research required.

References

ASTM (American Society for Testing and Materials) (2004) STM E1943–98 (2004) Standard guide for remediation of groundwater by natural attenuation at petroleum release sites

Bekins BA, Brennan JC, Tillitt DE, Cozzarelli IM, Illig JM, Martinović-Weigelt D (2020) Biological effects of hydrocarbon degradation intermediates: is the total petroleum hydrocarbon analytical method adequate for risk assessment? Environ Sci Technol 54(18):11396–11404. https://doi.org/10.1021/acs.est.0c02220

Bojan OK, Irianni-Renno M, Hanson AJ, Chen H, Young RB, De Long SK, Borch T, Sale TC, McKenna AM, Blotevogel J (2021) Discovery of oxygenated hydrocarbon biodegradation products at a late-stage petroleum release site. Energy Fuels 35(20):16713–16723. https://doi.org/10.1021/acs.energyfuels.1c02642

CL:AIRE (2014) An illustrated handbook of LNAPL transport and fate in the subsurface. CL:AIRE, London. ISBN 978-1-905046-24-9. Download at www.claire.co.uk/LNAPL

Corey AT (1994) Mechanics of immiscible fluids in porous media. Water Resources Publications, Highlands Ranch

Hadley PW, Newell CJ (2012) Groundwater remediation: the next 30 years. Groundwater 50(5):669–678

ITRC (2018) LNAPL site management: LCSM evolution, decision process, and remedial technologies. LNAPL-3. Washington D.C.: interstate technology and regulatory council. LNAPL Update Team. https://lnapl-3.itrcweb.com

ITRC (Interstate Technology & Regulatory Council) (2021) Sustainable resilient remediation SRR-1. Washington, D.C.: Interstate Technology and Regulatory Council, SRR Team. www.itrcweb.org

Lundegard PD, Johnson PC (2006) Source zone natural attenuation at petroleum hydrocarbon spill sites—II: application to a former oil field. Ground Water Monit Remediat 26(4):93–106. https://doi.org/10.1111/j.1745-6592.2006.00115.x

Lundy D, Gogel T (1988) Capabilities and limitations of wells for detecting and monitoring liquid phase hydrocarbons. In: Proceedings of the second national outdoor action conference on aquifer restoration, ground water monitoring and geophysical methods vol 1

O'Reilly KT, Sihota N, Mohler RE, Zemo DA, Ahn S, Magaw RI, Devine CE (2021) Orbitrap ESI-MS evaluation of solvent extractable organics from a crude oil release site. J Contam Hydrol 242:103855. https://doi.org/10.1016/j.jconhyd.2021.103855

Rincón JG (2020) Evaluating high-resolution site characterisation tools and multiphase modelling to predict LNAPL distribution and mobility (Doctoral dissertation)

Rivett MO, Wealthall GP, Dearden RA, McAlary TA (2011) Review of unsaturated-zone transport and attenuation of volatile organic compound (VOC) plumes leached from shallow source zones. J Contam Hydrol 123(3–4):130–156

Rivett MO, Sweeney R (2019) An introduction to natural source zone depletion at LNAPL sites. CL:AIRE Technical Bulletin TB20. https://www.claire.co.uk/component/phocadownload/cat egory/17-technical-bulletins?download=681:tb-20-an-introduction-to-natural-source-zone-dep letion-at-lnapl-sites

Smith JW, Davis GB, DeVaull GE, Garg S, Newell CJ, Rivett MO (2022) Natural source zone depletion (NSZD): from process understanding to effective implementation at LNAPL-impacted sites. Q J Eng Geol Hydrogeol

U.S. Environmental Protection Agency (USEPA) (1999a) 25 years of the safe drinking water act: history and trends. EPA 816-R-99-007

U.S. EPA (USEPA) (1999b) Remedial technology fact sheet. Monitored natural attenuation of petroleum hydrocarbons, EPA/600/F-98/021 May 1999. https://frtr.gov/matrix/documents/ Monitored-Natural-Attenuation/1999-Monitored-Natural-Attenuation-of-Petroleum-Hydroc arbons-Remedial-Technology-Fact-Sheet.PDF

U.S. Environmental Protection Agency (USEPA) (2011) Infographic: over 500,000 releases from underground storage tanks (USTs) cleaned up. https://www.epa.gov/system/files/documents/ 2021-12/500k-cleanups-11-19-21.pdf. Accessed on 17 Mar 2022

United Church of Christ Commision for Racial Justice (1987) Toxic waste and race in the United States, A National Report on the Racial and Socio-Economic Characteristics of Communities with Hazardous Waste Sites. https://www.nrc.gov/docs/ML1310/ML13109A339.pdf

Vila J, Jiménez-Volkerink SN, Grifoll M (2019) Biodegradability of recalcitrant aromatic compounds. In: Moo-Young M (ed) Comprehensive biotechnology, 3rd edn. Elsevier: Pergamon

Wargo J (1998) Our children's toxic legacy: how science and law fail to protect us from pesticides. Yale University Press

Wargo JP (2009) Green intelligence: creating environments that protect human health. Yale University Press, p 400

Wiedemeier TH, Rifai HS, Newell CJ, Wilson JT (1999) Natural attenuation of fuels and chlorinated solvents in the subsurface. Wiley, New York

Xia Y, Li H, Boufadel MC, Sharifi Y (2010) Hydrodynamic factors affecting the persistence of the Exxon Valdez oil in a shallow bedrock beach. Water Resour Res 46(10)

Chapter 2
Historical Development of Constitutive Relations for Addressing Subsurface LNAPL Contamination

Robert J. Lenhard and Jonás García-Rincón⊙

Abstract An overview of the historical development of *k-S-P* relations (constitutive relations) to model potential flow of light nonaqueous phase liquids (LNAPLs) in the subsurface is presented in this chapter. The focus is advancements proposed by Parker, Lenhard, and colleagues over time. Discussion includes constitutive relations for incorporation in numerical multiphase flow models as well as constitutive relations for predicting subsurface LNAPL volumes and transmissivities from fluid levels measured in monitoring groundwater wells. LNAPL saturation distributions are given for various subsurface properties and layering sequences.

Keywords Capillary pressure · LNAPL saturation · LNAPL-specific volume · Multiphase modeling · Relative permeability

2.1 Introduction

The potential contamination of groundwater by hydrocarbon fuels, such as gasoline (petrol), diesel, and heating oils, has existed since these products were produced. Hydrocarbon fuels, which exist as light nonaqueous phase liquids (LNAPLs), enter the subsurface by leaks in storage containers, pipes, and spills. Once in the subsurface, hydrocarbon compounds can potentially partition in water and contaminate groundwater resources. Typically, hydrocarbon fuels consist of many compounds each with different capabilities (properties) for partitioning into water, subsurface gas, and onto solid matter. Groundwater becomes contaminated by (1) dissolved fuel compounds moving downward by gravity in the vadose zone as water recharges the surface aquifer; (2) volatized fuel compounds partitioning in vadose zone water and

R. J. Lenhard (✉)
San Antonio, TX, USA
e-mail: rj.lenhard@yahoo.com

J. García-Rincón
Legion Drilling and Numac, 11–13 Metters Place, Wetherill Park, NSW 2164, Australia

© The Author(s) 2024
J. García-Rincón et al. (eds.), *Advances in the Characterisation and Remediation of Sites Contaminated with Petroleum Hydrocarbons*, Environmental Contamination Remediation and Management, https://doi.org/10.1007/978-3-031-34447-3_2

groundwater; and (3) direct partitioning into groundwater as the fuels collect at the top of the water-saturated region or get entrapped in it. In this chapter, we briefly discuss the efforts to model subsurface LNAPL behavior including permeability (k)-saturation (S)-pressure (P) relations; prediction of subsurface LNAPL saturations, volumes, and transmissivities based on LNAPL and water levels in nearby groundwater wells; and conclude with suggestions for some improvements in LNAPL predictive models. A focus is on constitutive (k-S-P) relations developed by Parker, Lenhard, and colleagues.

2.2   Recognition of Health Effects from LNAPLs in the Subsurface

Initially, there was little concern of health effects from LNAPLs in the subsurface because the health effects from exposure to some of the compounds were not well understood, except for introduced alkyl leads, e.g., tetraethyl lead in the 1920s (Lewis 1985; Oudijk 2007). Lead was known to have adverse human health effects for some time. Only after relatively high occurrences of sickness and cancer were reported, the health effects from exposure to LNAPL compounds became better understood. For example, benzene, toluene, and xylenes, which are components of LNAPLs, were eventually recognized as either toxic or carcinogenic to humans (IARC 1987; ATSDR 1995; EPA 2007). As the health effects from compounds in LNAPLs were better understood, more stringent regulations were introduced to minimize potential soil and groundwater contamination. Additionally, there were requirements to clean up contaminated sites to minimize the adverse health effects to humans, limit risks to the environment, and reduce other negative impacts from subsurface LNAPL contamination.

2.3   Predicting Subsurface LNAPL Behavior: Early Developments

To plan the remediation of LNAPL-contaminated sites, cost-effective approaches and designs are needed. Practitioners need to know the volume and distribution of LNAPL in the subsurface. Initially, the thickness of LNAPL in boreholes (groundwater wells) was used to estimate the LNAPL thickness in the subsurface (de Pastrovich et al. 1979; Hall et al. 1984). The LNAPL thickness in the subsurface was then used to calculate the subsurface LNAPL-specific volume (LNAPL volume per horizontal surface area), which is an important factor for designing remediation systems. Hampton and Miller (1988) conducted laboratory investigations of the relationship between the LNAPL thickness in a borehole and that in subsurface media and concluded that the LNAPL thickness in the subsurface is not a good measurement of the LNAPL-specific volume because other factors were involved such as porous media pore-size distributions and pore-scale physics that govern LNAPL saturation distributions. The pore-scale physics of LNAPLs in the subsurface that could potentially result in groundwater contamination, however, were well-known (van Dam 1967; Schwille 1967) when these early models were proposed. Because available computers at the time were limited, remediation practitioners used relatively simple expressions to help with remediation strategies and designs.

Abriola and Pinder (1985a,b) were among the first to develop a numerical model to address subsurface LNAPL contamination, which was able to solve the governing equations for multiphase (air, LNAPL, water) flow. Relatively soon afterward, several other multiphase flow numerical models were developed (e.g., Faust 1985; Osborne and Sykes 1986; Baehr and Corapcioglu 1987; Kuppusamy et al. 1987; Kaluarachchi and Parker 1989). To predict the potential flow of air, LNAPL, and water in incompressible porous media, a set of governing equations is

$$\phi \frac{\partial}{\partial t}(S_a \rho_a) = \nabla \left[\frac{kk_{ra}\rho_a}{\eta_a}(\nabla P_a + \rho_a g \nabla z) \right] \qquad (2.1)$$

$$\phi \frac{\partial}{\partial t}(S_o \rho_o) = \nabla \left[\frac{kk_{ro}\rho_o}{\eta_o}(\nabla P_o + \rho_o g \nabla z) \right] \qquad (2.2)$$

$$\phi \frac{\partial}{\partial t}(S_w \rho_w) = \nabla \left[\frac{kk_{rw}\rho_w}{\eta_w}(\nabla P_w + \rho_w g \nabla z) \right] \qquad (2.3)$$

where
  ϕ is the porosity (assumed to be constant for each media type),
  S_a, S_o, S_w are the air, LNAPL, and water saturations, respectively,
  ρ_a, ρ_o, ρ_w are the air, LNAPL, and water mass densities, respectively,
  k is the intrinsic permeability tensor,
  k_{ra}, k_{ro}, k_{rw} are the air, LNAPL, and water relative permeabilities, respectively,
  η_a, η_o, η_w are the air, LNAPL, and water dynamic viscosities, respectively,
  P_a, P_o, P_w are the air, LNAPL, and water fluid pressures, respectively,
  g is the scalar gravitational constant,
  t is time, and,
  z is elevation.

The equations (Eqs. 2.1–2.3) need to be solved simultaneously because the dependent saturation variable in each equation is a function of the independent pressure variable of other equations as well as the relative permeabilities. Consequently, the equations are relatively difficult to solve. Hassanizadeh and Gray (1990) proposed another approach and governing equations to describe multiphase flow based on macroscale thermodynamic theory. Their equations, which were based on fundamental laws of physics, were for only two-fluid systems, but Hassanizadeh and Gray (1990) stated that it would be straightforward to extend the theory to more than two fluids.

A simplifying assumption for solving Eqs. (2.1–2.3) can be made by assuming air to be at atmospheric pressure whenever there is interconnected air (continuous air phase). Richards (1931) made such assumption when considering water movement through soils. Under this assumption, then Eq. (2.1) can be neglected. This assumption is likely realistic for soils near the interface with the atmosphere, where air can freely move from soil pores to the atmosphere and vice versa. However deeper in the subsurface, continuous air may not be atmospheric because of time-dependent atmospheric conditions. The air pressure may vary with elevation. If there are clay or cemented layers, then there may be significant differences in the air pressures on opposite sides of the layers. Air flow has been observed at breathing boreholes which means an air pressure gradient exists. Furthermore, if vacuum extraction of LNAPL vapors is employed as a remediation option, then Eq. (2.1) is needed for any computer simulations (predictions). Employing the Richards (1931) assumption may not always be valid; it would depend on subsurface conditions and what is being simulated.

Important elements for predicting multiphase flow are the physical relations, particularly the relations describing how saturations and relative permeabilities vary as a function of the fluid pressures. Typically, the relations between fluid saturations and pressures are empirically parameterized and the parameters are utilized to predict fluid relative permeabilities. For two-fluid systems (air–water or LNAPL–water), only one set of parameters is needed. For three-fluid systems (air–LNAPL–water), there can be multiple sets of parameters.

Leverett (1941) proposed that the total-liquid saturation in multiphase systems, with gas and liquid phases, is a function of the capillary pressure at gas–liquid interfaces in which gas is the nonwetting fluid. Therefore, the total-liquid saturation (sum of water and LNAPL) would be a function of the air–LNAPL capillary pressure for air–LNAPL–water systems in porous media where water is the wetting fluid and LNAPL has intermediate wettability between air and water. From the petroleum industry, the water saturation is commonly assumed to be a function of the oil–water (LNAPL–water) capillary pressure in which water is a strongly wetting fluid. For three-fluid systems, the total-liquid saturation can be predicted from the air–LNAPL capillary pressure using one set of parameters and the water saturation can be predicted from the LNAPL–water capillary pressure using another set of parameters. Both sets of parameters need to reflect the same pore-size distribution for the predictions to be accurate and avoid numerical convergence issues as the two-fluid (air–water) system transitions into a three-fluid (air–LNAPL–water) system and vice versa.

To calculate the LNAPL saturation, the water saturation is subtracted from the total-liquid saturation. Commonly, water is assumed to be the wetting fluid, air is the nonwetting fluid, and LNAPL has intermediate wettability between water and air. Under this condition and neglecting fluid films, water will preferentially occupy the smallest pore spaces, air will occupy the largest continuous pore spaces, and LNAPL will occupy intermediate-sized pore spaces. However, this wettability sequence does not hold for all porous media or LNAPLs. In some cases, porous media can have LNAPL-wet, mixed-wet, or fractional-wet wettability instead of a strongly water-wet wettability (Anderson 1987). The distribution of fluids within the pore spaces will be governed by the wettability. Hence, the calculation of fluids saturations as a function of capillary pressures will depend on the wettability. In the following discussions in this chapter, we assume the porous media to be strongly water wet.

2.4 The Parker et al. (1987) Nonhysteretic Model

Parker et al. (1987) proposed a parametric formulation to describe nonhysteretic k-S-P relations for air–LNAPL–water systems in porous media where water is the wetting fluid, LNAPL has intermediate wettability, and air is the nonwetting fluid. The formulation can be calibrated relatively easily using more readily available air–water S-P measurements or published parameters. The scaling approach yields closed-form expressions of k-S-P relations that could be employed to predict behavior in two-fluid air–water, air–LNAPL, or LNAPL–water systems, or three-fluid air–LNAPL–water systems.

For S-P relations, Parker et al. (1987) proposed a scaled $S^*(h^*)$ functional as

$$S^*(h^*) = \overline{S}_w^{aw}(\beta_{aw}h_{aw}) \tag{2.4}$$

$$S^*(h^*) = \overline{S}_o^{ao}(\beta_{ao}h_{ao}) \tag{2.5}$$

$$S^*(h^*) = \overline{S}_w^{ow}(\beta_{ow}h_{ow}) \tag{2.6}$$

$$S^*(h^*) = \overline{S}_w^{aow}(\beta_{ow}h_{ow}) \tag{2.7}$$

$$S^*(h^*) = \overline{S}_t^{aow}(\beta_{ao}h_{ao}) \tag{2.8}$$

where superscripts aw, ao, ow, and aow refer to air–water, air–LNAPL, LNAPL–water, and air–LNAPL–water fluid systems, respectively; β_{aw}, β_{ao}, and β_{ow} refer to air–water, air–LNAPL, and LNAPL–water scaling factors, respectively; h_{aw}, h_{ao}, h_{ow} are air–water, air–LNAPL, and LNAPL–water capillary heads, respectively [$h_{aw} = (P_a - P_w)/\rho_w g$; $h_{ao} = (P_a - P_o)/\rho_w g$; $h_{ow} = (P_o - P_w)/\rho_w g$]; and $\overline{S}_w$, $\overline{S}_o$, and $\overline{S}_t$ are the effective water, LNAPL, and total-liquid saturations, respectively, defined by

$$\overline{S}_j = (S_j - S_m)/(1 - S_m) \tag{2.9}$$

in which S_j ($j = w, o, t$) are actual saturations and S_m is the minimum or irreducible wetting fluid saturation. To mathematically describe the relations, Parker et al. (1987) adapted the van Genuchten (1980) *S-P* equation to predict the effective saturations. However, the Brooks-Corey (1966) mathematical equation or other models for describing *S-P* relations also could be used.

The fluid pair scaling factors (β_{aw}, β_{ao}, and β_{ow}) are obtained by regressing air–LNAPL and LNAPL–water *S-P* data, if available, against air–water *S-P* data; however, obtaining such data may not be available or too time-consuming and expensive to conduct. Lenhard and Parker (1987a) overcame this issue by showing that the scaling factors can be estimated from ratios of interfacial tensions as

$$\beta_{aw} = 1 \tag{2.10}$$

$$\beta_{ao} = \sigma_{aw}/\sigma_{ao} \tag{2.11}$$

$$\beta_{ow} = \sigma_{aw}/\sigma_{ow} \tag{2.12}$$

where σ_{aw}, σ_{ao}, and σ_{ow} are air–water, air–LNAPL, and LNAPL–water interfacial tensions, respectively. By defining $\beta_{aw} = 1$, Parker et al. (1987) made the *S-P* parameters to be based on air–water *S-P* relations. This made it considerably easier to obtain parameters to describe the $S^*(h^*)$ relations because one could use available air–water *S-P* data or published air–water parameters and ratios of interfacial tensions, which are more easily measured than air–LNAPL and LNAPL–water *S-P* data. Consequently, investigators can conduct numerical simulations of multiphase flow without the need and cost for *S-P* data from different fluid systems. Enforcing

$$\beta_{aw} = 1/\beta_{ao} + 1/\beta_{ow} \tag{2.13}$$

which is the same as $\sigma_{aw} = \sigma_{ao} + \sigma_{ow}$ ensures the *S-P* relations of air–water and air–LNAPL–water systems will match as an air–water system transitions into an air–LNAPL–water system and vice versa (Lenhard et al. 2002), which is important for convergence of numerical models as fluid systems change. Lenhard and Parker (1987a) noted that when measuring interfacial tensions, all fluids need to be in equilibrium with each other, particularly water. The condition given by Eq. (2.13) indicates that LNAPL will spread on water surfaces. A similar scaling approach was advanced by Leverett (1941) when he was extrapolating *S-P* data for one rock sample to similar rock types with different intrinsic permeabilities and porosities (slightly different pore structure between rock types) in petroleum reservoirs. Parker et al. (1987) extrapolated *S-P* data for different fluid systems in a porous medium with constant rigid pore structure —not between different porous media.

For nonhysteretic *k-S* relations, Parker et al. (1987) followed van Genuchten (1980) by employing the Mualem (1976) model for air–water systems, but extended it for air–LNAPL–water fluid systems in porous media to yield closed-form equations for predicting three-fluid relative permeabilities as

$$k_{rw} = \overline{S}_w^{0.5} \left\{ 1 - \left[1 - \overline{S}_w^{1/m} \right]^m \right\}^2 \tag{2.14}$$

$$k_{ro} = (\overline{S}_t - \overline{S}_w)^{0.5} \left\{ \left[1 - \overline{S}_w^{1/m} \right]^m - \left[1 - \overline{S}_t^{1/m} \right]^m \right\}^2 \tag{2.15}$$

$$k_{ra} = \overline{S}_a^{0.5} \left(1 - \overline{S}_t^{1/m} \right)^{2m} \tag{2.16}$$

in which the k_{rw}, k_{ro}, and k_{ra} are the water, LNAPL, and air relative permeabilities, respectively, and m is a van Genuchten (1980) S-P parameter $\left(m = 1 - \frac{1}{n} \right)$. Closed-form expressions for relative permeabilities are in Parker and Lenhard (1989) when the Brooks-Corey (1966) S-P model is used in Burdine's (1953) relatively permeability integral.

The approach of Parker et al. (1987) allowed investigators to model air–water and air–LNAPL–water flow in the subsurface using numerical models that can be easily calibrated. Only air–water van Genuchten parameters α, n; the irreducible water saturation (S_m); and the interfacial tensions σ_{ao}, σ_{ow} are needed for the k-S-P relations when the van Genuchten (1980) and Mualem (1976) models are used for water-wet porous media. The nonhysteretic k-S-P relations were initially tested against air–LNAPL–water experimental data by Lenhard et al. (1988) in a multiphase numerical model. When the Brooks-Corey (1966) and Burdine (1953) models are used for water-wet porous media, only the air–water Brooks–Corey parameters h_d (displacement head) and λ (pore-size index); the irreducible water saturation (S_m); and the interfacial tensions σ_{ao}, σ_{ow} are needed for the k-S-P relations.

2.5 Hysteretic Model

Shortly after Parker et al. (1987) proposed the nonhysteretic k-S-P relations, Parker and Lenhard (1987) and Lenhard and Parker (1987b) proposed hysteretic k-S-P relations, which accounts for contact angle, pore geometry ("ink bottle"), and nonwetting fluid entrapment effects in the both vadose and liquid-saturated zones when fluids change from drying-to-wetting saturation paths and vice versa. The entrapment of LNAPL in water and entrapment of air in LNAPL and water are considered. Nonwetting fluid entrapment occupies pore space that would otherwise be occupied by a relative wetting fluid and displaces the relative wetting fluid into larger pore spaces. For example, entrapped LNAPL by water will displace water into larger pore spaces at the same actual water saturation and, thereby, increase the conductance rate of water. The effect increases the water relative permeability and decreases the LNAPL relative permeability at given actual water and LNAPL saturations for wetting fluid imbibition saturation paths than for wetting fluid drying saturation paths. To account for these effects, Parker and Lenhard (1987) and Lenhard and Parker (1987b) introduced the term "apparent" fluid saturations where the apparent two-phase (air–water)

water saturation is the sum of the effective entrapped air saturation in water and the effective water saturation; and the apparent three-phase (air–LNAPL–water) water saturation is the sum of the effective entrapped air and LNAPL saturations in water and the effective water saturation. The apparent saturations better index the pore sizes in which the continuous fluids reside than effective or actual saturations for predicting relative permeabilities. The hysteretic k-S-P relations were initially tested against transient air–LNAPL–water experimental data by Lenhard et al. (1995) in a multiphase numerical model, which yielded good results.

2.6 Predicting LNAPL Saturations, Volumes, and Transmissivity from Well Levels

To further help develop LNAPL remediation strategies and forecast three-dimensional LNAPL movement, Parker and Lenhard (1989) vertically integrated k-S-P relations and incorporated them into a two-dimensional numerical model (ARMOS) (Kaluarachchi et al. 1990). The earlier Parker et al. (1987) nonhysteretic k-S-P relations provided the framework. The model, ARMOS, was later modified to include LNAPL entrapment in both the unsaturated and liquid-saturated zones (Kaluarachchi and Parker 1992). The calculation of the LNAPL entrapment volume in a vertical slice of the unsaturated zone is based on the change in the air–LNAPL level in wells from rising water tables. The LNAPL entrapment volume in a vertical slice of the liquid-saturated zone is based on the change in the LNAPL–water level in wells from rising water tables. Parker et al. (1994) used ARMOS to model LNAPL migration and direct LNAPL recovery at LNAPL spill sites from knowledge of fluid levels in wells. Waddill and Parker (1997a) later proposed semi-analytical algorithms to predict LNAPL trapping and recovery from wells in diverse soils. The algorithms were incorporated in a version of ARMOS and tested against experimental data (Waddill and Parker, 1997b). Later, the approach of Parker and Lenhard (1989) was incorporated in American Petroleum Institute's (API) LDRM model (API 2007).

Parker and Lenhard (1989) developed the vertically integrated nonhysteretic k-S-P relations for incorporation into multiphase numerical models to assist remediation practitioners. However, not all LNAPL remediation practitioners have access to numerical models so Lenhard and Parker (1990) published equations for determining the LNAPL volume in a vertical slice of the subsurface from fluid level elevations in wells that can be used in simple computer programs. At the same time, Farr et al. (1990) also published equations to determine LNAPL volume in a vertical slice of the subsurface. Both publications assume vertical equilibrium conditions and use the Brooks-Corey (1966) S-P model for developing their equations. The difference in the approaches is Lenhard and Parker (1990) utilized elevations from the fluid levels in the wells to determine LNAPL and water pressures, and Farr et al. (1990) utilized depths from the surface to determine LNAPL and water pressures. The air phase is assumed to be atmospheric in both models.

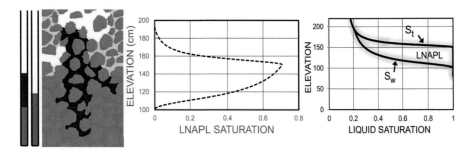

Fig. 2.1 A cartoon showing determining of total-liquid (S_t), water (S_w), and LNAPL saturations from wells adjacent to LNAPL-contaminated soils

Using the approach of Lenhard and Parker (1990) and Farr et al (1990) to determine LNAPL and water pressures, and hence air–LNAPL and LNAPL–water capillary pressures, from elevations (depths) of the air–LNAPL and LNAPL–water interfaces in the wells, the subsurface LNAPL saturations adjacent to wells can be calculated. Figure 2.1 shows an example of information that can be obtained. From the air–LNAPL and LNAPL–water capillary pressures, the total-liquid saturations (S_t) and the water saturations (S_w), respectively, can be calculated, which by difference yields the LNAPL saturations ($S_o = S_t - S_w$).

The shape of the Fig. 2.1 LNAPL saturation-elevation profile is similar to the "shark fin" shape description (Frollini and Petitta 2018) reported by researchers/ practitioners. However, the shape depends on porous media type. For example, Fig. 2.2 shows the predicted LNAPL saturations as a function of elevation for the scenario where the air–LNAPL interface in a well is at an elevation of 150 cm above a datum and the LNAPL–water interface is at an elevation of 100 cm above the same datum for a LNAPL thickness in the well of 50 cm. If an accompanying well is only screened in the water-saturated zone, then the air–water interface would occur at an elevation of 135.5 cm above the datum, which can be determined from Eq. 2.5 in Lenhard and Parker (1990). Hydrostatic conditions are assumed so the LNAPL and water pressures vary linearly with elevation. Furthermore, nonhysteretic S-P relations are assumed, i.e., entrapped and residual LNAPL saturations are not considered. The datum elevation for determining the LNAPL saturations can be arbitrarily set. The predicted S_t and S_w, from which the LNAPL saturations are determined in Fig. 2.2, are obtained using the van Genuchten (1980) model, but other S-P models could be used (e.g., Brooks and Corey 1966; Alfaro Soto et al. 2019). The left diagram in Fig. 2.2 is for a loamy sand soil and the right diagram is for a clay loam soil. The shape of the predicted LNAPL saturation-elevation distribution is noticeably different, especially the LNAPL saturation values at each elevation. The LNAPL saturations in the clay loam soil extend to higher elevations than those in the loamy sand soil, but not to a large extent. The LNAPL saturations in Fig. 2.2 become zero at an elevation of 188 cm for the loamy sand soil and become zero at an elevation of 193 cm for the clay loam soil. The LNAPL saturation predictions in Figs. 2.1 and

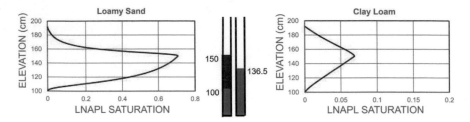

Fig. 2.2 Predicted LNAPL saturation-elevation distributions of a loamy sand (left) and a clay loam (right) soil for a scenario where the air–LNAPL interface in a well is at an elevation of 150 cm and the LNAPL–water interface is at 100 cm (center)

2.2 assumed that all the LNAPL is mobile or free as originally proposed by Lenhard and Parker (1990).

2.7 Incorporating Free, Residual, and Entrapped LNAPL Fractions

Later, Lenhard et al. (2004) proposed further adaptions to the k-S-P relations proposed by Parker et al (1987). Understanding that k_{ro} will be greater than 0 whenever the effective S_o ($\overline{S}_o$) is greater than 0, unless k_{ro} is arbitrarily set to zero at some $\overline{S}_o$, Lenhard et al. (2004) proposed an approach for methodically setting $k_{ro} = 0$ at variable S_o by considering a residual LNAPL saturation. When $k_{ro} > 0$ because $\overline{S}_o$ > 0, then LNAPL will continually drain from the vadose zone until vertical S-P equilibrium conditions are established. Setting $k_{ro} = 0$ arbitrarily at some fixed S_o is not realistic because the mobility of LNAPL at different water saturations will not be the same. For example, setting $k_{ro.} = 0$ at $\overline{S}_o = 0.15$ might be reasonable when the effective water saturation ($\overline{S}_w$) = 0.2, because the LNAPL will reside in small pores; but, not when $\overline{S}_w = 0.7$, because the LNAPL will reside in larger pores. The LNAPL may be immobile when $\overline{S}_w = 0.2$, but much of the LNAPL will likely be mobile at $\overline{S}_w = 0.7$, so setting $k_{ro} = 0$ at some arbitrary LNAPL saturation is not reasonable for all water saturations.

Lenhard et al. (2004) considered LNAPL to exist in three forms: free, entrapped, and residual. Free LNAPL is continuous and can move freely in porous media when a LNAPL total-pressure gradient exists. Entrapped LNAPL is water-occluded LNAPL that results from water displacing LNAPL from pore spaces and can potentially exist everywhere in the subsurface (i.e., above and below the water-saturated zone). Entrapped LNAPL is discontinuous and immobile; it will not move under water and LNAPL total-pressure gradients typical in porous media. Residual LNAPL is immobile LNAPL that is not water occluded, which may exist as LNAPL wedges and films. There may be LNAPL films connecting some residual LNAPL (wedges) to free LNAPL, but the movement of LNAPL through the films is likely to be very slow and

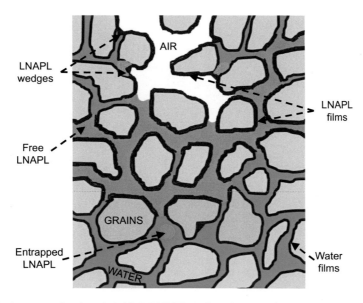

Fig. 2.3 A cartoon showing air (white), LNAPL (red), and water (blue) in subsurface pore spaces

cause the residual LNAPL to behave as if practically immobile under most LNAPL total-pressure gradients. Figure 2.3 shows a cartoon of air, LNAPL, and water in water-wet, subsurface, porous media as envisioned by Lenhard et al. (2004). Air is shown as white and exists when the air–LNAPL capillary pressure is greater than 0. LNAPL is shown as red with LNAPL films slightly darker red. The LNAPL films are attached to water films in pores containing air that previously were LNAPL filled and attached to water films in LNAPL-filled pores. Water-occupying pores is lighter blue. Water films wetting the solid particle grains (dark gray) is darker blue.

Figure 2.4 shows expanded sections of Fig. 2.3 which displays the Lenhard et al. (2004) concepts of free, entrapped, and residual LNAPL. Free LNAPL (Fig. 2.4a–c) is LNAPL continuous throughout the pore spaces that can move freely in response to LNAPL total-pressure gradients. It does not include LNAPL films (Fig. 2.4a,c) on water films (Fig. 2.4a,c) which theoretically will not move freely; the LNAPL films are assumed to be relatively immobile. Entrapped LNAPL (Fig. 2.4c) as explained earlier is water-occluded LNAPL which will not move in response to water and LNAPL total-pressure gradients. The water-occluded LNAPL is discontinuous and exists as blobs in pore bodies and LNAPL wedges that do not drain prior to water-filling pores during increasing water saturations. A similar situation may occur with water in smaller pores surrounded by larger pores that fill with LNAPL before the water in the small pores drain resulting in isolated water (Fig. 2.4b). Residual LNAPL is idealized as LNAPL wedges and films in the three-phase air–LNAPL–water zones (Fig. 2.4a). It also includes the relatively immobile LNAPL wedges and films (Fig. 2.4c) in the two-phase LNAPL–water zones.

Fig. 2.4 A cartoon showing the concepts of free, entrapped, and residual LNAPL from Fig. 2.3

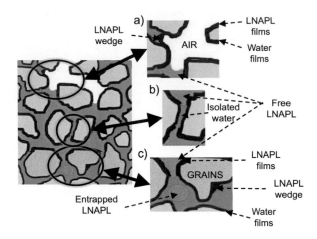

The concepts of free, entrapped, and residual LNAPL advanced by Lenhard et al. (2004) were incorporated in numerical codes by White et al. (2004) and Oostrom et al. (2005) and were shown to be important for predicting LNAPL behavior in the vadose zone. Additional models to incorporate a residual LNAPL saturation in k-S-P relationships were proposed by Wipfler and van der Zee (2001) and Van Geel and Roy (2002). Oostrom et al. (2005) tested both the Van Geel and Roy (2002) and Lenhard et al. (2004) formulations against transient three-fluid column experiments.

2.8   Recent Developments. The Lenhard et al. (2017) Model

Effects of entrapped and residual LNAPL (Lenhard et al. 2004) were considered by Lenhard et al. (2017) when predicting subsurface LNAPL saturations and transmissivity from LNAPL and water levels in wells. In the Lenhard et al. (2017) model, historical air–LNAPL and LNAPL–water interface levels in wells were considered. Specially, the highest air–LNAPL interface elevation and the lowest LNAPL–water interface elevation were included. The highest air–LNAPL interface elevation in an observation well reflects the highest elevations in the subsurface that may contain residual LNAPL. The lowest LNAPL–water interface elevation reflects the lowest elevations that may contain entrapped LNAPL. Both entrapped and residual LNAPL will affect the LNAPL relative permeability at a location because they reduce the amount of free or mobile LNAPL at that elevation. The LNAPL relative permeability distribution governs the LNAPL transmissivity.

For the same configuration of air–LNAPL and LNAPL–water interface elevations in a well as Fig. 2.2 (current levels) and assuming the air–LNAPL interface elevation at one time was 50 cm higher (i.e., historical elevation is 200 cm) and the LNAPL–water interface elevation was 50 cm lower (i.e., historical elevation is 50 cm), predicted free, entrapped, and residual LNAPL saturations are shown in

Fig. 2.5. Again, the van Genuchten (1980) model was used to predict the *S-P* relations consistent with the Lenhard et al. (2017) model. In comparison to Fig. 2.2, the upper free LNAPL elevation for the loamy sand soil in Fig. 2.5 is predicted to be about 20 cm lower. This results because residual LNAPL occurs, which is not mobile (free), and some minor entrapped LNAPL exists as water displaced LNAPL from those pores when the water table was raised (i.e., as the LNAPL–water interface in the well was raised from the historical level). The residual LNAPL for the loamy sand soil in Fig. 2.5 is predicted to occur about 50 cm higher than the free LNAPL in Fig. 2.2, because the historical air–NAPL level is 50 cm higher. The lower free LNAPL elevation is the same for the loamy sand soil in Figs. 2.2 and 2.5 because the lower free LNAPL occurrence is a function of the current LNAPL–water level (i.e., 100-cm elevation) and not historical levels. However, the lower entrapped LNAPL elevation for the loamy sand soil in Fig. 2.5 is predicted to be about 50 cm lower than the free LNAPL in Fig. 2.2 because the historical LNAPL–water level is 50 cm lower. There is no predicted entrapped LNAPL in Fig. 2.2 because it was not considered in the nonhysteretic *S-P* relations. For Fig. 2.5, the maximum possible entrapped LNAPL was set at 0.15, and the maximum possible residual LNAPL was set at 0.15. These are additional parameters in Lenhard et al. (2004, 2017). The LNAPL distributions in Fig. 2.5 can help practitioners design remediation programs focused on free, entrapped, and residual LNAPL and not only on free LNAPL as in Fig. 2.2.

The predicted free LNAPL volume in a vertical slice of the loamy sand soil per cm^2 of surface area in Fig. 2.2 is predicted to be 11.7 cm^3/cm^2, whereas the free LNAPL volume is predicted only to be 9.9 cm^3/cm^2 in Fig. 2.5. The difference is because of considering residual and entrapped LNAPL in Fig. 2.5. The difference also has a significant effect on predicted LNAPL transmissivities using the methodology in Lenhard et al. (2017). When entrapped and residual LNAPL are not considered, the predicted free LNAPL transmissivity is 6200 cm^2/day for the loamy sand soil with an assumed water-saturated hydraulic conductivity of 350 cm/day. When entrapped and residual LNAPL are considered, the predicted free LNAPL transmissivity is 4225 cm^2/day—a difference of over 30%. If a calculation is conducted that accounts for only LNAPL under positive LNAPL pressures, then the predicted free LNAPL

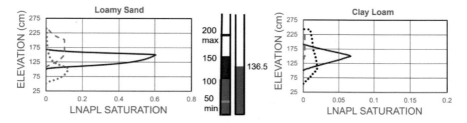

Fig. 2.5 Predicted free (red solid line), entrapped (black dotted line), and residual (green dashed line) of a loamy sand (left) and a clay loam (right) soil where the current air–LNAPL and LNAPL–water interface elevations in a well are the same as in Fig. 2.2 with the historical highest air–LNAPL level 50 cm higher than current and the historical lowest LNAPL–water level 50 cm lower than current

transmissivity is only 4830 cm²/day for the loamy sand in Fig. 2.2 and 3360 cm²/day in Fig. 2.5. Because only LNAPL under positive pressure will enter a borehole, estimations of LNAPL transmissivity should be based on LNAPL under positive pressure for planning direct LNAPL recovery from wells.

2.9   Layered Porous Media

When the contaminated subsurface can be represented as homogeneous, the subsurface LNAPL distributions can have a "shark fin" shape. For layered porous media, however, the subsurface LNAPL distribution shape can be more varied. Consider a scenario where two coarse layers are separated by a fine layer. Assume a LNAPL-contaminated subsurface with a 20-cm fine layer between two coarse layers that may be over 100 cm thick. In nearby monitoring wells, the air–LNAPL interface in one well occurs at an elevation of 160 cm above the datum and the LNAPL–water interface in the same well occurs at an elevation of 100 cm above the datum for a LNAPL thickness of 60 cm in the well. In an accompanying well screened only in the water-saturated zone, the air–water interface occurs at an elevation of 142 cm above the datum. Assuming hydrostatic conditions, nonhysteretic S-P relations, and the van Genuchten (1980) model to predict S_t and S_w, the subsurface LNAPL saturation distribution can be predicted. Figure 2.6 shows the results for the fine layer occurring between the 120- and 140-cm elevations. Because the fine layer is below the air–LNAPL interface in the well and above the elevation of the LNAPL–water interface, the fine layer will be completely saturated with LNAPL and water. The fine layer will have a higher water saturation than the coarse layers because of its smaller pore sizes. The contrast in pore sizes between the coarse and fine layers can produce significant differences in LNAPL saturations at the boundaries between the layers. The resulting shape of the subsurface LNAPL distribution will consequently be very different than a "shark fin" shape.

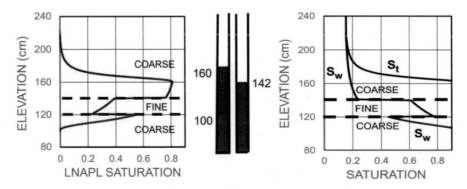

Fig. 2.6 Predicted LNAPL, total-liquid (S_t), and water (S_w) saturations for a scenario involving a fine layer between two coarse layers with 60 cm of LNAPL inside a nearby monitoring well

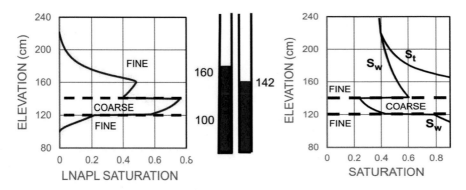

Fig. 2.7 Predicted LNAPL, total-liquid (S_t), and water (S_w) saturations for a scenario involving a coarse layer between two fine layers with 60 cm of LNAPL inside a nearby monitoring well

The shape of the subsurface LNAPL distribution curve will depend on whether there are contrasting pore sizes between potential layers, the location of the layers relative to the air–LNAPL and LNAPL–water interface elevations in nearby wells, and the thickness of LNAPL in the wells. As another example, consider a scenario where two fine layers are separated by a coarse layer. Assume a LNAPL-contaminated subsurface with a 20-cm coarse layer between two fine layers that may be over 100 cm thick. The opposite configuration as for Fig. 2.6. Further, assume the same conditions and fluid levels in wells as for Fig. 2.6. Using the same calculation protocols, Fig. 2.7 shows the results of the predicted total liquid, S_t, and water, S_w, saturations. For Fig. 2.7, the coarse layer is completely liquid-saturated for the same reason as the fine layer in Fig. 2.6, but the water saturation is low because of its larger pore sizes. The low water saturation yields a high LNAPL saturation for the coarse layer. The contrasting sizes of pores in the coarse and fine layers produce a significant difference in LNAPL saturations at the boundaries of the layers—just like in Fig. 2.6, but in the opposite direction. The shape of the subsurface LNAPL distribution curve can be quite complex for layered porous media and not a simple "shark fin" shape. Furthermore, an abrupt change in pore sizes at layer boundaries can result in LNAPL moving laterally versus downward similarly to a capillary break of constructed soil covers over waste pits. The subsurface LNAPL distribution depends on many factors.

2.10   Summary and Steps Forward

The historical development of k-S-P relations described in this chapter largely follows advancements proposed by Parker, Lenhard, and colleagues. Many of their proposed concepts and equations are incorporated in today's multiphase numerical models or code developers have used some of their concepts in building their models. However, further advancements are still required to help accurately predict the subsurface behavior of LNAPL, especially the effects of microbiological action and potential changes in fluid properties over time.

A key element of k-S-P models is that they mimic physics of multiphase flow, even if the models may be empirical. The various forms of LNAPL in the subsurface need to be addressed. Careful estimation of model parameters is needed also to produce good predictions, but this is not routinely done by contaminated land practitioners nor typically required by regulators. Improved and more accessible measurement methods for two-fluid and three-fluid systems should be developed and adopted.

Similarly, the effects of heterogeneous subsurface conditions should be acknowledged by investigators and considered when conducting numerical simulations or using simple predictive models. This may be incorporated into models specifically developed to represent heterogeneous environments with layered systems or multimodal pore-size porous media (see Chap. 3).

Some models to predict LNAPL volumes, distribution, and transmissivity may not account for certain processes, which may have not been considered relevant enough for practical purposes in the past. Additional research may be needed to elucidate the relevance of mechanisms like those associated to natural source zone depletion (e.g., entrapment of gases caused by methanogenic outgassing or presence of microorganisms and extracellular polymeric substances) on LNAPL distribution predictions. More information on natural source; zone depletion phenomena is presented in Chap. 5. Further research is also needed for systems not exhibiting strongly water-wet characteristics.

Good understanding of multiphase flow principles is paramount to develop sound conceptual site models and, therefore, should be an essential skill of regulators and practitioners managing LNAPL-contaminated sites. Multiphase models are currently underutilized and may help to analyze LNAPL remediation options and design cost-effective solutions, including mass recovery systems (see Chap. 12) and any other in-situ remediation strategies targeting the LNAPL source zone.

References

Abriola LM, Pinder GF (1985a) A multiphase approach to the modeling of porous media contamination by organic compounds-part I: equation development. Water Resour Res 21(1):11–18

Abriola LM, Pinder GF (1985b) A multiphase approach to the modeling of porous media contamination by organic compounds-part 2: numerical simulation. Water Resour Res 21(1):19–26

Anderson WG (1987) Wettability literature survey - Part 4: effects of wettability on capillary pressure. J Pet Technol 39:1283–1300. https://doi.org/10.2118/15271-PA

API (American Petroleum Institute) (2007) Volume 1: distribution and recovery of petroleum hydrocarbon liquids in porous media. API Publication 4760. Washington D.C, pp 53

ATSDR (Agency for Toxic Substances and Disease Registry) (1995) Toxicological profile for gasoline. U.S. Department of Health and Human Services. Division of Toxicology/Toxicology Information Branch 1600 Clifton Road NE, E-29. https://www.atsdr.cdc.gov/toxprofiles/tp72.pdf

Baehr AL, Corapcioglu MY (1987) A compositional multiphase model for groundwater contamination by petroleum products, 2 numerical solution. Water Resour Res 23:201–213

Brooks RH, Corey AT (1966) Properties of porous media affecting fluid flow. J Irrig Drain Div Am Soc Civ Eng 92:61–68

Burdine HT (1953) Relative permeability calculations from pore-size distribution data. Trans Soc Pet Eng AIME 198:71–77

de Pastrovich TL, Baradat Y, Barthel R, Chiarelli A, Fussell DR (1979) Protection of groundwater from oil pollution. CONCAWE, Report 3/79. Den Haag, Netherlands, pp 61

EPA (U.S. Environmental Protection Agency) (2007) Agency for toxic substances and disease registry (ATSDR), August 2007. Toxicological Profile for Benzene. https://www.epa.gov/sites/default/files/2014-03/documents/benzene_toxicological_profile_tp3_3v.pdf

Farr AM, Houghtalen RJ, McWhorter DB (1990) Volume estimation of light nonaqueous phase liquids in porous media. Ground Water 28(1):48–56

Faust CR (1985) Transport of immiscible fluids within and below the unsaturated zone: a numerical model. Water Resour Res 21:587–596

Frollini E, Petitta M (2018) Free LNAPL volume estimation by pancake model and vertical equilibrium model: comparison of results, limitations, and critical points. Geofluids, Article ID 8234167:13. https://doi.org/10.1155/2018/8234167

Hall RA, Blake SB, Champlin SCJ (1984) Determination of hydrocarbon thicknesses in sediments using borehole data. In: Proceedings of the fourth national symposium on aquifer restoration and ground water monitoring. National Water Well Assoc., Worrthington, OH, pp 300–304

Hampton DR, Miller PDG (1988) Laboratory investigation of the relationship between actual and apparent product thickness in sands. In: Proceedings petroleum hydrocarbons and organic chemicals in ground water: prevention, detection and restoration. National Water Well Assoc., Dublin, OH, pp 157–181

Hassanizadeh SM, Gray WG (1990) Mechanics and thermodynamics of multiphase flow in porous media including interphase boundaries. Adv Water Resour 13(4):169–186

IARC (International Agency for Research and Cancer) (1987) Summaries and evaluations: Benzene (Group 1). Lyon, International Agency for Research on Cancer, p 120 (IARC Monographs on the Carcinogenicity of Chemicals to Humans, Supplement 7. http://www.inchem.org/documents/iarc/suppl7/benzene.html

Kaluarachchi JJ, Parker JC (1989) Improving the efficiency of finite element method in modeling multiphase flow. Water Resour Res 25:43–54

Kaluarachchi JJ, Parker JC (1992) Multiphase flow in porous media with a simplified model for oil entrapment. Trans Porous Media 7:1–13

Kaluarachchi JJ, Parker JC, Lenhard RJ (1990) A numerical model for areal migration of water and light hydrocarbon in unconfined aquifers. Ad Water Resour. 13(1):29–40

Kuppusamy T, Sheng J, Parker JC, Lenhard RJ (1987) Finite element analysis of multiphase immiscible flow through soils. Water Resour Res 23:625–631

Lenhard RJ, Parker JC (1987a) Measurement and prediction of saturation-pressure relationships in three-phase porous media systems. J Contam Hydrol 1:407–424

Lenhard RJ, Parker JC (1987b) A model for hysteretic constitutive relations governing multiphase flow: 2 permeability-saturation relations. Water Resour Res 23:2197–2206

Lenhard RJ, Parker JC (1990) Estimation of free hydrocarbon volume for fluid levels in monitoring wells. Ground Water 28(1):57–67

Lenhard RJ, Oostrom M, White MD (1995) Modeling fluid flow and transport in variably saturated porous media with the STOMP simulator. 2. Verification and validation exercises. Adv Water Resour 18(6):365–373

Lenhard RJ, Oostrom M, Dane JH (2002) Chapter 7.5: prediction of capillary pressure-relative permeability relations. In: Methods of soil analysis. Part 1: physical methods. In: Dane JH and Topp GC (eds) Soil Science of America. Madison, WI. pp 1591–1607

Lenhard RJ, Oostrom M, Dane JH (2004) A constitutive model for air-NAPL-water flow in the vadose zone accounting for immobile, non-occluded (residual) NAPL in strongly water-wet porous media. J Contam Hydrol 73:283–304

Lenhard RJ, Rayner JL, Davis GB (2017) A practical tool for estimating subsurface LNAPL distributions and transmissivity using current and historical fluid levels in groundwater wells: effects of entrapped and residual LNAPL. J Contam Hydrol 205:1–11

Lenhard RJ, Dane JH, Parker JC, Kaluarachchi JJ (1988) Measurement and simulation of one-dimensional transient three-phase flow for monotonic liquid drainage. Water Resour Res 24:853–863

Leverett MC (1941) Capillary behavior in porous solids. Trans Soc Pet Eng AIME 142:152–169

Lewis J (1985) Lead poisoning: a historical perspective. EPA J 11(4):15–18. (U.S. Environmental Protection Agency) https://archive.epa.gov/epa/aboutepa/lead-poisoning-historical-perspective.html

Mualem Y (1976) A new model for predicting the hydraulic conductivity of unsaturated porous media. Water Resour Res 12:513–522

Osborne M, Sykes JF (1986) Numerical modeling of immiscible organic transport at the Hyde Park landfill. Water Resour Res 22:25–33

Oostrom M, White MD, Lenhard RJ, Van Geel J, Wietsma TW (2005) A comparison of models describing residual NAPL formation in the vadose zone. Vadose Zone J 4:163–174

Oudijk G (2007) The use of alkyl leads in gasoline age-dating investigations: new insights, common investigative techniques, limitations, and recommended practices. Environ Claims J 19(1–2):68–87. https://doi.org/10.1080/10406020601158329

Parker JC, Lenhard RJ (1987) A model for hysteretic constitutive relations governing multiphase flow: 1 saturation-pressure relations. Water Resour Res 23:2187–2196

Parker JC, Lenhard RJ (1989) Vertical integration of three-phase flow equations for analysis of light hydrocarbon plume movement. Trans Porous Media 5:187–206

Parker JC, Lenhard RJ, Kuppusamy T (1987) A parametric model for constitutive properties governing multiphase flow in porous media. Water Resour Res 23(4):618–624

Parker JC, Zhu JL, Johnson TG, Kremesec VJ, Hockman EL (1994) Modeling free product migration and recovery at hydrocarbon spill sites. Ground Water 32(1):119–128

Richards LA (1931) Capillary conduction of liquids through porous mediums. Physics 1:318–333. https://doi.org/10.1063/1.1745010

Schwille F (1967) Petroleum contamination of the subsoil—a hydrological problem. In: Hepple P (ed) The joint problems of the oil and water industries. Inst. Petrol, London, pp 23–54

Soto MAA, Lenhard R, Chang HK, van Genuchten MT (2019) Determination of specific LNAPL volumes in soils having a multimodal pore-size distribution. J Environ Mgt 237:576–584

van Dam J (1967) The migration of hydrocarbons in water-bearing stratum. In: Hepple P (ed) The joint problems of the oil and water industries. Inst. Petrol, London, pp 55–96

Van Geel PJ, Roy SD (2002) A proposed model to include residual NAPL saturation in a hysteretic capillary pressure-saturation relationship. J Contam Hydrol 58:79–110

van Genuchten MT (1980) A closed-form equation for predicting the hydraulic conductivity of unsaturated soils. Soil Sci Soc Am J 44:892–898 https://doi.org/10.2136/sssaj1980.03615995004400050002x

Waddill DW, Parker JC (1997a) A semianalytical model or predict recovery of light, nonaqueous phase liquids from unconfined aquifers. Ground Water 35(2):280–290

Waddill DW, Parker JC (1997b) Recovery of light, non-aqueous phase liquid from porous media: laboratory experiments and model validation. J Contam Hydrol 27:127–155

White MD, Oostrom M, Lenhard RJ (2004) A practical model for mobile, residual, and entrapped NAPL in water-wet porous media. Ground Water 42(5):734–746

Wipfler EL, van der Zee SEATM (2001) A set of constitutive relationships accounting for residual NAPL in the unsaturated zone. J Contam Hydrol 50:53–77

Chapter 3
Estimating LNAPL Volumes in Unimodal and Multimodal Subsurface Pore Systems

Miguel A. Alfaro Soto, Martinus Th. van Genuchten, Robert J. Lenhard, and Hung K. Chang

Abstract In this chapter, we compare modeling approaches for determining the specific volume of light nonaqueous phase liquids (LNAPLs) in the subsurface on top of the water-saturated zone following spills or leaks at or near the soil surface. We employ both unimodal and multimodal pore-size capillary pressure–saturation functions in our predictions. Hydrologic properties, both fluid properties and pore-size distribution, are important for accurate predictions. Before presenting our results, we discuss fluid interfacial tensions, fluid wettability, and pore structures. Data from the literature are used in our calculations to show that the use of multimodal unsaturated soil hydraulic properties can lead to significantly different subsurface LNAPL-specific volume predictions compared to when unimodal formulations are used. The differences can be significant even when the capillary pressure–saturation curves appear similar over the capillary pressure range relevant to the hypothetical LNAPL contamination scenarios. Consequently, the multimodal pore structure of porous media needs to be addressed to avoid potentially erroneous estimates of the resources and time needed to remediate LNAPLs from contaminated areas.

Keywords Capillary pressure-saturation relations · LNAPL-specific volume · Multimodal pore-size distribution · Pore structure · Subsurface LNAPL distribution

M. A. Alfaro Soto (✉) · H. K. Chang
Department of Applied Geology and Centro de Estudos Ambientais, UNESP—São Paulo State University, Rio Claro, SP, Brazil
e-mail: alfaro.soto@unesp.br

H. K. Chang
e-mail: chang.hung-kiang@unesp.br

M. Th. van Genuchten
Department of Nuclear Engineering, Federal University of Rio de Janeiro, UFRJ, Rio de Janeiro, RJ, Brazil
e-mail: vangenuchten@nuclear.ufrj.br

R. J. Lenhard
San Antonio, TX, USA

© The Author(s) 2024
J. García-Rincón et al. (eds.), *Advances in the Characterisation and Remediation of Sites Contaminated with Petroleum Hydrocarbons*, Environmental Contamination Remediation and Management, https://doi.org/10.1007/978-3-031-34447-3_3

3.1 Introduction

Soil hydraulic properties are very important for predicting the subsurface behavior of fluids, including light nonaqueous phase liquids (LNAPLs) that may accumulate above the water-saturated zone and potentially result in groundwater contamination. Accounting for pore-size distributions and fluid properties is critical. Relatively small differences in either may produce erroneous predictions. Models that predict subsurface LNAPL behavior commonly use capillary pressure–saturation (S-P) relations to describe fluid saturations and to estimate fluid relative permeabilities. S-P relationships and the size distribution of pores govern subsurface LNAPL volumes. Two of the most common S-P functions are those by Brooks and Corey (1964) and van Genuchten (1980), both of which were developed initially for porous media having normal or lognormal pore-size distributions.

Integrating LNAPL saturations over a vertical slice of the subsurface yields subsurface LNAPL-specific volumes, which are important for planning remedial cleanup activities (Wickramanayaque et al. 1991). Governmental regulatory agencies commonly require removing the LNAPL mass to the maximum extent practicable. Both the volume and architecture of the LNAPL body influence the performance of any remediation effort. Several authors have proposed methods to calculate subsurface LNAPL-specific volumes following accidental spills on the soil surface or by leaking underground storage tanks, mostly by utilizing fluid levels measured in monitoring wells (e.g., Charbeneau et al. 2000; Farr et al. 1990; Lenhard and Parker 1990; Lenhard et al. 2017; Sleep et al. 2000). Closed-form analytical equations are obtained when the Brooks and Corey S-P model is used to predict subsurface LNAPL-specific volumes (Farr et al. 1990; Lenhard and Parker 1990). Numerical integration, on the other hand, is needed when the van Genuchten S-P model is employed to predict subsurface LNAPL volumes (Parker and Lenhard 1989). Throughout this chapter, we will use the terms soils and porous media interchangeably, noting that pedogenesis processes may substantially influence hydraulic properties (Fogg and Zhang 2016).

An unresolved issue still is that all porous media do not necessarily have normal or lognormal pore-size distributions. Many natural porous media have multimodal pore-size distributions as shown by Peters and Klavetter (1988), Durner (1994), Romano et al. (2011), Wijaya and Leong (2016), Madi et al. (2018), among many others. The focus of this chapter is comparing predictions of subsurface LNAPL distributions in porous media when unimodal and multimodal pore-size distributions are assumed. We present our results after discussing the literature and data used for our calculations.

3.2 Pore Structures

Soils are made up of mostly mineral particles of different sizes (e.g., sand, silt, and clay). In their natural state, these components are often aggregated by means of a series of physical, chemical, and/or biological processes to create structures of different shapes and sizes. The aggregates commonly possess internal voids or pores of varying sizes, leading to a dual-porosity porous medium. Such a medium typically has larger (interaggregate) pores between the aggregates and smaller (intra-aggregate) pores within the aggregates. Pore sizes may range from macropores (larger than 50 nm) to mesopores (between 2 and 50 nm) to micropores (smaller than 2 nm) according to the International Union of Pure and Applied Chemistry (IUPAC).

Pore sizes in soils can be measured by various techniques as described by Dane and Topp (2002), such as mercury porosimetry, thermal porosimetry, nitrogen sorption, microscopy or x-ray tomography. Soils typically have a normal or lognormal pore-size distribution, generally leading to unimodal characteristics since most soils are made up of well-graded particles. However, many soils have more complex pore systems with bimodal or multimodal pore-size distributions (Peters and Klavetter 1988; Romano et al. 2011; Wijaya and Leong 2016; Madi et al. 2018).

Pore sizes govern the ability of porous media to retain fluids against fluid pressures. For example, fine-textured soils with their higher percentages of clay (> 50%) can drain water at low capillary pressures (0.01 MPa) similar as coarser-textured soils. This behavior is more related to the pore-size distribution than the grain size of the porous media. Bimodal or multimodal pore-size distributions can be found also in structured soils and fractured rocks containing large pores (fractures) interspersed with smaller and less permeable micropores of the soil or rock matrix (Peters and Klavetter 1988; Gerke and van Genuchten 1993). Multimodal pore-size distributions occur because of irregular particle-size distributions, the creation of secondary porosity as a result of genetic or other processes (such as physical or chemical aggregation, or biological processes), and/or the effects of glaciation (moraine) or solifluction. Many tropical soils (such as Ultisols and Oxisols) and volcanic soils (Andisols) frequently exhibit multimodal pore-size distributions (Spohrer et al. 2006; Alfaro Soto et al. 2008; Rudiyanto et al. 2013; Seki et al. 2021).

Pore sizes and their geometry affect S-P relations. Soils with unimodal and multimodal pore-size distributions often have very dissimilar S-P relationships. Mathematical functions used to describe unimodal pore-size distributions, hence, may be inadequate for describing multimodal pore-size distributions without adjustments. The standard Brooks and Corey (1964) and van Genuchten (1980) S-P models are applicable to media having normal or lognormal pore-size distributions, but may be inappropriate for predicting LNAPL distributions in porous media having multimodal pore- or particle-size distributions.

3.3   Water and LNAPL Saturations

In this section, several fluid and soil properties such as interfacial tensions, wettability, and capillary pressures are described briefly since they are important for calculating subsurface fluid saturation distributions. Additional information can be found in Chap. 2. Our discussion refers to continuous fluid phases (no fluid entrapment or fluid films).

When more than one fluid phase exists in a porous medium, interfaces will develop between the immiscible fluid phases. At the interfaces, there will likely be an imbalance of attractive forces causing the interfaces to contract, thus forming an interfacial tension (commonly called surface tension when between a gas and a liquid). A difference in fluid pressure across the interfaces will cause them to have a curvature with the convex side having the greater fluid pressure. The degree of curvature depends on the magnitude of the pressure difference between the immiscible fluid phases.

Wettability is the preference of a fluid to spread or adhere to a surface. Generally, the surface is considered to be a solid (porous media grains and/or aggregates). Several different types of wettability may exist, such as fractional wettability and mixed wettability (Tiab and Donaldson 1996), but these will not be discussed in this chapter. We will focus on porous media having a singular or constant wettability. Fluids that wet solid surfaces in preference to other fluids will make a contact angle less than 90 degrees when measured from the solid surface, and are then referred to as the wetting fluid. When a fluid spreads completely on a solid surface, the contact angle is 0 degrees. Different levels of wetting have been recognized. Strong wetting is when the contact angle is less than 30 degrees, moderate wetting when the contact angle is 30°–75°, and neutral wetting when the contact angle is 75°–105°. The other fluid in a two-fluid system is then referred to as the nonwetting fluid.

In porous media containing three fluids (e.g., air, a LNAPL, and water), wettability preference can refer also to which fluid will spread on another fluid (Parker et al. 1987; Tiab and Donaldson 1996). The fluid that wets the solids is the wetting fluid. The fluid that wets the wetting fluid is typically referred to as having intermediate wettability. The fluid wetting the intermediate wetting fluid is referred to as the nonwetting fluid. Neglecting fluid films, wettability dictates which fluid will preferentially occupy what pore sizes. The wetting fluid will generally occupy the smallest pores and the nonwetting fluid the largest pores. Fluids with intermediate wettability will occupy the intermediate-sized pores.

The fluid pressure difference across the interface of immiscible fluid phases is called the capillary pressure. If the capillary pressure is 0, then the interface will not be curved (such as a LNAPL on top of water in a jar). If the capillary pressure is greater than 0, then the interfaces will be curved with the relative wetting fluid on the concave side of the curvature. For example, water will be on the concave side of a curved air–water interface in a capillary tube inserted in a pool of water. Water would be the wetting fluid at a sub-atmospheric pressure, and air would be the nonwetting fluid at atmospheric pressure. The capillary pressure across interfaces is related to the interfacial tension between the relative wetting and nonwetting fluids and the radii of curvature of the interfaces in the orthogonal directions. This relationship is called the Laplace equation of capillarity (Corey 1994):

$$P_c = \sigma \left(\frac{1}{R_1} + \frac{1}{R_2} \right) \tag{3.1}$$

where σ is the interfacial tension of the fluid pair at the interface, and R_1 and R_2 are the radii of curvature of the interface in orthogonal directions.

The radii of curvature of the interfaces (R) can be related to pore radii by the contact angle the wetting fluid makes with the porous media solids such as

$$R = r/\cos \gamma \tag{3.2}$$

where r is the pore radius and γ the contact angle. If the cross-sectional area across a fluid interface is assumed to be circular, in which case $R_1 = R_2$, the Laplace equation reduces to

$$P_c = 2\sigma \cos \gamma /r \tag{3.3}$$

which is commonly given in textbooks. The above shows that the capillary pressure can be used to index pore sizes in which fluid interfaces occur, and to determine fluid saturations.

3.4   Capillary Pressure–Saturation Curves

If two immiscible fluids are present in a porous medium, fluid interfaces only occur between the two fluids. For an air–water fluid system, the fluid interfaces are between air and water. The degree of curvature of the air–water interfaces (i.e., the air–water capillary pressure) can index what pore sizes are occupied by which fluids (the Laplace equation of capillarity and wettability). The capillary pressure will then index the fluid saturations. Air–LNAPL–water fluid systems, where water is the wetting fluid and LNAPL has intermediate wettability between air and water, will have two types of fluid interfaces: one set of interfaces between the LNAPL and water and one set between air and the LNAPL. The radii of curvature of the two sets of interfaces (i.e., the capillary pressures) will be different because the fluids will occupy different pore sizes. Under this wettability (water is the wetting fluid), the LNAPL–water capillary pressure has long been known to index the water saturation. The air–LNAPL capillary pressure then indexes the total-liquid (LNAPL plus water) saturation as advanced by Leverett (1941). The difference between total-liquid saturation and water saturation is the LNAPL saturation.

The relationships between capillary pressure and saturation can be measured by either increasing or decreasing the capillary pressure or the fluid saturation and measuring the changes in fluid saturation or capillary pressure. When the air–water (for a two-fluid system) or LNAPL–water (for a three-fluid system) capillary pressure in water-wet porous media is increased from the completely water-saturated state, the resulting saturation path is called a drying (or drainage) process with regard to the wetting fluid. When the air–water (two-fluid system) or LNAPL–water (three-fluid system) capillary pressure is decreased, it is a wetting (or imbibition) process with regard to the wetting fluid. An example of an S-P curve for an air–water system is shown in Fig. 3.1 for a unimodal pore-size distribution when water is the wetting fluid. For the water drainage curve beginning at $S_w = 1$, the water saturation, at which there is no further water drainage when the air–water capillary pressure is increased, is shown as S_{wr}. Commonly, S_{wr} is called the residual (or irreducible) water saturation. For the water wetting (imbibition) curve beginning at S_{wr}, air can be occluded by water during water imbibition (i.e., air can become discontinuous by being surrounded by water). The maximum occluded nonwetting fluid saturation (i.e., air) at a capillary pressure of 0 is shown as S_{nr}. In the petroleum industry, the occluded nonwetting fluid is also called residual saturation, but its use generally is restricted to two-fluid systems (i.e., brine oil reservoirs). In the hydrology literature, S_{nr} is also called a residual saturation. In air–LNAPL–water systems, however, LNAPL can become occluded by water and, also, can become essentially irreducible in pore wedges and bypassed pores following LNAPL drainage in the vadose zone. To distinguish between these two processes, Parker and Lenhard (1987) called discontinuous LNAPL occluded by water as entrapped LNAPL, and referred to essentially irreducible LNAPL in the vadose zone as residual LNAPL, which is similar to water residual saturation (i.e., irreducible saturations as the capillary pressures are increased). One can have be entrapped LNAPL in both the vadose and the water-saturated zones.

The water drainage curve beginning at $S_w = 1$ in Fig. 3.1 and the water wetting curve beginning at $S_w = S_{wr}$ have been called main curves. Any wetting fluid drainage curve initiated from $S_w < 1$ or wetting process initiated from $S_w > S_{wr}$ is called a scanning curve. We note here also that S-P relationships are not unique. They are usually subject to hysteresis, which means that different S-P relations will occur depending on whether the fluids are draining or wetting. Nonwetting fluid entrapment, contact angle changes, and pore-size contrasts (the "ink-bottle effect" as described by Hillel 1980) are some reasons for hysteresis in S-P relations.

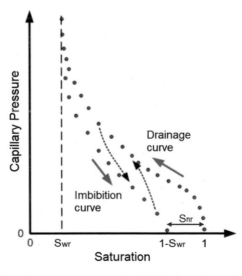

Fig. 3.1 Hypothetical capillary pressure–saturation (S-P) curve when water is the wetting fluid

Mathematical models are often used to describe the S-P relationships. Common models are those by Brooks and Corey (1964), van Genuchten (1980), and Kosugi (1996). Here, we limit ourselves to van Genuchten's formulation given by

$$\overline{S} = \left[1 + (\alpha h_c)^n\right]^{-m} \tag{3.4}$$

in which α, n, and m are quasi-empirical shape parameters, h_c is the capillary pressure head, and $\overline{S}$ is the effective wetting-fluid saturation given by:

$$\overline{S} = (S - S_r)/(1 - S_r) \tag{3.5}$$

where S is the actual wetting-fluid saturation and S_r the residual (or irreducible) wetting-fluid saturation. Equation (3.5) is generic for any fluid system and assumes that no entrapped (residual) nonwetting fluid is present (e.g., entrapped air in an air–water system).

For air–LNAPL–water fluid systems where water is the wetting fluid and LNAPL has intermediate wettability between air and water, two S-P relations are needed. One is the LNAPL–water S-P relationship, which relates the LNAPL–water capillary pressure to the water saturation. The second is the air–LNAPL S-P relationship, which relates the air–LNAPL capillary pressure to the total-liquid saturation. Both are needed to determine LNAPL saturations. When not considering entrapment of air and LNAPL, the effective water and effective total-liquid saturations in three-fluid air–LNAPL–water systems can be determined similarly to Eq. (3.5). The effective water saturation ($\overline{S}_w$) can be determined by replacing S and S_r in Eq. (3.5) by water saturation (S_w) and the residual water saturation ($\overline{S}_{wr}$), respectively, as

$$\overline{S}_w = (S_w - S_{wr})/(1 - S_{wr}) \tag{3.6}$$

The effective total-liquid saturation ($\overline{S}_t$) can be determined by replacing S and S_r in Eq. (3.5) by the total-liquid saturation (S_t), which is the sum of the water and LNAPL saturations, and the residual water saturation (S_{wr}), respectively, as

$$\overline{S}_t = (S_t - S_{wr})/(1 - S_{wr}) \tag{3.7}$$

Figure 3.1 is typical for porous media having unimodal pore-size distributions. Alfaro Soto et al. (2008) measured the air–water capillary pressure–water content relations of soils from the state of São Paulo, Brazil that showed characteristics of bimodal or trimodal pore-size distributions (Fig. 3.2a). The curves are different from the unimodal curve shown in Fig. 3.1. Many formulations for multimodal S-P curves have been proposed over the years (Ragab et al. 1981; Smettem and Kirkby 1990; Othmer et al. 1991; Mallants et al. 1997; Alfaro Soto et al. 2008; Rudiyanto et al. 2013; Li et al. 2014). One of the best-known models was formulated by Durner (1994), based on van Genuchten's (1980) unimodal S-P model:

$$\overline{S} = \sum_{i=1}^{j} w_i \left[1 + (\alpha_i h_c)^{n_i}\right]^{-m_i} \tag{3.8}$$

where j represents the number of subsystems composing the total pore distribution; w_i are subcurve weight factors varying between 0 and 1, which sum to unity; and α_i, n_i, and m_i are the van Genuchten parameters of each subsystem. Similar combinations can be constructed using Brooks and Corey (1964) type subfunctions, including a mixture of different types of subfunctions (e.g., Zhang and Chen 2005; Dexter et al. 2008; Li 2014; Seki 2021). The fitted curves in Fig. 3.2a were obtained using multiple van Genuchten (1980) subcurves as expressed by Eq. (3.8).

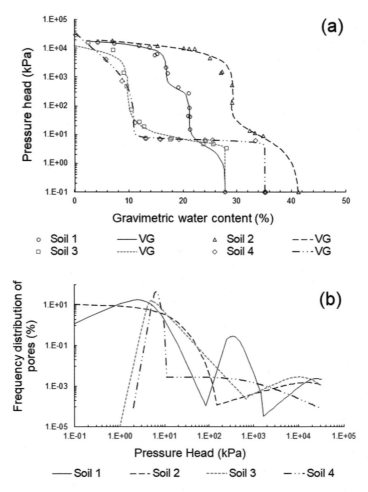

Fig. 3.2 Multimodal characteristics of soils from the state of São Paul, Brazil: **a** Capillary pressure–saturation curves, **b** pore-size distribution curves

The differences between the various curves in Fig. 3.2a are more clearly shown by their equivalent lognormal pore-size distributions given by $P(h_c) = dS(h_c)/dlog(h)$ (Fig. 3.2b). The plots reflect the bimodal and trimodal structures of the soils. Figure 3.2a highlights the limitations of standard unimodal models (Fig. 3.1) when describing S-P relationships of aggregated soils or fractured rocks.

To investigate how the predictions of subsurface LNAPL distributions may differ using unimodal and multimodal S-P formulations, we considered two lateritic soils with multimodal pore structures. Alfaro Soto et al. (2019) extracted data for these two soils (Soils A and F) from Mallants et al. (1997). Figure 3.3 shows the air–water capillary pressure data and best-fit unimodal and multimodal S-P curves for both soils. The unimodal S-P curves are based on Eq. (3.4), and the multimodal curves (bi- and trimodal) are based on Eq. (3.8). Figure 3.4 further shows the lognormal pore-size distributions, obtained again using $P(S) = dS(h)/dlog(h)$, for the soils in Fig. 3.3 as a function of $pF = \log(h_c)$ with h_c in cm. The two figures exhibit clear differences between the unimodal and multimodal curves.

Figure 3.3a shows that the unimodal and multimodal S-P curves of the sandy loam soil are very similar (Fig. 3.3a) for capillary heads less than about 0.55 m ($pF = 1.75$). Between capillary heads of about 0.55–1.6 m ($pF = 1.75$–2.2), the multimodal S-P

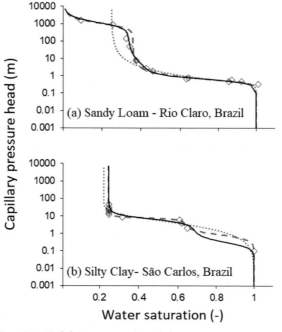

Fig. 3.3 Measured capillary pressure head–saturation data and optimized (best-fit) curves assuming unimodal, bimodal, and trimodal pore-size distributions for **a** A sandy loam soil from Rio Claro, Brazil, and **b** a silty clay soil from São Carlos, Brazil (Alfaro Soto et al. 2019)

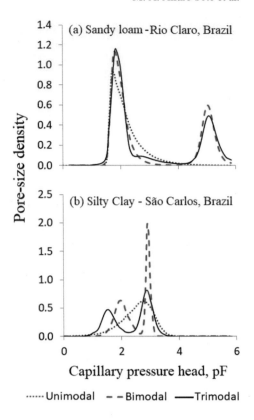

Fig. 3.4 Relative pore-size distributions assuming unimodal, bimodal, and trimodal pore systems for (**a**) A sandy loam from Rio Claro, Brazil and (**b**) a silty clay from São Carlos, Brazil (Alfaro Soto et al. 2019)

curves are slightly to the left of the unimodal *S-P* curve, indicating a somewhat greater volume of relatively large pores for the multimodal pore-size predictions within the range of corresponding pore sizes. The unimodal and multimodal *S-P* curves match again at a capillary head of around 1.6 m (pF = 2.2). At this point, the volume of pores larger than sizes corresponding to a capillary head of about 1.6 m is the same for the unimodal and multimodal *S-P* curves. Figure 3.4a shows the calculated pore-size density distributions of the sandy loam soil for the unimodal and bimodal sols, with the predicted distributions for the multimodal *S-P* formulation showing a greater density of pores with sizes corresponding to capillary heads of 0.55–1.6 m (pF = 1.75–2,2). Hence, a larger volume of pores is predicted (Fig. 3.4a) with the multimodal *S-P* curves in Fig. 3.3a for capillary heads between 0.55 and 1.6 m (Fig. 3.3a).

The smaller pores between capillary pressure heads of about 1.6 m (pF = 2.2) and 1000 m (pF = 5) occupy less volume when predicted with the unimodal *S-P* curve as compared to the multimodal *S-P* curves. For the range of pore sizes corresponding to capillary heads between 1.6 and 100 m (pF = 2.2–4.0), Fig. 3.4a shows a higher pore density for the unimodal *S-P* curve, but much less than the pore densities corresponding to capillary heads between 0.55 and 1.6 m. At a pF of about 4.0 in Fig. 3.4a, the pore densities predicted for the unimodal *S-P* curve show very low or negligible values to the end of the analyses, while the densities for the multimodal *S-P* curves increase again at a pF of 4.5, thus starting to form a second section of higher pore densities. After a pF of approximately 5, the pore densities of the multimodal distributions then begin decreasing until a pF of 6 (Fig. 3.4a), which is the end of the analyses. Still, based on the vertical shape of the unimodal *S-P* curve in Fig. 3.3a between capillary heads of 100 and 10,000 m, many pores smaller than those corresponding to a capillary head of 10,000 m are assumed to be present. For the multimodal *S-P* curves, almost all pores are larger than a size corresponding to a capillary head of 10,000 m since the associated water saturation value approaches zero.

For the silty clay soil, the unimodal and multimodal *S-P* curves are mostly very similar (Fig. 3.3b), except for capillary heads between 0.1 m and about 10 m (pF = 1–3). The multimodal *S-P* curves predicted larger volume of pore sizes within the range corresponding to capillary heads of 0.1 to 10 m (pF = 1–3), with the trimodal curve predicting a higher volume of these pores than the bimodal curve. Figure 3.4b shows that the predictions from the unimodal and multimodal *S-P* curves for the silty clay soil are very different for the pF range 1–3, clearly reflecting the different unimodal and multimodal pore structures. Within the pF range 1–3 (Fig. 3.4b), the multimodal curves show higher pore densities than the unimodal curve, especially in the larger pore sizes (i.e., lower pF values). Therefore, the multimodal *S-P* curves predict a higher volume of larger pores than the unimodal *S-P* curve here.

As noted by Mallants et al. (1997), the sandy loam soil showed a system of secondary (interaggregate) pores, and the silty clay soil a system of smaller primary (intra-aggregate) pores. The question is whether the differences in using unimodal pore-size *S-P* curves will yield dissimilar predictions of fluid behavior and subsurface LNAPL distributions, when using multimodal pore-size *S-P* curves. Whereas the unimodal and multimodal *S-P* curves are similar in some respects, will the differences yield dissimilar predictions, which will be investigated in the following section. This is addressed next.

3.5 Estimating LNAPL Saturations and Volumes from In-Well Thickness

The *S-P* relationships of porous media are useful for predicting the subsurface behavior of LNAPLs and for assessing subsurface LNAPL distributions to aid in designing LNAPL remedial operations. The relationships can be used in numerical multiphase flow and transport models for predicting transient fluid behavior. Assuming vertical equilibrium conditions, fluid levels in nearby monitoring wells can be employed to predict subsurface LNAPL distributions from *S-P* relationships. The predicted LNAPL and water saturations could then be utilized to predict LNAPL relative permeabilities and transmissivity as discussed in Chaps. 2 and 12.

Early approaches to calculate subsurface LNAPL volumes assumed that the LNAPL thickness in the subsurface is equal to that in the monitoring well. Later investigations, however, showed that the LNAPL thickness in a well is not a direct measurement of the LNAPL thickness in the subsurface (de Pastrovich et al. 1979; Mercer and Cohen 1990). Furthermore, using simple ratios of the LNAPL thickness in the subsurface to the LNAPL thickness in a well do not provide accurate predictions of actual LNAPL subsurface volumes (e.g., Hampton and Miller 1988).

Parker and Lenhard (1989), Farr et al. (1990) and Lenhard and Parker (1990) were the first to develop equations for calculating the volume of LNAPL in a vertical slice (the LNAPL-specific volume) of water-wet porous media using capillary pressure–saturation relationships, fluid and porous media properties, and the LNAPL thickness in a well. Parker and Lenhard (1989) also proposed equations for estimating the subsurface LNAPL transmissivity. The approach assumes good contact between the fluids in the subsurface and the well, and that the fluids are vertically static. Parker and Lenhard (1989), Farr et al. (1990), and Lenhard and Parker (1990) assumed all of the LNAPL to be mobile and continuous (i.e., nonhysteretic conditions). Later, API (2007), Jeong and Charbeneau (2014), Lenhard et al, (2017), and Wadill and Parker (1997) developed models accounting for immobile LNAPL in the unsaturated vadose and liquid-saturated zones.

The early and later models assumed applicability of unimodal pore-size distribution S-P models, such as the Brooks and Cory (1964) and van Genuchten (1980) equations. To show potential differences in the predictions of LNAPL saturation distributions when unimodal versus multimodal pore-size S-P formulations are used, Alfaro Soto et al. (2019) modified the API (2007) predictive model to include multimodal pore-size S-P formulation. They incorporated Durner's model for the multimodal pore-size S-P formulations. For further details of the modifications, readers are referred to Alfaro Soto et al. (2019).

Figure 3.5 shows predicted air, LNAPL, and water saturations as determined from fluid levels in a nearby monitoring well using the API (2007) model assuming a unimodal pore-size S-P formulation and the modified API (2007) model with a multimodal pore-size S-P formulation for a hypothetical soil with a heterogeneous pore system (i.e., the multimodal pore-size distribution). Figure 3.5a shows the hypothetical fluid levels in the monitoring well, and Fig. 3.5b, c show predicted fluid saturations using the unimodal pore-size S-P formulation and multimodal pore-size S-P formulation, respectively. As noted earlier, porous media with heterogeneous pore systems containing secondary porosity or multiple systems of pores may not be correctly represented by sigmoidal-type capillary pressure curves.

Alfaro Soto et al. (2019) calculated fluid saturations for a hypothetical scenario as a function of elevation using the API (2007) model for 2 m of LNAPL in nearby wells. They assumed air–water, air–LNAPL and LNAPL–water interfacial tensions of 65, 25, and 15 mN m^{-1}, respectively, and a LNAPL specific gravity of 0.8. In the API (2007) model, they used parameters for both unimodal and multimodal pore-size S-P formulations obtained by curve fitting to measured data (Fig. 3.3). Figure 3.6 shows the predicted fluid distributions for the sandy loam soil, while predictions for the silty clay soil are shown in Fig. 3.7. In both Figs. 3.6 and 3.7, the upper plots (a)

Fig. 3.5   a Schematic of the elevation of fluids in a monitoring well. The elevations z_{gs}, z_{ao}, z_{ow}, and z_{aw} correspond to a datum located at or below z_{ow}, the height of the air–LNAPL interface in the well, the height of the LNAPL–water interface in the well, and the height of the water table without a LNAPL, respectively, b_o represents the thickness of the LNAPL layer in the well, and h_o and h_w are the distance between the reference level and z_{ao} and z_{aw}, respectively (after Farr et al. 1990); **b** representative distribution of LNAPL in a unimodal porosity medium; **c** representative distribution of LNAPL in a trimodal porosity medium. (Adapted from Alfaro Soto et al. 2019)

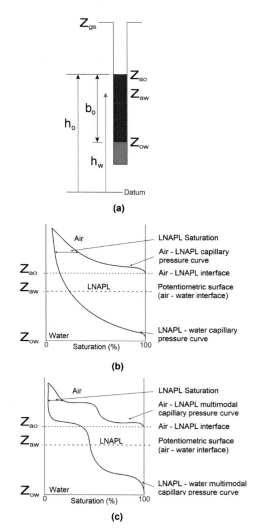

reflect predictions for the unimodal pore-size S-P formulation, and the lower plots (b) reflect the multimodal pore-size S-P formulation. The dashed lines in Figs. 3.6 and 3.7 furthermore indicate the elevation of the water table (i.e., where the water pressure is atmospheric) in a well screened only in the water-saturated zone.

Examination of Figs. 3.6 and 3.7 shows that the elevations of the LNAPL–water and air–LNAPL interfaces in the well are approximately 0.38 and 2.4 m, respectively. Using these values, the elevation of the air–water interface in a well screened only in the water-saturated zone can be calculated following Lenhard and Parker (1990) as

$$z_{aw} = (1 - \rho_{ro})z_{ow} + \rho_{ro}z_{ao} \qquad (3.9)$$

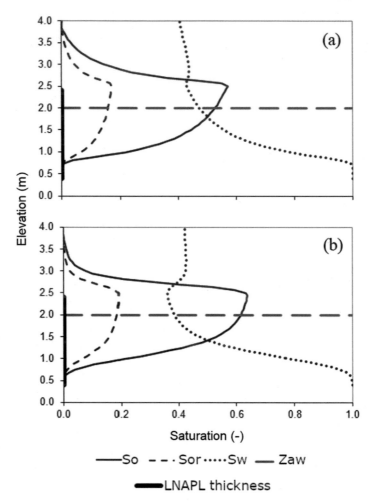

Fig. 3.6 Calculated LNAPL, water, and immobile LNAPL saturation profiles for the sandy loam soil in Figs. 3.3 and 3.4 using **a** Unimodal and **b** multimodal pore-size distributions based on the fitted optimized S-P curves (modified from Alfaro Soto et al. 2019)

where z_{aw}, z_{ow}, and z_{ao} are the elevations of the air–water, LNAPL–water, and air–LNAPL interfaces in wells, respectively, and ρ_{ro} is the LNAPL-specific gravity. From Eq. (3.9), the elevation of the air–water interface in the well is predicted to be 2.00, which agrees with Fig. 3.6 and 3.7 and confirms the estimates of the LNAPL–water and air–LNAPL interface elevations in the well.

LNAPL–water and air–LNAPL capillary heads can be calculated as a function of elevation and fluid levels in a well for equilibrium (static) conditions following Lenhard and Parker (1990) as

$$h_{ow} = (1 - \rho_{ro})(z - z_{ow}) \tag{3.10}$$

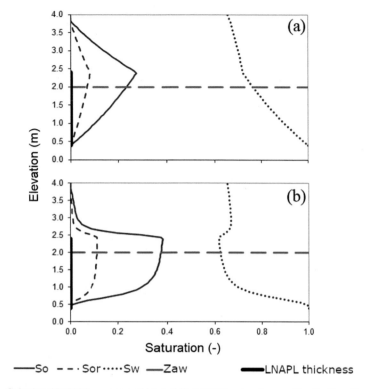

Fig. 3.7 Calculated LNAPL, water, and immobile LNAPL saturation profiles for the silty clay soil in Figs. 3.3 and 3.4 using **a** Unimodal and **b** multimodal pore-size distributions based on the fitted optimized *S-P* curves (modified from Alfaro Soto et al. 2019)

$$h_{\mathrm{ao}} = \rho_{\mathrm{ro}}(z - z_{\mathrm{ao}}) \tag{3.11}$$

where h_{ow} is the LNAPL–water capillary head, h_{ao} is the air–LNAPL capillary head, and z is elevation. The LNAPL–water and air–LNAPL capillary heads can then be scaled to yield equivalent air–water capillary heads (Parker et al. 1987). The scaling factors as estimated in API (2007) follow Lenhard and Parker (1987). An alternative scaling format is presented in Lenhard et al. (2017). The scaled LNAPL–water capillary heads estimate water saturations and the scaled air–LNAPL capillary heads total-liquid saturations using an air–water *S-P* curve, such as in Fig. 3.3.

Figure 3.6 shows that the unimodal and multimodal pore-size *S-P* formulations yield different predictions of subsurface fluid saturations above the water-saturated zone for the sandy loam soil. We assume the predicted fluid distributions in Figs. 3.6 and 3.7 are equilibrium (static) distributions so calculated scaled capillary heads can predict fluid distributions using Fig. 3.3. At an elevation of 2.5 m in Fig. 3.6a for the unimodal pore-size saturation distribution, which is only slightly higher than the LNAPL free surface, the LNAPL saturation is about 0.56 and the water saturation is about 0.44 for a total-liquid saturation of 1. The air–LNAPL capillary head at the 2.5 m elevation is 0.08 m [using Eq. (3.11)]. The scaled air–LNAPL capillary head is between 0.2 to 0.3 m, depending on the scaling format used. Based on the unimodal S-P curve in Fig. 3.3, the total-liquid saturation (S_t) should be 1 for these scaled air–LNAPL capillary heads. The LNAPL–water capillary head at 2.5-m elevation is 0.42 m [using Eq. (3.10)]. The scaled LNAPL–water capillary head is between 1.1 and 1.6 m. For this scaled LNAPL–water capillary head range, the water saturation (S_w) should be between about 0.46 and 0.56 using Fig. 3.3. Therefore, the LNAPL saturation should be between 0.44 (i.e., $S_t - S_w = 1 - 0.56$) and 0.54 (i.e., $S_t - S_w = 1 - 0.46$) using the scaled capillary pressures and Fig. 3.3, which is close to that estimated from Fig. 3.6a.

At 50 cm higher (elevation 3.0 m) in Fig. 3.6a, the LNAPL saturation is about 0.15 and the water saturation is about the same as for the 2.5-m elevation (i.e., 0.44) for a total-liquid saturation of about 0.59. It is not clear why the water saturation is not predicted to be lower in Fig. 3.6a at the 3.0 m-elevation because the LNAPL–water capillary head is greater than at the 2.5-m elevation. The air–LNAPL capillary head at the 3.0 m elevation is 0.48 m. The scaled air–LNAPL capillary head is between 0.8 and 1.25 m. For this scaled air–LNAPL capillary head range, the total-liquid saturation should be between about 0.52 and 0.72. The LNAPL–water capillary head at 3.0-m elevation is 0.52 m. The scaled LNAPL–water capillary head is between 1.4 and 2.25 m. For this scaled LNAPL–water capillary head range, the water saturation should be between about 0.42 and 0.54. The total-liquid, LNAPL, and water saturations from using the scaled capillary pressures and Fig. 3.3 agree with those estimated from Fig. 3.6a.

For the multimodal pore-size saturation distribution (Fig. 3.6b), the calculated air–LNAPL and LNAPL–water capillary heads and scaled capillary heads at the 2.5- and 3.0-m elevations are the same as for the unimodal pore-size saturation distribution (Fig. 3.6a) because z_{ow} and z_{ao} are unchanged. At an elevation of 2.5 m in Fig. 3.6b for the multimodal pore-size distribution, the LNAPL saturation is about 0.63 and the water saturation about 0.37 for a total-liquid saturation of 1. The scaled air–LNAPL capillary head, which is equal to an equivalent air–water capillary head, is between 0.2 and 0.3 m. For this scaled air–LNAPL capillary head range, the total-liquid saturation should be 1 using Fig. 3.3. The scaled LNAPL–water capillary head is between 1.1 and 1.6 m. For this scaled LNAPL–water capillary head range, the water saturation should be between about 0.44 and 0.5.

At 50 cm higher (elevation 3.0 m), the LNAPL saturation is about 0.08 and the water saturation is about 0.42 for a total-liquid saturation of about 0.5. The air–LNAPL capillary head at the 3.0-m elevation is 0.48 m. The scaled air–LNAPL capillary head is between 0.8 and 1.25 m. For this scaled air–LNAPL capillary head range, the total-liquid saturation should be between about 0.48 and 0.72. The LNAPL–water capillary head at 3.0-m elevation is 0.52 m. The scaled LNAPL–water capillary head is between 1.4 and 2.25 m. For this scaled LNAPL–water capillary head range, the water saturation should be between about 0.42 and 0.5. The total-liquid, LNAPL, and water saturations from using the scaled capillary pressures and Fig. 3.3 agree with those estimated from Fig. 3.6b.

There is reasonable agreement between the unimodal and multimodal pore-size saturation distributions in Fig. 3.6 and saturations estimated from calculated scaled capillary heads using Fig. 3.3. This because the S-P curves are relatively similar for air–water (scaled) capillary heads less than a few meters. However, because the multimodal S-P curve in Fig. 3.3 is slightly to the left of the unimodal S-P curve for the scaled capillary head range of 0.55–1.6 m, water saturations are less for the multimodal pore-size saturation distribution than the unimodal pore-size saturation distribution, which results in a larger LNAPL saturation for the multimodal pore-size saturation distribution when total-liquid saturation is 1. The slight differences in the unimodal and multimodal S-P curves can yield differences in predictions of subsurface LNAPL distributions, especially if the differences involve larger pores. When LNAPL accumulates above the water-saturated zone, LNAPL will always occupy the larger pores if it has intermediate wettability between water and air.

The differences between the unimodal and multimodal predictions in Fig. 3.6 can be explained also in another way. For equilibrium (static) conditions, the air–LNAPL capillary head is related to the LNAPL–water capillary head at a given elevation using Eqs. (3.10) and (3.11), i.e., for static z_{ow} and z_{ao}. Consider Fig. 3.3 at an air–water capillary head of 1 m where the unimodal and multimodal S-P curves begin to show differences. From the scaling format of Parker et al. (1987), the air–water capillary head is the scaled capillary head, which can be used to calculate the LNAPL–water capillary head (h_{ow}) by dividing by the LNAPL–water scaling factor. This leads to a value of 0.23 m for h_{ow} when the Lenhard and Parker (1987) and API (2007) LNAPL–water scaling factor is used, and a value of 0.37 m for h_{ow} when the Lenhard et al. (2017) LNAPL–water scaling factor is used. Using these values in Eq. (3.11) gives either an elevation (z) of 1.53 m for the Lenhard and Parker (1987) and API (2007) LNAPL–water scaling factor or an elevation (z) of 2.25 m for Lenhard et al. (2017) LNAPL–water scaling. Both of these elevations are below the air–LNAPL interface elevation in the well at 2.4 m; therefore, the sandy loam soil would be total-liquid saturated at these elevations. The same results are obtained when both elevations are used in Eq. (3.10), i.e., negative air–LNAPL capillary heads result, which indicate that the total-liquid saturation would be 1. For the scaled LNAPL–water capillary head of 1 m, water saturation would be approximately 0.6 for the unimodal S-P curve and approximately 0.54 for the multimodal S-P curves, which shows that higher LNAPL saturations result when using multimodal S-P relations as shown in Fig. 3.6. Even though the unimodal and multimodal S-P curves are generally similar for about 5 m above the water-saturated zone (Fig. 3.3), the slight differences for the larger pores still produced marked differences in the subsurface LNAPL saturation predictions. This may be important for planning remedial operations.

Figure 3.7 shows that the unimodal and multimodal pore-size S-P formulations yield significantly different saturation distributions for the silty clay soil. The calculated air–LNAPL and LNAPL–water capillary heads and scaled capillary heads at the 2.5- and 3.0-m elevations for Fig. 3.7 are the same as for Fig. 3.6, because, z_{ow} and z_{ao} are at the same elevations. For the unimodal pore-size distribution (Fig. 3.7a), the LNAPL saturation at the 2.5-m elevation is about 0.25 and the water saturation about 0.72 for a total-liquid saturation of 0.97. The scaled air–LNAPL capillary head is between 0.2 and 0.3 m. For this scaled air–LNAPL capillary head range, the total-liquid saturation should be close to 1 (i.e., 0.98–1) using Fig. 3.3. The scaled LNAPL–water capillary head is between 1.1 and 1.6 m. For this scaled LNAPL–water capillary head range, water saturation should be between about 0.78 and 0.82.

At a point 50 cm higher (elevation 3.0 m), the LNAPL saturation is about 0.13 and the water saturation about 0.7 for a total-liquid saturation of about 0.83. The air–LNAPL capillary head at 3.0-m elevation is 0.48 m. The scaled air–LNAPL capillary head is between 0.8 and 1.25 m. For this scaled air–LNAPL capillary head range, the total-liquid saturation should be between about 0.8 and 0.9. The LNAPL–water capillary head at 3.0-m elevation is 0.52 m. The scaled LNAPL–water capillary head is between 1.4 and 2.25 m. For this scaled LNAPL–water capillary head range, water saturation should be between about 0.7 and 0.74. The total-liquid, LNAPL and water saturations from using the scaled capillary pressures and Fig. 3.3 agree with those estimated from Fig. 3.7a.

For the multimodal pore-size saturation distribution (Fig. 3.7b), the LNAPL saturation at the 2.5-m elevation is about 0.14 and the water saturation is about 0.69, giving a total-liquid saturation of 0.83. The scaled air–LNAPL capillary head is between 0.2 and 0.3 m. For this scaled air–LNAPL capillary head range, the total-liquid saturation should be between about 0.87 and 0.91 using Fig. 3.3. The scaled LNAPL–water capillary head is between 1.1 and 1.6 m. For this scaled LNAPL–water capillary head range, the water saturation will be between about 0.67 and 0.7. The total-liquid, LNAPL, and water saturations from using the scaled capillary pressures and Fig. 3.3 are relatively close to those estimated using Fig. 3.7b.

At 50 cm higher (elevation 3.0 m), the LNAPL saturation is about 0.03 and the water saturation about 0.66, for a total-liquid saturation of about 0.69. The air–LNAPL capillary head at the 3.0-m elevation is 0.48 m. The scaled air–LNAPL capillary head is between 0.8 and 1.25 m. For this scaled air–LNAPL capillary head range, the total-liquid saturation should be between about 0.69 and 0.74. The LNAPL–water capillary head at 3.0-m elevation is 0.52 m. The scaled LNAPL–water capillary head is between 1.4 and 2.25 m. For this scaled LNAPL–water capillary head range, the water saturation is then between about 0.66 and 0.58. The total-liquid, LNAPL, and water saturations from using the scaled capillary pressures and Fig. 3.3 agree with those estimated using Fig. 3.7b.

There is again a reasonable agreement between the unimodal and multimodal pore-size saturation distributions in Fig. 3.7 and saturations estimated from calculated scaled capillary heads using Fig. 3.3. The differences between the unimodal and multimodal S-P curves in Fig. 3.3 show that the multimodal pore-size distribution predicts a larger volume of larger pores. The larger pores will contain more LNAPL and produce the differences in predicted LNAPL distributions seen in Figs. 3.6 and 3.7. When predicting subsurface LNAPL distributions, the size and volume of larger pores have a significant effect on the predictions. Significantly greater amounts of LNAPL can be predicted for relatively small differences in S-P curves. It is important to understand the potential distribution of pore sizes, especially larger pores. The multimodal pore-size S-P formulations were able to capture these effects of the larger pores in the sandy loam and silty clay soils when predicting subsurface LNAPL distributions above the water-saturated zone.

It should be noted that the symbol S_{or} in Figs. 3.6 and 3.7 denotes relatively immobile LNAPL saturation in the vadose and liquid-saturated zones present in the form of entrapped LNAPL, LNAPL films, and LNAPL wedges. As discussed in Chap. 2, the model by Lenhard et al. (2017) differentiates between relatively immobile LNAPL not occluded by water (residual LNAPL) and LNAPL occluded by water (entrapped LNAPL). The current version of the API (2007) model does not differentiate between the residual and entrapped LNAPL fractions and defines both as "residual LNAPL".

3.6 Conclusions

Knowing the hydraulic properties of porous media is very important for predicting subsurface LNAPL distributions. Correctly characterizing the pore-size distribution, especially the larger pores, is essential for accurately predicting LNAPL saturations. Appropriate *S-P* formulations hence must be used to capture multimodal pore-size structures. Our investigations show that unimodal and multimodal *S-P* formations yield different subsurface LNAPL saturations when LNAPL collects above the water-saturated zone, even when the *S-P* curves appear to be similar at scaled capillary pressures less than a couple of meters, especially for the sandy loam soil. If the larger pores are not properly characterized, then significant differences in predicted LNAPL saturations will be obtained. Our results suggest that the efficiency of certain technologies for recovering LNAPL released to groundwater will depend on the pore structure (i.e., unimodal versus multimodal distributions) of the subsurface. Erroneous predictions of potential LNAPL recoveries using conventional techniques hence may result if the pore system is not defined correctly.

Although more research is needed to identify the influential parameters in the distribution of multimodal pores, it is necessary to recognize that in practice, during the soil characterization and field contamination stages, efforts should be made to identify potential pore structures. Soil aggregate development, minerality, and degree of weathering are important factors. Additionally, the implementation of multimodal formulations is recommended to fully capture the effects of subsurface heterogeneity on the LNAPL distribution. Multimodal pore-size distributions can also affect relative permeability and LNAPL transmissivity estimates, which will impact potential LNAPL recoverability and such processes as vapor migration, mass partitioning, and natural source-zone depletion.

References

Alfaro Soto MA, Lenhard R, Chang HK, van Genuchten MT (2019) Determination of specific LNAPL volumes in soils having a multimodal pore-size distribution. J Environ Manag 237:576–584. https://doi.org/10.1016/j.jenvman.2019.02.077

Alfaro Soto MA, Chang HK, Vilar OM (2008) Evaluation of fractal scaling of sole Brazilian soils. Rev Bras Geociências 8:253–262 (in Portuguese)

API- American Petroleum Institute (2007) light non-aqueous phase liquid distribution and recovery model (LDRM), Publication 4760. https://www.api.org/oil-and-natural-gas/environment/clean-water/ground-water/lnapl/ldrm. Accessed 01 Sept 2021

Brooks RH, Corey AT (1964) Hydraulic properties of porous media, Hydrol. Paper No 3, Colorado State Univ, Fort Collins, CO

Corey AT (1994) Mechanics of immiscible fluids in porous media. Water Resources Publications, Highlands Ranch, Colorado

Charbeneau RJ, Johns RT, Lake LW, McAdams MJ III (2000) Free-product recovery of petroleum hydrocarbon liquids. Groundwater Monit Remediation 20:147–158

Dane JH, Topp GC (eds) (2002) Methods of soil analysis, Part 4 Physical methods. Soil Science of America, Madison, Wisconsin

de Pastrovich TL, Baradat Y, Barthel R, Chiarelli A, Fussel DR (1979) Protection of ground water from oil pollution. CONCAWE, The Hague, p 61

Dexter AR, Czyz EA, Richard G, Reszkowska A (2008) A user-friendly water retention function that takes account of the textural and structural pore spaces in soil. Geoderma 143:243–253. https://doi.org/10.1016/j.geoderma.2007.11.010

Durner W (1994) Hydraulic conductivity estimation for soils with heterogeneous pore structure. Water Resour Res 30:211–223

Farr AM, Houghtalen RJ, McWhorter DB (1990) Volume estimation of light nonaqueous phase liquids in porous media. Ground Water 28:48–56

Fogg GE, Zhang Y (2016) Debates—Stochastic subsurface hydrology from theory to practice: A geologic perspective. Water Resour Res 52(12):9235–9245

Gerke HH, van Genuchten MT (1993) A dual-porosity model for simulating the preferential movement of water and solutes in structured porous media. Water Resour Res 29:305–319

Hampton DR, Miller PDG (1988) Laboratory investigation of the relationship between actual and apparent product thickness in sands. In: Conference petroleum hydrocarbons and organic chemicals in ground water: prevention, detection and restoration. National Water Well Assoc, pp 157–181

Hillel D (1980) Fundamentals of soil physics. Academic Press, New York, pp 153–154

Jeong J, Charbeneau RJ (2014) An analytical model for predicting LNAPL distribution and recovery from multi-layered soils. J Contam Hydrol 156:52–61

Kosugi K (1996) Lognormal distribution model for unsaturated soil hydraulic properties. Water Resour Res 32:2697–2703

Lenhard RJ, Parker JC (1987) Measurement and prediction of saturation-pressure relationships in three-phase porous media systems. J Contam Hydrol 1:407–424

Lenhard RJ, Parker JC (1990) Estimation of free hydrocarbon volume from fluid levels in monitoring wells. Ground Water 28:57–67

Lenhard RJ, Rayner JL, Davis GB (2017) A practical tool for estimating subsurface LNAPL distributions and transmissivity using current and historical fluid levels in groundwater wells: effects of entrapped and residual LNAPL. J Contam Hydrol 205:1–11

Leverett MC (1941) Capillary behavior in porous solids. Trans AIME 142:151–169

Li X, Li J, Zhang L (2014) Predicting bimodal soil–water characteristic curves and permeability functions using physically based parameters. Comput Geotech 57:85–96

Madi R, Rooij GH, Mielenz H, Mai J (2018) Parametric soil water retention models: a critical evaluation of expressions of the full moisture range. Hydrol Earth Syst Sci 22:1193–1219

Mallants D, Tseng PH, Toride N, Timmerman A, Feyen J (1997) Evaluation of multimodal hydraulic functions in characterizing a heterogeneous field soil. J Hydrol 95:172–199

Mercer JW, Cohen RM (1990) A review of immiscible fluids in the subsurface properties, models, characterization, and remediation. J Contam Hydrol 6:107–163

Othmer H, Diekkruger B, Kutilek M (1991) Bimodal porosity and unsaturated hydraulic conductivity. Soil Sci 152:139–149

Parker JC, Lenhard RJ (1987) A model for hysteretic constitutive relations governing multiphase flow: 1 saturation-pressure relations. Water Resour Res 23:2187–2196

Parker JC, Lenhard RJ (1989) Vertical integration of three phase flow equations for analysis of light hydrocarbon plume movement. Transp Porous Media 5:187–206

Parker JC, Lenhard RJ, Kuppusamy T (1987) A parametric model for constitutive properties governing multiphase flow in porous media. Water Resour Res 23:618–624

Peters RR, Klavetter EA (1988) A continuum model for water movement in an unsaturated fractured rock mass. Water Resour Res 24:416–430

Ragab R, Feyen J, Hillel D (1981) Comparative study of numerical and laboratory methods for determining the hydraulic conductivity function of a sand. Soil Sci 134:375–388

Romano N, Nasta P, Severino G, Hopmans JW (2011) Using bimodal lognormal functions to describe soil hydraulic properties. Soil Sci Soc Am J 75:468–480

Rudiyanto Toride N, Sakai M, van Genuchten MT (2013) Estimating the unsaturated hydraulic conductivity of Andisols using the evaporation method. J Jpn Soc Phys 125:3–15. https://js-soilphysics.com/downloads/pdf/125003.pdf

Seki K, Toride N, van Genuchten MT (2021) Closed-form hydraulic conductivity equations for multimodal unsaturated soil hydraulic properties. Vadose Zone J 21. https://doi.org/10.1002/vzj2.20168

Sleep BE, Sehayek L, Chien CC (2000) A modeling and experimental study of light nonaqueous phase liquid (LNAPL) accumulation in wells and LNAPL recovery from wells. Water Resour Res 36:3535–3545

Smettem KRJ, Kirkby C (1990) Measuring the hydraulic properties of a stable aggregate soil. J Hydrol 177:1–13

Spohrer K, Herrmann L, Ingwersen J, Stahr K (2006) Applicability of uni- and bimodal retention functions for water flow modeling in a tropical Acrisol. Vadose Zone J 5:48–58

Tiab D, Donaldson EC (1996) Petrophysics: theory and practice of measuring reservoir rocks and fluid transport properties. Gulf Publishing Company, Houston, Texas, p 706

van Genuchten MT (1980) A closed-form equation for predicting the hydraulic conductivity of unsaturated soils. Soil Sci Soc Am J 44:892–898

Wijaya M, Leong EC (2016) Equation for unimodal and bimodal soil-water characteristic curves. Soils Found 56:291–300. https://doi.org/10.1016/j.sandf.2016.02.011

Wickramanayake GB, Gupta N, Hinchee RE, Nielsen BJ (1991) Free petroleum hydrocarbon volume estimates from monitoring well data. J Environ Eng 117:686–691

Zhang L, Chen Q (2005) Predicting bimodal soil-water characteristic curves. J Geotech Geoenviron Eng 131:666–670. https://doi.org/10.1061/(ASCE)1090-0241(2005)131:5(666)

Chapter 4
The Application of Sequence Stratigraphy to the Investigation and Remediation of LNAPL-Contaminated Sites

Junaid Sadeque and Ryan C. Samuels

Abstract The heterogenous and complex nature of soil and groundwater systems leads to significant technical challenges in remediating Light Non-Aqueous Phase Liquid (LNAPL) contamination and achieving cleanup goals within reasonable timeframes. Therefore, implementing a correlation approach that adequately addresses subsurface heterogeneity between sampling locations is critical for effective management of LNAPL-contaminated sites. In traditional LNAPL remedial investigations in clastic environments, correlation has typically been conducted using 'lithostratigraphy', which connects like-lithologies without recognizing the heterogeneity of subsurface sediments between boreholes. The resulting 'layer cake' stratigraphy of the subsurface is often too simplistic and inadequate for developing effective remedial strategies. In contrast, the sequence stratigraphic approach (supported by facies analysis) provides a more realistic subsurface correlation based on the predictable distribution of sediments in different depositional environments. The three-dimensional geologic framework derived from sequence stratigraphy can be used to map the heterogeneity between coarse- and fine-grained units across multiple scales and beyond the existing site data set. Moreover, this framework can be integrated with site hydrologic and chemical data to identify sediments with high- and low-fluid

The original version of this chapter has been revised: The author's figures and text corrections have been updated. The correction to this chapter can be found at https://doi.org/10.1007/978-3-031-34447-3_19

Supplementary Information The online version contains supplementary material available at https://doi.org/10.1007/978-3-031-34447-3_4.

J. Sadeque (✉) · R. C. Samuels
AECOM, Los Angeles, CA, USA
e-mail: Junaid.Sadeque@aecom.com

R. C. Samuels
e-mail: Ryan.Samuels@aecom.com

transmissive properties. This chapter demonstrates how a sequence-stratigraphy-based conceptual site model can be used to identify preferential LNAPL migration pathways and inform effective remedial decision-making.

Keywords Facies models · Geological heterogeneity · LNAPL conceptual site model · Sequence stratigraphy · Site investigation

4.1 Introduction

4.1.1 The Challenge of Subsurface Heterogeneity on LNAPL Remediation

Subsurface heterogeneity presents significant technical challenges in remediating Light Non-Aqueous Phase Liquid (LNAPL) contamination and achieving cleanup goals within reasonable timeframes. This is evident from a significant number of theoretical and field studies that analyze the impact of sediment heterogeneity on subsurface fluid flow and contaminant transport (e.g., Freeze 1975; Smith and Schwartz 1980; LeBlanc et al. 1991).

In sedimentary systems, LNAPL migration is largely controlled by the spatial distribution of preferential migration pathways and capillary barriers, with the influence of factors such as fluid properties and natural attenuation processes being discussed in Chaps. 2, 3, 5, and 12. In water-wet systems, LNAPL and air preferentially favor larger pores, since LNAPL imbibition into the pore space requires sufficient driving force to exceed the existing capillary forces (CL:AIRE 2014; Farr et al. 1990; Lenhard and Parker 1990; Huntley et al. 1994). As a consequence, LNAPL typically follows the path of least resistance, resulting in an irregular, heterogeneous distribution of LNAPL mass in the subsurface (e.g., Waddill and Parker 1997). Therefore, correctly relating vertical variability of sediments in boreholes with the geometries and spatial distribution of the underlying geological features is essential for investigating LNAPL-impacted sites.

Typically, subsurface geological correlation in remedial investigations is based on connecting 'like-lithology' between boreholes, without due consideration for the dynamics of their depositional processes. This simplistic approach often leads to miscorrelations of aquifers and aquitards.

On the other hand, facies analysis and sequence stratigraphy are presently used in the oil and gas industry as standard practices of subsurface correlation to characterize and predict petroleum reservoirs. This approach is based on determining depositional processes through geologic time rather than only relying on lithologic similarities. While sequence stratigraphy provides the overarching three-dimensional stratigraphic framework, facies analysis addresses the internal heterogeneity within that framework. The same stratigraphic principles can be repurposed to characterize aquifer heterogeneity and reduce uncertainty between and beyond sampling locations. Integration of site hydrologic and chemical data within the three-dimensional geologic framework derived from sequence stratigraphy (supported by facies analysis) provides an efficient tool for contamination investigation and remedial decision-making. This chapter elaborates on how these stratigraphic principles can be applied to clastic aquifers and aquitards and illustrates, with a case study, the application of this method in developing remedial strategies at LNAPL-impacted sites.

4.1.2   Application of Facies Models for Predicting Subsurface Heterogeneity

In sedimentary aquifers, groundwater flow and mass transport/storage are largely controlled by the geometry and spatial distributions of high- and low-permeability sediments. However, while traditional site characterization tools—i.e., borehole core description, geophysical logging tools, Cone Penetrometer Test (CPT), etc.—are good for determining vertical lithologic variability, their ability to define the heterogeneity between boreholes is limited (Fig. 4.1). Therefore, implementation of a scientifically defensible correlation method that adequately resolves the spatial uncertainty between boreholes is necessary for reducing data gaps and designing effective site remedies. This can be achieved by focusing on the understanding of the depositional processes controlling sediment distribution at the site.

Facies analysis is an established method for understanding geologic depositional processes (Posamentier and Walker 2006). The term 'facies' is used either descriptively, for a certain volume of sediment identified by their geological features

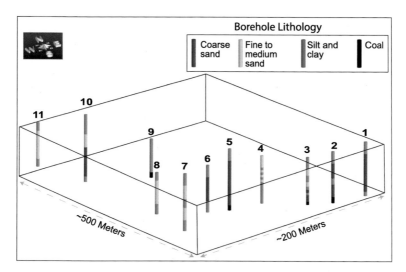

Fig. 4.1 Borehole core descriptions of driller's logs from multiple well locations at a hypothetical remedial site. The boreholes show considerable vertical variation in lithology. How are the different lithologies interconnected between the boreholes? This scenario reveals the challenge of accurately delineating flow versus confining zones in three dimensions

(e.g., grain size, sorting, sedimentary structures, fossils, etc.), or interpretatively for the inferred depositional environment (e.g., channels, deltas, alluvial fans, etc.) of that sediment volume (Anderton 1985). The latter is also informally dubbed as 'depositional facies' and used for developing facies models.

Facies models are general summaries of specific depositional environments (geological settings for sedimentary features) derived from a synthesis of empirical geological observations of ancient and modern sedimentary features, as well as laboratory flume experiments. As such, a facies model can be treated as a norm or conceptual template for future observations in that particular depositional environment. It provides the stratigrapher a sense of a reasonable scale of correlation within a reliable three-dimensional stratigraphic framework. Because facies models are derived from responses to their geological processes, they also serve as tools for hydrodynamical interpretation and prediction. It is the duty of the stratigrapher to compare their data with several tentative facies models and identify the one that satisfies most of the observations.

A detailed discussion of different depositional environments and facies models (Walker and James 1992; Posamentier and Walker 2006; James and Dalrymple 2010) is beyond the scope of this chapter. However, any established facies model for a particular depositional environment should guide the interpretation of borehole data by predicting the following critical sedimentological and stratigraphic features (among others):

1. Vertical grain size trends of sediments;
2. Grain size sorting, textures, and sedimentary structures;
3. Distribution of high-permeability and low-permeability units;
4. Sand-body geometry and dimensions (width-thickness ratios);
5. Clay deposit geometry and dimensions (width-thickness ratios);
6. Hydrodynamic conditions of deposition.

Once the understanding of vertical facies changes in boreholes is complete, the spatial distribution of those facies can be determined by the application of Walther's Law (Walther 1894). This law states that depositional environments that were lateral to one another will stack vertically, forming facies successions, reflecting successive changes in the environment (Middleton 1973). In other words, the vertical succession of beds/facies in boreholes mirrors the original lateral distribution in time as shown in Fig. 4.2, unless interrupted by erosion/unconformity.

Application of appropriate facies models in combination with Walther's Law and inferred facies geometry and dimensions is a powerful tool in interpreting a site's heterogeneity from borehole data. Figure 4.3 depicts an example of application of the fluvial facies model to the borehole dataset considered in Fig. 4.1. Observe that the vertical variation in lithology between boreholes previously posing a three-dimensional jigsaw puzzle can be adequately resolved only when seen as lithologies of different sub-environments (e.g., point bars, crevasse splays, levees, and oxbow lake deposits) of a fluvial facies model. By applying appropriate width-thickness ratios and depositional trends of these sub-environments based on the facies model, the complex inter-relationship of high- and low-permeability units can be reliably established.

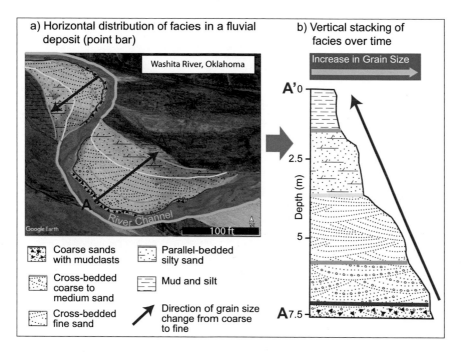

Fig. 4.2 Sedimentary environments that started side-by-side will end up overlapping one another over time due to progradation (deposition moving forward) and retrogradation (deposition retreating backward). The example illustrates the lateral migration (progradation) of a point bar reflected in vertical succession according to Walther's Law. See Electronic Supplementary Material for a video illustrating this concept (ESM_1)

The determination of high- and low-permeability zones of a site based on this facies model approach provides the ability to predict potential flow pathways of LNAPL migration that is scalable to the remedial problem at hand. For example, considering the hypothetical scenario in Fig. 4.3, one can infer the behavior and migration path of LNAPL along any section of the site (e.g., point bar deposits represented by wells 1, 2, and 3) near the water table by considering the lateral facies changes between fine-grained and coarse-grained sediments (Fig. 4.4).

4.2   Lithostratigraphy Versus Chronostratigraphy

4.2.1   The Pitfalls of Traditional Correlation Methods

A good understanding of the facies concept, although essential, does not alone ensure correct stratigraphic correlation. One must also consider the spatial facies variability over time. This section outlines the major pitfalls of traditional correlation methods, and how a chronostratigraphic approach (i.e., correlation based on timing of deposition) can provide a safeguard against miscorrelation of facies.

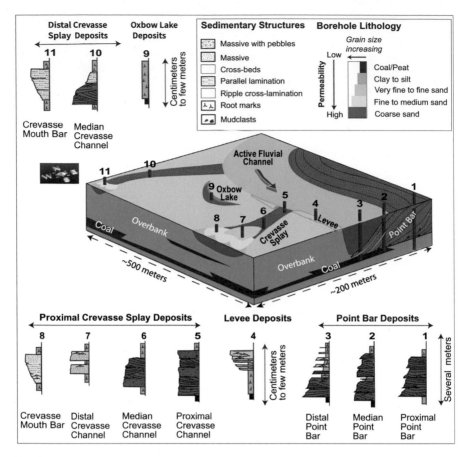

Fig. 4.3 Three-dimensional facies model of a meandering river system applied to the borehole data shown in Fig. 4.1. Borehole lithology and grain size distribution vary widely within short distances due to the complex relationship of different depositional sub-environments (e.g., point bars, crevasse splays, levees, and oxbow lake deposits). Without proper placement of each borehole lithology within the context of the facies model, accurate prediction of subsurface heterogeneity (transport versus storage zones) would be impossible

In the environmental industry, lithostratigraphic correlation has traditionally been used to resolve geological uncertainty between boreholes. This correlation method largely relies upon connecting 'sand-with-sand' and 'clay-with-clay' as long as the sedimentological characteristics between boreholes are comparable (Liu et al. 2017; Levitt et al. 2019). This method assumes a homogeneous horizontal continuity of depositional units between boreholes and does not rationally account for lateral variations in lithology due to facies changes (i.e., does not honor Walther's Law). This approach often results in a vertically stacked, unrealistic 'layer-cake' correlation of high-permeability (flow) and low-permeability (confining) units (Fig. 4.5a) and

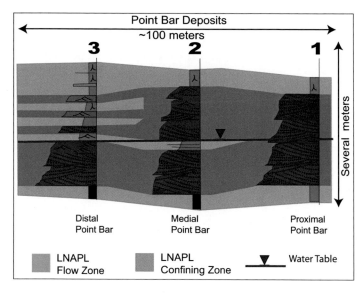

Fig. 4.4 Determination of potential LNAPL flow and confining zones based on correlation of depositional facies. See Fig. 4.3 for context

provides little correlation predictability beyond existing data points even where the depositional facies is broadly understood.

Over large distances and/or complex geologic settings, the lithostratigraphic approach is typically subject to miscorrelation and fails to address the heterogeneity of subsurface systems. Also, over relatively short distances (e.g., less than a hundred meters), traditional lithostratigraphy without the application of facies concepts often fails to identify the internal heterogeneity (confining versus flow units) within an aquifer system.

Most traditional computer-generated stratigraphic autocorrelation techniques used in the environmental industry are similarly based on lithostratigraphic correlation principles. Such autocorrelations are generally conducted by mathematical quantification of visual correlation based on statistical rules governing correlation lengths (e.g., input of kriging semi-variogram anisotropy). While this approach of correlation is reproducible, honors spatial observations, and mitigates the subjectivity of interpretation, it still fails to adequately address lateral facies variability according to true depositional patterns because of the simplistic underlying assumptions of lithostratigraphy in the algorithm. Consequently, integration of geological information is paramount to improve stochastic analyses (Fogg and Zhang 2016) as reflected by the development of methods such as multiple-point statistics based on trained images (Mariethoz and Caers 2015).

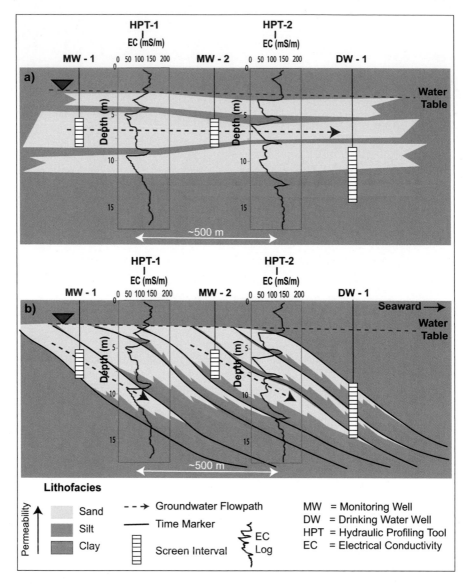

Fig. 4.5 Two alternative correlations between two well logs in a deltaic depositional environment along stratigraphic dip. Panel **a** depicts a lithostratigraphic ('layer-cake') correlation approach. Note the assumption of unimpeded fluid-flow direction (red arrow) resulting from this simplistic model. Panel **b** shows a chronostratigraphic correlation of the same dataset. Observe how the understanding of subsurface conditions has been significantly altered (modified after Ainsworth et al. 1999)

4.2.2   Chronostratigraphy—The Preferred Approach to Stratigraphic Correlation

Unlike lithostratigraphy, chronostratigraphy relies on correlation of facies within a stratigraphic framework defined by correct identification or inference of time-significant surfaces (time markers). The recognition that stratigraphic units may be defined by chronostratigraphically significant surfaces has been established in the North American Stratigraphic Code as allostratigraphy (North American Commission on Stratigraphic Nomenclature 1983, Rev. 2005). Identification of these time markers is not dependent upon biostratigraphy or absolute age-dating. They are derived from the study of outcrops (Van Wagoner et al. 1990; Posamentier and Allen 1999), seismic analogs (Vail et al. 1977), and laboratory flume experiments (Paola et al. 2009) that reveal the contemporaneous three-dimensional facies relationship of deposits within a particular depositional environment. Figure 4.5 shows how this chronostratigraphic approach can significantly improve the correlation of a dataset earlier addressed by lithostratigraphic correlation. In contrast to the scenario of a lithostratigraphic approach (Fig. 4.5a), the correct identification of lateral facies changes between basinward (seaward) dipping time markers (Fig. 4.5b) reveals internal heterogeneity and compartmentalization of the aquifer system that had remained previously undetected from lithostratigraphic correlation. This preferred approach of interpreting pre-existing lithologic data significantly alters the understanding of subsurface hydrostratigraphy and may also impact understanding of LNAPL migration at the site. Moreover, chronostratigraphy supported by knowledge of established facies models provides a stratigrapher the ability to predict facies heterogeneity beyond the data points and helps develop important investigation and remedial strategies for a site even from a limited amount of data.

4.3   Sequence Stratigraphy—A New Paradigm for the Environmental Industry

Sequence stratigraphy is a chronostratigraphic method of correlation pioneered by the petroleum industry to accurately define subsurface heterogeneity between boreholes and to predict reservoirs. The approach largely relies on understanding allocyclic controls (cyclic geological processes external to the considered sedimentary system—e.g., a basin) of sediment deposition. Global sea level variations, climatic changes, and tectonic subsidence/uplift are the major allocyclic processes considered (Fig. 4.6). Such processes often tend to generate predictable 'sequences' defined by unconformities. Allocyclic processes exhibit a larger spatial and temporal continuity than autocyclic processes (processes local to the basin) that result in cyclic deposition without a predetermined frequency. Understanding both the allocyclic and autocyclic processes that impacted the geology of a site is important for contamination remediation purposes. While allocyclic processes give the 'big picture' of regional

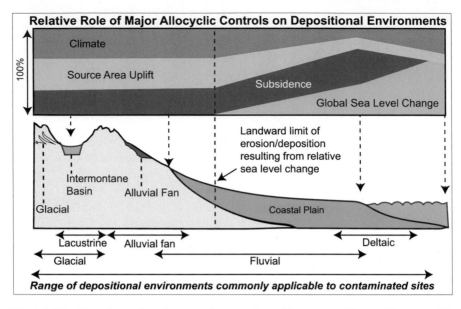

Fig. 4.6 The allocyclic stratigraphic controls on continental deposits are illustrated in terms of the ratios between the effects of climate, global sea level change, basin subsidence, and source area uplift after Shanley and McCabe (1994). Also, note that contaminated sites occur in a whole range of depositional environments

heterogeneity, autocyclic processes often provide the understanding of heterogeneity at the remediation scale. As shown in Fig. 4.6, the impact of global sea level change diminishes significantly in a landward direction. Rather, stratigraphic architecture in continental settings beyond the coastline is a function of climate, sediment supply, and tectonic subsidence/uplift (Shanley and McCabe 1994). Since the majority of contaminated sites are located in continental settings (continental-fluvial, glacial, lacustrine, and alluvial fan environments), application of sequence stratigraphy to those sites requires certain adaptations of the original concepts, which are largely dependent on sea level changes.

Table 4.1 shows some of the major allocyclic and autocyclic processes that interplay to control the three-dimensional distribution, stacking and hierarchy of facies geometries in different depositional environments, otherwise known as facies architecture (Miall 1988).

The most characteristic effect of the allocyclic processes, as opposed to autocyclic processes, is that they operate simultaneously in different basins in a similar, predictable way. Thus, they help to correlate strata over long distances and often, even from one basin to another. Based on this knowledge, the application of sequence stratigraphy has been historically focused toward recording and predicting the lateral shifts of facies within marine, coastal, and marginal fluvial depositional settings in response to global sea level changes.

Widely recognized since the 1980s as a new paradigm of stratigraphic analysis by both the petroleum industry and academic practitioners, the various definitions of

Table 4.1 Key allocyclic and autocyclic processes in different depositional environments

Depositional environment	Key processes controlling stratigraphic architecture	
	Regional (allocyclic)	Local (autocyclic)
Coastal (fluvial, fan delta, tidal estuary barrier bars, beach ridges) and shallow marine (delta, shoreface)	Global sea level curve applicable to the coastline Timing of major tectonic events (uplift/subsidence) impacting the region Paleoclimate (arid, humid, temperate, etc.) for the time interval of concern	Channel avulsion and delta lobe switching, tidal range, flooding events, storm events, slope, basin geometry, etc.
Continental fluvial	Tectonic events (e.g., uplift in the hinterland, basin subsidence), paleoclimate (arid–humid climatic cycles), base level (lake level) changes	Floodplain aggradation, avulsion frequency, slope change, flooding events, local structural controls (e.g., surface expression of faults) etc.
Alluvial fan and fan delta	Paleoclimate (arid–humid climatic cycles, glacial–interglacial), tectonic events (extensional or compressional tectonic regime)	Channel avulsion and fan lobe switching, flooding events, timing of local tectonic events (e.g., fault-block rotation, hanging wall generation) etc.
Continental glacial till (terminal, lateral, and ground moraine)	Global sea level, paleoclimate (glacial–interglacial periods), tectonic events (including glacial isostatic loading and rebound)	Compaction, local reworking, local water level, the depositional surface, point of sediment input relative to the glacier margin, etc.

sequence stratigraphy (Posamentier and Allen 1999; Galloway 1989; Embrey 2009) rely on two simple principles:

1. All deposition is the result of an interplay between accommodation (i.e., the space available for potential sediment accumulation) and sediment supply controlled by base level changes (Fig. 4.7); connectivity/heterogeneity of high-permeability sediments is a function of the accommodation (A) versus supply (S) in a depositional system.

Base level at the continental margin is typically equated to the relative sea level position, but it can also be related to local equilibrium surfaces such as the water surface of lakes and/or the graded profile of rivers in continental environments. With other parameters (e.g., sediment accumulation and tectonics) remaining constant, sea level may result in different accommodation scenarios. When A < S (e.g., during a sea level or base level fall), there is higher connectivity between aquifer sand bodies (channels/delta mouth bar/fan deposits), both vertically and horizontally. On

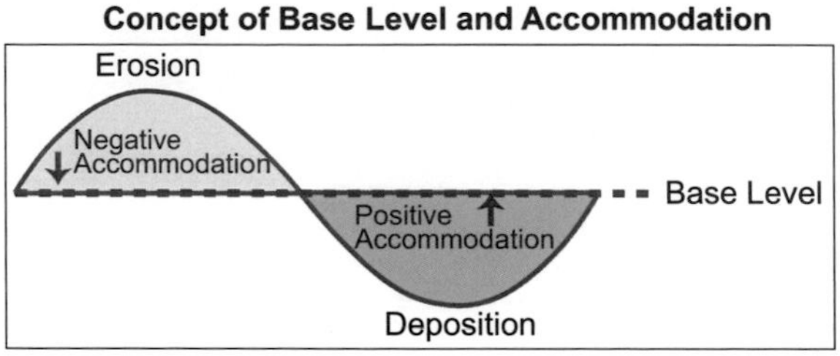

Concept of Base Level and Accommodation

Fig. 4.7 Joseph Barrell pioneered the concept of accommodation in relation to an equilibrium base level (Barrell 1917). Sediments tend to accumulate in the space available below the base level ('positive accommodation'). Sediments above the base level ('negative accommodation') tend to erode until reaching the base level

the other hand, when A > S (e.g., during a sea level or base level rise), there is lower connectivity and more heterogeneity between the sand bodies.

2. Depositional environments vary in a predictable sequence of facies architecture under particular settings of A/S in different systems tracts, which are defined as tracts (lands) of connected deposits accumulating during one phase of relative sea level/accommodation cycle and preserved between specific primary chronostratigraphic surfaces. Figure 4.8 illustrates the significant features of each systems tract predicted under a particular A/S scenario.

Over the years, several professionals (Sugarman and Miller 1997; Ehman and Edwards 2014; Meyer et al. 2016; Shultz et al. 2017; Samuels and Sadeque 2021) have successfully applied the principles of sequence stratigraphy to address hydrogeological and environmental issues. However, since contaminated sites often also include continental depositional environments or tectonically-controlled settings (Fig. 4.6), the application of sequence stratigraphy is not always straightforward. Therefore, a 'fit for purpose' adaptation of the methodology is recommended and is discussed further in the following section.

4.4 Methodology

4.4.1 Application of Sequence Stratigraphic Principles at Contaminated Sites

One of the biggest challenges for stratigraphic correlation in the environmental industry is the lack of reliable datasets. Sediment cores are seldom intact and preserved, while field-descriptions of lithology from driller's logs are frequently limited and sometimes biased. Moreover, many sites are shallow—typically less

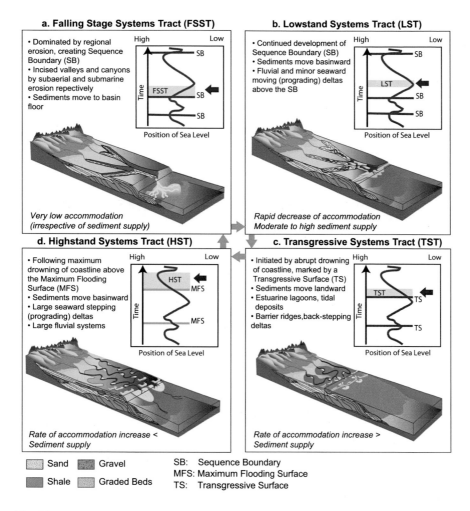

a. Falling Stage Systems Tract (FSST)

• Dominated by regional erosion, creating Sequence Boundary (SB)
• Incised valleys and canyons by subaerial and submarine erosion repectively
• Sediments move to basin floor

Very low accommodation (irrespective of sediment supply)

b. Lowstand Systems Tract (LST)

• Continued development of Sequence Boundary (SB)
• Sediments move basinward
• Fluvial and minor seaward moving (prograding) deltas above the SB

Rapid decrease of accommodation Moderate to high sediment supply

d. Highstand Systems Tract (HST)

• Following maximum drowning of coastline above the Maximum Flooding Surface (MFS)
• Sediments move basinward
• Large seaward stepping (prograding) deltas
• Large fluvial systems

Rate of accommodation increase < Sediment supply

c. Transgressive Systems Tract (TST)

• Initiated by abrupt drowning of coastline, marked by a Transgressive Surface (TS)
• Sediments move landward
• Estuarine lagoons, tidal deposits
• Barrier ridges, back-stepping deltas

Rate of accommodation increase > Sediment supply

Sand   Gravel
Shale Graded Beds

SB: Sequence Boundary
MFS: Maximum Flooding Surface
TS: Transgressive Surface

Fig. 4.8 Illustration after Kendall (2006) depicting predictable lateral shift of facies under varying accommodation versus supply scenarios dictated by the position of sea level. A sequence is defined as a full cycle of depositions between two sequence boundaries (generated by a Falling Stage Systems Tract—FSST—and/or Lowstand Systems Tract—LST). See Glossary of Terms for further explanations of different systems tracts and their stratigraphic markers. See Electronic Supplementary Material for a video illustrating these concepts (ESM_2)

than 150 m below ground surface (m bgs) and, in the case of LNAPL-contaminated sites, often less than 20 m bgs—and limited in scale (several thousand square meters or less).

Considering these limitations, this section proposes a practical, iterative approach to implement sequence stratigraphic principles and facies architecture analysis at contaminated sites. This methodology is applicable to any contaminated site within shallow marine and continental depositional environments.

4.4.2 Evaluation of Geologic Setting and Accommodation

Having a thorough understanding of regional geological setting is key to evaluating site datasets in terms of allocyclic factors (Table 4.1) and making more informed correlation decisions. This regional understanding can be achieved through the study of surface and subsurface geological maps, global sea level curves (Haq and Schutter 2008), and reports on tectonic and climatic history, among others.

Within this regional context, a knowledge of site-specific local depositional environments from previous reports and academic publications helps to understand the autocyclic controls on facies architecture governing local accommodation (Table 4.1). Geological investigations of contaminated sites initiate with an in-depth research for existing reports and papers that define the structural and depositional history of the targeted subsurface interval for the site. In the United States, regional and local geological information can be found in many public sources such as the United States Geological Survey (USGS), state geological surveys, universities, research institutions, local governments, published literature—e.g., Geological Society of America (GSA), American Geophysical Union (AGU), American Association of Petroleum Geologists (AAPG), and Society for Sedimentary Geology (SEPM)— and state database websites. Also, application of modern depositional analogs using Geographic Information System (GIS) and remote sensing data, as well as outcrop investigations (nearby gravel pits, cut banks, road cuts, quarries), can be useful for estimating width/thickness ratios of relevant clay deposit/sand-body geometries (e.g., Fielding and Crane 1987; Reynolds 1999; Gibling 2006) and to provide additional insights for subsurface heterogeneity as exemplified in Fig. 4.9. This strategy is particularly helpful in instances where good spatial and/or vertical data coverage at the site is lacking.

4.4.3 Analysis of Lithologic Data

Grain size and sorting are two lithologic characteristics that can significantly influence a unit's pore space geometry and permeability. In general, coarse-grained, well-sorted units exhibit higher permeability, while finer-grained and/or poorly sorted units display lower permeabilities. Therefore, defining changes in grain size and sorting within the existing dataset is critical for defining preferential migration pathways through both the vadose and saturated zones.

Borehole logs, geophysical logs (e.g., gamma, resistivity, neutron-porosity), CPT logs, Hydraulic Profiling Tool (HPT) logs, Electric Conductivity (EC) logs, and several other site characterization tools discussed in Chap. 7 are commonly used to provide valuable information for the classification of sediments at contaminated sites. Borehole core description ideally provides the 'ground truth' for accurate grain size and sorting information of sedimentary deposits. Therefore, great care must be taken to acquire high-quality data. Although often underutilized in the environmental

a. Identifying Suitable Outcrop in the Vicinity of the Site Area

b. Detailed Mapping of Bounding Discontinuities

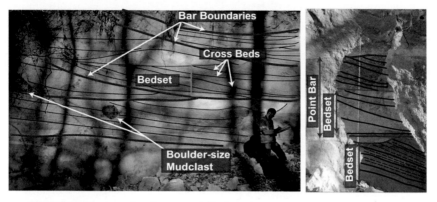

c. Studying Modern Analogs and Estimating Dimensions for Site data

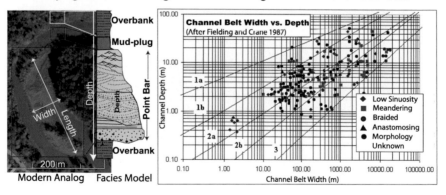

Fig. 4.9 Panel demonstrating how ancient and modern analog data can be used to enhance geologic correlation at a site. **a** Shows an outcrop of a meandering river system located close to a contaminated site. The relict river had migrated from right to left until it was eventually abandoned and 'plugged' with fine-grained sediments. **b** Depicts the method of identifying bedset boundaries (black markers) and bar boundaries (blue markers) for detailed correlation of the fluvial deposits by recognizing bounding discontinuities. Note the boulder-sized (0.3–0.6 m thick) rip-up clasts within the same outcrop, which reveal that thick clay deposits captured within nearby site boring logs may represent localized features rather than laterally continuous floodplain deposits. **c** shows how integration of outcrop measurements with Google Earth imagery provides a basis for determining appropriate width/thickness ratios of facies beneath the site for guiding correlation lengths between boreholes (Fielding and Crane 1987). See Electronic Supplementary Material for an uninterpreted version of Fig. 4.9b) (ESM_3)

industry, the Wentworth scale (Wentworth 1922; Fig. 4.10a) of grain size measurement is preferred over the Unified Soil Classifications System (USCS) classification (Fig. 4.10b) because it provides more precise grain size measurements for understanding depositional processes from borehole data. Moreover, once a grain size distribution is measured in the Wentworth scale, it can easily be broken down into the USCS classification for engineering purposes. On the other hand, if the data is lumped under the USCS classification, it is more difficult to be converted into the Wentworth scale.

Grain size measured in the Wentworth scale can be hand-plotted as a grain size log that reveals accurate grain size trends of the core data (Fig. 4.11a). However, processing previously obtained data in USCS classification can be also utilized for stratigraphic work, typically by color-coding the grain size and texture data (Fig. 4.11b) that are lumped within the USCS classes.

Despite the utility of borehole logs for accurate field measurements of lithology, they are limited by discrete sampling intervals and inherent logging biases. Therefore, a best-practice strategy to achieve a continuous, unbiased picture of subsurface conditions is to calibrate borehole logs with geophysical logs (e.g., gamma log, Fig. 4.11c).

Since good data quality is the first precondition for accurate and meaningful stratigraphic correlation, emphasis should be given in having multiple data types for individual boreholes (e.g., borehole core description supported by HPT, CPT, gamma, and resistivity logs).

4.4.4 Facies Architecture Analysis

Once the grain size and textural data (including sorting) have been processed, the identified grain size trends (Fig. 4.11) and lithologic compositions are compared with relevant facies models (see discussion in Sect. 4.1.2) for different depositional environments (Fig. 4.12) recognized during regional research (Fig. 4.13). Their spatial relationships are then determined using Walther's Law.

The hierarchy of three-dimensional architectural components (elements) of the depositional facies (e.g., shoreface, delta mouth bar, fluvial channel bars, alluvial fan sheet flood deposits, etc.) are further analyzed by recognizing the internal stacking patterns, geometry, and relationship of bounding surfaces shown by the data or its analog (Fig. 4.9). This determination of hierarchy in facies architecture ultimately helps to distinguish the larger stratigraphic architecture of sequences (i.e., hierarchy and stacking patterns of their building blocks—i.e., parasequences) resulting from allocyclicity and is critical for predicting aquifer heterogeneity across multiple scales. Figure 4.14a shows how the three-dimensional hierarchy of facies is established on the basis of an order of bounding surfaces from beds to elements. Figure 4.14b shows how stratigraphic architecture is understood in relation to stacking pattern

a. Wentworth Scale

Size (mm)	Symbol	Group	Grain Size Name
250	C	GRAVEL	Cobbles
65			
	P		Pebbles
4			
	G		Granules
2			
	Vcs	SAND	Very coarse sand
1			
	Cs		Coarse sand
0.5			
	Ms		Medium sand
0.25			
	Fs		Fine sand
0.125			
	Vfs		Very fine sand
0.625			
	Csi	SILT	Coarse silt
0.031			
	Msi		Medium silt
0.0156			
	Fsi		Fine silt
0.0078			
	Vfsi		Very fine silt
<0.0039			
	Cl	CLAY	Clay

b. USCS Scale

Size (mm)			Symbol	Group	Grain Size Name
More than 50% coarse grains > 0.075 mm	50% of coarse fraction > 4.75mm	Clean gravel <5% smaller than 0.075mm	GW	GRAVEL	Well-graded gravel, fine to coarse
			GP		Poorly graded gravel
		Gravel with >12% fines	GM		Silty gravel
			GC		Clayey gravel
	≥ 50% of coarse fraction	Clean sand	SW	SAND	Well-graded sand, fine to coarse
			SP		Poorly graded sand
		Sand with >12% fines	SM		Silty sand
			SC		Clayey sand
50% Fine grains ≥ 0.075 mm	Silt and clay liquid limit < 50	Inorganic	ML	SILT AND CLAY	Silt
			CL		Clay, low plasticity (lean clay)
		Organic	OL		Organic silt
	Silt and clay liquid limit ≥ 50	Inorganic	MH		Silt, high plasticity
			CH		Clay, high plasticity (fat clay)
		Organic	OH		Organic clay
	Highly Organic		Pt	PEAT	Peat

Fig. 4.10 A Comparison between the Wentworth scale and the USCS classification. Observe how grain size is combined with texture in the USCS classification. This method may be convenient for measurements by sieve. On the other hand, the Wentworth system is purely based on grain size differences measurable in the field by a hand lens and grain size comparator available in the market while giving a higher resolution of grain size variation

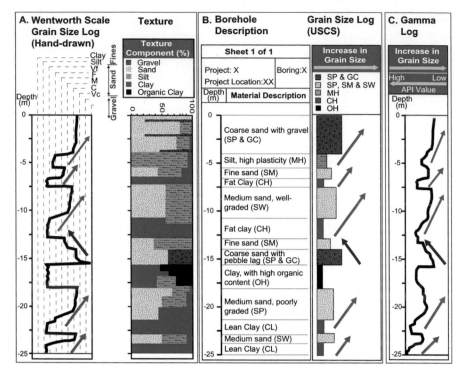

Fig. 4.11 Processing vertical grain size and texture from borehole using Wentworth and USCS data for calibration with geophysical logs. Blue arrows show coarsening-upward and red arrows show fining-upward trends in the vertical profile. **a** Separates grain size data from texture information and identifying vertical grain size trends (fining-upward/coarsening-upward) in Wentworth scale in hand-drawn log. **b** Color-codes grain size and texture to represent data collected in USCS scale. **c** Illustrates the calibration of gamma logs to borehole grain size data to use as proxy for continuous grain size logs. API (American Petroleum Institute) value is the unit of natural gamma radiation of sediments

and hierarchy of large-scale architectural units (elements and above) under allocyclic controls (e.g., sea level changes).

A key take-away from this approach is to recognize that it can address heterogeneity across multiple scales. Because scaling is often a challenge encountered by groundwater modelers and remedial practitioners, this approach can serve as an invaluable tool for developing more effective remedial strategies.

4.4.5 Correlation Between Boreholes

The key to successful correlation is to correctly identify time-significant surfaces (time markers) before populating the facies architecture between those surfaces.

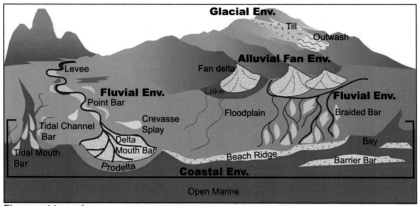

Figure not to scale

Fig. 4.12 Illustration showing major depositional environments (abbreviated as 'Env.') and their sub-environments. Note that while not all depositional processes are contemporaneous, the geology of a site may be dictated by multiple depositional environments over time

Determination of the relevant accommodation scenarios based on this approach has direct bearing on understanding the aquifer heterogeneity of a site.

4.4.5.1   Marine and Coastal Environments

If the target remedial zone of a contaminated site was deposited in a marine or coastal setting, correlation of the identified facies and propagation of the facies architecture should be conducted within a conventional depositional sequence stratigraphic framework based on the accommodation scenario and recognizable systems tracts (Fig. 4.8). The defining time-significant surfaces include marine flooding surfaces (e.g., mud or shale markers), wave ravinement surfaces, and regional erosional boundaries (sequence boundaries). Standard concepts and the basic methodology of sequence stratigraphy are described in numerous articles and books (Van Wagoner et al. 1990; Emery and Meyers 1996; Posamentier and Allen 1999; Catuneanu 2006; Embrey 2009). Once several cross-sections are developed (preferably both in dip and strike directions of a depositional system) and a three-dimensional understanding of the subsurface heterogeneity is attained (Fig. 4.15), this framework should be integrated with hydrogeologic and chemical data to interpret the hydrostratigraphic units for the site.

4.4.5.2   Continental and Tectonically-Controlled Settings

For a site in which the target depth interval of aquifer is recognized to have been deposited in a fluvial, alluvial fan, aeolian (loess), glacial, or structurally-controlled

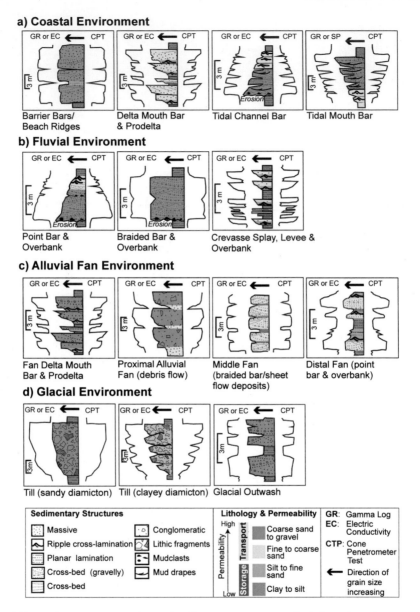

Fig. 4.13 Vertical grain size profiles with sedimentary structures typical of a variety of depositional environments. Black arrow indicates direction of grain size coarsening observed in boreholes. Responses of grain size trends are also captured in GR, EC, and CPT logs. Note that different facies may exhibit similar log motifs, emphasizing the need to place them within the context of the regional geology before correlation

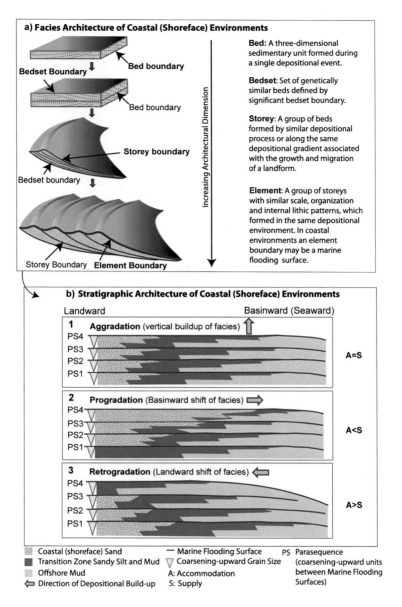

Fig. 4.14 a Example of facies architecture hierarchy in a coastal setting of shoreface deposits. The scale of architectural units increases from top to bottom (conceptualized after Vakaralov and Ainsworth 2013). **b** Conceptual cross-sections (adapted from Van Wagoner et al. 1990) showing stratigraphic architecture of coastal deposits based on large-scale stacking patterns related to sea level changes. Parasequences developed between marine flooding surfaces show aggradational, progradational, or retrogradational stacking patterns depending upon the accommodation versus sediment supply scenario of the system

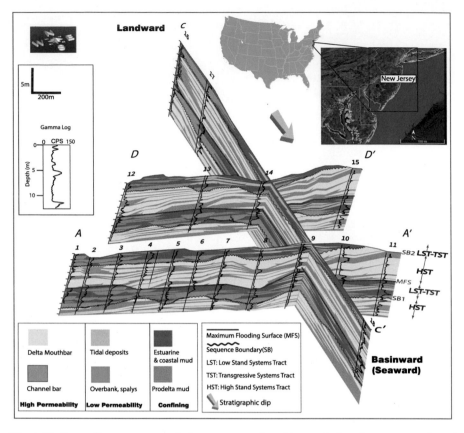

Fig. 4.15 Fence diagram composed of sequence stratigraphic correlations from a contaminated site that lies within the New Jersey coastal plain (USA). Hydrostratigraphic units were determined by integrating site hydrologic data into this three-dimensional stratigraphic framework (Sadeque 2020)

setting, we recommend the application of allostratigraphic methods ('descriptive sequence stratigraphy' after Brookfield and Martini 1999) to correlate and address the critical issues with the application of depositional sequence stratigraphy in the continental environments (summarized in Sect. 4.3 and Table 4.1).

This correlation approach, also informally designated 'Environmental Sequence Stratigraphy' (ESS) within the environmental industry (Shultz et al. 2017; Samuels and Sadeque 2021), primarily relies on defining and naming mappable discontinuity-bounded depositional successions without placing particular emphasis on the type of discontinuity used as a fundamental stratigraphic break.

The following are some of the common markers of mappable discontinuity that have time-significance for correlation:

1. Regional flooding surfaces, e.g., created by cyclic flooding events in continental fluvial settings, clay layers resulting from major channel migration and avulsion events, sheet-flood episodes in alluvial fan settings, and flooding surfaces related to glacial outwashes and glacial catastrophic events indicated by mud/shale markers and erosional features.
2. Major facies dislocations indicated by abrupt 'shallow facies' over 'deep facies', e.g., glacial over non-glacial facies and vice-versa.

3. Significant paleosol horizons and coal/peat horizons (in continental fluvial environments), caliche horizons (in alluvial fan environments), loess horizons (fine-grained aeolian deposits), and outwash (sandur) plain boundaries (in glacial environments).
4. Tectonically-driven incised valley boundaries, autocyclic fluvial erosions, bypass or erosional supersurface (in aeolian environment), and various glacial incisions, etc.

Pervasive markers that can be correlated over the entire region probably represent the allocyclic controls of tectonics and/or climate. Other markers local to the specific site may reflect autocyclic processes. But even if the distinction between these two types of markers is not always conclusive due to incomplete dataset, this approach of utilizing mappable discontinuities (Figs. 4.9b and 4.16) in combination with facies architecture analysis leads to a predictable time-significant framework of correlation.

4.4.6   Integration with Hydrogeology and Chemistry Data

Once the stratigraphic framework has been established, it should then be integrated with hydrogeologic and chemical data to guide the reinterpretation of site hydrostratigraphic units. Screen intervals, groundwater levels, and chemistry data (preferably color-coded by level of contamination) are just some examples of relevant data that should be added to the cross-sections to aid in the reinterpretation within the context of the stratigraphy. There must be agreement between interpreted stratigraphic pathways, potentiometric trends (vertical gradients and mapped trends), and analytical (isoconcentration) contours. The new understanding of subsurface conditions derived from this synergy should be used to redefine preferential groundwater flow paths and redraw hydrogeological contours that are more representative of the subsurface conditions of a site.

Figure 4.17 provides a generalized workflow for stratigraphic correlation as described in this chapter. Identifying depositional facies for the site data (by comparing with facies models and depositional analogs) and developing a stratigraphic framework for the depositional facies (using stratigraphic markers) are the two critical parts of the correlation process. Incorporating groundwater and chemistry data, as well as showing well screen intervals in the correlation provides a holistic view of the subsurface needed for developing an optimal remedial strategy at a site. Observe that following flow paths 1–5 lead to correlation by only facies analysis, whereas the flow paths of I–II and Ia–IIa help develop the sequence stratigraphic framework. The input of step 5 and step III together provide a facies-based sequence stratigraphic correlation. After correlation is complete, steps 7–8 are taken to address any data gaps and provide feedback for future iterations of the process.

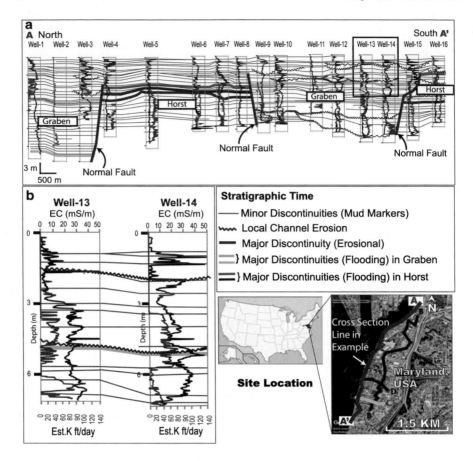

Fig. 4.16 Example of correlation by identification of bounding discontinuities in a structurally-controlled contaminated site in the Washington DC area using EC logs and conductivity (K) logs. **a** Correlation utilizing both major and minor bounding discontinues (time markers) in a horst and graben system. The faults were identified from regional research and 'slicken-sides' were identified using cores from the site. **b** Up-close excerpt of correlation showing the time markers in detail

4.5   Case Study: Using Sequence Stratigraphy to Inform Remedial Decision-Making at a Geologically Complex LNAPL-Impacted Site

This case study aims to demonstrate how sequence stratigraphy can enhance the Conceptual Site Model (CSM) of a LNAPL-impacted site by providing a more reliable understanding of subsurface conditions and heterogeneity needed for remediation.

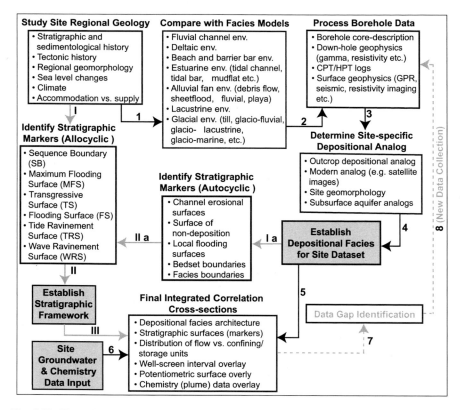

Fig. 4.17 Flow diagram for stratigraphic correlation proposed in this chapter. Grey boxes denote critical information required for the application of sequence stratigraphy to a contaminated site. Black and grey arrows indicate simultaneous steps taken to generate a sequence stratigraphic framework and conduct facies analysis within that framework

4.5.1 Site Background

An industrial site located in the West Coast Basin of the Los Angeles County Coastal Plain in the United States (Fig. 4.18) has been an active industrial facility for almost a century. The site is geologically complex and is impacted by various petroleum products. The remedial objective for the site is to implement aerobic bioremediation via biosparging at targeted zones with impacts to prevent the off-site migration of LNAPL.

Traditional studies of the subsurface geology of the region for groundwater evaluations (Poland et al. 1956; California Dept. of Water Resources 1961; Zielbauer et al. 1962) offered subsurface stratigraphic correlations by tying together deposits of similar lithologies between boreholes without considering facies heterogeneity. The resulting 'layer-cake' stratigraphy led to dividing the hydrostratigraphy of the

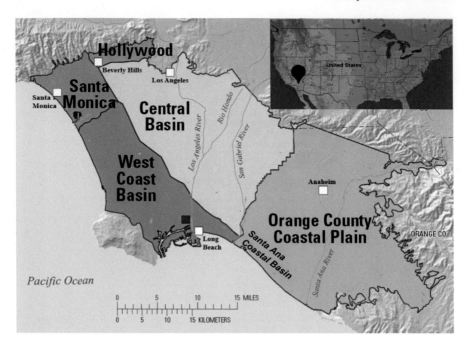

Fig. 4.18 Approximate geographical location of the site (shown in blue box) in the West Coast Basin of California, USA. *Modified from* U.S. Geological Survey. Public domain. *Credit* Miranda Fram

Los Angeles (LA) Basin into five major aquifers (Fig. 4.19). The present site investigation focuses on the 'Semi-perched aquifer', which consists of sands and gravels of 'Recent Alluvium' and varies between 0 and 20 m thick. The Semi-perched aquifer lies above the Gaspur aquifer, which lies at approximately 29–36 m below Mean Sea Level (MSL) and is composed of fluvial sediments of the lower Recent Alluvium.

Prior High-Resolution Site Characterization (HRSC) studies, including CPT with Laser-Induced Fluorescence (LIF), Membrane Interface Probe (MIP) and soil and groundwater sampling and analysis, confirmed that contamination beneath the site is concentrated within the Gaspur and Semi-perched aquifers. However, understanding of preferential contaminant migration pathways based on previous correlations proved to be inadequate for precise targeting of biosparging zones for LNAPL remediation (Fig. 4.20). This is because efficient biosparging requires adequate transmissivity of aquifer units in the target treatment zones to ensure airflow critical for aerobic biodegradation by indigenous bacteria.

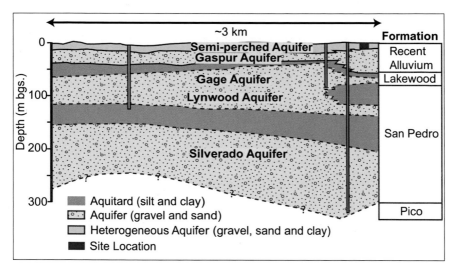

Fig. 4.19 Generalized hydrostratigraphic framework of the site area based on a lithostratigraphic approach. The blue box denotes the stratigraphic interval of interest for the present case study. Modified figure reproduced with permission from Ehman and Cramer (1997)

4.5.2   Application of Sequence Stratigraphy

The approach detailed in Sect. 4.4 was applied to the site to update the CSM. The objective of the sequence stratigraphic analysis was to develop a high-resolution stratigraphic framework to better define subsurface heterogeneity, characterize preferential LNAPL migration pathways, and aid in the design of targeted biosparging wells.

4.5.2.1   Evaluation of Geologic Setting–Regional Geology

The regional geology of the LA Basin (primarily consisting of the Central Basin, West Coast Basin, and the Santa Ana Coastal Basin) is described in terms of four primary structural blocks that are bound by fault zones. The site region is located on the southwestern structural block, within the LA Basin bounded by the Newport-Inglewood Fault (NIF) to the east, and the Palos Verdes Fault (PVF) to the west as shown in Fig. 4.21.

The LA Basin was formed during a phase of accelerated subsidence and deposition that began in the late Miocene and continued without significant interruption through the early Pleistocene. The basin was filled with thick successions of unconsolidated, stratified, and laterally discontinuous continental deposits derived from the highland areas and marine sediments to form the current broad plain. The deposits within the basin were deformed by tectonic events (folding and faulting), incised by rivers, and backfilled and buried by alluvium.

Sequence stratigraphic investigations based on seismic investigations (Ponti et al. 2007; Ehman and Edwards 2014) linked sea level changes to accommodation in the

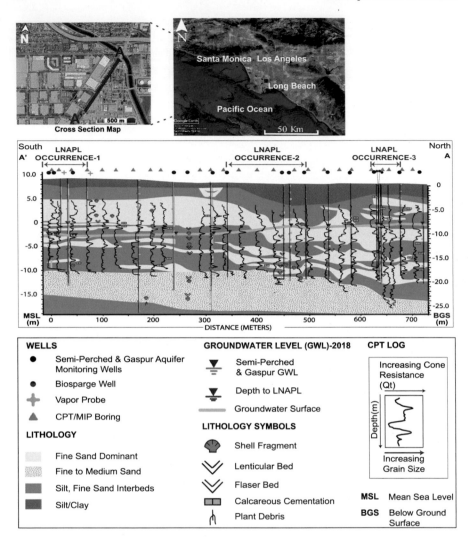

Fig. 4.20 Stratigraphic correlation at the site using a lithostratigraphic approach in the semi-perched aquifer interval. Observe that while the correlation provides a general understanding of sand and clay distribution beneath the site, it does not always provide the precise resolution of heterogeneity required for targeting optimal placement of biosparging tools in high-permeability intervals

LA Basin (Fig. 4.22) and developed a robust regional sequence stratigraphic framework (Fig. 4.23). Comparison of the traditional hydrostratigraphic units relative to this sequence stratigraphic framework shows that the aquifers and aquitards possess complex internal heterogeneity on a regional scale not recognized by the traditional lithostratigraphic correlation (Fig. 4.19).

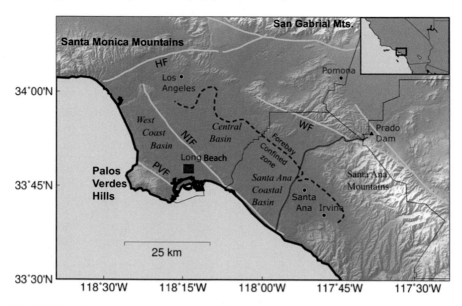

Fig. 4.21 Location and tectonic setting of the Los Angeles Central and West Coast Basins and the Santa Ana Coastal Basin (Riel et al. 2018). The blue box represents the approximate site location. The solid gold lines represent major faults in the area, including the Newport-Inglewood Fault (NIF), the Whittier Fault (WF), the Palos Verdes Fault (PVF), and the Hollywood Fault (HF). The solid blue line corresponds to the Santa Ana River. The dashed red line indicates the approximate boundary between the forebay and confined areas of the groundwater system

4.5.2.2   Evaluation of Site-Specific Geology

The geological investigation for the site includes sediments from the shallow depth intervals of the Upper Gaspur to Semi-perched aquifer (i.e., the primary focus of this case study) as shown by the blue box in Fig. 4.23. Regional investigations (e.g., Ehman and Edwards 2014; Ponti et al. 2007) showed that both the Gaspur and Semi-perched aquifers fall within the Dominguez sequence, based on the updated sequence stratigraphic nomenclature (Ehman and Edwards, 2014). As shown in Fig. 4.22, the initiation of the Dominguez sequence corresponds to a significant sea level fall less than 15 thousand years (Ka) ago. The relatively low sea level (low accommodation) indicates that the Dominquez sequence likely developed as an incised valley fill (Fig. 4.8a). Such incised valleys are typically formed by fluvial systems that extend basinward and erode into underlying strata in response to a significant relative fall in sea level (Fig. 4.24a). This hypothesis of the incised valley is supported by published literature near the study area (Fig. 4.24b) which shows the extent of the incised valley developed during the last significant sea level fall in the region, impacting the distribution of aquifers and aquitards at the case study locality.

From the above considerations, it is predicted that the aquifer and aquitard sediments of this interval will conform to the established sequence stratigraphic model

Fig. 4.22 Regional
sequence stratigraphic
framework for the site. The
sea level curve reflects
fluctuation between warm
and cold climates over time,
and the related development
of stratigraphic sequences
demarcated by sequence
boundaries (after Bassinot
et al. 1994; figure modified
after Ponti et al. 2007)

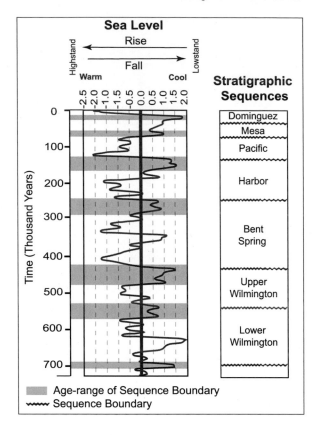

of an incised valley (Fig. 4.25) after Shanley and McCabe (1994). In Fig. 4.25, the position of sea level for each phase (1–4) is shown in red segments on the corresponding sea level curves. The first phase in the evolution of an incised valley is signified by the basinward propagation of an erosional surface (sequence boundary) accompanied by largely coastal sediment bypass (FSST). The second phase ('Time-2') is dominated by stacking of fluvial channel deposits in a low accommodation scenario of the LST. A tide-dominated estuarine setting ensues in the third stage ('Time-3') with rapidly rising sea level, following a transgressive surface shown as a blue line (transgressive systems tract or TST). Finally, in the fourth phase ('Time-4') with the turn-around of the sea level following a maximum flooding surface or MFS (pink line), a fluvio-deltaic setting of the highstand is emplaced (highstand system tract or HST). However, since LNAPL investigation at the site is primarily targeted within the Semi-perched zone of Recent Alluvium above the Gaspur aquifer, the stratigraphic correlation at the site is focused on interpreting aquifer heterogeneity within the upper units of the Dominguez sequence.

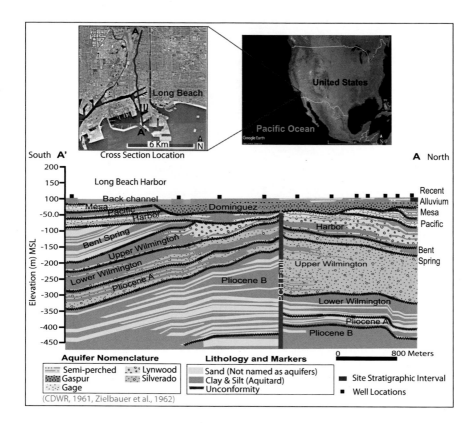

Fig. 4.23 Detailed regional sequence stratigraphic correlation based on well logs and seismic data reflecting the impact of sea level changes on sediments (modified after Ehman and Edwards (2014)). Note that the aquifer nomenclature derived by a 'layer cake' lithostratigraphic approach (shown in the legend) does not necessarily correspond to the sequence stratigraphic divisions encountered in the subsurface. Reproduced with permission from Pacific Section SEPM

4.5.2.3   Analysis of Site Lithologic Data

The subsurface geological data for the CSM consisted of borehole and CPT/MIP data down to approximately 27 m bgs. Pre-existing grain size data from borehole core descriptions were represented in USCS scale (Fig. 4.10b). Apart from grain size information, other important sedimentary and biologic features necessary for detailed facies interpretations (i.e., lenticular beds, mud clasts, shell fragments, plant material, etc.) were also meticulously recorded. The grain size data was color-coded to reveal vertical grain size and texture trends in the boreholes (as exemplified in Fig. 4.11b).

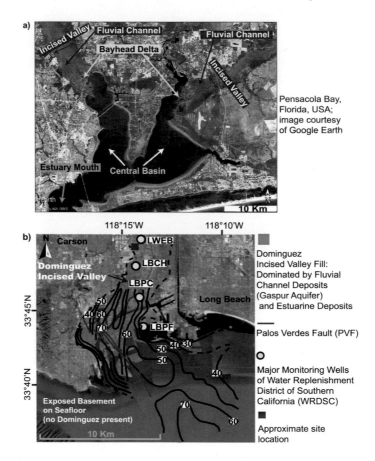

Fig. 4.24 **a** Modern incised valley example from the coast of Florida showing its various geomorphological components (modified from Google Earth image). Incised valleys can be up to several tens of meters thick and range in width from a kilometer to many tens of kilometers. **b** Seismic facies map of Dominguez sequence (consisting of the Gaspur and the Semi-perched aquifers) modified from Ehman and Edwards (2014). The contours in the figure represent the depth profile of the incised valley in meters. Reproduced with permission from Pacific Section SEPM

CPT logs (complemented with borehole core information since CPT provides information on soil behavior type, which is not necessarily the same as sediment classification in terms of grain size) were primarily used for facies analysis at individual boreholes. Where the reliability of the secondary grain size data was in question, more emphasis was given to the CPT data for shallow depths (~30 m bgs).

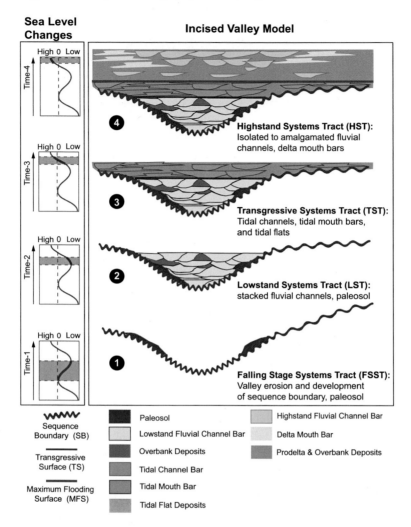

Fig. 4.25 Idealized evolution of an incised valley-fill (IVF) in response to relative sea level changes after Shanley and McCabe (1994). The figure is modified after Shanley and McCabe (1994). The brighter colors of depositional facies in this figure represent higher transmissivity compared to the darker colored depositional facies. Although no dimension is implied in the figure, studies (e.g., Wang et al. 2019) show that the thickness of IV typically ranges from 10 to 100 m, and the width of the IVF ranges from 10^3 to 10^5 m

4.5.2.4   Revised Geologic Framework

Identification of Facies

A total of eight different depositional facies were identified based on borehole core data and CPT responses with respect to the regional setting. These depositional facies are described under three facies associations below. Since the wells were very closely spaced (few meters apart), the lateral continuity of each facies of the individual facies associations between the wells could be estimated from direct observations of log responses.

(A)   *Fluvial Facies Association*

This depositional facies association consists of fluvial point bars, splays (crevasse splay and levee) and overbank fines (Fig. 4.26).

1. **Grain size/lithology**. Grain size of point bars ranges from very coarse to medium sand, with a common presence of pebble lags and mud clasts at the erosional base. The associated splay deposits are generally finer-grained, ranging from very fine sand to silt. Overbank deposits consist of clay and silt with a high proportion of carbonaceous organic matters.
2. **Log signatures**. The CPT log signature of the fluvial point bar deposits typically exhibits a fining-upward motif (Fig. 4.26c), with a sharp base of high cone resistance (Qt) values and a gradational top of low Qt values. Locally, these units may appear 'blocky' which may represent downstream accretion, as opposed to the fining-upward, lateral accretion of channel bars. Splay (crevasse splay and levee) deposits are represented by thin, coarsening-upward CPT signatures associated with channel bars. Splays are marked by a gradational base of low Qt values and sharp tops of high Qt values. Overbank deposits associated with the point bar deposits show significantly low Qt values in CPT logs.
3. **Thickness, geometry, and lateral continuity of flow units**. Typically, channel point bars range in thickness from 0.50 to 1.5 m. These deposits represent single story channels with a low lateral continuity (30–150 m). The point bar deposits typically represent lenticular geometries accreting perpendicular to the flow direction of the fluvial system. Splay deposits are up to 1 m thick and show longer range of lateral continuity than point bars at the site (30–250 m).
4. **Transmissivity**. Grain size of the point bars indicates moderate to high transmissivity (conductive layers). Splays and levees represent low to moderate transmissivity, while the muddy overbank deposits are likely to have very low transmissivity (confining layers).

(B)   *Bay-Head Delta Facies Association*

This facies association consists of bay-head delta mouth bars and prodelta deposits (Fig. 4.27). Such deposits typically occur in a bay/estuarine setting protected from

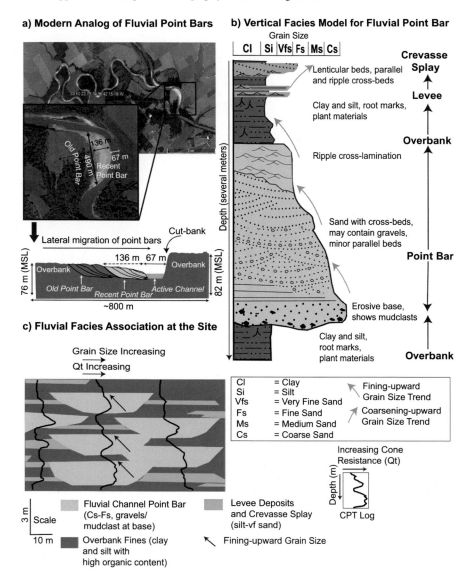

Fig. 4.26 **a** Modern analog of fluvial point bars with dimensions (modified from Google Earth image). **b** Vertical grain size profile of a typical point bar and associated facies conceptualized after Miall (2014). **c** Identification of fining-upward channel bar successions in site CPT data. The CPT information was calibrated with borehole core description for interpretation

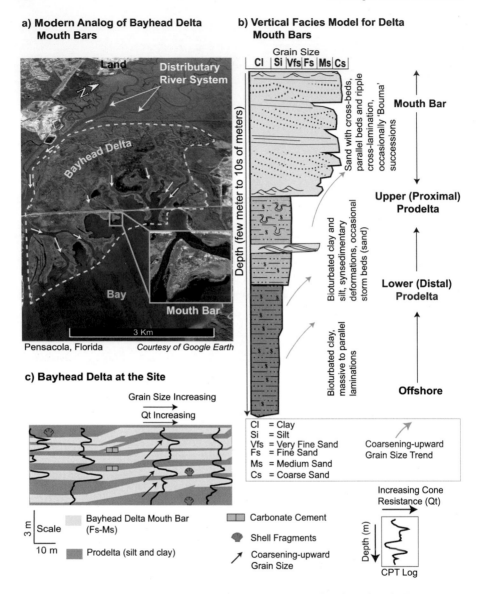

Fig. 4.27 a Modern analog of bay-head delta mouth bar (modified from Google Earth image).
b Vertical grain size profile of a typical delta mouth bar and associated facies conceptualized after
Bhattacharya and Walker (1991). **c** Identification of coarsening-upward mouth bar successions in
site CPT data. The CPT information was calibrated with borehole core description for interpretation

the open sea by barrier spits. However, these spits typically have low preservation
potentiality, and therefore, may not be present in the sediment record.

1. **Grain size/lithology**. Grain size of mouth bar deposits predominantly ranges from fine to medium sand with various proportions of clay and silt (5–20%). Coarse sand is observed locally. Gravel deposits are absent to rare and indications of carbonate cementation from early diagenesis are common. Prodelta deposits largely consist of silty mud, with occasional presence of shell fragments.
2. **Log signatures**. The bay-head mouth bar deposits appear 'coarsening-upward' to 'spiky' (as indicated by low Qt at the base and high Qt at the top) in CPT logs with locally 'blocky' responses (Fig. 4.27c). The associated prodelta deposits show uniformly low Qt values.
3. **Thickness, geometry, and lateral continuity of flow units**. Individual mouth bar units vary from 0.50 to more than 1.5 m thick. Amalgamated mouth bars locally reach a thickness of 2.4–3 m. The mouth bar units generally show high lateral continuity that ranges from 182 to more than 300 m in length. The mouth bars represent a shingled geometry of depositional units that follow the dip of the flow direction of the fluvial system.
4. **Transmissivity**. Mouth bars are inferred to be moderately transmissive (conductive layers), below the range of channel point bar deposits. Prodelta deposits are considered to have very low transmissivity (confining layers).

(C) *Tidal Deposits Facies Association*

This facies association consists of tidal channel bars (Fig. 4.28) and tidal mouth bars with estuarine mudflat deposits (Fig. 4.29).

1. **Grain size/lithology**. Grain size of tidal channel bars ranges from very fine to fine sand with a high content of mud. Tidal mouth bars range from very fine muddy sand to well-sorted fine/medium sand. Common sedimentological features of both tidal channels and tidal bars are mud-draped lenticular beds and flaser beds. Shell fragments are ubiquitous and carbonate cementation can occur locally. Rip-up clasts may be observed in tidal channels. Grain size of estuarine mudflat deposits varies from clay to silty clay. Abundant shell fragments and organic materials are present. The clay shows high plasticity ('fat clay').
2. **Log signatures**. The CPT log signature of the tidal channel deposits typically exhibits a fining-upward motif, with a sharp base of high Qt values and a gradational top of low Qt values (Fig. 4.28c). The tidal mouth bars show a gradational coarsening-upward motif (Fig. 4.29c) with high Qt values at the top and low Qt values at the base and have a more prominent 'spikey' appearance. The estuarine mudflats facies show low Qt values in general.
3. **Thickness, geometry, and lateral continuity of flow units**. The tidal channel bar deposits represent dimensions similar to the fluvial point bars at the site (0.50–1.5 m long, 100–150 m wide lenticular geometry). The tidal mouth bars are generally thinner than bay-head delta mouth bars at the site, attaining a thickness of 0.30–1 m. Amalgamation of tidal mouth bars is not identified. These tidal mouth bars represent a narrow, often 'spikey', shingled geometry of depositional units that follow the dip of the flow direction of the fluvial system. These units are estimated to have correlation lengths ranging from < 150–300 m.

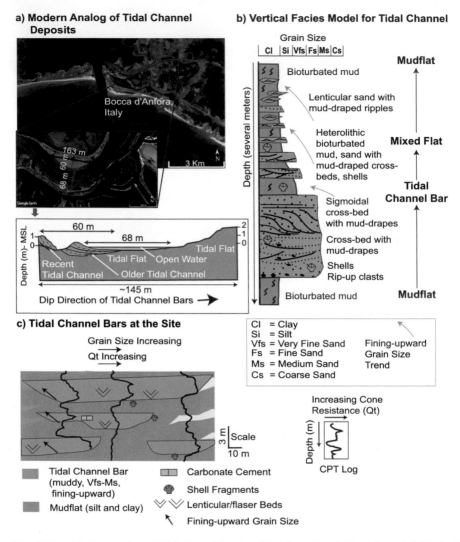

Fig. 4.28 **a** Modern analog of tidal channel bar (modified from Google Earth image). **b** Vertical grain size profile of a typical tidal channel bar conceptualized after Van Wagoner et al. (1990). **c** Identification of fining-upward muddy tidal channel bar successions in site CPT data. The CPT information was calibrated with borehole core description for interpretation

4. **Transmissivity**. Tidal channel bars and tidal mouth bars are considered to be conductive but are inferred to have lower transmissivity than fluvial point bars and bay-head delta mouth bars. This is because of flow inhibition caused by the common presence of intervening mud laminae in both. The estuarine mudflat deposits are considered to have very low transmissivity (confining layers).

Table 4.2 summarizes the main characteristics of the depositional facies identified at the site and their implications for LNAPL migration.

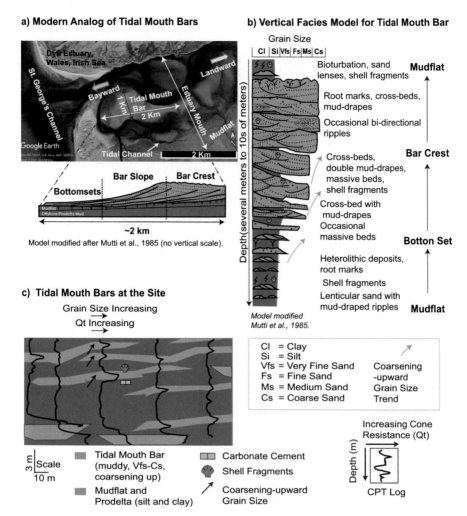

Fig. 4.29 a Modern analog of tidal mouth bar (modified from Google Earth image). The cross-sectional profile of an ideal tidal mouth bar is modified after Mutti et al. (1985). **b** Vertical grain size profile of a typical tidal mouth bar and associated facies after Mutti et al. (1985). **c** Identification of coarsening-upward mouth bar successions in site CPT data. The CPT information was calibrated with borehole core description for interpretation

Correlation and Interpretation of Borehole Data

Re-correlation of the previous cross-section (Fig. 4.20) for the site using sequence stratigraphic time markers reveals a complex stratigraphic architecture at the site (Fig. 4.30) controlled by sea level-driven accommodation changes (Figs. 4.22, 4.23, and 4.24b). As discussed in Sect. 4.5.2.2, the target depth interval of this case study

Table 4.2 Transmissivity and connectivity of various depositional facies at the site

Depositional facies	Transmissivity	Connectivity	Implication for LNAPL
Fluvial point bar	High	Low	Flow unit
Delta mouth bar	Moderate to high	High	Flow unit
Tidal channel bar	Low to moderate	Low	Flow unit (local)
Tidal bar	Low to moderate	Low to moderate	Flow unit (local)
Splay and levee	Low to moderate	Low	Flow unit (local)
Overbank fines	Very low	Moderate	Confining unit
Prodelta mud	Very low	Very high	Confining unit
Estuarine mud	Very low	Very high	Confining unit

(Semi-perched aquifer) covers only the shallow part of the Dominguez incised valley and does not include the basal fluvial channel deposits that comprise the Gaspur aquifer. The following is a brief description of the stratigraphic interpretation from bottom to top (Fig. 4.30) that establishes the sequence stratigraphic framework for the site.

The base of the stratigraphic interval covered in the cross-section starts above a probable transgressive surface (TS, Fig. 4.8c for illustration of concept). This surface is indicated by the signature of a regionally persistent clay-marker observed in cores and well logs that can be tied to a significant rise of sea level following the deposition of Dominguez basal fluvial channel deposits (Fig. 4.22). As the rate of accommodation generated from rising sea level exceeded sediment supply, a TST developed. This resulted in the deposition of a muddy tidal/estuarine environment and caused a landward (northward) shift of facies (see Fig. 4.8c and Fig. 4.14b-3 for concept).

Figure 4.31 shows the incorporation of depositional facies described in Sect. 4.5.2.4 to the stratigraphic framework developed in Fig. 4.30. The depositional package from −20 m above mean sea level (amsl) to −6 m amsl in this scenario is predominated by the Tidal Deposits Facies Association, as well as northward retro-grading ('back-stepping') sediments of the Bay-head Delta Facies Association, and Fluvial Facies Association. This depositional scenario conforms to the evolution of the incised valley fill as illustrated in Time-3 in Fig. 4.25. As the generation of accommodation started to slow down over time and sedimentation outpaced accom-modation, the Bay-head Delta Facies Association and Fluvial Facies Association started to prograde seaward (see concept illustrated in Fig. 4.14b-2). This significant 'turnaround' (e.g., seaward—southward—shift of facies) commenced following a maximum flooding of the coastline, marked by a regional mud layer (e.g., MFS) as shown in Figs. 4.30 and 4.31 (pink line), and led to the development of the HST (see concept illustrated in Fig. 4.8d). The HST is depicted from −6 to 9 m amsl and shows that the Fluvial Facies Association slowly overtopped the HST Bay-Head

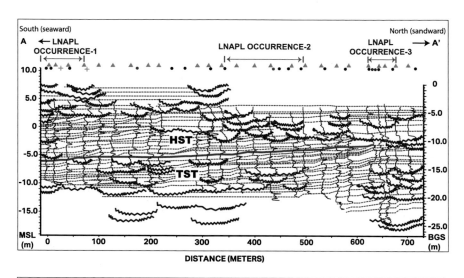

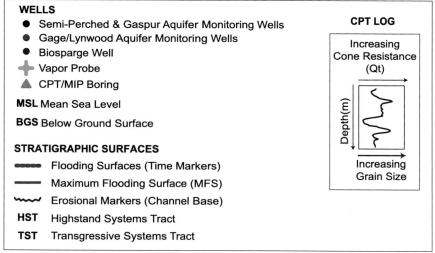

Fig. 4.30 Development of stratigraphic framework for the site based on identification of flooding surfaces and erosional markers from borehole grain size logs and CPT logs. An MFS shown in pink is identified as an important stratigraphic marker signifying the progradation of HST deposits over the predominantly tidal deposits of the underlying TST

Delta Facies Association, eventually pushing it farther seaward (outside the visual window of the example cross-section) over time. This depositional scenario conforms to the evolution of an incised valley fill illustrated in 'Time-4' in Fig. 4.25. The stratigraphic model for the site within the context of the regional understanding is shown in Fig. 4.32, and helps to demonstrate how the sequence stratigraphic correlation

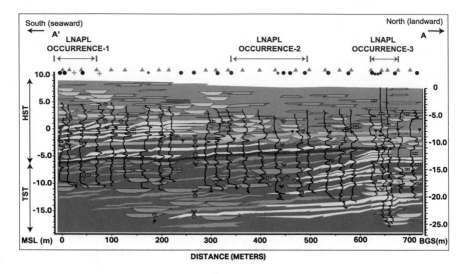

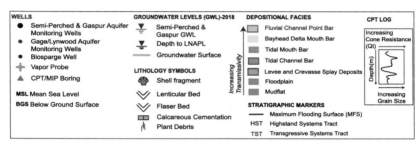

Fig. 4.31 Detailed stratigraphic correlation using site CPT logs and lithologic information. Ground-water level from 2018 is shown. Observe the southward progradation of deltaic and fluvial deposits above the MFS shown in pink. Deposits below the MFS largely consist of estuarine tidal deposits and back-stepping (southward-stepping) fluvio-deltaic deposits. See Electronic Supplementary Material for an uninterpreted version of Fig. 4.31) (ESM_3)

was used to reveal distinct, predictive stratigraphic packages of depositional facies through the target interval.

4.5.2.5   Implications for Biosparging

The sequence stratigraphic framework derived from the correlation and interpretation of borehole data was integrated with hydrologic and chemical data to characterize preferential LNAPL migration pathways and aid in the design of targeted biosparging wells.

LNAPL was detected at the site by in-situ tooling (LIF detector response) and observations in peaks of Photoionization Detector (PID) values, as well as charac-terization through laboratory analysis in forensic evaluation of samples. The current

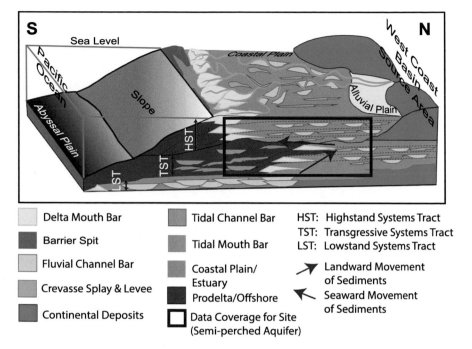

Fig. 4.32 A three-dimensional depositional model for the site derived from sequence stratigraphic correlation of site data and regional understanding. Inferred barrier spits in the model have not been identified in the dataset, likely because of poor preservation potential. The red box indicates the approximate lateral and vertical extent of the site dataset within the established stratigraphic framework. Note that the focus of this investigation was on the Semi-perched aquifer, which only covers the TST and HST deposits of the Dominguez sequence. The data does not include the LST channel bars (Gaspur aquifer) of the Dominguez incised valley

water table in the area lies at a depth about -0.5 m amsl to the north and about -2.5 m amsl to the south. An overlay of LNAPL plumes on the cross-section (Fig. 4.33) shows that the peak responses correlate to zones 1–3 m below the current water table. Impacts were not detected outside of the depth intervals of 2 to -12 m amsl.

The vertical distribution of LNAPL was dictated by the regional water table which declined steadily between the 1940s and the 1970s due to pumping activities in the region, followed by a near complete rebound from the 1970s to the present (Fig. 4.34). LNAPL released prior to 1970 extended down to 18–21 m below the current water table and into more permeable sands encountered at those depths. Since the rebound of the groundwater table, much of the deeper LNAPL became confined within more permeable lower sands (tidal channel bars and tidal mouth bars) interbedded with silt and clay. The mobile LNAPL not trapped beneath a low-permeability layer was redistributed vertically within the saturated portion of the soil column, following the rising water level and creating smear zones containing LNAPL of variable pore-fluid saturation in portions of the Semi-perched aquifer. Screen intervals of some

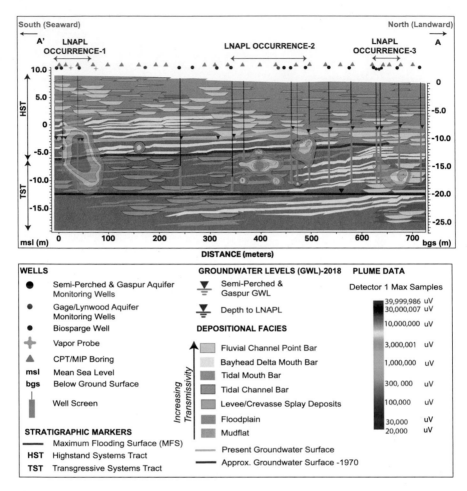

Fig. 4.33 LNAPL plumes superimposed upon the cross-section shown in Fig. 4.31. See text for detailed interpretation of LNAPL occurrences with respect to stratigraphy. Also, observe the approximate 1970 groundwater level, which helps to explain the presence of LNAPL at levels significantly lower than the present-day groundwater level

monitoring and recovery wells designed for a historically deeper water table were submerged by the rising groundwater level.

Three major zones of LNAPL ('LNAPL Occurrences 1–3') were observed along the cross-section (Fig. 4.33). Note that these plume shapes were derived prior to the stratigraphic work, and their geometry is largely a product of statistical interpolation (kriging) of chemical data, rather than geological constraints on fluid flow. However, the refined stratigraphic understanding shows that the plumes were located between the upper units of the TST and the lower units of the HST facies associations.

LNAPL Occurrence-1 appeared as a large, vertically elongated LNAPL plume in the section confined to southernmost seaward part of the section between 1.5 and

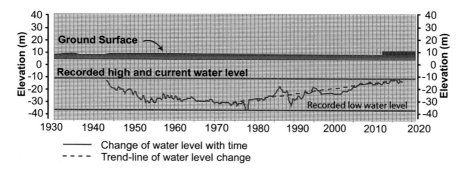

Fig. 4.34 Anthropogenic influences on groundwater at the site since the 1940s. Historic regional groundwater levels in the site area were significantly deeper than current levels. This was due to groundwater pumping between the 1940s and 1970s that lowered the local water table by as much as 18–21 m below current levels

−14 m amsl. Transmissive units in this zone of occurrence are predominated by upper TST tidal mouth bars and lower HST bay-head delta mouth bars, along with minor tidal channel bar deposits.

LNAPL Occurrence-2 showed two individual plumes largely confined within the upper TST in the landward direction to the north between −3 to −4 m amsl and −8 to −11 m amsl. Transmissive units in this zone of occurrence include tidal mouth bars and tidal channel bars.

LNAPL Occurrence-3 was represented by the landward-most part of the upper TST to the north between −5 and −11.5 m amsl. Transmissive units in this zone of occurrence are predominated by fluvial point bars and back-stepping bay-head delta mouth bars.

As evident from the sequence stratigraphic cross-section, the occurrence, connectivity, and elevations of conductive and confining layers frequently had vertical shifts exceeding 1.5 m even with locations only a few meters apart because of the seaward dipping nature of the strata. This observation was previously unrecognized in a lithostratigraphic correlation (Fig. 4.20). Level of impacts from LNAPL concentrations was also regularly as diverse. Considering this heterogeneity, over 60 biosparging wells were installed over a distance of approximately 800 m at the site location, where design of each well was developed based on the revised stratigraphic framework and then finalized based upon each individual zone's specific geological characteristics described in Sect. 4.5.2.4 and summarized in Table 4.2. Since the conductive layers through which LNAPL would be transported at the site appear to be predominated by thin shingles of tidal mouth bars and bay-head delta mouth bars (0.5–1.5 m thick), the screening intervals for the aerobic biosparging wells were set accordingly (in 0.5–1.5 m intervals). Thus, the revised sequence stratigraphic CSM proved essential for targeting optimal well placement/depth and aiding in the design of a more effective remedial strategy.

4.6 Summary

Subsurface heterogeneity, particularly in complex geological settings, imparts significant technical challenges in remediating LNAPL contamination and achieving
cleanup goals within reasonable timeframes. Therefore, implementing a correlation
approach that adequately addresses subsurface heterogeneity and reduces uncertainty between sampling locations (especially where data quality and coverage may
be limited) is critical for effective remedial decision-making.

In traditional LNAPL remedial investigations, correlation has typically been
conducted using 'lithostratigraphy', which attempts to connect sedimentary units
of similar lithology, often without any scientific basis, and without recognizing the
heterogeneity of subsurface sediments between boreholes. The resulting 'layer cake'
stratigraphy of the subsurface is often too simplistic and inadequate for developing
effective remedial strategies.

In contrast, the sequence stratigraphic approach (with the aid of facies models)
provides a more realistic subsurface correlation based on predictable distribution
of sediments in different depositional environments, and is supported by outcrop
observations, seismic studies, and various laboratory flume experiments.

The key factors in sequence stratigraphy are accommodation and supply of sediments. Classical sequence stratigraphy largely focuses on understanding the heterogeneity of marine and coastal environments controlled by relative sea level changes.
However, since contaminated sites are not limited to locations in marine and coastal
environments, the same principles of accommodation and supply can also be extended
to encompass continental environments (e.g., fluvial, alluvial fan, and glacial) where
tectonics and climate play a more significant role in controlling the patterns of deposition. In these settings, autocyclic geological processes often prove to be as important
as allocyclic geological processes for understanding and predicting the distribution
of sediments.

The three-dimensional geologic framework derived from this inclusive correlation approach can be used to delineate the spatial heterogeneity of sedimentary
units across multiple scales and to provide the theoretical validity for predicting
the distribution of aquifers and aquitards beyond the existing site data set. Moreover, integration of site hydrologic and chemical data within this framework can
result in precise identification of preferential contamination flow pathways. Thus,
a sequence-stratigraphy-based conceptual site model is recommended for guiding
remedial decision-making at LNAPL-impacted sites.

4.7 Future Directions

This chapter proposed a practical workflow for applying sequence stratigraphic techniques to continental settings. However, more theoretical studies in continental environments are necessary to further streamline a workflow that can better deal with contamination in systems such as alluvial fan and glacial settings.

Although sequence stratigraphy is gradually emerging as a best practice for environmental site management, integration of sequence stratigraphy into groundwater models remains a challenge. More research is necessary to develop a coherent methodology for integrating stratigraphic information with hydrogeology and stochastic frameworks of plume investigation through a multidisciplinary approach.

We recommend optimal spatial coverage of detailed core description (preferably using the Wentworth scale for grain size) from boreholes in combination with geophysical logs (e.g., gamma, resistivity, neutron etc.) or logs from direct-push profiling methods such as CPT and HPT (or others discussed in Chap. 7) for detailed stratigraphic correlation at a site. Application of surface geological techniques, such as shallow seismic investigation, resistivity imaging, and GPR data acquisition can further enhance high-resolution sequence stratigraphy. Complementing state-of-the-art-techniques of contamination detection (as those discussed in Chap. 8) with sequence stratigraphic correlation will provide more efficient methods for LNAPL investigation in the future.

Acknowledgements The authors wish to thank Jay Francisco for his assistance in obtaining client-permission for sharing the case study presented in this chapter. They are especially grateful to Assaf Reese, Katherine Carr, Sarah Price, and Danielle Cebra for their outstanding support in acquiring and compiling site data, which was essential for developing this chapter. Ben Campanaro provided valuable feedback and comments to improve the manuscript. Last, but not the least, the authors would like to thank Gina DiSimone and Maryanne Cleary for their support in shaping the chapter through graphic and formatting support.

References

Ainsworth RB, Sanlung M, Duivenvoorden STC (1999) Correlation techniques, perforation strategies, and recovery factors: an integrated 3-D reservoir modeling study, Sirikit Field, Thailand. AAPG Bull 83:1535–1551

Anderton R (1985) Clastic facies models and facies analysis. Geol Soc Lon Spec Pubs 18(1):31–47

Barrell J (1917) Rhythms and the measurements of geologic time. Geol Soc of Am Bull 28:745–904

Bassinot F, Labeyrie L, Vincent E, Quidelleur X, Shackleton N, Lancelot Y (1994) The astronomical theory of climate and the age of the Brunhes-Matuyama magnetic reversal. Earth Planet Sci Lett 126:91–108

Bhattacharya J, Walker R (1991) River- and wave-dominated depositional systems of the upper cretaceous Dunvegan formation, northwestern Alberta. Can Petrol Geol Bull 39(2):165–191

Brookfield M, Martini I (1999) Facies architecture and sequence stratigraphy in glacially influenced basins: basic problems and water-level/glacier input-point controls (with an example from the quaternary of Ontario, Canada). Sed Geol 123:183–197

California Dept. of Water Resources (1961) Planned Utilization of the Ground Water Basins of the Coastal Plain of Los Angeles County, Appendix A. California Department of Water Resources Bulletin 104:191

Catuneanu O (2006) Principles of sequence stratigraphy. Elsevier, Amsterdam

CL:AIRE (2014) An illustrated handbook of LNAPL transport and fate in the subsurface. London, UK. www.claire.co.uk/LNAPL

Ehman KD, Edwards B (2014) Sequence stratigraphic framework of upper pliocene to holocene sediments of the Los Angeles Basin, California: Pacific Section. SEPM Book 112, Studies on Pacific Region Stratigraphy 47

Ehman KD, Cramer RS (1997) Application of sequence stratigraphy to evaluate groundwater resources. In Kendall DR (ed) Proceedings of the American association of water resources symposium, conjunctive use of water resources: aquifer storage and recovery, American Water Resources Association, Herndon, Virginia, TPS-97-2: 221–230

Embrey AF (2009) Practical sequence stratigraphy. Reservoir 36:1–6

Emery D, Meyers KJ (1996) Sequence stratigraphy. Blackwell, Oxford

Farr A, Houghtalen R, McWhorter D (1990) Volume estimation of light nonaqueous phase liquids in porous media. Ground Water 28(1):48–56

Fielding CR, Crane RC (1987) An application of statistical modelling to the prediction of hydrocarbon recovery factors in fluvial reservoir sequences. SEPM Spec Pub 39:321–327

Fogg GE, Zhang Y (2016) Debates—stochastic subsurface hydrology from theory to practice: a geologic perspective. Water Resour Res 52(12):9235–9245

Freeze RA (1975) A stochastic-conceptual analysis of one-dimensional groundwater flow in nonuniform homogeneous media. J Water Resour Res 11(5):725–741

Galloway W (1989) Sequence stratigraphy; sequences and systems tract development. AAPG Bull 73(2):143–154

Gibling M (2006) Width and thickness of fluvial channel bodies and valley fills in the geological record: a literature compilation and classification. J Sed Res 76(5):731–770

Haq B, Schutter S (2008) A chronology of Palezoic sea-level changes. Science 322(5898):64–88

Huntley D, Wallace J, Hawk R (1994) Nonaqueous phase hydrocarbon in a fine-grained sandstone: 2. Effect of local sediment variability on the estimation of hydrocarbon volumes. Ground Water 32(5):778–783

James NP, Dalrymple RW (2010) Facies models 4. St. John's; Nfld: Geol Assoc of Can

Kendall C (2006) SEPMSTRAT.org. http://sepmstrata.org

LeBlanc DR, Garabedian SP, Hess KM, Gelhar LW, Quadri RD, Stollenwerk KG, Wood WW (1991) Large-scale natural gradient tracer test in sand and gravel, Cape Cod, Massachusetts: 1. Experimental design and observed tracer movement. Water Resour Res 27(5):895–910

Lenhard R, Parker J (1990) Estimation of free hydrocarbon volume from fluid levels in monitoring wells. Ground Water 28(1):57–67

Levitt JP, Degnan JR, Flanagan SM, Jurgens BC (2019) Arsenic variability and groundwater age in three water supply wells in southeast New Hampshire. Geoscience Frontiers 10(5):1669–1683. ISSN 1674-9871

Liu JP, DeMaster DJ, Nguyen TT, Saito Y, Nguyen VL, Ta TKO, Li X (2017) Stratigraphic formation of the Mekong River delta and its recent shoreline changes. Oceanography 30(3):72–83

Mariethoz G, Caers J (2015) Multiple-point geostatistics: stochastic modeling with training images. Wiley-Blackwell

Meyer J, Parker B, Arnaud E, Runkel A (2016) Combining high resolution vertical gradients and sequence stratigraphy to delineate hydrogeologic units for a contaminated sedimentary rock aquifer system. J Hydrol 534:505–523

Miall AD (1988) Architectural-element analysis: a new method of facies analysis applied to fluvial deposits. Earth Sci Rev 22(4):61–308

Miall AD (2014) Fluvial style. In: Fluvial depositional systems (14). Springer International Publishing, p 2–68

Middleton G (1973) Walther's Law of the correlation of facies. Bull Geol Soc of Am 84(3):979–988

Mutti E, Rosell J, Allen G, Fonnesu F, Sgavetti M (1985) The Eocene Baronia tidedominated delta shelf system in the Ager Basin. In: Mila MD, Rosell J (Eds.) Excursion Guidebook: VI Eur. Ref. Mtg. I.A.S. 13. Excursion, Lerida, Spain, 579–600

North American Commission on Stratigraphic Nomenclature (1983, Rev. 2005) North American Stratigraphic Code. AAPG Bull 89(11):1547–1591

Paola C, Straub K, Mohrig D, Reinhardt L (2009) The "unreasonable effectiveness" of stratigraphic and geomorphic experiments. Earth Sci Rev 97(1–4):1–43. ISSN 0012-8252

Poland JF, Piper AM (1956) Ground-water geology of the coastal zone; Long Beach-Santa Ana Area, California. US Geol Survey Water-Supply Paper 1109

Ponti D, Ehman K, Edwards B, Tinsley J, Hildenbrand T, Hillhouse J, Hanson J, Randall T, McDougall K, Powell C, Wan E, Land M, Mahan S, Sarna-Wojcicki, A (2007) A 3-dimensional model of waterbearing sequences in the Dominguez Gap Region, Long Beach, California. U.S. Geol Survey Open-File Report 2007 1013:34

Posamentier HW, Walker RG (2006) Facies models revisited. Tulsa, Okla: SEPM

Posamentier HW, Allen G (1999) Siliciclastic sequence stratigraphy; concepts and applications. SEPM 7

Reynolds AD (1999) Dimensions of Paralic Sandstone Bodies: AAPG Bull 83:11–229

Riel B, Simons M, Ponti D, Agram P, Jolivet R (2018) Quantifying ground deformation in the Los Angeles and Santa Ana Coastal Basins due to groundwater withdrawal. Water Resour Res 54(5):3557–3582

Sadeque J (2020) Role of sequence stratigraphy in remediation geology for the puchack well field superfund site, NJ. AECOM report for U.S. EPA

Samuels R, Sadeque J (2021) The application of environmental sequence stratigraphy at department of defense sites-technical report for NAVFAC EXWC

Shultz M, Cramer RS, Plank C, Levine H, Ehman KD (2017) Best practices for environmental site management: a practical guide for applying environmental sequence stratigraphy to improve conceptual site models. U.S. Environmental Protection Agency, Washington, DC, EPA/600/R-17/293

Shanley KW, McCabe P (1994) Perspectives on the sequence stratigraphy of continental strata. AAPG Bull 78(4):544–568

Smith L, Schwartz F (1980) Mass transport 1. A stochastic analysis or macroscopic dispersion. Water Resour Res 16(2):303–313

Sugarman P, Miller K (1997) Correlation of Miocene sequences and hydrogeologic units New Jersey Coastal Plain. Sed Geol 108(1–4):3–18

Vail P, Todd R, Sangree J (1977) Seismic stratigraphy and global changes of sea level; Part 5, chronostratigraphic significance of seismic reflections. AAPG Mem 26:99–116

Vakarelov BK, Ainsworth RB (2013) A hierarchical approach to architectural classification in marginal-marine systems: Bridging the gap between sedimentology and sequence stratigraphy. AAPG Bull 97(7):1121–1161

Van Wagoner J, Mitchum R, Campion K, Rahmanian V (1990) Silicislastic sequence stratigraphy in well logs, cores, and outcrops: concepts for high-resolution correlation of time and facies. AAPG Methods Explor Series 7:55

Waddill DW, Parker JC (1997) Simulated recovery of light, nonaqueous phase liquid from unconfined heterogeneous aquifers. Groundwater 35(6):938–947

Walker RG, James NP (1992) Facies Models: Response to Sea Level Change 2nd Edition. St. John's; Nfld: Geol Assoc of Can

Walther J (1894) Einleitung in die Geologie als historische Wissenschaft. Lithogenesis Der Gegenwart Jena: g. Fischer 3:535–1055

Wang R, Colombera L, Mountney NP, (2019) Quantitative analysis of the stratigraphic architecture of incised-valley fills: a global comparison of quaternary systems. Earth-Science Rev 200

Wentworth C (1922) A scale of grade and class terms for clastic sediments. J Geol 30(5):377–392

Zielbauer E, Kues H, Burnham W, Keene A (1962) Coastal basins barrier and replenishment investigation—Dominguez gap barrier project geologic investigation. Report by Los Angeles County
Flood Control District

Chapter 5
Natural Source Zone Depletion of Petroleum Hydrocarbon NAPL

Kayvan Karimi Askarani, Tom Sale, and Tom Palaia

Abstract In the last decade, it has become widely recognized that petroleum found in soil and groundwater in the form of non-aqueous phase liquid (NAPL) is depleted by naturally occurring microbial communities. Losses of petroleum NAPL via natural processes are referred to as natural source zone depletion (NSZD). The natural loss rates of petroleum NAPL are large enough that they can often be the primary component of a site management strategy. Losses of NAPL through NSZD processes provide by-products such as CO_2, CH_4, and heat. As such, based on consumption of O_2, production of CO_2 and CH_4, generation of heat, or changes in petroleum NAPL chemical composition over time, a variety of methods have been developed to measure NSZD rates. Each method has advantages and limitations. Therefore, care is needed to select the method that best fits site conditions and site- and project-specific data quality objectives.

Keywords Biodegradation · NSZD rate · Petroleum remediation · Soil gas efflux · Subsurface temperature

K. Karimi Askarani (✉) · T. Sale
Civil and Environmental Engineering Department, Colorado State University, Fort Collins, CO 80523, USA
e-mail: KKarimiAskarani@gsi-net.com

T. Sale
e-mail: TSale@engr.colostate.edu

K. Karimi Askarani
GSI Environmental Inc., 2211 Norfolk, Suite 1000, Houston, TX 77098, USA

T. Palaia
Jacobs Solutions, Greenwood Village, CO 80111, USA
e-mail: Tom.palaia@jacobs.com

© The Author(s) 2024
J. García-Rincón et al. (eds.), *Advances in the Characterisation and Remediation of Sites Contaminated with Petroleum Hydrocarbons*, Environmental Contamination Remediation and Management, https://doi.org/10.1007/978-3-031-34447-3_5

5.1   Overview of NSZD Process

Inadvertent releases of subsurface petroleum hydrocarbons in the form of light non-aqueous phase liquids (LNAPLs) and non-chlorinated dense non-aqueous phase liquids (DNAPLs), collectively termed "petroleum NAPL," are a common occurrence in the industrial world. Natural mechanisms in the subsurface depleting petroleum NAPL in the source zone are referred to as natural source zone depletion (NSZD) (ITRC 2009, 2018). NSZD begins right after a release. Volatilization and dissolution occur, but biodegradation eventually predominates as intrinsic microorganisms acclimate and use the petroleum NAPL as a growth substrate (ITRC 2018).

Significant NSZD is frequently observed to occur at petroleum NAPL release sites through a combination of direct-contact biodegradation of the petroleum NAPL body itself, biodegradation of solubilized hydrocarbons at the oil/water interface, volatilization of petroleum NAPL constituents followed by biodegradation in the vadose zone (as further discussed in Chap. 6), and, to a lesser extent, dissolution of petroleum NAPL constituents and aqueous biodegradation in the saturated zone (Johnson et al. 2006; ITRC 2018; CRC CARE 2018).

Evolution of biodegradation reactions in the petroleum NAPL source zone is dependent on available electron acceptors (Baedecker et al. 2011). Biodegradation of petroleum NAPL constituents occurs across the entire thickness of the smear zone (i.e., from unsaturated, to partially, to fully saturated) by naturally occurring microorganisms (ITRC 2018). Biodegradation occurs via a multitude of mechanisms in both aerobic and anaerobic conditions. Aerobic biodegradation occurs where ample oxygen (O_2) is present. Over time, depletion of O_2, manganese (Mn^{+4}), nitrate (NO_3), and iron (Fe^{+3}) leads to sulfate reduction and methanogenesis becoming the primary drivers of biodegradation (Bekins et al. 2005; Atekwana and Atekwana 2010; Molins et al. 2010; Irianni Renno et al. 2016; Garg et al. 2017; Smith et al. 2022). Following Fig. 5.1, methanogenesis is central to NSZD. While methanogenesis can be slow, it has the remarkable advantages of not being limited by the availability of an electron acceptor (API 2017). Methanogenesis is a multi-step syntrophic process that is intricate, and, as a metabolic phenomenon, two groups of organisms work together to degrade complex organics (Garg et al. 2017).

As illustrated in Fig. 5.1, under anaerobic conditions in Region 3, the consumption of hydrocarbons by methanogenic organisms produces methane (CH_4) and carbon dioxide (CO_2). Gas bubbles form from local exceedances of aqueous-phase gas solubilities in the saturated zone due to produced CH_4 and CO_2. Buoyancy forces can overcome capillary forces when the gas bubbles become sufficiently large leading to upward ebullition and outgassing of gases including CH_4 and CO_2 along with volatile organic compounds (VOCs) into the saturated zone (Amos et al. 2005; Ramirez et al. 2015; Garg et al. 2017). In the overlying vadose zone, the upward advection–diffusion-driven flux of CH_4 is encountered by a downward flux of O_2 from atmosphere (Region 2). Where the CH_4 and O_2 meet in the vadose zone, CH_4 is converted to CO_2 by methanotrophs and generates heat via the exothermic oxidation reaction (Stockwell 2015; Irianni Renno et al. 2016; Garg et al. 2017). Additionally,

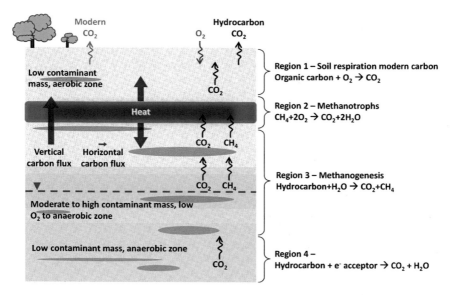

Fig. 5.1 Conceptual model of primary NSZD process (modified from Amos et al. 2005; Irianni-Renno et al. 2016; Karimi Askarani et al. 2018). The primary heat source occurs in Region 2, and heat generation through other processes (e.g., methanogenesis) is negligible

downward flux of O_2 may be consumed periodically by VOCs instead of CH_4 (Davis et al. 2005; Sookhak Lari et al. 2019).

Direct-contact biodegradation occurs in the immediate proximity to the petroleum NAPL, within pores with oil where air-phase porosity is present (e.g., top of an LNAPL body). By-product gases from this reaction are directly outgassed to the vadose zone and do not enter the aqueous phase. Research conducted at the Bemidji crude oil site attributed over 80% of the observed carbon efflux to direct-contact biodegradation and outgassing phenomena (Ng et al. 2015; ITRC 2018).

5.2 Measuring NSZD Rates

Measurement of NSZD rates is central to developing a conceptual understanding of petroleum NAPL sites, developing rational remedies, and tracking progress to clean up. First, it is critical to establish the purpose for which NSZD data will be used (i.e., establish data quality objectives). Initially, site characterization and conceptual site model (CSM) development may be the primary driver for measuring NSZD. With time, motives for resolving NSZD rates can move to support remedy selection decision and monitoring progress to clean up. Moreover, selection of an NSZD measurement scope and method relies on various other aspects of the petroleum NAPL CSM. Elements of the CSM, such as extent of NAPL, depth to groundwater, water table fluctuation, moisture content, lithology, and others, can impact the NSZD measurement method and rates. Key elements of the CSM and how they relate to NSZD measurement are described in detail in CRC CARE (2018).

Temporal and geospatial dynamics in the subsurface affect fluxes of O_2, CO_2, CH_4, and the associated heat. Incidentally, NSZD rates measured using these parameters may be variable in time and space. For instance, NSZD rates have been found to increase with temperature (Kulkarni et al. 2022b). NSZD rates can also vary from point to point with differences in the petroleum NAPL vertical distribution, condition (i.e., unconfined or confined), and proximity to the periphery of the NAPL body (CRC CARE 2018). For instance, as shown in Fig. 5.2, consumption of O_2 or production of CO_2 is low with a flat rate of change in concentration over depth within the vadose zone in the background location. Due to the absence of CH_4 and VOCs, no NSZD-related heat is generated at this location. In contrast, over the petroleum NAPL footprint where NSZD is occurring, there is significant O_2 consumption and CO_2 production with a rapid change in concentration over depth. Also, the presence of CH_4 and VOC provides strong heat signals in the subsurface. Hence, understanding the NAPL distribution is a crucial first step in specifying the locations for NSZD measurements.

The simplified conceptualization of processes governing NSZD (Fig. 5.1) shows the possible fluxes of gasses and heat generated for the methanogenesis-dominated

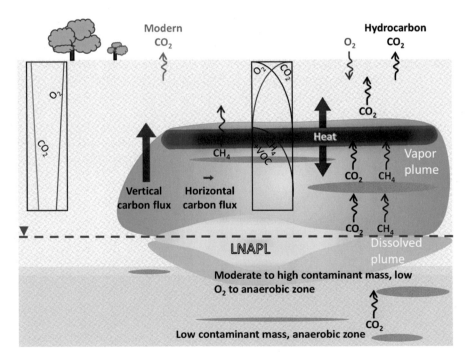

Fig. 5.2 Hypothetical soil gas concentration profiles and biogenic heat at background and high NSZD rate area (modified from CL: AIRE 2014; CRC CARE 2018)

case. Accordingly, following CRC CARE (2020) the most frequent methods for quantifying NSZD rates from the biogases focus on:

(1) vertical profiles of CO_2, O_2, CH_4, and VOCs concentrations in the vadose zone
(2) near surface CO_2 efflux, and/or
(3) temperature gradients in the vadose zone.

In addition, the NSZD rate can be quantified via observed changes in petroleum NAPL chemical composition over time (Ng et al. 2014 and 2015; DeVaull et al. 2020). This chapter includes a description of the chemical compositional change method but excludes discussion of dissolved-phase methods. Details of those generally well-established methods are covered in detail in various other available literature (CRC CARE 2018; ITRC 2009; NRC 2000).

In the following sections, methods for quantifying NSZD rates are summarized to provide a basis for quantifying NSZD rates including:

(i)   Concentration gradient
(ii)   Passive flux trap
(iii) Dynamic closed chamber
(iv)   Soil temperature
(v)   Chemical composition change.

Table 5.1 summarizes key attributes of these NSZD rate measurement methods (Tracy 2015; CRC CARE 2018; Karimi Askarani 2019). They are discussed in more detail below.

Overall, these data can be used to (i) demonstrate that NSZD is occurring; (ii) quantify rates of NSZD; (iii) incorporate NSZD into the CSM, and (iv) evaluate NSZD as a component of the remedy. As mentioned above, NSZD rate measurements inherently involve both temporal and geospatial variability in results and are affected by site conditions in the CSM. As such, site-specific characteristics and project goals should be carefully considered when selecting, designing, and implementing NSZD measurement methods (API 2017; ITRC 2018; CRC CARE 2018).

5.2.1 Soil Gas Methods

High levels of CH_4 and CO_2 above petroleum NAPL bodies and, uptake of O_2, provide important lines of evidence for NSZD. Here, CH_4 and CO_2 produced through methanogenic processes within the petroleum NAPL smear zone are transported via diffusion and advection upward through the vadose zone. Where CH_4 encounters O_2, it is oxidized to CO_2. Hence, NSZD rates can be quantified by measuring the flux of the gasses. Notably, background correction is required for all soil gas flux-based methods to isolate the gas flux associated with NSZD (J_{NSZD}) from natural soil respiration ($J_{Background}$). Following Sihota et al. (2011):

$$J_{NSZD} = J_{Total} - J_{Background} \tag{5.1}$$

Table 5.1 Summary of NSZD rate measurement characteristics (based on Tracy 2015; API 2017; CRC CARE 2018; Karimi Askarani 2019)

Characteristics	NSZD rate measurement method					
	Soil gas			Temperature		Other
	Concentration gradient	Dynamic closed chamber	Passive flux trap	Background-corrected	Non-background-corrected	
Measurement parameter(s)	CO_2, O_2, CH_4, VOC	CO_2, CH_4	CO_2	Soil temperature	Soil temperature	Chemical concentration in NAPL
Intrusiveness of method	Intrusive	Minimal[e]	Minimal[i]	Intrusive	Intrusive	Minimal to intrusive with soil cores
Period of measurement	Instantaneous	Typically, instantaneous	Time averaged integral value over days to weeks[j, k]	Continuous or instantaneous	Continuous or instantaneous	Time integrated over period of years to decades
Time to results	Days	Real-time field values	Weeks[i, j, k]	Real time	Real time	Weeks
Level of effort required	High[a]	Moderate	Low	Moderate	Moderate	Moderate to high with soil cores

(continued)

Table 5.1 (continued)

Characteristics	NSZD rate measurement method					
	Soil gas			Temperature		Other
	Concentration gradient	Dynamic closed chamber	Passive flux trap	Background-corrected	Non-background-corrected	
NSZD mechanism measured	Volatilization and biodegradation[a]	Volatilization and biodegradation, assuming VOC and CH$_4$ oxidation[f]	Volatilization and biodegradation, assuming VOC and CH$_4$ oxidation[f]	Volatilization and biodegradation, assuming VOC and CH$_4$ oxidation[l, m]	Volatilization and biodegradation, assuming VOC and CH$_4$ oxidation[l, m]	Dissolution, volatilization, and biodegradation
Transport process quantified	Diffusion of soil gas[a]	Advection and diffusion of soil gas[f]	Advection and diffusion of soil gas[f, k]	Conduction of heat from oxidation[l, m]	Conduction of heat from oxidation[l, m]	Not applicable
Background correction for non-NSZD-related processes	Required[a]	Required[d, g]	Required[d, g, k]	Required	Not required	Not required
Method of background correction	Background location, soil gas ^{14}C[p, g, r]	Background location, soil gas ^{14}C[p, q, r]	^{14}C[k]	Background location	Not required	Not required
Influence of barometric pumping	May be subject to barometric pumping[b]	May be subject to barometric pumping[h]	None (captures variation)[k]	Not applicable	Not applicable	Not applicable

(continued)

Table 5.1 (continued)

Characteristics	NSZD rate measurement method					
	Soil gas			Temperature		Other
	Concentration gradient	Dynamic closed chamber	Passive flux trap	Background-corrected	Non-background-corrected	Petroleum NAPL chemical composition
Influence of surface wind	Low[b, c]	Potential influence	Potential influence	Not applicable	Not applicable	Not applicable
Influence of precipitation and/or soil moisture	High[a, b]	Moderate to High	Moderate to High[j]	Moderate[m]	Minimal[o]	None
Influence of artificial ground surfaces	Minimal[d]	High	High	High in case of different surface at background area[l]	None[o]	None
Influence of heterogeneous subsurface	High[d]	Low	Low	Moderate[l]	Low[o]	None

[a]Johnson et al. (2006), [b]Maier and Schack-Kirchner (2014), [c]Poulsen and Møldrup (2006), [d]Coffin et al. (2008), [e]LI-COR (2010), [f]Molins et al. (2010), [g]Sihota et al. (2011), [h]Wyatt et al. (1995), [i]McCoy (2012), [j]Zimbron et al. (2014), [k]McCoy et al. (2015), [l]Stockwell (2015), [m]Warren et al. (2015), [n]Luo et al. (2013), [o]Karimi Askarani et al. (2020), [p]Reynolds (2019), [q]Crann et al. (2017), [r]Murseli et al. (2019)

After background correction, the gas flux (J_{NSZD}) can be converted to an NSZD rate using stoichiometry following CRC CARE (2018) as:

$$R_{NSZD} = \left[\frac{J_{NSZD} m_r MW}{10^6} \right] \times \frac{86,400 \text{ s}}{d} \qquad (5.2)$$

where R_{NSZD} is the total hydrocarbon degraded or NSZD rate (g/m^2/d), J_{NSZD} is the background-corrected soil gas flux measurement (micromoles per square meter per second (μmol/m^2/s)), m_r is the stoichiometric molar ratio of hydrocarbon degraded to CO_2 produced (unitless), and MW is the molecular weight of a representative hydrocarbon in the petroleum NAPL (g/mol).

It is important to note that the equations above use CO_2 efflux as the basis for the NSZD rate calculation. If site conditions do not allow for complete conversion of all CH_4 and VOCs emitted from the NSZD processes to be converted to CO_2, (i.e., O_2 diffusion into the subsurface is slowed by semi- or impervious ground cover or excessive soil moisture), the use of these equations will result in an underestimate of the NSZD rate. To accurately quantify the NSZD rate in this case, additional measurements of CH_4 and/or VOC efflux are required, and equations specific to CH_4 stoichiometry and VOC mass loss must be derived for the site. The total NSZD rate for the site will then equal the summation of the stoichiometric contributions from both CO_2, CH_4, and VOC mass loss.

5.2.1.1 Concentration Gradient Method

In the vadose zone, O_2 and CO_2 concentration profiles change due to underlying NSZD processes. Using background-corrected soil gas concentration gradients (i.e., slope of the concentration versus depth profile) of either O_2 or CO_2, and assuming that diffusion is the primary transport process, Fick's equation can be used to estimate gas fluxes and correspondingly NSZD rates (Johnson et al. 2006; Lundegard and Johnson 2006; ITRC 2009). The gradient method is based on Fick's first law of diffusion and used to estimate the gas flux as:

$$J_{diff} = D_v^{eff} \left(\frac{dc}{dz} \right) \qquad (5.3)$$

where J_{diff} is the steady-state diffusive flux (g/m^2/s), $\frac{dc}{dz}$ is the soil gas concentration gradient (g/m^3/m), and D_v^{eff} is the effective vapor diffusion coefficient (m^2/s).

The key assumptions of the gradient method include (i) diffusion is the primary governing process for gas flux, (ii) the vadose zone is homogeneous and isotropic with respect to diffusion coefficients, and (iii) it is a steady-state condition. The gradient method can be applied to any soil gas parameter, but is most commonly used with CO_2. If the concentration gradient of another gas such as CH_4 or O_2 is used, then the stoichiometry in the NSZD rate Eq. 5.2 must be modified accordingly.

Measurement Procedure—As illustrated in Fig. 5.3, application of the concentration gradient method includes the following:

(1) install multilevel vapor sampling probes at petroleum NAPL-impacted and background locations,
(2) measure O_2, CO_2, CH_4, and VOC concentrations in the sampling probes,
(3) immediately after the soil gas concentration measurements, conduct an in-situ tracer test to estimate an effective diffusion coefficient (Johnson et al. 1998),
(4) calculate concentration gradients (i.e., slope of the concentration profile) at background and petroleum NAPL-impacted locations to resolve background-corrected gradients, and
(5) estimate a point in time NSZD rate based on the fluxes of gases.

Details of the key steps involved in measuring soil gas gradients and estimating NSZD rates using the concentration gradient method are provided in CRC CARE 2018.

Challenges, Considerations, and Feasibility—The effective diffusion coefficient (D_v^{eff}) varies significantly with changes in soil water content and geology (Tillman and Smith 2005; Wealthall et al. 2010; Kulkarni et al. 2020). This will affect the derived NSZD rate from measured soil gas concentration profiles. Therefore, it is recommended to perform monitoring during dry weather and perform synoptic soil gas concentration measurements and diffusivity tracer tests (API 2017). Moreover, given a non-uniform vadose zone, D_v^{eff} and soil gas concentration profiles should be

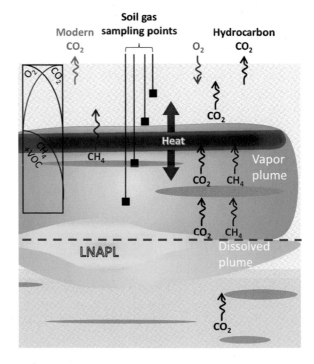

Fig. 5.3 Schematic of the concentration gradient method implementation (following Johnson et al. 2006)

measured within each unique geologic area and depth interval where gradients are present at the site. As the gradient method offers instantaneous flux on the period of measurement, additional measurements at different times of the year are required to ascertain the variability (API 2017). Gradient method can also be costly due to the high level of effort required for installation and collection of samples (CRC CARE 2018). Considering the aforementioned factors, the concentration gradient method is suitable for the sites with relatively uniform, non-stratified vadose zones (>1.5 m below root zone) (CRC CARE 2018).

5.2.1.2 Passive Flux Trap Method

Historically, flux traps were configured as chemical traps for collecting and measuring CO_2 gas venting from the subsurface to the atmosphere (Humfeld 1930; Edwards 1982; Rochette and Huchinson 2005). The historical flux trap method was modified for NSZD monitoring using a polyvinyl chloride (PVC) pipe at grade with two caustic sorbent elements (Zimbron et al. 2014; McCoy et al. 2015). As seen in Fig. 5.4, the CO_2 leaving the subsurface is absorbed by the bottom element and is converted into solid phase carbonate. The top element is used to collect atmospheric CO_2 to prevent it from reaching the bottom element. The absorbent elements are analyzed using an ASTM International method to quantify CO_2 efflux from the subsurface (McCoy et al. 2015).

Following CRC CARE (2018), the key assumptions inherent to this method include (i) CO_2 migrates vertically into receiver pipe and (ii) trap results are time-integrated over a typical multiday to several week period.

Measurement Procedure—The passive flux trap method is comprised of the following general steps:

(1) deploy the passive flux traps and leave them on-site for approximately two weeks or as needed to meet logistical needs or sorbent limitations,
(2) retrieve the traps and return them to a specialty laboratory,
(3) perform carbon and radiocarbon (^{14}C–CO_2) analysis,
(4) evaluate the background (i.e., modern) and NSZD (i.e., fossil fuel) fractions of carbon, and
(5) calculate the NSZD rate.

Details of the key aspects of deploying passive CO_2 flux traps and using their results to estimate NSZD rates are provided in CRC CARE 2018.

Challenges, Considerations, and Feasibility—Passive flux traps provide integral measurement of CO_2 efflux over the period of deployment. Based on challenges associated with passive flux trap method, this method is best suited for unpaved sites during dry periods (API 2017). Passive flux traps are unable to obtain a representative measurement of CO_2 efflux when installed on top of or through an impervious ground cover (e.g., an asphalt paved or concrete surface). As such, the method is not well suited for sites with large impervious areas or areas with highly compacted, low-permeability surface soil (API 2017; CRC CARE 2018). Wind is another potential

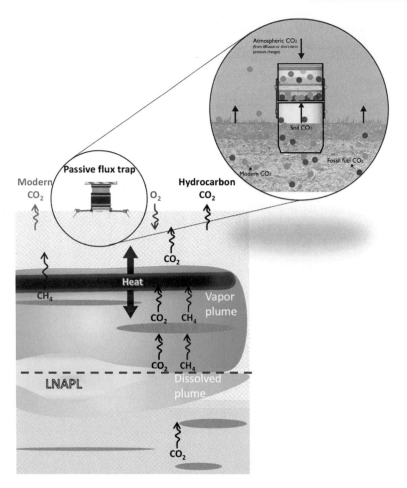

Fig. 5.4 Schematic of the passive flux trap method implementation (following www.soilgasflux. com)

challenge for this method as it might cause positive or negative air pressures that could bias CO_2 efflux through the traps (Tracy 2015). Therefore, it may be necessary to monitor wind speeds at sites with excessive winds and consider the correction for wind effects. Noteworthy, E-Flux redesigned the CO_2 traps to provide more accurate result with respect to this source of error (E-Flux, LLC 2015). In the case of precipitation during deployment, preferential flow can develop due to the rain cover preventing wetting of underlying soil (Johnson et al. 2006; Maier and Schack-Kirchner 2014). Hence, it is recommended to evaluate deployment duration and schedule a time to avoid heavy rainfall events.

5.2.1.3   Dynamic Closed Chamber Method

Chamber methods have been used to assess shallow soil respiration for more than 80 years (Norman et al. 1997; Rochette and Hutchinson 2005). Recently, dynamic closed chambers (DCC) have been applied for NSZD monitoring (Sihota et al. 2011). As shown in Fig. 5.5, a soil gas flux measurement chamber is placed on a PVC collar (e.g., LI-COR 2010). Soil gas accumulating in the chamber is circulated by a small pump between it and an infrared gas analyzer (IRGA) where CO_2 concentrations are measured approximately every 2 s. The rate of accumulation of CO_2 over the total measurement period (roughly 90 s) inside a known chamber volume is used to estimate the CO_2 efflux. A pressure equilibration device is fitted to the chamber to minimize measurement artifacts that could be caused by a pressure differential between the chamber and its surroundings. For the DCC method, it is prudent to collect total CO_2 efflux measurements in a background (unimpacted) location for use in comparison with those collected over the petroleum NAPL footprint.

Following CRC CARE (2018), the key assumptions of this method include (i) gas flux is vertically uniform in the subsurface, and (ii) the discrete CO_2 measurement (or multiple measurements) is representative of site conditions.

Measurement Procedure—Implementation of the DCC NSZD measurement method consists of the following general steps:

(1) install collars and allow re-equilibration,
(2) perform soil gas efflux survey using the portable chamber and IRGA,
(3) (optional) collect soil gas samples and send them to laboratory for analysis of ^{14}C,
(4) perform data validation,
(5) quantify the NSZD fraction of the measured total CO_2 efflux using background and/or ^{14}C–CO_2 results, and
(6) calculate NSZD rate.

Details of the key steps in designing and implementing a DCC-based CO_2 efflux monitoring program and estimating site-wide NSZD rates using the DCC method are provided in CRC CARE 2018.

Challenges, Considerations, and Feasibility—Considering the challenges, the DCC method is best suited to unpaved sites during dry periods with a relatively uniform background gas efflux. Diurnal and seasonal fluctuations in background CO_2 efflux, shallow soil water content, wind, impervious and compacted ground cover have a significant impact on the representativeness of DCC measurements. Hence, it is necessary to avoid rainfall events and impervious areas and monitor wind speed to correct the results for elevated wind speeds (API 2017). Since the DCC measures short-term CO_2 efflux subjected to diurnal and seasonal fluctuations, multiple measurements are required to understand the range of plausible NSZD rates. Using DCC in the areas with active carbon cycling in the root zone is also challenging.

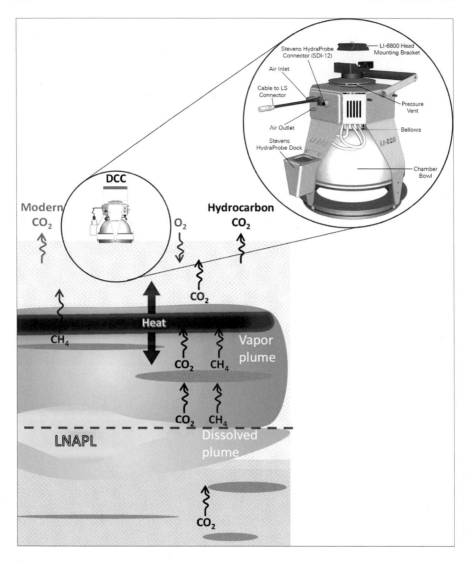

Fig. 5.5 Schematic of the dynamic closed chamber method implementation (following www.licor. com)

Background CO_2 efflux associated with root zones and natural organic matter respiration are best resolved using a background location. For sites with more complex background levels of CO_2 efflux, ^{14}C analysis can be used or DCC measurements can be collected during colder time of the year, when root zone/respiration activity is at a minimum (Sihota and Mayer 2012; API 2017; Crann et al. 2016; Reynolds 2019; Wozney et al. 2021).

5.2.2 Soil Temperature Methods

As mentioned above, observed elevated soil temperatures in the vadose hydrocarbon oxidation zone overly the petroleum NAPL footprint can be used to estimate NSZD rates (Sweeney and Ririe 2014; Stockwell 2015; Warren and Bekins 2015; Sale et al. 2018). Quantifying the amount of heat generated by NSZD in terms of J/s/m^2 or W/m^2 requires an estimate of the thermal gradient and the enthalpy of oxidation (Sale et al. 2018; Karimi Askarani et al. 2018; Kulkarni et al. 2020). Two methods of using soil temperatures to estimate NSZD rates are described in this chapter section.

5.2.2.1 Background-Corrected Method

Background-corrected approaches for quantifying NSZD rates using subsurface temperatures have been advanced by Sweeney and Ririe (2014), Warren et al. (2015), Sale et al. (2018), and Karimi Askarani et al. (2018). In this method, soil temperature measurements taken at discrete depths by thermocouples (Fig. 5.6) are used to obtain vertical vadose zone soil temperature profiles in both background and atop petroleum NAPL-impacted zones. Temperature at any depth in the impacted area is a function of heat generated through NSZD processes and heat from other sources (e.g., surface heating and cooling). Abiding by the assumption that all heat sources/sinks, except for the heat generated through NSZD processes, are similar in background and impacted locations, temperatures associated with NSZD can be simply estimated as (Stockwell 2015; Sale et al. 2018; Karimi Askarani et al. 2018):

where T_{NSZD} is the component of temperatures associated with NSZD (°C), T_{Imp} is the measured temperatures at an impacted location (°C), T_{Back} is the measured temperatures at the background location (°C), i is a fixed time (s), and z is a fixed vertical position (m).

$$T_{NSZD}\big|_z^i = T_{Imp}\big|_z^i - T_{Back}\big|_z^i \qquad (5.4)$$

Based on Fourier's first law, the heat flux associated with the NSZD process is expressed by (Hillel 1982):

$$q = -K\frac{\partial T}{\partial z} \qquad (5.5)$$

where q is the heat flux due to conduction (W/m^2), K is the thermal conductivity (W/°C/m), and $\frac{\partial T}{\partial z}$ is the change in temperature with respect to distance in the vertical direction (°C/m).

As shown in Fig. 5.6, performing an energy balance on a one-dimensional vertical reference volume, NSZD heat flux is obtained by (Karimi Askarani et al. 2018):

$$\dot{E}_{NSZD} = \dot{E}_{Top} - \dot{E}_{Bottom} + \frac{dE_{Sto}}{dt} \qquad (5.6)$$

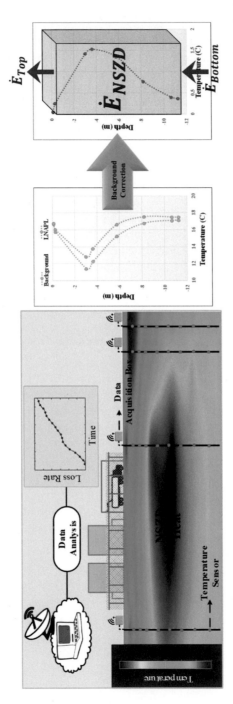

Fig. 5.6 Conceptual model and implementation schematic of the background-corrected soil temperature NSZD measurement method. The colors in the subsurface represent interpolated background corrected temperatures measured by multilevel thermocouples in the subsurface through a transect

where $\dot{E}_{\text{Top}}$ is the energy flux at the top of the reference volume (W/m²), $\dot{E}_{\text{Bottom}}$ is the energy flux at the bottom of the reference volume (W/m²), $\dot{E}_{\text{NSZD}}$ is the energy produced through NSZD over the height of energy balance element (W/m²), and E_{Sto} is the stored energy over the height of energy balance element (J/m²). The background-corrected methods developed by others neglect the energy storage $\left(\frac{dE_{\text{Sto}}}{dt}\right)$ in Eq. 5.6 to simplify the calculations.

Equation 5.6 is used to determine the NSZD rate using the following equation:

$$R_{\text{NSZD}} = \frac{-\dot{E}_{\text{NSZD}}\text{MW}}{\Delta H_r} \tag{5.7}$$

where the R_{NSZD} is in (g/m²/s), ΔH_r is the enthalpy released during oxidation of the NAPL (6770 kJ/mol for complete mineralization of 1 mol of decane), and MW is the molecular weight of a representative hydrocarbon in the petroleum NAPL (g/mol).

Following Karimi Askarani et al (2018), the key assumptions of the soil temperature method include (i) all factors controlling surface heating and cooling at impacted and background locations are sufficiently similar, (ii) soil thermal properties are constant through time with uniform unique values, (iii) heat conduction is the dominant mechanism of heat transfer, and (iv) horizontal transport of heat is considered negligible.

Measurement Procedure—Implementation of the background-corrected soil temperature method includes the following general steps:

(1) install soil temperature monitoring devices in both background and petroleum NAPL-impacted locations and record temperatures at various depth intervals within the hydrocarbon oxidation zone at least daily,
(2) quantify soil thermal properties (thermal conductivity),
(3) perform background correction to isolate NSZD-related heat, and
(4) calculate NSZD rate.

Challenges, Considerations, and Feasibility—Considering the following challenges with the background-corrected method, this approach is suitable for long-term NSZD monitoring at select key locations atop petroleum NAPL at sites with uniform hydrogeology, uniform ground cover, absence of heat sinks/sources (e.g., fluid-filled utility lines), and a water table deeper than 1.5 m, and background areas (absent of petroleum NAPL) that are representative of petroleum NAPL-impacted locations. Infiltration of precipitation can not only affect soil thermal properties by changes in soil water content but also play a role as a heat sink through cold seasons. Hence, it is recommended to modify soil thermal property values to incorporate these intermittent effects, if significant. In addition, backfilling and sealing the borehole equipped with thermometers can limit the infiltration. Soil heterogeneity may also result in a variable thermal gradient—heterogeneity can be addressed by using volumetric average of derived thermal property values from each unique lithology in the NSZD calculation (Warren and Bekins 2018). In the case of presence of external heat sources/sinks (e.g., pipelines), temperature monitoring should be performed at least 10 m away from external heat sources/sinks to avoid confounding background correction. Ideally, in-situ, site-specific, thermal soil properties should be measured

(Karimi Askarani et al. 2021; Kulkarni et al. 2022a). Notably, given periodic temperature data (e.g., daily) continuous NSZD rates can be estimated and averaged over time to obtain time-integrated and instantaneous NSZD rate.

5.2.2.2 Non-background-Corrected Method

To estimate the NSZD rate using the background-corrected method, background locations need to be largely similar to the petroleum NAPL-impacted locations in terms of all factors controlling surface heating and cooling through time (Karimi Askarani and Sale 2020). Flawed background correction can lead to fictitious negative NSZD rates and or elevated NSZD rates in areas where no petroleum NAPL is present. Alternatively, Karimi Askarani and Sale (2020) advanced a computational method to transform subsurface temperatures to NSZD rate without background correction, known as "*single stick*" in the literature. In this method, soil temperature measurements at discrete depths are only needed from petroleum NAPL-impacted areas. As shown in Fig. 5.7, at an impacted location, temperature at any point is a function of surface heating and cooling (q_s), subsurface heat source associated with NSZD process (q_{ss}; i.e., primarily being generated via the conversion of CH_4 and O_2 to CO_2 and H_2O in the vadose zone) and position of subsurface heat source (x'). Simplistically, given temperatures at two points, heat transport equations for each point can be written with q_s and q_{ss} as unknowns. The system of equations can be solved for q_s and q_{ss}.

In more detail, given a media with two primary heat sources (q_s and q_{ss}), independent soil thermal properties of position and direction, and negligible horizontal heat fluxes, the governing conductive heat transfer equation (Eq. 5.8, Carslaw and Jaeger (1959); Jury and Horton (2004); Hillel (2013)) is stated as:

Fig. 5.7 Conceptual model of non-background-corrected method (following Karimi Askarani 2019)

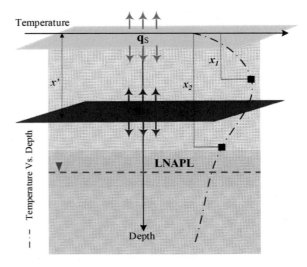

$$\frac{\partial^2 T}{\partial x^2} = \frac{1}{\kappa}\frac{\partial T}{\partial x} \tag{5.8}$$

where T is temperature (°C), t(s) is time, x(m) is spatial coordinate, and $\kappa = K/\rho c$ is thermal diffusivity (m^2/s).

Resolving Eq. 5.8 for temperatures associated with q_s and q_{ss} yields:

$$T(x,t) - T_0 = \frac{2q_s}{K}\left\{\sqrt{\frac{\kappa t}{\pi}}\exp\left(-\frac{x^2}{4\kappa t}\right) - \frac{x}{2}\mathrm{erfc}\left(\frac{x}{\sqrt{4\kappa t}}\right)\right\}$$
$$+ \frac{q_{ss}}{\rho c}\left\{\sqrt{\frac{t}{\pi\kappa}}\exp\left(-\frac{(x - x\prime)^2}{4\kappa t}\right) - \frac{|x - x\prime|}{2k}\mathrm{erfc}\left(\frac{|x - x\prime|}{\sqrt{4\kappa t}}\right)\right\}$$
$$+ \frac{q_{ss}}{\rho c}\left\{\sqrt{\frac{t}{\pi\kappa}}\exp\left(-\frac{(x + x\prime)^2}{4\kappa t}\right) - \frac{(x + x\prime)}{2k}\mathrm{erfc}\left(\frac{(x + x\prime)}{\sqrt{4\kappa t}}\right)\right\} \tag{5.9}$$

where T_0 is initial temperature (°C), c (J/kg/°C) is heat capacity, and ρ (kg/m^3) is sediment density. Given measured subsurface temperature ($T(x, t)$), Eq. 5.9 can be solved in a two-equation, two-unknown system for q_s and q_{ss}, following an iterative approach to determine $x\prime$. Derived calculations and procedures are explained in detail in Karimi and Sale (2020).

Following Karimi Askarani and Sale (2020), the key assumptions in this method are (i) surface and subsurface heat sources are represented as planar features, (ii) all processes controlling surface heating and cooling are included in surface heating and cooling term (q_s), and (iii) the position of subsurface heat source ($x\prime$) varies spatiotemporally.

Measurement Procedure—Implementation of the non-background-corrected soil temperature method to estimate NSZD rates consists of the following general steps:

(1) install the soil temperature monitoring devices atop petroleum NAPL-impacted locations and record temperatures at various depth intervals within the hydrocarbon oxidation zone at least daily,
(2) quantify soil thermal properties (thermal conductivity),
(3) compute the NSZD-related heat, and
(4) calculate NSZD rate.

Challenges, Considerations, and Feasibility—This advanced computational approach is suitable for long-term NSZD monitoring at select key locations atop petroleum NAPL sites with a water table depth greater than 1.5 m with the majority of the area absent of heat sinks/sources (e.g., fluid-filled utility lines). This method is applicable to sites with paved/impervious ground surface and unavailable background locations as there is no need for background correction. Infiltration of precipitation, heterogeneity, and presence of external heat sources are the challenges that need to be considered and addressed as explained in the background-corrected method. This method is computationally complex and is not publicly available. A promising aspect of the soil temperature methods is being able to obtain continuous NSZD rates that can be averaged over time to obtain a time-integrated NSZD rate.

5.2.3   Petroleum NAPL Chemical Composition Change

The chemical composition of petroleum NAPL changes over time as the weathering processes of dissolution, volatilization, and biodegradation constantly collectively act to deplete it. Therefore, the composition of petroleum NAPL can be evaluated over time as a direct measurement of petroleum NAPL losses. Biodegradation is identified as a significant contaminants of concern (COC) mass loss mechanism from the petroleum NAPL (Ng et al. 2014, 2015). Chemical analysis of petroleum NAPL can be performed to evaluate both bulk petroleum NAPL losses and losses of individual chemicals. Monitoring of losses of individual chemicals of concern is of high relevance to NSZD monitoring because none of the other methods are chemical-specific. Being chemical-specific, the results from the chemical composition change method can be used to evaluate the effectiveness of NSZD to reduce risks associated with toxicity to human health and the environment.

COC-specific NSZD rates in petroleum NAPL ($R_{COC\text{-}NAPL}$) can be calculated following CRC CARE (2018):

$$R_{COC\text{-}NAPL} = \frac{dm_{COC}}{dt} \rho_{NAPL} D_{NAPL} \tag{5.10}$$

where $R_{COC\text{-}NAPL}$ is the COC-specific NSZD rate (g/m^2/d), dm_{COC} is the change in chemical content in the petroleum NAPL normalized for a conservative marker in the NAPL (g/g), t is time (d), ρ_{NAPL} is the petroleum NAPL density (g/m^3), and D_{NAPL} is the specific volume of in-situ petroleum NAPL (m^3/m^2).

It is important to note that Eq. 5.10 is presented in a way that results in a unit of measurement common to the other NSZD measurement methods (i.e., g/m^2/d). Use of the chemical compositional change method can be used in a way that simply tracks reductions in $\frac{dm_{COC}}{dt}$ for specific chemicals of concern. For this reason, the routine of measurement of ρ_{NAPL} and D_{NAPL} is considered optional.

The key assumptions in using petroleum NAPL compositional change to estimate a COC-specific NSZD rate include (i) a time-series regression of the chemical content of the petroleum NAPL is representative of its NSZD rate, (ii) in-well petroleum NAPL samples collected after purging are representative of the petroleum NAPL in the surrounding formation, and (iii) $R_{COC\text{-}NAPL}$ is inclusive of losses due to dissolution, volatilization, waterborne biodegradation, and direct-contact oil biodegradation (CRC CARE 2018).

Measurement Procedure—Use of the petroleum NAPL compositional change method consists of the following general steps:

(1) well purging and petroleum NAPL sampling from in-well or soil samples at various time periods (e.g., several years apart),
(2) laboratory analysis of petroleum NAPL for COCs, conservative markers, and fluid density,
(3) calculation of COC and marker loss in petroleum NAPL and normalization of COC change in mass fraction over time,
(4) (optional) laboratory analysis of intact soil cores for capillarity and pore fluid saturation,
(5) (optional) estimation of petroleum NAPL-specific volume, and
(6) (optional) calculate the NSZD rate.

Details of the key aspects of using chemical compositional change in petroleum NAPL as a way to estimate chemical-specific NSZD rates are provided in CRC CARE 2018.

Recently, Devaull et al. (2020) advanced a quantitative analysis method that is consistent with constituent marker methods (Douglas et al. 1996). It consists of prescribing steps for (i) selecting the most appropriate conserved marker constituents based on measured composition data, (ii) doing parameter transformations, and (iii) performing regression analyses on time-series data sets.

Challenges, Considerations, and Feasibility—To accommodate for an anticipated high variability in the chemical quality of the petroleum NAPL, an adequate number of samples is necessary to observe the range in COC-specific NSZD rates. Calculation of the NSZD rate using multiple conservative markers is recommended that may be prudent to assess the variability in rate estimates (CRC CARE 2018). A possible constraint to this approach is the long data records (e.g., multiple years to decade) that are needed to get to resolve an NSZD rate. The petroleum NAPL compositional change method is suitable at the sites with a need for detailed understanding of NSZD related to rates of attenuation of a COC(s) that drive decision making on the remedial efforts. Overall, further research is needed to improve the calculation of depletion rates using petroleum NAPL composition data.

5.2.4 Emerging Science and Future Vision

5.2.4.1 Emerging Science

As presented in this chapter, there are ample available NSZD measurement methods; the results can be combined with other site characterization data collected for the CSM to improve understanding and remedial application of NSZD. All methods are credible but have advantages and disadvantages under certain site conditions. Regardless, all methods point to the widespread occurrence of petroleum NAPL NSZD at rates that often exceed depletions achievable with mechanically engineered remedies. As addressed in detail in Chap. 13, a current challenge is to determine the multiple lines of evidence needed to support a decision document that incorporates NSZD into a remedy that is regulatory compliant and protective to human health and the environment. As was the case in the early years of the natural attenuation remedy, the regulatory response to the emergence of NSZD science has been timid, and practitioners slow in bringing NSZD into both the CSM and remedial action at petroleum NAPL sites. Without doubt, however, NSZD is a part of the remedial solution for many (if not all) sites with petroleum liquids in soil and groundwater.

The future of NSZD-related questions being addressed today includes:

- What are the key factors governing NSZD processes and how might they be optimized to increase rates?
- What is the best approach to tracking the performance of NSZD-based remedies through time?
- How can NSZD be used to meet the modern challenges of resiliency, sustainability, and environmental and social governance?

Nascent tools for resolving the above questions are in development. These include but are not limited to cryogenic coring (Kiaalhosseini et al. 2016; Trost et al. 2018), molecular biological tools (further discussed in Chap. 10), and three-dimensional arrays of real-time sensors (e.g., temperature, ORP, pH, conductivity, and water level) that are linked to a cloud-based data storage, analytics, visualization, and reporting platform (Sale et al. 2021a, b). As this journey continues, more will be developed to help us with modern management of subsurface petroleum NAPL.

5.2.4.2   Future Vision

Applying new technologies, such as Internet of Things (IoT) and sensors, to the field of vapor, soil, and groundwater monitoring makes long-term monitoring of contaminated sites more feasible. They offer benefits in terms of real-time data enabling faster responses to adverse conditions, lower cost, improved safety, and a smaller environmental impact (Karimi Askarani and Gallo 2020; Blotevogel et al. 2021). For example, multilevel subsurface sensor-based monitoring is evolving at a rapid rate and providing ever-growing larger valuable data sets of temperature, soil oxidation reduction potential (ORP), and water levels at contaminated sites (Karimi Askarani and Sale 2021). Eventually, through cloud computing, these data sets will be managed and transformed into real-time actionable outputs such as NSZD rates and petroleum NAPL source zone depletion timeframe estimates. While IoT and cloud computing provide connection of sensors, platforms, and objects for data transmission and data transformation, synergistically, artificial intelligence and machine learning will capture the interrelationships of all this data and provide a better understanding of NSZD processes and establish the appropriate application of NSZD in managing petroleum NAPL sites (Karimi Askarani and Sale 2021).

5.2.5 Summary and Conclusion

Hundreds of thousands of sites are affected by historical releases of petroleum NAPL, including crude, fuels, lubricants, heating oil, creosote, and petrochemical wastes. In the past decade, NSZD—primarily the naturally occurring process of direct-contact oil biodegradation—has garnered considerable scientific and regulatory interest. This increased interest arises from the growing recognition that natural processes can lead to petroleum NAPL mass losses from the subsurface at rates rivaling the depletion rates achieved with mechanically engineered remedies. In this chapter, advanced methods used to measure NSZD rates based on gas flux, biogenic heat, and petroleum NAPL chemical composition are documented. The challenges and considerations with each method are provided to highlight site conditions that might control NSZD rates and help practitioners select the best approach to fit their CSM.

References

Amos RT, Mayer KU, Bekins BA, Delin GN, Williams RL (2005) Use of dissolved and vapor-phase gases to investigate methanogenic degradation of petroleum hydrocarbon contamination in the subsurface. Water Resources Res 41(2)

API (2017) Quantification of vapor phase-related natural source zone depletion processes, Publication No. 4784

Askarani KK, Sale TC (2020) Thermal estimation of natural source zone depletion rates without background correction. Water Res 169:115245. https://doi.org/10.1016/j.watres.2019.115245

Atekwana EA, Atekwana EA (2010) Geophysical signatures of microbial activity at hydrocarbon contaminated sites: a review. Surv Geophys 31(2):247–283

Baedecker MJ, Eganhouse RP, Bekins BA, Delin GN (2011) Loss of volatile hydrocarbons from an LNAPL oil source. J Contam Hydrol 126(3–4):140–152

Bekins BA, Hostettler FD, Herkelrath WN, Delin GN, Warren E, Essaid HI (2005) Progression of methanogenic degradation of crude oil in the subsurface. Environ Geosci 12(2):139–152

Blotevogel J, Karimi Askarani K, Hanson A, Gallo S, Carling B, Mowder C, Spain J, Hartten A, Sale T (2021) Real-time remediation performance monitoring with ORP sensors. Groundwater Monit Remediation 41(3):27–28. https://doi.org/10.1111/gwmr.12479

Carslaw HS, Jaeger JC (1959) Conduction of heat in solids, 2nd edn. Clarendon Press, Oxford, UK

Coffin RB, Pohlman JW, Grabowski KS, Knies DL, Plummer RE, Magee RW, Boyd TJ (2008) Radiocarbon and stable carbon isotope analysis to confirm petroleum natural attenuation in the vadose zone. Environ Forensics 9(1):75–84

Contaminated Land: Applications in Real Environments (CL:AIRE) (2014) An illustrated handbook of LNAPL transport and fate in the subsurface, CL:AIRE, London

Crann CA, Murseli S, St-Jean G, Zhao X, Clark ID, Kieser WE (2017) First status report on radiocarbon sample preparation at the A.E. Lalonde AMS Laboratory (Ottawa, Canada). Radiocarbon 59(3):695–704. https://doi.org/10.1017/RDC.2016.55

CRC CARE (2018) Technical measurement guidance for LNAPL natural source zone depletion, CRC CARE Technical Report no. 44, CRC for Contamination Assessment and Remediation of the Environment, Newcastle, Australia

CRC CARE (2020) The role of natural source zone depletion in the management of light non-aqueous phase liquid (LNAPL) contaminated sites, CRC CARE Technical Report no. 46, CRC for Contamination Assessment and Remediation of the Environment, Newcastle, Australia

Davis GB, Rayner JL, Trefry MG, Fisher SJ, Patterson BM (2005) Measurement and modeling of temporal variations in hydrocarbon vapor behavior in a layered soil profile. Vadose Zone J 4(2):225–239

DeVaull GE, Rhodes IA, Hinojosa E, Bruce CL (2020) Petroleum NAPL depletion estimates and selection of marker constituents from compositional analysis. Groundwater Monit Remediation 40(4):44–53

Douglas GS, Bence AE, Prince RC, McMillen SJ, Butler EL (1996) Environmental stability of selected petroleum hydrocarbon source and weathering ratios. Environ Sci Technol 30(7):2332–2339

Edwards NT (1982) The use of soda-lime for measuring respiration rates in terrestrial systems. Pedobiologia; (German Democratic Republic) 23

E-Flux, LLC (2015) Technical memo 1504.2: wind effects on soil gas flux measurements at ground level, Last revision 10 June 2015

Garg S, Newell CJ, Kulkarni PR, King DC, Adamson DT, Renno MI, Sale T (2017) Overview of natural source zone depletion: processes, controlling factors, and composition change. Groundwater Monit Remediation 37(3):62–81

Hillel D (1982) Introduction to soil physics. Academic Press Inc, San Diego, CA

Hillel D (2013) Fundamentals of soil physics. Academic Press, New York

Humfeld H (1930) A method for measuring carbon dioxide evolution from soil. Soil Sci 30(1):1–12

Interstate Technology & Regulatory Council, LNAPL Update Team, Washington, USA

Irianni-Renno M, Akhbari D, Olson MR, Byrne AP, Lefèvre E, Zimbron J, Lyverse M, Sale TC, Susan K (2016) Comparison of bacterial and archaeal communities in depth-resolved zones in an LNAPL body. Appl Microbiol Biotechnol 100(7):3347–3360

ITRC (2009) Evaluating LNAPL remedial technologies for achieving project goals, LNAPL-2, Interstate Technology & Regulatory Council, LNAPLs Team, Washington, USA

ITRC (2018) Light non-aqueous phase liquid (LNAPL) site management: LCSM evolution, decision process and remedial technologies, LNAPL-3, Interstate Technology & Regulatory Council, LNAPL Update Team, Washington, USA

Johnson P, Lundegard P, Liu Z (2006) Source zone natural attenuation at petroleum hydrocarbon spill sites—I: site-specific assessment approach. Groundwater Monit Remediation 26(4):82–92

Jury WA, Horton R (2004) Soil physics. Wiley, New York

Karimi Askarani K, Gallo S (2020) The role of emerging technologies in monitoring groundwater-soil systems. In: AGU fall meeting abstracts, vol 2020, pp H062-0006

Karimi Askarani K (2019) Thermal monitoring of natural source zone depletion (Doctoral dissertation, Colorado State University)

Karimi Askarani K, Gallo S, Kirkman AJ, Sale TC (2021) Method to estimate thermal conductivity of subsurface media. Groundwater Monit Remediation 41(1):99–105. https://doi.org/10.1111/gwmr.12419

Karimi Askarani K, Stockwell EB, Piontek KR, Sale TC (2018) Thermal monitoring of natural source zone depletion. Groundwater Monit Remediation 38(3):43–52. https://doi.org/10.1111/gwmr.12286

Kiaalhosseini S, Johnson RL, Rogers RC, Renno MI, Lyverse M, Sale TC (2016) Cryogenic core collection (C3) from unconsolidated subsurface media. Groundwater Monit Remediation 36(4):41–49

Kulkarni PR, Newell CJ, King DC, Molofsky LJ, Garg S (2020) Application of four measurement techniques to understand natural source zone depletion processes at an LNAPL site. Groundwater Monit Remediation 40(3):75–88

Kulkarni PR, Uhlir G, Newell CJ, Walker K, McHugh T (2022a) In-situ method to determine soil thermal conductivity at sites using thermal monitoring to quantify natural source zone depletion. J Hydrogeol Hydrol Eng 11:1

Kulkarni PR, Walker KL, Newell CJ, Askarani KK, Li Y, McHugh TE (2022b). Natural source zone depletion (NSZD) insights from over 15 years of research and measurements: a multi-site study. In: Water research, p 119170

Lari KS, Davis GB, Rayner JL, Bastow TP, Puzon GJ (2019) Natural source zone depletion of LNAPL: a critical review supporting modelling approaches. Water Res 157:630–646

LI-COR (2010) LI-8100A automated soil CO_2 flux system & LI-8150 multiplexer instruction manual. LI-COR, Inc., Lincoln, NE

Lundegard PD, Johnson PC (2006) Source zone natural attenuation at petroleum hydrocarbon spill sites—II: application to a former oil field. Groundwater Monit Remediation 26(4):93–106

Luo H, Dahlen PR, Johnson PC, Peargin T (2013) Proof-of-concept study of an aerobic vapor migration barrier beneath a building at a petroleum hydrocarbon-impacted site. Environ Sci Technol 47(4):1977–1984

Maier M, Schack-Kirchner H (2014) Using the gradient method to determine soil gas flux: a review. Agric Forest Meteorol 192–193:78–95

McCoy K (2012) Resolving natural losses of LNAPL using CO2 traps. Doctoral dissertation, Colorado State University. Libraries

McCoy K, Zimbron J, Sale T, Lyverse M (2015) Measurement of natural losses of LNAPL using CO_2 traps. Groundwater 53(4):658–667

Molins S, Mayer KU, Amos RT, Bekins BA (2010) Vadose zone attenuation of organic compounds at a crude oil spill site—interactions between biogeochemical reactions and multi-component gas transport. J Contam Hydrol 112(1):15–29

Murseli S, Middlestead P, St-Jean G, Zhao X, Jean C, Crann CA, Kieser WE, Clark ID (2019) The preparation of water (DIC, DOC) and gas (CO_2, CH_4) samples for radiocarbon analysis at AEL-AMS, Ottawa, Canada. Radiocarbon 61(5):1563–1571. https://doi.org/10.1017/RDC.201 9.14

National Research Council (NRC) (2000) Natural attenuation for groundwater remediation. committee on intrinsic remediation, Water Science and Technology Board, Board on Radioactive Waste Management, Commission on Geosciences, Environment, and Resources, National Academy Press, Washington DC

Ng GHC, Bekins BA, Cozzarelli IM, Baedecker MJ, Bennett PC, Amos RT (2014) A mass balance approach to investigating geochemical controls on secondary water quality impacts at a crude oil spill site near Bemidji, MN. J Contam Hydrol 164:1–15

Ng GHC, Bekins BA, Cozzarelli IM, Baedecker MJ, Bennett PC, Amos RT, Herkelrath WN (2015) Reactive transport modeling of geochemical controls on secondary water quality impacts at a crude oil spill site near Bemidji, MN. Water Resources Res 51(6):4156–4183

Norman JM, Kucharik CJ, Gower ST, Baldocchi DD, Crill PM, Rayment M, Savage K, Striegl RG (1997) A comparison of six methods for measuring soil-surface carbon dioxide fluxes. J Geophys Res: Atmospheres 102(D24):28771–28777

NRC (2000) Natural attenuation for groundwater remediation, Committee on Intrinsic Remediation, Water Science and Technology Board, Board on Radioactive Waste Management, Commission on Geosciences, Environment, and Resources, National Academy Press, Washington, USA

Poulsen TG, Møldrup P (2006) Evaluating effects of wind-induced pressure fluctuations on soil-atmosphere gas exchange at a landfill using stochastic modeling. Waste Manage Res 24(5):473–481

Ramirez JA, Baird AJ, Coulthard TJ, Waddington JM (2015) Ebullition of methane from peatlands: does peat act as a signal shredder? Geophys Res Lett 42(9):3371–3379

Reynolds L (2019) Soil $^{14}CO_2$ source apportionment for biodegradation in contaminates oils in permafrost climates: a novel technique for rapid sample collection by barium carbonate precipitation. A thesis submitted in partial fulfillment of the requirements for the degree of Master of Science. University of Ottawa, Carleton Geoscience Center, March

Rochette P, Hutchinson GL (2005) Measurement of soil respiration in situ: chamber techniques

Sale TC, Ham JM, Gallo WS, Askarani KK, Ferrie ZS, Scalia IV J (2021b) U.S. Patent No. 10,901,117

Sale TC, Stockwell EB, Newell CJ, Kulkarni PR (2018) U.S. Patent No. 10,094,719

Sale T, Gallo S, Askarani KK, Irianni-Renno M, Lyverse M, Hopkins H, Blotevogel J, Burge S (2021b) Real-time soil and groundwater monitoring via spatial and temporal resolution of biogeochemical potentials. J Hazard Mater 408:124403. https://doi.org/10.1016/j.jhazmat.2020. 124403

Sihota NJ, Mayer KU (2012) Characterizing vadose zone hydrocarbon biodegradation using carbon dioxide effluxes, isotopes, and reactive transport modeling. Vadose Zone J 11(4)

Sihota NJ, Singurindy O, Mayer KU (2011) CO_2-efflux measurements for evaluating source zone natural attenuation rates in a petroleum hydrocarbon contaminated aquifer. Environ Sci Technol 45(2):482–488

Smith JW, Davis GB, DeVaull GE, Garg S, Newell CJ, Rivett MO (2022) Natural Source Zone Depletion (NSZD): from process understanding to effective implementation at LNAPL-impacted sites. Quarterly J Eng Geol Hydrogeol

Stockwell EB (2015) Continuous NAPL loss rates using subsurface temperatures (Doctoral dissertation, Colorado State University)

Sweeney RE, Ririe GT (2014) Temperature as a tool to evaluate aerobic biodegradation in hydrocarbon contaminated soil. Groundwater Monit Remediation 34(3):41–50

Tillman FD, Smith JA (2005) Vapor transport in the unsaturated zone. Wiley

Tracy MK (2015) Method comparison for analysis of LNAPL natural source zone depletion using CO_2 fluxes (Doctoral dissertation, Colorado State University)

Trost JJ, Christy TM, Bekins BA (2018) A direct-push freezing core barrel for sampling unconsolidated subsurface sediments and adjacent pore fluids. Vadose Zone J 17(1):1–10

Warren E, Bekins BA (2015) Relating subsurface temperature changes to microbial activity at a crude oil-contaminated site. J Contam Hydrol 182:183–193

Wealthall GP, Rivett MO, Dearden RA (2010) Transport and attenuation of dissolved-phase volatile organic compounds (VOCs) in the unsaturated zone. In: British geological survey, groundwater pollution programme, Internal Report IR/09/037

Wozney A, Clark ID, Mayer KU (2021) Quantifying natural source zone depletion at petroleum hydrocarbon contaminated sites: a comparison of 14C methods. J Contam Hydrol 240:103795

Wyatt DE, Richers DM, Pirkle RJ (1995) Barometric pumping effects on soil gas studies for geological and environmental characterization. Environ Geol 25(4):243–250

Zimbron JA, Sale TC, Lyverse M, Chevron USA Inc and Colorado State University Research Foundation (2014) Gas flux measurement using traps. U.S. Patent 8,714,034

Chapter 6
Petroleum Vapor Intrusion

Iason Verginelli

Abstract Petroleum vapor intrusion (PVI) is the process by which volatile petroleum hydrocarbons released from contaminated geological materials or groundwater migrate through the vadose zone into overlying buildings. PVI science showed that petroleum hydrocarbons are subjected to natural attenuation processes in the source zone and during the vapor transport through the vadose zone. Specifically, in the presence of oxygen, aerobic biodegradation typically reduces or eliminates the potential for PVI. This behavior justifies the different approach usually adopted for addressing PVI compared to less biodegradable compounds such as chlorinated solvents. In some countries, it was introduced the concept of vertical exclusion distance criteria, i.e., source to building distances above which PVI does not normally pose a concern. For buildings where the vertical separation distance does not meet screening criteria, additional assessment of the potential for PVI is necessary. These further investigations can be based on modeling of vapor intrusion, soil gas sampling, indoor measurements or preferably a combination of these to derive multiple lines of evidence. The data collected are then used for a risk assessment of the vapor intrusion pathway. This chapter provides an overview of state-of-the-science methodologies, models, benefits and drawbacks of current approaches, and recommendations for improvement.

Keywords Indoor air · LNAPL · Risk assessment · Soil gas sampling · Vapor intrusion modeling

I. Verginelli (✉)
Laboratory of Environmental Engineering, Department of Civil Engineering and Computer Science Engineering, University of Rome Tor Vergata, Via del Politecnico 1, Rome 00133, Italy
e-mail: verginelli@ing.uniroma2.it

© The Author(s) 2024   139
J. García-Rincón et al. (eds.), *Advances in the Characterisation and Remediation
of Sites Contaminated with Petroleum Hydrocarbons*, Environmental Contamination
Remediation and Management, https://doi.org/10.1007/978-3-031-34447-3_6

6.1 Introduction

Petroleum vapor intrusion (PVI) is the term used to describe the migration of volatile petroleum hydrocarbons (PHCs) released in the subsurface from the vadose zone into overlying buildings (Fig. 6.1). As discussed in Chap. 1, petroleum contamination can occur at various types of sites including refineries, gasoline or diesel underground storage tanks (USTs), commercial and home heating oil in aboveground storage tanks (ASTs), pipelines or oil exploration and production (E&P) sites (ITRC 2014).

PHCs in the form of light non-aqueous phase liquids (LNAPLs) tend to migrate downward in the vadose zone under the force of gravity (Rivett et al. 2011). During the vertical percolation, the LNAPL is partially retained in the pores of the formation as a relatively immobile phase due to the establishment of capillary forces (ITRC 2009) and the presence of dead-end pores as discussed in Chaps. 1 and 2. If the quantity of the release is significant, LNAPLs can reach the capillary fringe and

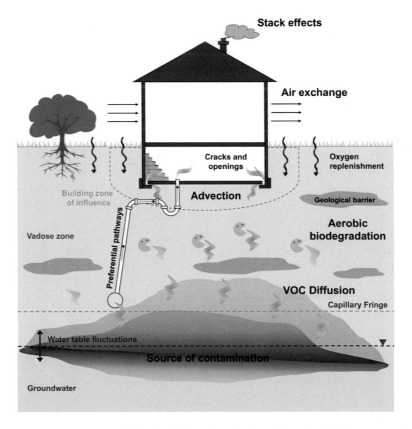

Fig. 6.1 Conceptual site model for PVI. Note that the figure is a simplified representation of the main processes involved in PVI. Subsurface environments and contaminant distributions are inherently complex and affected by dynamics such as water table fluctuations and weather effects

partially penetrate the saturated zone (U.S. EPA 1995). The LNAPL constituents in the saturated zone dissolve in groundwater generating a dissolved-phase plume downward to the source. Furthermore, the volatile constituents of LNAPL volatilize from the contaminated soil or groundwater and migrate in the subsurface mainly via diffusion and may potentially enter into the overlying buildings posing potential threats to safety (e.g., fire or explosion potential from petroleum vapors or methane) and human health (e.g., exposure to benzene from gasoline) (U.S. EPA 2015a). PHCs can be composed of hundreds of individual aromatic and aliphatic compounds. Petroleum contamination is typically assessed in terms of total petroleum hydrocarbons (TPH) with a specific focus on some individual compounds of major concern for vapor intrusion, such as BTEX (benzene, toluene, ethylbenzene, and xylenes), naphthalene, and methane (ITRC 2014). Due to its chemico-physical and toxicological properties, benzene is usually the risk driver of vapor intrusion although in some cases (e.g., gasoline contamination), C5–C8 aliphatics and C9–C12 aliphatics can contribute significantly to the overall PVI risks (Brewer et al. 2013). Furthermore, fuel additives and metabolites such as methyl tert-butyl ether (MtBE), ter-butyl alcohol (TBA), or ethylene dibromide (EDB) may also pose a risk to human health (U.S. EPA 2015a).

Advances in PVI science showed that PHCs are subjected to natural attenuation processes in the source zone and during the vapor transport through the vadose zone (ITRC 2018). It is well known that microorganisms can oxidize PHCs to carbon dioxide while utilizing electron acceptors such as molecular oxygen (Flintoft 2003; Fuchs et al. 2011) and that these microorganisms are practically ubiquitous across a wide range of subsurface conditions. In general, aerobic biodegradation rates are relatively rapid with respect to the rates of physical transport by diffusion and advection (U.S. EPA 2015a), leading to vapors attenuation by several orders of magnitude within a few meters. This behavior justifies the different approach usually adopted for addressing PVI compared to less biodegradable compounds such as chlorinated solvents.

This chapter provides an overview of the fate and transport of petroleum vapors in the subsurface and the methods available for the assessment of PVI.

6.2 Fate and Transport of Petroleum Vapors in the Subsurface

6.2.1 Natural Source Zone Depletion (NSZD)

As discussed in Chaps. 1, 5, 9, and 13, LNAPL constituents in the source zone undergo a series of naturally occurring processes that can lead to a progressive reduction of the source mass (DeVaull et al. 2020). In the early 90s, the primary mechanism attributed to the natural attenuation of LNAPL was the dissolution of the soluble constituents in groundwater and the subsequent biodegradation in the plume (Garg et al. 2017; ITRC 2018). Later, it was found that volatilization and methanogenic biodegradation occurring within the LNAPL body and adjacent vadose and saturated zones were the main drivers of the progressive depletion of the source (Garg et al. 2017). The combination of the sorption of LNAPL constituents onto subsurface solids, dissolution into pore water, volatilization into the vadose zone, and biodegradation within the LNAPL body is usually indicated as natural source zone depletion or NSZD (API 2017; ITRC 2018). An increasing number of studies have demonstrated that significant NSZD occurs in most sites impacted by PHCs, with measured depletion rates ranging from thousands to tens of thousands of liters per hectare per year (e.g., McCoy et al. 2014; Garg et al. 2017).

6.2.2 Phase Partitioning

PHCs can be present in the subsurface as separate (LNAPL), solid (sorbed to organic matter or geological materials), liquid (dissolved in water), or gas phases. Partitioning equations can be used to calculate the chemical concentrations in these different phases.

For instance, in the presence of LNAPL, the soil gas concentration of each constituent of interest, C_{sg} (g/m^3), can be estimated using the Raoult's law (Eq. 6.1):

$$C_{sg} = S_e \cdot H \tag{6.1}$$

where H (–) is the dimensionless Henry's law constant of the contaminant of concern and S_e (mg/L), the effective contaminant solubility in LNAPL mixture (Eq. 6.2):

$$S_{e,i} = X_i \cdot S_i \tag{6.2}$$

with X_i (mol/mol) representing the mole fraction of compound i in LNAPL mixture, S_i (mg/L), the aqueous solubility of the pure-phase compound and $S_{e,i}$ (mg/L) the effective solubility of compound i in LNAPL mixture.

When LNAPL is not present, linear equilibrium partitioning can be used. For a groundwater source, the soil gas concentration, C_{sg} (g/m^3), can be derived from the liquid-phase concentration, C_w (mg/L), through Henry's law (Eq. 6.3):

$$C_{sg} = C_w \cdot H \tag{6.3}$$

In the case of vapors originating from the soil, the concentration in the vapor phase can be calculated from the total soil concentration, C_{soil} (mg/kg), assuming again a linear equilibrium partitioning (ASTM 2000) as in Eq. 6.4:

$$C_{sg} = C_{soil} \cdot K_{as} \tag{6.4}$$

Table 6.1 Chemico-physical properties of some compounds typically of interest for vapor intrusion

Compound	Molecular weight	Henry's constant	Diffusion in air	Diffusion in water	K_{oc}	Solubility
	(g/mol)	(–)	(m²/h)	(m²/h)	(L/kg)	(mg/L)
Benzene	78.10	0.228	3.18E-02	3.50E-06	62	1743
Toluene	92.10	0.272	3.13E-02	3.10E-06	182	526
Ethylbenzene	106.20	0.323	2.70E-02	2.81E-06	363	169
Xylenes	106.20	0.314	3.13E-02	2.81E-06	240	180
Methane	16.04	29	7.02E-02	6.16E-06	90	23
Naphthalene	128.00	0.02	2.12E-02	2.70E-06	1549	31

with K_{as} (kg/L) being given by Eq. 6.5:

$$K_{as} = \frac{\rho_s \cdot H}{\theta_w + K_{oc} \cdot f_{oc} \cdot \rho_s + H \cdot \theta_a} \tag{6.5}$$

where ρ_s (kg/L) is the soil bulk density, θ_w (cm³/cm³) the moisture content of the soil, θ_a (cm³/cm³) the air-filled porosity, K_{oc} (L/kg) the organic carbon to water partition coefficient, and f_{oc} (g/g) the organic carbon fraction of the soil.

Table 6.1 reports the chemico-physical properties of petroleum compounds typically of interest for vapor intrusion.

An example of phase partitioning for benzene is depicted in Figs. 6.2 and 6.3. The two figures show the soil gas concentrations estimated for benzene as a function of groundwater and soil concentrations, respectively. For groundwater, the soil gas concentrations were estimated using the equations described before at the water table interface and above the capillary fringe (calculated with the attenuation factor AF_{cap} discussed in the next section for a sandy soil with the soil properties reported in Table 6.2). For the soil contamination, the soil gas concentrations were estimated for a sandy soil (see Table 6.2), assuming various contents of the f_{oc}.

This example shows that, depending on the extent of the contamination, the concentration of benzene in soil gas varies in the order of magnitude ranging from less than 1 to tens of g/m³. Note that in the case of a diesel or gasoline contamination, the BTEXs constitute only a small fraction of TPH (fraction of percent to few percent in mass), and thus, the effective solubility and saturation concentration in the soil of benzene (and consequently the maximum soil gas concentrations) can be lower (see Eq. 6.2) than the upper-bound values reported in these figures.

6.2.3 Molecular Diffusion

Diffusion is typically the dominant transport mechanism of vapors in the vadose zone (ITRC 2018). Molecular diffusion is the movement of a chemical from an area of

Fig. 6.2 Soil gas
concentration for benzene
estimated as a function of
groundwater concentration at
the water table interface and
above the capillary fringe
assuming a sandy soil and a
groundwater depth at 3 m
below ground surface

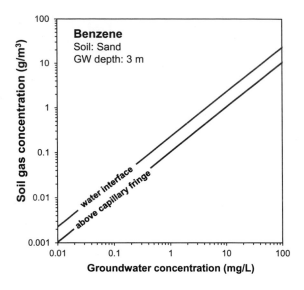

Fig. 6.3 Soil gas
concentrations for benzene
estimated as a function of
soil concentration and f_{oc} for
benzene assuming a sandy
soil

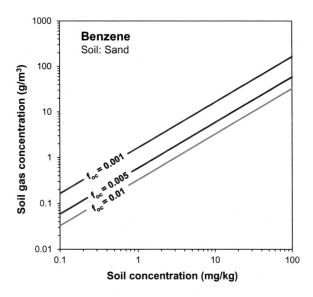

higher concentration to an area of lower concentration. Diffusion occurs in both the
aqueous and gas phases. The diffusive mass flux is directly proportional to the soil
vapor concentration gradient. Thus, the higher the soil vapor concentrations in the
source zone, the higher the flux.

Table 6.2 Soil properties (U.S. EPA 2017)

SCS soil type	ρ_s	θ_T	θ_w	$\theta_{w,cap}$	θ_r	h_{cap}	m	α	Permeability to vapor flow (k_v)
	kg/L	(–)	(–)	(–)	(–)	cm	(–)	(1/cm)	(cm^2)
Sand	1.66	0.375	0.054	0.253	0.053	17	0.685	0.035	9.91E-08
Loamy Sand	1.62	0.390	0.076	0.303	0.049	19	0.427	0.035	1.55E-08
Sandy Loam	1.62	0.387	0.103	0.320	0.039	25	0.310	0.027	5.34E-09
Sandy Clay Loam	1.63	0.384	0.146	0.333	0.063	26	0.248	0.021	1.75E-09
Loam	1.59	0.399	0.148	0.332	0.061	38	0.321	0.011	1.58E-09
Silt Loam	1.49	0.439	0.180	0.349	0.065	68	0.399	0.005	2.25E-09
Clay Loam	1.48	0.442	0.168	0.375	0.079	47	0.294	0.016	1.09E-09
Silty Clay Loam	1.37	0.482	0.198	0.399	0.090	134	0.343	0.008	1.43E-09
Silty Clay	1.38	0.481	0.216	0.424	0.111	192	0.243	0.016	1.25E-09
Silt	1.35	0.489	0.167	0.382	0.050	163	0.404	0.007	5.60E-09
Sandy Clay	1.63	0.385	0.197	0.355	0.117	30	0.172	0.033	1.46E-09
Clay	1.43	0.459	0.215	0.412	0.098	82	0.202	0.015	1.86E-09

The diffusive mass flux of a vapor in the soil, J_{diff} (g/m^2/h), is described by Fick's law:

$$J_{diff} = -D^{eff}\frac{dC}{dz} \tag{6.6}$$

where dC/dz (g/m^4) is the concentration gradient, and D^{eff} (m^2/h) the effective diffusion coefficient of the constituent in the porous medium.

The moisture content in the formation strongly affects the rate of the diffusive mass flux through the vadose zone. Diffusion coefficients in water are indeed about three to four orders of magnitude lower than the diffusion coefficients in air (see Table 6.1). Thus, as the moisture content in the formation increases, the diffusive flux decreases (ITRC 2018).

Another factor that can influence the diffusion coefficients is the subsurface temperature. In this case, as the temperature increases, the diffusive flux increases (Unnithan et al. 2021). Note that an increase in subsurface temperature also increases the Henry's law constant and the vapor pressure, thus affecting the phase partitioning of the contaminant in the subsurface.

Several equations were derived to relate the effective diffusion coefficient to the free-air diffusion coefficient of the compound and soil characteristics (Tillman and Weaver 2005). Typically, in vapor intrusion studies, the empirical equation derived by Millington and Quirk (1961) as reported by Johnson and Ettinger (1991) is used (Eq. 6.7):

$$D^{\text{eff}} = D_{\text{air}} \cdot \frac{\theta_a^{10/3}}{\theta_T^2} + \frac{D_{\text{wat}}}{H} \cdot \frac{\theta_w^{10/3}}{\theta_T^2} \qquad (6.7)$$

where D_{air} (m²/h) and D_{wat} (m²/h) are the diffusion coefficients in air and water, respectively, and θ_T (cm³/cm³) is the soil porosity.

Considering that the vertical moisture content profile through the soil is not constant, the overall diffusion coefficient in the vadose zone can be calculated by discretizing the system in n layers as suggested by Johnson and Ettinger (1991):

$$D_{\text{tot}}^{\text{eff}} = \frac{L}{\sum_i^n \frac{d_i}{D_i^{\text{eff}}}} \qquad (6.8)$$

where L (m) is the depth of source zone from the building foundations, d_i (m) the thickness of the ith layer, and D^{eff} (m²/h) the associated diffusion coefficient calculated considering the moisture content of this layer.

In cases involving homogenous soil, the moisture vertical profile can be estimated using the van Genuchten (1980) equation (Eq. 6.9):

$$S_w(z) = S_{\text{wr}} + (1 - S_{\text{wr}}) \cdot \left[\frac{1}{1 + (\alpha \cdot z)^m} \right]^m \qquad (6.9)$$

with:

$$S_w = \frac{\theta_w}{\theta_T} \qquad (6.10)$$

$$S_{wr} = \frac{\theta_r}{\theta_T} \qquad (6.11)$$

$$m = 1 - 1/n \qquad (6.12)$$

where z (cm) is the distance from the water table, θ_r (cm³/cm³) is the residual soil water content, and α (1/cm), m (–), and n (–) are the van Genuchten curve shape parameters (Table 6.2).

Therefore, in the capillary fringe, the diffusive vapor flux is relatively low compared to the one expected in the vadose zone. To estimate the attenuation factor in the capillary fringe, AF_{cap} (–), i.e., the ratio of the volatile organic compound (VOC) concentration at the top of the capillary fringe, C_{cap} (g/m³), to the VOC concentration in the soil gas in correspondence of the source in groundwater, C_{source} (g/m³), a two-layer model can be applied (U.S. EPA 2017). For instance, the attenuation factor in the capillary fringe can be estimated using Eqs. 6.13 and 6.14 (Verginelli and Baciocchi 2014):

$$AF_{\text{cap}} = \frac{C_{\text{cap}}}{C_{\text{source}}} \approx \left(1 - \frac{h_{\text{cap}}}{L}\right) \cdot \frac{D_{\text{tot}}^{\text{eff}}}{D_{\text{soil}}^{\text{eff}}} \qquad (6.13)$$

with:

$$D_{\text{tot}}^{\text{eff}} = \frac{L}{\frac{h_{\text{cap}}}{D_{\text{cap}}^{\text{eff}}} + \frac{L - h_{\text{cap}}}{D_{\text{soil}}^{\text{eff}}}} \tag{6.14}$$

where h_{cap} (m) is the thickness of the capillary fringe, D_{cap} (m^2/h) and D_{soil} (m^2/h) are, respectively, the effective diffusion coefficients in the capillary fringe and in the vadose zone calculated with Eq. 6.7 with the moisture content in the vadose zone and in the capillary fringe specific of the type of sediment considered in the site (see Table 6.2).

Note that this two-layer approach provides a conservative estimate of the attenuation through the capillary fringe. Indeed, by considering the vertical moisture profile obtained with the van Genuchten (1980) equation, the attenuation can result up to two orders of magnitude higher than the one calculated with the two-layer model approach (Hers et al. 2003; Shen et al. 2013; Yao et al. 2017, 2019).

6.2.4 Advection and Bubble-Facilitated Transport (Ebullition)

Advection is the transport mechanism by which soil gas moves due to pressure differences. The advective flux, J_{adv} (g/m^2/h), from a source zone with a known C_{sg} (g/m^3) can be estimated with Eq. 6.15:

$$J_{\text{adv}} = C_{\text{sg}} \cdot v \tag{6.15}$$

where v (m/h) is the advective velocity that can be calculated according to Darcy's law (Eq. 6.16):

$$v = \frac{k_{\text{v}}}{\mu_g} \cdot \frac{\Delta P}{L} \tag{6.16}$$

where k_{v} (m^2) is the formation permeability to vapor flow, μ_g (Pa h) is the vapor viscosity, and ΔP (Pa) is the pressure difference along a distance L (m).

In open ground conditions, the pressure differences can be generated by barometric pumping caused by ambient pressure and temperature variation and are usually limited to shallow depths (McHugh and McAlary 2009; Eklund 2016).

Pressure gradients between the air inside a building and the subsurface can be caused by several processes such as wind loading on the building, heating, ventilation, and air conditioning (HVAC) operation or the stack effect caused by heating of building air to temperatures higher than outdoor air (ITRC 2014). Thus, in the zone very close to a basement or a foundation, advective transport is likely to be the most significant contribution to PVI, as soil gases are generally swept into the building through foundation cracks due to the indoor–outdoor building pressure differential (U.S. EPA 2015b). Typically, pressure differentials between the building and the subsurface are relatively small (a few Pascals), so the building zone of influence of the pressure fields associated with the building-induced advective flow on soil gas flow is usually less than 1 m, vertically and horizontally (ITRC 2014; U.S. EPA 2015b; Ma et al. 2020a).

Advection may become important also at depth due to water table fluctuation (Tillman and Weaver 2007; Illangasekare et al. 2014; Liu et al. 2021) and near to the LNAPL source zone when the rate of gas production from methanogenesis is high (Ma et al. 2012; Yao et al. 2015). For instance, Molins et al. (2010) estimated that at the Bemidji site, advection was responsible for approximately 15% of the net flux of methane. Additionally, if methanogenesis is occurring in the saturated zone, then, after the groundwater becomes super-saturated with gas, bubble formation can occur, leading to gas transport to the vadose zone (Amos et al. 2005; Amos and Mayer 2006). Bubble formation is termed degassing. Bubble transport of gas from groundwater to the vadose zone is termed ebullition and occurs episodically (Sihota et al. 2013) along fractures in the formation. In this case, as shown by Soucy and Mumford (2017) and Ma et al. (2019), the mass flux of VOC transport could be up to two orders of magnitude higher than that of diffusive VOC transport.

6.2.5   Biodegradation During Vapor Transport

In the unsaturated zone, PHCs are readily degraded to carbon dioxide (CO_2) in the presence of oxygen (O_2) by subsurface microorganisms.

The mineralization reaction under aerobic conditions of a generic hydrocarbon C_nH_m can be written as in Eq. 6.17:

$$C_nH_m + \left(n + \frac{m}{4}\right)O_2 \rightarrow \frac{m}{2}H_2O + nCO_2 \tag{6.17}$$

From Eq. 6.17 and as indicated by Eq. 6.18, the mass ratio of O_2 consumption to the generic hydrocarbon C_nH_m mineralized is equal to:

$$\gamma = \left(n + \frac{m}{4}\right) \cdot MW_{O_2} / MW_{C_nH_m} \tag{6.18}$$

For many hydrocarbons of interest for vapor intrusion (e.g. benzene), this mass ratio is approximately 3 $g_{O_2}/g_{C_nH_m}$ (ITRC 2018).

Anaerobic degradation with other electron acceptors (e.g., nitrate or sulfate) can also occur with PHCs, but is usually neglected as there is no ready source for replenishment of these electron acceptors (ITRC 2014). Note that while aerobic biodegradation is the primary mechanism in the unsaturated zone, in the LNAPL source zone, as discussed earlier, PHCs typically degrade under methanogenic conditions with consequent production of methane and carbon dioxide (ITRC 2018).

There have been extensive compilations of rates of aerobic degradation for PHCs (e.g., DeVaull et al. 1997; Hers et al. 2000; Ririe et al. 2002; Davis et al. 2009a; DeVaull 2011). Typically, in PVI studies, first-order, water-phase aerobic degradation rates are considered. For instance, a compilation of first-order water phase biodegradation rate statistics from laboratory and field studies was reported by DeVaull (2011). Table 6.3 reports an extract of this study for some VOCs typically of concern. For more details, readers are directed to the original reference (DeVaull 2011).

Assuming a diffusion-dominated transport and first-order biodegradation, the attenuation factor due to biodegradation, AF_{bio} (–), i.e., the ratio of the VOC concentration at the top of the aerobic zone, C_{ss} (g/m^3), to the VOC concentration at the aerobic to the anaerobic interface, C_a (g/m^3), can be calculated as in Eq. 6.19 (ITRC 2014; Verginelli and Baciocchi 2021):

$$AF_{bio} = \frac{C_{ss}}{C_a} = \exp\left(-\frac{L_a}{L_R}\right) \qquad (6.19)$$

where L_a (m) is the thickness of the aerobic zone, and L_R (m) is the diffusive reaction length as defined by Eq. 6.20:

$$L_R = \sqrt{\frac{D^{eff} \cdot H}{\lambda \cdot \theta_w}} \qquad (6.20)$$

where λ (1/h) is the first-order reaction rate constant, θ_w (cm^3/cm^3) is the moisture content of the formation, D^{eff} (m^2/h) is the effective diffusion coefficient in the vadose zone, and H (–) is the dimensionless Henry's law constant.

The biodegradation attenuation factors calculated with the above equation as a function of the reaction length (L_R) and the thickness of the aerobic zone (L_a) are shown in Fig. 6.4. The reaction lengths expected in a sandy soil (see Table 6.2) for petroleum vapors of interest considering the median value of literature biodegradation rate constants (DeVaull 2011) are also reported in Fig. 6.4 as a reference. It can be

Table 6.3 First-order water phase biodegradation rates under aerobic conditions of some petroleum compounds typically of interest for vapor intrusion (DeVaull 2011 as reported by ITRC 2014)

Compound	First-order water phase biodegradation rates (h^{-1}) under aerobic conditions		
	Median value	Interquartile (1st to 3rd quartiles)	Data range (minimum to maximum)
Benzene	0.27	0.087–0.78	0.028–3
Toluene	0.72	0.19–1.4	0.028–77
Ethylbenzene	0.79	0.31–1.4	0.072–6.6
Xylenes	0.27	0.089–0.64	0.045–14
Methane	88	50–100	0.31–190
Naphthalene	0.12	0.054–5	0.021–9.8

Fig. 6.4 Biodegradation attenuation factor calculated as a function of the reaction length (L_R) and thickness of the aerobic zone (L_a). As reference in the is are reported the reaction length that can be expected in sandy soil for petroleum vapors of interest considering the median value of literature biodegradation rate constants (DeVaull 2011)

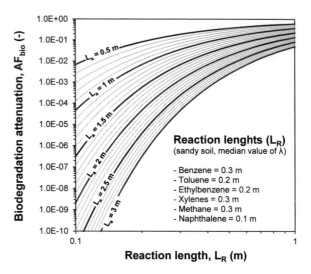

noticed that for the typical reaction lengths expected for the compounds of concern, a few meters of clean aerobic soil can ensure an attenuation in the vapor concentrations of several orders of magnitude.

The thickness of the aerobic zone, L_a (m), at the center of the building can be calculated using the expression derived by Verginelli et al. (2016a) to account for the building footprint (see Eqs. 6.21–6.23):

$$L_a = L - \frac{L}{\pi}\arccos\left[1 + \frac{\cos(\pi \cdot w_a) - 1}{\Omega}\right] \qquad (6.21)$$

with:

$$\Omega = \frac{1}{\cosh^2\left(\frac{\pi \cdot L_{\text{slab}}}{4L}\right)} \qquad (6.22)$$

$$w_a = \frac{\frac{L_b/L_{R,i}}{1+L_b/L_{R,i}}}{1 + \frac{D_{\text{ox}}\left(C_{\text{ox}}^{\text{atm}} - C_{\text{ox}}^{\text{min}}\right)}{\sum \gamma_i \cdot D_i^{\text{eff}} \cdot C_i^s}\left(1 - \frac{d_f}{d_s - L_b}\right)} \qquad (6.23)$$

where $C_{\text{ox}}^{\text{atm}}$ (g/m^3) is the atmospheric oxygen concentration (e.g., 21% v/v), $C_{\text{ox}}^{\text{min}}$ (g/m^3) is the minimum oxygen concentration (e.g., 1% v/v), C_i^s (g/m^3) is the petroleum vapor source concentration, γ_i (g/g) is the stoichiometric mass of oxygen consumed per mass of hydrocarbon i reacted, d_s (m) is the vertical source distance from open ground, d_f (m) is the slab depth from open ground, L_b (m) is the anaerobic zone thickness, $L_{R,i}$ (m) is the diffusive reaction length of hydrocarbon i and D (m^2/h) is the effective porous medium diffusion coefficients for the different species. Note that in the case of a hydrocarbons mixture, all the degradable compounds present in the source should be included.

6.2.6   Entry into the Building: Traditional and Preferential Pathways

Vapor intrusion can occur through different entry points in the building floors, walls, foundations, or through preferential pathways. The traditional vapor intrusion pathway refers to the entry of the VOCs from the subsurface through cracks, openings, and gaps in the basement (U.S. EPA 2015b). The intruded vapors into the building can mix and dilute with indoor air due to HVAC systems or windows opening. Dilution of sub-slab soil vapor concentrations is hence characterized by the building ventilation rate that is typically expressed as air exchanges per hour (AER).

For these scenarios, the sub-slab to indoor air attenuation factor, AF_{ss} (–), i.e., the ratio of the VOC concentration in indoor air, C_{indoor} (g/m^3), to the VOC concentration in the sub-slab, C_{ss} (g/m^3), can be expressed as the ratio of the soil gas entry rate, Q_{soil} (m^3/h), to the building ventilation rate, $Q_{building}$ (m^3/h), as it is shown in Eq. 6.24:

$$AF_{ss} = \frac{C_{indoor}}{C_{ss}} = \frac{Q_{soil}}{Q_{building}} = \frac{Q_{soil}}{AER \cdot V_{building}} \qquad (6.24)$$

where AER (1/h) is the air exchange rate, and $V_{building}$ (m^3) is the building volume.

Alternatively, for a screening purpose, U.S. EPA (2015b) proposed an empirical sub-slab to indoor air attenuation factor of 0.03 that represents the upper-bound value of empirical datasets. Note that the representativeness of this empirical attenuation factor proposed by U.S. EPA has been the subject of much debate (Song et al. 2011; Brewer et al. 2014; Yao et al. 2018; Lahvis and Ettinger 2021).

The sub-slab attenuation factor (AF_{ss}) calculated with Eq. 6.24 as a function of the air exchange rate (AER) and the building volume ($V_{building}$), assuming a soil gas entry rate of 10 L/min (i.e., the upper bound of the values of Q_{soil} indicated by U.S. EPA 2002) is shown in Fig. 6.5. From Fig. 6.5, it can be observed that the 0.03 value of AF_{ss} is very conservative as it can be considered representative of an air exchange rate of 0.18 h^{-1} (i.e., the 10th percentile of the values reported by U.S. EPA 2018), a soil gas entry rate of 10 L/min (i.e., the upper bound of the values of Q_{soil} indicated by U.S. EPA 2002) and a building volume of 100 m^3 (i.e., a building area of 40 m^2 considering a building mixing height of 2.5 m for a slab-on-grade scenario).

Instead, preferential pathways are specific migration routes that can cause higher contaminant flux into a building compared to the average transport through the formation (Nielsen and Hvidberg 2017; Beckley and McHugh 2020; Unnithan et al. 2021).

Fig. 6.5 Sub-slab attenuation factor (AF_{ss} = $Q_{soil}/Q_{building}$) as a function of the building volume ($V_{building}$) and of the air exchange rate (AER) for different values of the soil gas entry rate (Q_{soil})

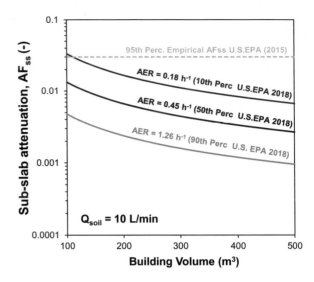

For instance, sewer pipes and other utility conduits (e.g., fiber optics, cable television, and telephone cables) are preferential pathways that can result in vapor intrusion when VOC vapors migrate through the interior of the conduits into buildings (McHugh et al. 2017; Roghani et al. 2018). According to the data collected by Beckley and McHugh (2020) from more than 30 sites across the United States, the magnitude of vapor attenuation from the sewer into the buildings is often large (> 1000× attenuation) although building-specific plumbing faults can result in much lower vapor attenuation (<100×). The authors concluded that, in general, sites with a higher risk for sewer vapor intrusion are those with direct interaction between the subsurface VOC source (e.g., groundwater) and the sewer line.

Although less investigated, a high permeability region in the vadose zone (e.g., gravel layers or fractured rocks) or the tree roots system can also act as preferential pathways for the vapor migration in the subsurface (Unnithan et al. 2021).

6.3 PVI Assessment

6.3.1 Vertical and Lateral Exclusion Distance

Source to building distance criteria can be applied to screen the vapor intrusion pathway. If the building is closer to the source than the screening distance, then further evaluations are recommended (Ma et al. 2020a). For instance, a source to building distance of 30 m (100 ft) has been used in different countries for both lateral and vertical distance-based screening since the early 2000s.

For petroleum vapor intrusion, shorter distances can be adopted due to PHCs biodegradation within the vadose zone (Fig. 6.6). For instance, Davis et al. (2009b) analyzed more than two hundred vapor samples estimating that 1.5 m (5 ft) and 10 m (30 ft) thickness of clean soil is sufficient to attenuate PHC vapors from dissolved-phase and LNAPL sources to non-detectable levels, respectively. Later, McHugh et al. (2010) proposed a separation distance of 3 m (10 ft) for petroleum vapors resulting from dissolved-phase groundwater sources and a separation distance of 10 m (30 ft) for LNAPL vapor sources. Lahvis et al. (2013) have estimated a separation distance at petroleum UST sites of 4 m (13 ft) for LNAPL sources, whereas for dissolved-phase vapor sources, the probability of detecting benzene vapor above the screening level of 30 $\mu g/m^3$ was found to be below 5%. Similar values were also reported by U.S. EPA (2013) where, depending on the method adopted for the data interpretation (i.e., vertical distance method or clean soil method), screening distances of 0 to 1.6 m (5.4 ft) for dissolved-phase sources (benzene groundwater concentration below 1 mg/L and benzene soil gas screening level of 100 $\mu g/m^3$) and 4.1–4.6 m (13.5–15 ft) for LNAPL sources at UST sites and 5.5–6.1 m (18–20 ft) for LNAPL sources at non-UST sites were defined. CRC CARE (2013) has also determined vertical screening distances of 1.5 m for dissolved phase and 3–5.6 m for LNAPL sources (note that these values are the ones reported in the document not considering the 1.5-fold uncertainty factor). These empirical screening distances are also supported by mathematical modeling. For instance, results from numerical (Hers et al. 2000; Abreu and Johnson 2006; Abreu et al. 2009; Hers et al. 2014) and analytical models (DeVaull 2007; Yao et al. 2014; Verginelli and Baciocchi 2014) were consistent with the empirical exclusion distance values reported above, showing that, in nearly all cases, source to building vertical separation distances greater than 2 m or 5 m are sufficient to attenuate to acceptable risk-based levels PHC vapors from dissolved-phase or LNAPL sources, respectively. For the lateral source to building exclusion distance, Verginelli et al. (2016b) estimated that 6 and 9 m lateral distances are sufficient to attenuate petroleum vapors below risk-based values for groundwater and soil sources, respectively. Note that these screening criteria were derived assuming that preferential pathways are not present.

A key aspect to be evaluated for the applicability of these distance criteria is the establishment of oxygenated zones beneath large buildings (Abreu et al. 2013). A building slab can potentially act as a surface cap reducing the migration of oxygen into the soil beneath the building and, consequently, limiting the attenuation of vapor concentrations due to aerobic biodegradation (Knight and Davis 2013). In most PVI conceptual site models, atmospheric oxygen at the ground surface beyond the building perimeter is considered the primary source of oxygen in the subsurface (Abreu and Johnson 2006; Abreu et al. 2009). Verginelli et al. (2016a) developed an analytical solution (Eq. 6.25) to predict the critical slab size, $L_{slab,c}$ (m), normalized to the source to building vertical distance, L (m), above which the development of an "oxygen shadow" (i.e., anoxic zone) at the center of the impervious building is expected:

154   I. Verginelli

Fig. 6.6 Typical vertical and
lateral source to building
distance criteria adopted for
petroleum vapor intrusion
that ensure an acceptable risk
(e.g., a carcinogenic risk, R,
equal to 10⁻⁶). Note that
these screening criteria are
set assuming that preferential
pathways are not present

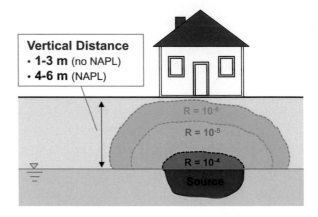

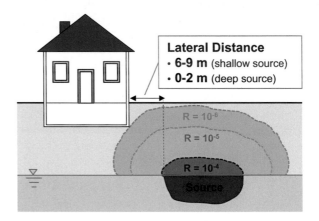

$$\frac{L_{\mathrm{slab,c}}}{L} = \frac{4}{\pi} \ln\left[\cot\left(\frac{\pi \cdot w_a}{4}\right)\right] \tag{6.25}$$

with w_a (−) defined as reported in Eq. 6.23.

The critical slab width to source depth ratio ($L_{\mathrm{slab,c}}/L$) calculated with the above equation as a function of the vapor source concentration for a slab-on-grade ($d_f = 0$) and two basement scenarios ($d_f = 0.25d_s$ and $d_f = 0.5d_s$) is shown in Fig. 6.7. The establishment of an oxygen shadow at the center of the building depends on the source depth and vapor source concentration. For instance, for a vapor source concentration of 100 g/m³ and a source depth of 10 m, an oxygen shadow at the center of buildings larger than 15 m (slab-on-grade, $L_{\mathrm{slab,c}}/L \approx 1.5$) or 5 m (basement, $L_{\mathrm{slab,c}}/L \approx 0.5$) is expected, i.e., highlighting that vertical screening criteria cannot be applied and further investigations might be needed to evaluate the sub-slab oxygen conditions. Note that the establishment of an oxygenated zone beneath impervious slabs is relatively insensitive to the type of formation as the oxygen replenishment from open ground linearly depends on the ratio of the diffusion coefficients of oxygen

Fig. 6.7 Critical normalized building slab size ($L_{slab,c}/L$) above which the development of an oxygen shadow is expected at the center of the building calculated with the solution derived by Verginelli et al. (2016a)

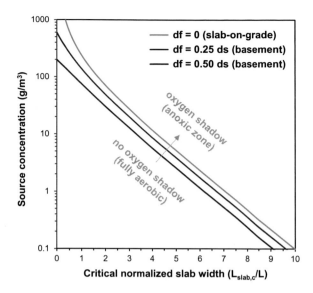

and vapor into the soil that for the different types of soil remains almost constant (Knight and Davis 2013; Verginelli et al. 2016a).

6.3.2 Analytical and Numerical Modeling

When the distance criteria discussed in the previous section are not satisfied or are not applicable (e.g., large buildings for which an oxygen shadow is expected), the indoor concentrations, C_{indoor} (g/m³), can be estimated through mathematical modeling based on the concentrations detected in the source (Eq. 6.26):

$$\begin{cases} C_{indoor} = C_{gw} \cdot H \cdot AF_{cap} \cdot AF_{bio} \cdot AF_{ss} \text{ (groundwater)} \\ C_{indoor} = C_{soil} \cdot K_{as} \cdot AF_{bio} \cdot AF_{ss} \text{ (soil)} \end{cases} \qquad (6.26)$$

The different parameters in Eq. 6.26 were already discussed in the previous sections.

As shown by Yao et al. (2013), Bekele et al. (2013) and Verginelli and Yao (2021), in the last decades, there have been numerous analytical (e.g., Johnson and Ettinger 1991; Parker 2003; DeVaull 2007; Davis et al. 2009a; Verginelli and Baciocchi 2014; Yao et al. 2014, 2015, 2016a; Verginelli et al. 2016a) and numerical models (Hers et al. 2000; Abreu and Johnson 2006; Abreu et al. 2009; Hers et al. 2014) to simulate the migration of subsurface volatile organic compounds into the building of concern and that can be used for the estimation the attenuation factors reported in the above equation. Among these, the Johnson and Ettinger (1991) model is the most widely used algorithm for assessing the vapors intrusion into enclosed spaces for non-biodegradable compounds and is incorporated in many risk assessment standards (e.g., ASTM E2081-00). For petroleum vapors, the BioVapor tool (DeVaull et al. 2010) is commonly used. BioVapor incorporates the 1-D analytical solution derived by DeVaull (2007) based on a mass continuity among upward diffusive flux, entry rate through the foundation and indoor exchange rate of hydrocarbons, and accounts for oxygen limited biodegradation. Alternatively, to evaluate the petroleum vapors and oxygen 2-D profile below the building foundations, the PVI2D tool (Verginelli et al. 2016c) can be used. PVI2D incorporates the analytical model developed by Yao et al. (2016a) and can be used to provide 2-D soil gas concentration profiles for both hydrocarbon and oxygen, based on coupled oxygen-hydrocarbon transport and reaction besides source-to-indoor air concentration attenuation factors. An example of the benzene and oxygen soil gas concentration profiles simulated using the PVI2D tool is shown in Fig. 6.8. In this example, a benzene vapor source concentrations of 50 g/m^3, a sandy soil (Table 6.2), and a biodegradation rate of 0.27 h^{-1} are assumed. As shown by Yao et al. (2016b), for homogenous site conditions, PVI2D provides soil gas concentration profiles and source-to-indoor air attenuation factors similar to the ones obtained using the more sophisticated numerical model developed by Abreu and Johnson (Abreu and Johnson 2006; Abreu et al. 2009).

For complex contamination scenarios involving transient transport, soil heterogeneities, preferential pathways, and non-uniform sources, numerical modeling is needed. In PVI field, the Abreu and Johnson's model is the reference model by the majority of practitioners and has been employed in almost all U.S. EPA's PVI technical documents (Abreu and Schuver 2012; Abreu et al. 2013) since its publication.

It is worth noting that, especially in potentially critical scenarios (e.g., NAPL sources), mathematical modeling should be used as one of the lines of evidence for PVI assessment in conjunction with site investigation data as part of a multiple-lines-of-evidence approach (U.S. EPA 2015b). Indeed, mathematical modeling may not reflect all of reality as subsurface heterogeneity (e.g., geological barriers, fractured soils, plants roots), preferential pathways, or barometric pumping cycling can hardly be schematized and predicted (Unnithan et al. 2021). Besides, for their simplicity, in most cases, practitioners prefer 1D models compared to more sophisticated 2D and 3D models leading to a further oversimplification of the processes that can affect the transport of petroleum vapors in the subsurface and the consequent intrusion into the building.

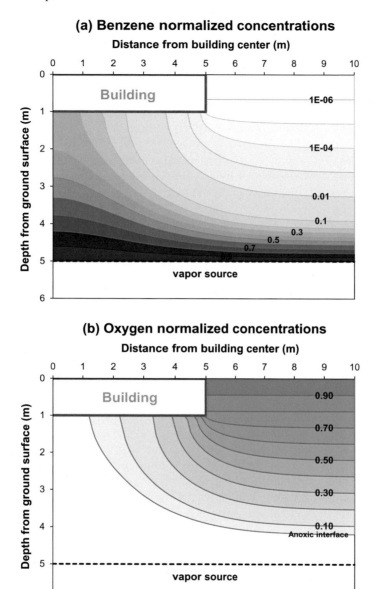

Fig. 6.8 Example of benzene and oxygen soil gas concentration profiles simulated with PVI2D (Verginelli et al. 2016c; Yao et al. 2016a) for a basement scenario. In this example, a benzene vapor source concentrations of 50 g/m^3, a sandy soil (Table 6.2), and a biodegradation rate of 0.27 h^{-1} are assumed. The hydrocarbons and oxygen soil gas concentrations are normalized to source concentration and atmosphere concentration, respectively

6.3.3   Soil Gas Sampling

Measurement of soil gas is a common approach for evaluating the vapor intrusion pathway. Soil gas data compared to soil matrix and groundwater data represent a direct measurement of the contaminant that can potentially migrate into indoor air. Soil gas sampling is usually carried out by installing temporary or permanent probes into the subsurface (U.S. EPA 2015b) as illustrated in Fig. 6.9. These probes are typically equipped with active sampling systems that pump out the air-filled porosity of the soil, and the vapor samples are collected using Tedlar® bags, passivated stainless-steel canisters, or sorbent tubes (e.g., automated thermal desorption tubes). The samples are then sent to the laboratory for the VOCs analyses. Over recent years, passive samplers have become an attractive alternative option for the monitoring of soil gas. Passive sampling is based on the molecular diffusion of the compounds from the subsurface to a collecting medium in response to a chemical potential difference (Górecki and Namiesnik 2002). Passive sampling has potential advantages over conventional active methods, including simpler protocols, smaller size for ease of shipping and handling, and lower costs (McAlary et al. 2014; Salim and Górecki 2019). A wide variety of "linear uptake" passive samplers have been designed and tested for soil gas, including Petrex tubes, Emflux® cartridges, Beacon B-Sure Samples™, Gore™ Modules, and PDMS membranes (Hodny et al. 2009; McAlary et al. 2014; Liang et al. 2020). Each of these methods provides results in units of the mass of contaminant adsorbed over the exposure duration and then the correlation between the mass adsorbed and the soil vapor concentration can be quantitatively assessed based on specific uptake rates. The uptake rate is the key parameter for a quantitative assessment of soil gas concentration by passive samplers, and it depends on the environmental conditions (e.g., subsurface temperature, humidity, and pressure) and on the geometry of the sampling device and the sorbent characteristics (McAlary et al. 2014). Alternatively, "equilibrium" passive samplers based on polymers were recently proposed. For instance, Gschwend et al. (2022) investigated the use of thin, low-density polyethylene (PE) films as the sampling absorbent for estimating the soil gas concentrations of BTEX and chlorinated solvents.

Soil gas samples can be differentiated by the location of the samples. Near-slab soil gas samples are collected outside a structure but within a short distance (a few meters) of the building's perimeter. Soil gas samples collected at higher distances from the perimeter of the building are referred to as exterior soil gas samples and can be useful for delineating the vapor plume, for screening areas for subsequent indoor sampling or for the evaluation of future vapor intrusion scenarios. Finally, sub-slab soil gas samples are collected from the subsurface immediately under the building foundation or slab.

As indicated by Eq. 6.27, based on the concentrations detected in soil gas, the vapor intrusion pathway is evaluated through the application of the sub-slab to indoor air attenuation factor, AF_{ss} (–), in the case of sub-slab sampling or considering also the biodegradation attenuation factor, AF_{bio} (–), in the case of soil gas data collected at some depth from the building zone:

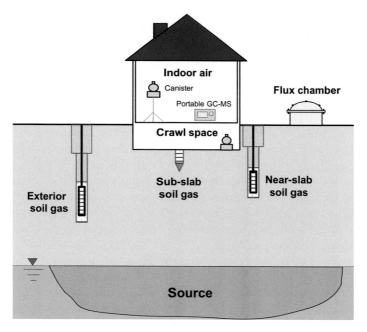

Fig. 6.9 Soil gas, flux chamber, and indoor air sampling

$$\begin{cases} C_{\text{indoor}} = C_{\text{sg}} \cdot \text{AF}_{\text{ss}} \text{ (sub-slab)} \\ C_{\text{indoor}} = C_{\text{sg}} \cdot \text{AF}_{\text{bio}} \cdot \text{AF}_{\text{ss}} \text{ (deep soil gas)} \end{cases} \qquad (6.27)$$

Alternatively, or in conjunction with the soil gas probes, flux chambers can be installed at the ground surface to estimate the emissions of VOCs from the subsurface. Flux chambers are inverted containers installed at the ground surface that can directly measure the emissions of VOCs from the subsurface to the atmosphere (Verginelli et al. 2018). These chambers can be used to estimate the outdoor volatilization or can be installed in buildings with exposed soil (e.g., pier-and-beam construction) or with uncracked concrete slab foundations with sealed expansion joints to quantify the vapor intrusion rate (Ma et al. 2020b). Further details on flux chambers can be found in Chap. 5.

6.3.4 Indoor Air Sampling

Measuring indoor air is the most direct approach for vapor intrusion to provide the indoor concentrations needed for risk assessment. Indoor air sampling is typically performed using passivated stainless-steel canisters with sampling periods of 24 and 8 h for residential and industrial buildings, respectively (U.S. EPA 2015a). The samples are then sent to the laboratory for the VOCs analyses providing a time-weighted indoor concentration for the contaminants of concern averaged over the sampling period. Alternatively, passive sampling (e.g., passive sorbent sampler) can be performed, allowing a long sample collection time (days to weeks) that can reduce the effects of short-term temporal variability by yielding a time-integrated average VOC concentration (McHugh et al. 2017).

The above methods provide information on the average inhalation exposure needed to calculate long-term chronic risks. To evaluate short-term acute risks or episodical potential threats to safety (e.g., explosive concentrations of petroleum vapors or methane), alternative methods were recently developed. These include high-frequency indoor gas chromatograph-mass spectrometer (GC–MS) measurements (e.g., Gorder and Dettenmaier 2011; Beckley et al. 2014) eventually coupled with concurrent measurement of dynamic controlling factors (e.g., Kram et al. 2019, 2020) or manipulation of indoor–outdoor pressure conditions (building pressure cycling method) to measure changes in indoor air VOC concentration under positive or negative indoor air pressure (e.g., McHugh et al. 2012). High-frequency indoor GC–MS measurements are typically performed with field-portable instruments designed for on-site analysis that allow measuring indoor air VOC concentrations in almost real-time (a few minutes) during the course of the field investigation (Gorder and Dettenmaier 2011). Continuous monitoring enables the collection of a large volume of contaminant concentration data over time (tens of analyses per day), thus allowing to evaluate short-term indoor air concentration variations and the presence of background VOCs sources (Beckley et al. 2014). By coupling this analysis with simultaneous monitoring with dedicated detectors of additional parameters such as sub-foundation pressure, wind speed and barometric pressure could help to determine the cause of the vapor intrusion (Kram et al. 2019, 2020).

Building pressure cycling (BPC) is a technique that manipulates building air pressure and ventilation to promote or inhibit the intrusion of vapors into the building using either blower doors or by manipulation of existing HVAC systems (McHugh et al. 2012). Indoor air VOC concentrations are then measured either using real-time monitoring or by traditional indoor sampling after three to five times the air volume of the building has been flushed (Lutes et al. 2019). BPC reduces the spatial and temporal variability of indoor air and allows to determine the reasonable maximum exposure (RME) under negative building pressure for the assessment of short-term risks and allows to determine the presence of background sources under positive pressure conditions (McHugh et al. 2012).

Regarding the latter aspect, for the evaluation and eventual management of the vapor intrusion pathway, particular caution should be paid to identify all the potential indoor and outdoor VOCs sources (usually indicated as background) that are not related to the subsurface source of vapors (Rago et al. 2021). As discussed by U.S. EPA (2015b), indoor air in many buildings can indeed contain detectable levels of a number of VOCs associated with the use and storage of consumer products (e.g., cleaners, air fresheners, scented candles, or other household products), combustion processes (e.g., smoking, cooking, and home heating) and releases from interior building materials (e.g., carpets, paint, and wood-finishing products). Furthermore, in urban centers, outdoor ambient concentrations of some VOCs (e.g., benzene) may exceed allowable indoor risk-based levels, as a result of vehicle traffic emissions. For the above reasons, as suggested by U.S. EPA (2015b), indoor air sampling should always be carried out in conjunction with other lines of evidence (e.g., soil gas sampling), eventually coupled with compound-specific isotope analysis (CSIA). CSIA (Chap. 11) provides an independent line of evidence to distinguish between vapor intrusion and indoor sources of VOCs by analyzing the isotope fractionation in indoor air and soil gas samples (Beckley et al. 2016). For PHCs, the isotopes to be used are ^{13}C and ^{2}H. The isotope composition of VOCs originating from the subsurface is often clearly different than that of pristine manufactured products acting as indoor sources of the same VOCs, and thus, this difference allows the differentiation between VOCs from indoor sources and those from true vapor intrusion sources (McHugh et al. 2011).

6.3.5 Risk Assessment

The measured or estimated indoor concentrations, C_{indoor} (g/m^3), for the different target compounds are then used in the risk assessment to establish the need for mitigation actions of the vapor intrusion pathway.

According to the risk-based corrective action (RBCA) procedure outlined in the ASTM (2000) standard, the carcinogenic and non-carcinogenic risks can be estimated as shown in Eq. 6.28:

$$\begin{cases} R = C_{indoor} \cdot IUR \cdot EC \text{ (carcinogenic effects)} \\ HQ = C_{indoor} \cdot \dfrac{EC}{RfC} \text{ (non - carcinogenic effects)} \end{cases} \qquad (6.28)$$

where R (–) is the carcinogenic risk, HQ (–) is the hazard quotient for toxic effects, and RfC (g/m^3) and IUR (m^3/g) are the reference concentration and the inhalation unit risk that represent the toxicological parameters for toxic and carcinogenic effects (see Table 6.4). EC (–) is the exposure factor that can be estimated as in Eq. 6.29:

$$EC = \frac{EF_{gi} \cdot EF \cdot ED}{BW \cdot AT \cdot 365 \, \frac{days}{year} \cdot 24 \, \frac{hours}{day}} \qquad (6.29)$$

where BW (kg) is the body weight, EF (d/y) is the annual exposure frequency, EF_{gi} (h/d) the daily exposure frequency, ED (y) is the exposure duration, and AT (y) is the averaging time (set equal to ED for the toxic effects).

The above equations can be rearranged as in Eq. 6.30 to calculate the acceptable indoor risk-based levels, $C_{indoor,acc}$ (g/m^3), that ensure the target risk, TR (–), and hazard quotient, THQ (–):

Table 6.4 Toxicological parameters (U.S. EPA 2020) and example of indoor risk-based concentration for some petroleum compounds of concern for petroleum vapor intrusion. The acceptable indoor risk-based levels ($C_{indoor,acc}$) were calculated using the following parameters: TR = 10^{-6}, THQ = 1, BW = 70 kg, EF = 350 d/y, EF_{gi} = 24 h/d, ED = 30 y, AT = ED for toxic effects, AT = 70 y for carcinogenic effects

Compound	RfC	IUR	$C_{indoor,acc}$
	(mg/m^3)	$(\mu g/m^3)^{-1}$	($\mu g/m^3$)
Benzene	0.03	7.80×10^{-6}	0.31
Toluene	5	–	5214
Ethylbenzene	1	2.50×10^{-6}	0.97
Xylenes	0.1	–	104
Naphthalene	0.003	3.40×10^{-5}	0.072

$$C_{indoor,acc} = \min \begin{cases} \dfrac{TR}{IUR \cdot EC} & \text{(carcinogenic effects)} \\ \dfrac{THQ \cdot RfC}{EC} & \text{(non - carcinogenic effects)} \end{cases} \quad (6.30)$$

Table 6.4 reports an example of the indoor risk-based concentrations calculated with the above equations for some petroleum compounds typically of concern for petroleum vapor intrusion, assuming a target cancer risk level of one per million (10^{-6}) and a target hazard quotient of 1 for non-carcinogenic effects. The assumed exposure factors were: BW = 70 kg, EF = 350 d/y, EF_{gi} = 24 h/d, ED = 30 y, AT = ED for toxic effects and AT = 70 y for carcinogenic effects.

As discussed earlier, benzene is usually the most critical compound for vapor intrusion although as shown by Brewer et al. (2013) when benzene is less than approximately 1% in the hydrocarbon mixture (i.e., when the concentration of TPH is more than 900 times that of benzene) TPH could drive vapor intrusion risk (for C5–C8 and C9–C12 aliphatics).

6.4 Conclusions

Although there were significant advances in PVI science over the last decades, knowledge gaps still exist. Regarding the vapor intrusion route, the importance of sewers or other utility tunnels as preferential pathways for VOC migration into buildings has received increased focus in recent years (Ma et al. 2020a), but their understanding, especially in the case of petroleum vapors, is very limited. Bubble-facilitated transport above NAPL sources is also attracting attention, but more field and laboratory studies are still required to understand the significance of this pathway. The role of pervious slabs for PVI is another aspect that could be addressed in the future as diffusive transport through permeable concrete slabs can enhance the oxygen availability in the subsurface (U.S. EPA 2015a), thus creating conditions favorable for the occurrence of aerobic biodegradation even below large buildings (Verginelli et al. 2016a). Finally, considering that most of the field experience was gained in the United States and, for some aspects (e.g., vertical distance criteria), also in Australia, future studies could be oriented to evaluate if the criteria and approaches adopted in these countries should be modulated to account for difference in types of building construction, climate, or geological settings.

Regarding soil gas and indoor sampling, in the last decade, there was increasing attention in developing and testing new methods to account for the dynamics of vapor intrusion and determine the presence of indoor background sources. These methods include real-time GC–MS analysis eventually coupled with concurrent measurement of dynamic controlling factors (e.g., sub-foundation pressure, wind speed, and barometric pressure), building pressure cycling methods to measure changes in indoor air VOC concentration under positive or negative indoor air pressure, quantitative passive soil gas sampling, and CSIA. The further development and refinement of these methods along with a better understanding of the fate and transport of petroleum vapors in the subsurface will trace the path for a better understanding and management of petroleum vapor intrusion.

References

Abreu LD, Johnson PC (2006) Simulating the effect of aerobic biodegradation on soil vapor intrusion into buildings: influence of degradation rate, source concentration, and depth. Environ Sci Technol 40(7):2304–2315. https://doi.org/10.1021/es051335p

Abreu LD, Ettinger R, McAlary T (2009) Simulated soil vapor intrusion attenuation factors including biodegradation for petroleum hydrocarbons. Ground Water Monit Remediat 29(1):105–117. https://doi.org/10.1111/j.1745-6592.2008.01219.x

Abreu LD, Schuver H (2012) Conceptual model scenarios for the vapor intrusion pathway. EPA 530-R-10-003. USEPA Office of Resource Conservation and Recovery, Washington, DC. www.epa.gov/sites/production/files/2015-09/documents/vi-cms-v11final-2-24-2012.pdf. Accessed Dec 2021

Abreu LD, Lutes CC, Nichols EM (2013) 3-D modeling of aerobic biodegradation of petroleum vapors: effect of building area size on oxygen concentration below the slab. Office of Underground Storage Tanks (OUST). EPA 510-R-13-002. http://www.epa.gov/oust/cat/pvi/building-size-modeling.pdf. Accessed Dec 2021

Amos RT, Mayer KU, Bekins BA, Delin GN, Williams RL (2005) Use of dissolved and vapor-phase gases to investigate methanogenic degradation of petroleum hydrocarbon contamination in the subsurface. Water Resour Res 41(2). https://doi.org/10.1029/2004WR003433

Amos RT, Mayer KU (2006) Investigating ebullition in a sand column using dissolved gas analysis and reactive transport modeling. Environ Sci Technol 40(17):5361–5367. https://doi.org/10.1021/es0602501

API (2017) Quantification of vapor phase-related natural source zone depletion processes. American Petroleum Institute, Publication No. 4784. Available at http://www.api.org

ASTM (2000) Standard guide for risk-based corrective action. ASTM, West Conshohocken, PA, E2081-00. Accessed Dec 2021

Beckley L, Gorder K, Dettenmaier E, Rivera-Duarte I, McHugh T (2014) On-site gas chromatography/mass spectrometry (GC/MS) analysis to streamline vapor intrusion investigations. Environ Forensics 15(3):234–243. https://doi.org/10.1080/15275922.2014.930941

Beckley L, McHugh T, Philp P (2016) Utility of compound-specific isotope analysis for vapor intrusion investigations. Ground Water Monit Remediat 36(4):31–40. https://doi.org/10.1111/gwmr.12185

Beckley L, McHugh T (2020) A conceptual model for vapor intrusion from groundwater through sewer lines. Sci Total Environ 698:134283. https://doi.org/10.1016/j.scitotenv.2019.134283

Bekele DN, Naidu R, Bowman M, Chadalavada S (2013) Vapor intrusion models for petroleum and chlorinated volatile organic compounds: opportunities for future improvements. Vadose Zone J 12(2). https://doi.org/10.2136/vzj2012.0048

Brewer R, Nagashima J, Kelley M, Heskett M, Rigby M (2013) Risk-based evaluation of total petroleum hydrocarbons in vapor intrusion studies. Int J Environ Res Public Health 10(6):2441–2467. https://doi.org/10.3390/ijerph10062441

Brewer R, Nagashima J, Rigby M, Schmidt M, O'Neill H (2014) Estimation of generic subslab attenuation factors for vapor intrusion investigations. Ground Water Monit Remediat 34(4):79–92. https://doi.org/10.1111/gwmr.12086

CRC CARE (2013) Petroleum hydrocarbon vapour intrusion assessment: Australian guidance. CRC CARE Technical Report no. 23, CRC for Contamination Assessment and Remediation of the Environment, Adelaide, Australia. www.crccare.com/publications/technical-reports. Accessed Dec 2021

Davis GB, Patterson BM, Trefry MG (2009a) Evidence for instantaneous oxygen-limited biodegradation of petroleum hydrocarbon vapors in the subsurface. Ground Water Monit Remediat 29(1):126–137. https://doi.org/10.1111/j.1745-6592.2008.01221.x

Davis GB, Patterson BM, Trefry MG (2009b) Biodegradation of petroleum hydrocarbon vapours. CRC CARE 2009b, Technical Report no. 12, CRC for Contamination Assessment and Remediation of the Environment, Adelaide, Australia. www.crccare.com/publications/technical-reports. Accessed Dec 2021

DeVaull GE, Ettinger RA, Salanitro JP, Gustafson JB (1997) Benzene, toluene, ethylbenzene, and xylenes (BTEX) degradation in vadose zone soils during vapor transport: first-order rate constants (No. CONF-971116-). Ground Water Publishing Co., Westerville, OH (United States)

DeVaull GE (2007) Indoor vapor intrusion with oxygen-limited biodegradation for a subsurface gasoline source. Environ Sci Technol 41(9):3241–3248. https://doi.org/10.1021/es060672a

DeVaull GE, McHugh TE, Newberry P (2010) Users' manual "Biovapor: a 1-D vapor intrusion model with oxygen-limited aerobic biodegradation"; American Petroleum Institute, Washington, DC. https://www.api.org/oil-and-natural-gas/environment/clean-water/ground-water/vapor-intrusion/biovapor. Accessed Dec 2021

DeVaull GE (2011) Biodegradation rates for petroleum hydrocarbons in aerobic soils: a summary of measured data. In: International symposium on bioremediation and sustainable environmental technologies. Reno, NV

DeVaull GE, Rhodes IA, Hinojosa E, Bruce CL (2020) Petroleum NAPL depletion estimates and selection of marker constituents from compositional analysis. Ground Water Monit Remediat 40(4):44–53. https://doi.org/10.1111/gwmr.12410

Eklund B (2016) Effect of environmental variables on vapor transport. In: Proceedings of the 26th annual international conference on soil, water, energy, and air, San Diego, CA

Flintoft L (2003) Boost for bacterial batteries. Nat Rev Microbiol 1(2):88–88. https://doi.org/10.1038/nrmicro758

Fuchs G, Boll M, Heider J (2011) Microbial degradation of aromatic compounds—from one strategy to four. Nat Rev Microbiol 9(11):803–816. https://doi.org/10.1038/nrmicro2652

Garg S, Newell CJ, Kulkarni PR, King DC, Adamson DT, Renno MI, Sale T (2017) Overview of natural source zone depletion: processes, controlling factors, and composition change. Ground Water Monit Rem 37(3):62–81. https://doi.org/10.1111/gwmr.12219

Górecki T, Namiesnik J (2002) Passive sampling. Trends Analyt Chem 21:276–291. https://doi.org/10.1016/S0165-9936(02)00407-7

Gorder KA, Dettenmaier EM (2011) Portable GC/MS methods to evaluate sources of cVOC contamination in indoor air. Ground Water Monit Remediat 31(4):113–119. https://doi.org/10.1111/j.1745-6592.2011.01357.x

Gschwend P, MacFarlane J, Jensen D, Soo J, Saparbaiuly G, Borrelli R, Vago F, Oldani A, Zaninetta L, Verginelli I, Baciocchi R (2022) In situ equilibrium polyethylene passive sampling of soil gas VOC concentrations: modeling, parameter determinations, and laboratory testing. Environ Sci Technol. https://doi.org/10.1021/acs.est.1c07045

Hers I, Atwater J, Li L, Zapf-Gilje R (2000) Evaluation of vadose zone biodegradation of BTX vapours. J Contam Hydrol 46(3–4):233–264. https://doi.org/10.1016/S0169-7722(00)00135-2

Hers I, Zapf-Gilje R, Johnson PC, Li L (2003) Evaluation of the Johnson and Ettinger model for prediction of indoor air quality. Ground Water Monit Rem 23(2):119–133. https://doi.org/10.1111/j.1745-6592.2003.tb00678.x

Hers I, Jourabchi P, Lahvis MA, Dahlen P, Luo EH, Johnson PC, Mayer KU (2014) Evaluation of seasonal factors on petroleum hydrocarbon vapor biodegradation and intrusion potential in a cold climate. Ground Water Monit Rem 34(4):60–78. https://doi.org/10.1111/gwmr.12085

Hodny JW, Whetzel Jr JE, Anderson II HS (2009) Quantitative passive soil gas and air sampling in vapor intrusion investigations. In: Proceedings of the AW&MA vapor intrusion 2009 conference

Illangasekare T, Petri B, Fucik R, Sauck C, Shannon L, Sakaki T, Smits K, Cihan A, Christ J, Schulte P (2014) Vapor intrusion from entrapped NAPL sources and groundwater plumes: process understanding and improved modeling tools for pathway assessment. Colorado School of Mines Golden. https://clu-in.org/download/issues/vi/VI-ER-1687-FR.pdf. Accessed Dec 2021

ITRC (2009) Evaluation of natural source zone depletion at sites with LNAPL. Interstate Technology and Regulatory Council, LNAPL Team, Washington, DC, April 2009. https://www.itrcweb.org/guidancedocuments/lnapl-1.pdf. Accessed Dec 2021

ITRC (2014) Petroleum vapor intrusion: fundamentals of screening, investigation, and management. Interstate Technology and Regulatory Council, Vapor Intrusion Team, Washington, DC, October 2014. https://projects.itrcweb.org/PetroleumVI-Guidance/. Accessed Dec 2021

ITRC (2018) Light non-aqueous phase liquids (LNAPL) document update, LNAPL-3. Interstate Technology & Regulatory Council, LNAPL Update Team, Washington, USA. https://lnapl-3.itrcweb.org/. Accessed Dec 2021

Johnson PC, Ettinger RA (1991) Heuristic model for predicting the intrusion rate of contaminant vapors into buildings. Environ Sci Technol 25(8):1445–1452. https://doi.org/10.1021/es00020a013

Knight JH, Davis GB (2013) A conservative vapour intrusion screening model of oxygen-limited hydrocarbon vapour biodegradation accounting for building footprint size. J Contam Hydrol 155:46–54. https://doi.org/10.1016/j.jconhyd.2013.09.005

Kram ML, Hartman B, Clite N (2019) Automated continuous monitoring and response to toxic subsurface vapors entering overlying buildings—selected observations, implications and considerations. Remed J 29(3):31–38. https://doi.org/10.1002/rem.21605

Kram ML, Hartman B, Frescura C, Negrao P, Egelton D (2020) Vapor intrusion risk evaluation using automated continuous chemical and physical parameter monitoring. Remed J 30(3):65–74. https://doi.org/10.1002/rem.21646

Lahvis MA, Hers I, Davis RV, Wright J, DeVaull GE (2013) Vapor intrusion screening at petroleum UST sites. Ground Water Monit Rem 33(2):53–67. https://doi.org/10.1111/gwmr.12005

Lahvis MA, Ettinger RA (2021) Improving risk-based screening at vapor intrusion sites in California. Ground Water Monit Rem 41(2):73–86. https://doi.org/10.1111/gwmr.12450

Liang C, Chang JS, Chen TW, Hou Y (2020) Passive membrane sampler for assessing VOCs contamination in unsaturated and saturated media. J Hazard Mater 401:123387. https://doi.org/10.1016/j.jhazmat.2020.123387

Liu X, Ma E, Zhang YK, Liang X (2021) An analytical model of vapor intrusion with fluctuated water table. J Hydrol 596:126085. https://doi.org/10.1016/j.jhydrol.2021.126085

Lutes CC, Holton CW, Truesdale R, Zimmerman JH, Schumacher B (2019) Key design elements of building pressure cycling for evaluating vapor intrusion—a literature review. Ground Water Monit Rem 39(1):66–72. https://doi.org/10.1111/gwmr.12310

Ma E, Zhang YK, Liang X, Yang J, Zhao Y, Liu X (2019) An analytical model of bubble-facilitated vapor intrusion. Water Res 165:114992. https://doi.org/10.1016/j.watres.2019.114992

Ma J, Rixey WG, DeVaull GE, Stafford BP, Alvarez PJ (2012) Methane bioattenuation and implications for explosion risk reduction along the groundwater to soil surface pathway above a plume of dissolved ethanol. Environ Sci Technol 46(11):6013–6019. https://doi.org/10.1021/es300715f

Ma J, McHugh T, Beckley L, Lahvis MA, DeVaull GE, Jiang L (2020a) Vapor intrusion investigations and decision-making: a critical review. Environ Sci Technol 54(12):7050–7069. https://doi.org/10.1021/acs.est.0c00225

Ma J, McHugh T, Eklund B (2020b) Flux chamber measurements should play a more important role in contaminated site management. Environ Sci Technol 54(19):11645–11647. https://doi.org/10.1021/acs.est.0c04078

McAlary T, Wang X, Unger A, Groenevelt H, Górecki T (2014) Quantitative passive soil vapor sampling for VOCs-part 1: theory. Environ Sci Process Impacts 16(3):482–490. https://doi.org/10.1039/C3EM00652B

McCoy K, Zimbron J, Sale T, Lyverse M (2014) Measurement of natural losses of LNAPL using CO_2 traps. Groundwater 53(4):658–667. https://doi.org/10.1111/gwat.12240

McHugh TE, McAlary T (2009) Important physical processes for vapor intrusion: a literature review. In: Proceedings of AWMA vapor intrusion conference, San Diego, CA

McHugh TE, Davis R, DeVaull GE, Hopkins H, Menatti J, Peargin T (2010) Evaluation of vapor attenuation at petroleum hydrocarbon sites: considerations for site screening and investigation. Soil Sediment Contam 19(6):725–745. https://doi.org/10.1080/15320383.2010.499923

McHugh TE, Kuder T, Fiorenza S, Gorder K, Dettenmaier E, Philp P (2011) Application of CSIA to distinguish between vapor intrusion and indoor sources of VOCs. Environ Sci Technol 45(14):5952–5958. https://doi.org/10.1021/es200988d

McHugh TE, Beckley L, Bailey D, Gorder K, Dettenmaier E, Rivera-Duarte I, MacGregor IC (2012) Evaluation of vapor intrusion using controlled building pressure. Environ Sci Technol 46(9):4792–4799. https://doi.org/10.1021/es204483g

McHugh TE, Loll P, Eklund B (2017) Recent advances in vapor intrusion site investigations. J Enviro Manage 204:783–792. https://doi.org/10.1016/j.jenvman.2017.02.015

Millington RJ, Quirk JP (1961) Permeability of porous solids. Trans Faraday Soc 57:1200–1207

Molins S, Mayer KU, Amos RT, Bekins BA (2010) Vadose zone attenuation of organic compounds at a crude oil spill site—interactions between biogeochemical reactions and multicomponent gas transport. J Contam Hydrol 112(1–4):15–29. https://doi.org/10.1016/j.jconhyd.2009.09.002

Nielsen KB, Hvidberg B (2017) Remediation techniques for mitigating vapor intrusion from sewer systems to indoor air. Remed J 27(3):67–73. https://doi.org/10.1002/rem.21520

Parker JC (2003) Modeling volatile chemical transport, biodecay, and emission to indoor air. Ground Water Monit Rem 23(1):107–120. https://doi.org/10.1111/J.1745-6592.2003.TB00789.X

Rago R, Rezendes A, Peters J, Chatterton K, Kammari A (2021) Indoor air background levels of volatile organic compounds and air-phase petroleum hydrocarbons in office buildings and schools. Ground Water Monit Rem 41(2):27–47. https://doi.org/10.1111/gwmr.12433

Ririe GT, Sweeney RE, Daugherty SJ (2002) A comparison of hydrocarbon vapor attenuation in the field with predictions from vapor diffusion models. Soil Sediment Contam 11(4):529–554. https://doi.org/10.1080/20025891107159

Rivett MO, Wealthall GP, Dearden RA, McAlary TA (2011) Review of unsaturated-zone transport and attenuation of volatile organic compound (VOC) plumes leached from shallow source zones. J Contam Hydrol 123(3–4):130–156. https://doi.org/10.1016/j.jconhyd.2010.12.013

Roghani M, Jacobs OP, Miller A, Willett EJ, Jacobs JA, Viteri CR, Pennell KG (2018) Occurrence of chlorinated volatile organic compounds (VOCs) in a sanitary sewer system: Implications for assessing vapor intrusion alternative pathways. Sci Total Environ 616:1149–1162. https://doi. org/10.1016/j.scitotenv.2017.10.205

Salim F, Górecki T (2019) Theory and modelling approaches to passive sampling. Environ Sci Process Impacts 21(10):1618–1641. https://doi.org/10.1039/C9EM00215D

Shen R, Pennell KG, Suuberg EM (2013) Influence of soil moisture on soil gas vapor concentration for vapor intrusion. Environ Eng Sci 30(10):628–637. https://doi.org/10.1089/ees.2013.0133

Sihota NJ, Mayer KU, Toso MA, Atwater JF (2013) Methane emissions and contaminant degradation rates at sites affected by accidental releases of denatured fuel-grade ethanol. J Contam Hydrol 151:1–15. https://doi.org/10.1016/j.jconhyd.2013.03.008

Song S, Ramacciotti FC, Schnorr BA, Bock M, Stubbs CM (2011) Evaluation of USEPA's empirical attenuation factor database. Air Waste and Management Association. Emissions Monitoring, 16–21 Feb 2011

Soucy NC, Mumford KG (2017) Bubble-facilitated VOC transport from LNAPL smear zones and its potential effect on vapor intrusion. Environ Sci Technol 51(5):2795–2802. https://doi.org/10. 1021/acs.est.6b06061

Tillman FD, Weaver JW (2005) Review of recent research on vapor intrusion. Washington, DC 20460: US Environmental Protection Agency, Office of Research and Development

Tillman FD, Weaver JW (2007) Temporal moisture content variability beneath and external to a building and the potential effects on vapor intrusion risk assessment. Sci Total Environ 379(1):1–15. https://doi.org/10.1016/j.scitotenv.2007.02.003

Unnithan A, Bekele D, Chadalavada S, Naidu R (2021) Insights into vapour intrusion phenomena: current outlook and preferential pathway scenario. Sci Total Environ 796:148885. https://doi. org/10.1016/j.scitotenv.2021.148885

U.S. EPA (1995) Light nonaqueous-phase liquids. EPA/540/S-95/500. EPA Groundwater Issue. http://nepis.epa.gov/Exe/ZyPURL.cgi?Dockey=10002DXR.txt. Accessed Dec 2021

U.S. EPA (2002) OSWER draft guidance for evaluating the vapor intrusion to indoor air pathway from groundwater and soils (Subsurface Vapor Intrusion Guidance). Office of Solid Waste and Emergency Response, Washington, D.C. EPA530-D-02-004. https://nepis.epa.gov/Exe/ZyP URL.cgi?Dockey=P1008OTB.TXT. Accessed Dec 2021

U.S. EPA (2013) Evaluation of empirical data and modeling studies to support soil vapor intrusion screening criteria for petroleum hydrocarbon compounds. EPA 510-R-13-001. Washington, DC: U.S. Environmental Protection Agency, Office of Solid Waste and Emergency Response. https://www.epa.gov/sites/default/files/2014-09/documents/pvi_database_report.pdf. Accessed Dec 2021

U.S. EPA (2015a) Technical guide for addressing petroleum vapor intrusion at leaking underground storage tank sites. Office of Underground Storage Tanks (OUST). EPA 510-R-15-001, 2015a. http://www.epa.gov/oust/cat/pvi/pvi-guide-final-6-10-15.pdf. Accessed Dec 2021

U.S. EPA (2015b) OSWER technical guide for assessing and mitigating the vapor intrusion pathway from subsurface vapor sources to indoor air. OSWER Publication 9200.2-154. U.S. Environmental Protection Agency, Office of Solid Waste and Emergency Response, June 2015. https://www.epa.gov/sites/production/files/2015-09/documents/oswer-vapor-intrus ion-technical-guide-final.pdf. Accessed Dec 2021

U.S. EPA (2017) Documentation for EPA's implementation of the Johnson and Ettinger model to evaluate site specific vapor intrusion into buildings. Office of Superfund Remediation and Technology Innovation, Washington, DC. https://semspub.epa.gov/work/HQ/100000489.pdf. Accessed Dec 2021

U.S. EPA (2018) Update for Chapter 19 of the exposure factors handbook. EPA/600/R-18/121F. https://cfpub.epa.gov/ncea/risk/recordisplay.cfm?deid=340635. Accessed Dec 2021

U.S. EPA (2020) Toxicity and chemical/physical properties for Regional screening level (RSL) of chemical contaminants at superfund sites. U.S. Environmental Protection Agency, Region 9 May 2020. http://www.epa.gov/region9/superfund/prg/. Accessed Dec 2021

Van Genuchten MT (1980) A closed-form equation for predicting the hydraulic conductivity of unsaturated soils. Soil Sci Soc Am J 44 (5):892–898. https://doi.org/10.2136/sssaj1980.036159 95004400050002x

Verginelli I, Baciocchi R (2021) Refinement of the gradient method for the estimation of natural source zone depletion at petroleum contaminated sites. J Contam Hydrol 241:103807. https://doi.org/10.1016/j.jconhyd.2021.103807

Verginelli I, Baciocchi R (2014) Vapor intrusion screening model for the evaluation of risk-based vertical exclusion distances at petroleum contaminated sites. Environ Sci Technol 48(22):13263–13272. https://doi.org/10.1021/es503723g

Verginelli I, Yao Y, Wang Y, Ma J, Suuberg EM (2016a) Estimating the oxygenated zone beneath building foundations for petroleum vapor intrusion assessment. J Hazard Mater 312:84–96. https://doi.org/10.1016/j.jhazmat.2016.03.037

Verginelli I, Capobianco O, Baciocchi R (2016b) Role of the source to building lateral separation distance in petroleum vapor intrusion. J Contam Hydrol 189:58–67. https://doi.org/10.1016/j.jconhyd.2016.03.009

Verginelli I, Yao Y, Suuberg EM (2016c) An excel®-based visualization tool of two-dimensional soil gas concentration profiles in petroleum vapor intrusion. Ground Water Monit Rem 36(2):94–100. https://doi.org/10.1111/gwmr.12162

Verginelli I, Pecoraro R, Baciocchi R (2018) Using dynamic flux chambers to estimate the natural attenuation rates in the subsurface at petroleum contaminated sites. Sci Total Environ 619:470–479. https://doi.org/10.1016/j.scitotenv.2017.11.100

Verginelli I, Yao Y (2021) A review of recent vapor intrusion modeling work. Ground Water Monit Rem 2:138–144. https://doi.org/10.1111/gwmr.12455

Yao Y, Shen R, Pennell KG, Suuberg EM (2013) A review of vapor intrusion models. Environ Sci Technol 47(6):2457–2470. https://doi.org/10.1021/es302714g

Yao Y, Yang F, Suuberg EM, Provoost J, Liu W (2014) Estimation of contaminant subslab concentration in petroleum vapor intrusion. J Hazard Mater 279:336–347. https://doi.org/10.1016/j.jhazmat.2014.05.065

Yao Y, Wu Y, Wang Y, Verginelli I, Zeng T, Suuberg EM, Jiang L, Wen Y, Ma J (2015) A petroleum vapor intrusion model involving upward advective soil gas flow due to methane generation. Environ Sci Technol 49(19):11577–11585. https://doi.org/10.1021/acs.est.5b01314

Yao Y, Verginelli I, Suuberg EM (2016a) A two-dimensional analytical model of petroleum vapor intrusion. Water Resour Res 52(2):1528–1539. https://doi.org/10.1002/2015WR018320

Yao Y, Wang Y, Verginelli I, Suuberg EM, Ye J (2016b) Comparison between PVI2D and Abreu–Johnson's model for petroleum vapor intrusion assessment. Vadose Zone J 15(11):1–11. https://doi.org/10.2136/vzj2016.07.0063

Yao Y, Verginelli I, Suuberg EM (2017) A two-dimensional analytical model of vapor intrusion involving vertical heterogeneity. Water Resour Res 53(5):4499–4513. https://doi.org/10.1002/2016WR020317

Yao Y, Verginelli I, Suuberg EM, Eklund B (2018) Examining the use of USEPA's generic attenuation factor in determining groundwater screening levels for vapor intrusion. Ground Water Monit Rem 38(2):79–89. https://doi.org/10.1111/gwmr.12276

Yao Y, Mao F, Xiao Y, Luo J (2019) Modeling capillary fringe effect on petroleum vapor intrusion from groundwater contamination. Water Res 150:111–119. https://doi.org/10.1016/j.watres.2018.11.038

Chapter 7
High-Resolution Characterization of the Shallow Unconsolidated Subsurface Using Direct Push, Nuclear Magnetic Resonance, and Groundwater Tracing Technologies

Gaisheng Liu, John F. Devlin, Peter Dietrich, and James J. Butler Jr.

Abstract Groundwater protection and contaminated site remediation efforts continue to be hampered by the difficulty in characterizing physical properties in the subsurface at a resolution that is sufficiently high for practical investigations. For example, conventional well-based field methods, such as pumping tests, have proven to be of limited effectiveness for obtaining information, such as the transmissive and storage characteristics of a formation and the rate at which groundwater flows, across different layers in a heterogeneous aquifer system. In this chapter, we describe a series of developments that are intended to improve our discipline's capability for high-resolution characterization of subsurface conditions in shallow, unconsolidated settings. These developments include high-resolution methods for hydraulic conductivity (K) characterization based on direct push (DP) technology (e.g., DP electrical conductivity probe, DP permeameter, DP injection logger, Hydraulic Profiling Tool (HPT), and High-Resolution K tool), K and porosity characterization by nuclear magnetic resonance (NMR), and groundwater flux characterization by monitoring the movement of thermal or chemical tracers through distributed temperature sensing

G. Liu (✉) · J. J. Butler Jr.
Kansas Geological Survey, University of Kansas, 1930 Constant Ave., Lawrence, KS 66047-3724, USA
e-mail: gliu@ku.edu

J. J. Butler Jr.
e-mail: jbutler@ku.edu

J. F. Devlin
Geology Department, University of Kansas, Lindley Hall, Room 217, 1475 Jayhawk Blvd., Lawrence, KS 66045-7613, USA
e-mail: jfdevlin@ku.edu

P. Dietrich
Department Monitoring and Exploration Technologies, Helmholtz Centre for Environmental Research-UFZ, Permoserstr. 15, Leipzig 04318, Germany
e-mail: peter.dietrich@ufz.de

© The Author(s) 2024
J. García-Rincón et al. (eds.), *Advances in the Characterisation and Remediation of Sites Contaminated with Petroleum Hydrocarbons*, Environmental Contamination Remediation and Management, https://doi.org/10.1007/978-3-031-34447-3_7

(DTS) equipment or the point velocity probe (PVP). Each of these approaches is illustrated using field or laboratory examples, and a brief discussion is provided on their advantages, limitations, as well as suggestions for future developments.

Keywords Direct push injection logging · Groundwater velocity · Hydraulic conductivity · Hydrostratigraphic characterization · Nuclear magnetic resonance profiling

7.1 Introduction

Groundwater protection and contaminated site remediation remain a difficult challenge for society due to our inability to characterize subsurface conditions at a sufficiently high level of resolution and accuracy. The success of efforts to protect or remediate a site requires a good understanding of how contaminants move in the saturated subsurface. Contaminant transport is primarily controlled by two physical mechanisms, advection, the movement of dissolved solutes with flowing groundwater, and diffusion, movement of contaminants from areas of high to low concentrations as a result of solute molecular thermal motion. Except in low-permeability materials, such as clays, where flow velocity is very small, advection typically plays a much more significant role than diffusion on the physical transport of contaminants in groundwater.

Groundwater flow can be mathematically related to hydraulic conductivity K [L/T] and gradient i [dimensionless], through Darcy's Law (Eq. 7.1), whose physical interpretation is illustrated in Fig. 7.1:

$$Q = -K \times i \times A, \tag{7.1}$$

where Q [L^3/T] is the flow rate across area A [L^2] of a porous medium. In groundwater hydrology, it is more common to use the Darcy flux, q [L/T], which is defined as in Eq. 7.2:

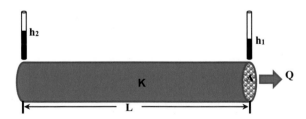

Fig. 7.1 Schematic of Darcy's Law flow calculation (Liu and Butler 2019). Groundwater flow rate is proportional to K, which can vary over several orders of magnitudes in a heterogeneous system. Definition of other notations is given in text

$$q = Q/A = -K \times i. \tag{7.2}$$

The hydraulic gradient is computed following Eq. 7.3:

$$i = (h_1 - h_2)/L, \tag{7.3}$$

where h_1 and h_2 are the hydraulic heads at the ends of the computational domain and L is the distance between the points where h_1 and h_2 are measured. Note that the Darcy flux describes the average flow rate across the total cross-sectional area, which is not the same as the seepage velocity in the pore space. The seepage velocity can be related to the Darcy flux through Eq. 7.4:

$$v = q/\phi_e = -K \times i/\phi_e, \tag{7.4}$$

where ϕ_e is the effective porosity (dimensionless), i.e., the proportion of pore space (relative to total aquifer volume) through which water can move under the hydraulic gradient. Because groundwater cannot move readily through extremely small pores (e.g., those in clays), dead-end pores, or pores that do not connect to the actual flow paths, the effective porosity is less than the total porosity (Zheng and Bennett 2002).

Groundwater flux (e.g., seepage velocity defined in Eq. 7.4) is a dynamic quantity whose value and direction will change whenever there is a change in K (e.g., borehole construction, dissolution or precipitation of solids) or i (e.g., pumping, episodic recharge events, increase or decrease of boundary heads). Compared to K, direct measurement of groundwater flux is more difficult in the field. As a result, the majority of current contaminant site characterization efforts focus on K (Boggs et al. 1992; Dagan and Neuman 1997; Fogg et al. 2000), although direct flux characterization has received increasing attention in recent years (Devlin 2020). In this chapter, we will discuss the development of high-resolution approaches for K, porosity, and flux characterization.

Conventional K characterization approaches, such as pumping tests, slug tests, or flowmeter profiling, have proven to be of limited effectiveness for obtaining information at the level of detail that is needed to accurately quantify groundwater and contaminant movement in heterogeneous media (Butler 2005; Liu and Butler 2019). There are two major drawbacks associated with the conventional approaches. The first is their dependence on existing wells, which are often sparsely distributed and not necessarily located at places of greatest interest. The second is the many theoretical and procedural limitations of well-based approaches (e.g., in-well hydraulics and short-circuiting flow through filter packs or leaky in-well packers). To address these concerns, a series of high-resolution characterization methods for unconsolidated settings based on direct push (DP) technology have been developed (Butler et al. 2002; Schulmeister et al. 2003; Dietrich and Leven 2005; McCall et al. 2005; Butler et al. 2007; Dietrich et al. 2008; Liu et al. 2009, 2012; Maliva 2016; Liu et al. 2019; McCall and Christy 2020). Compared to the conventional approaches, these

DP methods can provide K measurements at a much higher level of detail, and allow the measurements to be acquired in virtually any location.

Nuclear magnetic resonance (NMR), a widely used borehole logging technique in the petroleum industry for characterizing hydrocarbon reservoirs, has been adapted for hydrological applications in recent years (Walsh et al. 2011, 2013). By measuring the responses of water molecules to perturbed magnetic fields, NMR provides information about the total amount of water (i.e., porosity in saturated zones), as well as the distribution of pore sizes in the formation. That information can then be used to estimate K (Dlubac et al. 2013; Walsh et al. 2013; Knight et al. 2016; Kendrick et al. 2021). Furthermore, based on the NMR responses of water molecules in different sizes of pores, porosity can be divided into relatively mobile and immobile portions. As discussed earlier, the effective porosity, which is similar to the mobile porosity determined by NMR, is the key parameter for calculating pore water velocity from Darcy's flux. The immobile porosity, on the other hand, provides important information that is needed to characterize the diffusion of contaminants from low-K sediments back into high-K zones at sites where contaminant concentrations in those zones have been largely reduced by remediation efforts. Currently, the most common approach for determining porosity is to collect core samples in the field, which are then sent to a laboratory for measurements. Due to the high costs associated with this method, porosity measurements are typically kept at a minimum in most site characterization investigations. Thus, the NMR approach has great potential as a site characterization tool to obtain information on porosity as well as K estimates.

There has been a growing interest in measuring groundwater flux directly in the field as the remediation community switches from contaminant concentration-based decision-making to one that is based on contaminant mass discharge (Suthersan et al. 2010; Devlin 2020). The contaminant mass discharge, a product of groundwater flux and contaminant concentration, requires the groundwater flux to be reliably measured or estimated. Different approaches are available for obtaining groundwater flux estimates (Labaky et al. 2007; Bayless et al. 2011; Devlin 2020). These approaches can be divided into two general categories, (1) Darcy's law calculation where both K and i need to be estimated independently, and (2) tracer-based approaches that provide a direct estimate of groundwater flux assuming a quantifiable relationship between the flux and tracer responses. For the Darcy's law approach, the calculated flux is typically an average value over a relatively large area. As a result, the resolution is often not suitable for contaminant remediation activities. For tracer-based approaches, heat and solutes are two common tracers and responses are monitored by measuring subsurface temperature and solute concentrations (or concentration surrogates). By injecting and monitoring a small amount of tracer with an electrical conductivity different from the ambient formation fluid, centimeter scale groundwater velocity measurements can be made by a point velocity probe (PVP; Labaky et al. 2007). Initial applications of the PVP focused on deployment of the probe in unconsolidated formations, so that the probe was in direct contact with porous media and the impacts of borehole or well construction were minimized. Subsequent developments extended the PVP approach to wells and open boreholes to allow for measurements at as many depths as needed (Osorno et al. 2018), and in fractured rocks (Heyer et al. 2021).

A major limitation of tracer-based approaches is that only a few measurements can be acquired at pre-selected locations at a single time. Over the last decade, technological developments in fiber-optic distributed temperature sensing (DTS) have led to a significant improvement in using heat to measure and monitor various hydrological processes (Selker et al. 2006; Lowry et al. 2007; Moffett et al. 2008; Henderson et al. 2009; Leaf et al. 2012; Striegl and Loheide 2012; Liu et al. 2013; Read et al. 2013; Bakker et al. 2015; Maldaner et al. 2019; Munn et al. 2020; Simon et al. 2021). Due to the high-resolution temporal (sub-minute) and spatial (cm scale with wrapped DTS cable) temperature measurements by DTS, continuous information on groundwater flux distribution can potentially be obtained by tracking heat movement along the entire vertical interval of the borehole. However, a significant challenge for heat-based approaches is that density-driven buoyancy flow can affect the measurements, especially when the borehole is open and/or there is a significant amount of annular space between the measuring device and well casing. In some cases, separating the impact of thermal conduction from heat advection can also be a challenge if the rate of groundwater flux is small.

Butler (2005) provided an overview of the major methods for shallow subsurface characterization of K at that time. Many developments have occurred since then, particularly with the four approaches described in the previous paragraphs (DP, NMR, PVP, and DTS). The purpose of this chapter is to provide an overview of those developments. Specifically, we will discuss the characterization of K by DP approaches, the characterization of K and porosity by NMR, and the characterization of groundwater flux by PVP and DTS. For DP and NMR approaches, we will demonstrate their performance with data collected at a long-term research site of the Kansas Geological Survey (KGS). The DTS flux approach will be demonstrated in a laboratory sand tank while the PVP approach will be demonstrated with data from a sand aquifer at Borden, Canada. We conclude the chapter by summarizing the advantages and disadvantages of these approaches and offering some suggestions on future work to address key limitations.

7.2 Characterization of Hydraulic Conductivity by Direct Push Approaches

7.2.1 Direct Push Technology

Direct push is a drilling method that advances a rod string into unconsolidated materials using hydraulic rams supplemented with vehicle weight and, depending on the application, high-frequency percussion hammers (Fig. 7.2). In contrast to conventional rotary drilling, no material is removed from the subsurface. Instead, the sediments in the immediate vicinity of the rod string are pushed aside during advancement. The major advantages of DP technology include minimal site distur- bance, practical elimination of drill cuttings, reduced exposure of field personnel to hazardous materials at contaminated sites, lower time and costs (costs reduced by 40–60% compared to rotary drilling; personal communication, Wes McCall, Geoprobe Systems, Nov. 11 2021), and high mobility of the equipment (DP rigs can work under many field conditions). The main disadvantage is that DP is limited to unconsolidated sediments with penetration depths typically less than 30 m. The shoving aside and thus repacking of near-rod materials, which may lead to a change in their hydraulic properties, can sometimes have adverse impacts on field results, especially in finer-grained sediments.

DP technology has been used for a wide array of subsurface investigations ranging from characterization of aquifer properties (Butler et al. 2002; Lunne et al. 2002; Butler 2005; Lessoff et al. 2010; Liu et al. 2012; Liu and Butler 2019; McCall and Christy 2020), groundwater and soil sampling (Artiola et al. 2004; Schulmeister et al. 2004; Nielsen and Nielsen 2006; U.S. EPA 2016), subsurface injection of remedial reagents (Stroo and Ward 2010), and installing wells and monitoring devices for long-term water resources management (ITRC 2006; Liu et al. 2016). In the area of hydrogeologic characterization, a series of innovative approaches have been devel- oped for DP applications to allow information about K to be obtained at a resolution, accuracy and speed that was previously not possible (McCall et al. 2002; Butler et al. 2007; Dietrich et al. 2008; Liu et al. 2009, 2019; McCall and Christy 2020). In this section, we focus on using the DP electrical conductivity (EC) probe, DP permeameter (DPP), DP injection logger (DPIL), Hydraulic Profiling Tool (HPT), and High-Resolution K (HRK) tool to characterize vertical K profiles in unconsol- idated settings. Results from the Larned field site of the KGS are used to illustrate and compare the performance of these four approaches.

7.2.2 Larned Research Site

The Larned Research Site (LRS), which was established by the KGS for studying stream-aquifer interactions and riparian zone processes in the early 2000s, is located adjacent to a US Geological Survey (USGS) stream-gaging station on the Arkansas River near the city of Larned in west-central Kansas (Butler et al. 2007; Fig. 7.3). The shallow unconsolidated subsurface at the site consists of three major units: the Arkansas River alluvial aquifer, a clay-dominated confining layer in the middle, and the High Plains Aquifer (HPA) resting on Pennsylvanian bedrock (shale and limestone) at the bottom (Butler et al. 2004). The alluvial aquifer contains many intermittent clay lenses, as indicated by the abrupt increase of EC values particularly in the lower two-thirds of the interval. The upper portions of the HPA also contain intermittent clay lenses where the EC values are elevated. The thickness of each hydrogeological unit varies significantly across the site, with the depth to water around 3–5 m. Due to extensive pumping from the HPA, groundwater levels have

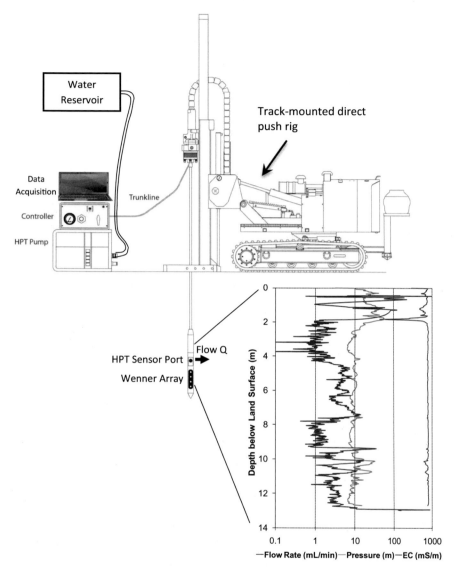

Fig. 7.2 A direct push rig and an example HPT profile (image of DP rig courtesy of Geoprobe Systems)

been declining since the late 1950s in western Kansas where the upstream reaches the Arkansas River are located. As a result, the Arkansas River near the LRS has been dry for much of the last few decades except during wet years or after intensive storms.

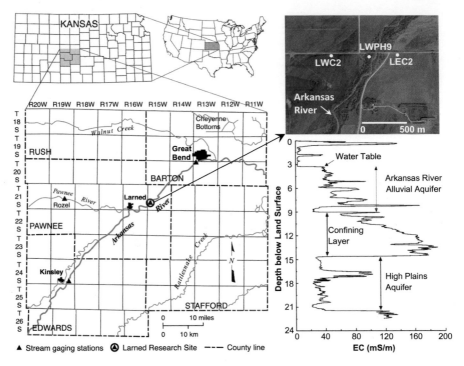

Fig. 7.3 Overview of the Larned Research Site. The image on the upper right is from Google Earth (accessed on 10/7/2013). The generalized stratigraphy on the lower right was determined from DP EC (Wenner array) profiling (Butler et al. 2004). Low EC values indicate sand and gravel, while high values indicate clay at this site. Notice the jump in electrical conductivity at the water table where the probe moved from very dry to saturated sands and gravels

7.2.3   Direct Push Electrical Conductivity

The EC of subsurface saturated zones is primarily determined by the pore-fluid chemistry, clay content (electrically conductive; not all clays are electrically conductive), and total porosity. At freshwater sites where clay mineralogy is consistent and variations in pore-fluid chemistry are negligible, such as the LRS, EC is largely a measure of the clay content (porosity variations typically have a much less significant impact than clay content) and can thus be used as a qualitative indicator of K. Because no water injection is needed during probe advancement and measurement, EC profiling can be performed with significantly less effort and more rapidly than other DP approaches.

Figure 7.4 displays three EC logs collected across the site. At LWC2, the EC values are generally the highest in the upper alluvial aquifer, indicating that the K of the alluvial zone is lower at this location than at the other two locations. The EC values are much higher for the middle confining clay layer in all profiles. The thickness of the confining layer is similar between LWC2 and LWPH9 (interval extending from depths of 10–15 m), but larger at LEC2 (interval extending from 10 to 19 m). The EC variations between the profiles indicate that the clay distribution is highly heterogeneous across the site. Below the middle confining layer, the EC values decrease to around 30 mS/m in the HPA. The thickness of the HPA is similar between LWC2 and LWPH9, but smaller at LEC2. The EC values show an abrupt increase once the probe enters the shale below the HPA.

Although EC provides a good indicator of relative K when electrically conductive clay is a dominant factor in subsurface lithology, only a small number of studies have explored the use of EC for K estimation in unconsolidated clayey formations (see the review paper by Purvance and Andricevic 2000 and cited literature). This is largely because K is very difficult to measure in clay-rich formations where K is low and hydraulic tests (typically used to obtain independent K estimates) would take a long time to complete in the field. As a result, current K characterization efforts are mainly focused on moderate to high-K zones (Bohling et al. 2012). Nonetheless, future work is needed to exploit the potential of EC as a means to estimate K in unconsolidated materials having clay as a major constituent. On the other hand, in zones that contain little clays such as the HPA (Fig. 7.4), where the pore size distribution likely has the most significant impact on aquifer permeability, the EC does not provide an effective measure of K (Schulmeister et al. 2003).

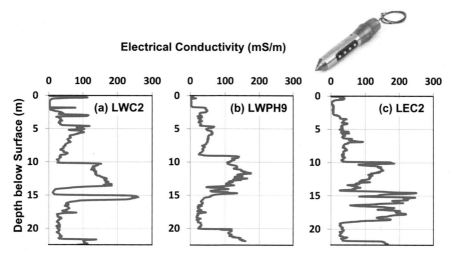

Fig. 7.4 DP EC logs at three locations at the LRS: **a** LWC2, **b** LWPH9, and **c** LEC2. See Fig. 7.3 for the locations of LWC2, LWPH9, and LEC2. An example DP EC probe with the Wenner sensor array is shown at the upper right (image of EC probe courtesy of Geoprobe Systems)

7.2.4 Direct Push Permeameter

The DP permeameter (DPP) is a tool that can be used to obtain reliable K estimates on a scale of relevance for most contaminant transport investigations (Stienstra and van Deen 1994; Lowry et al. 1999; Butler et al. 2007; Liu et al. 2008). It consists of a short cylindrical screen attached to the lower end of a DP rod and two pressure transducers inset into the rod above the screen (Fig. 7.5). During DPP advancement, water is continuously injected through the injection screen to prevent clogging. After the measurement depth is reached, advancement ceases and water injection is stopped to allow the heads in the aquifer to recover to background conditions. After the aquifer heads recover, a series of short-term injection tests are performed, and K is estimated from the spherical form of Darcy's Law (Eq. 7.5) using the injection rate and the injection-induced pressure responses at the two transducers (Butler et al. 2007),

$$ K_{\text{DPP}} = \frac{Q}{4\pi(\Delta h_1 - \Delta h_2)} \left(\frac{1}{r_1} - \frac{1}{r_2} \right), \tag{7.5}$$

where Q is injection rate, Δh_1 and Δh_2 are the injection-induced pressure head changes at pressure transducers PT1 and PT2, and r_1 and r_2 are the distances from PT1 and PT2 to the injection screen, respectively (Fig. 7.5). K_{DPP} is a weighted average over the interval between the screen and the farthest transducer; material outside of that interval has little influence on the estimated K (Liu et al. 2008). The major advantage of the DPP is that it only requires steady-shape conditions (constant hydraulic gradient with time) as opposed to the steady-state condition (constant head with time) required by most other pumping-based approaches (e.g., flowmeter tests). The steady-shape analysis allows for shorter test times in the field. In addition, the resulting K estimates are not sensitive to the low-K skin that can potentially develop due to material compaction during tool advancement. Background water level fluctuations due to regional stresses (e.g., well pumping, stream stage changes) also have a minimal impact on DPP results because the fluctuations are largely cancelled out using the steady-shape analysis.

Figure 7.5 shows an example of DPP test data collected at the LRS (location LWPH9) at a test depth of 6 m. Three injection tests were performed. The first test used an injection rate of $\approx$800 mL/min, which was started at a time of 120 s. The injection-induced pressure head difference (i.e., the pressure head difference between the two pressure transducers during injection minus the pressure head difference prior to injection) was 0.07 m. The injection-induced pressure head difference in the first test determined how the injection rate was adjusted in the next test. Because 0.07 m is a relatively small value, the injection rate was raised to $\approx$1600 mL/min in the second test. In the third test, the injection rate was adjusted back to the level of the first test to check if the injection-induced pressure returned to the level observed during the first test. For the three tests shown in Fig. 7.5, the calculated K values were 7.4, 7.7, and 7.4 m/d, respectively. The consistency between the K estimates from different tests is a good indication of the reliability of DPP results; a lack of consistency could

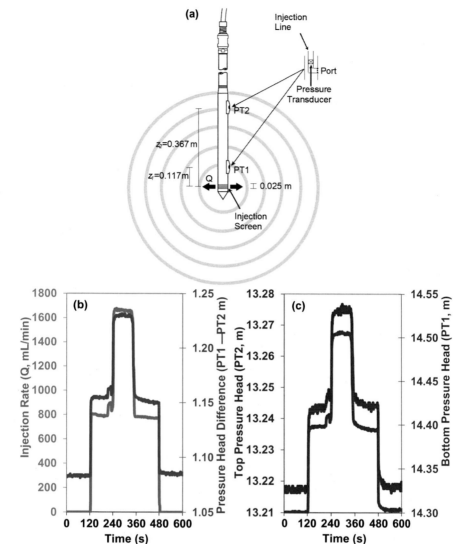

Fig. 7.5 **a** Schematic of the DPP, **b** injection rate and injection-induced pressure head difference between the top and bottom transducers for a DPP test at a depth of 6 m below land surface at the LRS (location LWPH9), **c** measured pressure heads at the top and bottom transducer for this series of tests. In **a**, the injection lines at the pressure transducer ports were added during the development of the HRK tool

be due to sensor instability (e.g., air bubbles in the injection water can cause unstable sensor readings) or formation alteration during DPP tests (e.g., vertical channeling along the probe surface).

Figure 7.6 shows the DPP K results for the shallow alluvial aquifer and the deeper HPA at LWC2, LWPH9, and LEC2. No DPP tests were performed in the middle low-K clay layer due to time constraints (with current flow instrumentation, one DPP test in this layer would take days to weeks to complete). Consistent with the EC profiles, K shows a much greater variability with depth in the alluvial aquifer than in the HPA. Except for a few thin layers, the K in the alluvial aquifer is lower than that in the HPA. The average DPP K value for the alluvial aquifer is 25 m/d, 25 m/d, and 70 m/d at LWC2, LWPH9, and LEC2, respectively, while the average DPP K value for the HPA is above 200 m/d at all three locations.

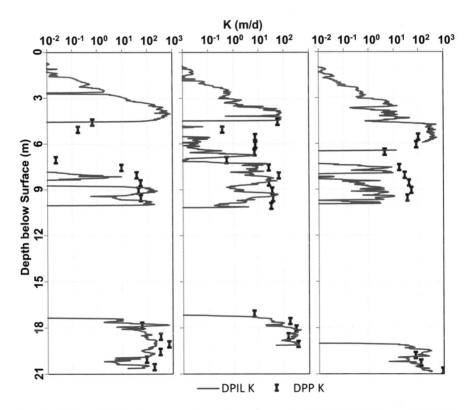

Fig. 7.6 DPP and DPIL K estimates at three locations at the LRS: LWC2 (left graph), LWPH9 (middle), and LEC2 (right). DPP values are calculated using the spherical form of Darcy's Law. The calibrated a and b are 5.8 and -12.4 for computing DPIL K (b changes with the unit of DPIL flow/pressure head ratio; see Eq. 7.6). Neither DPP nor DPIL K estimates were obtained in most clay layers. For DPIL, the injection pressure is so large that it exceeds the pressure sensor limit in these layers. For DPP, both the time for recovery of heads to background conditions and for the injection-induced pressure difference to stabilize exceed practical time constraints in the field

The current DPP tool provides a spatial resolution of $\approx$0.4 m (i.e., the distance between the screen and farthest transducer; Fig. 7.5). Although a higher resolution is possible by advancing the tool a smaller distance between measurements and analyzing the set of tests together with a numerical model (Liu et al. 2008), practical time constraints (a DPP test sequence requires 10–15 min in moderate to high-K intervals) usually limit the measurement spacing to 0.4 m or larger in most cases (Butler et al. 2007; Knight et al. 2016). As such, the DPP does not provide information about conditions between test intervals. In addition, it can produce anomalous results in highly stratified formations if no information about the stratification is available (Liu et al. 2008; Zschornack et al. 2013).

7.2.5   Direct Push Injection Logger

The DP injection logger (DPIL) is a tool that can be used to rapidly obtain high-resolution information about relative variations in K (Dietrich et al. 2008). The original design of the tool consists of a short cylindrical screen attached to the lower end of a DP rod immediately behind the drive point (Fig. 7.7a). After the tool is advanced to a depth at which a K measurement is desired, advancement ceases. Water is then injected using a sequence of different rates while the injection rate and pressure are measured at the surface. Line losses between the surface and downhole injection screen can be estimated analytically (assuming laminar flow conditions) or from a regression analysis between pressures at different injection rates. The line losses are removed from the total injection pressure measured at the surface, and the ratio of injection rate over the corrected injection pressure head can then be used for K estimation. In Dietrich et al. (2008), the DPIL ratio was converted into K estimates by regression with K estimates from nearby slug tests. In the original tool design, the scale of each measurement is the vertical length of the injection screen (0.025 m). Because K measurements were made after halting probe advancement, the original DPIL profiling procedure is referred to as the discontinuous advancement mode. DPIL can also be performed while the probe is advanced into the subsurface; this profiling procedure is referred to as continuous DPIL (Liu et al. 2012).

7.2.6   Hydraulic Profiling Tool

The Hydraulic Profiling Tool (HPT) is a DP tool developed by Geoprobe Systems that combines continuous DPIL with EC profiling (Geoprobe 2007). Profiling K variations with the HPT has now been established as a standard practice by the American Society for Testing and Materials (ASTM 2016). Compared to the original DPIL, HPT consists of a single screened port inset into a DP rod at some distance above the drive point, with an EC sensor array between the drive point and the screened port (Fig. 7.7b). As the tool is advanced, water is injected continuously

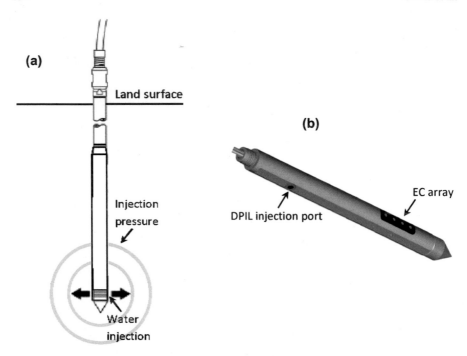

Fig. 7.7 **a** Schematic of the original DPIL with a screen at the lower end of the probe rod (after Dietrich et al. 2008), **b** artistic rendering of the HPT (continuous DPIL probe combined with an EC Wenner array; image courtesy of Geoprobe Systems)

through the screen. The injection rate is monitored at the surface, while the pressure is measured directly behind the screen, essentially eliminating the need for a line loss correction between the pressure sensor and injection screen. The distance between the injection port and the bottom drive point (0.36 m) largely reduces the impact of the mechanical stress produced by the advancement of the drive point during continuous profiling. Because halting probe advancement is not required for injection pressure measurement (except for rod changes or pressure dissipation tests; during dissipation tests, probe advancement is halted, water injection is stopped, and the recovery of pressure to the background level is monitored; dissipation tests provide an estimate of the hydrostatic pressure at different depths and two such tests are usually sufficient per profile), HPT profiling is very rapid in the field, and up to six profiles of 12 m in length can be obtained in a day under good conditions (Bohling et al. 2012). HPT produces a K estimate every 0.015 m, thus providing unprecedentedly high-resolution information about K in the saturated zone across the entire interval traversed during probe advancement.

Figure 7.8 shows the injection rates and pressures measured by HPT at LWC2, LWPH9, and LEC2. The injection rate was set to $\approx$250 mL/min in the permeable zones (the injection rate decreased in lower K materials because of the high injection pressures). For the middle clay layer, the injection rate became zero when the injection pressure exceeded the upper limit of the pressure sensor (shown as gaps in the injection pressure curves). For all three profiles, the injection pressure is smallest in the HPA and in the lower portion of the alluvial aquifer, indicating K is highest at those depths. As discussed earlier, the ratio of injection rate over injection pressure head, which is also plotted in Fig. 7.8, can be used for K estimation by relating those ratios to nearby K estimates obtained from other approaches.

The major advantage of the discontinuous DPIL mode is that the quality of K measurement is relatively high as the injection rate can be varied at each depth interval to assess if consistent K estimates are obtained across multiple rates. However, no information is available for the intervals between the discontinuous measurements.

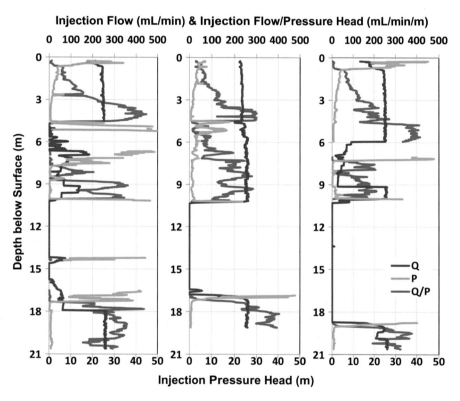

Fig. 7.8 Continuous DPIL logs at three locations at the LRS: LWC2 (left graph), LWPH9 (middle), and LEC2 (right). Logs are obtained using the HPT manufactured by Geoprobe Systems. The injection rate (red) and rate/pressure head (blue) reference the top axis, and the injection pressure head (green) references the bottom axis. The injection pressure head was calculated as the total injection pressure head measured by the sensor minus the hydrostatic pressure head (estimated by pressure dissipation tests at different depths). For the middle clay layer, the injection rate became zero when the injection pressure exceeded the upper end of the pressure range of the sensor (shown as gaps in the pressure curve)

The frequent suspension of advancement for measurements also makes the discontinuous DPIL profiling less time-efficient than HPT. As a result, HPT is preferred by most practitioners.

The major advantages of HPT are the speed and resolution at which K information can be obtained. However, the current HPT has two limitations that are important for field applications. The first is that HPT has both a lower and upper limit for estimation of K. In high-K zones, the injection-induced pressures are small and may become indistinguishable from the HPT system noise. In low-K zones, on the other hand, the injection-induced pressures are high and may exceed the upper limit of the pressure sensor. In addition, it is difficult, if not impossible, to assess if vertical channeling along the probe surface or formation alteration is occurring as a result of the elevated injection pressures. McCall and Christy (2010) estimated the K range that can be reasonably assessed by HPT to be 0.03–25 m/d under typical conditions. Note that both the upper and lower K limits can be extended by adjusting the injection flow rates; however, this requires modifications of the current flow control and measurement system (Bohling et al. 2012; Liu et al. 2012, 2018). The second limitation of HPT is that it only provides a relative indicator of K. To obtain improved K estimates, additional calibration data are needed to develop a site-specific relationship between HPT measurements and K. Dietrich et al. (2008) and McCall and Christy (2010) used K values from nearby slug tests and performed regression analyses to develop empirical relations for converting the DPIL ratios into K. Zhao and Illman (2022) combined HPT with inverse modeling and hydraulic tomography to improve the estimation of K in a highly heterogeneous site. Liu et al. (2009) used DPP tests, which were performed at the same location as the DPIL, to transform the DPIL ratios into K estimates; the resulting approach is referred to as the High-Resolution K (HRK) tool and is discussed in the following paragraphs. Borden et al. (2021) developed a physically based equation to derive K values from HPT data and an empirical hydraulic efficiency factor.

7.2.7 High-Resolution K (HRK) Tool

As mentioned above, the DPP provides reliable K estimates at a relatively coarse resolution (0.4 m), while the continuous DPIL can provide a measure of the relative variations of K at a much finer resolution (0.015 m). The HRK tool was developed to better realize the potential of these approaches by coupling the DPP and DPIL into a single probe (Liu et al. 2009). The tool has a similar appearance to the DPP (i.e., two pressure transducer ports above the bottom injection screen; Fig. 7.9). The difference is that water is injected through both pressure transducer ports during tool advancement (water also injected through the bottom screen to prevent clogging); the DPP only injected water through the bottom screen during advancement. By injecting water through the transducer ports and measuring injection responses during tool advancement, the HRK tool essentially functions in the continuous DPIL mode. Next, at selected depth intervals, tool advancement ceases and the DPP injection tests are performed. Because the DPIL and DPP measurements are collocated, the DPP data can be used to directly transform the DPIL ratios into K without the need to compare measurements at different support scales as in the regression-based approaches. In the HRK analysis, the DPIL K is calculated using a power-law relation as shown in Eq. 7.6 (Liu et al. 2009):

$$K = 10^b(\text{DPIL})^a, \text{ or } \log_{10}(K) = a \log_{10}(\text{DPIL}) + b, \tag{7.6}$$

where DPIL is the ratio of injection rate (mL/min) over pressure head (m); a and b are empirical coefficients whose values are determined through calibration. Specifically, the DPIL K values calculated based on (7.6) are used as input to a numerical model that simulates the DPP test responses. The values of a and b are adjusted such that the simulated DPP responses match the observed DPP data in the field. Instead of calibrating a and b for each HRK profile individually, a single set of a and b is used to achieve the best overall match for all profiles at a site (Liu et al. 2009).

Figure 7.6 shows the DPIL and DPP K estimates from HRK profiling at LWC2, LWPH9, and LEC2; K estimates were not available for the middle clay layer as discussed earlier. Overall, the high-resolution DPIL K estimates are consistent with the DPP values calculated from the spherical form of Darcy's Law. As a result of the high resolution, the HRK profiles provide quantitative K estimates for thin layers that would be difficult to identify with other approaches, including the DPP when used alone.

7.2.8   Summary of Direct Push Approaches

As no flow injection is involved, DP EC profiling can be used to obtain a significant amount of information about the shallow unconsolidated subsurface in a short time. However, because of the insensitivity to K variations in sediments with little clay, it is commonly applied as a site screening tool to develop a general understanding of hydrostratigraphy and identify zones for further interrogation by more quantitative approaches such as DPIL, DPP, or HRK. For zones with electrically conductive clays as a major constituent, such as the alluvial sediments at the LRS, EC could potentially be converted into semi-quantitative K estimates; further work is needed to establish the relations between EC and K in these settings.

Among the various hydraulically-based DP approaches, continuous DPIL (i.e., HPT) is one of the most powerful approaches and can be used to obtain relative K variations at many locations across a site in a time-efficient manner. However, the relations for converting DPIL data into quantitative K estimates are typically site-specific and independent data are needed for DPIL calibration. Therefore, for sites with significant heterogeneity in K, we recommend a combination of DPIL and HRK, with the DPIL profiles covering the entire site and the HRK profiles performed at a

(a)

(b)

Fig. 7.9 Prototype HRK tool: **a** Picture showing the bottom injection screen with the two pressure transducers inset into the rod above the screen, and **b** expanded view of the pressure transducer screen. Water is injected through the bottom screen and both transducer ports during probe advancement

few strategic locations for detailed DPIL calibration. In addition, due to the practical limits of the instrumentation, the current HPT has a rather restrictive range for K measurement. The DPP (or the DPP component of HRK), on the other hand, can be applied in most unconsolidated settings except those with low-K (e.g., < 0.001 m/d) where the time required to complete each test may be hours to days and is too long for practical investigations.

7.3   Characterization of Hydraulic Conductivity and Porosity by Nuclear Magnetic Resonance Profiling

7.3.1   Nuclear Magnetic Resonance

Over the past decade, significant progress has been made in adapting proton nuclear magnetic resonance (NMR) profiling, a widely used borehole logging technique in the petroleum industry, to various applications in near-surface hydrology (Walsh et al. 2011, 2013; Knight et al. 2016; Krejci et al. 2018; Kendrick et al. 2021). This approach involves measuring the response of hydrogen atoms to a series of magnetic perturbations (Fig. 7.10). That response is a function of, among other things, water-filled porosity and the pore-size distribution of subsurface materials. The current NMR tools can provide K and porosity estimates that are averaged over a 0.10- or 0.25-m vertical interval, although a higher resolution can potentially be achieved by advancing the tools at a finer interval and analyzing the entire profile data with global optimization techniques.

Three steps are typically involved during borehole NMR measurements (Dunn et al. 2002; Walsh et al. 2013). First, after the tool is moved to a measurement depth, the nuclear spins of hydrogen atoms in the surrounding materials are allowed to align (equilibrate) with a static magnetic field (B_0) created by permanent magnets in the tool. Second, an alternating magnetic perturbation (B_1) is applied at the resonant frequency, during which the hydrogen atoms absorb energy and precess toward the plane that is perpendicular to the static field. Third, the magnetic perturbation is turned off and the excited hydrogen atoms relax back to equilibrium with the static field. During the relaxation, hydrogen atoms will emit magnetic energy that can be detected by the induction coil in the tool. The amplitude of initial magnetization (M_0 in Fig. 7.10) at the time of the magnetic perturbations being turned off is a function of the total water content (i.e., porosity in the saturated zone), while the decay of the NMR signal is dependent on the pore size distribution among other factors (e.g., surface geochemistry of grains). Note that the resonant frequency of the measurement volume is not a single value, but a range of values due to the magnetic gradient caused by the permanent magnets on the tool. As a result, a sequence of B_1 perturbations at different frequencies (i.e., repeats of the second and third steps) is typically applied to improve the NMR data signal-to-noise ratio (Walsh et al. 2013).

The relaxation of NMR magnetization is usually described by an exponential function (Fig. 7.10). The relaxation rate is fast (short relaxation time T_2) in small pores and slow in large pores (long relaxation time). The actual porous media are composed of a network of pores with different sizes, so a distribution of T_2 values (typically uniformly spaced on a log scale) are predefined for the NMR data analysis. The amplitude of initial magnetization for each predefined T_2 is estimated by a least squares fit between the NMR relaxation data and the exponential functions for all T_2 values.

The initial magnetization provides the measurement of total porosity in saturated zones. K can be estimated from the NMR-determined porosity and relaxation time

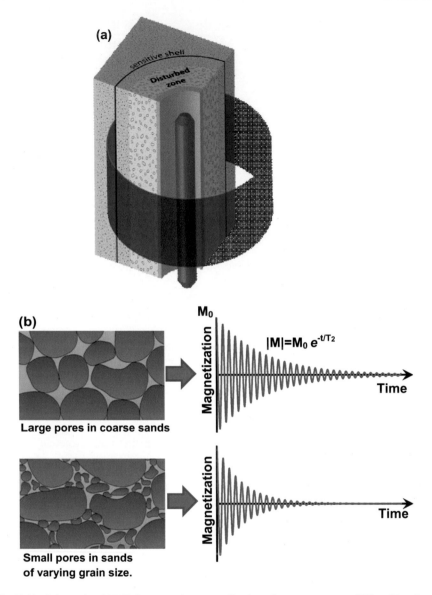

Fig. 7.10 Schematic of NMR for groundwater applications (images courtesy of Vista Clara Inc.): **a** The measurement domain is a thin shell around the tool that is suspended in a borehole (after Walsh et al. 2013), and **b** the relaxation of hydrogen atoms is affected by the pore size distribution. The yellow color in **a** indicates a disturbed zone from borehole drilling; the NMR measurement domain is typically outside the disturbed zone. There are two key NMR relaxation characteristics shown in **b**: first, the initial magnetization (M_0) depends on the total amount of hydrogen atoms (total water content), and second, the relaxation rate is fast in small pores and slow in large pores

distribution. Different empirical relations between K and NMR signals have been used in the petroleum industry (Dlubac et al. 2013; Kendrick et al. 2021). One of the most commonly used of those approaches for hydrological applications is the Schlumberger Doll Research (SDR) equation (Eq. 7.7) (Knight et al. 2016),

$$K_{NMR} = b\phi^m (T_{2ML})^n, \tag{7.7}$$

where b (m/s^3), m, and n are empirically determined constants; ϕ is the porosity determined from the NMR initial magnetization; T_{2ML} is the arithmetic mean of log relaxation times weighted by the amplitudes of initial magnetization. b is often referred to as the lithologic constant and contains information about all the parameters affecting permeability except porosity and relaxation time distribution (e.g., the surface-area-to-volume ratio of the pore space). Based on comparing the NMR and DP K data at three sites, Knight et al. (2016) suggested a universal set of b, m, and n could be used for unconsolidated sand and gravel aquifers ($b = 0.05$ to 0.12 m/s^3, $m = 1, n = 2$).

7.3.2 Nuclear Magnetic Resonance Application at Larned Research Site

Figure 7.11 shows the NMR relaxation data acquired at LWC2, LWPH9, and LEC2 at the LRS. Data were collected by a DP version of the Javelin tool manufactured by Vista Clara Inc, for which the measurement domain was a 0.5-m vertical shell at approximately 14.5 cm from the center of the probe (radial thickness of the shell is 2 mm). As the DP rods used for NMR tool deployment could not be advanced through the middle tight clay layer (a more powerful DP rig was not available during the field campaign), measurements were made in the upper alluvial aquifer only. At all three profiles, the lower portions of the alluvial aquifer have relatively higher amplitude of the initial magnetization with longer relaxation times (dark red colors), indicating that there are more large pores at those depths.

Figure 7.12 shows the NMR porosity and K estimates at LWC2, LWPH9, and LEC2. The sharp increase in the water-filled porosity around a depth of 4 m indicates the water table. The NMR K values are calculated using Eq. 7.7 with $b = 0.08$ m/ s^3, $m = 1$, and $n = 2$. Overall, the NMR and DPP K estimates are consistent with each other. For a few low-K zones, however, the NMR K values are significantly higher than the DPP values (i.e., depth 7 m at LWC2, depths 5 and 7 m at LWPH9). This is likely because the SDR coefficients in Eq. 7.7 are primarily calibrated for relatively permeable zones. A different set of SDR coefficients may be needed to estimate K from NMR data in low-K zones where the fine-grained materials may play an important role in affecting formation permeability.

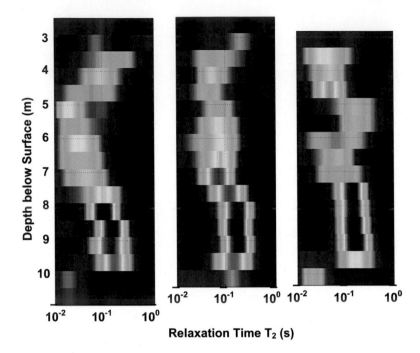

Fig. 7.11 NMR relaxation data at three locations at the LRS: LWC2 (left graph), LWPH9 (middle), and LEC2 (right). Data were acquired using a DP version of the Javelin tool (diameter 6 cm) manufactured by Vista Clara Inc. The color indicates the amplitude of magnetization fitted for each relaxation time $T2$ using the exponential decay function (Fig. 7.10). The red color means high initial magnetization amplitude (i.e., larger fraction of water for that $T2$), and the blue color means low initial magnetization amplitude (i.e., smaller fraction of water for that $T2$). The sum of the initial magnetization across all $T2$ represents the total water content at that depth; the relationship between the initial magnetization and water content is tool specific and is determined by laboratory calibration

The NMR tool used at the Larned site produced K and ϕ estimates that were averaged over a 0.5-m vertical interval; more recent NMR tools can provide measurements over a 0.25 or 0.10 m interval. Each measurement required about 5 min per interval. Higher resolution is possible by advancing the tool at a small interval (e.g., decimeter) and analyzing the entire profile data with global optimization techniques. As a subsurface characterization tool, NMR has two advantages over other approaches. First, it provides a direct measure of porosity in the saturated zone (and moisture content in the unsaturated zone), while most of the other approaches do not provide any information about porosity. Second, NMR can potentially be a very powerful tool in low-K zones, which are known to present a significant issue at many contaminated sites through the slow, persistent release of contaminants into more permeable zones.

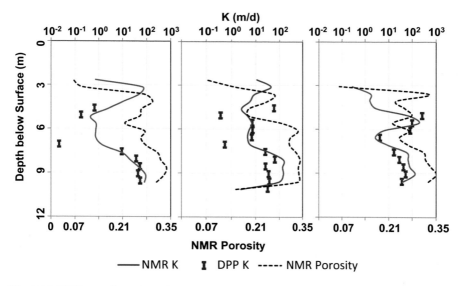

Fig. 7.12 NMR porosity and *K* estimates at three locations at the LRS: LWC2 (left graph), LWPH9 (middle), and LEC2 (right). The DPP *K* estimates are also plotted for comparison with NMR results. The DPP and NMR profiles are within 1 m of each other at each location

In low-*K* zones, hydraulic-based approaches, such as pumping and slug tests, are not as effective as in coarser materials because test durations are long (e.g., days to months or longer; Butler 2019). For the DP injection-based approaches (DPIL, DPP, HRK), a particular challenge is the significant amount of pressure generated by tool advancement when *K* is low (Liu et al. 2019). For example, the pressure generated by tool advancement may overwhelm DPIL injection pressure when HPT profiles are performed in silt and clay layers. The time to complete a DPP test in low-*K* settings can last from hours to days (typically too long for practical applications). NMR, on the other hand, can provide rapid measurements of formation properties in low-*K* zones because no flow injection is needed and the tool advancement pressure, if NMR is deployed by DP, has little impact on NMR responses. Future work is needed to further investigate the use of NMR for characterizing *K* and *ϕ* in silts and clays.

NMR has been increasingly used for environmental investigations in semi- and fully consolidated rock, including shale, limestone, sandstone, and fractured granite (personal communication, David Walsh, Vista Clara, Nov. 22 2021). Recent developments on combining NMR measurement with DP in unconsolidated formations have allowed NMR *K* and *ϕ* estimates to be made at a vertical resolution of under 10 cm (Vista Clara 2021).

7.4 Groundwater Velocity Characterization

7.4.1 Characterization of Velocity by Distributed Temperature Sensing

7.4.1.1 Distributed Temperature Sensing

Heat has been used extensively as a tracer to study groundwater and its interactions with other systems (Anderson 2005; Constantz 2008; Rau et al. 2014). As a temperature measurement technology, fiber-optic DTS was introduced to the hydrological community in the early 2000s (MacFarlane et al. 2002). After the mid-2000s, reductions in the cost of the instrumentation and increased data quality spurred a significant interest of using DTS to measure and monitor various hydrologic processes (Selker et al. 2006; Tyler et al. 2009). In fiber-optic DTS, a laser pulse of a certain duration (e.g., 10 ns) is sent down a fiber-optic cable. As the laser pulse travels along the fiber, it interacts with the fiber materials and produces backscattering. One group of backscattering light is known as Raman scattering. The intensity of the anti-Stokes Raman backscatter (wavelength shorter than that of the source laser) is dependent on the temperature of the cable, while the intensity of Stokes Raman backscatter (wavelength longer than that of the source laser) is largely insensitive to cable temperature. Therefore, the intensity ratio of anti-Stokes versus Stokes Raman backscatter can be used to estimate the temperature distribution along the length of the cable. Because the signal of Raman backscatter from each pulse is very weak, thousands of pulses per second are typically required to make a temperature measurement at each cable location.

One of the common DTS applications has been using heat as a tracer to investigate groundwater movement (e.g., Lowry et al. 2007; Leaf et al. 2012; Becker et al. 2013; Liu et al. 2013; Munn et al. 2020; Simon et al. 2021). For example, Lowry et al. (2007) used DTS-based temperatures to identify several discrete groundwater discharge zones along a wetland stream. Leaf et al. (2012) used DTS to monitor the vertical movement of heat in two open boreholes, which led to an improved understanding of flow processes in a hydrostratigraphic unit at the site. Becker et al. (2013) estimated the water infiltration rates from a surface recharge basin based on the propagation of the diurnal temperatures of the infiltrated water measured by DTS. Liu et al. (2013) found that the temperature responses to active heating of a probe in a well were consistent with the K profiles previously determined at the same location, and concluded that the temperature responses could be used to approximate the variations of groundwater flux at different depths in the well. The wrapping of the sensing cable around the probe was able to significantly increase the measurement resolution (to about 1.5 cm), which led to a much improved understanding of the hydrostratigraphic controls on groundwater flow processes at the site. Munn et al. (2020) used DTS with active heating to measure borehole flows in fractured rocks under different hydraulic conditions. Results indicated that flow measurements in fracture systems can be significantly affected by cross-flow between fractures along open boreholes.

7.4.1.2   Groundwater Flux Characterization Tool

To illustrate the potential of DTS as an approach for characterizing groundwater flux, the Groundwater Flux Characterization (GFC) tool developed by Liu et al. (2013) is assessed here in a controlled laboratory setting. The GFC tool was constructed by wrapping a DTS fiber-optic cable and resistance heating cable around a sealed hollow PVC pipe (Fig. 7.13). For groundwater flux profiling, the GFC tool is first deployed to a measurement interval in an existing well. DTS temperatures are then monitored for a period of time (30 min or more) until the thermal disturbance from tool deployment has dissipated. Once temperatures return to background conditions, heating begins by flowing an electric current through the resistance cable. The rate of groundwater flux in the surrounding aquifer is proportional to the average temperature increase during heating as expressed in Eq. 7.8:

$$\Delta T_{\text{ave}} = \frac{1}{t_1 - t_0} \int_{t_0}^{t_1} [T(t) - T_0] \mathrm{d}t, \qquad (7.8)$$

where T_0 is the temperature [K or °C] before heating starts at time t_0; t_1 is the time when heating ceases; and $T(t)$ is the temperature at time t during heating (Fig. 7.13). The heating duration $(t_1 - t_0)$ is typically between 5 and 10 h and kept constant for comparisons between measurement intervals. The larger the rate of groundwater flux, the faster the movement of heat away from the probe by groundwater advection, and the smaller the temperature increase computed by Eq. 7.8 during the heating tests.

Two key assumptions are invoked using Eq. 7.8 to estimate horizontal groundwater flux. The first is that vertical flow is negligible, as the vertical flow will cause heat to move vertically along the probe and the resulting temperature responses will be difficult to separate from the temperature responses produced by horizontal flux at different depths. This limitation can be potentially addressed by zoned heating (i.e., discrete sections of the tool are heated while the temperatures of the entire probe is monitored), which requires a significant modification of the heating component of the tool. The second assumption is that the vertical variations in the thermal conductivity of the materials near the test well are negligible, so that the temperature differences between depths are mainly a result of groundwater flux instead of thermal conduction. This assumption appears to be valid when the well is backfilled with an artificial filter pack at the time of well installation. For wells without artificial filter packs, caution is needed when using Eq. 7.8 to predict groundwater flux.

For the laboratory tests discussed here, a smaller version of the GFC probe was constructed (Knobbe et al. 2015). The wrapped PVC pipe was reduced to a length of 0.98 m, with the total lengths of fiber-optic and heating cables at 43 and 40 m, respectively. As a result, each DTS measurement (1 m of fiber-optic cable) is equivalent to a 0.02-m vertical interval on the probe. There are a couple of differences between the laboratory probe and the prototype tool developed by Liu et al. (2013). The wall thickness of the PVC pipe is smaller (about 50% less), thus reducing the

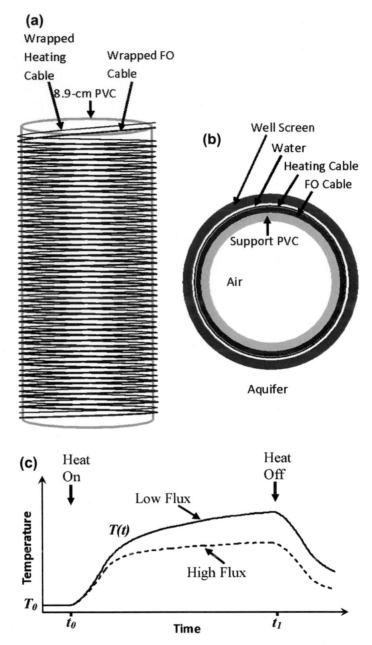

Fig. 7.13 **a** Schematic of the GFC tool, **b** planar schematic view of the GFC tool in a well, and **c** heat-induced temperature increase at different flux rates (assuming a constant rate of thermal conduction) (after Liu et al. 2013)

thermal mass and improving the sensitivity of the probe to heating. The heating cable is placed inside the fiber-optic cable (the original tool has the heating cable outside of the fiber-optic cable), diminishing the potential of heating-induced buoyancy flow in the annular space between the GFC tool and the well screen.

7.4.1.3 GFC Laboratory Test Results

Figure 7.14 shows the setup of the laboratory sand tank for testing the GFC probe (Knobbe et al. 2015). The inner dimensions of the rectangular box (steel container) are 1.83 m by 1.14 m by 1.14 m. Rigid foam insulation boards were installed inside the container to minimize the thermal interactions between the sand tank aquifer and the ambient surroundings. Commercial medium grade sand (Quikrete, Medium No. 1962, #20 - #50, 0.8–0.3 mm) was used for creating the synthetic aquifer. A 1.3 m long 10.2 cm inner diameter test well (schedule 40 PVC with screen slot width 0.025 cm) was installed at the center of the box. Two reservoirs, which were constructed using perforated PVC pipes as space retainers to provide additional support to the screened reservoir walls, were used to establish the flow field in the sands. Flow was from right to left (parallel to the longest side of the box) in the sands by pumping water from the left reservoir into the right one with a peristaltic pump. The sand aquifer has an average K of 218 m/d and an effective porosity of 33%.

Figure 7.15 shows the results of GFC tests in the sand tank at different flow rates. When the flow rate was zero, the DTS-measured temperature increase during heating was the largest at different depths. The average temperature increase over the heated section was 1.68 °C. As the flow rate increased, the temperature increase became smaller. The average temperature increase reduced to 1 °C when the Darcy's flow rate increased to 0.78 m/d. A repeat heating test was performed at a flow rate of 0.78 m/d for quality assurance, and the temperature responses were nearly identical to those from the original test. For the tested range of flow rates (0–0.78 m/d), the relation between the average temperature increase and flow rate appears to be approximately linear (Fig. 7.15b).

The GFC laboratory test results show the promise of the tool for characterizing vertical variations in groundwater velocity. Because of the significant impact of thermal conduction, the relationship between temperature response and groundwater velocity is expected to vary in different formations or wells with different construction specifications. Further tests of the approach under different field settings are needed before it can be widely applied as a velocity characterization tool. Heating-induced buoyancy effects should always be taken into account, especially when the annular space between the tool and well screen is large (e.g., larger than several millimeters).

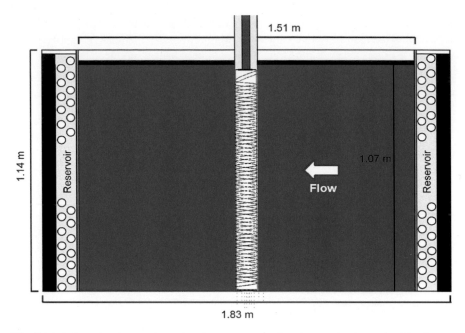

Fig. 7.14 Lab sand tank setup for testing the groundwater flux characterization probe. The diagram shows the cross-section of the tank along the flow direction (flow in the sands created by pumping water directly from the left reservoir to the right reservoir). The width of the tank (perpendicular to the cross-section) is 1.14 m. Perforated PVC pipes are used as space retainers in the reservoirs to provide additional support to the screened reservoir walls against the sands. Black color represents the impermeable, thermally-insulating foam boards that minimize the thermal interaction between the sand aquifer and room surroundings

7.4.2 Characterization of Groundwater Velocity by Point Velocity Probe

7.4.2.1 Point Velocity Probe

The point velocity probe (PVP) uses localized tracer tests to measure the magnitude and direction of groundwater velocity at the centimeter scale (Labaky et al. 2007). A PVP operates by injecting a small amount of tracer solution into the formation (typically less than 1 mL) and measuring the tracer movement around the probe surface and across two or more detectors (Fig. 7.16). As the detectors on a PVP are a pair of parallel electrical wires that measure EC, the injected tracer needs to have a significantly different EC from the formation fluid. The most common tracer used in PVP tests is a dilute saline solution (e.g., NaCL with concentration < 1 g/L), which provides an electrical conductivity signal well above background values in freshwater systems (Devlin 2020). In some cases where the formation fluid has high salinity, deionized water can be used as the tracer, and the decrease in the measured EC is used to quantify tracer breakthrough at the detectors.

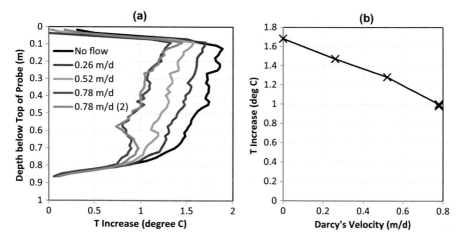

Fig. 7.15 The GFC temperatures from lab sand tank tests: **a** Time-averaged temperature increase during GFC heating at different flow velocities, and **b** temperature increase averaged along the length of the probe. The temperature data between depths of 0.18 and 0.63 m in (**a**) are used to compute the depth-averaged temperature in (**b**); small portions of the upper and lower heated sections are not included in the average temperature calculation due to vertical boundary impacts. In **a**, the temperature increase was larger at shallow depths; this might be an indication of impacts from heat-induced buoyancy or vertical variation in formation thermal conductivity (the more compacted sands at deeper locations likely have higher thermal conductivity)

Both the direction and magnitude of the ambient groundwater velocity at the PVP measurement interval can be determined from the tracer breakthrough data at the two detectors, d_1 and d_2, using Eqs. 7.9 and 7.10 (Fig. 7.17),

$$\alpha = \tan^{-1}\left(\frac{v_1\gamma_1(\cos\gamma_2 - 1) + v_2\gamma_2(1 - \cos\gamma_1)}{v_1\gamma_1\sin\gamma_2 - v_2\gamma_2\sin\gamma_1} \right), \tag{7.9}$$

$$v_g = \frac{v_1\gamma_1}{2(\cos\alpha - \cos(\alpha + \gamma_1))} \tag{7.10}$$

where v_w is the magnitude of the ambient groundwater seepage velocity prior to the installation of the PVP, α is the angle between ambient flow and the injection port i, γ_1 is the angle between i and d_1, γ_2 is the angle between i and d_2, and v_1 and v_2 are the apparent velocities determined from tracer breakthrough at d_1 and d_2. Note that after the installation of PVP, the flow field will be altered in the immediate vicinity of the probe as groundwater has to move around the probe surface. As a result, the tracer breakthrough responses from PVP detectors measure apparent velocities on the probe surface, not the ambient velocity itself. The apparent velocities at the detectors are computed from the breakthrough data as in Eq. 7.11 (Fig. 7.17),

$$v_1 = \frac{r\gamma_1}{\Delta t_1}, v_2 = \frac{r\gamma_2}{\Delta t_2}, \tag{7.11}$$

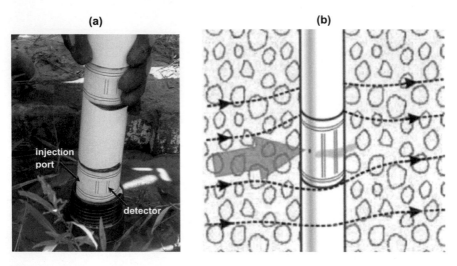

Fig. 7.16 Overview of the PVP: **a** Photo of a multilevel arrangement of probes showing injection ports and detectors, and **b** illustration of tracer movement around the probe surface during a test (Ozark Underground Laboratory 2021). Injection ports can be mounted in up to three locations on the probe surface 120° apart, to ensure detection of tracer solution no matter how the probe is oriented with respect to the ambient groundwater flow field (Gibson and Devlin 2018). Each detector consists of two parallel electric wires. The measured EC between the wires provides information on tracer solution breakthrough. In **b**, the blue-shaded arrow indicates the general groundwater flow before the probe is inserted into the formation; the dashed lines indicate the altered groundwater flow lines produced by probe installation. The red color indicates the movement of injected tracer around the probe surface. The diameter of the PVP can vary from 2.5 to 15 cm, depending on the application

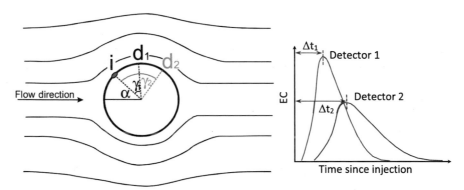

Fig. 7.17 Typical tracer breakthrough during a PVP test. The diagram on the left shows the position of injection port (i), detector 1 (d_1), and detector 2 (d_2) relative to the ambient groundwater flow: α is the angle between ambient flow and i, γ_1 is the angle between i and d_1, and γ_2 is the angle between i and d_2

where r is the radius of the probe (m), and Δt_1 and Δt_2 are the durations between the start of injection and arrival of tracer concentration peaks at detectors 1 and 2 (s).

The ambient velocity calculated using Eqs. 7.9–7.11 is the horizontal component of groundwater flow. If there is a vertical flow component, it can, in principle, be monitored by the breakthrough data at the vertical detectors. Limited experimental data indicate the vertical detectors function as designed. However, further work is needed to fully assess the accuracy of vertical velocity measurements.

Since the PVP was first introduced into the groundwater community, early applications of the approach have focused on direct measurement of groundwater velocity by installing the probe in dedicated boreholes completed in unconsolidated, non-cohesive sediments, where the sediments can collapse and be in direct contact with the probes. This practice eliminates biases from well construction issues on the measured flow. The dedicated borehole PVP has been used for characterizing the transience of a flow field at a site undergoing bioremediation of hydrocarbons (Schillig et al. 2011, 2016), estimating contaminant mass discharge across streambanks and streambeds (Rønde et al. 2017; Cremeans et al. 2018), and measuring groundwater flow in horizontal wells for passive in situ remediation (Cormican et al. 2021).

During recent years, the PVP has also been modified for measuring groundwater flow in wells (Osorno et al. 2018, 2022) and at the groundwater-surface water interface (GWSWI) (Cremeans and Devlin 2017). In the former case, due to the significant impacts of well construction, as well as the impacts of the PVP on annular flow, empirical relationships, in the form of calibration curves, are relied upon to convert the in-well velocity measurements to ambient groundwater velocity under different well and PVP construction parameters. The in-well PVP has been used for collecting hundreds of groundwater velocity measurements across an alluvial aquifer to identify contaminant preferential flow paths (sand and gravel lenses distributed discontinuously in tight clays), and characterizing groundwater flows in fractured rocks (Ozark Underground Laboratory 2021; Heyer et al. 2021).

In the case of GWSWI studies, the PVP was adapted by miniaturizing the probe and equipping it with a hyporheic shield to isolate vertical flow. Using this technology, Cremeans et al. (2018) mapped a stream bed to identify localized zones of high discharge, facilitating the characterization of contaminant discharge zones associated with a plume of chlorinated solvents. The adapted probe was also used to measure groundwater discharge at the base of a small lake near Bemidji, MN (French et al. 2021). Preliminary results suggested that there was rapid, upward flow into the base of a thick muck layer; the flow then followed a path along the lake bottom without much mixing with the bulk of the lake water body due to the muck layer acting as a flow barrier.

7.4.2.2 PVP Case Study

Many in situ remediation technologies rely on biological processes to break down contaminants, during which aquifer bioclogging can occur as a result of biomass accumulation in the pore space. Despite the wide recognition of its significance, few studies have investigated bioclogging and its impact on groundwater flow under field conditions. This is mostly due to the difficulty of making repeated K or groundwater velocity measurements at a scale that is sufficiently small for assessing aquifer property changes caused by biological processes. The development of the PVP provides an effective means of addressing this concern.

Schillig et al. (2011) presented a PVP case study for investigating the transience of groundwater flow in a sand aquifer undergoing benzene, toluene, ethylbenzene, and xylene bioremediation. Figure 7.18 shows the site setting. The unconfined sand aquifer under study was isolated by a clay aquitard on the bottom and two sheet pilings on the sides. Hydrocarbon was released into the aquifer between depths 2.5 and 4.0 m in the source zone. Five dedicated multilevel PVP stands, each equipped with four probes, were installed along a transect 13 m downstream of the hydrocarbon source. Five fully screened remediation wells were used to administer dissolved oxygen, using Oxygen Release Compound (ORC®) 1 m upstream of the PVP transect. Oxygen was released to the aquifer for stimulating aerobic biodegradation of the dissolved hydrocarbons. The ORC® releases oxygen with diminishing strength over time, so the wells were recharged with additional oxygen three times during the experiment in September 2005, February 2006, and September 2006.

Figure 7.19 shows the groundwater velocity changes at the PVP transect at the different times; August 2005 represents the pre-oxygen addition background condition for the experiment, so the differences plotted are zero (Fig. 7.19a). The plotted values are interpolated from the velocity changes at the 20 measurement locations (Fig. 7.18). After the first addition of oxygen in September 2005, groundwater velocity showed a clear decrease across the transect, with the largest reduction occurring near the lower right corner of the plot (Fig. 7.19b). Flow directions (not shown) remained largely similar to the background field, except in the lower right area where the direction of flow changes from northeast to northwest. Figure 7.19c shows the measured velocity three months after the second addition of oxygen in February 2006. Compared to October 2005, groundwater velocity in May 2006 increased significantly—possibly the result of seasonal changes of flow in the aquifer—and flow directions changed more significantly across a large portion of the transect. Most notably, the largest changes from background were found to occur in the locations coinciding with the highest concentrations of hydrocarbon, where bioactivity might be expected to be maximized. Following the May measurements, the ORC® was permitted to deplete to exhaustion, allowing the site to return to ambient conditions. The resilience of the system was tested with a final addition of oxygen in September 2006. One month later, in October 2006 (Fig. 7.19d), the measured groundwater appeared to undergo a large decrease in velocity in a fashion resembling the response a year before. Samples of aquifer material recovered in core and examined for microbial biomass near the PVP transect, and at a separate location removed from the biostimulation, showed significant increases in microbial numbers in the biostimulated zone, lending support to the notion that biomass accumulation might have been responsible for at least some of the observed velocity transience.

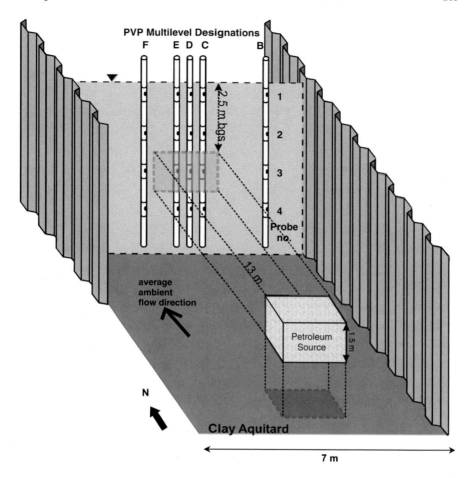

Fig. 7.18 Study site setting at Borden, Canada by Schillig et al. (2011). The aquifer is comprised of 7 m of well-sorted fine- to medium-grained sands bounded by sheet piling on the east and west and a clay aquitard underlying the sand. The depth to the water table fluctuates with precipitation and is generally within 1 m of ground surface. Groundwater generally flowed from the south to the north, with a hydrocarbon source 13 m upstream of a fence of PVP probes. Each PVP borehole had 4 measurement intervals with a vertical spacing of 0.8 m. Five fully screened remediation wells (not shown here) with dissolved oxygen release compound were installed 1 m upstream of the PVP fence to stimulate aerobic biodegradation of the released hydrocarbon. The objective of the study was to document changes to the groundwater velocity in response to microbial growth and activity during the aerobic biodegradation of the hydrocarbons

It should be pointed out that the plotted values in Fig. 7.19 provide point assessments of velocity responses, and the interpolations presented should not be considered true representations of the detailed velocity distribution. To obtain such a picture of the aquifer structure and velocity distribution, the density of sampling would have to be increased considerably—a testimonial to the challenges in aquifer characterization. Further effort is needed to determine the density of sampling needed to achieve

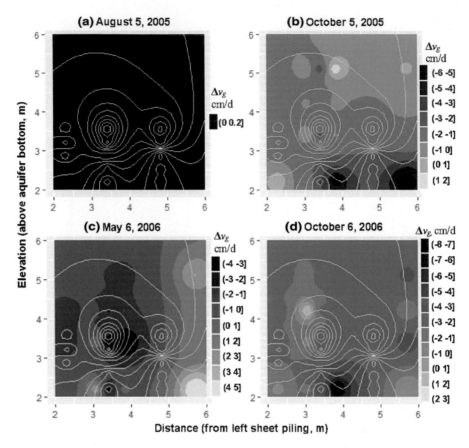

Fig. 7.19 Changes in velocity at the PVP transect at different times before and after adding oxygen to the hydrocarbon contaminated aquifer: **a** August 2005, difference from background is zero, **b** October 2005, following O_2 addition in September 2005, **c** May 2006, after established biodegradation, and **d** October 2006 re-addition of O_2 following a 2–3 month waning of the O_2 source (after Schillig et al. 2011). For comparison, the hydrocarbon concentrations measured in October 2006 (based on water samples from 36 multilevel sampling wells across the site) are plotted as contours on the maps (the two centers of the contours at elevation 3.6 m represent peak concentrations). The changes in groundwater velocity were believed to be at least partially caused by biological growth following the additions of oxygen at different time. Note the coincidence of the greatest changes in velocity with the most concentrated portion of the hydrocarbon plume where biodegradation was expected to be the most significant

practical uses for such data, for example the estimation of water flow across the transect, or contaminant mass discharges and the associated uncertainty. Nonetheless, the sampling density adopted for this project was sufficient to identify and quantify possible hydraulic responses to biostimulation.

7.5 Summary and Conclusions

In this chapter, we discussed high-resolution characterization of the physical properties in shallow, saturated, unconsolidated subsurface conditions using DP technology, NMR logging, DTS, and PVPs. These developments have led to a significantly improved ability to obtain information about subsurface properties (K and porosity) and groundwater velocity at a speed and resolution that has not previously been possible. These advances have allowed the discipline to gain important insights into the fundamental controls on groundwater flow and transport processes under field conditions (Schillig et al. 2011; Bohling et al. 2012; Fiori et al. 2013; Liu et al. 2013; Dogan et al. 2014; Knight et al. 2016; Munn et al. 2020).

The performance of these approaches was demonstrated using examples from field and laboratory settings. DP EC profiling could potentially provide a good indicator of relative K when electrically conductive clay is a dominant control on permeability, although more work is needed to explore this use of EC. The DPP is a tool that can provide reliable K estimates in moderately to highly permeable zones. It has several major advantages over conventional hydraulic testing approaches (such as slug and flowmeter tests) due to the steady-shape requirement for the analysis. In addition, the DPP is not impacted by low-K skins, which are known to be a major challenge for slug tests. The practical time constraints, however, limit the DPP vertical resolution to 0.4 m or larger under most field conditions. The continuous DPIL is a high speed and resolution (0.015 m) approach, but only characterizes relative variations in K. A general empirical model may be used to estimate K values within a narrow, but useful, K range (McCall and Christy 2010). Alternately, regressions with additional data, such as slug test K values from nearby locations, are used to convert the DPIL data into site-specific K estimates. The HRK tool was designed to exploit the advantages of the DPP and DPIL by coupling them into a single probe. Because DPP tests are collocated with DPIL logs, they can be used to directly transform the DPIL data into K without the need to compare measurements at different support scales as in the regression-based approaches.

In field investigations, the DPIL and HRK can be applied in a complementary fashion (Bohling et al. 2012). Because the continuous DPIL is rapid, many DPIL profiles can be performed across a site in a short period. The HRK tool, on the other hand, takes more time (due to halting tool advancement for DPP tests) and can be applied at far fewer locations. By coupling the HRK tool with DPIL, the DPIL K power-law relation determined from the HRK profiles can be applied to all the DPIL logs, resulting in a high-resolution characterization of K for the entire site at only a fraction of time and costs of other approaches (Bohling et al. 2012).

The current DP tools (including the DPP, DPIL, and HRK tool) do not perform well in low-K settings. The DPP tests take a long time to complete when K is low (e.g., < 0.001 m/d). For DPIL, it remains a challenge to inject water at the small rate (e.g., 1 mL/min) that is needed in low-K formations to avoid high injection pressure and formation alteration. The pressure generated from tool advancement can also significantly impact the measured injection pressure signal. In addition to the low-K

limit, the DPIL also has an upper limit as the injection-induced pressure responses become very small in highly permeable zones. Future work is needed to improve the range of reliable K estimates by the DPP, DPIL, and HRK tool.

Logging with NMR technology, a widely used borehole technique in the petroleum industry, has been adapted to various applications in near-surface hydrology (Walsh et al. 2011, 2013). Compared to the DP approaches, a major advantage of NMR is that it provides an estimate of porosity in addition to K. Because the NMR signal is directly a function of the total number of water molecules in the measurement zone, the accuracy of estimated porosity is much higher than that of K. Given the challenges of DP approaches in low-K settings, NMR logging holds great potential for characterizing K and porosity in low-K formations.

There has been a growing interest in measuring groundwater flux directly in the field as the remediation community switches from contaminant concentration-based decision-making to one based on contaminant mass discharge (Suthersan et al. 2010; Devlin 2020). In recent years, significant progress has been made in using heat and other tracers to measure groundwater flux. Liu et al. (2013) developed a GFC tool by wrapping a DTS fiber-optic cable and resistance heating cable around a sealed hollow PVC pipe. The GFC tool was assessed in a series of laboratory tests in a sand tank, and results showed a linear relationship between the heating-induced temperature increase and the ambient groundwater flux over the range of 0–0.78 m/d (Darcy's flux). One key assumption in using the GFC is that the vertical variations in the thermal conductivity of the materials near the test well are negligible, so that the impacts of thermal conduction can be ignored. However, when thermal conduction is important, the detailed characteristics of temperature change with time may be used to quantify formation thermal conductivity in addition to groundwater flux (Simon et al. 2021).

The PVP uses localized tracer tests to measure the magnitude and direction of groundwater velocity at the centimeter scale (Labaky et al. 2007). It operates by injecting a small amount of tracer solution into the formation and measuring the movement of tracer around the probe surface via multiple detectors. Both the direction and magnitude of the ambient groundwater velocity can be determined from the tracer breakthrough data at the two detectors. Early applications of the approach focused on direct measurement of groundwater velocity by installing the probe in dedicated boreholes, while recent adaption of the tool allows it to be used for flow measurement in wells, at the groundwater-surface water interface, and in horizontal wells (Ozark Underground Laboratory 2021). The PVP has been used in a variety of site characterization projects, and its high-resolution measurement can provide improved understanding of groundwater flow under transient conditions.

The approaches discussed in this chapter were primarily developed for use in intervals saturated with groundwater. When non-aqueous phase liquids (NAPLs) are present, additional research is needed to investigate the performance of these approaches as NAPL-water movement will not only depend on the properties of the NAPL and groundwater but also on their interactions. One of the most common approaches for characterizing the transmissivity of LNAPL (NAPL lighter than water) intervals is slug testing (Charbeneau et al. 2016; Butler 2019). However, this only provides an estimate on the mobility of LNAPLs as a free phase in the immediate vicinity of the test well. To characterize LNAPL movement throughout the groundwater system, detailed information about relative permeabilities, NAPL saturations and their spatial distribution is needed. Furthermore, a practical challenge for characterizing sites with LNAPL is the potential risk of vertically spreading the contaminants across different geologic units by well construction (Newell et al. 1995). This challenge can be largely overcome in shallow unconsolidated sediments with DP-based approaches such as the Membrane Interface Probe (MIP; ASTM 2018) and laser-induced fluorescence (LIF) probes presented in Chap. 8. As discussed in Chap. 9, NMR logging could be used to map the distribution of NAPL because of the abundance of hydrogen in petroleum NAPL, although the presence of NAPL can affect the utility of NMR for characterizing porosity and K. For PVP-based studies, preliminary testing has demonstrated that velocity measurements are sensitive to the presence of gases in the pore space (Cormican et al. 2021), which is analogous to the NAPL case since tracer responses are mainly controlled by pore water flow that is restricted when NAPLs are present. Additional research is needed to extend the preliminary work and investigate how the presence of NAPL affects the measured velocity under different subsurface conditions.

There has been an increasing recognition of the need for "focused" site remediation with remedial measures targeted at the processes and locations that are crucial to remediation success (ITRC 2004). Identifying these critical processes and their locations, however, remains a difficult challenge at many sites as the conventional approaches (e.g., the well-based pumping and slug tests) often fail to provide information at sufficient details to make those identifications. The high-resolution approaches discussed in this chapter hold great promise for addressing this challenge. Each of these approaches has its own advantages and limitations, and future studies are needed to refine their performance as discussed above. Finally, when site conditions are complex, a combination of different approaches may provide the best solution for site investigators to obtain the high-resolution data that are needed to develop and implement a successful site remediation program.

References

American Society for Testing and Materials (ASTM). New practice for detection of hydrocarbon liquids in soils by fluorescence with the optical imaging profiler using direct push methods. WK66935. In review. ASTM International, West Conshohocken, Pennsylvania. www.astm.org.

American Society of Testing and Materials (ASTM) (2016) Standard practice for direct push hydraulic logging for profiling variations of permeability in soils. D8037. ASTM International, West Conshohocken, Pennsylvania. www.astm.org

American Society for Testing and Materials (ASTM) (2018) Standard practice for volatile contaminant logging using a membrane interface probe (MIP) in unconsolidated formations with direct push methods. D5878. ASTM International, West Conshohocken, Pennsylvania. www.astm.org.

Anderson MP (2005) Heat as a Ground Water Tracer. Groundwater, 43:951–968. https://doi.org/10.1111/j.1745-6584.2005.00052.x

Artiola J, Pepper IL, Brusseau ML (2004) Environmental monitoring and characterization. Academic Press, Cambridge

Bakker M, Caljé R, Schaars F, van der Made KJ, de Haas S (2015) An active heat tracer experiment to determine groundwater velocities using fiber optic cables installed with direct push equipment. Water Resour Res 51:2760–2772. https://doi.org/10.1002/2014WR016632

Bayless ER, Mandell WA, Urisc JR (2011) Accuracy of flowmeters measuring horizontal groundwater flow in an unconsolidated aquifer simulator. Ground Water Monit R 31(2). http://doi.org/10.1111/j1745-6592.2010.01324.x

Becker MW, Bauer B, Hutchinson A (2013) Measuring artificial recharge with fiber optic distributed temperature sensing. Groundwater 51:670–678. https://doi.org/10.1111/j.1745-6584.2012.01006.x

Boggs JM, Young SC, Beard LM (1992) Field study of dispersion in a heterogeneous aquifer 1, overview and site description. Water Resour Res 28(12):3281–3291

Bohling GC, Liu G, Knobbe SJ, Reboulet EC, Hyndman DW, Dietrich P, Butler JJ Jr (2012) Geostatistical analysis of centimeter-scale hydraulic conductivity variations at the MADE site. Water Resour Res 48:W02525. https://doi.org/10.1029/2011WR010791

Borden RC, Cha KY, Liu G (2021) A physically based approach for estimating hydraulic conductivity from HPT pressure and flowrate. Groundwater 59(2):266–272

Butler JJ Jr (2005) Hydrogeological methods for estimation of hydraulic conductivity. In: Rubin Y, Hubbard S (eds) Hydrogeophysics. Springer, The Netherlands, pp 23–58

Butler JJ Jr (2019) The design, performance, and analysis of slug tests, 2nd edn. CRC Press, Boca Raton, p 266

Butler JJ Jr, Healey JM, McCall GW, Garnett EJ, Loheide SP II (2002) Hydraulic tests with direct-push equipment. Ground Water 40(1):25–36

Butler JJ Jr, Whittemore DO, Zhan X, Healey JM (2004) Analysis of two pumping tests at the O'Rourke Bridge site on the Arkansas River in Pawnee County, Kansas. Kansas Geological Survey Open-File Report 2004-32. Lawrence, Kansas

Butler JJ Jr, Dietrich P, Wittig V, Christy T (2007) Characterizing hydraulic conductivity with the direct-push permeameter. Groundwater 45(4):409–419. https://doi.org/10.1111/j.1745-6584.2007.00300.x

Charbeneau R, Kirkman A, Muthu R (2016) LNAPL transmissivity workbook: a tool for baildown test analysis—user guide. American Petroleum Institute Publication 4762

Constantz J (2008) Heat as a tracer to determine streambed water exchanges, Water Resour Res 44:W00D10. https://doi.org/10.1029/2008WR006996

Cormican A, Devlin JF, Osorno TC, Divine D (2021) Design, testing, and implementation of a real-time system for monitoring flow in horizontal wells. J Contam Hydrol 238. http://doi.org/10.1016/j.jconhyd.2021.103772

Cremeans MM, Devlin JF (2017) Validation of a new device to quantify groundwater-surface water exchange. J Contam Hydrol 206:75–80

Cremeans MM, Devlin JF, McKnight U, Bjerg P (2018) Application of new point measurement device to quantify groundwater-surface water interactions. J Contam Hydrol 211:85–93

Dagan G, Neuman SP (1997) Subsurface flow and transport: a stochastic approach. Cambridge University Press, Cambridge, UK

Devlin JF (2020) Groundwater velocity. The Groundwater Project, Guelph, Ontario, Canada, 64 p. ISBN: 978-1-77470-000-6, downloadable at https://gw-project.org/books/groundwater-velocity

Dietrich P, Leven C (2005) Direct push technologies. In: Kirsch R (ed) Groundwater geophysics. Springer, Berlin, pp 321–340

Dietrich P, Butler JJ Jr, Faiss K (2008) A rapid method for hydraulic profiling in unconsolidated formations. Ground Water 46(2):323–328. https://doi.org/10.1111/j.1745-6584.2007.00377.x

Dlubac K, Knight R, Song Y, Bachman N, Grau B, Cannia J, Williams J (2013) Use of NMR logging to obtain estimates of hydraulic conductivity in the high plains aquifer, Nebraska, USA. Water Resour Res 49(4):1871–1886. https://doi.org/10.1002/wrcr.20151

Dogan M, Van Dam RL, Liu G, Meerschaert MM, Butler JJ Jr, Bohling GC, Benson DA, Hyndman DW (2014) Predicting flow and transport in highly heterogeneous alluvial aquifers. Geophys Res Lett 41:7560–7565. https://doi.org/10.1002/2014GL061800

Dunn KJ, Bergman DJ, Latorraca GA (2002) Nuclear magnetic resonance—petrophysical and logging applications. Pergamon, Oxford

Fiori A, Dagan G, Jankovic I, Zarlenga A (2013) The plume spreading in the MADE transport experiment: could it be predicted by stochastic models? Water Resour Res 49:2497–2507. https://doi.org/10.1002/wrcr.20128

Fogg GE, Carle SF, Green C (2000) Connected network paradigm for the alluvial aquifer system. In: Zhang D, Winter CL (eds) Theory, modeling, and field investigation in hydrogeology. A special volume in honor of Shlomo P. Neuman's 60th birthday. Geological Society of America Special Papers 348, pp 25–42

French L, Heyer B, Osorno T, Jones M, Devlin JF (2021) Groundwater-lake water flow characterization using seepage velocity point measurements in Unnamed Lake, MN (abstract). In: Governor's conference on the future of water in Kansas (Virtual), Kansas

Geoprobe (2007) Geoprobe Hydraulic Profiling Tool (HPT) system. Standard operating procedure. Technical Bulletin No. MK3137. Kejr Inc., Salina, KS

Gibson B, Devlin JF (2018) Laboratory validation of a point velocity probe for measuring horizontal flow from any direction. J Contam Hydrol 208:10–16

Henderson R, Day-Lewis F, Harvey C (2009) Investigation of aquifer-estuary interaction using wavelet analysis of fiber-optic temperature data. Geophys Res Lett 36(6):1–6

Heyer BR, Osorno TC, Devlin JF (2021) Laboratory testing of real-time flux measurements in fractured media. J Hydrol 601:126639

ITRC (Interstate Technology & Regulatory Council) (2004) Remediation process optimization: identifying opportunities for enhanced and more efficient site remediation. RPO-1. Interstate Technology & Regulatory Council, Remediation Process Optimization Team, Washington, D.C. Available on the Internet at https://www.itrcweb.org

ITRC (Interstate Technology & Regulatory Council) (2006) The use of direct-push well technology for long-term environmental monitoring in groundwater investigations. Prepared by the Interstate Technology & Regulatory Council Sampling, Characterization and Monitoring Team. Available on the Internet at https://www.itrcweb.org

Kendrick AK, Knight R, Johnson CD, Liu G, Knobbe S, Hunt RJ, Butler JJ Jr (2021) Assessment of NMR logging for estimating hydraulic conductivity in glacial aquifers. Groundwater 59:31–48. https://doi.org/10.1111/gwat.13014

Knight R, Walsh DO, Butler JJ Jr, Grunewald E, Liu G, Parsekian AD, Reboulet EC, Knobbe S, Barrows M (2016) NMR logging to estimate hydraulic conductivity in unconsolidated aquifers. Groundwater 54(1):104–114. https://doi.org/10.1111/gwat.12324

Knobbe S, Liu G, Butler JJ Jr (2015) Laboratory investigation of distributed temperature sensing to characterize groundwater flux (abstract). Novel methods for subsurface characterization and monitoring: from theory to practice. Lawrence, KS

Krejci M, Lett M, Lloyd A, Hopper T, Neville T, Birt B (2018) Groundwater assessment in a coal measures sequence using borehole magnetic resonance. ASEG Ext Abstr 1:1–5. http://doi.org/10.1071/ASEG2018abT7_3H

Labaky W, Devlin JF, Gillham RW (2007) Probe for measuring groundwater velocity at the centimeter scale. Environ Sci Technol 41(24):8453–8458. https://doi.org/10.1021/es0716047

Leaf AT, Hart DJ, Bahr JM (2012) Active thermal tracer tests for improved hydrostratigraphic characterization. Groundwater 50(5):726–735. https://doi.org/10.1111/j.1745-6584.2012.00913.x

Lessoff SC, Schneidewind U, Leven C, Blum P, Dietrich P, Dagan G (2010) Spatial characterization of the hydraulic conductivity using direct-push injection logging. Water Resour Res 46:W12502. https://doi.org/10.1029/2009WR008949

Liu G, Butler JJ Jr (2019) Hydraulic conductivity characterization methods. www.enviro.wiki/index.php?title=Characterization_Methods_%E2%80%93_Hydraulic_Conductivity

Liu G, Bohling GC, Butler JJ Jr (2008) Simulation assessment of the direct-push permeameter for characterizing vertical variations in hydraulic conductivity. Water Resour Res 44:W02432. https://doi.org/10.1029/2007WR006078

Liu G, Butler JJ Jr, Bohling GC, Reboulet E, Knobbe S, Hyndman DW (2009) A new method for high-resolution characterization of hydraulic conductivity. Water Resour Res 45:W08202. https://doi.org/10.1029/2009WR008319

Liu G, Butler JJ Jr, Reboulet EC, Knobbe S (2012) Hydraulic conductivity profiling with direct push methods. Grundwasser 17(1). http://doi.org/10.1007/s00767-011-0182-9

Liu G, Knobbe S, Butler JJ Jr (2013) Resolving centimeter-scale flows in aquifers and their hydrostratigraphic controls. Geophys Res Lett 40:1098–1103. https://doi.org/10.1002/grl.50282

Liu G, Knobbe S, Reboulet EC, Whittemore DO, Händel F, Butler JJ Jr (2016) Field investigation of a new recharge approach for ASR projects in near-surface aquifers. Groundwater 54(3):425–433. https://doi.org/10.1111/gwat.12363

Liu G, Knobbe S, Borden R, Butler JJ Jr (2018) A low permeability extension of hydraulic profiling tool: challenges and solutions. Abstract (H51A-08) presented at 2018 AGU fall meeting, Washington D.C., 10–14 Dec 2018

Liu G, Borden R, Butler JJ Jr (2019) Simulation assessment of direct push injection logging for high resolution aquifer characterization. Groundwater 57(4):562–574. https://doi.org/10.1111/gwat.12826

Lowry W, Mason N, Chipman V, Kisiel K, Stockton J (1999) In-situ permeability measurements with direct push techniques: phase II topical report, 102 p. SEASF-TR-98-207 Report to DOE Federal Energy Technical Center

Lowry CS, Walker JF, Hunt RJ, Anderson MP (2007) Identifying spatial variability of ground-water discharge in a wetland stream using a distributed temperature sensor. Water Resour Res 43:W10408. https://doi.org/10.1029/2007WR006145

Lunne T, Powell JJ, Robertson PK (2002) Cone penetration testing in geotechnical practice. CRC Press, Boca Raton

MacFarlane AP, Förster A, Merriam D, Schrötter J, Healey JM (2002) Monitoring artificially stimulated fluid movement in the Cretaceous Dakota aquifer, western Kansas. Hydrogeol J 10(6):662–673

Maliva RG (2016) Direct-Push technology. In: Aquifer Characterization Techniques, Springer Hydrogeology, p 383–402. Springer, Cham

Maldaner CH, Munn JD, Coleman TI, Molson JW, Parker BL (2019) Groundwater flow quantification in fractured rock boreholes using active distributed temperature sensing under natural gradient conditions. Water Resour Res 55:3285–3306. https://doi.org/10.1029/2018WR024319

McCall GW, Christy TM (2010) Development of hydraulic conductivity estimate for the Hydraulic Profiling Tool (HPT). In: The 2010 North American environmental field conference and exposition. The Nielsen Environmental Field School, Las Cruces, NM, 12–15 Jan 2010

McCall GW, Christy TM (2020) The hydraulic profiling tool for hydrogeologic investigation of unconsolidated formations. Groundwater Monit R 40:89–103. https://doi.org/10.1111/gwmr.12399

McCall GW, Butler JJ Jr, Healey JM, Lanier AA, Sellwood SM, Garnett EJ (2002) A dual-tube direct-push method for vertical profiling of hydraulic conductivity in unconsolidated formations. Environ Eng Geosci 8(2):75–84

McCall GW, Nielsen DM, Farrington SP, Christy TM (2005) Use of direct-push technologies in environmental site characterization and ground-water monitoring. In: Nielsen DM (ed) The practical handbook of environmental site characterization and ground-water monitoring, 2nd edn. CRC Press, Boca Raton, pp 345–472

Moffett K, Tyler S, Torgersen T, Menon M, Selker JS, Gorelick S (2008) Processes controlling the thermal regime of saltmarsh channel beds. Environ Sci Technol 42(3):671–676

Munn JD, Maldaner CH, Coleman TI, Parker BL (2020) Measuring fracture flow changes in a bedrock aquifer due to open hole and pumped conditions using active distributed temperature sensing. Water Resour Res 56:e2020WR027229. http://doi.org/10.1029/2020WR027229

Newell CJ, Acree SD, Ross RR, Huling SS (1995) Light nonaqueous phase liquids. US EPA, Washington DC, p 36

Nielsen DM, Nielsen G (2006) The essential handbook of ground-water sampling. CRC Press, Boca Raton

Osorno T, Firdous R, Devlin JF (2018) An in-well point velocity probe for the rapid characterization of groundwater velocity at the centimeter-scale. J Hydrol 557:539–546. https://doi.org/10.1016/j.jhydrol.2017.12.033

Osorno TC, Devlin JF, Bohling GC (2022) Geostatistics of the Borden aquifer: high-resolution characterization using direct groundwater velocity measurements. Water Resour Res 58(3):e2020WR029034

Ozark Underground Laboratory (2021) Point velocity probe handbook. Protem, MO. https://www.ozarkundergroundlab.com

Purvance DT, Andricevic R (2000) On the electrical-hydraulic conductivity correlation in aquifers. Water Resour Res 36(10):2905–2913. https://doi.org/10.1029/2000WR900165

Rau GC, Andersen MS, McCallum AM, Roshan H, Acworth RI (2014) Heat as a tracer to quantify water flow in near-surface sediments. Earth-Sci Rev 129:40–58, ISSN 0012-8252. https://doi.org/10.1016/j.earscirev.2013.10.015

Read T, Bour O, Bense V, Le Borgne T, Goderniaux P, Klepikova MV, Hochreutener R, Lavenant N, Boschero V (2013) Characterizing groundwater flow and heat transport in fractured rock using fiber-optic distributed temperature sensing. Geophys Res Lett 40:2055–2059. https://doi.org/10.1002/grl.50397

Rønde V, McKnight US, Sonne AT, Balbarini N, Devlin JF, Bjerg PL (2017) Contaminant mass discharge to streams: comparing direct groundwater velocity measurements and multi-level groundwater sampling with an in-stream approach. J Contam Hydrol 206:43–54. https://doi.org/10.1016/j.jconhyd.2017.09.010

Schillig PC, Devlin JF, McGlashan M, Tsoflias G, Roberts JA (2011) Transient heterogeneity in an aquifer undergoing bioremediation of hydrocarbons. Ground Water 49(2):184–196

Schillig PC, Devlin JF, Rudolph D (2016) Upscaling point measurements of groundwater velocity for enhanced site characterization in a glacial outwash aquifer. Groundwater 54(3):394–405

Schulmeister MK, Butler JJ Jr, Healey JM, Zheng L, Wysocki DA, McCall GW (2003) Direct-push electrical conductivity logging for high-resolution hydrostratigraphic characterization. Ground Water Monit R 23(3):52–62

Schulmeister MK, Healey JM, Butler JJ Jr, McCall GW (2004) Direct-push geochemical profiling for assessment of inorganic chemical heterogeneity in aquifers. J Contam Hydrol 69(3–4):215–232

Selker JS, van de Giesen N, Westhoff M, Luxemburg W, Parlange MB (2006) Fiber optics opens window on stream dynamics. Geophys Res Lett 33:L24401. https://doi.org/10.1029/2006GL027979

Simon N, Bour O, Lavenant N, Porel G, Nauleau B, Pouladi B, Longuevergne L, Crave A (2021) Numerical and experimental validation of the applicability of active-DTS experiments to estimate thermal conductivity and groundwater flux in porous media. Water Resour Res 57:e2020WR028078. http://doi.org/10.1029/2020WR028078

Stienstra P, van Deen JK (1994) Field data collection techniques–unconventional sounding and sampling methods. In: Rengers N (ed) Engineering geology of quaternary sediments. Balkema, Rotterdam, pp 41–55

Striegl AM, Loheide SP II (2012) Heated distributed temperature sensing for field scale soil moisture monitoring. Groundwater 50(3):340–347. https://doi.org/10.1111/j.1745-6584.2012.00928.x

Stroo HF, Ward CH (2010) In situ remediation of chlorinated solvent plumes, vol 2. Springer Science & Business Media, Berlin

Suthersan S, Divine C, Quinnan J, Nichols E (2010) Flux-informed remediation decision making. Ground Water Monit R 30(1):36–45. https://doi.org/10.1111/j.1745-6592.2009.01274.x

Tyler SW, Selker JS, Hausner MB, Hatch CE, Torgersen T, Thodal CE, Schladow SG (2009) Environmental temperature sensing using Raman spectra DTS fiber-optic methods. Water Resour Res 45:W00D23. http://doi.org/10.1029/2008WR007052

U.S. Environmental Protection Agency (2016) Expedited site assessment tools for underground storage tank sites: chapter V—direct push technologies. EPA 510-B-16-004. Office of Underground Storage Tanks, Washington, D.C. https://www.epa.gov/ust/expedited-site-assessment-tools-underground-storage-tank-sites-guide-regulators

Vista Clara. https://www.vista-clara.com/2021/01/25/leveraging-direct-push-for-nmr-logging-measurements. Accessed 22 Dec 2021

Walsh DO, Grunewald E, Turner P, Hinnell A, Ferre P (2011) Practical limitations and applications of short dead time surface NMR. Near Surf Geophys 9:103–113. https://doi.org/10.3997/1873-0604.2010073

Walsh DO, Turner P, Grunewald E, Zhang H, Butler JJ Jr, Reboulet E, Knobbe S, Christy T, Lane JW, Johnson CD, Munday T, Fitzpatrick A (2013) A small-diameter NMR logging tool for groundwater investigations. Groundwater 51:914–926. https://doi.org/10.1111/gwat.12024

Zhao Z, Illman WA (2022) Improved high-resolution characterization of hydraulic conductivity through inverse modeling of HPT profiles and steady-state hydraulic tomography: Field and synthetic studies. J Hydrol 612:128124

Zheng C, Bennett GD (2002) Applied contaminant transport modeling, 2nd edn. Wiley, New York, 621 p

Zschornack L, Bohling GC, Butler JJ Jr, Dietrich P (2013) Hydraulic profiling with the direct-push permeameter: assessment of probe configuration and analysis methodology. J Hydrol 496:195–204

Chapter 8
High-Resolution Delineation of Petroleum NAPLs

Randy St. Germain

Abstract Previous chapters of this book demonstrate that a cohesive and well-supported conceptual site model (CSM) of non-aqueous phase liquid (NAPL) petroleum is commonly the cornerstone of successful risk analysis and/or remediation design. It is difficult to overstate however, the extent to which the heterogeneity of source term NAPL distribution confounds one's efforts to develop an accurate NAPL CSM. In most cases, only near-continuous measurements of NAPL in the soil are capable of adequately conceptualizing a site's complex NAPL distribution. Continuous NAPL logging, conducted at a significant number of locations across a petroleum release site, is necessary to better comprehend the chaotic nature of the NAPL's distribution. Applying high-resolution screening techniques sitewide is known as high-resolution site characterization (HRSC) and this chapter describes how the most commonly applied HRSC techniques can make the difficult task of logging continuously for petroleum NAPL, and its associated groundwater impacts, not only possible but fairly routine.

Keywords High-resolution site characterization · Laser-induced fluorescence · LNAPL conceptual site model · Petroleum hydrocarbons · Subsurface heterogeneity

8.1 History of Subsurface Petroleum Hydrocarbon Investigation

Regulations regarding petroleum releases in the subsurface were put in place because petroleum contains numerous toxic water soluble compounds including benzene, toluene, ethyl-benzene, and xylenes (BTEX) as well as polycyclic aromatic hydrocarbons (PAHs). These somewhat water soluble compounds partition out of the petroleum non-aqueous phase liquid (NAPL) into groundwater, making groundwater

R. St. Germain (✉)
Dakota Technologies, Inc., 2201 12th Street N., Suite A, Fargo, ND 58102, USA
e-mail: stgermain@dakotatechnologies.com

© The Author(s) 2024
J. García-Rincón et al. (eds.), *Advances in the Characterisation and Remediation of Sites Contaminated with Petroleum Hydrocarbons*, Environmental Contamination Remediation and Management, https://doi.org/10.1007/978-3-031-34447-3_8

213

the "canary in the coal mine" at sites where petroleum releases are suspected. If the groundwater is found to be impacted, a petroleum NAPL release has been confirmed.

The first course of action to be taken to determine if a release has occurred has typically been to grab a limited number of soil samples and install the ubiquitous set of monitoring wells, all intended to be placed strategically so as to intercept any dissolved phase contaminants that might indicate a petroleum NAPL release affecting the groundwater. There is little hand-wringing at this stage with regard to what methodologies are employed to obtain samples, because we are initially looking only to find any contaminants of concern (COCs), such as BTEX. We simply sample the groundwater, send it in, wait for the laboratory results, and any decision points are relatively straightforward. Only after finding COCs above a certain threshold in groundwater, confirming NAPL has been released, do we progress to "action levels". In other cases, NAPL's presence does not need to be inferred from groundwater contamination, because it is encountered in the monitoring wells, confirming the NAPL's presence directly.

Historically, the next step of the investigation was to conduct more discrete sampling and more monitoring wells to see "how big the plume is", kicking off quarterly sampling of those wells, and in general a repeat of previous steps in an ever-expanding fashion. This was often accompanied by a lot of head-scratching as to why some monitoring wells had NAPL, many did not, some contained high dissolved phase, some very nearby did not, NAPL came and went in wells, and so on. Even repeat sampling, at the very same locations but on different dates, produced concentrations that changed wildly. The petroleum appeared to be moving about dramatically with the passing of time.

What was not widely understood at the time was that the majority of sites have very complex NAPL and groundwater flow path architecture. What appeared to investigators as dramatic changes of NAPL and dissolved phase over time were, in reality, mirages caused by natural heterogeneity and/or seasonal or well-induced groundwater movement. For many NAPL release sites, the longer the investigation lasted, and the more wells and soil sampling that was conducted, the more bizarre and contradictory the conceptual site model (CSM) became.

Due to the fact that regulations were focused on the effects to groundwater, the focus for characterization was also on groundwater. These characterizations were often well-conducted, using an array of sophisticated discrete level monitoring wells and sampling systems. This amounted to little more than an advanced understanding of the symptoms of the petroleum's presence however, rather than the distribution of the root cause itself. This became evident when remedial designs based largely on groundwater data were initially thought to be a successful treatment due to promising declines in groundwater concentrations, but subsequently suffered rebounds in the dissolved phase. Our early focus on groundwater left us with relatively little knowledge about the source term NAPL, for which the applied remedy had little if any efficacy. Add onto that our underappreciation of geology's tremendous role (especially its inherent heterogeneity) in the early goings of the industry, and it is no wonder that many early attempts at site remediation were destined to be modestly effective at best, and often completely ineffective. It took a long time for the industry to realize that once we had established that NAPL was present in sufficient quantities that it was sourcing a dissolved phase problem, our continued use of methods that focused primarily on the dissolved phases of petroleum no longer made sense. Relying on means designed to measure water, such as monitoring wells that were never designed to measure NAPL, was not sensible.

A familiar adage states that "where there is smoke, there is fire", and NAPL and its dissolved phase have a similar relationship. Dissolved and vapor phase distribution is helped along by diffusion and groundwater flux, so they behave more like smoke which diffuses freely and is carried by these "winds". The petroleum NAPL is akin to the fire, moving discretely through soil pores, its path determined by the soil's grain size, pore sizes, structure, and geometry. And all the while the NAPL is effusing its telltale "smoke" (the dissolved and vapor phase) into adjacent soils and groundwater. NAPL follows a distribution pattern dictated by gravity in the vadose zone, its specific gravity in the saturated zone, and preferential pathways available to NAPL's viscosity and surface and interfacial tensions, distributed by the whims of geology.

Firefighters faced with finding a fire in a building filled with blinding smoke employ infrared cameras, which allow them to focus on the fire itself and see past the smoke. For similar reasons, investigators seriously interested in understanding petroleum product distribution in the subsurface need to adopt techniques that are highly NAPL-specific in their response, or that are responsive to all phases but continue to increase or otherwise recognizably change their response upon encountering NAPL. As the industry slowly turned its focus to the source term NAPL (not just inferring NAPLs presence by gauging NAPL in monitoring wells and monitoring groundwater) a few early adopter regulators were keen to put their focus back onto the NAPL, in particular relying on direct sensing technologies such as laser-induced fluorescence (LIF) to change their agency's entire mindsight with regard to petroleum release investigations (Stock 2011).

The term "NAPL body" in context of this chapter means the distribution of light non-aqueous phase liquids (LNAPL) and dense non-aqueous phase liquids (DNAPLs) in the subsurface. This includes both multi-component DNAPLs such as bunker fuel and creosotes as well as the single component chlorinated solvent DNAPLs that often comes to mind when the term DNAPL is mentioned. The term NAPL body as used here includes light staining, those nearly invisible deposits of NAPL that are not readily discernible as free product, discrete ganglia, pools, or droplets. For instance, when one soaks up a diesel spill in the mechanic's shop with a granular clay-based floor adsorbent, those particles no longer drip or even have a sheen, but they certainly contain diesel's relatively non-volatile NAPL within their pores, and this NAPL can still act as a source for dissolved phase when the sorbent granules are placed in contact with water. A microscope or other visual magnifier and an ultraviolet (UV) light might be needed to confirm that NAPL exists in fine-textured soils, but indeed it does, and more often than we assume based on examination of cores with the naked eye outdoors in daylight.

The term NAPL body used here expressly excludes the considerably larger plume of vapor and aqueous phase forms created as some fraction of compounds contained within the NAPL emanate away from the NAPL body. It excludes as well those compounds that sorb onto soil particles that they encounter during their convective and/or diffusive travels. This is not to say that sorbed phase contaminants of this nature are not critical sources to consider, especially in the case of more water soluble halogenated DNAPLs such as trichloroethylene (TCE) and the like. These DNAPLs often source enough dissolved phase contaminant, over long enough periods of time, that the remnant sorbed phase adsorbed in low-permeability soils can go on to act as strong source terms via back diffusion, long after the true DNAPL has been depleted (Brooks et al. 2020). This topic has been thoroughly explored, halogenated solvent fate and transport behavior in particular, but back diffusion occurs for petroleum NAPLs as well, just not at such extremes most likely due to attenuation mechanisms deserving of further research. The exclusion of the dissolved/sorbed body from the data set allows one to focus their attention on the originally released source term NAPL, that often is the driver for any continuing sourcing assuming it has not yet been depleted of its water soluble compounds.

The early portion of this chapter focuses on techniques that are capable of generating data that accurately represent this NAPL body either exclusively (NAPL only) and/or the combination of the NAPL body and the associated high dissolved or sorbed phase that is located in relatively close proximity to the NAPL body proper. Of these, only approaches that employ fluorescence spectroscopy can generate semi-quantitative and qualitative responses that are highly preferential to the NAPL body alone. Fluorescence spectroscopy is best able to capture the nuances of the distribution of NAPL body while simultaneously resisting influence from dissolved, sorbed, or vapor phases (there are major departures from this behavior which will be covered). For these reasons, we will explore fluorescence means in the far greater detail and explain how it generates data that accurately represents how the NAPL body has distributed itself and, in some cases, how it can indicate NAPL's chemistry has changed since its release.

Of these various fluorescence means, time-resolved LIF has the superior semi-quantitative, qualitative, false positive rejection, and monotonic behavior across a wide concentration range and wide variety of NAPL types. Therefore, we will focus on LIF but point the reader to other fluorescence-based means as appropriate.

8.2   High-Resolution Petroleum Hydrocarbon NAPL Screening

8.2.1   Capabilities Necessary to Delineate NAPL

Generating a CSM representative of petroleum NAPL requires tools responding to chemical constituents that are representative of only the NAPL or, at a minimum, tools that indicate NAPL's presence is nearby (meters), for instance high dissolved phase BTEX. The techniques should also be relatively immune to significant "false positive" responses generated by the soil particles because these materials an be misinterpreted as NAPL (by fluorescence systems). If the system used is not sufficiently immune to responding to false positives, then the system's response should at least allow for a false positive's recognition as such, in order to allow for their subsequent removal from the NAPL CSM.

Some primary capabilities desired to delineate petroleum NAPLs include:

- Near-continuous collection of measurements.
- Sufficient sensitivity to contaminants/phase of interest.
- Response that is resistant to poisoning/blinding upon encountering NAPL.
- Rejection of false positives (or recognition of them in hopes one can filter their positive response out of the NAPL CSM).
- Ability to access the required depths.

Secondary capabilities that can prove helpful include:

- Monotonic response with NAPL saturation.
- Speciation (insight into changes in the chemistry and/or class of NAPL).
- Speed—the more productive the tool the less it costs, not only in terms of characterization itself, but improved efficacy of remediation as a result of higher data density or more expansive characterization.
- Real-time visualization of results, so adaptive field campaigns can be conducted.
- Minimum production of investigation-derived waste and associated costs.

8.2.2 Choosing the Appropriate Method

Once it has been established that a site would benefit from high-resolution NAPL screening and there is an understanding of what parameters should be measured, a method or ideally set of methods that respond properly to the contaminant (and phase of that contaminant) that is driving your investigation can be selected (ITRC 2019).

8.3 High-Density Coring and Sampling (HDCS)

Traditional core sampling is capable of producing data that can certainly be classified as "high-resolution" and can generate a fully accurate NAPL CSM. However, the equipment, the personnel, and the procedures used for traditional soil sampling did not generally evolve with a focus on NAPL screening. Practitioners who are accustomed to traditional sampling approaches, such as sampling every meter or two or sampling only in targeted intervals (e.g., at the potentiometric surface), have to radically change their mindset if their goal is NAPL delineation. Should they hope to achieve the data density needed to overcome the chaos introduced by NAPL distribution's heterogeneous distribution that can seem at times to be a mirage, partitioners have to modify their budgets, adjust their expected production rates downward, and generally change the very culture of how they approach sampling. Nevertheless, with careful preparation and mindset it can be done and done well (Byker 2021).

8.3.1 Advantages and Disadvantages of HDCS

Some HDCS advantages are:

• Getting to the necessary depths

Drilling or direct-push systems are almost universally able to get to depth in some manner or another be it auger, direct-push, sonic, or other method.

• Availability

Equipment capable of obtaining subsurface materials is almost universally available around the world.

• Multiple lines of evidence (MLOE)

Even NAPL CSMs generated with direct sensing tools benefit from limited targeted validation sampling. Simply getting one's hands and eyes on the affected soil can sometimes shed light on why the direct sensing tool has responded the way it has. Geologist's impressions, screening tool responses, NAPL-indicative dyes, and other methods all contribute to a greater appreciation for core-scale NAPL distribution that is impossible to achieve with one method alone.

An approach perfected over time by numerous skilled researchers and consultants who validated LIF logs to develop confidence in or further the value of LIF-based NAPL CSMs (Ernest Mott-Smith et al. 2014) consists of measuring and recording MLOE, along with photos of cores alongside a measuring ruler. The MLOE data is written with markers onto plastic sheeting or even laminated printouts laid under the core on the processing table (McDonald et al. 2018). This makes note-taking a straightforward and less error prone affair and allows a single photo to capture the entire "data dashboard". The MLOE data are all generated from narrow fixed intervals so as to spatially align the MLOE with the core, a technique developed to combat severe localized heterogeneity of chlorinated DNAPL during validation studies of dye-based LIF (Einarson et al. 2016). Data dashboard style photography reduces data transcription mistakes (such as typos, mistakes transcribing notebooks to spreadsheets, sample jar labels rendered illegible from shipping damage). Of course, if one is to rely solely on the photography for record-keeping, duplicate photos with two cameras (or perhaps recording the data into a field notebook at the end) are wise. Data dashboard methods can even eliminate the need for further "visualization" of the data in spreadsheets and graphs because it is already spatially organized versus depth in the photo.

Many of the figures depicted in subsequent sections will be using such a data dashboard style to demonstrate the method's utility for spotting telltale signs of

NAPL's differing behavior from that of other petroleum contaminant phases, and how method can also reveal data misinterpretation mirages for what they are.

- Ability to triage

Screening visually or with handheld photo-ionization detector (PID), dyes, and other rapid screening approaches can be used to identify appropriate intervals for laboratory sample grabs or when to adaptively switch over to high-resolution mode where slower NAPL-discerning techniques would be fruitful. Triaging intervals according to soil core data that is generated with real-time benchtop screening tools allows for adaptive decision-making (deciding on the next sampling locations for instance) because there is no waiting due to shipping samples and laboratory analysis time.

- Familiarity with stakeholders

People trust data generated with "what they know" and are inherently suspicious of techniques they are unfamiliar with. This blind allegiance can at times be irrational and counter-productive, yet it remains that core sampling is often "the easier sell", especially with stakeholders who are entrenched with doing things the way they have always been done.

Some HDSC disadvantages are:

- Lower productivity

The sampling machinery is often capable of producing cores at a relatively rapid rate, although commonly slower than most direct sensing technologies. Judicious processing of those cores is an additional rate limiting step that greatly reduces production. This is especially true when the core processing is done at a screening density that achieves resolution high enough to assure that small NAPL features (cm to dm scale) are not missed.

- Laboratory costs

Laboratory analysis costs can become prohibitively high even at modest (e.g., 0.3 m) resolution. Great savings can be realized, while still achieving a robust NAPL CSM, by conducting a carefully orchestrated MLOE approach. Selecting occasional grabs for laboratory work at lower density, and using those data to develop an understanding of how these sparser "gold standard" laboratory results relate to the higher density screening methods used, bolsters our confidence in the high-density screening method data. That said, many laboratory methods fail to contribute significantly to the NAPL-only CSM, because they either respond to only select fractions of petroleum hydrocarbons or they respond to all phases contained within the soil, not just the NAPL expressly. It became clear midway through our industry's development that the regulatory framework had been placing far too much emphasis on "data quality" (laboratory data being the gold standard) to the detriment of data density. This led to the development of the Triad methodology (triadcentral.clu-in.org), an approach to decision-making during characterization that sought to strike a balance between data quality and quantity, giving recognition to the fact that data density was required to overcome the uncertainty caused by spatial heterogeneity of contaminant and hydrogeological properties.

• Skilled labor demands

Judicious processing of cores requires experience and a discipline akin to a military exercise. Assembling such a skilled team costs more than a group of inexperienced personnel. When planning for core screening projects, it is easy to delude ourselves with mental images of the sun shining, birds chirping, good lighting, mild temperatures, and plenty of rest. Reality is often far from this ideal with long hot or frigid days, perhaps some freezing conditions thrown in, making people grow short in patience and encouraging a desire to start taking shortcuts not long into the project. A proper work plan, a sound leader to guide the process, and a crew willing to "stick to the script" is vital. Ignoring the geology (or being lazy toward it) could result in missing a major clue as to how or why the NAPL body has distributed itself.

• Poor recovery and soil disturbance

Years spent validating LIF logs with physical coring has taught us that NAPL has a way of transporting and storing itself in variety of soil types, some of which are very difficult to sample effectively including large gravels, running sands, and the like. It is not uncommon to get full recovery cores all the way down to the interval where the LIF had indicated NAPL, only to have the very core where the NAPL was indicated come back up-hole with only partial or no recovery at all. Direct-push sensing on the other hand typically results in "100% recovery" top to bottom, regardless of the soil's sampling behavior (outside of refusal). There are methods less prone to recovery issues like sonic drilling but this is often highly disruptive to the contaminant, the soil, or both. Cryogenic coring techniques represent a promising alternative requiring further research and development.

• Investigation-derived waste (IDW)

Conducting HDSC alone to develop a NAPL CSM generates a large amount of IDW compared to direct sensing with limited validation sampling. Depending on the contaminant and regulations, the management and disposal of the large soil volumes necessary to generate a high-resolution model of the NAPL CSM can be expensive enough to be a factor and certainly not the most sustainable solution practitioners should aim for.

• Data handling

Copious amount of data can be generated from HDSC and recording it all is rather daunting and fraught with chance for errors. In addition, site data often gets strung out over many months or years, with change-over in consulting firms or even agencies in charge. In many cases, just a few sloppy errors (or ambiguous observations such as a lone nebulous term such as "impacted" written on a drilling log) can throw doubt onto a data set that had the opportunity to be very insightful but lost value due to poor or inarticulate archiving.

8.3.2 HDSC - Best Practices

- Dedicated workspace

Large table(s) placed in a location where rain, direct sun, and other distracting weather elements are minimized is important. Skilled NAPL delineation practitioners often cover their tables with white disposable plastic and write detailed notes on the table recording the location, depths, geology, NAPL observations, PID response, NAPL sensitive dye tests, and other pertinent data (McDonald et al. 2018).

- Splitting cores

Core liners should be cut lengthwise, followed with splitting of the cores using spatulas, drywall taping knives or similar straight edged tools. Whatever technique is used to split the core longitudinally, it is desirable that the technique retains the interior soil structure. For cores containing loose granular soils, an excellent alternative is to freeze the cores, let the outer liner thaw for a few minutes to reduce brittleness, lay them in a core jig for safety, then cut them completely in half, using stone cutting masonry blades with a circular saw.

Immediately after cutting the frozen cores, one should shave the thin layer of frozen mud off the face of the core to reveal the details of the soil and NAPL distribution. If delineating gasoline or other NAPLs rich in volatile organic compounds (VOCs), it is crucial to keep the core covered in metal foil (never plastic), keep the core cold, and process the cores with as little delay as possible after thawing. Core photography while still in the frozen state works very well, but high humidity environments can cause frost to build up too quickly, obscuring the surface. Freezing cores is obviously not for high throughput, but for cores from important depth intervals, where discovering the details of distribution of NAPL versus the geology is needed.

- Core Photography

Quality photographs record core information in a way that is irreplaceable. If conditions allow it, one should set up a geologist's UV mineral lamp in a dark room or trailer and take both visible and UV-induced photos. As shown in Fig. 8.1 it is even more effective to mount the camera and core in fixed locations so one can see both the visible and UV from the same perspective. Figure 8.1 contains photos of a 1.2-m long gasoline NAPL-impacted soil core photographed under various conditions. The cyan blue is gasoline NAPL fluorescence, the purple hue is reflected UV lamp visible color bleed, and the creamy orange is fluorescence of an indicator dye (indicating the most freely available NAPL). The bottom of the core is at right and the arched patterns in the fine laminar soil layers were caused by higher sampler friction on the outer soils. The narrow black lenses in the visible photograph at bottom are a seam of granulated activated carbon amendment.

Fig. 8.1 A soil core that has been frozen and cut in half lengthwise. From top to bottom are UV with indicator dye (orange), UV, and visible lighting conditions

Keep in mind that kerosene (jet fuel), aviation gasoline, bunker fuels, and other NAPLs are difficult if not impossible to document properly with UV-excited photographs because they simply do not fluoresce appreciably in the visible wavelengths.

- Subsampling

Composite sampling, by design, generates an "average" response across the soil column. Subsequently, it fails to provide any insight into the nature of the NAPL's distribution and/or degree of heterogeneity. The narrower the depth interval sampled, the more accurately it represents the extremes that NAPL is typically capable of achieving across short distances. Subsampling with Encore samplers (or syringe bodies with the needle attachment cut off) combined with inexpensive screening tools such as dyes, glove tests, PID, and others, which generate data consistently and rapidly for little cost, is highly desirable to learn about how NAPL distributes itself in the soil. A small fraction of these numerous samples are often co-sampled at the same horizon as other tests and are sent in for formal laboratory testing.

- Indicator dyes

Shake tests consist of adding hydrophobic indicator dyes and water to soil samples in a jar that is shaken briefly and then examined. The dyes change color only when they dissolve into an organic liquid and they are particularly good for testing for NAPL presence. Unlike laboratory or PID readings, dye tests are much more decisive because they are reacting to a physical transformation of the hydrophobic dyes that only become colorful when NAPL has solvated them, regardless of sorbed or dissolved phase concentrations in the soil core. Oil Red O and Sudan IV are famously used in the chlorinated NAPL sector (Cohen et al. 1992), and companies sell various colored dye "kits" that make the test easy to implement properly with relatively little experience. Buying scintillation vials and bulk dyes and doing it yourself is much cheaper and achieves excellent results (Einarson et al. 2018). Most NAPLs respond well, but NAPLs that are very dark or black (such as coal tars and tank bottoms) will almost certainly not respond due to the absorbance of light by carbon and other chromophores that quench or physically filter out any nuanced color changes. One should also be careful of assessing a "light positive" response caused not by dye dissolving into trace NAPL but is instead simply grains that are the same color as the dye when it contacts NAPL. In cases where even trace NAPL is of interest, it is best to have a vial of control soil placed next to the jar to which indicator dye has been added, so flecks of soil coincidentally shaded in the same color as the dye change are more easily appreciated and assessed as soil, not NAPL positives.

- PID

Inexpensive handheld PID devices are useful for screening of VOCs that are either in NAPL form or are emanating from dissolved or sorbed phases typically located close to source NAPL. PIDs are often used in headspace mode, where a sample is placed in a container and the headspace is measured. For NAPL screening, the abundance of VOCs may allow a more direct continuous "sniff" along the cores surface, insertion into a series of shallow divots, or cupped with a clean gloved hand to help assure the VOCs are only emanating from the core section being screened. In this way, rapid progress can be made along the core until the PID alerts the screening team to go from their relatively "coarse" sampling (along with geologic examination) to high-resolution subsampling.

- Glove stain test

Testing for NAPL by looking for staining of nitrile gloves with NAPL is an inexpensive and reliable test for NAPL, because only NAPL can move directly and rapidly into the nitrile polymer and stain it, unlike the vapor and dissolved phase, soil, and/or water. Exposing the glove to NAPL-free soils, then washing the glove with soap and water and rinsing, will result in a stain-free surface. But exposing the glove even to sheen traces of NAPL will result in NAPL collecting into the glove's polymer (a process called solid phase extraction). A glove also allows one to work the glove into dark, fine, and opaque soils and sediments, where visual detection or dyes are notoriously ineffective, allowing one to search around for any NAPL droplets. When used in this manner the glove is sampling far more soil than surface-based techniques, allowing for more exhaustive detection of any NAPL present. A light-hued glove, in particular light green, is a popular choice. It might seem like an amateurishly simple technique, but for many NAPLs the glove test can uncannily discern NAPL when dyes, photos, or the human eye cannot.

- Shake tests

Placing soil in a container with water and then shaking the mixture causes NAPLs to free themselves from the soil and float on the water's surface. This is a reliable technique for most NAPLs, but stiff cohesive soils like clays are often difficult to disperse enough to free up much NAPL. Letting the water settle for some time greatly aids detection of NAPLs versus trying to detect them under turbid conditions. One should be aware that exceptionally clear NAPLs remain challenging at lower saturations, and that some NAPLs collect on the glass as opposed to forming a visual sheen or layer at the water's surface.

- Concise language and terminology

Nothing is more frustrating than having gone through exhausting and expensive high-resolution core screening process, only to find that the processing team's terminology and decision-making was inconsistent. Vague simple terms like "impacted", "odor", and "saturated" should be more specific, such as "NAPL-impacted (sheen)", "aromatic odor", and "groundwater saturated". Degree of impacts should be kept on a simple coarse scale of perhaps zero to three or four. Handheld instrument responses can then be normalized to that same scale—allowing for all the data to be hung vertical on the same axes, using differing colors or symbols for each line of evidence. This normalization makes it easy to see when and where the various lines of evidence agree (building confidence) or maybe even highly disagree, perhaps pointing at a NAPL that is present but has dramatically different chemical properties than the target NAPL originally being delineated. For instance, a chemically intact gasoline NAPL will cause a high response to PID, glove, and dye while a clear mineral oil might cause a poor PID and glove response, but a vivid dye response.

8.3.3 HDCS Logging in Practice

Figure 8.2 represents an idealized scenario of a boring location continuously cored and screened for gasoline NAPL. Figure 8.2 incorporates a number of elements that are routinely encountered during a typical NAPL-specific screening exercise. As previously stated, the relationship between multiple lines of evidence can be exceedingly difficult to conceptualize unless they are all hung together and graphed vertically. MLOE "data dashboard" figures like Fig. 8.2 help us make sense of different data types which sometimes contradict or support each other. Various depth intervals (listed alphabetically at far right) have been selected for discussion so as to help the reader identify commonly encountered situations that often lead to misinterpretation or underappreciation of the meaning of data produced with high-resolution techniques.

Key elements to consider in Fig. 8.2 include:

- The four columns at far left (Soil Profile, Gasoline NAPL, and VOCs) represent the true soil profile in this scenario. If the HDCS techniques used to generate MLOE data perform ideally, the MLOE data at right (PID, Dye, Lab, and UV Photo) should match what is pictured in the idealized "reality" in the four columns on the left.
- The idealized columns at far left will be held static throughout the discussion of HDCS, MIP, and LIF methods in subsequent sections, so that one can appreciate the differences between the methods when they are applied to identical soil and NAPL conditions.

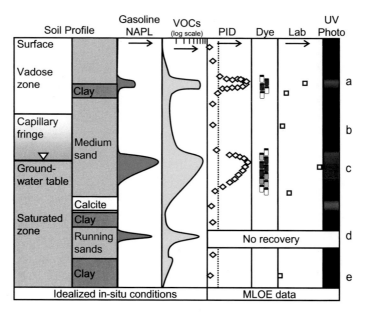

Fig. 8.2 Idealized scenario of NAPL and VOC distribution in the soil column with MLOE data resulting from HDCS methodology

- The VOCs column describes total VOCs (e.g., BTEX) versus depth and is plotted in log format. This is necessary because while VOCs dissolved in groundwater, soil pore gases, and sorbed phase near NAPL can reach significant levels, the concentration of these non-NAPL phases of VOCs is dwarfed by their relative abundance in NAPL proper. Notice in Fig. 8.2 how VOCs are encountered throughout the entire soil column shown, because VOC components are always more widely and homogeneously distributed than NAPL.
- The NAPL column contains various forms of NAPL—ranging from fairly saturated and obvious in a soil core to very light staining that might be difficult for a geologist to identify visually.
- The density of the sampling intervals shown here was arbitrarily chosen. We are not suggesting one typically sample 50 intervals at each location, but then again it does occur. Sampling density is a balance of budget, time, and goals.
- Notice the switch from coarse sampling intervals to narrow intervals (and NAPL-specific methods) took place only after the PID threshold was exceeded. Time and money were saved by screening at coarser intervals only with the PID.
- Intervals A and C were screened at high-resolution and included the more laborious NAPL-specific methods, the dye shake tests (Oil Red O in clear vials), and UV photos. Once the PID threshold dropped back below the action level, the coarser spacing was resumed. A glove staining test would probably not have provided an acceptable alternative to the dye test for fresh clear gasolines but would for many discolored or heavier NAPLs.

- Interval A contains perched NAPL sitting on clay according to the MLOE, but UV photography only indicated NAPL in the top half of the NAPL lens due to the clay's fine grains' ability to hide NAPL fluorescence relative to sand (Apitz et al. 1992) and/or the difficulty of NAPL to penetrate fine-textured soils.
- Both the dye testing and the UV photos indicated NAPL at intervals A and C. These positive NAPL responses were corroborated by substantially elevated concentrations in the validation laboratory samples grabbed at intervals that were indicated to contain NAPL.
- All four MLOE data streams in interval B agree that while VOCs were modestly present, no actual source term NAPL exists. The laboratory samples confirm that the screening tools are responding appropriately (i.e., not exhibiting falsely negatives for NAPL). This illustrates the utility of confirming both low and high responses, not just the highs.
- Looking closely at interval C's MLOE data versus NAPL saturation, it is clear that NAPL was encountered, but the depth at which it registered in the MLOE data is deeper than our model's "reality". Simple human error, compression of tooling, drilling issues, and a host of other factors often lead to improperly determining the depth of the core material and thus the depth to NAPL. These depth mismatches are a common thorn in the side of everyone who validates direct sensing with HDCS.
- Interval C contains the classic "shark's fin" NAPL saturation that appears in homogeneous sandy soils with a stable phreatic surface (Tomlinson et al. 2014). Partial NAPL saturations often result in "pink" red dye tests rather than bright red. A laboratory sample grab at the same interval as the slightly positive pink dye shake test is implemented confirms the modest level of NAPL impact indicated by the light pink (less than obvious) dye response.
- Interval D contains a significant lens of NAPL trapped below clay in running sand, but this was not discovered by HDCS due to lack of recovery of the running sands. This is a common occurrence because NAPLs often reside in difficult-to-sample soils.
- Interval E was submitted for laboratory analysis in order to generate data on the low end of the petroleum contamination spectrum, without which the behavior of the MLOE screening tools cannot be fully validated with respect to false negatives. False negatives are notoriously easy to ignore because investigators may think that certain depth intervals are NAPL-free, but until and unless it has been demonstrated it is premature to assume so.
- Notice that the fluorescence photo colors change with depth, not only between the false positive calcite and NAPL, but within various horizons of the gasoline NAPL itself. These color changes are the result of significant chemistry changes. It is common for perched gasoline near the soil surface to weather significantly, and that is illustrated in this model. Notice in the UV Photo column that the NAPL deposit at interval A is turquoise (weathered) in contrast to the bluer (inferring a more intact) gasoline NAPL at interval C. The top side of interval C also weathered to a turquoise color similar to interval A. How NAPL fluorescence colors vary and can change, and the significant role that colors can play in identifying NAPLs,

discerning false positives, and identifying weathering will be detailed in the LIF discussion sect. 8.6.3.1.

8.4   Direct Sensing of Petroleum NAPL

Early pioneers of geotechnical soil assessments developed simple probes that were pushed steadily down into the soil, without rotation, while measuring the probe tip's resistance. This cone penetration test (CPT) method was later enhanced with the addition of a friction sleeve and eventually pore water pressure sensors. CPT systems have gone on to revolutionize geotechnical characterization and the speed at which geotechnical data can be gathered. Researchers at the United States Army Corp of Engineers Waterways Experimental Station (WES) added a sapphire window to the side of their probe in the early 1990s to enable detection of petroleum hydrocarbon NAPLs by their fluorescence. Percussion delivered direct-push sampling platforms were developed and gained popularity in the late 80s and early 90s. Researchers and probing companies subsequently developed and manufactured sensor systems that allowed percussion delivered direct sensing of contaminants in situ. Like the geotechnical revolution prior, the ability to measure petroleum chemistry continuously with depth greatly advanced the field of subsurface petroleum hydrocarbon characterization.

There is a myriad of tools and sensors available for characterizing aqueous phase (groundwater) hydrocarbons, but this chapter focuses on two mainstream direct sensing technologies that are capable of characterizing the source term NAPL and/ or the high dissolved phase that indicates source term NAPL is close by. The next sections will discuss the Membrane Interface Probe (MIP) that measures the VOC fraction of NAPLs exclusively, which is pertinent to gasolines whose formulation is complex and forever changing with regulations and technology (Chin and Batterman 2012), and will more closely examine LIF which is essentially blind to the VOCs and responds instead to the PAHs (semi-VOCs and non-VOCs) that are mostly contained within the source term NAPL itself. These two major classes of petroleum logging tools both directly sense petroleum contamination but come at the problem from very different chemical perspectives.

8.5   Membrane Interface Probe (MIP)

The MIP is a VOC-sensitive direct-push delivered tool manufactured by Geoprobe Systems® (Salina, KS, USA). It screens for VOCs continuously and rapidly, much like a handheld VOC sensor, but with the major advantage of being combined with a durable steel tool that can be hammered into the ground where VOCs are measured

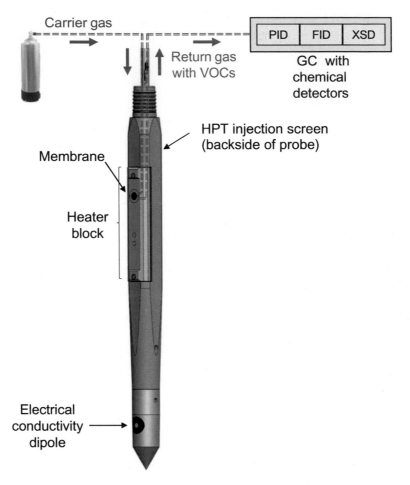

Fig. 8.3 MIP transports VOCs to uphole detectors via tubing with the aid of carrier gas flow (after figure courtesy of Geoprobe Systems)

in situ (Christy 1996; ITRC 2019). As illustrated in Fig. 8.3, carrier gas is introduced into small tubing (preferably polyether ether-ketone or PEEK) that delivers the gas down to a heated gas transfer port built into the side of the direct-push delivered probe. A durable metal-supported hydrophobic polymer matrix shields the carrier gas from intrusion by water, NAPL, and solid particles while allowing VOCs and semi-VOCs to pass though into the carrier gas stream, where they are subsequently carried back uphole and into a series of detectors usually housed in a gas chromatograph instrument. A heater block consisting of a resistive heater coil and a thermocouple holds the membrane's support fixture at the set temperature (normally 120 °C) which warms the membrane and surrounding formation. The high temperature hastens the diffusion of analytes across the membrane and into the port and also serves to drive off

VOCs or semi-VOCs that may cause a residual response when the probe is advanced downward into less impacted soils.

The VOC transfer rate across the membrane generally trends with VOC content of the formation, but factors such as grain size may influence the pressures outside the probe and the transfer rate of VOCs across the membrane (Costanza et al. 2002). As such, low-permeability materials such as clays are thought to enhance the VOC transfer across the membrane, while high-permeability materials decrease it (Costanza et al. 2002). The probe is typically advanced at 30-cm (1-ft) intervals. At the end of each interval, advancement is paused (generally for 45 s) while the MIP's heater block and membrane reach the desired temperature and heat the surrounding formation, after which the probe is again advanced to the next depth. VOCs are capable of crossing into the probe at all times, but maximum transfer typically occurs as the temperature of the membrane and adjoining formation approaches the heater block target temperature.

As mentioned, the sensing of the VOCs that cross into the probe does not actually take place inside the probe—although successful implementation of downhole halogen-specific detector (XSD) has accomplished down-hole detection of contaminants in membrane interface gas flow (Lieberman 2007). Typically the VOCs are transported uphole by an inert carrier gas (normally nitrogen) via tubing that is incorporated into a trunkline of carrier tubing and wires that have been pre-strung through the direct-push rods. Transport to the surface allows for the vapors to be introduced into a climate-controlled housing, where the analyte gas stream is dried of interfering water vapor (generally using Nafion™ tubing and sometimes a water trap) and passed through multiple detectors. The climate-controlled housing assures that the sensors and electronics are stable and able to respond properly under field conditions where ambient temperatures range widely.

A trio of sensors is typical and includes a PID, a flame ionization detector (FID), and an XSD, although alternative configurations have been used with the objective of improving the qualitative and quantitative interpretation of MIP results [e.g., (Bumberger et al. 2016)]. Proper interpretation of the various detector responses enables stakeholders to identify the main classes of VOCs the MIP is encountering at various depths, and gain insight into the relative concentration of VOCs in the formation. Data generated by multiple detectors can also function to identify false positives such as methane which is produced during biodegradation of hydrocarbons and natural organics such as buried wood and vegetation.

Before and after each MIP log, the MIP practitioner will run chemical response tests using compounds of concern at a known concentration to evaluate response consistency of the primary detectors and to calculate the gas trip time by measuring the time required for the detectors to respond after the membrane was exposed to the chemical standard.

Transporting the VOCs back to the surface for detection by sensitive and reliable detectors is advantageous but employing a permeable membrane and tubing for this purpose also creates challenges. Petroleum hydrocarbon NAPLs are an extremely complex assembly of molecules, with each individual compound having a unique volatility, as well as some degree of affinity for the membrane and/or the transport tubing's interior surface, depending on the molecule's size and structure. Consequently, many gases spend considerable time adsorbing to and then desorbing from, the membrane and interior surfaces of carrier tubing during upward transport. Generally speaking, the higher the molecular weight and boiling point of the molecule, the longer the tubing travel (retention) time, resulting in a behavior similar to that observed in chromatographic columns. The most volatile compounds experience a rapid and relatively unencumbered transport to the surface, while sequentially heavier VOCs and semi-VOCs arrive later and may remain stuck in the colder carrier tubing for hours. Some may never clear (cold trapping), in particular where the MIP was delivered through heavier (diesel or crude oil) NAPLs. This may even result in carrier lines needing to be discarded and replaced.

Water vapor also passes through the heated membrane and travels to the colder carrier tubing, where it can condense to form a film of water that can act as an additional stationary phase, exaggerating the chromatographic behavior. Low-permeability formations may enhance this problem by generating higher membrane exterior gas pressures (Adamson et al. 2014). Decreasing the heater block temperature down to 95–100 °C has been proposed to reduce water ingress into the probe and trunkline, although this adjustment may simultaneously contribute to the risk of blocking the carrier tubing by lowering the gas temperature. A heated trunkline version of MIP was introduced in 2009 to alleviate some of these issues, as well as to increase the contaminant transport rate through the trunkline, by maintaining an elevated temperature in the tubing. Heated carrier lines are made of stainless steel instead of PEEK, which makes them harder to manipulate in the field. One situation where the heated trunkline MIP appears to offer a great advantage is operating in sub-freezing weather, because it prevents icing of water vapor condensate in the carrier tubing that can block gas flow.

A common consequence of trunkline carryover is that MIP detector responses often remain elevated for some time after the probe has passed through significant contamination (NAPL) and advances into intervals with lower contaminant concentrations. This is because heavy loads of NAPL-sourced analyte gases are still adsorbed to the walls of the tubing and will continue to bleed out over time. If the probe is still being advanced during this carryover response, it appears to the observer that "dragdown" of the contaminant is occurring. The heavier the molecular weight of the contaminant encountered, and/or the greater the contaminant concentration encountered, the more severe the trunkline carryover issue. Gasolines vary in their formulation chemistry as well, with some gasoline chemistries exhibiting little carryover, while others can cause noticeable carryover if significant concentrations were encountered by the probe.

There are operational tricks that experienced MIP operators have to reduce the complications caused by carryover. If the carryover is significant, MIP operators often pause the advancement of the probe to allow the carrier line to clear so as to better define the contaminant distribution below highly contaminated intervals. During this pause, the temperature of the probe and the carrier gas flow may be increased to flush the residual contamination from the system (Adamson et al. 2014). If the petroleum loading of the carrier tubing is severe, the wait can take hours. MIP providers have also learned to approach the suspected location of LNAPL bodies (gasoline for example) from outside the suspected LNAPL limits, moving carefully in toward the potential source LNAPL. They can safely approach from the outside because the predominantly smaller and lighter molecules that are water soluble (the symptom of the LNAPL source) naturally also have a lower affinity for the carrier tubing's interior wall. Another option MIP operators have is to increase the setpoint temperature on the MIP probe from 120 to 140 °C, which will reduce the number of heat cycle events required to desorb analytes off of the MIP membrane and move them into the carry gas stream. This may improve the delineation of the contaminant profile especially when encountering LNAPL, although it may simultaneously increase the risk of water ingress and condensation. MIP trunklines might also include two carrier return lines, allowing the operator to switch to an uncontaminated one while the other clears.

Increasing the push rate and/or the carrier gas flow in highly contaminated intervals are other ways of reducing the diffusion rate of LNAPL constituents through the membrane and try to prevent saturation of the gas lines and chemical detectors. Using relatively short trunklines may also contribute to decrease the risk of cold trapping. MIP practitioners generally try in earnest to avoid getting into significant saturations of LNAPLs and generally attempt to avoid any encounters with medium weight LNAPLs like diesel. Encounters with even heavy NAPLs such as crudes will typically result in long-term or even permanent trunkline contamination, so use of MIP at sites suspected of involving NAPLs heavier than gasoline is often not advised.

The ensuing carryover prevents the rapid drop in response that should be observed as the probe moves out of the LNAPL down into non-LNAPL containing soils. Thus, the MIP often delineates the top of the LNAPL body adequately but is quickly blinded by the high response, rendering it unable to determine the bottom of the LNAPL impacts. This has led researchers to investigate the carryover phenomenon carefully and even investigate if perhaps the top and bottom of highly contaminated intervals might best be delineated using multidirectional MIP screening (Bumberger et al. 2012) and even methods that utilize a secondary carrier tube (Bumberger et al. 2016) that can be switched over to prevent carryover effects and which could also enable improved logging of low contaminant concentrations as achieved with low-level MIP systems.

8.5.1 MIP Logging in Practice

Figure 8.4 features the same idealized scenario used in Fig. 8.2 but with MIP direct sensing applied rather than HDCS to compare the MIP's response versus the true contaminant distribution defined by our idealized in situ condition model.

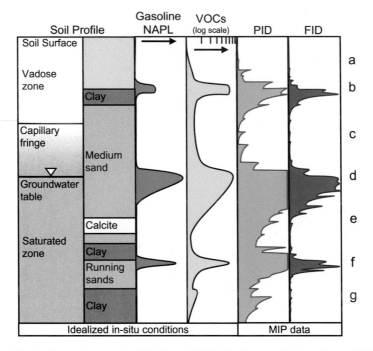

Fig. 8.4 Idealized scenario of vertical NAPL and VOC distribution with hypothetical PID and FID detector data from MIP direct sensing

Key elements of Fig. 8.4 include:

- The VOCs column represents BTEX and short-chain aliphatics, in other words, the volatile fraction of gasoline that MIP detects well. The idealized VOC profile includes VOCs in all phases, including the NAPL, dissolved (aqueous), volatilized (gas), and sorbed phases.
- While the MIP has a variety of detectors, this scenario shows only the PID and FID. In cases involving only petroleum (e.g., gasoline) without halogenated additives, the XSD would not give response, so it has been omitted here.
- The FID is superior to the PID when attempting to define the NAPL because it is less sensitive at lower concentrations but has a wider dynamic range and top end response. An FID response exceeding 1×10^7 μV is considered by many to be indicative of NAPL. The PID is less NAPL-predictive because it typically experiences greater carryover effects than the FID, thus limiting its ability to delineate NAPL-impacted intervals. The PID is also more sensitive and it may enter detector saturation territory well before true NAPL is encountered, although the detector and computer gain settings can be adjusted before the detectors reach a level that exceeds their maximum output signal.
- The (admittedly cartoonish) "spikes" rising out of the MIP response at semi-regular intervals are due to momentary increases of VOC flow crossing the

membrane during the ~ 45 s pause every 0.3 m. With the probe's downward motion paused, the membrane and the formation in the probe vicinity warm rapidly, creating optimal VOC transfer temperatures and causing the VOC transfer spike. This is followed by the rapid depletion of the VOCs at the membrane's immediate exterior.

- Interval A shows the MIP's positive and growing response to VOCs associated with NAPL's nearby presence.
- In interval B, the MIP passes into a modestly saturated lens of gasoline NAPL perched on a clay lens, evidenced by the sudden increase in MIP response. The PID saturates quickly while the FID increase better reflects the extent of the NAPL. In addition, the transition into clay resulted in more backpressure and the associated increase in the response versus the medium sands. The FID still has "room to run" so its response increases, while the PID rise is dampened by its maximum output signal using the default detector (high gain) settings.
- In interval C, the VOCs actually drop off quickly, but the PID response is affected by carryover in the tubing. The carryover shown in the FID is less pronounced because of its more NAPL-selective response.
- Interval D represents a scenario that MIP practitioners commonly try to avoid, contacting enough NAPL in the pore space to source a sustained high flow of VOCs. The PID response is saturated and the relatively massive increase in VOCs associated with high NAPL content is shrouded in this saturation state where the response simply cannot go any higher. The FID response better depicts the NAPL-impacted interval.
- The FID's high response and lengthy and sustained PID carryover in interval E is indication enough that substantial impacts (NAPL) were encountered. In many cases, the MIP logging might be paused at interval E to allow the trunkline to clear. In this case, probe advancement is continued for demonstrative purposes. Notice also that the VOC profile is tailing away slowly under the NAPL in interval D contributing to the high PID responses. The FID, again, shows better fidelity to NAPL.
- In interval F, the clay lens and subsequent transition into NAPL in running sands causes another increase in response. However, the NAPL's presence (especially for the PID) remains shrouded in the saturation and carryover from interval D, so recognition of this NAPL is uncertain.
- In interval G, higher adsorption capacity clay causes heightened MIP response, due to the increase in VOCs available to desorb into the membrane. As discussed, there is also evidence that MIP response varies with sediment type, not just concentration (Costanza et al. 2002; Adamson et al. 2014; Christy et al. 2015) and that less-permeable materials (e.g., clay) may generate increased partial pressures outside the membrane versus sands, which improve VOC transfer across the membrane. Accordingly, no matter what type of sensors are being evaluated it is difficult to design and conduct laboratory and field experiments that are capable of parsing the numerous parameters that contribute to MIP and other in situ sensor responses occurring deep underground.

In summary, MIP remains the premier tool for in-situ delineation of total petroleum VOC contamination and it is useful for certain volatile petroleum NAPL delineation tasks, but one must keep in mind how some factors, including the presence of NAPL and other highly contaminated intervals, may impact the MIP results. Figure 8.5 illustrates this point well and contains MIP data that was gathered at a petroleum (gasoline) filling station in 2018. Notice that the MIP's PID and FID responded to VOCs almost continuously with depth, while the XSD stayed low, assuring investigators that halogenated compounds were not encountered. The contribution from all VOC phases, combined with modest trunkline carryover (red lines), caused responses that looked like physical contamination "dragdown" outside the probe. Despite their diminutive size, the tiny NAPL deposits, in Fig. 8.5, sourced enough VOCs in their various forms to the surrounding soils to assure a robust MIP response across the entire span of VOC-affected soils, while they were only modestly above the minimum detection threshold for the LIF system as shown in the LIF log at right. Subsequent coring adjacent to these MIP and LIF locations (< 1 m) confirmed the presence of randomly distributed discrete NAPL weeps from pinhole-sized features in the silty soil. These weeps were photographed after they produced a sheen across the core's face in the UV-induced photographed at right. Further information on LIF is provided in Sect. 8.6.

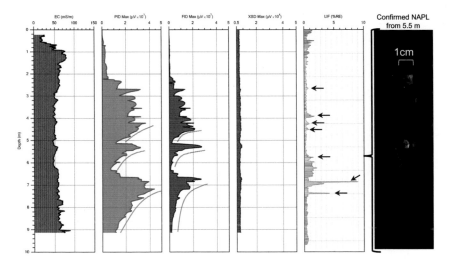

Fig. 8.5 Co-located MIP and LIF logs acquired at a gasoline release site

8.6   Laser-Induced Fluorescence (LIF)

LIF is an analytical technique applied across a host of disciplines, including drug discovery and genomics, not just contaminated site characterization. The technique consists of directing laser light at fluorescent molecules which absorb some of the light, which excites them, ultimately inducing them to emit light that indicates the molecule's presence. Most petroleum-based NAPLs such as fuels, oils, coal tars, and creosotes contain enough naturally fluorescent molecules such as PAHs (Berlman 1965) to allow for the NAPL's detection using LIF. LIF responds almost exclusively to fluorescent molecules that reside primarily in the NAPL and thus in-situ LIF measurements have not been found to be significantly responsive to gaseous or aqueous phase contaminants in the pore spaces.

As depicted in Fig. 8.6, LIF is typically deployed with direct-push tooling. As the tool is steadily pushed into the unconsolidated formation at 2 cm/s, brief (~1–2 ns) pulses of laser light are carried by a fiber optic down to sapphire window in the side of a steel probe (see Fig. 8.7), where it is directed out the sapphire window in the side of the probe, causing fluorescence to be emitted wherever NAPL-affected soil (or any other fluorescent material) is encountered. Laser light is directed at the formation soil that is pressing hard up against the outer surface of the sapphire window and while there is some slight penetration into the pore spaces, penetration is limited to approximately 1 cm in depth at most in favorable conditions (e.g., transparent coarse sand grains). The intensity of the fluorescence induced in the formation generally scales with NAPL impact, but is affected by other factors described in the following paragraphs. A portion of the fluorescence couples back into the probe through the window and is carried back up to the surface by a second fiber optic cable, where the returning fluorescence is dispersed according to color and detected versus time (temporally) with a photomultiplier tube. In this fashion, direct-push LIF provides rapid and cost-effective delineation of most petroleum NAPLs.

Production rates often exceed more than 10 locations per day (depending on probing conditions), yielding cost-effective high-density fluorescence data that can then be displayed as singular logs, series of logs taken across transects (fence diagrams), or 3D visualizations to give a quick idea of the distribution of the NAPL sitewide. LIF is often coupled with other downhole tools on the same tool string (e.g., electrical conductivity or a continuous direct-push injection logging module; see Chap. 7 for more information on these technologies) allowing the investigator to gain a better understanding of how the site geology is controlling the NAPL distribution. Cobbles, gravels, caliche, and other difficult to penetrate materials often limit the attainable depth and sometimes make direct-push logging impractical.

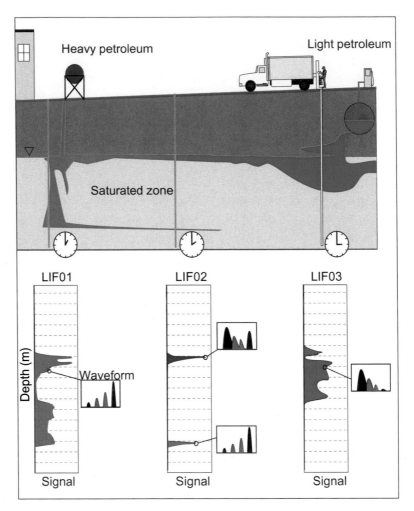

Fig. 8.6 A conceptual illustration of conducting LIF characterization of NAPL at a site where two differing NAPL bodies have co-mingled

8.6.1 History

Early development of continuous wave (non-lifetime) LIF technology began at the United States Army Corp of Engineers' Waterways Experimental Station (WES) in the early 1990s. It is with the Naval Research, Development, Test and Evaluation Division, they initially developed a system that relied on a downhole mercury lamp to excite the PAH compounds in situ and was delivered into the subsurface with cone penetrometer direct-push technology. The mercury lamp was later replaced by light from a 337 nm wavelength laser delivered via a fiber optic to the subsurface. The fluorescence was transmitted back to the surface with another

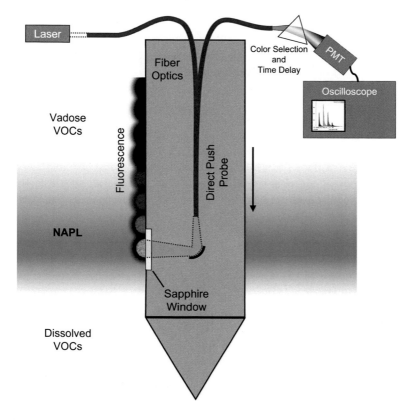

Fig. 8.7 Laser-induced fluorescence probe concept. PMT stands for photomultiplier tube

fiber optic to an optical multi-channel analyzer (OMA) which measured the spectral and intensity properties of the returning fluorescence (Lieberman et al. 1991). This system was one component of the Site Characterization and Analysis Penetrometer System (SCAPS) (Knowles and Lieberman 1995) and it was well suited to detecting medium-to heavy-weight fuels like diesel and crude, but performed less satisfactorily on gasoline and was nearly blind to jet fuel (kerosene), which was a major contaminant of concern of the U.S. Air Force (USAF).

The USAF coincidentally had been working for several years with North Dakota State University to develop a system that employed a pulsed tunable dye laser capable of emitting the shorter wavelength energies (260–290 nm wavelength) required to excite and detect the fluorescence of BTEX and naphthalenes in groundwater using fiber optic probes (Meidinger et al. 1993). At the request of the USAF, the system was modified to allow its use to include measuring NAPL on soils through the sapphire window of a CPT system. Research field trials showed promise and the system was commercialized as the Rapid Optical Screening Tool (ROST™) by Dakota Technologies, Inc. (Fargo, ND, USA) in 1994 (Nielsen et al. 1995). A subsequent generation system that utilizes a fixed wavelength xenon-chloride excimer laser (308nm) for excitation was developed by Dakota Technologies, Inc. in 2005 and is currently marketed and sold under the brand name Ultra-Violet Optical Screening Tool (UVOST®). In 2004, improvements on the core technology were made to optimize LIF's performance for coal tars, creosotes, and other heavy NAPLs, leading to a new generation of optical screening tool (OST) that was branded the Tar-specific Green Optical Screening Tool (TarGOST®) (St. Germain et al. 2006; Okin et al. 2006). In 2010, Dakota teamed with consultants to develop and test a concept that had been conceived of by U.S. Navy scientists but never put into practice. A fluorescent NAPL-indicator dye injection system was added to LIF in order to induce non-fluorescent NAPLs (including chlorinated solvent DNAPLs and monoaromatic NAPLs such as benzene and toluene) to fluoresce, and this model was branded as the Dye-enhanced Laser-Induced Fluorescence (DyeLIF™) system and has been offered to the industry since 2014 (St. Germain et al. 2014).

While numerous researchers over the years have proposed optimal laser and detection wavelength systems for in situ detection of various classes of NAPLs in situ in the last 30 years, Dakota Technologies, Inc. remains the only company to have commercialized time-resolved LIF systems through both sales of systems and characterization services. Unfortunately, optimizing a fluorescence system to detect the unique chemistry of any specific NAPL type is often accompanied by a significant sacrifice in that system's ability to detect other types of NAPL because each NAPL type requires unique excitation and fluorescence emission combinations for optimum performance (Kram et al. 2004; Kram and Keller 2004; Kram and Keller 2004). While a universally responsive system is technically feasible, the prohibitive cost and complexity of such a fully capable machine is currently beyond the industry's ability to financially support its development. Over the decades, government and industry funded research, in combination with field implementation at thousands of sites, has eventually led to the creation of a "family" of LIF systems that strike a balance between optimized detection of some particularly important classes of NAPL without overly sacrificing their ability to discern other classes of NAPLs and/or false positives.

8.6.2 LIF Family of Optical Screening Tools

There are currently three LIF systems and each strikes a balance between specialization and broad applicability. Figure 8.8 illustrates how the three types of LIF fit in with the wide variety of NAPLs commonly released into the subsurface and in need of delineation for risk assessment and/or remedy design.

1. UVOST: This time-resolved LIF system has excitation and emission capabilities optimized for use on light to medium weight petroleum fuels and oils. Specifically, the UVOST employs a 308-nm XeCl laser, which produces 1–2 ns duration pulses of light to excite PAHs (and false positive non-NAPL fluorophores should they exist). UVOST detects ultraviolet to blue color fluorescence versus time,

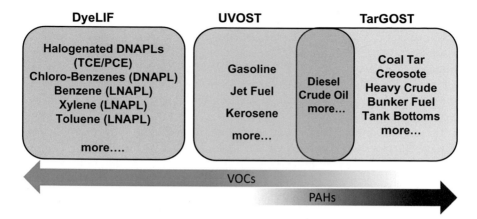

Fig. 8.8 Laser-induced fluorescence systems and their optimal NAPL detection performance

resulting in data called waveforms. Under optimal conditions, UVOST responds monotonically to LNAPLs that are commonly encountered at gasoline dispensing stations, military fuel depots, refineries, pipelines, and bulk handling facilities. Heavy molecular weight NAPLs ("heavies") such as coal tar or bunker at far right in the figure will likely be vastly under-reported (very weak response) with UVOST, but it is not likely for heavies to generate true false negatives (no signal whatsoever).

2. TarGOST: This LIF system uses a pulsed (1–2 ns duration) 532-nm wavelength laser to excite the very large PAHs in heavy NAPLs that all UV-induced fluorescence tools struggle to detect. Heavy NAPLs are often multi-component products with densities equal to or higher than water, and this NAPL group includes coal tars, creosote, tank bottoms, bunker fuel, and similarly dark highly recalcitrant NAPLs.

3. DyeLIF: This is a time-resolved fluorescence system optimized for detection of chlorinated solvent DNAPLs and other solvent NAPLs that are not naturally capable of fluorescing due to their lack of PAH content. DyeLIF is spectroscopically similar to TarGOST but is aided by injection of a fluorescent indicator dye that stains colorless NAPLs, causing them to fluoresce. The dye functions as an insurance policy that promotes fluorescence to occur in cases where non-fluorescent NAPLs do not fluoresce enough to allow for confident detection.

All three LIF systems shown in Fig. 8.8 are relatively blind to certain NAPLs (false negatives). UVOST, for example, which is designed to characterize light to medium weight fuels, is nearly unresponsive to many coal tars which are high-risk, multi-component DNAPLs. Vice versa for TarGOST, which is designed for coal tar delineation but is essentially blind to gasolines or kerosene. It is strongly encouraged that potential users of LIF data contact the developers of various LIF systems for guidance in the proper tool selection and benchtop LIF analysis of samples of target NAPLs prior to any fieldwork.

UVOST (and its predecessor ROST) were the first time-resolved LIF systems to be commercialized, and they are most strongly associated with the term "LIF". Since the bulk of petroleum NAPL releases can be addressed with UVOST, we'll primarily use UVOST data to discuss temporal-based fluorescence detection and analysis capabilities but they are shared across all three LIF systems. Later, the TarGOST and DyeLIF sections will focus on the unique attributes that set these systems apart from UVOST.

In addition to the family of LIF probes, it bears mention that Geoprobe Systems® recently commercialized the Optical Image Profiler (OIP). The OIP design is an improvement upon the GeoVIS, a U.S. Navy invention (U.S. Patent 6,115,061) co-developed with the Strategic Environmental Research and Development Program (SERDP). The GeoVIS consisted of a video microscope sapphire-windowed direct-push probe combined with a high-resolution piezocone sensor (U.S. Patents 6,208,940 and 6,236,941). It was therefore capable of yielding real-time sediment and contaminant images as well as some information on hydraulic head, hydraulic conductivity, sediment type, and effective porosity data as a function of depth (Kram 2008).

The OIP improved on the GeoVIS by employing a 275 nm wavelength light-emitting diode (LED) as an excitation light source, enabling it to serve as a logging tool that senses NAPL via UV-induced fluorescence. The UV and white light LEDs are combined with a color camera for detection and are housed inside a sapphire-windowed probe. Fluorescence images (or visible white light if desired) of the subsurface are acquired with every 15 mm increase in depth (McCall et al. 2018). Fluorescence is logged as a percent area of the sediment images (photographs) deemed by a digital filter algorithm to represent NAPL. The OIP is capable of effective screening for many common types of petroleum NAPL (e.g., diesel and crude oil). The probe advancement can be paused at any depth to collect in situ fluorescence and visible still photographs without the blurring effects observed during normal logging. There is also an OIP-G tool essentially using the OIP camera-based system but with a green laser diode for excitation combined with a visible filter for detecting visible fluorescence of heavy NAPLs.

The OIP's color camera's still images register some differences in fluorescence color (see Sect. 8.6.3.1) between certain NAPL types and there is an ongoing effort to process the color information for improved in situ determination of NAPL type. However, OIP and OIP-G's ability to determine NAPL types (and/or degree of weathering) or to recognize false positive fluorescence arising from natural organics (calcite for example) is currently limited, placing additional emphasis on the need for physical sampling and knowledge about the site's NAPL release history. While it has been reported that calcite false positives for OIP are minimal (ITRC 2019), this may be due to the image pixel filtering algorithm, which is designed to prevent any low-intensity, false positive fluorescence from meeting the requirements for being representative of NAPL. The consequence of setting this filtering algorithm's intensity and/or color threshold at levels necessary to remove most false positives is that the algorithm may also reject weak (but often very important) NAPL fluorescence, as is frequently observed in the case of detecting gasoline NAPL's weak fluorescence in fine-textured sediments for example. Another significant current limitation of the OIP system is the lack of ultraviolet sensitivity inherent in cameras, which reduces the response of OIP to kerosene and jet fuels or select gasoline formulations, whose fluorescence is composed almost entirely of ultraviolet light (Fig. 8.15). Future improvements in the OIP system may help to overcome this deficiency.

LIF and OIP logs can therefore differ significantly at many sites, but generally speaking many of the data interpretation concepts, best practices, validation techniques, and other elements that will be covered for LIF data will apply to OIP data as well.

8.6.3 NAPL Fluorescence

PAHs are the highly fluorescent compounds found in petrogenic (related to rock sourced) and pyrogenic (related to combustion) liquids. PAHs are characterized by multiple benzene ring structures, with the simplest of these being naphthalene and its two fused benzene rings. PAHs in purified form are typically hydrophobic crystalline solids, but they are highly soluble in organic solvents, including the complex mixtures of molecules that make up the bulk of most petroleum NAPLs. PAHs are a highly diverse family with hundreds of possible molecular weights and varying degree of substitutions (naphthalene versus its numerous methyl-naphthalenes forms for example).

Crude oils are a complex mixture of chemicals that includes PAHs. Crude oils are refined to select chemical ranges that give fuels and oils their unique properties, and this refining process also affects the PAH size distribution contained in fuels and oils. Kerosene (jet fuels) contain almost exclusively the smallest two-ring PAHs (naphthalenes), gasoline contains small to medium-sized PAHs, diesel fuels contain a modest percentage of many sizes of PAHs, while heavy molecular weight NAPLs such as bunker fuel contain relatively high PAH content, again of all shapes and sizes. These differences in PAH content result in stark differences in how they fluoresce because PAHs all have their own unique way of fluorescing depending on their structure and surroundings.

Fluorescence is an inherently sensitive technique for detecting PAHs, with sub microgram per liter (sub-ppb) detection limits for most PAHs when they are dissolved in a non-aqueous solvent such as methanol. While rigorous analyses of gasoline's PAH content are rarely undertaken, it has been reported that benzo(a)pyrene concentrations are in the single parts per million (ppm) range in multiple gasolines (Zoccolillo et al. 2000). Because gasolines contain well over 24 two- to six-ring PAHs (C_{10}–C_{22}) their combined presence, while it likely does not exceed even 1% mass, is more than sufficient to render gasoline modestly fluorescent. So even though discussions in our industry do not often focus on gasoline's PAH content in terms of risk, gasolines do contain enough PAHs to be adequately detected with UV-induced fluorescence techniques including LIF.

The vast majority of fluorescence from petroleum NAPL impacted soils is due primarily to the PAHs contained in the NAPL, not the PAHs in the aqueous or gas phase that fill pore spaces unoccupied by the NAPL. There are typically far less PAHs outside the NAPL because PAHs are hydrophobic and only semi-volatile, with decreasing aqueous solubility versus increasing molecular weight. One significant exception to LIF's preferential response to PAHs in NAPL is observed with heavy molecular weight NAPLs (bunker, coal tar, creosote) that often contain very high PAH concentrations and have poor solvent properties. Equilibrium can drive enough PAHs out of heavy NAPLs and into the aqueous phase that UV-induced fluorescence of PAHs in pore groundwater can rival or exceed the NAPL's fluorescence—and therefore are easily misinterpreted as LNAPL. This troublesome and significant UV-induced fluorescence "mirage" was one major driver behind the development of TarGOST, which is immune to this aqueous phase PAH false positive fluorescence.

For fluorescence to occur, it is first necessary that electrons in the PAH be driven into an excited state. Energy must be introduced to the PAH (a process called excitation) and the absorbance of light is one of a number of ways for this excitation to occur. In the case of time-resolved LIF, short duration (1–2 ns) flashes of laser light excite any PAH that is capable of absorbing the laser's color. These excited-state PAHs are unstable, and immediately seek to shed that excess energy in order to get back to their preferred ground state. One major mechanism of releasing this energy is for the PAH to emit that excess energy as light, which is usually of a longer wavelength (less energetic) than the light that was originally absorbed by the PAH (referred to as Stokes shift). There are three main properties of petroleum NAPL fluorescence relevant to LIF systems: color or wavelength (λ), intensity, and lifetime (τ).

8.6.3.1 Fluorescence Color

The fluorescence of PAHs varies in energy, depending on the size, structure, and degree of substitution of the PAHs that are fluorescing. The wavelength of PAH fluorescence (which we see as color when involving visible wavelengths) trends with their size and degree of substitution, with the smallest PAHs fluorescing in the UV range (highest energy) and transitioning to ever redder colors (lower energy) with increasing molecular weight. Naphthalenes emit UV light that humans cannot see, 3- and 4-ring PAHs emit visible indigo, violet, blue, and green, and so on all the way up to very large PAHs that fluoresce even into the near infrared. The color of fluorescence being observed is key to interpreting in situ LIF data and aids in identifying fuel types or discerning when false positives might be causing the fluorescence rather than NAPLs.

The color of light emitted by NAPLs containing a mixture of PAH sizes and shapes is also affected by the PAH concentration, not just the PAH sizes and structures, due to interaction between the various sized PAHs at high concentrations. In gasoline, kerosene and diesel for instance, the PAHs are dilute enough that there is little direct interaction between excited state PAHs. But, as a NAPL's PAH concentration increases to the point where PAHs are no longer separated from each other by the solvent molecules (heavy crude oils for instance), the smaller PAHs can transfer their excited state energy directly to larger PAHs rather than fluorescing (transfer in the opposite direction, from larger to smaller, is rare). The net result is less fluorescence yield from the smaller (bluer fluorescing) PAHs and a red-shift of fluorescence in favor of the larger PAHs.

When heavy crude is diluted with non-fluorescent hexane (a non-participant in the fluorescence process), the color of the fluorescence shifts to ever bluer colors with increasing dilution, because the increase in hexane content is simply creating distance between the PAHs. The PAH size distribution of the crude remains the same during dilution, but PAH interaction has been reduced.

Selective removal of PAHs based on size can also change a petroleum NAPL's color. Weathering selectively removes smaller PAHs and packs the PAHs closer together, causing a red-shift. Highly adsorptive soils and activated carbon amendments on the other hand can selectively strip out the larger PAHs because they are more prone to adsorption, causing blue-shifting the NAPL's fluorescence color.

8.6.3.2 Fluorescence Intensity

Under controlled conditions most refined fuels and crude oils fluorescence with an intensity (brightness) that scales with the saturation of the NAPL present in the pore spaces (Teramoto et al. 2019), but there are a host of factors that influence the intensity of a NAPL's fluorescence.

Major factors influencing fluorescence intensity include:

- Innate chemistry—some NAPLs like diesels and light crudes are simply created at the refinery with an optimum balance of solvent to PAHs that results in vivid fluorescence, while others are low in PAHs (gasoline) or too PAH-rich (bunker fuel).
- Soil matrix—particle size and the accompanying differences in soil pore spaces that house the NAPL.
- NAPL solvent—light, clear, lower viscosity NAPL solvent bodies are best, black opaque tars and sludges are the worst host NAPL to be detected using LIF.
- Energy transfer—the more excited state PAH energy transfer events that take place the lower the intensity of the fluorescence, because it is a lossy process that quenches fluorescence.
- Quenching—non-fluorescent molecules in NAPL can steal energy away from excited state PAHs causing a net decrease in fluorescence. Molecular oxygen is an efficient quencher of UV-induced fluorescence that commonly affects benchtop LIF of lighter fuels (crude oil and lighter) (Parmenter and Rau 1969).

8.6.3.3   Fluorescence Lifetime (Rate of Decay)

Exciting a population of PAHs in NAPL with a brief pulse of laser results in some of the PAHs in the NAPL fluorescing immediately (a short lifetime) while other PAHs may take hundreds of ns (long lifetimes). The time it takes for all the excited state PAHs to emit their fluorescence is dependent on each PAH's size, shape, and even local environment. Because petroleum NAPLs are a complex mixture of PAHs, a NAPL's fluorescence lifetime is the combined lifetimes of the entire fluorophore population's lifetime. Lifetimes even vary by wavelength. Diesel, for instance, typically exhibits short-lived UV fluorescence (350 nm wavelength) accompanied by much longer-lived blue-violet fluorescence (400–500 nm wavelength).

Major factors influencing a fluorescent NAPL's fluorescence lifetime include:

- NAPL's PAH content, sizes, and structures.
- NAPL's solvent qualities—good NAPL hosts (like diesel) encourage long lifetimes, bad NAPL hosts (tars and sludges) discourage long lifetimes.
- Energy transfer—low PAH/solvent ratios support long lifetimes while high PAH/solvent ratios support energy transfer which shortens lifetimes.
- Quenching—non-fluorescent quenching molecules remove energy from excited state PAHs and this shortens the observed lifetimes because the longer an excited state PAH waits to fluoresce, the higher the likelihood of contact and resulting quenching.

8.6.4   UVOST Waveforms

Fluorescence color, intensity, and lifetime change with the complex interplay between the NAPL's PAHs and the NAPL's bulk solvent properties and measuring them all is key if one desires to use fluorescence to screen for the presence, quantity, and type of any NAPL. But in order to take full advantage of all three, they must all be measured, and this takes several seconds or even minutes with traditional laboratory fluorescence instrumentation. Subsurface logging demands that fluorescence be gathered in the subsurface with a probe that is moving, so a method capable of simultaneously collecting fluorescence color, intensity, and lifetimes is required if one is to maximize knowledge about the fluorophore's identities. Measuring the fluorescence temporally at multiple wavelengths results in an excitation-emission matrix shown in Fig. 8.9 (left). This three dimensional measurement illustrates the complex nature of diesel fluorescence resulting from excitation with a 290-nm wavelength pulsed laser. It is beyond the scope of this discussion to delve into the nuances of the techniques used to accomplish the task of detecting and processing the comprehensive fluorescence response simultaneously and rapidly in UVOST, but the end result is a more compact two-dimensional multi-wavelength waveform that represents all three dimensions of fluorescence shown in Fig. 8.9 (right). Each completed UVOST log is a collection

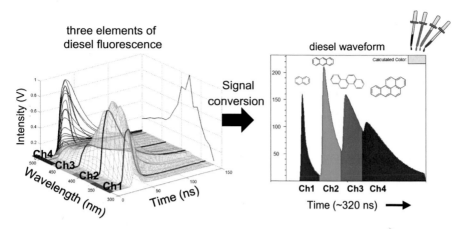

Fig. 8.9 Laser-induced fluorescence of diesel and associated time-resolved waveform

of hundreds or thousands of such waveforms, each encoded with the probe depth below surface at which it was collected.

Waveforms represent the three elements of fluorescence of PAHs in the following ways:

1. **Waveform Color**

- Color is represented by the waveform's four fluorescence decay pulses called "channels". The x-axis of Fig. 8.9 (waveform at right) represents 320 ns of time, although the axes labels are omitted so as to avoid confusion regarding its dual nature (because the x-axis represents both time and color). Four fluorescence pulses of differing wavelengths (colors) arrive at the detector at sequentially longer delay times and arrive as a train, so they bleed together to some degree.
- The four channels for the UVOST system represent bands of fluorescence centered at 350, 400, 450, 500 nm (from left to right)—these are ultraviolet, violet, indigo, and blue colors.
- Each of the four channels' contributions are represented by filling the area under the voltages with blue, green, orange, and red respectively—these fill colors are not the true colors of fluorescence being detected, but rather primary colors used to maximize contrast in the colorizing fluorescence data graphics in order to ease interpretation.
- Note the square color swatch in the upper right corner of the waveform's border— the relative intensity of the four channels defines the fill colors of the fluorescence response versus depth in the log. Because these fill colors are derived from each waveform, one can see similarities/changes in fluorescence "at a glance" with depth.
- Figure 8.10 illustrates how the shorter wavelength channels of UVOST (blue and green) generally represent fluorescence from 2- and 3-ring PAH fluorescence, the middle (green and orange) by 3- and 4-ring PAH fluorescence, and the rightmost

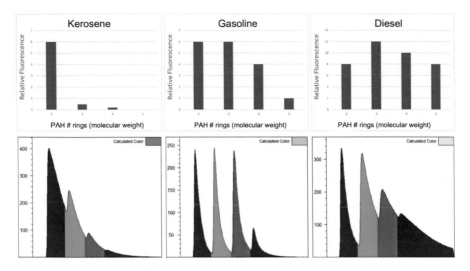

Fig. 8.10 NAPL fluorescence color dependence with PAH size and relative concentrations, and its effect on waveforms for common NAPL types

(orange and red) by 4-ring and larger PAH fluorescence—remember that the size and structure of a PAH determines the color of light it emits, with the blue end of the waveform dominated by small PAH fluorescence and the red end dominated by larger PAH fluorescence.

2. **Waveform Intensity**

- The y-axis on a waveform represents the intensity (brightness) of the fluorescence which is measured as a voltage generated by the photomultiplier tube detector which converts fluorescence flux to an electric current that is measured with a digital storage oscilloscope.
- The area under the curve of all four channels of the waveform is summed, divided by the area of all four channels for a reference emitter (RE) that was measured prior to logging.
- This normalization process allows the total fluorescence to be reported as a signal in %RE or total fluorescence of the LIF waveform versus the RE's waveform. The example waveform in Fig. 8.14 (right) had a strength of 203.6 %RE, so this diesel had about twice as bright total fluorescence intensity as the RE fluid measured prior to the diesel.
- For a homogeneous soil type and LNAPLs such as fuels, fluorescence is expected to increase monotonically with increasing concentration of fuels/oils in the soil pores. Figure 8.11 contains UVOST data obtained by measuring the fluorescence of diesel loaded onto moist coarse sand. Row three contains photos of those same sands (excited with a 308nm UVOST laser). Row four contains the resulting waveforms generated by each sample.

Diesel (mg/kg)	0	10^1	10^2	10^3	10^4	10^5
Signal (%RE)	0.5	1.1	9.2	50.5	132	414
308 nm photo	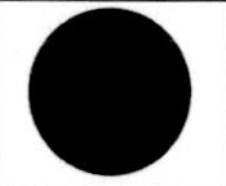				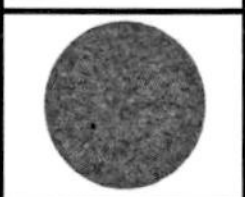	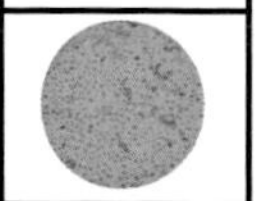
Waveform		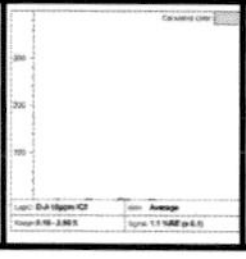	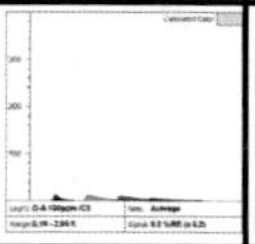	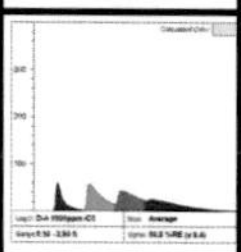	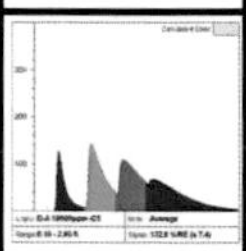	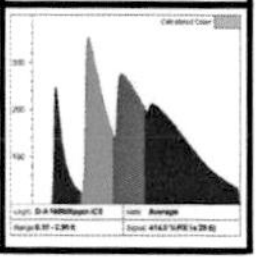

Fig. 8.11   Waveforms of diesel NAPL on 20–40 silica at decade concentration loadings

Fig. 8.12   Response of common NAPLs on 20–40 silica sand at decade series loadings

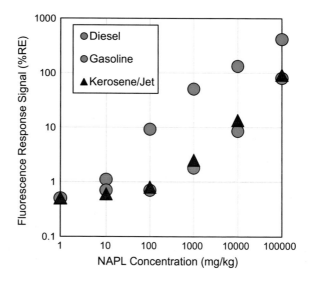

- Some fuel types fluoresce much more intensely than others. Diesel and crude fluoresce very well, while gasoline and kerosene (jet) fluoresce about 10–50% as intensely, as is depicted in Fig. 8.12, which shows the response for diesel, gasoline, and kerosene on UVOST.
- Porosity and grain sorting have a considerable influence on NAPL fluorescence (Alostaz et al. 2008) because coarse sediments allow neat NAPL pools to form up against the sapphire window and may even contain clear quartz grains whose transparency can increase the soil volume being interrogated. Fine-textured soils that assemble in a tightly organized matter more fully occupy the surface of the sapphire window and NAPL can hide in the soil's complex structure.

- Figure 8.13 illustrates sediment type's influence on fluorescence. Nine soils purchased as a kit from Midwest Geosciences Group (Indiana, USA) were loaded with 10,000 mg/kg (±100 mg/kg) of on-road diesel and 10% (±0.1%) of water, mixed well, left to rest 24 h, then examined under UV for homogeneity (proper mixing). Their fluorescence was measured, as well as clean sand at 10% moisture, using a UVOST system fitted with standard subsurface tooling. Figure 8.13 illustrates the dramatic difference in intensity due to sediment type alone (the clean sand/system background was 0.5% RE). Fortunately, NAPLs often distribute themselves within the more favorable soils in a geologic feature, if even the tiniest of pathways, and it is these NAPL-impacted seams and small pathways that the LIF responds to as the window slides past, if even for a brief moment. It should be mentioned that the waveform shapes (color and lifetimes) were essentially the same for each sediment, only their intensity (size) varied.

3. Waveform Lifetimes

- The waveform's x-axis spans 320 ns in time in order to capture NAPL's brief fluorescence pulses.
- Fluorescence lifetime is defined as the length of time it takes for the pulse of fluorescence to decrease to ~ 1/3 (1/e) of its peak intensity.
- The LIF system's software calculates the approximate lifetime and display them digitally.

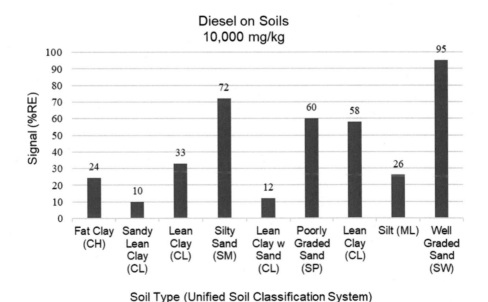

Fig. 8.13 UVOST response to diesel NAPL loaded at 10,000 mg/kg onto various soils (soils were pre-moistened to 10% water by weight)

- Notice that the lifetimes of some channels bleed into subsequent channels, which can influence the fill-colorization calculation—recent versions of LIF system software contribute the proper amount by using a relatively sophisticated allotment calculation.
- Long lifetimes often indicate ideal conditions, free from oxygen, with a solvent-rich fluorescence-friendly NAPL environment. Natural gas condensates, jet fuels, and diesels are ideal habitats for excited state PAHs and are known for their long lifetimes.
- Short lifetimes are indicative of excited state environments that are not friendly to excited state PAHs and one major cause is quenching by neighboring molecules that provide non-radiative pathways for excited state energy to leave the PAH, essentially snuffing out potential fluorescence before it can occur.
- Energy transfer causes a shortening of the lifetimes—the more closely packed the PAHs the greater the chance of energy transfer and quenching. This shortening begins to occur in heavy crude oils and gets steadily more extreme as the NAPLs get heavier (trending toward higher PAHs and less solvent) (Wang and Mullins 1994).
- Molecular oxygen, typically not present in situ at NAPL release sites due to anaerobic conditions, is a very efficient PAH excited state energy scavenger.
- Heavy NAPLs get spectacularly short lifetimes, even shorter lived than mineral fluorescence. This often allows for successful identification of heavies versus minerals.
- Heavies are immune to oxygen quenching due to their internal self-quenching which already quenches most susceptible PAHs regardless of molecular oxygen's presence.

8.6.4.1   Reference Emitter (RE)

Under normal circumstances, if one increases the excitation energy (the intensity of the laser light that is directed into the NAPL) one observes increasing fluorescence. Once excitation intensities get high enough however, the vast majority of PAHs are already driven into the excited state, so applying even more laser light intensity is fruitless. If we desire to consistently compare fluorescence data from one NAPL to another, or from one site to another, we can account for laser intensity variations in one of two ways:

1. Hold the intensity of laser excitation laser light delivered out the sapphire window to very consistent levels so that any changes in fluorescence can be attributed to the NAPL/soil matrix being measured, not to changes in the laser intensity.
2. Measure a fluorescent reference material's fluorescence emission just prior to logging and normalize the subsequent in situ fluorescence readings by the response of this RE.

The first approach is difficult to achieve under laboratory conditions, let alone field conditions. So early on in the commercialization of LIF (Bujewski and Rutherford

1997), it was recognized that a stable fluorescent material, capable of fluorescing with consistency across the range of wavelengths typical of fuels and oils, be formulated and made available to practitioners of LIF. This proprietary blend of petroleum oils and stabilizers was dubbed "RE" and has been used for UV normalizing LIF data since the late 1990s and RE remains in use today. RE calibration is conducted in a fashion similar to the tank of 100 ppm isobutylene gas used to calibrate handheld PIDs. While PID responses are reported as the sample's response relative to isobutylene's response in "PPM", LIF systems report total fluorescence intensity as a percentage of the RE's total fluorescence flux called Signal (%RE).

Figure 8.14 shows the cuvette holder filled with RE fluid (top) and the RE's waveform at left. The RE has a cross-section of 9992 pico-Volt-seconds (pVs) and a subsequently measured diesel on sand waveform (semi-transparent over RE's at right) had a cross-section of 23,082 pVs. The resulting signal for this diesel on sand was thus 231.0%RE, a little over twice the total fluorescence intensity as the RE fluid.

The RE also serves functions beyond normalization for laser energy. Firstly, there is a desired instrumental range of fluoresce intensities, because if the RE response is too small (from insufficient laser light) then subsequent in situ fluorescence will also be too dim and electronic noise might interfere. If too much laser light is delivered, the fluorescence of even modest NAPL saturation would overwhelm the detection system (detector saturation would occur), reducing the upper NAPL saturation range of the instrument. LIF system operators adjust the laser intensity into the fiber so as

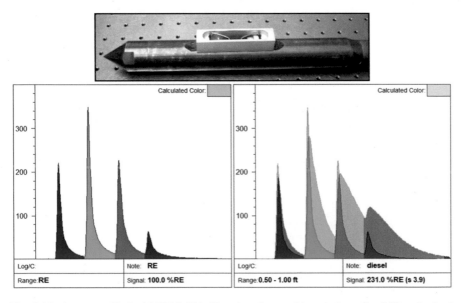

Fig. 8.14 A cuvette filled with RE fluid held against the sapphire window of an LIF probe (top). The RE fluid's waveform (bottom left) and the waveform produced by diesel on sand, with the RE waveform superimposed on top (bottom right)

102 %RE   47 %RE   90 %RE   77 %RE

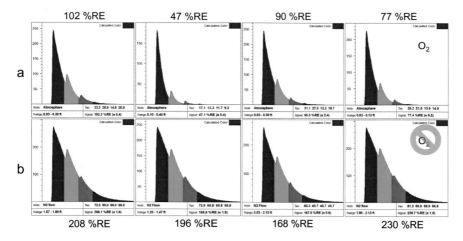

208 %RE   196 %RE   168 %RE   230 %RE

Fig. 8.15 Waveforms of kerosene/jet fuel NAPLs on sandy soil. Row A is at equilibrium with room air's oxygen, row B contains waveforms acquired after purging the NAPL of oxygen with nitrogen gas

to achieve an RE waveform that is of optimal intensity. The LIF software will not allow logging to proceed if the waveform is not within the proper range. Secondly, the LIF technician examines the relative response across the four channels of the waveform. If the ratio of the four channels relative to each other is not within the factory recommendations, the detection system needs to be adjusted to achieve the desired ratios. If not held within a few percent of their desired ratios, then subsequent waveforms acquired in situ of common recognizable NAPL types will not look familiar to data analysts. Imagine if one of the four channel were only 25% of the strength it typically exhibits. The result is that subsequent downhole encounters with NAPLs, even those of commonly recognized NAPLs, would be unrecognizable.

8.6.4.2   Waveforms of Common LNAPLs

Generally, similar fluorescence waveforms are emitted by families of fuels and oils with similar chemistries. This often allows LIF analysts to identify types of fuels and oils in situ, or at least whittle the possibilities down to one or two possibilities (Lu et al. 2014). But each fuel is a unique product depending greatly on the source crude and formulation, so even fuels of the same type do differ to some degree, sometimes surprisingly so, right out of the dispenser at the filling station.

Figure 8.15 contains waveforms of sandy material saturated with intact (unweathered) jet fuel and kerosene NAPLs. Notice the great dominance in the first channel, which is where naphthalenes fluoresce. Because the samples in row A were taken under ambient air conditions, molecular oxygen is present at ~ 20%, and molecular oxygen quenching is significant. The waveforms in row B are about twice as fluorescent and twice as long lived because the NAPL is relatively oxygen-free. Any benchtop studies of kerosene and jet fuel (measuring NAPLs recovered from wells for instance) should be interpreted with acknowledgment that the response is as much as a factor of two lower than what one will observe in situ because oxygen rapidly enters NAPL during normal handling in air. Interestingly, when kerosene weathers it does not red-shift to any significant degree like gasoline, rather its waveform retains the same shape and simply fades in intensity. This is because kerosenes contain insufficient numbers of PAHs of three or more rings to assume the task of fluorescing as the naphthalenes (and any other UV-fluorescing species) get preferentially washed out of the NAPL due to their higher semi-volatility and water solubility.

Figure 8.16 contains waveforms of three intact (fresh) gasoline NAPL saturated onto sandy soil at left. At far right is gasoline NAPL that had been saturated onto sandy soil and then allowed to weather indoors in a loosely lidded jar for seven years. Intact (unweathered) gasoline waveforms are typically shaped as shown, but exhibit variability due to their differing formulations. Oxygen quenching is a significant factor with gasoline as demonstrated by the two- to three-fold enhancement of both lifetimes and intensity observed in row B versus row A. Weathering has a significant impact on gasoline, demonstrated by the seven-year-old gasoline waveform at far right which is dramatically red-shifted. Notice, however, that oxygen quenching does not play a major role in the highly weathered sample. This is likely due to the reduction in gasoline NAPL volume, allowing energy transfer between PAHs to take place (a form of quenching) thus reducing oxygen's contribution to quenching of PAHs within the sticky condensed coating of residual NAPL on the sand grains. There has been no clear indication that octane has much effect on fluorescence across hundreds of gasolines tested. However, premium blends appear less susceptible to weathering and have longer lifetimes in general versus lower octanes, due to differing formulation at the refinery.

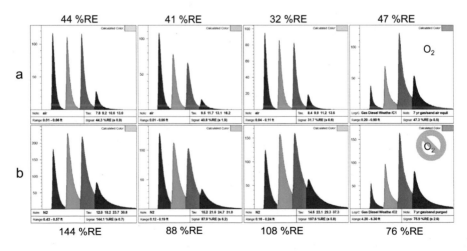

Fig. 8.16 Waveforms of gasoline NAPLs on sandy soil

Figure 8.17 contains a sampling of diesel NAPL waveforms. Diesel has the unique characteristic of having a short lifetime blue peak (two-ring PAH UV emission) that is about the same height as the longer lifetime green peak, followed by three- to four-fold longer lifetimes in the subsequent channels. The reduction in oxygen quenching after removal (row B) is dramatic, with lifetimes so long that they look rather silly. How much of this is due to molecular dynamics versus detector saturation is unclear, but it makes many diesels instantly recognizable. Notice though how the diesel at far left has a waveform easily mistaken for gasoline, demonstrating there are often exceptions to the general waveform shapes of common fuel NAPLs. Evidence of weathering in diesels is not nearly as obvious as it is for gasolines, as is demonstrated in the waveform at far right which was weathered for seven years in the laboratory along with the gasoline discussed above. Notice there is reduction of the smaller more readily weathered PAHs (bluer channels) however, which is consistent with weathering of all NAPLs. Diesels have the longest lifetimes of all the commonly encountered NAPLs, with the possible exception of natural gas condensates. Notice also how the decay rate at the right side of channels three and four for de-oxygenated samples in the bottom row looks somewhat odd with a hump-backed appearance. This is likely due to interaction between PAHs via energy transfer, photon cycling that involves larger PAHs absorbing smaller PAH fluorescence and subsequently emitting fluorescence at longer wavelengths, or other complex interplay between PAHs for the excitation energy. These processes can take time and can be repeated, causing the unnatural upward bulge rather than the usual decay just after peak emission in latter channels.

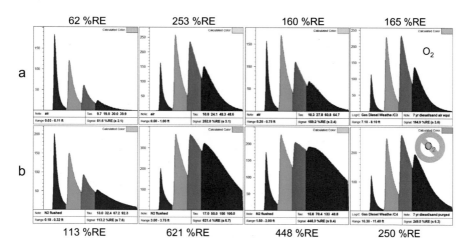

Fig. 8.17 Waveforms of diesel NAPLs on sandy soil

8.6.4.3 Waveforms of Medium and Heavy NAPLs

Figure 8.18 contains UVOST waveforms of heavier unrefined NAPLs that are beginning to show the effects of having too many PAHs versus solvent molecules, which begin to affect their fluorescence. Energy transfer and photon cycling effects begin to have an ever larger impact as the NAPLs get heavier in molecular weight (PAH content) from left to right. Recall that these processes are lossy (often do not result in fluorescence), resulting in far less fluorescence than the NAPL's robust PAH content would suggest. Dramatically shorter lifetimes are also the result when numerous larger PAHs are too close to the smaller PAHs that initially absorbed the UV laser excitation. This lifetime reduction gets steadily worse as NAPL density (relative PAH/solvent content) increases. These short lifetimes should be regarded as an obvious indication to the analyst that heavy NAPL may exist, and in great quantity, yet the heavy NAPL generates only a whisper of the signal %RE (relative to refined products that contain PAH/solvent ratios far friendlier to the fluorescence process). The reason we employ the word "may" above is because waveforms from false positives often display similar characteristics, making differentiation challenging. Identifying heavies and their relative saturation in soil with any UV-based fluorescence tool is fraught with the possibility of dismissing the heavies as harmless false positives or assigning harmless false positives as heavies. Heavy NAPLs should be investigated with TarGOST, the tool designed to respond appropriately to heavy NAPLs across a range of NAPL saturations.

- Crude oils respond well to UVOST LIF and as long as one is seeing modestly long lifetimes, such as the response in light crude (Fig. 8.18 far left), one can safely assume the relationship between Signal %RE and degree of NAPL saturation is likely to be monotonic, or at least nearly so. Light to medium crude waveforms are usually weighted slightly right of center (slightly red-shifted) on the waveform time axis and have modest duration lifetimes.
- The fill colors in the upper right corner of the waveforms in Fig 8.18 are all fading orange to red, this is another visual clue of the presence of heavier (or extremely weathered) NAPLs in a UVOST log, which will have their Signal %RE plot filled with red indicating zones where heavies (or false positives) are encountered.

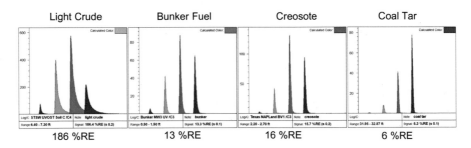

Fig. 8.18 Waveforms of light crude, bunker fuel, creosote, and coal tar on sandy soil

- Because their chemistry spans the transition from relatively ideal fluorescence behavior to extreme energy transfer effects (Wang and Mullins 1994), crude oils also spectacular variety of waveforms, depending greatly on their source and chemistry. They can be relatively blue-shifted and long-lived or tremendously red-shifted and short-lived, making crudes difficult to identify definitively as crude with LIF data alone.
- Crudes contain not only fluorophores, but chromophores as well, which are molecules that absorb wavelengths of light involved in LIF processes but do not fluoresce, so chromophores can cause significant quenching in crudes and other heavies.
- The fluorescence of NAPLs heavier than light crudes also begins to exhibit fluorescence intensity behavior that does not scale properly (monotonically) with the degree of NAPL pore saturation. Most heavy NAPLs will be grossly under-reported by any UV-excitation fluorescence system.
- Bunker fuel (second from left) is heavier in average molecular weight than light crude, higher in relative PAH content and contains less VOC solvents, so energy transfer red-shifts the color, lifetimes begin to shorten, and precipitous drop in total fluorescence occurs. UVOST is capable of absence/presence of bunker, but false positive interference and almost certain lack of monotonic behavior begins to complicate interpretation and limit utility of the data.
- Creosote (second from right) is in the same category of fluorescence as bunker fuels in that they often fluoresce at the same intensity across a wide range of NAPL saturation. From 100 ppm, all the way up to full saturation, they can yield approximately the same %RE response. While completely missing creosote is unlikely for UVOST, the logs at creosote sites give little indication of the relative NAPL content in the soil, only absence/presence.
- Coal tar (far right) is the most concentrated PAH NAPL encountered with LIF, with PAHs sometimes occupying over 50% by weight. This oversaturation of PAHs is instantly recognized in the coal tar waveform, which has the shortest lifetimes of any NAPL, combined with a fluorescence intensity that is also extremely low and sometimes not exceeding the natural soil's response. Coal tars can still hint that they are present however, by a red-shifting and short lifetime waveform even more extreme than soils or other false positives, but it takes a keen eye and discipline to spot them, especially if their presence is unexpected at the site.

8.6.4.4 Waveforms of False Positives

Soils can generate fluorescence that does not originate from the NAPLs being targeted. These non-NAPL positive responses range from weakly fluorescing to strong and even dominating fluorescence (higher than target NAPL at full saturation). Sources include organic materials such as phragmites roots, meadow mat, certain clays, calcareous sands, buried woody debris, or even aqueous phase PAHs associated with nearby heavy NAPLs like coal tar. Almost any man-made non-metallic materials fluoresce (concrete slurry, paper, sewage, biodiesel, plastic, fabrics, etc.)

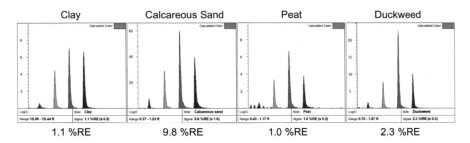

Fig. 8.19 Waveforms of commonly encountered false positives

so investigators should always be on the watch-out for waveforms that do not match the fuels/oils on site, especially shallow. Figure 8.19 contains a sampling of typical false positives encountered during UVOST logging.

Aspects to keep in mind about false positives include:

- Fluorescence intensity from false positives is sometimes not that high (< 5%RE), but those levels can actually be quite problematic when they occur in and among weakly fluorescing NAPL (e.g., gasoline) seams or weeps of NAPL found in small grained soils which creates NAPL and soil combinations that are not strongly fluorescing themselves.
- Almost all false positive soil waveforms differ from NAPLs due to their short lifetimes and the "trident spear" shape in the last three channels. This makes recognition of NAPL in and among these common false positives, and subsequent rejection of false positives from the NAPL CSM, possible and even routine.
- Clays (far left in Fig 8.19) often fluoresce in the few percent RE range. They can be confused for heavies and limited targeted sampling is often required to establish the true origins of this waveform.
- Calcareous sands (second from left) are common along coastal environments and can sometimes yield strong fluorescence signals. Probing in constantly fluorescing soils seems like an impossible environment to sensitively detect NAPL fluorescence in, but the longer lifetimes exhibited by LNAPLs typically make discernment of NAPL relatively easy to discern from fluorescing sand.
- Peat (second from right) is highly variable in its fluorescence, ranging from none to as high as 50%RE. Finding heavy NAPL contamination in peat lenses is one of the most challenging scenarios for LIF.
- Any living or recently decayed vegetation will exhibit fluorescence, but it is typically limited to a few %RE intensity and recognizable as non-NAPL in nature due to the familiar "trident spear" waveform. Aquatic vegetation false positives are commonly encountered at the sediment surface during offshore barge-mounted LIF projects on canals, rivers, and other water bodies adjacent to onshore NAPL release sources.

- The only long-lived "blue-dominant" false positive waveforms encountered in the 30 years of LIF have been volcanic soils in the Aleutian Islands and these fluoresced in the 0–3%RE range.
- One should always budget for modest level of targeted sampling from LIF locations and depths where potential false positives exist to determine what materials are causing unfamiliar fluorescence waveforms and their fluorescence must be removed from the data set if they are found not to be target NAPL.

An extremely problematic type of false positive mentioned previously is aqueous phase PAHs that occur in sandy or gravelly soils in the immediate vicinity of heavy NAPLs such as coal tar. The reason it is rare is that probing in/near very heavy NAPLs with UV is avoided by reputable LIF providers, who recognize the danger of misidentifying dissolved phase as LNAPL and subsequently reporting them as such. Figure 8.20 contains a benchtop demonstration of the danger of probing coal tar with UV fluorescence. The jar (A) contains coal tar and water and the tar is stuck to the jar's sides and bottom with a layer of water between the NAPL. After some time (days to weeks), the PAH-rich tar sources smaller more soluble PAHs to the water and these smaller two- to three-ring PAHs are readily excited by UVOST's UV laser. Because the larger PAHs are practically insoluble in water, there are not enough of them to promote energy transfer and its associated quenching. This leaves the smaller aqueous phase PAHs to fluoresce unencumbered, as demonstrated by

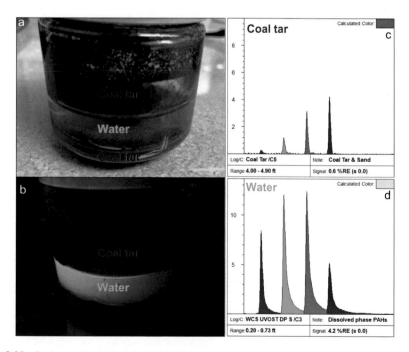

Fig. 8.20 Coal tar and aqueous phase PAHs in a jar at left accompanied by their inverse responses to UVOST LIF when saturated onto sand at right

the sky blue fluorescence in the jar (B). They fluoresce so well, in fact, that when this same water is used to saturated sand, its fluorescent waveform mimics partially saturated gasoline NAPL (4.2% RE) and would almost certainly be interpreted as such (D) by the analyst. The waveform of the source coal tar meanwhile (C) at total saturation on sand, yields a factor of seven lower response (only 0.6%RE) which is easily dismissed as insignificant (false positive), especially by inexperienced data analysts.

LNAPLs do not suffer from this aqueous phase PAH false positive because the LNAPL solvent body is quite capable of hosting hydrophobic PAHs. Heavy NAPLs on the other hand are at or near saturation with PAHs and the tremendous concentration gradient between the DNAPL and water is high enough to generate high dissolve phase PAHs (compared to waters adjacent to LNAPLs). Waters adjacent to LNAPLs do of course contain dissolved PAHs, but at a concentration too low to rival that of their LNAPL source NAPL when that water is loaded onto soil.

8.6.4.5   Weathering Effects on LIF Response

Weathering selectively removes the most soluble, most volatile, and most readily metabolized hydrocarbons first (including PAHs), which means the smaller PAHs are the first to go. This process selects an ever larger size population of the PAHs who continue to reside in the remaining NAPL. And because the larger remaining PAHs fluoresce in redder colors, weathering causes a red-shift in waveform color. At the same time, the loss of the NAPL "solvent body" (including the BTEX and aliphatic fluids that help solvate PAHs) makes for a less fluorescence-friendly liquid hydrocarbon environment. This loss of solvate often shortens the lifetime, so combined with the color shift, a weathered lighter fuel's waveform will begin to mimic heavier NAPLs, and eventually look more like an oil or even a tar than the original LNAPL. This makes sense to us because weathering processes are slowly converting the lighter LNAPL to a heavier NAPL, which in its final extreme stages contains only the large and "sticky" molecules that are the hallmark of heavy NAPLs.

Because the average molecular weight of gasoline is relatively light, it contains many volatile and relatively soluble molecules, making gasoline particularly susceptible to weathering. A simple but effective method for emulating mechanical weathering of gasoline is to introduce nitrogen gas at 300 ml/minute into a vial containing ~ 5 g of 87 octane gasoline saturated onto moist sand (10% water). Logging the UVOST LIF response continuously as the nitrogen flows allows one to observe the waveform changes that occur due to the chemical makeup changes over time. Figure 8.21 contains the UVOST waveforms taken over time, from pre-flow of nitrogen (air equilibrium) to 20 min after flow was begun (20 min N_2 flow). The voltage scale on the four waveforms is scaled independently to allow us to closely observe the color and lifetime changes, hiding the fact that they were not all the same height (voltage).

At air equilibrium, the waveform was heavily quenched by room air's 20% molecular oxygen, as evidenced by the short lifetimes and low %RE. Whatever PAHs were

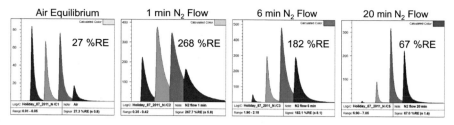

Fig. 8.21 Artificial weathering of gasoline via nitrogen gas flow

responsible for the fluorescence in channel two (green channel), they were particularly susceptible to having their excited state energy stolen by molecular oxygen. Just one minute after starting the flow of nitrogen, the bulk of the oxygen has been removed and both the lifetimes and intensity increased dramatically for this particular gasoline formulation. Removal of the most volatile of the PAHs (responsible for the blue, UV channel) was already being reduced at this point. Six minutes after flow started, continuing selective removal of both the more volatile PAHs and solvent body VOCs continued to red-shift the fluorescence and shorten the lifetimes. After 20 minutes of flow, the gasoline fluorescence has transformed markedly from its original (intact) fluorescence. This virtual rainbow of responses is commonly observed at gasoline release sites and helps reveal insights into the condition of gasoline NAPL with depth.

Mechanically weathering gasoline at such a rapid pace is of course a much faster process than that occurring in situ, but it serves as a convincing demonstration of the fluorescence red-shifting that weathering causes. Remember that jet/kerosene weathering is typically indicated by decreasing fluorescence but no color shift because there is an insufficient number of larger PAHs left to continue the fluorescing once the semi-volatile naphthalenes have been weathered out. In general, the heavier (and therefore more recalcitrant) a NAPL is, the less its fluorescence response changes with weathering.

8.6.4.6   Cluster Diagrams

As we have just seen, waveforms are a compact way of envisioning all main fluorescence properties, but their vast numbers make them difficult to comprehend site-wide. What was lacking for quite some time in LIF's history was a method of reducing waveforms down to a more compact form, where each and every waveform's contribution to the log could be visualized in one graphic, and this lead to the development of the cluster diagram.

Cluster diagrams begin with a waveform reduction process to each of the waveforms in a log, breaking every waveform down into its four individual fluorescence decays. The top row of Fig. 8.22 shows a waveform of typical fuels (kerosene, gasoline, and diesel) that have undergone the data reduction processing with individual decays shown after process along with waveform in black outline. The gasoline waveform (top center) shows the four channels' intensity factors (black arrows) and the four lifetimes (orange arrows) that were determined during the processing for that waveform. The four relative strengths represent relative color, so their relative left–right balance is used to calculate each waveform's bubble and their position in the color axis (x-axis). In this fashion, very blue-shifted waveforms like kerosene land at far left of the cluster diagram and very red-shifted NAPLs (highly weathered or heavy NAPLs) land very far to the right.

The average of each waveform's four channel lifetimes is used to determine where on the lifetime axis (y-axis) of the cluster diagram each waveform's bubble should be placed. The cluster data are normalized to place very short lifetimes of < 1 ns (like coal tar or soil fluorescence) toward the bottom, and long lifetimes like that of diesel near the top. Each waveform is thus transformed into a bubble that occupies its own unique position on the cluster diagram "map", with its position describing the color and lifetimes contained in the original waveform. The fluorescence intensity of the total fluorescence (%RE) is portrayed by sizing each waveform's bubble according to its signal %RE, up to a capping level that changes with each LIF family.

Figure 8.22 bottom row contains cluster diagrams for UVOST logs gathered during weathering studies of six kerosenes (left), ten gasolines (center), and six diesels (right). Cluster plots of weathering experiments for a number of the same fuel

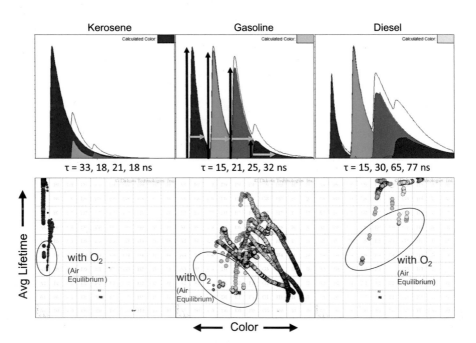

Fig. 8.22 Top: Three common fuel class waveforms reduced to independent decays. Bottom: cluster diagrams of artificial weathering for six kerosenes/jet fuels (left), ten gasolines (center), and six diesels (right)

types serves to illustrate where the certain fuel type's bubbles land on the clustering diagram, as well as how individual chemical formulation differences and weathering affects their positions. The samples were flushed with a gentle stream of nitrogen gas to displace oxygen and left to flow over a 20-min period. The changes in molecular oxygen content (and subsequent degree of quenching) combined with selective weathering of smaller semi-VOCs resulted in continuous "tracks" of bubbles for each NAPL species, each track representing shifts in color and lifetime that occurred in the fluorescence during this artificial weathering process. Recall that oxygen's presence shortens the lifetimes and this is evident in the bubbles circled in blue, showing the cluster positions for waveforms acquired while the fuels still contained oxygen. In the gasoline cluster (center), bubbles rose sharply on the lifetime axis during the initial oxygen removal (a process taking less than a minute to complete) and then lifetimes trend steadily downward again within the 20-min nitrogen flow that is causing mechanical weathering of semi-VOCs and the VOC solvent body. Notice also how the gasolines' bubbles spread to the right in color (red-shifting) as discussed in the previous section.

The gasoline samples' bubbles are cast across a wide range of both fluorescence color and lifetimes, while diesel and kerosene are more stable with regard to color shifts, creating tighter clusters, but shifted up and down by oxygen quenching effects. This same "vertical stacking" pattern for diesel is often observed for diesel NAPL located near the soil surface, where oxygen and weathering have the most influence due to earth tides. The general trend is that cluster bubbles for all LNAPLs generally head to the right (red-shift color) and down (shorter lifetimes) as the degree of weathering increases, but effect is most pronounced in gasolines because they vary most with original formulation at the refinery and with weathering. Figure 8.23 contains a UVOST cluster diagram map, with regions where fluorescence responses for various fuel classes reside highlighted. There is modest overlap between the different fuel classes, indicating that LIF data is definitive in some cases, but variations in NAPL chemistries due to source and weathering can create less than certain waveform interpretation in many instances.

8.6.5   Analysis and Interpretation of LIF Logs

8.6.5.1   LIF Logging in Practice

Over the last 30 years, the most common log interpretation mistake, by far, has been to focus solely on fluorescence intensity, without tempering the intensity-focused interpretation with the other valuable information contained in the log. Intensity alone simply cannot be trusted to represent the NAPL in situ, due to highly variable fluorescence intensity from NAPL to NAPL, weathering, and false positives sometimes out-fluorescing the NAPL, so adjusting our interpretation by including the color and lifetime in the analysis is a key part of LIF log interpretation.

Log interpretation will be now be demonstrated based on the same idealized scenario employed to illustrate the application of HDCS and MIP (Figs. 8.2 and

Fig. 8.23 UVOST LIF
cluster diagram showing the
regions occupied by common
NAPLs and false positives

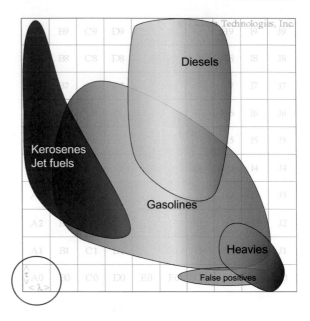

8.4, respectively), which will be assumed to be contaminated with gasoline NAPL (Fig. 8.24).

In Fig. 8.24, the Signal %RE x-axis approximately covers the range 0–150 %RE which most gasolines typically produce. Gasoline NAPL is complex in its fluorescence due to its propensity for weathering rapidly and more extremely than other commonly delineated petroleum NAPLs. The Visible Fluorescence column in Fig. 8.24 illustrates this, showing us how gasoline behaves like a chameleon with regard to its fluorescence color. Depending on the source and age it varies wildly, occasionally fluorescing nearly as blue (UV) as kerosene, while premium gasoline can come close to mimicking diesels. Weathering can cause red-shifting that generates waveforms that mimic light to medium crudes. Every gasoline release site's fluorescence looks different because every site has a unique history of gasoline formulations released, age and weathering processes. Figuring out the details of exactly what, when, and how past NAPL releases occurred is always difficult, but an LIF survey provides, at a minimum, an immensely useful guide to where gasoline NAPL is and insights into the chemical nature of the NAPL at thousands of data points in the subsurface.

Key observations for the gasoline NAPL profile in Fig. 8.24 include:

• Interval A

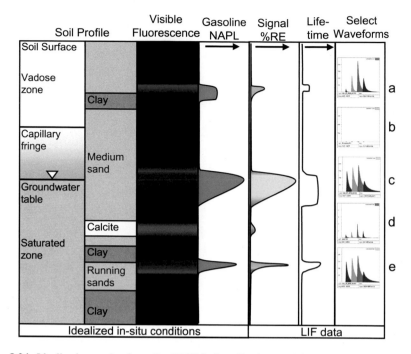

Fig. 8.24 Idealized scenario of gasoline NAPL in the soil column at left and resulting LIF (UVOST) data at right

The NAPL is perched near the surface, leaving it more vulnerable to aerobic and physical weathering, so the intensity is somewhat lower than it would be had the NAPL not been weathered. More dramatic signs of weathering are recognizable in the Visible Fluorescence column showing a turquoise color which is red-shifted from the deep blue–violet color typically exhibited by intact gasolines. The callout waveform here is classic highly weathered gasoline (which looks similar to light crude). The lifetimes are modestly short, consistent with gasoline that is weathered and may be exposed to oxygen. Notice that the signal fill color here is orange, so we might be tempted to call this false positive fluorescence, but the lifetimes are still long enough (noticeably up off the natural soil baseline) to assure us this is NAPL.

- Interval B

Despite the presence of VOCs, LIF fails to register any Signal (%RE) response above background because LIF is not capable of detecting gaseous, aqueous dissolved or sorbed VOCs like a MIP can.

- Interval C

The "shark's fin" shaped profile in the Signal %RE replicates the idealized Gasoline NAPL saturation profile nicely, but a close inspection sees color shifts occurring across this profile. The very top of the shark's fin exhibits several signs of weathering including the turquoise color in Visible Fluorescence, and both the lifetime and Signal %RE increases do not increase in concert with the gasoline NAPL saturation. Once down into the higher saturation of the gasoline NAPL, we return to a classic intact gasoline waveform with robust UV contribution in the 350 nm (blue) channel of the waveform, and the modestly long lifetimes that intact gasolines usually exhibit. The cobalt blue in the Visible Fluorescence is typical of intact gasoline's blue, indigo, and violet emission that human eyes can just barely see. This is rendered as a chartreuse signal fill color in the OST software, which is a fill color commonly generated for intact gasolines.

- Interval D

The orange fill color in the Signal %RE response here hints that it is very likely soil fluorescence. Due to the nature of this being a gasoline release (which we know from the waveforms above and below), there is a heightened chance that Interval D actually represents a thin lens of extremely weathered gasoline. However, there is a lack of increase in the lifetime column, indicating that it is unlikely to be extremely weathered gasoline and is instead mineral in nature (false positive). Either interpretation of this interval is low risk (compared to the more robust and confident NAPL intervals).

- Interval E

Notice how the chartreuse signal %RE fill color on top side of the Signal %RE response fades to orange at the bottom. Like Interval C, this trapped lens of gasoline shows signs of weathering, but in reverse. This interval's more weathered gasoline NAPL is located where groundwater dissolution is taking place along the NAPL body's bottom edge. The more intact gasoline NAPL in Interval E is that which is held up against the low-k clay that protects this NAPL deposit from weathering. The lifetimes support this as well, shortening in concert with the red-shifting of color due to weathering.

8.6.5.2   Field Logs

For purposes of providing a single comprehensive field log, Fig. 8.25 was created by stitching together three different UVOST field logs obtained at three NAPL release sites in the USA in 2018–2019. The potentiometric surface elevation indicated in this profile (at 10 m below ground level) truly was co-located in relation to the NAPL data shown in that same interval C.

Prior to discussing the fluorescence of this field log, let us introduce several elements of LIF field logs not yet covered. At left are the callouts, each numbered 1–5 in their upper corner. Each waveform has information below it describing the depth and signal %RE for that waveform. Callouts 2 and 5 were single waveforms,

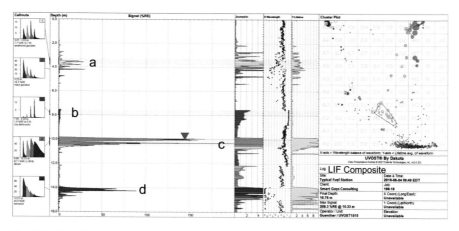

Fig. 8.25   Top: Composite UVOST field log containing typical responses for gasoline, diesel, false positive soils, and kerosene

while callouts 1, 3, and 4 were selected from a range of depths, so they contain the average of all waveforms that were acquired between those depths.

The cluster diagram for this log's waveforms is at far upper right. Small tags with the callout numbers are made next to bubbles representing the waveforms and, in the case of range callouts, a polygon roughly encircling those bubbles represented in the range. Next to the cluster in the vertical log section, you see the X: Wavelength column that contains each waveform bubble's x-axis cluster diagram position (A-J). Alongside, it is the Y: Lifetime log that indicates each waveform bubbles' y-axis position (0–10) on the cluster diagram. This allows for high definition examination of both color shifts and lifetime at all depths.

- Interval A

This first encounter of significant fluorescence is not a cartoonish hump like those used in the simplistic training models (Fig. 8.24). NAPL fluorescence is often observed as a bristly pattern of high and low fluorescence encounters which are accurately mimicking the highly heterogeneous NAPL distribution commonly observed in situ.

A core from this interval would yield discrete patchy mottling or distinct lamina of gasoline NAPL-impacted soils sandwiched between soils that contain little or no NAPL (or fine soils such as clay that partially hide the fluorescence). This insight into the existence of small-scale extremes in distribution is another aspect of LIF logs that often goes underappreciated but can be valuable for those considering potential NAPL recovery, NAPL transmissivity (García-Rincón et al. 2020), and associated remediation system design.

The presence of gasoline far above the water table may suggest a release source is perhaps nearby (some gasoline perched during its fall through the profile) and that the gasoline is perhaps sitting above a capillary barrier.

- Interval B

This interval contains what at first glance appears to be a trivially small Signal %RE. But absent evidence to the contrary, one can never rely on intensity alone to conclude a response is insignificant. Examination of the waveform, fill color, and lifetime suggest interval B is either a harmless false positive soil or organic. But a much more consequential interpretation is that it is instead a heavy NAPL such as coal tar which would yield a similarly red-shifted signal. We simply cannot rule this possibility out, so this interval should be validated at least once at this site or further assessed considering the site history and other site characteristics (e.g., the presence of fluorescing kaolinite or montmorillonite).

- Interval C

This NAPL is distributed in a shark's fin fashion poised on the water table. The waveforms, calculated fill color, long lifetime, and high center-right location on the cluster diagram all point to an intact diesel. A lens of low hydraulic conductivity soils probably exists at ~10.1 meters, causing the brief "cut out" in the Signal %RE response of this otherwise classic shark's fin profile that is typical of NAPL residing on groundwater surfaces in transmissive soils. Careful examination of the calculated fill color along with X: Wavelength and Y: Lifetime columns, indicates that there is weathering at the bottom of the fin, and those weathered NAPL responses are represented by the green to light orange bubbles in coordinates F4, F5, G5, G6, G7 of the cluster diagram. This is the vertical stacking behavior, a weathering behavior that is commonly observed with diesel (see Fig. 8.22 lower right).

One might be tempted to assume that the NAPL in interval C contains far more saturated NAPL than other intervals, and is therefore more consequential to the NAPL CSM than intervals A or D, but that is not necessarily true. This is another common interpretation error, that is, not considering the inherent fluorescence intensity differences between NAPL types and varying sediment types. Because gasoline is considered to have higher risk for sourcing BTEX, interval A might well be of more interest to stakeholders than interval C and in fact the gasoline in interval A could have saturation levels matching that of the diesel in interval C.

- Interval D

Interval D is trapped well below the water table, which is not uncommon due to water table fluctuations. Interval D's fluorescence is extremely blue-shifted, long-lived, and its waveform is consistent with kerosene. Callout tag 5 of the cluster diagram shows it is located in the extreme upper left, where no other common NAPL types are found with the exception of some natural gas condensate formulations. The few signs of weathering are at the top and bottom of this NAPL lens, and those waveforms are stringing down toward the bottom right of the cluster (coordinates B8, B7, and C6), consistent with weathering of NAPLs that drives them in this downward and slightly right (red) direction.

If one desires to do site-specific calibration studies in order to convert signal %RE to NAPL saturation or similar metric, spiking NAPL recovered from the site onto representative site soils in a decade dilution series (as was done to generate Fig. 8.12) will get one closest to realistic conversion from Signal %RE to estimated NAPL saturation.

8.6.5.3   Non-negative Least Squares (NNLS) Fitting

Our final topic with regard to fully utilizing the qualitative aspects of fluorescence inherent in an LIF data set's waveforms, involves processing the thousands of waveforms gathered across a NAPL release site through non-negative least squares (NNLS) fitting or similar techniques in order to infer, continuously with depth, where different NAPL types and/or false positives are found. This is typically conducted after all types of fluorescence observed at the site have been somehow vetted, leaving us confident that we know enough to determine what fluorescence we want to keep (target response) versus those type of fluorescence we would like to discard or otherwise ignore.

The waveforms that represent the main classes of fluorescence observed in the log (or entire site) are called the basis set, a collection of up to five waveforms in the current OST software. The basis set created for our composite log is shown in Fig. 8.26.

The processing involves application of a NNLS fitting algorithm to determine which waveforms in the basis set, and how much contribution of each waveform, need to be combined in order to generate a synthetic waveform that best fits every in situ waveform in the log. The result contains a separate log for each basis set waveform (fluorescence type), representing how much (in %RE) each basis set waveform contributed to achieve the best fit. Figure 8.27 contains the results for our composite study log. The signal scale is held to 25 %RE to allow us to see the results for all types of fluorescence, including the weaker ones (resulting in some NAPL types plotting off-scale to the right). Notice how the NNLS process successfully isolated each basis set fluorescence type according to where it was observed, with the exception of highly weathered diesel at 12–13 m below ground level, where the NNLS process chose mostly weathered gasoline instead. This was because weathered diesel was not available in the basis set and highly weathered gasoline and highly weathered diesel color and lifetime begin to converge, so NNLS chose the closest match in its basis set to fit the weathered diesel.

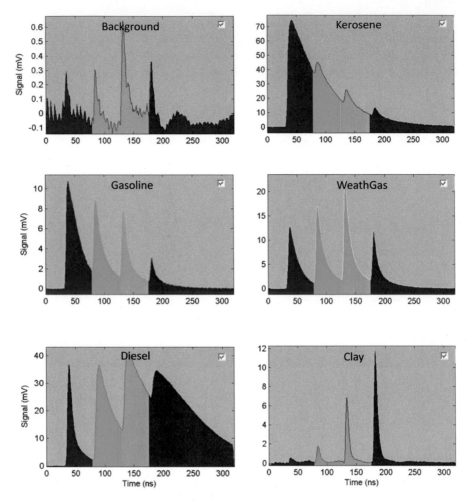

Fig. 8.26 Basis set of waveforms for NNLS processing of the composite log

NNLS processing is often applied to an entire set of LIF logs from one site characterization effort, which often contains dozens to hundreds of logs. These logs contain many thousands of waveforms, all from unique points in 3D space. NNLS processing enables stakeholders to separate mixed NAPL bodies up into their respective NAPL types, eliminate false positives from target NAPLs, and other complex situations. Figure 8.28 contains the fluorescent response of a mixed NAPL body at a gasoline filling station site that was separated into fluorescence types (corresponding to intact gasoline, weathered gasoline, and diesel) using NNLS processing (ITRC 2019).

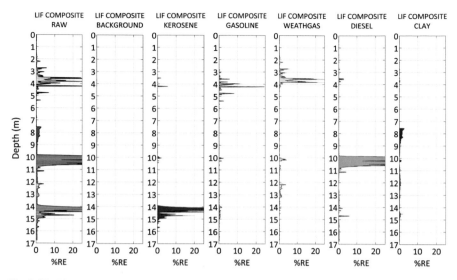

Fig. 8.27 The composite log at left along with the resulting basis set logs, each containing their contribution to the total fluorescence signal (in %RE) with depth

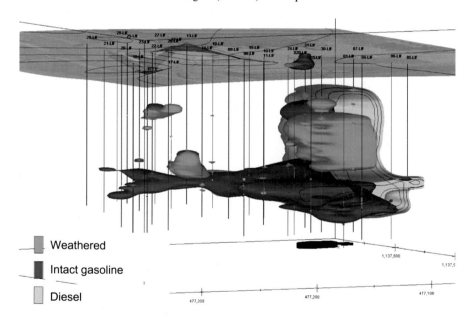

Fig. 8.28 3D visualization of a mixed NAPL body broken up into separate bodies using NNLS processing. Green corresponds to "weathered gasoline", blue to "intact gasoline", and yellow to "diesel"

8.6.5.4 Validation Sampling of LIF

Targeted validation of LIF logs with co-located core sampling is invaluable, but can also be misleading unless one is prepared to manage the effects of NAPL's often extremely heterogeneous distribution. Comparisons between LIF logs and physical samples taken adjacent to NAPL indications in LIF logs often show poor correlation, with the primary cause being localized heterogeneity. The simple act of acquiring samples from a separate adjacent borehole can result in mysterious depth offsets due to compression, partial recovery, and human errors. The key to conducting valid comparisons between the two is to implement controls by comparing multiple lines of evidence of the soil subsamples from the cores with benchtop LIF of those very same subsample intervals.

Figure 8.29 contains a "data dashboard" style assembly of data from validation of a fairly typical UVOST log that was acquired at a gasoline NAPL release site. Validation of this particular field log was requested by the client, so a core was retrieved from the same depth interval as the LIF log response using a direct-push sampler. The core was taken less than one meter away from the in situ UVOST log. Recovery was nearly 100%, the core was capped, shipped unfrozen to the LIF provider, frozen for 24 h, split open with a masonry saw and the exposed soil surface was scraped clean. The clean face was screened continuously with UVOST by sliding the sapphire window sensor along the core's face, the core was photographed in visible and UV lamp light, then sprayed with a proprietary NAPL indicator dye, and photographed under UV again. The in situ log is at far left of Fig. 8.29, the benchtop UVOST log is at center, and core photos are located at right. Hanging the multiple lines of evidence in a vertical fashion makes analysis a lot easier for most than staring at spreadsheets of data.

The first thing one notices in Fig. 8.29 is that the in situ LIF log at left is noticeably different than the ex situ LIF log adjacent to it. The NAPL in the in situ log appeared to be one half meter of depth deeper than the ex situ log's NAPL response. The cause for this depth mismatch is, as usual, a mystery. It could have been human error, NAPL heterogeneity in the soil, direct-push rod flex, or some combination. In addition, the ex situ LIF result is lower signal with more weathered looking (red-shifted) waveforms compared to in situ response. This difference is consistent with our experience for benchtop validation works, especially in the case of gasolines. The difference is likely due to both molecular oxygen quenching and sample handling induced weathering affecting the ex-situ LIF response.

The ex-situ LIF responses match up almost seamlessly with the fluoresce photography and dye testing. Only by conducting such "apples to apples" analyses on narrow intervals, with multiple NAPL-sensitive methods, are were able to successfully control for NAPL heterogeneity and/or any human or mechanical error. Notice how poorly the downhole data (green diamonds) correlates with all the other lines of evidence. This is because the green diamond data points were gathered from different soils than all the other lines of evidence. All other lines of evidence were gathered from the exact same soils, thus the up-hole data have good correlation.

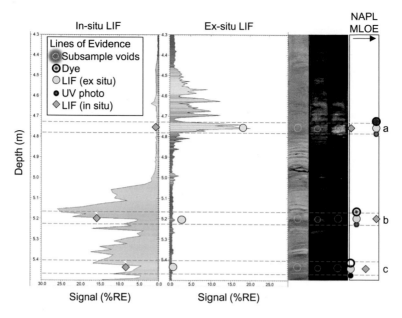

Fig. 8.29 Validation of a UVOST log with benchtop UVOST, photography, and emulated NAPL validation at narrow intervals

8.7   Tar-Specific Green Optical Screening Tool (TarGOST®)

In the UVOST discussion so far, we described the alarmingly low response generated by heavy (dense) NAPLs such as coal tars, creosotes, and bunker fuels. At best, UV-induced fluorescence intensity of heavies rises a bit at low NAPL saturations but then plateaus, showing no additional response with increasing saturation. At their worst the fluorescence response of heavies begin to decrease as NAPL saturation increases, making UV-induced fluorescence unsuitable for screening for heavies due to the high risk of false negatives. This low fluorescence response might be tolerable if it were not for false positive fluorescence emitted by organic and soils, which normally contribute only a subdued fluorescence relative to the brightly fluorescing lighter NAPLs such as refined fuels. But false positives are often able to emit enough fluorescence to make total fluorescence intensity alone a poor indicator of their presence. And in some case, the false positives fluoresce just as brightly, if not brighter, than heavy NAPLs, making it nearly impossible to screen for heavy NAPLs without grossly exaggerating their footprint. Finally, there is also the fatal flaw of the fluorescence of aqueous phase 2–3 ring PAHs that often reach high concentrations in pore waters adjacent to heavy NAPL (Fig. 8.20) and are easily mistaken for an LNAPL by UV-induced fluorescence sensors.

The TarGOST form of LIF was introduced in 2004 as a solution to these issues and was rigorously vetted for use on coal tar NAPLs at former manufactured gas plants (MGPs) (Coleman et al. 2006). TarGOST differs from UVOST in that it responds monotonically to the vast majority of heavy NAPLs. TarGOST waveforms also more capably discern differences between heavy NAPL chemistries and the false positives of organic materials that so often exist in quantity at wood treater (creosote) and former MGP sites (due to being built alongside water bodies). TarGOST is virtually unresponsive to dissolved phase PAH fluorescence, eliminating the risk that this false positive introduces to UV-induced fluorescence. One very rare exception are acid tars, which can source acid form (water soluble) PAHs to groundwater which TarGOST can respond to.

TarGOST's waveform callouts, cluster diagrams, and associated reporting are similar to that previously discussed. The excitation and fluorescence emission wavelengths were simply shifted to lower energies (longer wavelengths). This was done to specifically target directly the exceptionally large PAHs that dominate the fluorescence of heavies thus passing over to, some extent, the energy transfer quenching observed with UV-excitation fluorescence.

Heavy NAPLs are often DNAPLs, so the herding effect that gravity and LNAPL's buoyancy have on driving LNAPLs toward the groundwater's potentiometric surface does not apply to them. This causes heavies like coal tars to sink and crawl laterally, distributing themselves even more heterogeneously than LNAPLs. Compound this with the fact that heavy NAPL sites have had up to a century for the released NAPL to distribute themselves, and the architecture of DNAPL bodies is inevitably more complicated and chaotic than that of LNAPLs. There are occasionally very large and important sites, where addressing the nature and extent of any and all petroleum LNAPLs, coal tar and creosote (DNAPLs) in one logging event is considered critical and worth the extra expense. Dual-windowed subs are available for such occasions which allow both UVOST and TarGOST to occur simultaneously in the same logging event (Tomlinson et al. 2017). Dual LIF logging assures that any NAPL encountered will have at least one LIF system capable of responding monotonically, almost foolproof recognition of NAPL type and improved identification of false positives.

Figure 8.30 contains a fairly typical TarGOST log. Log interpretation remains much the same as for UVOST logs but with some key differences. One difference is the introduction of an excitation laser scatter channel, which replaces the left-most fluorescence channel, leaving the last three channels of the waveform to represent fluorescence. The laser scatter (A) and fluorescence (B) channels are processed separately and displayed in their own columns. Signal %RE (C) is calculated using a proprietary method that results in a response that scales monotonically with the presence of heavy NAPLs. Note that UVOST and TarGOST use different RE substances, so there is not a direct relationship in %RE values across TarGOST versus UVOST.

The cluster diagram bubble positions for TarGOST logs are based only on the three fluorescence channels (at right in the waveforms) to characterize the fluorescence colors and lifetimes as is done with UVOST's four channels. Bluer colors (lighter NAPLs) plot on the left of the x-axis and redder colors (heavier NAPLs) on the right, with average lifetimes plotted on the y-axis. Heavy NAPL fluorescence lifetimes observed with TarGOST are never as long-lived as UVOST LNAPL fluorescence, so the lifetime axis of the cluster is far more sensitive to even minor lifetime shifts than

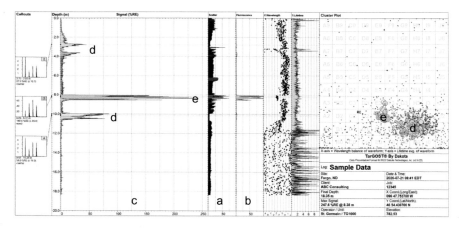

Fig. 8.30 TarGOST log consisting of former MGP coal tar NAPL in callouts 2 and 4, with a layer of woody debris in callout 3

the UVOST cluster diagram. Contained in both the upper and lower intervals (D) of the log in Fig. 8.30 is the coal tar DNAPL that was being targeted for characterization. The turquoise response in interval E was immediately recognized as not matching the known coal tar waveform, so that interval was sampled and found to be a layer of woody debris.

It is important to recognize that interpretation of this log based on the fluorescence intensity alone (column B) is folly because the false positive (wood) is dwarfing the coal tar NAPL's fluorescence. This is a common occurrence at wood treater and former MGP sites. In this particular case, lifetimes alone did not serve as a sufficient aid for differentiation of the two, but the fluorescence color combined with the lifetimes of the two were quite different. In other cases, the fluorescence color of target NAPL and false positive are identical, but lifetime differences allow for their differentiation and successful data filtering via NNLS techniques. In the rare cases (well over 500 sites have been investigated with TarGOST to date) that significant false positive organics and the target NAPL fluoresce are identical in both color and lifetime, interpretation of the TarGOST logs is limited in its specificity, which dramatically reduces our confidence in the LIF site model's veracity. It is difficult to overstate how often fluorescence of heavy NAPLs and natural organics are encountered at the same site and need to be separated from each other. In fact, it is more often the rule than the exception. This issue is effectively dealt with by applying NNLS fitting and other approaches (St. Germain 2011), but it is something that will almost certainly occur, so it is important that one budgets for some limited validation and processing, and work closely with your service provider to stay alert for its inevitable occurrence.

Because heavy NAPLs have such a high PAH to solvent NAPL ratio, self-quenching of fluorescence between PAHs is significant, causing generally short lifetimes in all heavy NAPLs. Molecular oxygen quenching of NAPL, a major influencer in benchtop UV-induced fluorescence measurements of LNAPLs discussed previously, has not been observed to be an issue for benchtop TarGOST fluorescence measurements of heavy NAPLs. In addition, weathering-induced loss of PAHs appears to have a negligible effect on the fluorescence color or lifetimes of heavies, because it is likely that there is such a surplus of PAHs of all sizes, that weathering's minor effects are not noticeable with regard to fluorescence.

TarGOST waveforms are capable of general discernment between NAPL classes, in a manner similar to UVOST. Figure 8.31 depicts cluster diagram bubbles from a suite of medium to heavy NAPLs, along with their waveforms.

The trend for TarGOST fluorescence, like UVOST, is for the smallest of these larger PAHs to fluoresce toward the bluer wavelength side (left) of the diagram and red-shift with increasing size of PAH (and degree of energy transfer) toward the right (redder wavelength) side. The heavier the NAPL, the shorter the lifetime typically, so they are located toward the bottom of the diagram (low in the lifetime

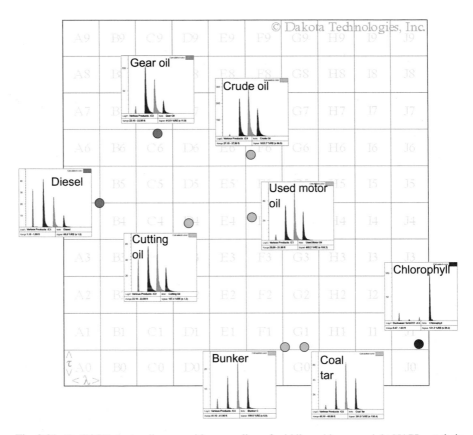

Fig. 8.31 TarGOST cluster diagram with a sampling of middle and heavy weight NAPLs at their positions on the cluster diagram

axis). Chlorophyll's commonly encountered fluorescence waveform was included here because probing through green plant material of any type, such as grass on soil or moss on the sediment surface during a barge project, commonly causes this readily recognizable waveform. Fluorescent orange or hot pink survey paint also generates a waveform that is very similar to that of chlorophyll.

8.8   Dye-Enhanced Laser-Induced Fluorescence (DyeLIF™)

The chemical structure of familiar contaminants such as TCE, as well as monoaromatic NAPLs such as benzene and toluene, does not support the energy absorbance and emission processes necessary to be detected using UVOST or TarGOST as it is in the case of PAHs. These NAPLs can produce a LIF response because they often contain enough PAH impurities from their manufacture, PAHs picked up during their use to clean or degrease, and may even contain naturally occurring fluorophores that were solvated after the NAPL's release into the environment. In the case of chlorinated solvents, these DNAPLs will also solvate a host of non-fluorescent molecules including optical chromophores and other non-fluorescent molecules that simply absorb the excitation laser light and/or quench the fluorescence of excited state fluorescent molecules. For this reason, logging tool developers and providers of LIF systems have generally avoided applying LIF toward these "unpredictable" chlorinated DNAPL releases, for fear of liability of reporting "clean" logs that were actually contaminated with DNAPL and the general protection of LIF's reputation as being reliable for indicating NAPLs.

In order to improve on the reliability of applying LIF toward non- or modestly fluorescing NAPLs (including monoaromatic NAPLs such as benzene and toluene), a viable technique was conceived in 2009 and eventually commercialized by a team of direct-push technology developers and consultants (Einarson et al. 2018; St. Germain 2014).

The DyeLIF version of LIF relies on injection of a fluorescent dye onto the soil, several centimeters ahead of the LIF sensor, so as to ensure a response from any NAPLs present. This approach is similar in fashion to the visual dye testing for the presence of NAPLs in soil samples using Oil Red O and Sudan IV dyes shake tests (Cohen et al. 1992). The DyeLIF system is basically a specialized form of LIF that retains many of the same time-resolved elements as the TarGOST and UVOST systems but with the expansion to address non-fluorescing NAPLs not normally detectable by UVOST and/or TarGOST.

Figure 8.32 illustrates the general concept behind the DyeLIF method, which involves injecting a NAPL-indicating dye below the LIF window (at approximately 60 mL/min) in order to render any NAPL fluorescent, which is detected by the LIF system following closely behind. Rather than the dye color changing from a black to red color like Oil Red O or Sudan IV chromatic dye testing, DyeLIF's indicator dye transitions from fluorescing very weakly to fluorescing orders of magnitude more intensely when it contacts a NAPL capable of solvating the dye. Its fluorescence color

276   R. St. Germain

also blue-shifts dramatically, along with an increase in its fluorescence lifetime. As
with other forms of LIF, this behavior is continuously recorded with depth by storing
the series of waveforms that had been generated.

The dye fluid injection process is monitored by sensors that allow for measurement
of the pressure required to inject the aqueous dye fluid into the soil as the probe is
advanced into the subsurface. This allows for estimation of the potentiometric surface
elevation and hydraulic conductivity using methods similar to those described in
Chap. 7 for the HPT.

The DyeLIF technology has been rigorously tested under an Environmental Security Technology Certification Program (ESTCP) funded validation study (Einarson
et al. 2016) and has been successfully applied at sites across the United States. In order
to improve confidence in the resulting fluorescence response logs, every DyeLIF

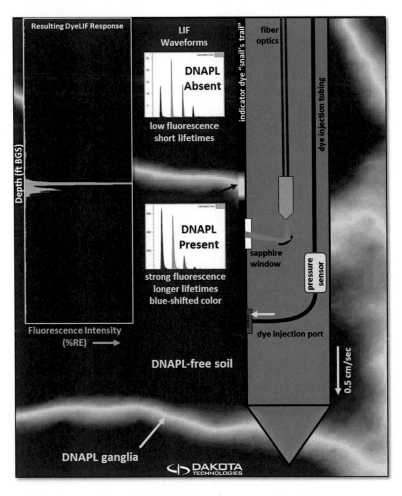

Fig. 8.32 Conceptual diagram of a DyeLIF probe passing through a DNAPL-impacted formation

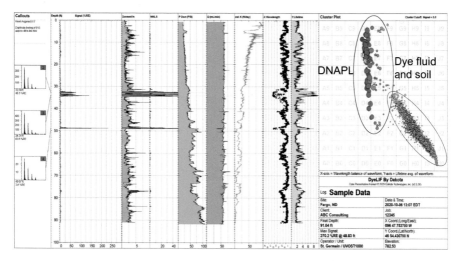

Fig. 8.33 Typical DyeLIF field log

project relies extensively on targeted validation sampling techniques. Recovered cores are screened using the MLOE benchtop screening approach developed for the ESTCP validation study in order to meet the challenges of validating in situ responses with coring in the face of extreme heterogeneity that DNAPLs are famous for. The validation approach has been steadily improved upon by innovative consultants who are using the DyeLIF to delineate DNAPL at their sites (Horst et al. 2018).

Figure 8.33 contains a typical DyeLIF field log. DyeLIF data is acquired at higher density (~ 0.5 cm spacing) than other LIF so as to more effectively detect even small DNAPL ganglia. Pressure and flow of the dye fluid (mostly water) injection is monitored and used to produce estimated hydraulic conductivity values. Results of NNLS analysis of the waveforms (processed after MLOE validation is finished) is plotted in red and indicates those responses considered to be highly confident for DNAPL fluorescence (i.e., all "false positives" have largely been removed via NNLS).

The proprietary dye that acts as the NAPL indicator for DyeLIF has been found to perform well with monoaromatic NAPLs such as benzene and toluene, chlorobenzenes, DCA, TCA, PCE, TCE, chloroform, Freon, and a host of other typically nonfluorescent NAPLs. DyeLIF has responded to a wide suite of halogenated solvents but bench-testing of site-specific NAPL (or testing of reagent grade solvents of the same variety in question) is always advised. Nitrobenzenes are a notable exception, having been found to quench the indicator dye fluorescence completely in benchtop testing.

DyeLIF has no response to dissolved, vapor, or adsorbed phase chlorinated solvent molecules, making it truly NAPL-specific. DyeLIF, like TarGOST, is relatively blind to the smaller 2–4-ring PAHs responsible for the bulk of the fluorescence for many light fuel NAPLs such as kerosene and gasoline. However, DyeLIF does respond well to the larger PAHs present in diesel and other heavier petroleum NAPLs, so the presence of these NAPLs at the site will likely confound DyeLIF's ability to provide a halogenated NAPL-specific response but instead a general NAPL response.

Proper grouting is crucial at sites where DyeLIF is deployed. Typically, the most confident method available (tremie from bottom up) is employed to prevent migration of DNAPL down DyeLIF or sampling boring preferential pathways. It is most common that the DyeLIF push rig is kept busy all day with one rig in order to maximize daily production for the DyeLIF system. Because co-sampling and grouting is going to be necessary no matter what, it is typical to dedicate a second (usually less expensive rig) to tremie grouting and/or validation sampling.

8.9 General Best Practices for LIF

As discussed in previous sections, LIF can provide a wealth of valuable information regarding NAPL nature and extent, but suffers enough interferences and exceptions to the norm that all LIF data deserves careful scrutiny.

- Because LIF technology provides real-time results, plans should be put in place to adapt to the results in real time so as to make optimal use of the investigation budget. Expect the unexpected with regard to both the NAPL architecture and chemistry.
- Ask the LIF vendor to test site NAPL samples if they are available in order to ascertain that the target NAPL has an appropriately intense fluorescence response and to check for monotonicity of the response with changing NAPL saturation (across three or four orders of magnitude).
- Budget for limited targeted validation sampling. This will enable you to validate your LIF data interpretation which will leave all stakeholders confident in their ability to take substantive action based on the LIF NAPL CSM. Remember that false positive fluorescence must be identified and removed so that the LIF CSM represents the NAPL CSM.
- Recognize and factor into your interpretation the fact that different NAPLs have inherently different fluorescence intensities. For example, applying diesel's highly fluorescent response characteristics to gasoline NAPL will cause one to grossly underestimate gasoline NAPL impacts.
- Be careful not to apply irrationally high signal %RE thresholds for indication of NAPL at gasoline sites such as 5, 10, or even 20 %RE because even the smallest signal %RE responses indicate gasoline NAPL (as long as the waveforms supports it being gasoline).
- Background fluorescence (the waveform generated by a clean sapphire window) should never exceed 1%RE, with < 0.5 %RE being an optimum value to strive to achieve—low background makes small NAPL fluorescence responses outside the window far easier to detect.
- With UVOST one should consider very low fluorescence (< 1%RE) to be representative of NAPL (i.e., "getting down into the weeds" to detect NAPL) if the color and lifetime are supportive. It might seem like overkill, or a practice that only researchers might worry about, but a goal with LIF is to bound the NAPL

body, to find its outer limits so the NAPL CSM footprint can be developed. That is not possible unless you are probing the outer fringes of the NAPL body, which by definition will eventually produce small enough signals so that one has to carefully scrutinize the LIF logs in order to detect them.

- Continue LIF logging well below the CSM's groundwater surface, being careful not buy into the outdated "floating pancake" theory that has traditionally resulted in missing substantial trapped NAPL mass. Going one or two rod lengths below the water table costs relatively little and often reveals NAPL impacts essential to building a sound NAPL CSM (ITRC 2019).
- When pushing next to a well containing NAPL as a "test" of LIF, do not expect to encounter a response on the first log, because the NAPL is almost certainly not in a uniform layer. NAPL heterogeneity is often extreme and many times it takes two or more logging events next to a well before one intersects the small NAPL yielding preferential pathway that is sourcing NAPL to the well.
- The probe advancement rate must be monitored for deviations to account for associated response variations. The push rate should not exceed that specified for the method. Advancement rates slower than the standard rate simply produce higher data densities with minimal risk to data quality but negatively impacting daily production while pushing too rapidly causes risk of missing small seams of NAPL and distorts the nuances of geology that are routinely revealed in LIF logs.
- Electrical conductivity and hydraulic profiling add-ons can be advanced along with fluorescence methods on the same tool, often providing important clues as to why the NAPL is located where it is and how geology is influencing that distribution.
- Do not wait long after an LIF project to critically review the LIF data, do it while all the lines of evidence are fresh in everyone's minds.
- Discuss with the LIF operator how hard they are to push before advancement becomes difficult, to reduce the likelihood of equipment breakage or becoming stuck. A little caution will likely produce more data in the long run due to repair time if one is too aggressive.
- Assure your LIF provider is following proper setup/alignment of the Shock-Protected Optical Compartment (SPOC) and evaluating for (and mitigating for) fogging of the SPOC which occurs inside sapphire window when warm probes are pushed into cold ground. Fogging interferes with the LIF response and can be evaluated easily by holding ice up against the outside of the window and examining the inside of the window for fogging.
- Following advancement of each LIF push, the operator must inspect the sapphire window for fogging, chips, cracks, or other problems and remedy them before continuing—an example would be NAPL getting inside the probe due to a leak in the probe seals, causing a constant elevated fluorescence response due to this leaked NAPL inside which masks any true NAPL outside during logging.
- Low viscosity NAPLs such as gasoline need to be applied to the window on a tiny "anthill" of moist sand piled onto the probe window, so as to prevent thin-sheeting and evaporation that produce artificially low fluorescent responses for bench testing of volatile NAPLs like gasolines.

- Benchtop NAPL analysis can also be done following the field work rather than prior, but this does deny the investigators the ability to factor their findings (what waveforms shapes to watch for) into real-time interpretation of the logs during the investigation.
- As is the case in any drilling and direct-push investigation of high-risk contaminants, one should avoid puncturing through to lower aquifers and causing cross-contamination when possible. That said, it is surprisingly often the case that what previously had been considered a continuous and competent geologic feature that has not been penetrated by the DNAPL indeed has been (either through unrecognized natural pathways or previous investigative work such as poorly designed or improperly installed monitoring wells).

8.10   Conclusions

It is useful to ask ourselves what aspects of applying HRSC to NAPL delineation (seeing the logs that reveal where the NAPL is) appeals to us. It may be that it frees us from our previously simplistic beliefs regarding NAPL behavior, beliefs that we were forced to adopt because of our inability to imagine its truly chaotic nature—thus we collectively simplified it in our minds early on in the development of our field of practice. This "slow to learn our lessons" approach is not intuitive, especially in light of the fact that people understood and dealt with geology's heterogeneity long ago. Karl von Terzaghi, the father of soil mechanics (a discipline which predates environmental soil work by over a century) described this challenge for structural engineering site characterization almost a century ago:

> Unfortunately, soils are made by nature and not by man, and the products of nature are always complex… As soon as we pass from steel and concrete to earth, the omnipotence of theory ceases to exist. Natural soil is never uniform. Its properties change from point to point while our **knowledge of its properties are limited to those few spots at which the samples have been collected**. In soil mechanics the accuracy of computed results never exceeds that of a crude estimate, and the principal function of theory consists in teaching us what and how to observe in the field.
>
> ~ Karl von Terzaghi, 1936

It is hard to imagine anyone having said it better with regard to NAPL characterization as well. But it did take an unexpectedly long time for in situ "screening tools" to prove themselves to be much more capable of properly characterizing complex NAPL bodies than soil sampling. Whatever the reasons, it was somehow difficult for us to move beyond coarse discrete sampling and move to high-resolution high-density screening level estimates. Numerous federally funded programs such as Triad, a federal/state interagency partnership in the U.S., were developed to address this innate hesitancy, but changing the groupthink of a very large regulation-driven industry turns out to be a slow process.

Consulting engineers have contributed greatly to HRSC's acceptance, not only by utilizing newly introduced tools and proving their worth in practice, but by offering advice and counsel on improvements of the tool and even conceiving of new techniques and tools for service providers to commercialize. Unfortunately, many consulting firms in smaller markets are staffed with personnel who lack the support and guidance to help them recognize the benefits of HRSC, so they remain comfortable "doing things the way it is always been done". Programs like Interstate Technology and Regulatory Council (ITRC) in the U.S. have been instrumental in bringing awareness to smaller agencies and engineering firms who do not always have the budget or manpower to hire and train internal experts who can stay abreast of the latest advances in theory and technology.

As we look back on the tools and methods discussed in this chapter, we can see a consistent pattern: petroleum NAPL's tremendous chemical complexity, combined with geology's complexity, is fraught with mirages and complexities that continually vex our ability to characterize a NAPL body's true architecture. Fortunately, we have compiled enough of an understanding of the data generated by HRSC tools such as MIP and LIF that pieces of the puzzle are beginning to fit together and reveal NAPLs' secrets. As we look to the future, it is clear that advances in technology will continue to drive innovation in every aspect of our lives, and the petroleum release characterization industry is no exception.

Assimilation of numerous sensors into the same probe is one area of advancement which allows more data to be gathered simultaneously in one logging event. But we must keep practicality in mind because, while tool manufacturers might succeed in building such "mega-probes", the service providers in the field (and subsequently their clients) will have to pay for those increasingly expensive devices. HRSC tool developers ply their trade at the intersection of two very different worlds. They are tasked with taking fairly sensitive and complex instruments and ruggedizing them enough to withstand, quite literally, a severe beating at the end of a very large jackhammer that shoves them 30 m or more into the ground. And they probe not just soft soils, but a variety of difficult probing soils, including conditions so difficult they inevitably prove to be too much for the probe, leading to damage and even loss. Another challenge for these newer mega-probes is that keeping all those sensors calibrated and functioning properly becomes, at some point, too much for even the most skilled and experienced operators. Their complexity makes mega-probes more prone to failure simply because there are simply more things that can go wrong, and the result is that you sometimes spend less time probing versus servicing the probe.

The majority of HRSC NAPL characterization projects are conducted in one to possible three mobilizations over a period of months at most, which results in a comprehensive sitewide snapshot of the NAPL's nature and distribution. This is, in fact, one of HRSC's benefits in that we are assessing the entire petroleum NAPL body in one generally continuous effort. This affords greater confidence in the NAPL CSM because the data are acquired under consistent conditions and using "machine vision" rather than a hodgepodge of reports generated using a variety of conventional methods over a decades' long string of sporadic visits. This snapshot in time introduces the opportunity for another area of advancement in HRSC, employing HRSC as a pre- and post-treatment investigation tool. One example of pre- and post-remedy utilization includes UVOST NAPL surveys being conducted at sites where activated carbon (AC) treatment is being injected to treat gasoline releases (technology further discussed in Chap. 16). The initial UVOST NAPL survey is used to guide the placement of the AC. If, after the prescribed time period for effective performance of the AC has passed and there is evidence that the AC's performance is not meeting objectives, follow-up UVOST logging is brought in to determine the cause. Perhaps the AC injection displaced the NAPL off-site, or the AC injection was designed without LIF so its design failed to place AC in contact with source term NAPL or VOCs. Interestingly, high concentrations of AC mixed with NAPL entirely eliminate the NAPL's fluorescence due to AC's black color and its ability to absorb the larger visible-wavelength fluorescing PAHs. However, partial AC contact with NAPL leaves a telltale sign of partial contact with NAPL by severely blue-shifting NAPL's fluorescence to UV only, because the very smallest PAHs (or highly substituted monoaromatics) are not sorbed effectively by AC. Post-treatment UVOST surveys successfully identify those parts of the NAPL body untouched by the AC injection, those NAPLs only modestly contacted by AC, and absence of NAPL fluorescence tells us that either the injection was successful (significant AC presence) or the NAPL is not there at all. Labeling of the AC with a fluorescent tag to allow LIF to respond to AC specifically is being explored, and if successful would aid in our understanding the AC's exact distribution during the post-treatment LIF NAPL survey.

Another promising use of pre- and post-treatment HRSC involves TarGOST LIF surveying of coal tar at former manufactured gas plants. One method of remediating these stubborn NAPLs is solvation of coal tar into the dissolved phase using surfactants or natural water-miscible solvents like d-limonene, which is followed up with collection well recovery or natural attenuation (EPRI 1993). Fortunately, solvation of coal tar is readily recognized in TarGOST logs due to dramatically blue-shifted and longer lived waveforms relative to intact coal tar's waveforms which are red-shifted and extremely short-lived. Thus a pre-treatment TarGOST survey can be done prior to remediation to guide d-limonene injection well and recovery well placement. After the solvation treatment and recovery period, co-located follow-up TarGOST logging precisely indicates those depth intervals that had total, partial, and ineffective removal of coal tar. This gives the practitioners HRSC information with which to go back in and "mop up" the remaining coal tar that the first round of treatment has missed.

In addition to direct-push logging methods discussed here, researchers and developers continue to pursue other methods of characterizing NAPL in the field, including electrical resistivity imaging to detect NAPL bodies (Halihan et al. 2017)—which is also discussed in 9—screening cores and discrete samples with handheld mid-infrared field instruments (Webster et al. 2016), NAPL indicating liners designed to indicate NAPL in boreholes (NAPL FLUTe), continuous NAPL profiling in shallow sediments using solid phase extraction based LIF samplers (EPRI 2007), or nuclear magnetic resonance logging tools for detecting and quantifying LNAPL in situ (Spurlin et al. 2019)—method that is discussed in Chaps. 7 and 9.

Some researchers are beginning to tackle measuring subsurface NAPL transport dynamics in the field, for instance, throughout the measurement of NAPL transmissivity using tracer dye dilution in wells (Pennington et al. 2016) or observing NAPL saturation changes over time and space with UV-transparent wells (Zimbron 2020).

It will be interesting to see where advances in NAPL HRSC methodology take us in the future. The latest generation of practitioners having grown up with advanced technology in the palm of their hand, so we can expect researchers and tool developers to develop capabilities and performance we cannot currently imagine. And, because technology is now so integral in our daily lives, it is likely that the adoption of these new approaches will proceed much more rapidly than when HRSC was first introduced 30 years ago, when personal computers for home use, the internet, and cell phones were just starting to go mainstream.

References

Adamson DT, Chapman S, Mahler N, Newell C, Parker B, Pitkin S, Rossi M, Singletary M (2014) Membrane interface probe protocol for contaminants in low-permeability zones. Groundwater 52:550–565. https://doi.org/10.1111/gwat.12085

Alostaz MD, Biggar K, Donahue R, Hall G (2008) Soil type effects on petroleum contamination characterization using ultraviolet induced fluorescence excitation-emission matrices (EEMs) and parallel factor analysis (PARAFAC). J Environ Eng Sci 7(6):661–675. https://doi.org/10.1139/S08-037

Apitz SE, Borbridge LM, Bracchi K, Lieberman SH (1992) The fluorescent response of fuels in soils: insights into fuel-soil interactions. SPIE Int Soc Opt Photonics 1716:139–147

Berlman IB (1965) Handbook of fluorescence spectra of aromatic molecules. Academic Press Inc., New York

Brooks MC, Yarney E, Huang J (2020) Strategies for managing risk due to back diffusion. Ground Water Monit Remediat 41(1):76–98. https://doi.org/10.1111/gwmr.12423

Bujewski G, Rutherford B (1997) The rapid optical screening tool (ROST™), laser-induced fluorescence (LIF) system for screening of petroleum hydrocarbons in subsurface soils, innovative technology verification report, Sandia National Laboratories

Bumberger J, Peisker K, Reiche N, Radny D, Dietrich P (2016) A triggered depth-dependent sampling system to overcome the carry-over effects of the membrane interface probe. Groundwater Monit R 36:54–61. https://doi.org/10.1111/gwmr.12163

Bumberger J, Radny D, Berndsen A, Goblirsch T, Flachowsky J, Dietrich P (2012) Carry-over effects of the membrane interface probe. Ground Water 50(4):578–84. https://doi.org/10.1111/j.1745-6584.2011.00879.x

Byker G (2021). http://naplansr.com/sediment-sample-collection-and-field-screening-to-inform-napl-mobility-and-migration-evaluations/

Chin JY, Batterman SA (2012) VOC composition of current motor vehicle fuels and vapors, and collinearity analyses for receptor modeling. Chemosphere 86(9):951–958. https://doi.org/10.1016/j.chemosphere.2011.11.017

Christy TM (1996) A drivable permeable membrane sensor for the detection of volatile compounds in soil. In: National ground water associations outdoor action conference. Las Vegas, Nevada

Christy TM, Pipp DA, McCall W (2015) Membrane interface probe protocol for contaminants in low-permeability zones. Groundwater 53:185–186. https://doi.org/10.1111/gwat.12305

Cohen RM, Bryda AP, Shaw ST, Spalding CP (1992) Evaluation of visual methods to detect NAPL in soil and water. Groundwater Monit Rem 12:132–141. https://doi.org/10.1111/j.1745-6592.1992.tb00072.x

Coleman A, Nakles D, McCabe M, DiGnazio F, Illangasekare T, St. Germain R (2006) Development of a characterization and assessment framework for coal tar at MGP sites. EPRI, Palo Alto, CA, p 1010137

Costanza J, Pennell K, Rossabi J, Riha B (2002) Effect of temperature and pressure on the MIP sample collection process. In: Proceedings of the third international conference on remediation of chlorinated and recalcitrant compounds, pp 367–372

Einarson M, Fure AD, St Germain D, Parker B, Chapman SW (2016) Direct-push optical screening tool for high-resolution, real-time mapping of chlorinated solvent DNAPL architecture. ESTCP project report ER-201121

Einarson M, Fure A, St. Germain R, Chapman S, Parker B (2018) DyeLIF™: a new direct-push laser-induced fluorescence sensor system for chlorinated solvent DNAPL and other non-naturally fluorescing NAPLs. Groundwater Monit Remediat 38(3):28–42

EPRI (1993) Solvent extraction for remediation of manufactured gas plant sites, electric power research institute (EPRI), 1993, Project 3072–02

EPRI (2007) A continuous in-situ dart profiling system for characterizing MGP coal tar and PAH impacts in sediments: a technology using laser-induced fluorescence in sediments Electric Power Research Institute (EPRI). https://www.epri.com/research/products/000000000001014749

Ernest Mott-Smith PE, Cal Butler PG, Ed Hicks PE (2014) Evaluation of TarGOST investigations for multiple hazardous waste sites in the southeast, black and Veatch. In: 5th International symposium and exhibition on the redevelopment of manufactured gas plant sites (MGP 2014)

García-Rincón J, Gatsios E, Rayner JL, McLaughlan RG, Davis GB (2020) Laser-induced fluorescence logging as a high-resolution characterisation tool to assess LNAPL mobility. Sci Total Environ 725:138480. ISSN 0048-9697. https://doi.org/10.1016/j.scitotenv.2020.138480

Halihan T, Sefa V, Sale T, Lyverse M (2017) Mechanism for detecting NAPL using electrical resistivity imaging. J Contam Hydrol 205:57–69. ISSN 0169-7722. https://doi.org/10.1016/j.jconhyd.2017.08.007. https://www.sciencedirect.com/science/article/pii/S0169772216302157

Horst J, Welty N, Stuetzle R, Wenzel R, St. Germain R (2018) Fluorescent dyes: a new weapon for conquering DNAPL characterization. Groundwater Monit Rem 38(1):19–25 . https://triadcentral.clu-in.org/over/index.cfm

ITRC (Interstate Technology & Regulatory Council) (2019) Implementing advanced site characterization tools. ASCT-1. Washington, D.C.: Interstate Technology & Regulatory Council, Advanced Site Characterization Tools Team. https://asct-1.itrcweb.org/

Knowles DS, Lieberman SH (1995) Field results from the SCAPS laser-induced fluorescence (LIF) sensor for in-situ subsurface detection of petroleum hydrocarbons. In: Proceedings of SPIE 2504, environmental monitoring and hazardous waste site remediation. https://doi.org/10.1117/12.224113

Kram ML, Keller AA (2004) Complex NAPL site characterization using fluorescence, part 2: analysis of soil matrix effects on the excitation/emission matrix. Soil Sediment Contam Int J 13(2):119–134

Kram ML, Keller AA (2004) Complex NAPL site characterization using fluorescence, part 3: detection capabilities for specific excitation sources. Soil Sediment Contam Int J 13(2):135–148

Kram ML, Keller AA, Massick SM, Laverman LE (2004) Complex NAPL site characterization using fluorescence, part 1: selection of excitation wavelength based on NAPL composition. Soil Sediment Contam Int J 13(2):103–118

Kram (2008). https://serdp-estcp.org/Program-Areas/Environmental-Restoration/Contaminated-Groundwater/ER-200421/ER-200421

Lieberman (2007) Direct-push chemical sensors for DNAPL and other VOCs, Dr. Stephen Lieberman, space and naval warfare systems command, ER-200109. https://www.serdp-estcp.org/Program-Areas/Environmental-Restoration/Contaminated-Groundwater/Monitoring/ER-200109

Lieberman SH, Theriault GA, Cooper SS, Malone PG, Olsen RS, Lurk PW (1991) Rapid subsurface in situ field screening of petroleum hydrocarbon contamination using laser-induced fluorescence over optical fibers. In: Proceedings of second international symposium on field screening methods for hazardous wastes and toxic chemicals, U.S. Environmental Protection Agency, Las Vegas, pp 57–63

Lu J, St. Germain RS, Andrews T (2014) NAPL source identification utilizing data from laser induced fluorescence (LIF) screening tools. In: Proceedings of the 2013 INEF conference on environmental forensics, pp 77–97

McCall W, Christy TM, Pipp DA et al (2018) Evaluation and application of the optical image profiler (OIP) a direct-push probe for photo-logging UV-induced fluorescence of petroleum hydrocarbons. Environ Earth Sci 77:374. https://doi.org/10.1007/s12665-018-7442-2

McDonald S, Prabhu C, Gbondo-Tugbawa S, Weissbard R, St. Germain R (2018) Confirming laser-induced fluorescence NAPL delineation in Newtown creek superfund site. https://www.battelle.org/docs/default-source/conferences/chlorinated-conference/proceedings/2018-chlori nated-conference-proceedings/h3-high-resolution-site-characterization/h3_1005_-683_mcd onald.pptx?sfvrsn=de138088_0

Meidinger R, St. Germain RW, Dohotariu V, Gillispie GD (1993) Fluorescence of aromatic hydrocarbons in aqueous solution. Field screening methods for hazardous wastes and toxic chemicals. Las Vegas, NV, pp 395–403

NAPL FLUTe. https://www.flut.com/napl-flute

Nielsen BJ, Gillispie G, Bohne DA, Lindstrom DR (1995) A new site characterization and monitoring technology. In: Vo-Dinh T (ed) Environmental monitoring and hazardous waste site remediation: proceedings of international society for optics and photonics (SPIE), vol 2504, pp 278–290

Okin MB, Carroll SM, Fisher WR, St. Germain RW (2006) Case study: confirmation of TarGOST laser-induced fluorescence DNAPL delineation with soil boring data. Land Contam Reclam 14(2):502–507(6)

Parmenter CS, Rau JD (1969) Fluorescence quenching in aromatic hydrocarbons by oxygen. J Chem Phys 51:2242–2246. https://doi.org/10.1063/1.1672322

Pennington A, Smith J, Koons B, Divine CE (2016) Comparative evaluation of single-well LNAPL tracer testing at five sites. Groundwater Monit R 36:45–58. https://doi.org/10.1111/gwmr.12155

Spurlin M, Barker B, Cross B, Divine C (2019) Nuclear magnetic resonance logging: example applications of an emerging tool for environmental investigations. Remediat J 29:63–73. https://doi.org/10.1002/rem.21590

St. Germain R (2011) Laser-induced fluorescence (LIF) primer. Appl NAPL Sci Rev Demystifying NAPL Sci Remediat Manag 1(9)

St. Germain R (2014) Eliminating natural organic interference from TarGOST® data using advanced waveform analysis, poster. In: The Fifth international symposium and exhibition on the redevelopment of manufactured gas plant sites (MGP 2014). https://www.dakotatechno logies.com/docs/default-source/presentations-and-papers/eliminating-natural-organic-interfere nce-from-targost-data-mgp14.pdf?sfvrsn=5b33c331_6

St. Germain R, Adamek S, Rudolph T (2006) In situ characterization of NAPL with TarGOST® at MGP sites. Land Contam Reclam 14:573–578. https://doi.org/10.2462/09670513.741

St. Germain RW, Einarson MD, Fure A, Chapman S, Parker B (2014) Dye based laser-induced fluorescence sensing of chlorinated solvent DNAPLs. In: Proceedings of the 3rd international symposium on cone penetration testing, Las Vegas, Nevada, USA

Stock P (2011) Using LIF to find LNAPL, L.U.S.T. Line: Bulletin 68. http://www.neiwpcc.org/lus tlineold/lustline_pdf/lustline_68.pdf

Teramoto E, Isler E, Polese L, Baessa M, Chang H (2019) LNAPL saturation derived from laser induced fluorescence method. Sci Total Environ. https://doi.org/10.1016/j.scitotenv.2019.05.262

Tomlinson D, Thornton S, Thomas A, Leharne S, Wealthall G (2014) Illustrated handbook of LNAPL transport and fate in the subsurface, p 15

Tomlinson DW, Wealthall GP, Thorson DM, Himmelheber DW, Cumberland HL, Beech JF, Adamek S, St. Germain R (2017) Combined high-resolution site characterization tool for forensic analysis and delineation of petroleum dense and light nonaqueous-phase liquid. In: The ninth international conference on remediation and management of contaminated sediments, 2017

Wang X, Mullins OC (1994) Fluorescence lifetime studies of crude oils. Appl Spectrosc 48(8):977–984. https://doi.org/10.1366/0003702944029703

Webster GT, Soriano-Disla JM, Kirk J, Janik LJ, Forrester ST, McLaughlin MJ, Stewart RJ (2016) Rapid prediction of total petroleum hydrocarbons in soil using a hand-held mid-infrared field instrument. Talanta 160:410–416. ISSN: 0039–9140. https://doi.org/10.1016/j.talanta.2016.07.044. (https://www.sciencedirect.com/science/article/pii/S0039914016305471)

Zimbron (2020) Methods, systems, and devices for measuring in situ saturations of petroleum and NAPL in soils, U.S. Patent 10,677,729 B2

Zoccolillo L, Babi D, Felli M (2000) Evaluation of polycyclic aromatic hydrocarbons in gasoline by HPLC and GC-MS. Chromatographia 52:373–376. https://doi.org/10.1007/BF02491036

Chapter 9
Biogeophysics for Optimized Characterization of Petroleum-Contaminated Sites

Estella A. Atekwana, Eliot A. Atekwana, Leonard O. Ohenhen, and Silvia Rossbach

Abstract Oil spills are common occurrences on land and in coastal environments. To remediate oil spills, the contaminated volume has to be defined, appropriate remedial measures should be undertaken, and evidence must be provided for the successful remediation. Geophysical techniques can aid site investigation and remediation efforts. The insulating properties of hydrocarbons make them ideal targets for employing a variety of geophysical techniques for their characterization. Nonetheless, the geophysical response of hydrocarbon-contaminated sites is non-unique and depends on factors such as: (1) the release history, (2) hydrocarbon distribution and partitioning into different phases (vapor, free, dissolved, entrapped, and residual phases) in the unsaturated and saturated zones, (3) seasonal hydrologic processes, (4) extent of biodegradation and (5) aquifer salinity and host lithology. Where the contaminants have been biodegraded, the geophysical response depends on the by-products of different terminal electron acceptor processes (TEAPs). In this chapter, we review the different pathways by which TEAPs mediate geophysical property changes. We provide select field case studies from hydrocarbon-contaminated sites across the globe, including different climatic regimes and water salinity conditions. We show that the geophysical response can be transient, hence, data interpretation should be guided by an understanding of the hydrobiogeochemical processes at each site.

E. A. Atekwana (✉) · E. A. Atekwana
Department of Earth and Planetary Sciences, University of California Davis, Davis, CA, USA
e-mail: eaatekwana@ucdavis.edu

E. A. Atekwana
e-mail: eatekwana@ucdavis.edu

L. O. Ohenhen
Department of Geosciences, Virginia Tech, Blacksburg, VA, USA
e-mail: ohleonard@vt.edu

S. Rossbach
Department of Biological Sciences, Western Michigan University, Kalamazoo, MI, USA
e-mail: silvia.rossbach@wmich.edu

© The Author(s) 2024   287
J. García-Rincón et al. (eds.), *Advances in the Characterisation and Remediation of Sites Contaminated with Petroleum Hydrocarbons*, Environmental Contamination Remediation and Management, https://doi.org/10.1007/978-3-031-34447-3_9

Keywords Biodegradation · Biogeophysics · Hydrogeophysics · LNAPL contamination · Near-surface geophysics

9.1 Introduction

Contamination of soil and sediments in the unsaturated and saturated zones by accidental oil spills (e.g., from pipeline breakage, filling stations, refineries, tankers, well blowouts, etc.) is a common occurrence. Since 2010, close to 100 major spills have been reported with an estimated minimum of 5.5 million barrels released into the environment (https://en.wikipedia.org/wiki/List_of_oil_spills#Complete, access 1/25/2022), making hydrocarbons, especially light non-aqueous phase liquids (LNAPL), a major source of groundwater contamination. Both natural attenuation and active remediation measures are used to mitigate and clean up hydrocarbon-contaminated environments (Brown and Ulrich 2014). Geophysical techniques play a crucial role in mitigation and cleanup efforts. Geophysical techniques are used to aid in contaminant delineation, estimate the impacted volume of the subsurface, characterize the site geology/lithology, delineate preferential contaminant flow pathways, as well as monitor the efficacy of the remediation process (Atekwana and Atekwana 2010). Geophysical techniques have the added advantage of being relatively inexpensive, minimally invasive to non-invasive, and provide continuous data coverage over spatial and temporal scales not captured by other more invasive techniques commonly applied at LNAPL-contaminated sites.

Geophysical techniques measure the contrast in the physical properties between a target of interest and the background. LNAPLs and other petroleum products are resistive, and hence a fresh oil spill should result in a resistive response as the LNAPLs replace more conductive pore water. Indeed, fresh spill experiments document the resistive signatures of LNAPLs (DeRyck et al. 1993). The resistive signature was the model used for investigating hydrocarbon-contaminated sites until the seminal work of Sauck et al. (1998), where the authors documented conductive responses and provided strong evidence linking geophysical signature changes to microbial processes at a jet fuel spill site. Today, a variety of geophysical techniques such as electrical resistivity imaging (ERI), electromagnetic (EM) low frequency methods, ground penetrating radar (GPR), induced polarization (IP), and self-potential (SP) have been extensively used to investigate hydrocarbon-contaminated field sites (Atekwana and Atekwana 2010; Abbas et al. 2017, 2018 and references therein). As in other bioreactor systems, the geophysical techniques measure the changes in physical properties resulting from the intermediate and end products of different microbial-mediated processes occurring during the breakdown of hydrocarbons. Although fresh spills occur regularly, in this chapter, we focus on the use of geophysical characterization of aged (several years to decades old) spills in the subsurface, as thousands of such sites exist globally, and need to be remediated. The choice of geophysical methodologies to be used at field sites requires an understanding of the geophysical properties measured by each methodology.

In this chapter, we provide an overview of the terminal electron acceptor processes (TEAPs) at hydrocarbon-contaminated sites. Then, we discuss pathways by which these TEAPs and their by-products mediate changes in physical properties of sedimentary systems. Finally, we provide select case studies of how the by-products of the TEAPs are expressed in geophysical signatures, and thus their utility in investigating LNAPL impacted sites in different environments.

9.1.1 Terminal Electron Acceptor Processes at LNAPL Impacted Sites

When bacteria degrade highly reduced petroleum hydrocarbons, they must contend with a large number of electrons. Most bacteria contain proteins and other compounds in their cell membranes that facilitate the transport of electrons, constituting an electron transport chain (Madigan et al. 2018). An electron transport chain allows bacteria to acquire energy, because as electrons are transported through the chain, protons are exported, and a membrane potential is created. The membrane potential can be converted into mechanical and chemical energy.

The greatest amount of energy is derived if oxygen is used as a terminal electron acceptor in aerobic respiration. Because of their much longer evolutionary history compared to animals and plants, many bacteria are endowed with proteins that can relay the electrons not only onto oxygen, but also onto other dissolved and solid terminal electron acceptors. These other electron acceptors include soluble nitrate and sulfate ions, as well as minerals containing oxidized iron or manganese. Based on thermodynamic considerations, the use of electron acceptors other than oxygen results in less energy gain for the bacteria. If there are no external electron acceptors available in the environment, then bacteria pass on the electrons to metabolic intermediates such as pyruvate (an internal electron acceptor) in a process called fermentation. Bacteria that ferment release acids, alcohols, and gasses. These fermentation products can be used as substrates by other microbial groups. A well-known example of these "syntrophic" associations are fermenting microorganisms that excrete carbon dioxide and hydrogen, which, in turn, can be used by methanogenic Archaea, which produce methane out of these substrates.

When electron-rich petroleum hydrocarbons are spilled into the subsurface and pool in an aquifer, the available electron acceptors are depleted by the metabolic activities of specific bacterial populations, preferentially in the order of their energy potential. For example, because the use of oxygen as a terminal electron acceptor results in the greatest energy gain for aerobic bacteria, aerobic bacteria will successfully outcompete all other microbial populations. When all oxygen is used up, and nitrate is available in the groundwater, then bacterial populations able to use nitrate as terminal electron acceptor are favored. This process will continue with $Mn(IV)$, $Fe(III)$, and sulfate, if they are present in the contaminated environment. When

all available external electron acceptors are depleted, then the only process left is
fermentation/methanogenesis.

Figure 9.1 shows the predominant TEAPS and their products that occur in a
confined aquifer contaminated by petroleum hydrocarbon. The conceptual model
shows sequential redox zonation from oxygen on the outside to methanic in the core of
the contaminant plume (Fig. 9.1a). In practice, however, the redox zonations are not so
discrete and there is often overlapping of different redox zonations (Fig. 9.1b) because
biodegradation is not necessarily driven by thermodynamics alone but also by the
spatial and temporal availability of electron acceptors and microbes in the impacted
subsurface (Meckenstock et al. 2015). Thus, it is not uncommon to have zones of iron/
manganese reduction occurring concurrently and overlapping with methanogenesis.
In addition, seasonal hydrologic recharge events may shift the contaminant mass as
well as deliver fresh electron acceptors to the microorganisms causing a switch from
one TEAP to another (e.g., switch from methanogenesis to iron reduction and vice
versa; Meckenstock et al. 2015; Teramoto et al. 2020; Beaver et al. 2021). While
dissolved gasses, such as oxygen, and ions such as nitrate and sulfate easily diffuse
to the bacterial membrane-bound reductases, an interesting problem arises from the
use of solid electron acceptors. Some bacteria are able to reduce Fe(III) and Mn(IV)
that occur in solid form as terminal electron acceptors—these bacteria can "breathe"
metals.

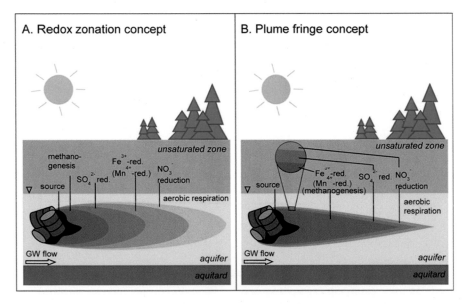

Fig. 9.1 a Discrete redox zonations often associated with microbial oxidation of hydrocarbons
based on redox ladder concept. **b** A new representation of the redox concept showing multiple
terminal electron acceptor processes occurring in the same region of the contaminant plume (From
Meckenstock et al. 2015 with permission)

Well-known and thoroughly studied iron-reducing bacteria are *Geobacter* and *Shewanella*. The iron-reducing bacteria have developed several solutions to the problem of dealing with insoluble electron acceptors. They can either (1) use diffusible compounds such as humic acids as electron shuttles, (2) be in close contact with the iron minerals, or (3) use extracellular appendages such as nanowires or an array of extracellular multiheme cytochromes to relay the electrons onto the metal ions in solid minerals (Lovley and Walker 2019). Interestingly, many methanogens are able to reduce iron, and it seems that they prefer to switch to the more energetic advantageous iron reduction as soon as they encounter minerals containing oxidized iron e.g., goethite, nontronite, illite or smectite (Prakash et al. 2019; Ferry 2020; Liu et al. 2011a, b; Zhang et al. 2012, 2013).

To put it succinctly, microorganisms mediate geochemical reactions by transferring electrons between compounds to meet their energy demands. Geophysical techniques could detect the intermediate and end products of these biogeochemical reactions (biosignatures) in the subsurface. In the next section, we will discuss different microbial-mediated physical property changes of soil and sediments and their resulting geophysical signatures.

9.1.2 By-Products of Microbial-Mediated Redox Processes Drive Geophysical Property Changes

Bacterial metabolic activities can influence the geophysical properties of the subsurface environment by changing the pore fluid chemistry through enhanced mineral dissolution, altering the composition and properties of solids (mediating precipitation of different mineral phases or changes of existing mineral phases) and influencing the overall microbial presence and their activities (microbial cells and biofilms).

9.1.2.1 Pathway 1: Pore Water Chemistry Changes

As shown in Table 9.1, several of the microbial generated metabolic products can change the pore fluid chemistry. As an example, the process of fermentation produces organic acids such as acetic acid or formic acid (Madigan et al. 2018). In addition, CO_2 is an end product of hydrocarbon mineralization. The produced CO_2 dissolves in water to produce carbonic acid. The acids dissolve in pore water, lowering the pH and enhancing the dissolution and leaching of aquifer minerals releasing ions (e.g., Ca^{2+}, Mg^{2+}, K^+, Si^{4+}) into solution (Sauck 2000; Atekwana et al. 2000; Atekwana and Atekwana 2010), causing an increase of ion concentration in the pore water. Similarly, redox reactions produce ions that directly go into solution. Dissimilatory nitrate reduction produces NH_4^+, while Mn^{4+} is reduced to Mn^{2+} and Fe^{3+} is reduced to Fe^{2+}. The dissolved ions and organic acids elevate the pore water electrical conductivity (specific conductance), causing a decrease in the bulk resistivity of the formation (e.g., Sauck et al. 1998; Atekwana et al. 2000; Sauck 2000; Atekwana and Slater 2009; Atekwana and Atekwana 2010 and references there in) as illustrated in Fig. 9.2. As a result, hydrocarbon impacted subsurface environments undergoing significant intrinsic bioremediation (or natural attenuation) are typically characterized by changes in pore fluid chemistry and elevated pore water conductivity (Sauck 2000), making electrical geophysical techniques (ERI, GPR, EMI, IP, or SP) ideal for their characterization and monitoring. In Sect. 9.3, we provide select case studies of geophysical signature changes driven by changes in pore fluid conductivity.

9.1.2.2   Pathway 2: Microbial-Mediated Mineral Precipitation

Bacteria are involved in the dissolution and precipitation of minerals (Hoffmann et al. 2021). Environmental conditions that are conducive for mineral precipitation include ion concentrations that are higher than their solubility and the presence of crystallization nuclei. Bacterial cell surfaces can provide both. Bacterial cell walls contain a variety of negatively charged groups, such as phosphate and carboxyl groups that are stabilized by divalent cations. These can serve as nucleation sites and high local concentrations of ions. In addition, metabolic activities of bacteria can influence the pH of the local environment, e.g., the pore water. For example, when bacteria degrade proteins or urea, ammonium ions are released, which may lead to a locally higher pH, which, in turn, can lead to the increased precipitation of minerals such as carbonates. An extensive list of minerals that have been found to be precipitated in association with bacteria is shown in the recent review by Hoffmann et al. (2021). The list includes a variety of carbonates, phosphates, silicates, sulfides, sulfates, and oxides. As one example, the product of iron reduction, Fe(II), has a much higher solubility at neutral pH than Fe(III). Dissolved Fe(II) atoms can adsorb to ferrihydrite [$(Fe^{3+})_2O_3 \cdot 0.5\ H_2O$] and induce the transformation of ferrihydrite to lepidocrocite, goethite, or magnetite (Hansel et al. 2003, 2005). Hansel et al. (2005) found that the transformation of ferrihydrite to goethite and lepidocrocite occurred at low Fe(II) concentrations (≤ 0.2 mM), while concentrations of ≥ 2 mM Fe(II) resulted in the formation of magnetite. The precipitation of magnetite can be used to infer the presence of iron reduction as a TEAP in the subsurface (Atekwana et al. 2014) which might be useful for the design of remediation programs. Nonetheless, although magnetite precipitation is typically associated with iron reducing bacteria such as *Geobacter* and *Shewanella*, there is now mounting evidence indicating that methanogens can switch their metabolism from methanogenesis to iron reduction causing the precipitation of magnetite (Beaver et al. 2021; Amiel et al. 2020; Shang et al. 2020; Sivan et al. 2016).

Magnetite is conductive and magnetic and thus the presence of magnetite can be detected by using electrical and magnetic geophysical techniques. In contrast, pyrite is conductive and not magnetic thus can only be detected by electrical geophysical techniques. Precipitated minerals like calcite are non-conductive and non-magnetic and can be detected by electrical geophysical techniques (e.g., Saneiyan et al. 2019).

Table 9.1 Common microbial metabolic processes in the subsurface, with electron acceptors used and metabolic products that can be found in the environment. The information in this table is compiled from Mann et al. (1988), Lee et al. (2011), Madigan et al. (2018), Yu and Leadbetter (2020), Hoffmann et al. (2021) and references therein

Process	Electron acceptors/ compounds used	Intermediate and end products	Bacterial-induced precipitation of secondary minerals
Aerobic respiration	Oxygen (O_2); reduced carbon: CH_2O	Water (H_2O), CO_2, CO_3^{2-}, HCO_3^-	Calcite: $CaCO_3$, dolomite: $CaMg(CO_3)_2$, kutnahorite: $CaMn(CO_3)_2$, siderite: $FeCO_3$, magnesite: $MgCO_3$, otavite: $CdCO_3$, strontianite: $SrCO_3$, rhodochrosite: $MnCO_3$, cerussite: $PbCO_3$, hydrozincite: $Zn_5(CO_3)_2(OH)_6$, dypingite: $Mg_5(CO_3)(OH)_2\cdot 5H_2O$, witherite: $BaCO_3$
Anaerobic respiration			
Nitrate respiration	NO_3^-	NO_2^-, NH_4^+	Gwihabaite: $(NH_4,K)NO_3$
Denitrification	NO_3^-	NO_2^-, NO, N_2O, N_2	
Manganese respiration	Mn^{4+}	Mn^{2+}	Hausmannite: Mn_3O_4, rhodochrosite: $MnCO_3$, rambergite: MnS
Iron respiration	Fe^{3+}	Fe^{2+}	Magnetite: Fe_3O_4, siderite: $FeCO_3$, vivianite: $Fe_3(PO_4)\cdot 2\,H_2O$, mackinawite: FeS, pyrite: FeS_2, greigite: Fe_3S_4, baricite: $(MgFe)_3(PO_4)_2\cdot 8H_2O$
Sulfate respiration	SO_4^{2-}	SO_3^{2-}, H_2S, HS^-	Mackinawite: FeS, pyrite: FeS_2, greigite: Fe_3S_4
Sulfur respiration	S^o	H_2S, HS^-	Mackinawite: FeS, pyrite: FeS_2, greigite: Fe_3S_4
Selenate respiration	SeO_4^{2-}	SeO_3^{2-}	
Arsenate respiration	AsO_4^{3-}	AsO_3^{3-}	
Methanogenesis	CO_2	CH_4	
Acetogenesis	CO_2	CH_3COOH (acetic acid)	
Fermentation	Organic-C	Gases (H_2, CO_2), acids (acetic, formic), alcohols (ethanol, butanol)	See carbonates under "aerobic respiration" above
Reductive dechlorination	Chlorobenzoate	Benzoate, HCl	
Nitrogen fixation	N_2	NH_3	
Ammonification	Organic-N	NH_3, NH_4^+	Gwihabaite: $(NH_4,K)NO_3$

(continued)

Table 9.1 (continued)

Process	Electron acceptors/ compounds used	Intermediate and end products	Bacterial-induced precipitation of secondary minerals
Hydrogen oxidation	H_2	H_2O	
Methane oxidation	CH_4	CO_2	See carbonates under "aerobic respiration" above
Ammonium oxidation	NH_4^+	NO_2^-	
Nitrite oxidation	NO_2^-	NO_3^-	Gwihabaite: $(NH_4,K)NO_3$
Anaerobic ammonia oxidation (Anamox)	$NH_4^+ + NO_2^-$	N_2	
Iron oxidation	Fe^{2+}	Fe^{3+}	Magnetite: Fe_3O_4, hematite: Fe_2O_3, ferrihydrite: $Fe_2O_3 \cdot 0.5H_2O$, goethite: α-FeO(OH), strengite: $FePO_4 \cdot 2H_2O$, Nontronite: $Na_{0.3}Fe^{3+}_2(Si,Al)_4O_{10}(OH)_2 \cdot nH_2O$, Greigite: Fe_3S_4
Manganese oxidation	Mn^{2+}	Mn^{4+}	Manganite: MnOOH, vernadite: MnO_2, hausmannite: Mn_3O_4, todorokite: $(Ca,Na,K)_x(Mn^{4+},Mn^{3+})_6O_{10} \cdot 3.5H_2O$, birnessite: $(Na,Ca,K)_x(Mn^{4+},Mn^{3+})_2O_4 \cdot 1.5H_2O$
Sulfur oxidation	H_2S, $S°$, $S_2O_3^{2-}$, SO_3^{2-}	SO_4^{2-}	Gypsum: $CaSO_4 \cdot 2H_2O$, celestite: $SrSO_4$, barite: $BaSO_4$

In essence, the ability of geophysical techniques to detect zones of microbial-mediated mineral precipitation depends on the physical property of the precipitated minerals and whether they occur in sufficient concentrations to be detected (Abdel Aal et al. 2014; Revil et al. 2015).

9.1.2.3 Pathway 3: Microbial Cells and Biofilms

Although bacteria have been studied for more than 100 years in liquid laboratory media (planktonic cells), it turns out that in nature, most bacteria are sessile and form biofilms or bioclusters (Flemming et al. 2016; Baveye 2021). Biofilms are defined as a community of microorganisms that stick to a surface in a self-produced matrix. The matrix consists of excreted biomolecules, including polysaccharides, proteins, lipids, and deoxyribonucleic acid (DNA). Biofilms survive adverse environmental conditions much better than planktonic cells. For example, biofilms are more resistant to antibiotics, can withstand drying out, and are more protected against

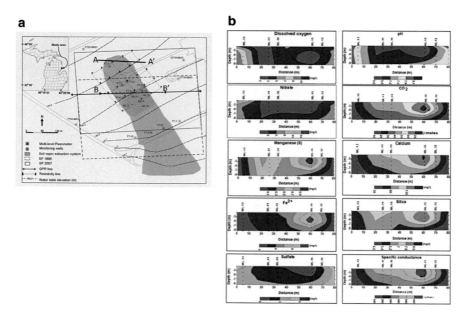

Fig. 9.2 Spatial distribution of subsurface terminal electron acceptors and weathering products in groundwater from the FT-2 site at the Wurtsmith Air Force Base, Oscoda, Michigan, USA, a site of active biodegradation: **a** location of site in Michigan, pink outlines the FT-2 plume. The plume resulted from JP4 jet fuel spilled as a result of fire training exercises from 1958 to 1991. **b** The geochemistry data were acquired in 2007. The core of the plume (well ML-15 and 12) shows lower dissolved oxygen and nitrate resulting from the utilization of oxygen and nitrates by indigenous microorganisms. Reduction of manganese(IV) by manganese reducing bacteria and iron(III) by iron reducing bacteria results in elevated concentrations of manganese(II) and iron(II) respectively, whereas utilization of sulfate by sulfate reducing bacteria results in the sulfate depletion. The terminal electron acceptor processes are accompanied by acid production, lower pH and elevated concentrations of CO_2, accelerating the weathering and dissolution of calcium and silica, increasing the ionic content expressed as an increase in the specific conductance. Details on the geochemical analysis can be found in Che-Alota et al. (2009). Figure modified from Atekwana and Atekwana (2010) with permission

grazing by protozoa. These biofilms, consisting of microbial cells and the matrix they produce, can clog pore spaces in sediments and aquifer matrices and therefore change the geophysical properties of the subsurface contaminated with hydrocarbons (e.g., Fig. 9.3). In both natural and engineered systems, biofilms are a key factor in clogging of sediment pore spaces and fluid flow pathways (Baveye et al. 1998). Studies have shown that biofilm development may significantly reduce the porosity (by 50–90%) and permeability (by 95–99%) of porous media (Bouwer et al. 2000; Dunsmore et al. 2004) and thus may negatively impact the efficacy of remediation programs. Therefore, determining where these biofilms are forming in the subsurface from different amendment strategies is an important consideration.

Fig. 9.3 Scanning electron micrograph image of biofilm growth in the pores of sediments with the potential to clog the pores of sediments, altering the hydraulic conductivity. Figure adapted from Sharma et al. (2021) with permission

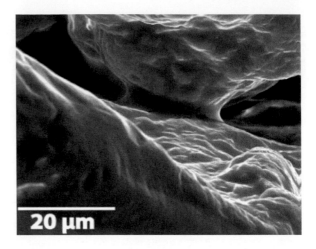

9.2 Geophysical Methods

While geophysical techniques can be used for investigating the subsurface lithology and characterization of fractured medial to determine preferential contaminant flow paths, in this section, we provide a summary only of surface geophysical methods that are commonly used for the determination of the presence of LNAPL-contaminated sediments and for the monitoring of the progress of bioremediation. Admittedly, a number of direct push logging tools are also used but are beyond the scope of this chapter. A more detailed overview of geophysical methods can be found in most environmental and near surface geophysics textbooks (e.g., Reynolds 2011; Rubin and Hubbard 2005; Binley and Slater 2020).

9.2.1 Electrical Methods

Electrical methods measure voltages associated with electrical current flow in the subsurface to elucidate subsurface geoelectrical properties. Electrical methods can measure currents injected (or induced) into the subsurface (ERI, EMI, GPR, IP) while others like SP and Magnetotelluric (MT) can measure naturally occurring electrical and electromagnetic fields.

9.2.1.1 Electrical Resistivity

Electrical resistivity is one of the most versatile and commonly used electrical methods applied in the study of hydrocarbon-contaminated environments. The fundamental principle of the electrical resistivity method is based on the ease or difficulty of charges to flow through a material. Electric current flow in the subsurface occurs in three ways: (1) electronic conduction in materials containing free electrons such as metals, (2) electrolytic (ionic) conduction by ions in the pore waters, and (3) surface conduction (interfacial conduction) occurring in the electrical double layer at the interfaces of minerals in contact with the pore water (important for clays and disseminated metals). In the absence of metallic minerals, electrolytic conduction is by far the most common mechanism by which current flows in subsurface geologic media.

In practice, the electrical resistivity method involves the injection of direct current into the ground through electrodes connected to an artificial source of current and determining the apparent resistivity by measuring the potential at other electrodes in the vicinity of the current (Telford et al. 1990; Loke 2001). Apparent resistivity is the resistivity that would yield the measured relationship between applied current and the potential difference for a particular arrangement and spacing of electrodes of an electrically homogeneous and isotropic half-space. The measured apparent resistivity depends on the physical properties of the material and is obtained as the product of measured resistance (R) and the geometric factor (K) for a given electrode configuration (Eq. 9.1), and its unit is ohm·m.

$$\rho_a = RK, R = \frac{\delta V}{I} \qquad (9.1)$$

where ρ_a is the apparent resistivity, R is the resistance, K is the geometric factor dependent on electrode configuration, δV is the potential difference, and I is the induced current.

There are several electrode configurations used for several application purposes, such as Wenner, Schlumberger, Wenner-Schlumberger, pole-pole, dipole–dipole arrays, or a combination of these. A detailed analysis of the advantages and disadvantages of each electrode configuration is discussed by Loke (2001). The apparent resistivity values obtained from field measurements are not "true" or actual resistivity, but the true resistivity is estimated by carrying out an inversion of the apparent resistivity using available software packages. Some commonly used inversion software are RES2DINV (Loke 2001), AGI EarthImager, and the open sourced ResIPy (Blanchy et al. 2020). In most cases the electrical resistivity is dependent on the porosity, the degree of saturation, and the conductivity of the fluid filling the pores. Thus, it is expected that the high resistivity of the LNAPLs replacing the more conductive pore waters will result in a resistive response. However, biodegradation and the production of additional ions will enhance electrolytic conductivity and will result in an expected conductive response. Nevertheless, the response is also a function of the host mineralogy/geology. For example, clays, certain remediation amendments, and salt water are very conductive and so even a relatively conductive plume may still result in a resistive response.

9.2.1.2   Ground Penetrating Radar

The GPR method is a non-destructive method that uses pulsed electromagnetic waves in the MHz-GHz range to image the subsurface. The GPR system comprises a signal generator, transmitting and receiving antennas, and a control console to display the signal generated (Reynolds 2011). The GPR antenna transmits a high frequency (typically 25 MHz to 1 GHz) electromagnetic (EM) wave that propagates into the subsurface. When the propagating EM wave encounters a boundary or interface with contrasting electromagnetic properties [magnetic permeability (μ), dielectric permittivity (ε), or electrical conductivity (σ)], some of the transmitted energy is reflected to the surface. Since in most materials the magnetic permeability does not vary much, the dielectric permittivity is the property that determines the EM wave velocity in most materials (Eq. 9.2).

$$V = 1/\sqrt{\mu\varepsilon} \qquad (9.2)$$

Commonly the dielectric behavior is characterized in terms of the relative dielectric permittivity (also known as the dielectric constant):

$$\kappa = \varepsilon/\varepsilon_0 \qquad (9.3)$$

where ε_0 is the permittivity of a vacuum (8.8542×10^{-12} F/m). Neglecting the effects of magnetic permeability (μ) since near surface sediments and soils are typically non-magnetic, the EM wave velocity can be rewritten as:

$$V = c/\sqrt{\kappa} \qquad (9.4)$$

where c is the velocity of light in a vacuum (0.3 m/ns). Water has a significantly higher dielectric constant ($\kappa = 80$) than soils and sediments ($\kappa = 4$–10) and LNAPLs ($\kappa \sim 2$), hence GPR techniques are commonly used for providing information on water content (saturation) (Davis and Annan 1989).

The reflected waves across interfaces and the two-way travel times are recorded by a receiver antenna. The amplitude of the reflection depends on the magnitude of the contrast of electromagnetic properties across the boundary. The attenuation factor (α) and the depth of penetration is dependent on the electrical conductivity (σ), magnetic permeability (μ), and dielectric permittivity (ε) of the media through which the signal is propagating (Reynolds 2011) as shown in Eq. (9.5).

$$\alpha \approx \frac{\sigma}{2}\sqrt{\frac{\mu}{\varepsilon}} \qquad (9.5)$$

From Eq. (9.5), it is evident that attenuation is related to conductivity; the higher the conductivity is, the greater the attenuation and the shallower the depth of penetration will be (Pettersson and Nobes 2003). GPR signals are greatly attenuated in highly conductive materials such as clayey soils and sediments and by high salinity pore waters, limiting their penetration and use in such environments. At hydrocarbon-contaminated sites where biodegradation increases the pore water conductivity, the amplitude of the reflected energy is attenuated resulting in low signal amplitudes (muted reflections, or shadow zones) coincident with the zone of contamination (Sauck et al. 1998; Sauck 2000).

9.2.1.3 Electromagnetic Induction

EM methods are commonly used at hydrocarbon-contaminated sites to map shallow LNAPL contamination, as well as locate underground utilities such as buried pipes including underground storage tanks (e.g., Atekwana et al. 2002). The basic principles of shallow EM techniques can be found in McNeill (1980, 1991). Briefly, the EM method uses the response of the ground to the propagation of incident alternating electromagnetic waves that are made up of two orthogonal components consisting of an electric field (E) and a magnetic field (H) in a plane perpendicular to the direction of the EM wave propagation. A transmitter coil is used to generate the primary electromagnetic field, which propagates above and below ground. When a subsurface conductor is encountered, the primary magnetic field induces eddy currents in the conductor. The eddy currents in turn induce a secondary magnetic field which is measured by the receiver coil but is delayed in phase. The receiver coil measures a resultant field which is made up of the primary (from the transmitter coil) and secondary magnetic field generated by the subsurface conductor. The phase lag between the primary and secondary fields is used to make useful deductions about the subsurface conductor. In general, the depth of penetration of the EM wave is a function of the frequency and conductivity. The skin depth Z_s is defined as the depth at which the amplitude of the plane wave is reduced to $1/e$ or 37% of its initial amplitude. This depth of penetration of an EM wave is given by:

$$Z_s = 503.8\sqrt{f\sigma} \tag{9.6}$$

where f is the frequency of the EM wave in Hz and σ is the conductivity (Reynolds 2011). The depth of penetration of EM systems is also a function of the distance between the receiver and transmitter coils as well as the orientation of the coils.

In ground conductivity meters such as the EM-31 manufactured by Geonics Ltd, which is commonly used at hydrocarbon-contaminated sites, the receiver coil measures both the magnitude and phase of the secondary field and provides two readings: the quadrature component, which is related linearly to the apparent conductivity is measured in units of millisiemens per meter (mS/m) and the in-phase component, which is a measure of the metal content in parts per thousands (ppt) provides useful information for determining if the source of the anomaly is a metal (e.g., buried metallic pipe). The maximum depth of penetration for the EM-31 signals is ~6 m.

9.2.1.4   Induced Polarization

Although direct current (DC) resistivity methods are most used to characterize LNAPL sites, they constitute a bulk measurement, responsive to both electrolyte and solid–fluid interface (surface) chemistry, and therefore are unable to differentiate between the relative contributions of electrolytic versus interface conductivity or discriminate between clay-rich sediments and more saline pore water. The IP technique, which can be measured either in the time domain or frequency domain, is an extension of the DC-resistivity method. It is more sensitive to changes in the electrochemistry of the pore water–mineral interface (e.g., Lesmes and Frye 2001) and allows for the complex electrical properties (electromigration and polarization) to be measured (e.g., Binley and Kemna 2005; Revil et al. 2012). Because biodegradation potentially modifies the mineral surface properties and because microorganisms typically are attached to the mineral grains, the induced polarization technique, particularly the spectral induced polarization (SIP, a multifrequency measurement) has been documented as being sensitive to the presence of NAPLs and also suitable for investigating the effects of bio-physicochemical changes of electrical properties in hydrocarbon-contaminated sediments (Abdel Aal et al. 2004, 2006; Orozco et al. 2012; Johansson et at. 2015). In the SIP technique, the impedance magnitude $|\sigma|$ and the phase shift φ (between a measured sinusoidal voltage and an induced sinusoidal current) is measured over a range of frequencies, typically between 100 MHz and 1000 Hz.

From the magnitude and phase measurements, the real ($\sigma' = |\sigma| \cos \varphi$) and imaginary ($\sigma'' = |\sigma| \sin \varphi$) parts of the sample complex conductivity (σ^*) are calculated:

$$\sigma^*(\omega) = \sigma'(\omega) + i\sigma''(\omega) \tag{9.7}$$

where, the in-phase (real; σ') conductivity component represents electromigration in the subsurface and is sensitive to changes in fluid chemistry, whereas the out-of-phase (imaginary, σ'') conductivity represents the charge polarization which at low frequencies (<1000 Hz) results primarily from the polarization of ions in the electrical double layer (EDL) at the mineral–fluid interface and i is the square root of $\sqrt{-1}$ (Lesmes and Frye 2001). Conduction and polarization at the fluid–grain interface are a function of surface area, pore size geometry, surface charge density, and surface ionic mobility (Lesmes and Frye 2001). The real conductivity is what is typically measured in electrical resistivity imaging. Several laboratory experiments have clearly established that σ'' is more sensitive to microbial activity and presence of biofilms than σ' (Abdel Aal et al. 2004; Davis et al. 2006; Zhang et al. 2014; Mellage et al. 2018, 2019; Kimak et al. 2019).

Most modern electrical resistivity instruments are equipped to make time domain IP (TDIP) measurements during ERI surveys. TDIP is acquired simultaneously with resistivity measurements by measuring the transient decay of the voltage after the current is shut-off, typically in the form of the integral of the decay curve over a defined time window.

For TDIP applications, the parameters measured are the apparent chargeability (m_a) calculated as:

$$m_a = \frac{V_s}{V_p} \tag{9.8}$$

where, m_a is the apparent chargeability (with unit mV/V) at time t, V_s is the secondary voltage measured at time t after the current is off, V_p is the primary voltage measured when the current is on. It should be noted that V_s is only significant in polarizable subsurface environments. At low frequencies (below 10 Hz), the measured parameters in the FDIP and TDIP are proportional and related by the equation:

$$m_a = -\kappa\varphi \tag{9.9}$$

where, κ is the constant of proportionality, which can be experimentally derived. Because FDIP measurements are more challenging to acquire in the field, TDIP chargeability values acquired during electrical resistivity surveys can be transformed to phase angle measurements using Eq. (9.9) (Binley and Kemna 2005).

9.2.1.5 Self Potential

The SP method involves the measurement of differences in natural electric potentials developed in the Earth. The potentials measured can range from a few millivolts (mV) to greater than 1 V (Reynolds 2011). The sign of the potential (negative or positive) is diagnostic of the source generating the potential. SP anomalies can be generated in the Earth through the following processes:

(1) Electrokinetic (electrofiltration, streaming) potentials result from electrolytes flowing through a capillary tube or a porous medium. These types of potentials are transient and are typically generated during groundwater flow associated with recharge events, dam seepage, groundwater pumping, or discharge zones.
(2) Electrochemical (diffusion or liquid junction) potentials result from local differences in the mobilities of electrolytes (anions and cations) due to differences in concentrations. These potentials are small (in the tens of mV range) and can occur at contaminated sites where differences in the concentrations of ions within and outside the plume exist. The electrochemical potentials can also be generated at shale–sand interfaces and are the major source of SP anomalies measured in well logging.
(3) Thermoelectric potential results from temperature gradients such as in geothermal areas. In addition, microbial activity in hydrocarbon-contaminated environments may result in thermal anomalies (e.g., Warren and Bekins 2018) that could result in temperature gradients between contaminated and background regions generating thermoelectric potentials. This type of potential is also transient and exists as long as the thermal differences exist.

(4) Mineral potentials are typically very large (100 s mV), associated with massive sulfide ore bodies, and used in the exploration of these deposits. While their origin is not well understood, they are hypothesized to be caused by the existence of geobatteries (Sato and Mooney 1960), where an ore body straddles the water table serving as an electronic conductor for electron transport from anodic reactions below the water table to cathodic reactions occurring above the water table. Naudet et al. (2004) and Revil et al. (2010) have documented the existence of large SP anomalies at contaminated sites resulting from biogeobatteries associated with redox reactions. Here, the electronic conductors are biofilms, conductive bacterial appendages (nanowires), or bio-metallic minerals within the water table fluctuation zone that move electrons from reducing environments below the water table to oxidizing environments above the water table.

At hydrocarbon-contaminated sites, the cause of SP may be related to electrokinetic potentials, electrochemical potentials, thermoelectric potential or mineral potentials or a combination of these processes.

Self-potential measurements are easy to make. The equipment required for measuring SP anomalies consists of a pair of non-polarizable electrodes (e.g., Cu/$CuSO_4$ porous pots) connected by insulated cable to a high input impedance (~100 MΩ) voltmeter, which is used to read the potential with a 0.1 or 1 mV resolution. Field surveys are commonly done in two ways: gradient surveys and fixed base station mode. In gradient surveys, two electrodes (with constant spacing between them—typically 5–10 m) are moved successively along a survey line and the potential difference between the two electrodes is divided by the spacing between the electrodes expressed in mV/m. Electrodes are moved along the line in leapfrog fashion. In the fixed electrode configuration, one electrode is fixed at a location outside the target zone referred to as the base station and the potential difference between the base station and roving electrode is measured. SP measurements are susceptible to cultural noise associated with stray currents such as those from electric trains, or geological noise from changing soil conditions (wet to dry, forested to grasslands, elevation difference) and telluric currents; hence care must be taken in the processing to identify and eliminate noise.

This unsophisticated methodology is also reflected in the results, which are often interpreted qualitatively and provide general information about anomaly shape and amplitude. However, more advanced processing and inversion can be used to retrieve the current source densities and quantitative information obtained on the redox potentials at the site (e.g., Naudet et al. 2004; Revil et al. 2010; Giampaolo et al. 2014; Abbas et al. 2017, 2018).

9.2.2 Magnetic Method

Magnetic methods obtain information related to the intensity, direction, gradient of the magnetic field of the Earth. In the context of application to hydrocarbon contamination, magnetic methods exploit variations in the magnetic mineralogy of materials. Some common magnetic minerals which may cause variation in the measured magnetic properties include magnetite, maghemite, hematite, pyrrhotite, and greigite. One of the magnetic methods most applied in investigating hydrocarbon contamination is magnetic susceptibility.

Magnetic susceptibility is defined as the degree to which a sediment or mineral can be magnetized when a magnetic field is applied. Magnetic susceptibility (χ) is the ratio of applied magnetization (M) to the applied magnetic field (H):

$$M = \chi H \qquad (9.10)$$

where, M and H have the same units (A/m), and χ is the volume-specific dimensionless magnetic susceptibility.

Magnetic susceptibility measurements can be made on cores retrieved from hydrocarbon-contaminated sediments or by downhole measurements at field sites. In downhole measurements, a magnetic susceptibility probe is lowered down a borehole that is free of any metallic materials (including well casings). The role of magnetite cannot be overstated for hydrocarbon-contaminated sites, as even minor precipitates of magnetite can produce large magnetic susceptibility measurements. This enhanced magnetite production has been observed at many hydrocarbon-contaminated sites (Mewafy et al. 2011; Rijal et al. 2010, 2012; Atekwana et al. 2014; Lund et al. 2017) and is a key feature in the application of magnetic methods in hydrocarbon-contamination monitoring.

9.3 Geophysical Applications and Case Studies

A variety of geophysical techniques have been successfully applied at hydrocarbon-contaminated sites to infer the presence of fresh and biodegraded hydrocarbons, as well as to monitor their remediation. Although numerous laboratory experiments have been conducted to determine the efficacy of geophysical techniques to detect both fresh and biodegraded hydrocarbon-contaminated media (Atekwana and Atekwana 2010), in this chapter, we focus on field characterization and provide example case studies from different environments (fresh to saline aquifers), geographic and climatic regions (desert and cold). We have presented, in Sect. 9.1 above, the need to understand the processes occurring at hydrocarbon-contaminated sites, as this will determine the appropriate geophysical techniques to use.

9.3.1 Geophysical Signatures of Changes in Pore Fluid Conductivity

9.3.1.1 Conductive Response in Contaminated Freshwater Aquifers

An examination of the published peer review literature indicates that the conductive plume model (Sauck et al. 1998; Atekwana et al. 2000) is the dominant response observed at the majority of aged contaminated sites irrespective of geographic location. Hence, electrical techniques, which are sensitive to changes in pore fluid conductivity are best for characterizing aged hydrocarbon-contaminated sites (e.g., Werkema et al. 2003). The GPR and ERT are the two most commonly applied techniques due to their relative ease of use even for the non-practitioner. We discuss two case studies where electrical resistivity and GPR were used to investigate aged LNAPL plumes.

The first case study is a site at the decommissioned Wurtsmith Air Force Base in Oscoda, Michigan, USA, near the shores of Lake Huron. This site was the location of the seminal work of Sauck et al. (1998) that provided strong evidence linking the measured geophysical signatures to microbial degradation of the hydrocarbons and resulted in the development of the conductive plume model by Sauck (2000). Hydrocarbon contamination resulted from more than three decades of bi-weekly fire training exercises with JP-4 as the dominant contaminant resulting in what is known as the FT-02 plume. By the 1990s, the free product plume was 0.3 m thick. Much of the hydrocarbon contamination was restricted to the upper parts of the saturated zone and within the capillary fringe zone (McGuire et al. 2000; Skubal et al. 2001). The contaminant plume at FT-02 was approximately 75 m wide and extended 30 m upgradient and about 450 m downgradient to the southeast of the source area (Che-Alota et al. 2009; Skubal et al. 2001) (Fig. 9.2a). Chemical analyses of the groundwater showed elevated benzene, toluene, ethylbenzene, and xylenes (BTEX) concentrations in addition to elevated groundwater conductivities. The site geology consists mostly of clean well-sorted fine to medium sands coarsening downward to gravel of eolian origin approximately 20 m thick, underlain by a confining unit that is a brown to gray lacustrine silty clay unit 6–30 m thick (Bermejo et al. 1997; Sauck et al. 1998) with the water table at ~3–5 m below ground surface (bgs). The geophysical results shown in Fig. 9.4 were acquired in 2003 and presented in Che-Alota et al. (2009) and Atekwana and Slater (2009).

The electrical resistivity survey was conducted using a Syscal R2 resistivity meter manufactured by IRIS with 72 electrodes using a dipole–dipole array and a 10 m electrode spacing. The apparent resistivity measurements were inverted using RES2DINV (Loke and Barker 1996) to retrieve the true resistivities at the site. The GPR survey was acquired using a Geophysical Survey System Inc. (GSSI) Sir 10 A + system with 100 MHz bistatic antennae recording for a total of 400 ns. A fixed 1.4 m separation between RX-TX pairs was used and gains were automatically set at the beginning of the line and the data were processed using the RADAN software. Extensive geochemistry was acquired over the site and the data are presented in Fig. 9.2. The zone of contamination is characterized by a zone of low resistivity (<

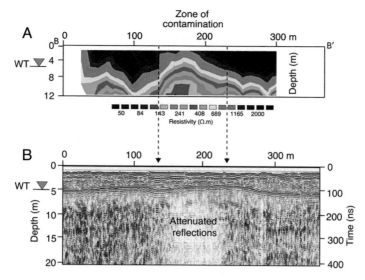

Fig. 9.4 **a** Electrical resistivity and **b** GPR image of the FT-02 plume at the decommissioned Wurtsmith Air Force Base in Oscoda Michigan, USA (modified from Atekwana and Slater 2009 with permission). The zone of attenuated GPR reflections is coincident with a region of low resistivity of the resistivity profiles that is coincident with the location of the plume. Profile location is shown on Fig. 9.2a as B–B′

150 $\Omega \cdot m$) and attenuated GPR reflections (see arrows in Fig. 9.4). The geochemistry of groundwater indicates that biodegradation is active at the site. The biodegradation by-products, e.g., Fe^{2+}, Mn^{2+} and weathering products (e.g., silica and calcium) have significantly perturbed the groundwater geochemistry at the site resulting in elevated ion chemistry within the plume (~4 × background) that is detected by the geophysical measurements. The attenuated GPR reflections result from the elevated conductivity within the plume.

The second case study we present is a decommissioned refinery near Paris, France (Abbas et al. 2018). Operations at the refinery started in the 1920s. Here the contamination consists of LNAPLs and other hydrocarbon products. The free phase plume had a thickness range of 0.1–1 m. The geology consists of mostly fine to medium-grained sands coarsening to gravel with a thickness ranging between 8 and 25 m (Abbas et al. 2018). This layer hosts an alluvial aquifer at a depth ranging from 4 to 11 m. A chalk aquifer underlies the alluvial aquifer at a depth ranging from 11 to 25 m. Extensive geochemical data exist for the site and were reported in Abbas et al. (2018). The electrical resistivity data were acquired with a Syscal Pro system using a Wenner and dipole–dipole arrays with a 2 m electrode spacing. The acquired apparent resistivity data were inverted using the RES2DINV (Loke and Barker 1996). The GPR data were acquired using a Mala Geoscience's GPR system with a 250 MHz antenna. The groundwater geochemistry indicates biodegradation was active at the site with low TEA concentrations (e.g., oxygen, nitrate, sulfate) and elevated total dissolved solids.

The electrical resistivity profile (Fig. 9.5a) and GPR profile (Fig. 9.5b) over the clean areas (0–75 m X position) showed high electrical resistivity (>150 Ω·m) that markedly decreased and strong GPR reflections that were attenuated within the contaminated zone (X position 75 m to end of profile). The zone of attenuation started at ~2 m beneath the surface. Note that the water table was approximately 8 m beneath the surface, hence this attenuation began in the unsaturated zone extending into the saturated zone. The resistivity profile showed a region of low resistivity (< 50 Ω·m) coincident with the region of attenuated GPR reflections consistent with a conductive response related to elevated ion concentration.

Both case studies presented suggest that in freshwater aquifers where biodegradation of the LNAPL is active, the zone of LNAPL contamination can be characterized by low resistivity and attenuated GPR reflections. These responses have been reported in many freshwater aquifers globally and across many climatic zones including dry, desert environments in Iraq (Al-Menshed and Thabit 2018), tropical environments such as in Brazil (e.g., Lago et al. 2009; Moreira et al. 2019, 2021), temperate environments such as in Serbia (Burazer and Burazer 2017), France (Blondel et al. 2014), and warm temperate–subtropical environments such as Jiangsu Province, China (e.g., Shao et al. 2021).

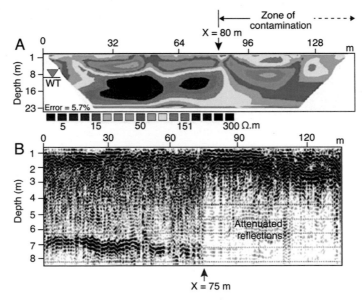

Fig. 9.5 **a** Electrical resistivity, **b** GPR image from a decommissioned refinery near Paris, France. The zone of attenuated GPR reflections is coincident with a region of low resistivity which is coincident with the location of the plume (modified from Abbas et al. 2018 with permission)

9.3.1.2 Characterizing LNAPL Sites Using Spectral Induced Polarization (SIP) and Time Domain Induced Polarization (TDIP)

While the electrical resistivity imaging provides information on bulk electrical property changes that are largely controlled by the electrolytic conductivity, it is unable to discriminate between fluid and lithologic effects. In contrast, SIP allows for the measurement of additional parameters such as the real (σ') and imaginary (σ'') components of the complex conductivity ($\sigma*$), which can provide information on physicochemical processes occurring at the mineral grain–pore fluid interface. As such it can provide additional information that can allow for the discrimination of fluid conductivity and lithology (Slater and Lesmes 2002), as well as discerning microbial effects (Abdel Aal et al. 2004, 2006; Davis et al. 2006; Atekwana and Slater 2009; Orozco et al. 2012, 2021; Mewafy et al. 2013; Johansson et al. 2015). A large body of literature exists examining SIP response to soil-hydrocarbon mixtures in laboratory settings (Vanhala 1997; Schmutz et al. 2010; Abdel Aal et al. 2014; Deng et al. 2018 and references there in). These studies all suggest that SIP is excellent for detecting the presence of hydrocarbons in porous media. Despite this large advantage of SIP over ERT, only a limited number of studies exist of SIP measurements of hydrocarbon contamination in field settings (e.g., Flores-Orozco et al. 2012; Blondel et al. 2014; Maurya et al. 2018). This is largely due to the limited availability of field SIP instrumentation and problems of noise from electromagnetic coupling at higher frequencies.

In Fig. 9.6, we present a case study by Orozco et al. (2021) of SIP measurements at a kerosene-contaminated site downgradient of a military base in Decimomannu, Italy. The subsurface of the site is characterized by a backfill layer with a thickness ranging between 0.2 and 1 m and underlain by recent alluvium composed mostly of gravels and sands which extend to a maximum depth between 4 and 6 m. The sand and gravel layers are underlain by "Hazelnut clay" lenses, composed of sandy-gravelly clays of hazelnut color with a thickness of 1–1.5 m. Total Petroleum Hydrocarbon (TPH) concentrations in the subsurface ranged from 100 to 18,000 µg/L. Fluid electrical conductivities were highest at the clean locations (~1860 µS/cm) compared to contaminated locations (~1200 µS/cm). The results presented include the real, imaginary, and phase response at 15 Hz (detailed results are presented in Orozco et al. 2021). The results showed that except for the phase (φ) response, the real (σ') and imaginary (σ'') components of the complex conductivity for the contaminated sediments (Fig. 9.6b–d) were higher than those in clean sediments (Fig. 9.6a) and higher in the dissolved phase (Fig. 9.6c) than in the free phase (Fig. 9.6b). This is clearly demonstrated in the correlation plot of Fig. 9.6d. The authors explain the increase in real and imaginary conductivity of the contaminated locations compared to the clean location as resulting from: (1) the disconnection of the electrical double layer as a result of the immiscible oils in the pores; and (2) the increase of ions from the bulk electrolyte into the electrical double layer associated with the non-miscible oils in the pore water.

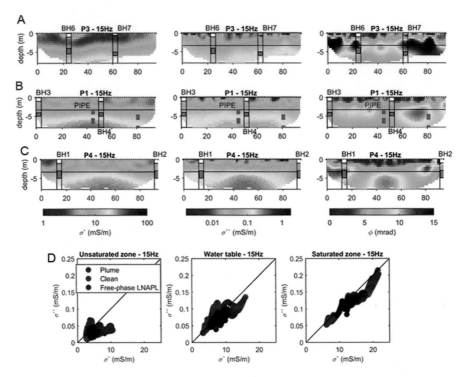

Fig. 9.6 Spectral induced polarization (SIP) results from a kerosene impacted site near Decimo-mannu, Italy. The results shown are the 15 Hz data of the real (σ'), imaginary (σ''), and phase (φ) from **a** clean, **b** free phase, **c** dissolved phase, and **d** correlation of the real (σ') and imaginary (σ'') component of the complex conductivity from pixel values extracted from the SIP data at different depths: between 1 and 2 m (unsaturated zone), 2.80 and 3.80 m (around the water table), and 4.5 and 5.5 m (saturated zone). The SIP parameters are extracted from the clean sediments, free-phase LNAPL, and dissolved phase plume. The black line is the linear increase related to the model σ' = 100 σ'', which is used to indicate regions dominated by electrolytic conduction (below line) and surface conduction (along the line). The horizontal line at 3.3 m depth indicates the position of the groundwater table at the time of the survey. BH1–BH7 are lithologic boreholes superimposed on the SIP profiles, with the boxes indicating: the backfill materials on the top (white), the recent alluvial (no color), the Hazelnut clays (gray), and the ancient alluvial sediments (no color, at the bottom). Modified from Orozco et al. (2021) with permission

    Although the above study is unable to use SIP to estimate the volume of TPH in the soil, a study by Deng et al. (2018) documents in a sandbox experiment that SIP can map an oil plume and estimate the volumetric oil content with an efficacy rate of 70–80% with the efficacy rate higher for the imaginary conductivity component.

    In the second example, we describe the use of TDIP measurements as a surrogate for FDIP measurements. Here TDIP measurements were acquired during ERI surveys of the FT-02 plume in 2007 (the location of the profile A-A' is shown on Fig. 9.2) at the Wurtsmith Air Force Base, Oscoda MI, described in Sect. 9.3.1.1. The TDIP survey was conducted using an IRIS Syscal Pro with an axial dipole–dipole array with 3 m electrode spacing. The TDIP measurements were transformed to produce phase angle, real, imaginary and responses following Binley and Kemna (2005). The results from the IP inversion are presented on Fig. 9.7. Boundaries of the plume as determined by geochemical data are marked on the figures by vertical black lines. The phase angle data (Fig. 9.7a) show a decrease in the phase angle (values range between -1 and -2.5 mRads) within a region coincident with the plume (approximately 35–95 m horizontal distance) compared to regions outside the plume (-4 to -5 mRads). At ~40 m there is an elliptical zone of anomalously high phase angles (6 to -7 mRads) that is probably due to noise from buried infrastructure. The data showed that a zone of higher real and imaginary conductivity occurring below the water table (5 m) is coincident with the plume (Figs. 9.7b, c). Outside of the plume boundary, there is no significant variation in both the real and imaginary conductivity values below the water table. We interpret the higher real and imaginary conductivities within the zone of contamination as related to increase in ion concentrations resulting from enhanced mineral weathering due to acids produced during biodegradation. Studies by Che-Alota et al. (2009) at this field site indicated that when compared to the 1996 and 2003 data, the bulk electrical resistivity within the contaminated zone had reverted to near background conditions in 2007 due to contaminant mass removal from a soil vapor extraction system installed in 2001. However, we note that the TDIP profile still showed the effects of biodegradation consistent with groundwater geochemical data (e.g., elevated CO_2, Fe^{2+}, Mn^{2+}, and decreased sulfate and nitrate; Fig. 9.2b), perhaps related to the higher sensitivity of IP. This may suggest that IP may be more diagnostic to microbial processes when compared to the DC resistivity.

Other TDIP studies from hydrocarbon-contaminated sites have shown mixed results. For example, Blondel et al. (2014) reported a clear decrease in bulk electrical resistivity related to biodegradation of an oil spill site but no measurable changes in the normalized changeability or quadrature conductivity. On the other hand, in other studies, Deceuster and Kaufmann (2012) and Abbas et al. (2018) reported both low bulk electrical resistivity anomalies and high-chargeability anomalies coincident with the contaminated aquifer. The above discussion points to the fact that, sometimes the electrical response can be ambiguous. It is therefore important to constrain geophysical interpretations with groundwater geochemistry data.

9.3.2 Resistive Response in Saline Aquifers

Although many aged hydrocarbon-contaminated sites show a conductive electrical resistivity response associated with biodegradation, not all sites are conductive. We provide examples from a coastal environment where a resistive response is observed. Figure 9.8a shows the results from an electrical resistivity survey conducted at a former perfumery plant in Jinghai, southwest of Tianjin, China, with a benzene and ethylbenzene spill history. Details of this study are presented in Xia et al. (2021). The site geology consists of a ~1.5 m thick top layer of backfill underlain by ~1 m of silt, clay and silty clay layers. Underlying this, at a depth of 13–16.5 m, is a dense

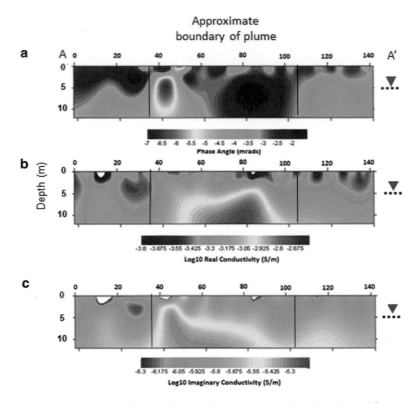

Fig. 9.7 **a** Phase angle, **b** Real conductivity, and **c** Imaginary conductivity obtained from transformation of time domain induced polarization measurement at the FT-02plume site, Wurtsmith Air Force Base, Oscoda, Michigan, USA. Plume boundary is marked by vertical black lines. Location of profile is shown as A–A' on Fig. 9.2a. Water table is marked by a dashed line and inverted triangle

silt layer forming an impermeable boundary. Contaminant concentrations obtained from groundwater samples showed their highest concentrations in the range of 2592 μg/l for benzene and 92,800 μg/l for ethylbenzene. Soil sampling showed that most of the contaminants were hosted in the silt and clay layers. The water table is ~4 m bgs and fluctuates annually within 1.9–4.1 m bgs. The groundwater has high salinity in the range of 4–10 g/L. The electrical resistivity data were acquired using a combination of a gradient and Wenner arrays with an electrode spacing of 2 m. The data were inverted using RES2DINV to obtain the resistivity structure of the subsurface (Loke and Barker 1996). The locations of borings with contaminant concentration are superimposed on the resistivity profile. The resistivity profile shows two layers, an upper resistive layer that is 5 m thick with ~12–20 Ω·m resistivity and a lower conductive layer with resistivity values <6 Ω·m. The highest resistivity values (>30 Ω·m) in the shallow subsurface (upper 2.5 m) are associated with high concentrations of benzene and ethylbenzene.

Despite the long history of contamination at this site, the LNAPL contamination showed a resistive response. This response may be the result of lack of significant

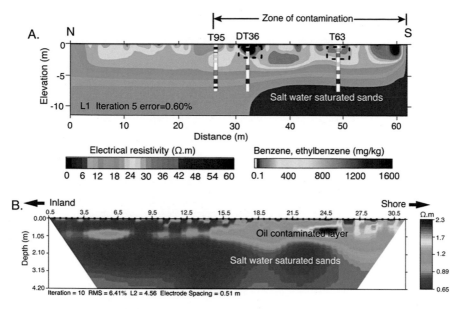

Fig. 9.8 a Geophysical signatures of LNAPL contamination in saline aquifers showing the electrical resistivity tomography profile from a former perfumery plant in Jinghai, southwest of Tianjin, China. T95, DT36, and T63 are soil borings with benzene and ethylbenzene concentrations superimposed on the resistivity profile (modified from Xia et al. 2021 with permission). High concentrations of benzene and ethylbenzene are coincident with regions of high resistivity and the region of high resistivity is associated with the oil-contaminated zone. **b** Electrical resistivity profile from Grande Terre Barrier Island, Louisiana, USA acquired during the 2010 BP Deep Horizon Oil spill in the Gulf of Mexico (Ross 2013). Note that the region of high resistivity is associated with the oil-contaminated zone

degradation or the fact that the spill is hosted in a more conductive host lithology (silts and clays) as well as conductive groundwater. Hence any conductive signature arising from biodegradation might be masked by the conductive clays or saline groundwater.

The second example we present is from the Grande Terre Barrier Island off the southeastern coast of Louisiana, USA (Fig. 9.8b). Details of the study can be found in Heenan et al. (2015). The contamination resulted from the BP Deep Horizon spill which occurred on April 20, 2010, spilling 4.1 M barrels of oil in the Gulf of Mexico. Although dispersants were used to disperse the oil, the spill resulted in the contamination of coastal communities and barrier islands. The electrical resistivity data were acquired using a dipole–dipole array with 0.5 m spacing approximately 4 months after the spill. The subsurface geology consisted of fine to medium-grained sands saturated with salt water. The resistivity profile presented in Fig. 9.8b shows that the resistivity is generally very low (<2.5 Ω·m) due to the water salinity. However, a layer of relatively higher resistivity (~1.5–2.3 Ω·m) occurring at shallow depths (~1–2 m) is spatially correlated to the impacted layer based on soil borings and thickens toward the shoreline. In this example, the high resistivity response resulted from a fresh spill which is resistive and hosted in an aquifer with very conductive salt water. It is important to point out that continued monitoring of the site over an 18-month period documented the attenuation of the resistive anomaly as a result of biodegradation (Heenan et al. 2015).

Although we have presented only two examples of LNAPL spills and biodegradation in saline aquifers in coastal regions, another good example is a spill in the Niger Delta area of Nigeria. A resistive response was the dominant response observed in the electrical resistivity profiles (Raji et al. 2018; Uchegbulam and Ayolabi 2014). The resistive response probably resulted from the fact that this is an area of active oil exploration with repeated spills from pipeline breakage. In addition to the resistive response observed in saline aquifers, a resistive response can also be obtained when the spills are hosted in conductive lithologies such as clays and shales as demonstrated by a study in Argentina by Osella et al. (2002).

9.3.3   Example from Cold, Permafrost Environments

Pettersson and Nobes (2003) provide an example of the use of geophysical techniques (EM-31 and GPR) to investigate hydrocarbon contamination (mostly JP4 and JP5) at Scott Base, located on Ross Island in the Ross Sea region in Antarctica. Scott Base has been occupied since 1957 and anthropogenic activities have resulted in hydrocarbon contamination at the base. Figure 9.9 shows the pre-melt EM survey results from Scott Base presented as apparent conductivity. The red colors represent high conductivities and blue colors represent low conductivities.

Areas identified and labeled as 1, 2, 3, and 4 represent regions of low conductivity (high resistivity) ranging from 1 to 3 mS/m which are interpreted to be the result of the hydrocarbon contamination. Hydrocarbon contamination was confirmed by dug pits that showed oil slicks on water. Laboratory analysis confirmed the presence of hydrocarbons with TPH concentrations ranging from 1860 to 5560 mg/kg. Although studies from other permafrost regions have documented the potential for hydrocarbon degradation in subzero temperatures (e.g., Rike et al. 2003; Børresen et al. 2003), albeit at a slower rate due to low temperatures and low nutrients, at Scott Base, the LNAPL contamination results in a resistive response. It is possible that the high resistivity (low conductivity) response at Scott Base may be due to the occurrence of repeated fresh spills, the lack of extensive biodegradation or the fact that the region is underlain by soils with high salt concentration due to the proximity of the site to the sea (Pettersson and Nobes 2003).

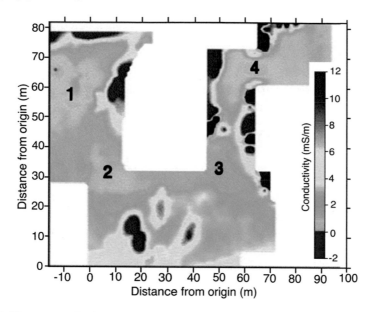

Fig. 9.9 Electromagnetic signatures of LNAPL contamination in a cold, permafrost region. 1–4 represent locations of low electrical conductivity coincident with regions of LNAPL contamination (Modified from Pettersson and Nobes 2003 with permission)

9.3.4 Geophysical Signatures of Microbial-Mediated Mineral Precipitation

We discussed in Sect. 9.1.2 how microorganisms can change the pore water chemistry resulting in the precipitation of different mineral phases as documented in Table 9.1. Different minerals can be precipitated depending on the TEAPs and can therefore be used at hydrocarbon-contaminated sites to delineate zones of contamination, as well as infer the TEAPs at the site. The geophysical technique used to detect zones of bio-mediated mineral precipitation depends on the physical property of the biomineral (Atekwana and Adel Aal 2015). For example, the precipitation of magnetite resulting from the activity of iron-reducing bacteria can be used to infer zones where iron reduction is occurring. Magnetite is both magnetic and conductive, requiring the use of either electrical or magnetic techniques for its detection (e.g., Mewafy et al. 2013; Atekwana et al. 2014). In fact, laboratory investigations by Porsch et al. (2010) suggest that magnetic properties such as magnetic susceptibility (MS) can be used to assist in the delineation of hydrocarbon contamination in the environment. We present, in Fig. 9.9, examples of the use of MS to delineate zones of enhanced MS at three different hydrocarbon-contaminated sites.

Figure 9.10a shows the magnetic susceptibility measurements from cores retrieved from an abandoned refinery site in Carson City, Michigan, USA. The refinery at the site was in operation for more than 60 years and was decommissioned in the 1990s.

Historical releases from storage tanks and pipelines resulted in the contamination of the subsurface aquifer. This site has been the focus of several geophysical (Atekwana et al. 2000, 2004; Werkema et al. 2003; Abdel Aal et al. 2006), microbial (Allen et al. 2007), and geochemical (Atekwana et al. 2004) studies. These studies confirmed active intrinsic bioremediation at the site. In situ resistivity measurements down boreholes identified a zone of enhanced conductivity coincident with the zone of contamination and enhanced microbial activity. This conductivity enhancement was also coincident with the water table fluctuation zone (WTFZ). The magnetic susceptibility values displayed in Fig. 9.10a were obtained from cores retrieved from the site in 2007 and measured in the laboratory using a benchtop MS meter. The results showed a decrease in MS from the surface down to an elevation of 226 m. This elevation is also coincident with the top of the hydrocarbon smear zone formed as a result of seasonal water table fluctuations. At 225.5 m, the MS increased with variable excursions to an elevation of 224 m below which it decreased to the end of the measured core. This zone of enhanced MS was coincident with the zone of contamination and the zone of enhanced conductivity documented by Werkema et al. (2003). In fact, Werkema et al. (2003) suggested that the pore water conductivity in this zone was around five times the conductivity at the uncontaminated location. We can infer that the MS was the result of magnetite precipitation resulting from microbial iron reduction within the WTFZ.

Figure 9.10b shows *in situ* downhole MS measurements from a decommissioned refinery site in, Montana, USA. The primary contaminant was crude oil. The depth to groundwater was approximately 3.6–4 m below ground level with groundwater

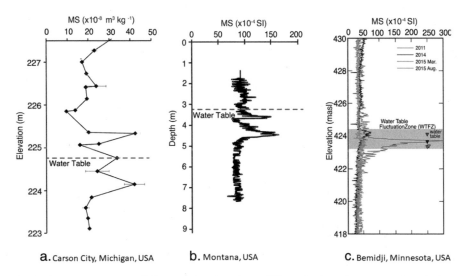

Fig. 9.10 Magnetic susceptibility measured at three different hydrocarbon-contaminated sites. **a** Carson City, Michigan, USA; **b** Montana, USA; **c** Bemidji, Minnesota, USA. All three sites show positive excursions in the magnetic susceptibility within the water table fluctuation zone. **c** Adapted from Lund et al. (2017) with permission

specific conductance at ~15,000 μS/cm making the aquifer at this site very saline. MS logging involves lowering a MS probe/sensor into the borehole while making MS measurements. The MS logging tool can be operated in uncased, or PVC cased wells that have a diameter of 2″ (~50 mm) or bigger. The results showed stable MS values from the surface down to a depth of 3 m at which point significant excursions occur reaching a value of 150 SI × 10^{-4}. This zone of enhanced MS was ~1.5 m thick and extended to a depth of 5 m. Below 5 m the values stabilized at ~100 SI × 10^{-4}. Geochemical data at the site suggested biodegradation was occurring at the site and the plume was methanic.

9.3.5 Geophysical Investigations at Bemidji, Minnesota, USA

Lund et al. (2017) and Atekwana et al. (2014) document similar MS responses from a site near Bemidji, Minnesota, USA, where a crude oil pipeline ruptured, releasing 1,700,000 L of crude oil into the environment. After initial clean up, at least 400,000 L of hydrocarbon remained in the subsurface and the site has been used as a natural laboratory for investigating natural attenuation processes associated with hydrocarbon contamination. The site is administered by the United States Geological Survey. The site has therefore been the focus of numerous investigations including geophysical (Mewafy et al. 2011, 2013; Heenan et al. 2017), microbial (Bekins et al. 2001; Beaver et al. 2016, 2021), and geochemical studies (e.g., Cozzarelli et al. 2010). The site geology consists of ~20 m-thick moderately calcareous silty sand and outwash glacial deposits overlying clayey till of unknown thickness (Bennett et al. 1993). Figure 9.10c shows the MS measurements recorded at the site from 2011 to 2015. The initial MS results were reported in Atekwana et al. (2014). The data showed a marked increase (positive excursion) of MS within the zone of water table fluctuation which was not observed at uncontaminated locations. The MS decreases significantly within the saturated zone below the WTFZ. Microbial data from the site reported in Beaver et al. (2016, 2021) documented that this zone of positive MS excursions coincides with the methanic zone where active methanogenesis is occurring. The zone of iron reduction was above this enhanced MS layer representing a paradox. To explain this enigma, Beaver et al. (2021) suggested methanogens may switch their metabolism from methanogenesis to iron reduction resulting in the precipitation of magnetite.

Atekwana et al. (2014) suggested that MS measurements could be used as a low cost, rapid monitoring tool for assessing the extent of hydrocarbon contamination, and to delineate zones where magnetic mineral precipitation due to iron reduction was occurring. Nonetheless, continued measurements over a 4-year period (2011–2015) documented significant attenuation of the enhanced MS across the WTFZ by ~90%, suggesting that the MS signals were transient (Lund et al. 2017). Although the reduction in MS signal magnitude may result from dissolution of the magnetite (e.g., Ameen et al. 2014), recent studies by Ohenhen et al. (2022) provided evidence suggesting microbial-mediated anaerobic conversion of magnetite to maghemite.

The MS data across all three sites presented in the above case studies show MS excursions within the WTFZ consistent with observations reported at other hydrocarbon-contaminated sites (e.g., Rijal et al. 2010, 2012) and suggest that this interface is biogeochemically active, representing a hotspot of microbial activity. We also note that studies at a hydrocarbon-contaminated site in Iran by Ayoubi et al. (2020) further document the use of MS in predicting hydrocarbon levels of the impacted subsurface volume.

9.3.6   Temporal (Time-Lapse) Geophysical Investigations of Hydrocarbon-Contaminated Sites

One of the major advantages of the application of geophysical techniques to the characterization of LNAPL-contaminated sites is its ability to provide high resolution spatio-temporal images of the subsurface that can be used for monitoring natural attenuation as well as for the monitoring of active bioremediation. We provide some case studies below on the use of geophysics for monitoring LNAPL sites.

Sauck et al. (1998) provided one of the first documented case studies on the use of the self-potential method for mapping a LNAPL plume. In this study, self-potential measurements were first acquired in 1996 over the FT-02 plume at the WAFB site as described above in Sect. 9.3.1.1 (Fig. 9.11a). In 2007, the SP survey was repeated (Fig. 9.11b) and results were presented in Che-Alota et al. (2009). Both data sets were acquired at the same time of the year. The 1996 data (Fig. 9.11a) show a positive NW–SE trending positive SP anomaly reaching a maximum of ~24 mV coincident with the approximate plume boundary as delineated from hydrochemistry and GPR surveys (Sauck et al. 1998). This positive anomaly is in contrast with more negative SP values (around -12 to -30 mV) characterizing the background. Although the site was undergoing intrinsic bioremediation with large negative redox potentials, the source mechanism of the SP anomaly is attributed to diffusion potentials as the values are too small to result from a biogeobattery. In 2007, the repeated survey showed a similar positive anomaly characterizing the plume with two distinct differences: (1) the strength of the anomaly was attenuated and the maximum values are in the range of 3–6 mV; and (2) there is a region of negative SP anomalies (-18 to -34 mV) in the central part of the plume coincident with a soil vapor extraction (SVE) system that was installed in 2001 (Che-Alota et al. 2009). Thus, the decrease in the magnitude of the SP anomaly over the plume is attributed to ongoing natural attenuation processes as well as the reduction of the contaminant mass from the SVE system. Coincidentally, as detailed in Che-Alota et al. (2009), the geophysical anomalies (GPR, ER) observed in 1996 were significantly attenuated by 2007 with both the resistivity and GPR reverting to background conditions.

Giampaolo et al. (2014) used the SP method to monitor a crude oil-contaminated site in Trecate, Italy. In 1994, an oil well blowout from the TR24 ENI-Agip operated

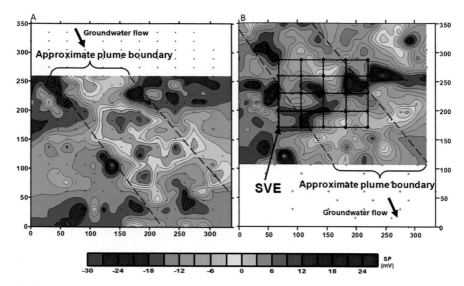

Fig. 9.11 **a** SP anomaly map of a portion of the FT-02 measured in 1996 and **b** in 2007. The grid showing the location of a soil vapor extraction system installed at the site in 2001 is superimposed on the 2007 SP map (modified from Che-Alota et al. 2009 with permission). The 2007 map shows significant attenuation in the magnitude of the positive SP anomaly compared to the 1996 map

exploration well resulted in 15,000 m³ of middleweight crude oil released to the environment contaminating soils and groundwater. Although cleanup efforts removed most of the contaminants, some of the crude oil infiltrated into the subsurface reaching the groundwater. The site geology is characterized by a thick deposit of Holocene glaciofluvial and fluvial deposits made up of poorly sorted silty sand and gravel in extensive lenses, typical of braided streams sediments (Giampaolo et al. 2014). The site has been the focus of several geophysical investigations (GPR, IP, ERT, SP) including Cassiani et al. (2014) and Godio et al. (2010). A time-lapse SP survey was conducted over a 12-month period (March 2010, October 2010, and March 2011), covering both contaminated and uncontaminated regions (Giampaolo et al. 2014). The corrected self-potential maps are shown in Fig. 9.12a–c and document significant time-lapse differences. In general, the SP anomalies are positive over the contaminated region for the March 2010 and 2011 data ranging from 10 to 65 mV, whereas the October data show a bipolar anomaly ranging from − 15 to 25 mV with the positive pole in the north and negative pole in the south. The authors relate the positive anomaly observed in the March 2010 and 2011 data to a decreased value of the electrokinetic coupling coefficient in the contaminated vadose zone because of biodegradation, whereas the bipolar anomaly in October is related to a difference in the redox conditions occurring in the northern and southern parts of the area. The small SP anomalies at this site rule out biogeobatteries as the source of the anomalies due to absence of electronic conductors.

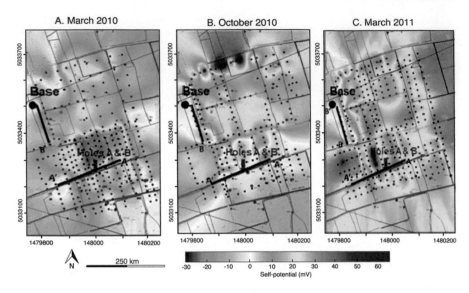

Fig. 9.12 Self-potential maps obtained during March 2010 (**a**), October 2010 (**b**), and March 2011 (**c**) surveys from a crude oil-contaminated site in Trecate, Italy. Note the differences in the time-lapse images with the pronounced bipolar anomaly recorded in the October survey (modified from Giampaolo et al. 2014 with permission)

Incidentally, time-lapse SP monitoring of the Bemidji, MN plume described in Sect. 9.3.4 above by Heenan et al. (2015) using downhole SP probes imaged a strong dipolar anomaly with a negative anomaly above the WTFZ and a positive anomaly below. The observed SP dipole is centered around the zone of highest magnetic susceptibility within the WTFZ. Thus, the microbially-mediated magnetite layer serves as an electronic conductor, transporting electrons from reducing conditions (iron reduction and methanogenesis) below the saturated zone into the unsaturated zone where oxidizing conditions (iron oxidation) occur. This geobattery response appears to be transient in nature, probably driven by hydrobiogeochemical processes.

The above case studies illustrate that SP techniques can be used to not only map the extent of the LNAPL plume but can also map the dynamics of the plume driven by redox and hydrologic processes including monitoring the attenuation of the plume due to active and passive remediation. SP also has the advantage that it can be used to estimate the redox potential distribution (Eh) at organic-rich contaminated sites (e.g., Naudet et al. 2003, 2004; Abbas et al. 2017).

9.3.7 Other Emergent Geophysical Techniques

We would be amiss if we did not include other emergent geophysical techniques that can improve the delineation and quantification of LNAPL contamination.

9.3.7.1 Nuclear Magnetic Resonance (NMR)

Chapter 7 provides the principles of NMR technique and its application to determination of porosity and permeability. Here we provide an example of the use of NMR in the detection of NAPLs, as this technique has been extensively used in petroleum exploration to estimate hydrocarbon and water volume. In this field example, Fay et al. (2017) measured the spin–spin relaxation (T_2 relaxation) times and the diffusion coefficient (D) (D results not shown here) to investigate hydrocarbon impacted sediments at a site in Pine Ridge, South Dakota, USA. Oil field operations at the Pine Ridge, South Dakota site resulted in releases of petroleum products into the subsurface from multiple leaky underground storage tanks. Hydrocarbon contamination was first discovered at the site in 1992. Significant free product thickness of up to 4.3 m thick was observed in wells in 1995 when remediation efforts were initiated. At the time of the survey in 2016, 13–42 cm of free products was still present in the monitoring wells. The subsurface geology is characterized by lacustrine and fluvial sediments consisting of silts, silty sand with siltstone lenses. The water table elevation ranges from 7.8 to 11.1 m below ground surface with ~3.5 m of water table fluctuation. Supporting laboratory measurements were acquired to constrain the field data interpretation. Javelin and Dart NMR probes were used for data acquisition in two different bore holes.

Figure 9.13a shows the Javelin probe T_2 relaxation results. On this figure, warmer colors (yellow) indicate increased fluid volume at a particular decay time. The dashed white line on the plots is the T_2 at 0.1 s and is used to distinguish between the water and contaminant signal. Based on laboratory simulations, the contaminant in silt is expected to plot to the right of this line. The blue water symbol on the plots is the water table elevation at time of measurement and the grey water table symbols are the minimum and maximum water table elevations over six years. The results show bimodal T_2 distribution at several depths which is more pronounced at depths below the water table (below 8.4 m for MW-4 and 7.5 m for MW-16). Fay et al. (2017) interpreted the peak at long T_2 as resulting from the presence of contaminants. In both wells, $T_2 > 0.1$ s is observed at all measurements below the water table and within the smear zone. Although the authors did not consider the effects of biodegradation, we note that the water table fluctuation zone is the zone of most intense biogeochemical activity (often concomitant with mineral precipitation such as magnetite as well as biofilm formation) with maximal changes in geophysical signatures (e.g., Werkema et al. 2003; Atekwana et al. 2014; Sharma et al. 2021). Figure 9.13b shows a plot of the fluid content estimated from the T_2 logs in MW-4 and MW-16. The blue line is the total fluid content, and the red line is attributed to contaminant. In MW-4, the total fluid content ranges from 12 vol% in the unsaturated zone to a maximum of 28 vol% within the saturated zone. Similarly, the contaminant volume is lower in the unsaturated zone (1 vol%) and increases to a maximum of 5 vol% within the saturated zone. In MW-16, the contaminant volume increases to 9.5 vol%.

The observations documented in the Fay et al. (2017) study suggest that NMR is an effective tool that can be used to detect the presence of contaminants and quantify their volume in the sediments. Although not discussed here, the NMR tool can also be

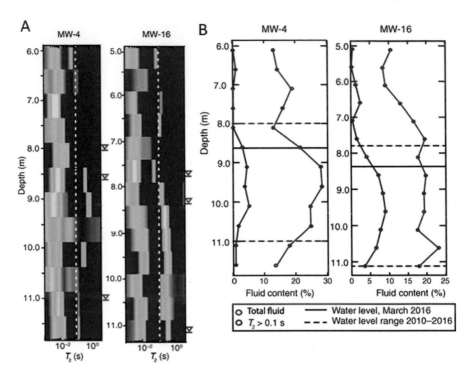

Fig. 9.13 **a** Nuclear magnetic resonance measurements of the T_2 logs measured with the Javelin in 15 cm (6 in) wells MW-4 and MW-16. The color shows the distribution of T_2 values, with warmer colors (yellow) indicating higher signal amplitudes. The depth of the water table in March 2016 is indicated in blue; the gray symbols show the range of water levels observed between 2010 and 2016. The dashed white line shows the $T_2 = 0.1$ s line. **b** Total fluid content (blue) and fluid content with long T_2 (red). MW-4 and MW-16 are the monitoring wells. Solid black line is the water table elevation at time of measurement and dashed black line shows the range of water levels measured from 2010 to 2016. Figure modified from Fay et al. (2017) with permission

used to detect in situ biofilm formation expected to be associated with biodegradation especially within the water table fluctuation zone (e.g., Kirkland et al. 2015).

9.4 Conclusions and Key Take-Aways

We have provided case studies from environments across the globe that show that geophysical investigations can be used to detect fresh spills, as well as effectively used to assess the biogeochemical changes occurring from intrinsic and engineered bioremediation at contaminated hydrocarbon spill sites. In fresh or new spills, the geophysical response is driven by the displacement of (and mixing with) conductive pore water by the resistive LNAPL resulting in a resistive response imaged by electrical geophysical techniques. However, with time, biodegradation causes more conductive pore waters from ionic metabolic by-products resulting in a conductive response which is by far the most common response observed at hydrocarbon sites. This makes electrical geophysical techniques (ERI and GPR) the most applied techniques. Nonetheless, not all biodegraded sites show a conductive response as the response is also dictated by the background geology and fluids properties. Where clays and saline waters characterize the background conditions, then even conductive pore waters resulting from biodegradation would result in a resistive response. Thus, knowledge of the background conditions is important for the investigation and interpretation of the geophysical signatures.

The geophysical response is also driven by microbial-mediated redox processes that result in the precipitation of minerals. The choice of geophysical technique to use to delineate these "hot zones" of microbial activity depends on the contrast in physical properties between the biomineral and the mineralogy of the background geology. We provided case studies showing that magnetite which can be precipitated during microbial-mediated iron reduction is both magnetic and conductive and can be imaged using magnetic susceptibility or SIP techniques. Interestingly, the magnetic susceptibility response was strongest within the water table fluctuation zone pointing to this zone as the most biogeochemically active. In fact, it is within this zone that studies have documented the strongest geophysical responses.

The geophysical detection and monitoring of hydrocarbon contamination are not trivial, as results are often mixed and data interpretation must proceed with caution. For example, the application of the same geophysical technique at different locations at the same site can produce dramatically different results due to the variability of the contaminant mass distribution and the activity of indigenous microorganisms. As depicted by the magnetic susceptibility response in Fig. 9.10, the largest geophysical response may occur within the water table fluctuation zone, where free phase LNAPLs were found, resulting in an abundance of electron donors. This is not surprising as enhanced biogeochemical cycling driven by dynamic environmental conditions results from the mixing of electron donors and acceptors stimulating microbial respiration and shifts in microbial activity, carbon turnover, and elemental cycling. However, the magnitude of the geophysical response decreased downgradient away from the core of the plume where free phase LNAPLs existed to distal ends of the plume where the plume was in the dissolved phase, as well as within the saturated zone.

As previously suggested by Atekwana and Atekwana (2010), the geophysical response of hydrocarbon-contaminated media depends on several factors including: (1) the type of the contaminant present (crude oil, jet fuel, diesel fuel, etc.), (2) their distribution and partitioning into different phases (vapor phase, residual and entrapped phases, free phase, dissolved phase) in the unsaturated and saturated zone, (3) dynamic hydrologic processes (e.g., advective transport, seasonal recharge), (4) release history (e.g., continuous release over a long time versus a single release), (5) the saturation and wetting (oil or water wetted) history of the contaminated media, and (6) biological processes. We add to this list (1) background geology (e.g., clay vs sand), (2) pore water salinity, (3) hydrobiogeochemical processes, and (4) time of year when measurements are made. In conclusion, geophysical technologies offer clear benefits in the characterization of hydrocarbon-contaminated sites especially when optimized by biogeochemistry data. They can therefore be used for identifying contaminated areas, remediation monitoring, and post-remediation monitoring.

Looking to the future, more studies are needed in the following areas:

1. Improved understanding of how hydrobiogeochemical processes drive the geophysical response. For example, the transient magnetic susceptibility is curiously related to the hydrology and water levels at the site and appears to be at a maximum during periods of high-water level but declines during drought periods. Thus, a mechanistic understanding is needed to explain the transient magnetic susceptibility signatures.
2. Relating the geophysical signatures to terminal electron acceptor processes. At present, only responses related to iron reduction can be conclusively determined.
3. Estimating the volumetric content of the LNAPL from the geophysical signal. There is some encouraging evidence provided by Deng et al. (2018) that SIP can be used to estimate the volumetric content of the LNAPL. However, this study was a controlled spill in a well-characterized sandbox experiment. To this last point, we need more field investigations that address this point as well as determine the threshold volume that can be detected by geophysical methodologies, especially in complex geology. NMR may hold such potential as it can both detect and quantify the LNAPL volume.
4. We need more innovations to make SIP instruments more readily usable in field settings like current ERI technologies.

Acknowledgements We acknowledge the different funding agencies [the US National Science Foundation (EAR-1742938; OCE-1049301; EAR-0651433), Chevron Energy Technology Company (Grant # CW852844)] whose financial support helped develop many of the ideas discussed in the chapter. We also acknowledge the different authors whose work we have presented in the case studies. Ryan Joyce and Jay Nolan helped to acquire TDIP data shown on Figure 9.7.

References

Abbas M, Jardani A, Brigaud L, Dupont JP, Soueid AA, Revil A, Begassat P (2017) Redox potential distribution of an organic-rich contaminated site obtained by the inversion of self-potential data. J Hydrol 554:111–127. https://doi.org/10.1016/j.jhydrol.2017.08.053

Abbas M, Jardani A, Machour N, Dupont JP (2018) Geophysical and geochemical characterisation of a site impacted by hydrocarbon contamination undergoing biodegradation. Near Surf Geophys 16(2):176–192

Abdel Aal GZ, Atekwana EA (2014) Spectral induced polarization (SIP) response of biodegraded oil in porous media. Geophys J Int 196(2):804–817

Abdel Aal GZA, Atekwana EA, Slater LD (2004) Effects of microbial processes on electrolytic and interfacial electrical properties of unconsolidated sediments. Geophys Res Lett 31(12):1–4

Abdel Aal GZA, Slater LD, Atekwana EA (2006) Induced-polarization measurements on unconsolidated sediments from a site of active hydrocarbon biodegradation. Geophysics 71(2):H13–H24

Abdel Aal GZ, Atekwana EA, Revil A (2014) Geophysical signatures of disseminated iron minerals: a proxy for understanding subsurface biophysicochemical processes. J Geophys Res Biogeosci 119(9):1831–1849

Allen JP, Atekwana EA, Atekwana EA, Duris JW, Werkema DD, Rossbach S (2007) The microbial community structure in petroleum-contaminated sediments corresponds to geophysical signatures. Appl Environ Microbiol 73(9):2860–2870

Al-Menshed FH, Thabit JM (2018) Comparison between VES and 2D imaging techniques for delineating subsurface plume of hydrocarbon contaminated water southeast of Karbala City. Iraq. Arab J Geosci 11(7):1–9

Ameen NN, Klueglein N, Appel E, Petrovský E, Kappler A, Leven C (2014) Effect of hydrocarbon-contaminated fluctuating groundwater on magnetic properties of shallow sediments. Stud Geophys Geodaet 58(3):442–460

Amiel N, Shaar R, Sivan O (2020) The effect of early diagenesis in methanic sediments on sedimentary magnetic properties: Case study from the SE Mediterranean continental shelf. Front Earth Sci 8:283

Atekwana EA, Abdel Aal GZA (2015) Iron biomineralization controls on geophysical signatures of hydrocarbon contaminated sediments. J Earth Sci 26(6):835–843. https://doi.org/10.1007/s12 583-015-0611-2

Atekwana EA, Atekwana EA (2010) Geophysical signatures of microbial activity at hydrocarbon contaminated sites: a review. Surv Geophys 31(2):247–283

Atekwana EA, Slater LD (2009) Biogeophysics: a new frontier in Earth science research. Rev Geophys 47(4):147

Atekwana EA, Sauck WA, Werkema DD Jr (2000) Investigations of geoelectrical signatures at a hydrocarbon contaminated site. J Appl Geophys 44(2–3):167–180

Atekwana EA, Sauck WA, Abdel Aal GZ, Werkema DD (2002) Geophysical investigations of vadose zone conductivity anomalies at a hydrocarbon contaminated site: implications for the assessment of intrinsic bioremediation. J Environ Eng Geophys 7:103–110

Atekwana EA, Werkema DD Jr, Duris JW, Rossbach S, Atekwana EA, Sauck WA, Cassidy DP, Means J, Legall FD (2004) In-situ apparent conductivity measurements and microbial population distribution at a hydrocarbon-contaminated site. Geophysics 69(1):56–63

Atekwana EA, Mewafy FM, Abdel Aal G, Werkema DD Jr, Revil A, Slater LD (2014) High-resolution magnetic susceptibility measurements for investigating magnetic mineral formation during microbial mediated iron reduction. J Geophys Res Biogeosci 119(1):80–94

Ayoubi S, Khademi H, Shirvani M, Gyasi-Agyei Y (2020) Using magnetic susceptibility for predicting hydrocarbon pollution levels in a petroleum refinery compound in Isfahan Province. Iran. J Appl Geophys 172:8. https://doi.org/10.1016/j.jappgeo.2019.103906

Baveye P, Vandevivere P, Hoyle BL, DeLeo PC, de Lozada DS (1998) Environmental impact and mechanisms of the biological clogging of saturated soils and aquifer materials. Crit Rev Environ Sci Technol 28(2):123–191

Baveye P (2021) Soil "biofilms":"Bioclusters" would be a much better descriptor. Spanish J Soil Sci 11(1)

Beaver CL, Williams AE, Atekwana EA, Mewafy FM, Abdel Aal G, Slater LD, Rossbach S (2016) Microbial communities associated with zones of elevated magnetic susceptibility in hydrocarbon-contaminated sediments. Geomicrobiol J 33(5):441–452

Beaver CL, Atekwana EA, Bekins BA, Ntarlagiannis D, Slater LD, Rossbach S (2021) Methanogens and their syntrophic partners dominate zones of enhanced magnetic susceptibility at a petroleum contaminated site. Front Earth Sci 9:156

Bekins BA, Cozzarelli IM, Godsy EM, Warren E, Essaid HI, Tuccillo ME (2001) Progression of natural attenuation processes at a crude oil spill site: II. Controls on spatial distribution of microbial populations. J Contam Hydrol 53(3–4):387–406

Bennett PC, Siegel DE, Baedecker MJ, Hult MF (1993) Crude oil in a shallow sand and gravel aquifer—I. Hydrogeology and inorganic geochemistry. Appl Geochem 8(6):529–549

Bermejo JL, Sauck WA, Atekwana EA (1997) Geophysical discovery of a new LNAPL plume at the former Wurtsmith AFB, Oscoda, Michigan. Groundwater Monit Remed 17(4):131–137

Binley A, Kemna A (2005) DC resistivity and induced polarization methods. In: Rubin Y, Hubbard SS (eds) Hydrogeophysics: water science and technology library, vol 50. Springer, Dordrecht. https://doi.org/10.1007/1-4020-3102-5_5

Binley A, Slater L (2020) Resistivity and induced polarization: theory and applications to the near-surface earth. Cambridge University Press, Cambridge

Blanchy G, Saneiyan S, Boyd J, McLachlan P, Binley A (2020) ResIPy, an intuitive open source software for complex geoelectrical inversion/modeling. Comput Geosci 137:104423

Blondel A, Schmutz M, Franceschi M, Tichané F, Carles M (2014) Temporal evolution of the geoelectrical response on a hydrocarbon contaminated site. J Appl Geophys 103:161–171

Børresen M, Breedveld GD, Rike AG (2003) Assessment of the biodegradation potential of hydrocarbons in contaminated soil from a permafrost site. Cold Reg Sci Technol 37(2):137–149

Bouwer EJ, Rijnaarts HHM, Cunningham AB, Gerlach R (2000) Biofilms in porous media. In: Bryers J (ed) Biofilms II: process analysis and applications. Wiley-Liss Inc., Hoboken, pp 123–158

Brown LD, Ulrich AC (2014) Bioremediation of oil spills on land. In: Fingas M (ed) Handbook of oil spill science and technology. John Wiley and Sons, Hoboken, NJ

Burazer M, Burazer N (2017) Geophysical and geochemical investigation of hydrocarbon subsurface contamination. In: Proceedings of the August 31st to September 2nd 15th international conference on environmental science and technology, Rhodes, Greece

Cassiani G, Binley A, Kemna A, Wehrer M, Orozco AF, Deiana R, Boaga J, Rossi M, Dietrich P, Werban U, Zschornack L, Godio A, Gandomi AJ, Deidda GP (2014) Noninvasive characterization of the Trecate (Italy) crude-oil contaminated site: links between contamination and geophysical signals. Environ Sci Pollut Res 14:1–18. https://doi.org/10.1007/s11356-014-2494-7

Che-Alota V, Atekwana EA, Atekwana EA, Sauck WA, Werkema DD Jr (2009) Temporal geophysical signatures from contaminant-mass remediation. Geophysics 74(4):B113–B123

Cozzarelli IM, Bekins BA, Eganhouse RP, Warren E, Essaid HI (2010) In situ measurements of volatile aromatic hydrocarbon biodegradation rates in groundwater. J Contam Hydrol 111(1–4):48–64

Davis JL, Annan AP (1989) Ground-penetrating radar for high-resolution mapping of soil and rock stratigraphy. Geophys Prospect 37:531–551

Davis CA, Atekwana E, Slater LD, Rossbach S, Mormile MR (2006) Microbial growth and biofilm formation in geologic media is detected with complex conductivity measurements. Geophys Res Lett 33:L18403

Deceuster J, Kaufmann O (2012) Improving the delineation of hydrocarbon-impacted soils and water through induced polarization (IP) tomographies: a field study at an industrial waste land. J Contam Hydrol 136–137:25–42

Deng YP, Shia XQ, Revil A, Wu J, Ghorbani A (2018) Complex conductivity of oil-contaminated clayey soils. J Hydrol 561:930–942

DeRyck SM, Redman JD, Annan AP (1993) Geophysical monitoring of a controlled kerosene spill. In: Proceedings of the symposium on the application of geophysics to engineering and environmental problems (SAGEEP), San Diego, CA, pp 5–19

Dunsmore BC, Bass CJ, Lappin-Scott HM (2004) A novel approach to investigate biofilm accumulation and bacterial transport in porous matrices. Environ Microbiol 6(2):183–187

Fay EL, Knight RJ, Grunewald ED (2017) A field study of nuclear magnetic resonance logging to quantify petroleum contamination in subsurface sediments. Geophysics 82:EN81–EN92. https://doi.org/10.1190/geo2016-0369.1

Ferry JG (2020) Methanosarcina acetivorans: a model for mechanistic understanding of aceticlastic and reverse methanogenesis. Front Microbiol 11:1806

Flemming HC, Wingender J, Szewzyk U, Steinberg P, Rice SA, Kjelleberg S (2016) Biofilms: an emergent form of bacterial life. Nat Rev Microbiol 14(9):563–575

Giampaolo V, Rizzo E, Titov K, Konosavsky P, Laletina D, Maineult A, Lapenna V (2014) Self-potential monitoring of a crude oil-contaminated site (Trecate, Italy). Environ Sci Pollut Res Int 21(15):8932–8947. https://doi.org/10.1007/s11356-013-2159-y

Godio A, Arato A, Stocco S (2010) Geophysical characterization of a nonaqueous-phase liquid-contaminated site. Environ Geoscience 17(4):141–216

Hansel CM, Benner SG, Neiss J, Dohnalkova A, Kukkadapu RK, Fendorf S (2003) Secondary mineralization pathways induced by dissimilatory iron reduction of ferrihydrite under advective flow. Geochim Cosmochim Acta 67(16):2977–2992

Hansel CM, Benner SG, Fendorf S (2005) Competing Fe(II)-induced mineralization pathways of ferrihydrite. Environ Sci Technol 39(18):7147–7153

Heenan J, Slater LD, Ntarlagiannis D, Atekwana EA, Fathepure BZ, Dalvi S, Ross C, Werkema DD, Atekwana EA (2015) Electrical resistivity imaging for long-term autonomous monitoring of hydrocarbon degradation: lessons from the Deepwater Horizon oil spill. Geophysics 80(1):B1–B11

Heenan JW, Ntarlagiannis D, Slater LD, Beaver CL, Rossbach S, Revil A, Atekwana EA, Bekins B (2017) Field-scale observations of a transient geobattery resulting from natural attenuation of a crude oil spill. J Geophys Res Biogeosci 122(4):918–929

Hoffmann TD, Reeksting BJ, Gebhard S (2021) Bacteria-induced mineral precipitation: a mechanistic review. Microbiology 167:001049. https://doi.org/10.1099/mic.0.001049

Johansson S, Fiandaca G, Dahlin T (2015) Influence of non-aqueous phase liquid configuration on induced polarization parameters: conceptual models applied to a time-domain field case study. J Appl Geophys 123:295–309

Kimak C, Ntarlagiannis D, Slater LD, Atekwana EA, Beaver CL, Rossbach S et al (2019) Geophysical monitoring of hydrocarbon biodegradation in highly conductive environments. J Geophys Res Biogeosci 124(2):353–366. https://doi.org/10.1029/2018JG004561

Kirkland CM, Herrling MP, Hiebert R, Bender AT, Grunewald E, Walsh DO, Codd SL (2015) In situ detection of subsurface biofilm using low-field NMR: a field study. Environ Sci Technol 49:11045–11052

Knight RJ, Nur A (1987) The dielectric constant of sandstones, 60 khz to 4 mhz. Geophysics 52(5):644–654. https://doi.org/10.1190/1.1442332

Lago AL, Elis VR, Borges WR, Penner GC (2009) Geophysical investigation using resistivity and GPR methods: a case study of a lubricant oil waste disposal area in the city of Ribeirão Preto, São Paulo. Brazil. Environ Geol 58(2):407–417

Lee JH, Kennedy DW, Dohnalkova A, Moore DA, Nachimuthu P, Reed SB, Fredrickson JK (2011) Manganese sulfide formation via concomitant microbial manganese oxide and thiosulfate reduction. Environ Microbiol 13(12):3275–3288

Lesmes P, Frye M (2001) Influence of pore fluid chemistry on the complex conductivity and induced polarization responses of Berea sandstone. J Geophys Res 106:4079–4090

Liu D, Dong H, Bishop ME, Wang H, Agrawal A, Tritschler S, Eberl DD, Xie S (2011a) Reduction of structural Fe(III) in nontronite by methanogen Methanosarcina barkeri. Geochim Cosmochim Acta 75(4):1057–1071

Liu D, Wang H, Dong H, Qiu X, Dong X, Cravotta CA III (2011b) Mineral transformations associated with goethite reduction by Methanosarcina barkeri. Chem Geol 288(1–2):53–60

Loke MH, Barker RD (1996) Rapid least-squares inversion of apparent resistivity pseudosections by a quasi-Newton method1. Geophys Prospect 44(1):131–152

Loke MH (2001) Electrical imaging surveys for environmental and engineering studies, a practical guide to 2-D and 3-D surveys: RES2DINV Manual, IRIS Instruments

Lovley DR, Walker DJ (2019) Geobacter protein nanowires. Front Microbiol 10:2078

Lund AL, Slater LD, Atekwana EA, Ntarlagiannis D, Cozzarelli I, Bekins BA (2017) Evidence of coupled carbon and iron cycling at a hydrocarbon-contaminated site from time lapse magnetic susceptibility. Environ Sci Technol 51(19):11244–11249

Madigan MT, Bender KS, Buckley DH, Sattley WM, Stahl DA (2018) Brock biology of microorganisms, 15th edn. Pearson Education, New York City

Mann S, Sparks NHC, Scott GHE, De Vrind-De Jong EW (1988) Oxidation of manganese and formation of Mn$_3$O$_4$ (hausmannite) by spore coats of a marine *Bacillus* sp. Appl Environ Microbiol 54(8):2140–2143

Maurya PK, Balbarini N, Møller I, Rønde V, Christiansen AV, Bjerg PL, Auken E, Fiandaca G (2018) Subsurface imaging of water electrical conductivity, hydraulic permeability and lithology at contaminated sites by induced polarization. Geophys J Int 213(2):770–785

McGuire JT, Smith EW, Long DT, Hyndman DW, Haack SK, Klug MJ, Velbel MA (2000) Temporal variations in parameters reflecting terminal-electron-accepting processes in an aquifer contaminated with waste fuel and chlorinated solvents. Chem Geol 169(3–4):471–485

McNeill JD (1991) Advances in electromagnetic methods for groundwater studies. Geoexploration 27:65–80. https://doi.org/10.1016/0016-7142(91)90015-5

McNeill JD (1980) Electromagnetic terrain conductivity measurements at low induction numbers. Geonics Limited, Mississauga, p 15

Meckenstock RU, Elsner M, Griebler C, Lueders T, Stumpp C, Aamand J, Agathos SN, Albrechtsen HJ, Bastiaens L, Bjerg PL, Boon N (2015) Biodegradation: updating the concepts of control for microbial cleanup in contaminated aquifers. Environ Sci Technol 49(12):7073–7081

Mellage A, Smeaton CM, Furman A, Atekwana EA, Rezanezhad F, Van Cappellen P (2018) Linking spectral induced polarization (SIP) and subsurface microbial processes: results from sand column incubation experiments. Environ Sci Technol 52(4):2081–2090. https://doi.org/10.1021/acs.est.7b04420

Mellage A, Smeaton CM, Furman A, Atekwana EA, Rezanezhad F, Van Cappellen P (2019) Bacterial stern layer diffusion: experimental determination with spectral induced polarization (SIP) and sensitivity to nitrite toxicity. Near Surf Geophys Spec Issue Recent Dev Induced Polar 17(6):623–635. https://doi.org/10.1002/nsg.12058

Mewafy FM, Atekwana EA, Werkema DD Jr, Slater LD, Ntarlagiannis D, Revil A, Skold M, Delin GN (2011) Magnetic susceptibility as a proxy for investigating microbially mediated iron reduction. Geophys Res Lett 38(21):1577

Mewafy FM, Werkema DD Jr, Atekwana EA, Slater LD, Aal GA, Revil A, Ntarlagiannis D (2013) Evidence that bio-metallic mineral precipitation enhances the complex conductivity response at a hydrocarbon contaminated site. J Appl Geophys 98:113–123

Moreira CA, Junqueira PG, Casagrande MFS, Targa DA (2019) Geophysical study in a diesel contaminated area due to a railway accident in Cerquilho (SP, Brazil). Brazil J Geophys 37(4):529–543

Moreira CA, Helene LPI, Hartwig ME, Lourenço R, do Nascimento MMPF, Targa DA (2021) Geophysical and structural survey in the diagnosis of leaks at a fuel station in a uranium mine in decommissioning phase (Poços de Caldas, Brazil). Pure Appl Geophys 178(9):3489–3504

Naudet V, Revil A, Bottero JY, Bégassat P (2003) Relationship between self-potential (SP) signals and redox conditions in contaminated groundwater. Geophys Res Lett 30(21):96

Naudet V, Revil A, Rizzo E, Bottero JY, Bégassat P (2004) Groundwater redox conditions and conductivity in a contaminant plume from geoelectrical investigations. Hydrol Earth Syst Sci 8(1):8–22

Ohenhen LO, Feinberg JM, Slater LD, Ntarlagiannis D, Cozzarelli IM, Rios-Sanchez M, Isaacson CW, Stricker A, Atekwana EA (2022) Microbially induced anaerobic oxidation of magnetite to maghemite in a hydrocarbon-contaminated aquifer. J Geophys Res Biogeosci 127:e2021JG006560. https://doi.org/10.1029/2021JG006560

Orozco AF, Kemna A, Oberdörster C, Zschornack L, Leven C, Dietrich P, Weiss H (2012) Delineation of subsurface hydrocarbon contamination at a former hydrogenation plant using spectral induced polarization imaging. J Contam Hydrol 136:131–144

Orozco AF, Ciampi P, Katona T, Censini M, Papini MP, Deidda GP, Cassiani G (2021) Delineation of hydrocarbon contaminants with multi-frequency complex conductivity imaging. Sci Total Environ 768:144997

Osella A, de la Vega M, Lascano E (2002) Characterization of a contaminant plume due to a hydrocarbon spill using geoelectrical methods. J Environ Eng Geophys 7(2):78–87

Pettersson JK, Nobes DC (2003) Environmental geophysics at Scott Base: ground penetrating radar and electromagnetic induction as tools for mapping contaminated ground at Antarctic research bases. Cold Reg Sci Technol 37(2):187–195

Porsch K, Dippon U, Rijal ML, Appel E, Kappler A (2010) In-situ magnetic susceptibility measurements as a tool to follow geomicrobiological transformation of Fe minerals. Environ Sci Technol 44(10):3846–3852

Prakash D, Chauhan SS, Ferry JG (2019) Life on the thermodynamic edge: respiratory growth of an acetotrophic methanogen. Sci Adv 5(8):eaaw9059

Raji WO, Obadare IG, Odukoya MA, Johnson LM (2018) Electrical resistivity mapping of oil spills in a coastal environment of Lagos. Nigeria. Arab J Geosci 11(7):1–9

Revil A, Mendonça CA, Atekwana EA, Kulessa B, Hubbard SS, Bohlen KJ (2010) Understanding biogeobatteries: where geophysics meets microbiology. J Geophys Res Biogeosci 115(G1):496

Revil A, Karaoulis M, Johnson T, Kemna A (2012) Review: some low-frequency electrical methods for subsurface characterization and monitoring in hydrogeology. Hydrogeol J 20:617–658. https://doi.org/10.1007/s10040-011-0819-x

Revil A, Abdel Aal GZ, Atekwana EA, Mao D, Florsch N (2015) Induced polarization response of porous media with metallic particles—Part 2: comparison with a broad database of experimental data. Geophysics 80(5):D539–D552

Reynolds JM (2011) An introduction to applied and environmental geophysics, 2nd edn. Wiley-Blackwell, Hoboken

Rijal ML, Appel E, Petrovský E, Blaha U (2010) Change of magnetic properties due to fluctuations of hydrocarbon contaminated groundwater in unconsolidated sediments. Environ Pollut 158(5):1756–1762

Rijal ML, Porsch K, Appel E, Kappler A (2012) Magnetic signature of hydrocarbon-contaminated soils and sediments at the former oil field Hänigsen, Germany. Stud Geophys Geodaet 56(3):889–908

Rike AG, Haugen KB, Børresen M, Engene B, Kolstad P (2003) In situ biodegradation of petroleum hydrocarbons in frozen arctic soils. Cold Reg Sci Technol 37(2):97–120

Ross CS (2013) Geophysical and geochemical characterization and delineation of a crude oil spill in a highly saline environment. MS Thesis, Oklahoma State University, p 74

Rubin Y, Hubbard SS (eds) (2005) Hydrogeophysics, vol 50. Springer Science & Business Media, New York

Saneiyan S, Ntarlagiannis D, Ohan J, Lee J, Colwell F, Burns S (2019) Induced polarization as a monitoring tool for in-situ microbial induced carbonate precipitation (MICP) processes. Ecol Eng 127:36–47

Sato M, Mooney HM (1960) The electrochemical mechanism of sulfide self-potentials. Geophysics 25(1):226–249

Sauck WA (2000) A model for the resistivity structure of LNAPL plumes and their environs in sandy sediments. J Appl Geophys 44(2–3):151–165

Sauck WA, Atekwana EA, Nash MS (1998) High electrical conductivities associated with an LNAPL plume imaged by integrated geophysical techniques. J Environ Eng Geophys 2:203–212

Schmutz M, Revil A, Vaudellet P, Batzle M, Femenía Viñao P, Werkema DD (2010) Influence of oil saturation upon spectral induced polarization of oil-bearing sands. Geophys J Int 183(1):211–224. https://doi.org/10.1111/j.1365-246X.2010.04751.x

Shang H, Daye M, Sivan O, Borlina CS, Tamura N, Weiss BP, Bosak T (2020) Formation of zerovalent iron in iron-reducing cultures of Methanosarcina barkeri. Environ Sci Technol 54(12):7354–7365

Shao S, Guo X, Gao C, Liu H (2021) Quantitative relationship between the resistivity distribution of the by-product plume and the hydrocarbon degradation in an aged hydrocarbon contaminated site. J Hydrol 596:1–13

Sharma S, Jaiswal P, Raj R, Atekwana EA (2021) In-situ biofilm detection in field settings using multichannel seismic. J Appl Geophys 193:104423

Sivan O, Shusta SS, Valentine DL (2016) Methanogens rapidly transition from methane production to iron reduction. Geobiology 14(2):190–203

Skubal KL, Barcelona MJ, Adriaens P (2001) An assessment of natural biotransformation of petroleum hydrocarbons and chlorinated solvents at an aquifer plume transect. J Contam Hydrol 49(1–2):151–169

Slater LD, Lesmes D (2002) IP interpretation in environmental investigations. Geophysics 67(1):77–88

Telford WM, Geldart LP, Sheriff RE (1990) Applied geophysics, 2nd edn. Cambridge University Press, Cambridge

Teramoto EH, Vogt C, Baessa MPM, Polese L, Soriano AU, Chang HK, Richnow HH (2020) Dynamics of hydrocarbon mineralization characterized by isotopic analysis at a jet-fuel-contaminated site in subtropical climate. J Contam Hydrol 234:103684

Uchegbulam O, Ayolabi EA (2014) Application of electrical resistivity imaging in investigating groundwater pollution in Sapele Area, Nigeria. J Water Resour Protect 6(14):1369

Vanhala H (1997) Mapping oil-contaminated sand and till with the spectral induced polarization (SIP) method. Geophys Prospect 45(2):303–326

Warren E, Bekins BA (2018) Relative contributions of microbial and infrastructure heat at a crude oil-contaminated site. J Contam Hydrol 211:94–103

Werkema DD Jr, Atekwana EA, Endres AL, Sauck WA, Cassidy DP (2003) Investigating the geoelectrical response of hydrocarbon contamination undergoing biodegradation. Geophys Res Lett 30(12):58

Xia T, Dong Y, Mao D, Meng J (2021) Delineation of LNAPL contaminant plumes at a former perfumery plant using electrical resistivity tomography. Hydrogeol J 29(3):1189–1201

Yu H, Leadbetter JR (2020) Bacterial chemolithoautotrophy via manganese oxidation. Nature 583(7816):453–458

Zhang J, Dong H, Liu D, Fischer TB, Wang S, Huang L (2012) Microbial reduction of Fe(III) in illite–smectite minerals by methanogen Methanosarcina mazei. Chem Geol 292:35–44

Zhang J, Dong H, Liu D, Agrawal A (2013) Microbial reduction of Fe (III) in smectite minerals by thermophilic methanogen Methanothermobacter thermautotrophicus. Geochim Cosmochim Acta 106:203–215

Zhang C, Revil A, Fujita Y, Munakata-Marr J, Redden G (2014) Quadrature conductivity: a quantitative indicator of bacterial abundance in porous media. Geophysics 6(79):D363–D375

Chapter 10
Molecular Biological Tools Used in Assessment and Remediation of Petroleum Hydrocarbons in Soil and Groundwater

Dora M. Taggart and Trent A. Key

Abstract Molecular biological tools (MBTs) are used to assess and characterize the microbiology and associated biological processes at contaminated sites, including ecological roles, phylogenetic diversity, and metabolic and co-metabolic capabilities related to contaminant biodegradation and biotransformation. MBTs have changed our approach to the assessment and remediation of petroleum hydrocarbons in the environment. In contaminated-site management, several MBTs and stable isotope analysis methods can be used to measure the presence, abundance, and activity of petroleum hydrocarbon-degrading microorganisms and transformation pathways: (1) quantitative polymerase chain reaction (qPCR) to quantify microorganisms and genes encoding enzymes for biodegradation or biotransformation, (2) DNA sequencing to comprehensively identify the microorganisms presence and microbial community structure, (3) stable isotope probing (SIP) provides conclusive evidence of biodegradation through the use of ^{13}C-labeled surrogate compounds of petroleum hydrocarbons that microbes use as carbon and/or energy sources, and (4) compound-specific isotope analysis (CSIA) measures the ratio of naturally occurring stable isotopes within a contaminant to indicate (bio)degradation. This chapter discusses the use, benefits, and limitations of MBTs. Several real-world case studies are provided to demonstrate how to investigate the biodegradation of petroleum hydrocarbon constituents under varying field conditions to better characterize governing biogeochemical processes and to better inform remedial decisions.

Keywords Bioremediation · DNA sequencing · Petroleum hydrocarbons · qPCR · Stable isotope probing

D. M. Taggart (✉)
Microbial Insights, Inc, Knoxville, TN, USA
e-mail: DTaggart@microbe.com

T. A. Key
ExxonMobil Environmental and Property Solutions Company, Spring, TX, USA
e-mail: trent.a.key1@exxonmobil.com

© The Author(s) 2024
J. García-Rincón et al. (eds.), *Advances in the Characterisation and Remediation of Sites Contaminated with Petroleum Hydrocarbons*, Environmental Contamination Remediation and Management, https://doi.org/10.1007/978-3-031-34447-3_10

10.1   Introduction

Molecular biological tools (MBTs) are a category of environmental molecular diagnostics (Busch-Harris et al. 2008), which provide essential information for evaluating ongoing and potential degradation of petroleum hydrocarbons (PHCs) by microorganisms. MBTs provide critical information for effective site characterization, remedy selection, performance monitoring, and site closure (Fig. 10.1).

MBTs measure the organisms and genes involved in contaminant biodegradation and provide evidence for understanding changes in rates of abiotic and biotic degradative processes over background. MBTs complement conventional hydrogeological, chemical, and geochemical data, as part of a multiple lines of evidence approach to provide a comprehensive picture of the concentrations of contaminants and associated products, the redox status, electron acceptors, electron donors, microbiology, and degradation of contaminants. This multiple lines of evidence approach is critically important for selecting the appropriate remediation strategy, monitoring remediation effectiveness, and supporting transition to monitored natural source zone depletion (NSZD) and monitored natural attenuation (MNA). Natural attenuation processes are further discussed in Chaps. 5, 9, and 13. Details of active bioremediation approaches are presented in Chap. 14.

Here, we review MBT integration into site management and use case studies to highlight the critical evidence MBTs provide to decision-makers.

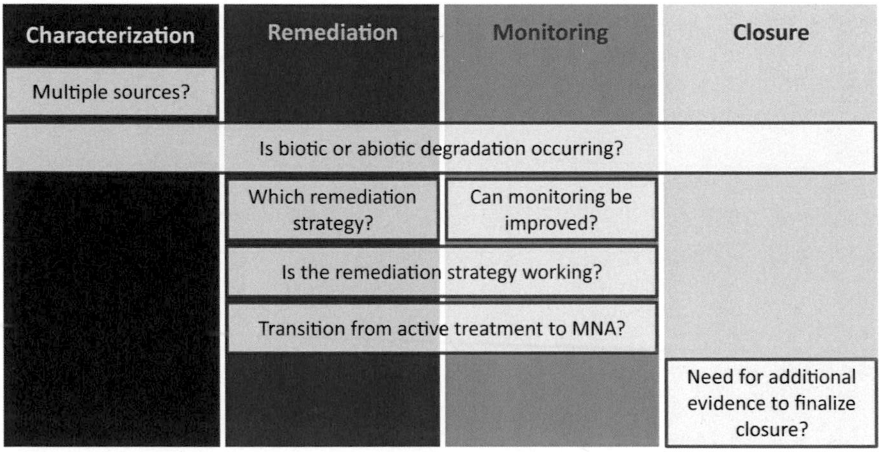

Fig. 10.1 Questions MBTs can help answer. MNA refers to monitored natural attenuation. Adapted from Interstate Technology and Regulatory Council (ITRC 2013)

10.2   MBTs Used in PHC Investigation and Remediation

Several MBTs can be employed at contaminated sites—quantitative polymerase chain reaction (qPCR), qPCR arrays, reverse transcription-qPCR (RT-qPCR), stable isotope analysis methods such as stable isotope probing (SIP) and compound-specific isotope analysis (CSIA), or DNA sequencing (amplicon sequencing or metagenome sequencing)—to improve conceptual site models or aid in the remediation design or decision-making, as described below. The samples to be assayed are typically obtained using grab sampling or passive sampling approaches like in-situ microcosms (ISMs).

For nucleic acid-based MBTs, including qPCR, RT-qPCR, and DNA sequencing, groundwater sampling methods can significantly influence results (Alleman et al. 2005; Ritalahti et al. 2010). Therefore, the sampling protocol should be defined and maintained for the duration of the monitoring effort. At a minimum, sampling should use low-flow purging methods, and, immediately after sampling, the samples should be placed in coolers with ice packs and/or blue ice to ensure refrigeration at 4 °C until arrival at the analytical laboratory (Lebrón et al. 2011).

10.2.1   In-Situ Microcosms (ISMs)

ISMs sample subsurface conditions to assess biodegradation associated with natural attenuation (e.g., MNA) and enhance bioremediation, specifically biostimulation (i.e., addition of rate-limiting nutrients, electron acceptors, and electron donors). These data can be used to estimate amendment effectiveness prior to pilot- or full-scale biostimulation approaches (Taggart and Clark 2021). ISMs typically consist of a two- or three-unit assembly, one unit for assessing biodegradation associated with MNA and the second and third units for assessing biodegradation associated with various amendments for biostimulation. Each unit commonly consists of a length of slotted PVC pipe that hoses samplers to quantify geochemical parameters—including electron acceptors (nitrate and sulfate), dissolved gases (methane, ethene, and ethane), and chloride—concentration of PHCs and associated products, and microorganism genera and genes by a nucleic acid-based MBT (e.g., qPCR). The analyses are conducted after the ISM is incubated in a monitoring well for 30–60 days and can be used in combination with SIP or CSIA (Sects. 10.2.4 and 10.2.5 below; see also Chap. 11 for further details on CSIA). Successful ISM studies can provide in-situ biodegradation data and could be done to complement bench-, pilot-, or full-scale biostimulation design and implementation.

10.2.2   Quantitative Polymerase Chain Reaction (qPCR)

qPCR is used to quantify taxonomic groups (genera, species, and strains) and genes which encode enzymes capable of degrading PHCs in soil or groundwater. qPCR is now used instead of counting colonies on agar plates and other culture-based methods that recover a lower percentage of microbes. qPCR is accurate, sensitive to low microbial concentrations (detection limit as low as 10 cells/ml), and precise in detecting genes over a large dynamic range (~seven orders of magnitude) (Forootan et al. 2017).

Thus, qPCR determines the presence of target microbes and genes, their abundance (expressed as cells/mL), changes in their abundance over time or space, and whether they are present in sufficient quantities for bioremediation. At baseline, qPCR is used to assess microbial abundances pretreatment. In response to treatments, the growth of the microorganisms, measured as an increase in genera/species/strains or specific biodegradation genes, provides evidence supporting treatment effectiveness. Interpreting qPCR over space can be conceptualized as higher abundances of the microbes and contaminant-degrading genes in groundwater or soil impacted by PHC compared to background groundwater or soil, which provides direct evidence for biodegradation.

A single qPCR quantifies a single gene. However, PHCs at contaminated sites are typically complex mixtures of aliphatic, cyclic, and heterocyclic compounds that are potentially biodegraded via multiple genes in multiple anaerobic and aerobic pathways. Thus, the quantification of multiple genes can improve the assessment of biodegradation capacity at a site, depending on site-specific scenarios.

Although the microbes and genes quantified need to have been previously characterized and published, assays are commercially available for the quantification of numerous genes involved in degradation of contaminants of concern (COC). The current commercially available number of different gene targets with potential applicability to PHC-impacted sites is > 80, including more than 12 functional genes associated with polycyclic aromatic hydrocarbon (PAH) biodegradation (Table 10.1).

qPCR assays can be multiplexed within a single-tube reaction containing multiple primer pairs and probes. However, these reactions can be prone to mis-priming, artifact amplification products, and erroneous results from cross-reactions. To overcome this risk, qPCR arrays can be used that consist of numerous individual reaction wells (64 per square) of nanoliter volume (33 nL); each reaction contains a single primer pair and probe and is monitored for reaction kinetics individually. Analysis of a broad spectrum of target genes simultaneously is particularly useful for assessing mixtures of contaminants or when evaluating the potential for multiple biodegradation pathways or treatments.

Table 10.1 Genes and microorganisms involved in the biodegradation of PHCs and other compounds. Abbreviations correspond to the naming convention used by most commercial laboratories. It is not an academic gene labeling convention

Contaminant class	Redox	Abbreviation	Target	Description
Benzene, Toluene, Ethylbenzene, Xylene (BTEX)	Anaerobic	BSS	Benzylsuccinate Synthase	Mediates the first step in the anaerobic biodegradation of toluene and in some cases ethylbenzene and xylenes
		ABC	Anaerobic Benzene Carboxylase	Anaerobic biodegradation of benzene
		GMET	*Geobacter metallireducens* Functional Genes	Targets functional genes including a predicted oxidoreductase specifically required for anaerobic benzene metabolism
		BCR	Benzoyl Coenzyme A Reductase	Reduces the benzene ring structure of the central intermediate Benzyl-CoA
	Aerobic	TOD	Toluene and Benzene Dioxygenases	Capable of degrading toluene, benzene, and chlorobenzene along with co-oxidation of a variety of compounds including ethylbenzene, p-xylene, and m-xylene
		RMO	Ring hydroxylating Toluene monooxygenase (toluene-3 and 4 monooxygenases)	Catalyze the initial oxidation and sometimes second oxidation steps in aerobic BTEX biodegradation
		RDEG	Ring hydroxylating Toluene monooxygenase (toluene-2 monooxygenases)	Catalyze the initial oxidation and sometimes second oxidation steps in aerobic BTEX biodegradation

(continued)

Table 10.1 (continued)

Contaminant class	Redox	Abbreviation	Target	Description
		PHE	Phenol Hydroxylase	Catalyze the continued oxidation of phenols produced by RMOs
		TOL	Toluene/Xylene Monooxygenase	Biodegradation of toluene by attacking at the methyl group
		EDO	Ethylbenzene Dioxygenase	Responsible for aerobic biodegradation of alkylbenzenes including ethylbenzene and isopropylbenzene or cumene
		TCBO	Trichlorobenzene and Biphenyl/Isopropylbenzene Dioxygenases	Includes benzene and isopropylbenzene dioxygenases
Methyl Tert-Butyl Ether (MTBE) and Tert-Butyl Alcohol (TBA)	Aerobic	PM1	*Methylibium petroleiphilum* PM1	Capable of utilizing MTBE and TBA as growth-supporting substrates
		TBA	Tert-Butyl Alcohol Monooxygenase	Catalyzes the continued biodegradation of TBA, an intermediate produced during aerobic MTBE and ETBE biodegradation
		ETHB	P450 cytochrome monooxygenase	Initiates aerobic biodegradation of ethyl tert-butyl ether (ETBE) and is capable of co-oxidation of MTBE and tert-amyl methyl ether (TAME)

(continued)

Table 10.1 (continued)

Contaminant class	Redox	Abbreviation	Target	Description
Diesel, Naphthalene, and PAHs	Anaerobic	ANC	Anaerobic Naphthalene Carboxylase	Only known pathway for anaerobic biodegradation of naphthalene
		MNSSA	Naphthylmethylsuccinate Synthase	Initiates anaerobic biodegradation of 2-methylnaphthalene by catalyzing the addition of fumarate onto the methyl group
	Aerobic	NAH	Naphthalene Dioxygenase	Initiates aerobic metabolism of naphthalene by incorporating both atoms of molecular oxygen into the ring. Also capable of catalyzing oxidation of anthracene, phenanthrene, acenaphthylene, acenaphthene, and fluorine
		NidA	Naphthalene-inducible Dioxygenase	Capable of mineralizing naphthalene and degrading some higher molecular weight PAHs including pyrene and benzo[a]pyrene
		PHN	Phenanthrene Dioxygenase	Phenathrene/naphthalene dioxydenases capable of degrading phenathrene and naphthalene but have broad specificity
		ARHA	Acenaphthylene Dioxygenase	Capable of catalyzing the degradation of acenaphthene, acenaphthylene, naphthalene, phenanthrene, anthracene and fluoranthene

(continued)

Table 10.1 (continued)

Contaminant class	Redox	Abbreviation	Target	Description
		NAHM	Marine Naphthalene and PAH Dioxygenases (*nahAc, phnA*)	Quantifies naphthalene dioxygenase genes from marine organisms that are capable of catalyzing the degradation of naphthalene, 2-methylnaphthalene, phenanthrene fluoranthene, and pyrene
		PHE	Phenol Hydroxylase	Involved in aerobic biodegradation of BTEX and catalyze the continued oxidation of phenols produced by toluene monooxygenases and indicate the potential for aerobic BTEX biodegradation
TPH-Alkanes and Crude Oil	Anaerobic	ASSA	Alkylsuccinate Synthase	Initiates anaerobic biodegradation of alkanes with chain lengths from C6 to at least C18
	Aerobic	ALKB	Alkane Monooxygenase	Initiates the aerobic biodegradation of n-alkanes with carbon lengths from C5 to C16
		ALMA	Alkane Monooxygenase	Catalyzes the aerobic biodegradation of C20–C32 alkanes by some *Alcanivorax* species that are dominant in marine systems
		CAR	Carbazole Dioxygenase	Catalyzes the oxidation of carbazole and other high-molecular-weight aromatics such as dibenzofuran

(continued)

Table 10.1 (continued)

Contaminant class	Redox	Abbreviation	Target	Description
Dioxin, Dibenzofuran	Aerobic	dbfA	Dibenzofuran dioxygenase	Catalyzes the first step in the aerobic degradation of dibenzofuran. Can also degrade dibenzo-p-dioxin at lower levels
Dibenzothiophene	Aerobic	dbtA	Dibenzothiophene Dioxygenase from *Burkholderia* spp.	Catalyzes the degradation of the intermediate dibenzothiophene
Styrene	Aerobic	STY	Styrene Monooxygenase	Catalyzes the epoxidation of styrene to styrene oxide
Nitrogen, Nitrate, Nitrite, Ammonia	Anaerobic	NIF	Nitrogen-fixing Bacteria via nitrogenase (*nifD*)	Nitrogen fixation converts nitrogen gas into ammonia which can be assimilated by organisms
		DNF	Denitrifying Bacteria via dissimilatory nitrite reductase (*nirS* and *nirK*)	Responsible for converting nitrite to nitric oxide
		ADNF	Archaeal Denitrifying Bacteria via nitrite reductase (*nirS* and *nirK*)	Responsible for converting nitrite to nitric oxide in archaeal organisms
		Anammox	Anaerobic ammonia oxidation from *Brocadia, Kuenenia, Scalindua, Anammoxyglobus, and Jettenia* spp.	Responsible for converting nitrite and ammonia directly into molecular nitrogen
		AMXNIRS	Anammox Nitrite Reductase (*nirS*)	Responsible for reducing nitrite to nitric oxide inside the anammoxosome
		AMXNIRK	Anammox Nitrite Reductase (*nirK*)	Responsible for reducing nitrite to nitric oxide inside the anammoxosome

(continued)

Table 10.1 (continued)

Contaminant class	Redox	Abbreviation	Target	Description
	Aerobic	AMO	Ammonia oxidizing bacteria via ammonia monooxygenase (*amoA*)	Responsible for converting ammonia to hydroxyl amine which is then converted to nitrite by hydroxylamine oxidoreductase
		AOA	Ammonia oxidizing archaea via ammonia monooxygenase (*amoA*)	Responsible for converting ammonia to hydroxyl amine in archaeal organisms
		NOB	Nitrite oxidizing bacteria (*Nitrospira* spp.)	Responsible for converting nitrite to Nitrate
		NOR	Nitrite oxidizing bacteria via nitrite oxidoreductase	Responsible for converting nitrite to Nitrate
Prokaryotic Groups	Variable	AAB	Acetic Acid Bacteria via alcohol dehydrogenase (*adhA*) from *Acetobacter*, *Gluconobacter*, and *Komagataeibacter* spp.	Catalyzes the oxidation of ethanol to acetic acid which can be a potential cause of corrosion
		AGN	Acetogens	Acetogenic bacteria are strict anaerobes that produce acetate from the conversion of H_2 and CO_2, CO, or formate
		AMGN	Acetoclastic Methanogens (*Methanosarcina* spp.)	Acetoclastic methanogens dismutate acetate to form methane

(continued)

Table 10.1 (continued)

Contaminant class	Redox	Abbreviation	Target	Description
		APS	Sulfate-reducing bacteria via Adenosine 5' phosphosulfate reductase	Anaerobic hydrocarbon oxidation/biogeochemical reduction
		ARG	*Archaeoglobus*	Targets a genus of sulfate-reducing archaea
		ARC	Total Archaea	Quantifies total archaea
		BCE	Exopolysaccharide from *Burkholderia* spp.	Gene involved in the production of exopolysaccharide (EPS) and biofilm formation by some *Burkholderia* spp.
		CLAD	*Cladosporium*	*Cladosporium resinae* is a common fuel contaminant that it has been described as the "kerosene fungus." *C. resinae* grows on hydrocarbons including alkanes to produce organic acids
		DCS	*Deinococcus*	Genus of bacteria considered very efficient primary biofilm formers and therefore have been implicated in slime formation and biofouling
		EBAC	Total Eubacteria	Quantifies total Eubacteria

(continued)

Table 10.1 (continued)

Contaminant class	Redox	Abbreviation	Target	Description
		FEOB	Iron Oxidizing Bacteria (*Gallionella*, *Leptothrix*, *Mariprofundus*, and *Sphaerotilus* spp.)	Iron oxidizing bacteria are a group of microorganisms implicated in biogeochemical cycling of iron which may be relevant to anaerobic hydrocarbon biodegradation
		FER	Fermenting Bacteria (fermentative *Firmicutes*)	Anaerobic bacteria that produce organic acids and hydrogen. Organic acid and hydrogen production can be relevant to understand hydrocarbon biodegradation, downstream methanogenesis, and natural source-zone depletion processes
		GEO	*Geobacter*	Iron-reducing bacteria reduce insoluble ferric iron, which occurs in anaerobic hydrocarbon biodegradation using ferric iron as an electron acceptor. This assay targets a common iron-reducing bacteria, *Geobacter*
		GLK	Glycerol-utilizing acetogens via glycerol kinase	Microbial degradation of glycerol, a byproduct of biodiesel production from fats, leads to the generation of VFAs (lactic and propionic acid)
		IRB/SRB	Iron- and sulfate-reducing bacteria	Quantifies iron- and sulfate-reducing *Deltaproteobacteria*

(continued)

Table 10.1 (continued)

Contaminant class	Redox	Abbreviation	Target	Description
		IRA	Iron-reducing archaea (*Ferroglobus* and *Geoglobus*)	Quantifies two genera of iron-reducing archaea
		IRB	Iron-reducing bacteria	Iron-reducing bacteria reduce insoluble ferric iron to soluble ferrous iron, which occurs in anaerobic hydrocarbon biodegradation using ferric iron as an electron acceptor
		LAB	Lactic Acid Bacteria via lactate dehydrogenase	Quantifies bacteria that produce lactic acid. Lactic acid produced by these organisms can contribute to corrosion. However, some strains can also produce expolysaccharides
		MGN	Methanogens via methyl coenzyme reductase (*mcrA/mrtA*)	Methanogens utilize hydrogen for growth and can be relevant to understand hydrocarbon biodegradation and natural source-zone depletion processes. Methanogens also compete with halorespiring bacteria for available hydrogen

(continued)

Table 10.1 (continued)

Contaminant class	Redox	Abbreviation	Target	Description
		MnOB	Manganese Oxidizing Bacteria via	Manganese oxidizing bacteria are a group of microorganisms implicated in biogeochemical cycling of manganese which may be relevant to anaerobic hydrocarbon biodegradation
		MOB	Methanotrophs	Targets methane oxidizing bacteria
		PMMO	Particulate Methane Monooxygenase (*pmoA*)	Targets the *pmoA* gene of methane oxidizing bacteria which oxidizes methane to form methanol. *pmoA* is capable of cometabolizing trichloroethylene
		MTS	*Meiothermus*	*Meiothermus* spp. are efficient primary biofilm formers and frequently implicated in slime formation and biofouling
		SHEW	*Shewanella*	Anaerobic bacteria, which can utilize hydrogen as an energy source, reduce ferric iron and sulfide to ferrous iron and sulfide indicating that it can play a role in anaerobic hydrocarbon biodegradation and attenuation

(continued)

Table 10.1 (continued)

Contaminant class	Redox	Abbreviation	Target	Description
		SOB	Sulfur Oxidizing Bacteria	Often aerobic bacteria oxidize sulfide or elemental sulfur producing sulfuric acid. Commonly implicated in the corrosion of concrete
		SRA	Sulfate-reducing archaea	Sulfate-reducing archaea consume hydrogen, produce hydrogen sulfide and have been implicated in anaerobic hydrocarbon biodegradation and attenuation
		SSPH	*Sporomusa sphaeroides*	Genus of anaerobic, acetic acid-producing bacteria (homoacetogens)
		TFUN	Total Fungi	Quantifies total fungi. Fungi are capable of degrading organic materials, hydrocarbons, foods, and wood and produce organic acids
Biosurfactant Producers	Variable	SurG	Glycolipid Biosurfactants	Targets the *rhlA* and *rhlC* genes involved in the production of mono- and di-rhamnolipids in *Pseudomonas* spp. as well as the *treS*, and *treY* genes involved in the production of trehalose in Rhodococcus spp. These glycolipid biosurfactants may be implicated in the bioremediation of petroleum hydrocarbons

(continued)

Table 10.1 (continued)

Contaminant class	Redox	Abbreviation	Target	Description
		SurP	Lipopeptide Biosurfactants	Targets the *SrfAC*, *licC*, *aprE* genes involved in the production of *Surfactin, lichenysin,* and *Subtilisin* in *Bacillus* spp. as well as the *visC* gene involved in the production of viscosin in *Pseudomonas* spp. These lipopeptide biosurfactants may be implicated in the bioremediation of petroleum hydrocarbons
		SurL	Liposaccharide Biosurfactants	Targets the *weeA* and *alnB* genes involved in the production of emulsan and alasan in *Acinetobacter* spp. These liposaccharide biosurfactants may be implicated in the bioremediation of petroleum hydrocarbons
		SurT	Trehalose Biosurfactants	Targets a specific group of glycolipid biosurfactants that are produced by *Rhodococcus* and *Mycobacterium* spp. and are involved in the uptake of low polarity hydrocarbons. The SurT assays quantifies two genes responsible for the production of these biosurfactants

10.2.3 Reverse Transcription Quantitative Polymerase Chain Reaction (RT-qPCR)

RT-qPCR is a variation of qPCR that quantifies RNA transcripts expressed from genes. RT-qPCR not only confirms the presence of a target microorganism(s) or functional gene(s) (i.e., gene that produces an enzyme that degrades a contaminant) but also demonstrates microbes are active for biodegradation. For example, *Methylibium petroleiphilum* strain PM1 quantified by qPCR in the moderate range (10^3–10^4 cells/mL) often does not correlate with aerobic methyl tert-butyl ether (MTBE) biodegradation; rather moderate strain concentrations have been associated with low to moderate MTBE biodegradation (i.e., data in tertiles: low, moderate, and high) (Taggart and Clark 2021). Similarly, in aerobic biodegradation of BTEX compounds (benzene, toluene, ethylbenzene, and xylene), qPCR quantification of the genes encoding toluene dioxygenase and phenol hydroxylase can be high, but the genes' expression can be low (Taggart and Clark 2021). Thus, site-specific scenarios, such as those described above, may indicate that RNA analysis should be performed.

10.2.4 Stable Isotope Probing (SIP)

SIP is an MBT which assesses biodegradation through the use of a synthetic stable isotope probe; i.e., the COC is synthesized with carbons replaced by ^{13}C. This ^{13}C-labeled stable isotope probe is absorbed to a matrix (e.g., the activated carbon in commercially available Bio-Sep® beads in an ISM) that is placed in monitoring wells. After 30–45 days, the matrix is retrieved, and the probe's biodegradation by microbes from the aquifer that colonized the matrix is assessed by quantifying the ^{13}C-labeled COC remaining and the quantity of ^{13}C incorporated into microbial biomass—phospholipid fatty acids (PLFA) or nucleic acids (DNA and RNA)—and $^{13}CO_2$ (i.e., dissolved inorganic carbon: carbon dioxide, bicarbonate, and carbonate). PLFA is a major component of bacterial cell membranes, and for contaminants microbes used as carbon sources, the incorporation of ^{13}C from the stable isotope probe into PLFA indicates that the microbes have grown new cells (biomass). Additionally, carbon can be oxidized to CO_2 (mineralization) through PHCs biodegradation. Therefore, the method can also be used to assess biodegradation by microbes that use the probe for energy, as a carbon source, or co-metabolite (i.e., not incorporated into biomass). The incorporation of ^{13}C from the probe into either microbial biomass or CO_2 provides conclusive evidence of biodegradation of PHCs in the subsurface.

Evidence of biodegradation from SIP is most commonly used together with changes in contaminant concentrations and contaminant-degrading microbes (determined by qPCR or RT-PCR) for supporting MNA as a site management strategy. An advantage of SIP is that the microbes or genes involved in the biodegradation do not need be known.

10.2.5   Compound-Specific Isotope Analysis (CSIA)

As further discussed in Chap. 11, CSIA measures the ratio of naturally occurring stable isotopes (e.g., $^{13}C/^{12}C$, $^{2}H/^{1}H$) of a contaminant as an indicator of abiotic degradation or biodegradation. Each element in a compound has a distinct isotopic ratio that is a function of the starting material, manufacturing process, and relative mechanisms of attenuation (e.g., biotic/abiotic degradation, volatilization). Isotopic ratios change (e.g., isotopic fractionation) as compounds degrade due to the slightly faster reaction of the lighter isotope, e.g., ^{12}C reacts more quickly than ^{13}C, resulting in the parent compound becoming enriched with ^{13}C.

Isotope enrichment factors found at sites are compared with those from literature and reference databases that have been previously described for each contaminant. Reference databases also provide links to literature and can generate plots, such as contaminant-degradation (mole fractions), dual-isotope, and modified-Kuder plots.

For compounds that become isotopically enriched, CSIA can identify the occurrence and extent of degradation. In addition to the extent of degradation, CSIA can potentially provide information on contaminant sources and whether plumes consist of comingled contaminants.

However, CSIA is less sensitive in some scenarios. For compounds that exhibit lower isotopic fractionation, such as high-molecular-weight PAHs, fractionation often cannot be concluded until the parent contaminant is biodegraded by > 50–80%. Isotopic enrichment discriminates phase transfer processes poorly compared to biodegradation, although for low-molecular-weight volatile compounds, such as gasoline range hydrocarbons, heavier isotopes are preferentially vaporized (Imfeld et al. 2014). Lastly, contaminants dissolving into the groundwater (e.g., from NAPL) can in some circumstances mask isotope fractionation.

10.2.6   DNA Sequencing

DNA sequencing is used to study mixed communities of organisms by the analysis of their collective DNA. Communities can be analyzed using amplicon sequencing (i.e., 16S rRNA gene sequencing) or metagenome sequencing (i.e., metagenomics) to comprehensively survey the genera/species and genes at a site capable of contaminant biodegradation. Both DNA sequencing approaches provide insights to the large unculturable portion of microbial communities (Mason et al. 2012).

In amplicon sequencing, one gene—typically the 16S ribosomal RNA (rRNA) gene—is sequenced in essentially all the microorganisms in a sample. Primers are used to sequence from conserved regions (DNA sequences that are identical across taxonomic groups of interest) of the target gene and across hypervariable regions (DNA sequences that are different across taxonomic groups of interest) of the gene to capture variation in the target gene of multiple microorganisms. The presence of COC-degrading functional genes can be inferred for taxa (genera or species) with characterized and published genomes. For example, the 16S rRNA gene sequence is used to identify the presence and abundance of *Methylibium* spp., which contains a functional gene that degrades MTBE under aerobic conditions.

In contrast, metagenome sequencing attempts to identify the sequence of all genes of all organisms present in appreciable numbers in a sample. The DNA from a sample is converted to a library of fragments that is sequenced using high-throughput next generation sequencing (NGS). All sequences are aligned, annotated, and analyzed through comparison with several databases of genomes, genes, and proteins to assign and identify microbial community taxonomy, the genes, and to infer gene functions. Metagenome sequencing can identify genes with previously unknown sequences and genes, such as acetylene hydratase, encoded by diverse sequences across genera.

When the contaminant-degrading genera and genes are not known, DNA sequencing can be used to understand how microbial community members change across a site spatially and temporally, and during or after remediation. Additionally, samples can be grouped by microbial composition and correlated with contaminant chemical compositions and geochemical data. Metagenomics can be useful when the function of species within a genus differs. For example, all *Geobacter* reduce iron, but only certain *Geobacter* species reduce sulfur (Rickard 2012). Thus, these functional differences within the genus *Geobacter* may not be distinguished based on amplicon sequencing (e.g., 16S RNA gene sequencing). As compared to metagenome sequencing, amplicon sequencing can identify more microbial species in a sample due to a greater depth of sequencing and with lower risk for false positives which can result from genome reassembly algorithms (Tovo et al. 2020). Although amplicon sequencing can also identify genera and species, it is only based on known genomes. Lastly, metagenome sequencing can identify functional genes encoding enzymes involved in biodegradation in genomes of unknown microorganisms.

Amplicon sequencing of samples from the Deepwater Horizon oil spill was performed to support the assessment of the fate and intrinsic biodegradation of hydrocarbons related to the release. Sequencing data were used to identify hydrocarbon-degrading bacteria, such as *Oleispira antarctica, Thalassolituus oleivorans,* and *Oleiphilus messinensis,* which were enriched within the PHC plume and aided in the biodegradation and attenuation of the oil spill (Hazen et al. 2010).

Both amplicon and metagenome sequencing are useful for understanding how microbial community compositions correlate with contaminant concentrations and geochemistry and change across a site spatially and temporally, and during or after remediation.

10.3 Selection of MBTs

10.3.1 QPCR Versus RT-QPCR

qPCR is commonly recommended as the standard baseline tool to assess the presence, absence, and concentration of COC-degradation-associated microorganisms and/or functional genes due to the quantitative nature and ease of handling DNA compared to RNA (used in RT-qPCR). qPCR is most useful when sampling across steep concentration gradients of COC and/or geochemistry. RT-qPCR is recommended in cases where quantification by qPCR has not correlated with the expected activity. For example, qPCR quantification of the DNA encoding toluene dioxygenase and phenol hydroxylase genes can be high, but the expression of the genes can be low (Taggart and Clark 2021). Thus, COC and/or geochemistry concentrations not correlating with gene concentrations may indicate that RNA analysis should be performed.

10.3.2 SIP Versus CSIA

SIP cannot detect abiotic degradation, so if abiotic degradation is occurring through the addition of zero valent iron (ZVI) or by the presence of naturally occurring minerals such as iron sulfides (FeS, pyrite), iron oxides (e.g., magnetite), green rust, or iron carbonate, then CSIA should be used. If the contaminants are used as both a carbon and energy source, then SIP may be the appropriate choice since some compounds exhibit lower isotopic fractionation during biodegradation (Mancini et al. 2003). SIP may be a better choice if the contaminants are BTEX and MTBE or tert-butyl alcohol (TBA) because it can be successfully employed in the presence of light non-aqueous phase liquid (LNAPL) or high COC concentrations that can confound CSIA results. If the plume is long and dilute, two-dimensional CSIA with carbon and hydrogen isotopes should be considered (Bouchard et al. 2018).

10.3.3 QPCR Versus qPCR Arrays Versus DNA Sequencing

If genera and genes that degrade a COC are known, such as those involved in aerobic and anaerobic degradation of PAH, they can be quantified by qPCR and reported as gene copies/mL. Individual genes can be quantified using qPCR or a qPCR array (See Sect. 10.2.1). If the genera and genes are not known or the goal is to comprehensively understand the microbial communities and how they change spatially (comparing different sampling wells), temporally (comparing samples of a well at different times), with remediation, COC composition, and geochemical data, then DNA sequencing should be used.

10.4   Case Studies

10.4.1   Transition to MNA at a Former Retail Gasoline Station

At a former gasoline station, mass reduction of contaminants by dual-phase extraction (DPE) and soil vapor extraction (SVE) reduced BTEX and MTBE concentrations to asymptotic levels where the mechanical remediation techniques were no longer effective. In assessing the potential for MNA or the need for bioremediation, most monitoring well data indicated decreasing concentrations of BTEX and MTBE, but the trend was inconclusive in one area. Therefore, more evidence was needed to evaluate the option of enhanced bioremediation prior to transition to MNA. qPCR arrays and SIP were used for this purpose (Fig. 10.2).

A suite of genes involved in aerobic and anaerobic biodegradation of BTEX, PAHs, and other PHCs was quantified using a qPCR array (QuantArray®-Petro). Gene concentrations were compared between sampling wells with decreasing trends and those showing no trend (as defined by Mann–Kendall analysis), which showed that genes involved in BTEX (aerobic and anaerobic) and MTBE biodegradation were at higher concentrations (cells/mL) in the monitoring wells where the BTEX and MTBE were decreasing compared to the wells where there was no trend.

These concentrations of BTEX-, MTBE-, and TBA-degrading microorganisms and genes were compared to the Microbial Insights (MI) qPCR Database of almost one hundred thousand samples from sites from around the world (Microbial Insights

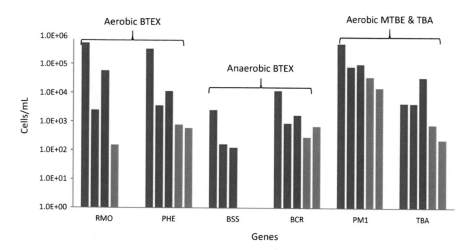

Fig. 10.2 Comparison of the concentrations of genes (cells/mL) involved in biodegradation of BTEX, MTBE, and TBA between monitoring wells showing decreasing COC trends (black bars) or no trend (white bars). RMO, toluene/benzene monooxygenases; PHE, phenol hydroxylase; BSS, benzylsuccinate synthase; BCR, benzoyl coenzyme A reductase; PM1, *Methylibium petroleiphilum* strain PM1 rRNA gene; TBA, *tert*-butyl alcohol hydroxylase

Inc. 2021). In wells where the COCs were decreasing, PHE and RMO concentrations in most samples exceeded 70–95% of the samples in the database (Fig. 10.3). Conversely, in samples where low concentrations (ppb) of COCs were lingering and not decreasing, these gene concentrations were below the median of samples in the MI Database. Additionally, the benzylsuccinate synthase (BSS) gene, involved in anaerobic biodegradation of toluene and xylene, was detected only in monitoring wells with decreasing COC trends, and the concentration was relatively low (~35th percentile) in two of the three wells. Furthermore, the MTBE-utilizing strain PM1 and the TBA hydroxylase gene were higher than the median in all monitoring wells, but PM1 concentrations in wells with decreasing trends were in the upper quartile (75–95th percentiles) and TBA hydroxylase concentrations were high (>90th percentile).

MBT data generated using a qPCR array provided an additional line of evidence to support transitioning from physical/chemical remediation (DPE and SVE) to MNA. However, the concentrations of genes involved in aerobic BTEX biodegradation were low in one area of the dissolved plume, indicating the potential for aerobic BTEX biodegradation was being limited by existing conditions in that portion of the site.

SIP using ^{13}C-benzene and ^{13}C-MTBE as probes was performed to obtain conclusive evidence of BTEX and MTBE biodegradation. The probes were adsorbed to powdered activated carbon (PAC) within Bio-Sep® beads (25% Nomex and 75% PAC). After incubation of the beads in the monitoring wells for 30 days, they were retrieved and analyzed by gas chromatography with a flame ionized detector (GC-FID) for the incorporation of ^{13}C into biomass, i.e., PLFA. The SIP results

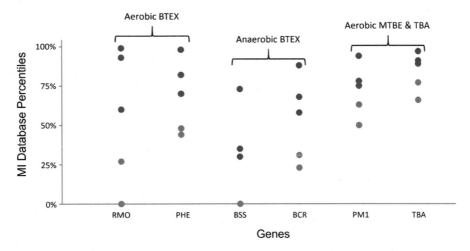

Fig. 10.3 Percentiles of gene concentration in the Microbial Insights Database for the concentrations of genes involved in biodegradation of BTEX, MTBE, or TBA in monitoring wells with decreasing COC trends (black circles) or no trend (white circles). Genes involved in degrading BTEX or MTBE and TBA: RMO, toluene/benzene monooxygenases; PHE, phenol hydroxylase; BSS, benzylsuccinate synthase; BCR, benzoyl coenzyme A reductase; PM1, *Methylibium petroleiphilum* strain PM1 rRNA gene; TBA, *tert*-butyl alcohol hydroxylase

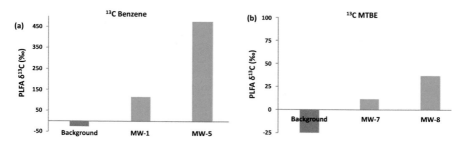

Fig. 10.4 Incorporation of ^{13}C from ^{13}C-benzene and ^{13}C-MTBE stable isotope probes (SIP) into biomass (PLFA). The δ^{13}C value for most natural substrate/carbon sources is between $-$ 20 and $-$ 30‰ (or per mille). Thus, background PLFA δ^{13}C values are around $-$ 25‰ under natural conditions (white bars). For both benzene (**a**) and MTBE (**b**), the enrichment of ^{13}C in PLFA (black bars) compared to background conclusively demonstrates in-situ biodegradation of these contaminants occurring under existing conditions in the aquifer

conclusively demonstrated that ^{13}C-benzene and ^{13}C-MTBE were being biodegraded (Fig. 10.4).

In conclusion, in areas where the contaminant concentrations were trending downward, qPCR array data, along with historical COC and geochemical data, demonstrated high concentrations of BTEX- and MTBE-degrading microbes. Additionally, in these downward-trending areas, SIP conclusively demonstrated that benzene and MTBE were being degraded in situ. Further, the results were supported the transition from mechanical remediation to MNA at the site. The use of MBTs to aid site decision-making increased efficiency, reduced costs, and supported earlier site closure.

10.4.2 Oxygen Addition at a Former Retail Gasoline Station

Oxygen was added to the subsurface at a former gasoline station to enhance biodegradation of COCs (BTEX and MTBE) and decrease the time to closure. To assess the effectiveness of the oxygen addition, microorganisms capable of degrading the COCs were quantified before injection (baseline) and periodically after injection (performance monitoring). At baseline, a variety of genes responsible for aerobic BTEX biodegradation were present at relatively low to moderate concentrations in the MI qPCR Database (10^2–10^4 cells/bead); after oxygen addition, these genes increased by nearly 3 orders of magnitude (OoMs), demonstrating that the oxygen injection stimulated the growth of the aerobic BTEX degraders (Fig. 10.5).

The microorganisms were sampled from monitoring wells (MW-1 and MW-2, two source area wells) using Bio-Trap® samplers, and the genes involved in degrading the COCs were quantified using qPCR array (QuantArray®-Petro). After injection of the oxygen-releasing material, concentrations of ring hydroxylating toluene monooxygenase (RDEG, RMO) and phenol hydroxylase (PHE) genes increased as

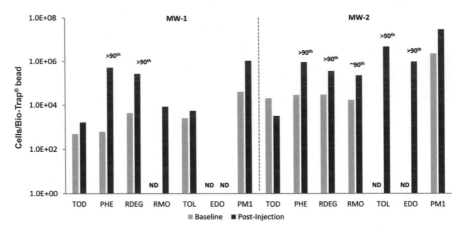

Fig. 10.5 Concentrations of BTEX-degrading genes and an MTBE-degrading strain (PM1) at baseline (white bars) and after injection of an oxygen-releasing product (black bars). Percentiles for sample concentrations in the MI Database are indicated above the bars

high as 2–3 orders of magnitude (OoM) in MW-1. Likewise, PHE, RDEG, and RMO concentrations increased more than one OoM in MW-2, demonstrating the growth of aerobic BTEX-degrading microbes. In both monitoring wells after injection, PHE and RDEG concentrations were > 90th percentile in the MI Database. Furthermore, other aromatic oxygenase genes not detected (ND) before injection increased dramatically (up to $> 10^6$ cells/bead) following injection (MW-2; TOL, EDO).

In response to the oxygen addition, genes associated with aerobic BTEX biodegradation degraders increased nearly three OoMs and the MTBE-degrading strain PM1 (*M. petroleiphilum*) increased more than one OoM, which provided a line of evidence that the enhanced bioremediation was successfully implemented and occurring. Thus, qPCR analyses were included in the performance monitoring program at this site to increase performance certainty.

10.4.3 Continuation of MNA at a Pipeline Release in a Remote Area

A former pipeline release in a remote area impacted the groundwater with BTEX, which appeared to decrease based on Mann–Kendall trend analysis. However, to reduce uncertainty due to heterogeneous subsurface conditions, additional evidence of BTEX biodegradation was needed to support the continuation of MNA. qPCR was used to quantify BSS and anaerobic benzene carboxylase (ABC) genes as part of the routine groundwater monitoring of the BTEX plume as well as background wells. The BSS gene can initiate anaerobic biodegradation of toluene (Biegert et al. 1996; Heider et al. 2016) and xylenes (Achong et al. 2001; Kniemeyer et al. 2003; Gieg and

Toth 2020; Rabus et al. 2016), and ABC is suggested to initiate anaerobic benzene biodegradation (Abu Laban et al. 2010). During ongoing groundwater monitoring, samples from two background wells (MW-6 and MW-7) and three BTEX-impacted wells (MW-8, MW-9, and MW-10) were analyzed for the concentrations of BSS and ABC genes by qPCR, which indicated the growth of anaerobic BTEX degraders within the dissolved plume (Fig. 10.6).

BSS genes for anaerobic toluene biodegradation were detected in only one of the background wells (MW-6) and only during two of the six sampling events. Conversely, BSS genes were detected in all BTEX-impacted wells (MW-8, MW-9, and MW-10) during all but one sampling event. Moreover, BSS gene concentrations (cells/bead) in impacted wells were high. BSS concentrations in MW-8 were higher than in 80–95% of samples in the MI Database, and in MW-9 and MW-10 were above the median, ranging between the 60 and 95th percentiles. Finally, ABC

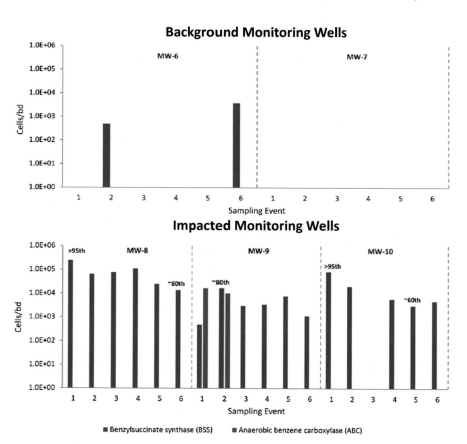

Fig. 10.6 Comparison of the concentrations of genes associated with anaerobic BTEX biodegradation (cells/bead) between background (upper panel) and impacted (lower panel) monitoring wells. Missing bars indicate that the sample was assayed but the gene was not detected. Percentiles for sample concentrations in the MI Database are indicated above the bars

genes for anaerobic benzene biodegradation were occasionally detected in impacted monitoring well MW-9 at high concentrations.

The qPCR data indicated high concentrations of genes associated with anaerobic BTEX biodegradation in the dissolved plume, which, along with converging trends in contaminant concentrations and geochemistry, was a key factor in continuing MNA as an efficient and risk-based remediation approach.

Due to subsurface conditions and fluctuations (e.g., changes in depth to groundwater, potential influx of nutrients from nearby agricultural application), 16S rRNA gene sequencing (i.e., amplicon sequencing) was used to gain further evidence of PHC biodegradation along the plume. These phylogenetic data were analyzed by Principal Coordinate Analysis (PCoA) to visualize similarities in the microbial communities across the site and over time. On PCoA graphs, similar microbial communities cluster together and communities differing most from each other plot farthest apart. In the PCoA graph of communities sampled from monitoring wells at the pipeline-release site (Fig. 10.7), clustering is seen for the background wells (MW6 and MW7) on the right of the PCoA plot, and on the left, the microbial communities of the impacted wells (MW8, MW9, and MW10) cluster together, demonstrating their relatedness and their difference versus the background wells.

The microbial communities of background well MW7 and impacted well MW10 were compared. The background well's (MW7) top genus was the facultative aerobe, *Dechloromonas* (~29% of total reads), while anaerobes like sulfate-reducing *Desulfobulbus* and potentially iron-reducing *Geobacter* were less abundant (~2%). Conversely, in the impacted well, *Geobacter* was the top genus (~34%), likely due to the anaerobic conditions in the dissolved plume, while *Dechloromonas* was much less abundant (~4%).

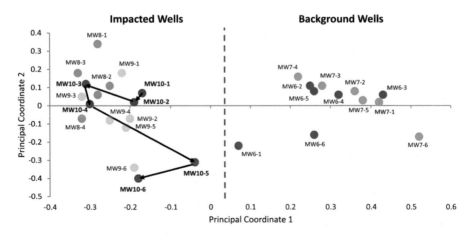

Fig. 10.7 Principal coordinate analysis (PCoA) of the 16S rRNA gene amplicon sequencing data. The microbial communities of impacted and background monitoring wells (MW) are plotted. The numbers following the dash in the monitoring well designation indicate sequential sampling events. The black arrows trace the sequential sampling of MW10

Changes in microbial communities over time can also be seen in the PCoA plot. For example, the first four samples from MW10 form a cluster while the fifth and sixth samples are separated, suggesting a shift in the microbial community with time (Fig. 10.7). The changes in microbial communities from sampling event one to six in MW10, shown in the PCoA plot, are shown as a stacked bar chart in Fig. 10.8. During the first three sampling events (MW10-1 to MW10-3), the community was diverse, but relatively stable in structure at the genus level from one sampling event to the next; *Geobacter* was the most abundant genus, comprising ~ 20–40% of total reads. *Rhodoferax* were consistently ~ 10% of total reads. At MW10-4, the microbial community composition changed more substantially: *Geobacter* remained the top genus (~35% of total reads), but the iron-reducing genus *Albidiferax* increased to over 21% of total reads, and *Rhodoferax* decreased while *Polaromonas* increased. By MW10-5, marked changes were evident, potentially resulting from fluctuating subsurface conditions: anaerobic *Geobacter* decreased from over 30 to ~ 3% of total reads, while aerobic and facultative microorganisms (*Pseudomonas, Hydrogenophaga, Dechloromonas,* and *Polaromonas*) were detected in a substantial proportion. Finally, by the last sampling event (MW10-6), *Geobacter* rebounded from MW10-5, *Rhodoferax* was the most abundant genus, and facultative anaerobic sulfur oxidizers (*Sulfuritalea* and *Sulfuricurvum)* were detected.

In conclusion, DNA sequencing revealed that microbial communities in the impacted areas were markedly different than background communities. Moreover, the community composition at the impacted wells changed over time, likely as a consequence of the fluctuations in subsurface conditions that were observed. *Geobacter* spp. were typically detected at the highest relative abundances in monitoring wells within the dissolved plume. Although at least one species of *Geobacter* is capable of anaerobic BTEX biodegradation, most cannot; therefore, *Geobacter* presence even in high relative proportions does not indicate the potential for BTEX biodegradation. However, supplementing the amplicon sequencing data with qPCR data, which indicated a 4–5 OoM increase in the concentration of genes involved in anaerobic biodegradation of BTEX (BSS and ABC genes) in impacted wells compared to background wells, provided greater clarity to the occurrence of anaerobic BTEX biodegradation across the plume. Lastly, decreasing concentrations of BTEX and qPCR and amplicon sequencing data provided additional evidence that biodegradation was a mechanism of attenuation at the site and supported ongoing implementation of MNA.

10.5 Cost Considerations in MBT Selection

Wilson et al. (Wilson et al. 2004) obtained information on remediation costs for petroleum-impacted sites from the Clu-In Database (http://cluin.org/), maintained by the U.S. EPA, and used it to identify average and median remediation cost per site. The mean (median) costs (in USD) for these cleanups, adjusted for cumulative inflation annually from 2004 to 2021, was $580,516 ($286,717) for all the combined

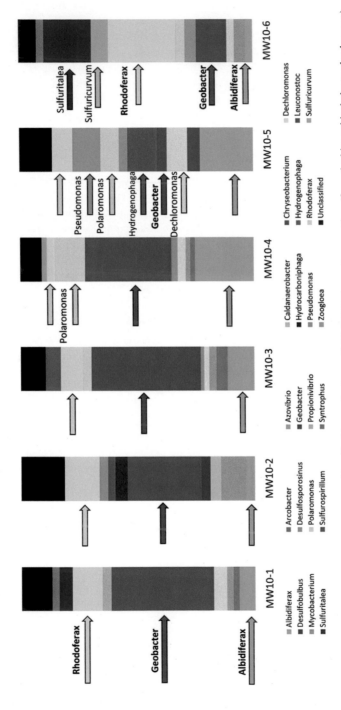

Fig. 10.8 Stacked bar chart of the topmost genera identified in monitoring well 10 (MW10) samples over time (from the hierarchical cluster dendrogram)

112 sites, $352,784 ($261,000) for the service station sites, $1,488,557 ($348,000) for the public drinking water sites, and $2,765,150 ($791,700) for the industrial sites. The average cost for application of MBTs at a PHC-contaminated site is roughly $10,000 per event for stable isotope probing and qPCR arrays on 5 monitoring locations.

10.6   Summary

MBTs provide additional lines of evidence during site characterization, remedy selection, performance monitoring, and site closure. Although each MBT provides valuable information, MBTs should be selected based on specific questions as outlined in Fig. 10.1 that need to be developed depending on site-specific needs and conditions and then addressed. Additionally, MBTs should be analyzed in conjunction with other traditional site data, such as hydrogeologic, contaminant, and geochemical data, to increase certainty associated with biogeochemical processes governing contaminant attenuation.

10.7   Future Directions

The information gained from MBTs is increasingly being used by practitioners around the world (Taggart and Clark 2021). However, the link between the abundance of microorganisms with genetic potential to degrade compounds and their rates of degradation is an important frontier of research. Pushing forward to refine rates of degradation, catalyzed by knowledge exchanged between disciplines (e.g., molecular biologists and geologists), will make MBT data of higher value in optimizing corrective action plans.

Additionally, metabolomics is a new tool for the environmental restoration industry that may become an integral part of site screening and design. In environmental restoration, metabolomics measures small (molecular weight < 1000) water-soluble compounds that are microbial metabolic products, such as corrinoids, electron-shuttling compounds, signaling compounds for organismal communication, and anabolic and catabolic pathway intermediates—providing insights into microbial activity and function that complement nucleic acid-targeted qPCR data. Metabolomics can target the quantification of specific indicator molecules or can identify patterns in complex metabolome datasets that correlate with specific microbial processes (e.g., sulfate reduction) or overall activity of the microbiome. A cost-effective and user-friendly MBT with these capabilities may further reduce site assessment and remediation costs and be an asset to the industry.

References

Abu Laban N, Selesi D, Rattei T, Tischler P, Meckenstock RU (2010) Identification of enzymes involved in anaerobic benzene degradation by a strictly anaerobic iron-reducing enrichment culture. Environ Microbiol 12(10):2783–2796

Achong GR, Rodriguez AM, Spormann AM (2001) Benzylsuccinate synthase of Azoarcus sp. strain T: cloning, sequencing, transcriptional organization, and its role in anaerobic toluene and m-xylene mineralization. J Bacteriol 183(23):6763–6770

Alleman B, Chandler D, Cole J, Edwards E, Fields M, Haas P, Halden R, Hashsham S, Hazan T, Johnson P (2005) SERDP and ESTCP expert panel workshop on research and development needs for the environmental remediation application of molecular biological tools. Deputy Director of Defense Research and Engineering Arlington VA Strategic

Biegert T, Fuchs G, Heider J (1996) Evidence that anaerobic oxidation of toluene in the denitrifying bacterium *Thauera aromatica* is initiated by formation of benzylsuccinate from toluene and fumarate. Eur J Biochem 238(3):661–668

Bouchard D, Marchesi M, Madsen EL, DeRito CM, Thomson NR, Aravena R, Barker JF, Buscheck T, Kolhatkar R, Daniels EJ, Hunkeler D (2018) Diagnostic tools to assess mass removal processes during pulsed air sparging of a petroleum hydrocarbon source zone. Groundwater Monit Rem 38(4):29–44

Busch-Harris J, Sublette K, Roberts KP, Landrum C, Peacock AD, Davis G, Ogles D, Holmes WE, Harris D, Ota C, Yang X, Kolhatkar A (2008) Bio-traps coupled with molecular biological methods and stable isotope probing demonstrate the in situ biodegradation potential of MTBE and TBA in gasoline-contaminated aquifers. Ground Water Monit Rem 28(4):47–62

Forootan A, Sjoback R, Bjorkman J, Sjogreen B, Linz L, Kubista M (2017) Methods to determine limit of detection and limit of quantification in quantitative real-time PCR (qPCR). Biomol Detect Quantif 12:1–6

Gieg LM, Toth CRA (2020) Signature metabolite analysis to determine in situ anaerobic hydrocarbon biodegradation. In: Boll M (ed) Anaerobic utilization of hydrocarbons, oils, and lipids. Springer International Publishing, Cham, pp 361–390

Hazen TC, Dubinsky EA, DeSantis TZ, Andersen GL, Piceno YM, Singh N, Jansson JK, Probst A, Borglin SE, Fortney JL, Stringfellow WT, Bill M, Conrad ME, Tom LM, Chavarria KL, Alusi TR, Lamendella R, Joyner DC, Spier C, Baelum J, Auer M, Zemla ML, Chakraborty R, Sonnenthal EL, D'Haeseleer P, Holman HY, Osman S, Lu Z, Van Nostrand JD, Deng Y et al (2010) Deep-sea oil plume enriches indigenous oil-degrading bacteria. Science 330(6001):204–208

Heider J, Szaleniec M, Martins BM, Seyhan D, Buckel W, Golding BT (2016) Structure and function of benzylsuccinate synthase and related fumarate-adding glycyl radical enzymes. J Mol Microbiol Biotechnol 26(1–3):29–44

Imfeld G et al. (2014) Carbon and hydrogen isotope fractionation of benzene and toluene during hydrophobic sorption in multistep batch experiments. Chemosphere 107:454–461

ITRC (2013) Technical and regulatory guidance: environmental molecular diagnostics, new site characterization and remediation enhancement tools. EMD-2. Interstate Technology and Regulatory Council, Environmental Molecular Diagnostics Team, Washington, D.C. Retrieved from https://clu-in.org/download/contaminantfocus/mtbe/MTBE-ITRC-EMD2.pdf.

Kniemeyer O, Fischer T, Wilkes H, Glockner FO, Widdel F (2003) Anaerobic degradation of ethylbenzene by a new type of marine sulfate-reducing bacterium. Appl Environ Microbiol 69(2):760–768

Lebrón CA, Petrovskis E, Loffler F, Henn K (2011) Application of nucleic acid-based tools for monitoring monitored natural attenuation (MNA), biostimulation and bioaugmentation at chlorinated solvent sites. Naval Facilities Engineering Command Port Hueneme CA Engineering Service Center

Mancini SA, Ulrich AC, Lacrampe-Couloume G, Sleep B, Edwards EA, Lollar BS (2003) Carbon and hydrogen isotopic fractionation during anaerobic biodegradation of benzene. Appl Environ Microbiol 69(1):191–198

Mason OU, Hazen TC, Borglin S, Chain PSG, Dubinsky EA, Fortney JL, Han J, Holman H-YN, Hultman J, Lamendella R, Mackelprang R, Malfatti S, Tom LM, Tringe SG, Woyke T, Zhou J, Rubin EM, Jansson JK (2012) Metagenome, metatranscriptome and single-cell sequencing reveal microbial response to Deepwater Horizon oil spill. ISME J 6(9):1715–1727

Microbial Insights Inc. (2021) The MI environmental microbiology database v2.0. Retrieved from https://microbe.com/mi-database-2/

Rabus R, Boll M, Heider J, Meckenstock RU, Buckel W, Einsle O, Ermler U, Golding BT, Gunsalus RP, Kroneck PM, Kruger M, Lueders T, Martins BM, Musat F, Richnow HH, Schink B, Seifert J, Szaleniec M, Treude T, Ullmann GM, Vogt C, von Bergen M, Wilkes H (2016) Anaerobic microbial degradation of hydrocarbons: from enzymatic reactions to the environment. J Mol Microbiol Biotechnol 26(1–3):5–28

Rickard D (2012) Sedimentary iron biogeochemistry (chapter 3). In: Rickard D (ed) Developments in sedimentology, vol 65. Elsevier, pp 85–119

Ritalahti KM, Hatt JK, Lugmayr V, Henn K, Petrovskis EA, Ogles DM, Davis GA, Yeager CM, Lebrón CA, Löffler FE (2010) Comparing on-site to off-site biomass collection for dehalococcoides biomarker gene quantification to predict in situ chlorinated ethene detoxification potential. Environ Sci Technol 44(13):5127–5133

Taggart DM, Clark K (2021) Lessons learned from 20 years of molecular biological tools in petroleum hydrocarbon remediation. Rem J (ahead of print)

Tovo A, Menzel P, Krogh A, Cosentino Lagomarsino M, Suweis S (2020) Taxonomic classification method for metagenomics based on core protein families with Core-Kaiju. Nucleic Acids Res 48(16):e93

Wilson BH, Hattan G, Kuhn J, McKay R, Wilson JT (2004) Costs and issues related to remediation of petroleum-contaminated sites. Presented at NGWA remediation conference, New Orleans, LA

Chapter 11
Compound-Specific Isotope Analysis (CSIA) to Assess Remediation Performance at Petroleum Hydrocarbon-Contaminated Sites

Daniel Bouchard, Julie Sueker, and Patrick Höhener

Abstract Compound-specific isotope analysis (CSIA) is an advanced characterization tool increasingly used by field practitioners to demonstrate degradation of compounds such as benzene, toluene, ethylbenzene, and xylene (BTEX) in petroleum hydrocarbon-contaminated aquifer systems. Formerly used to demonstrate occurrence of in situ biodegradation of BTEX during natural attenuation in groundwater, CSIA underwent substantial research and development to confidently be applied in the frame of engineered remediation efforts. Due to the feasibility to demonstrate destruction of contaminants by tracking the change in isotopic composition caused by either biotic or abiotic processes, mass destruction process initiated by the remediation treatment can be distinguished from other co-occurring non-destructive mass removal process(es) such as sorption and dilution. For this reason, CSIA has become a valuable characterization tool to directly assess the performance of the remediation treatment on specifically selected contaminants. This chapter presents the principles of CSIA application to assess performance of in situ remediation treatments applied to BTEX-contaminated sites. The information introduced herein on CSIA is presented from the perspective of supporting field practitioners in their intention to implement the tool at field sites.

Keywords BTEX · CSIA · In situ remediation · Remediation performance · Stable isotope

D. Bouchard (✉)
GHD, Montreal, Canada
e-mail: daniel.bouchard@GHD.com

J. Sueker
Arcadis U.S., Inc., Broomfield, USA
e-mail: julie.sueker@arcadis.com

P. Höhener
Aix-Marseille University, Marseille, France
e-mail: patrick.hohener@univ-amu.fr

11.1 Introduction

Over the years, diverse remediation technologies and approaches have been developed to remediate petroleum hydrocarbon (PHC)-contaminated sites, often targeting benzene, toluene, ethylbenzene, and xylene (BTEX) compounds due to their toxicity and physico-chemical properties. The knowledge and the experience gained were applied to remediation technologies increasingly more effective and for which diverse practical guidance documents were issued to assist field application (USACE 2014; NAVFAC 2013; EPA 2017). With respect to known BTEX chemical and physical properties and site-specific conditions, these remediation technologies are implemented, individually or combined, to stimulate either destructive mass removal processes (e.g., biodegradation or chemical oxidation) or non-destructive mass removal processes (e.g., volatilization). Regardless of the selected remediation technology, the fate of BTEX compounds and treatment cost-effectiveness are nonetheless closely related to successful field implementation of the selected remedial technology. Proper establishment of the intended mass removal process thus requires specific attention to ensure optimal performance of the remediation technology. However, although a remediation technology can be designed to stimulate a specific mass removal process, additional mass removal process(es) may co-occur, often inevitably. To demonstrate that the intended mass removal process is occurring as initially designed, advanced characterization tools capable of discriminating co-occurring processes are required.

CSIA is a process-based advanced characterization tool that tracks the stable isotope composition of selected volatile organic compounds (VOCs) present in groundwater. In essence, the change in stable isotope composition of selected VOCs is monitored during the treatment period to reveal valuable information about the attenuation process that concentration analysis cannot provide. Application of CSIA as a characterization tool has been studied for more than two decades, the subject of numerous scientific review publications (Kuntze et al. 2019; Thullner et al. 2012; Vogt et al. 2016; Elsner 2010), described in textbooks (Aelion et al. 2010; Jochmann and Schmidt 2012), and summarized by environmental agencies (EPA 2008) and coalition groups (ITRC 2011) for field practitioners. While most of the work cited above concerns CSIA application in a context of natural attenuation and biodegradation of contaminants, it is only recently that exhaustive development efforts have been directed toward engineered in situ remediation (Bouchard et al. 2018a). With CSIA being able to discriminate between co-occurring mass removal processes, the application of CSIA in engineered in situ remediation thus represents a great asset in assessing treatment performance. By demonstrating the occurrence of the intended mass removal process for newly implemented treatment or following modification actions, CSIA supports decision-making for timely system optimization, resulting in remediation cost efficiency. In this chapter, a methodology to deploy CSIA specifically to assess BTEX remediation treatment performance is presented with focus on groundwater assessments. In the attempt to remain concise and field application oriented, this chapter does not aim to provide a complete review of theoretical knowledge behind CSIA application or a thorough discussion of all parameters commonly included in exhaustive performance monitoring [as for instance suggested in technical guidance documents such as EPA (2017)], but rather focuses on key concepts allowing field practitioners to confidently utilize CSIA at their field sites.

11.2 CSIA Principles

11.2.1 Background and Concepts

Stable carbon (C) and hydrogen (H) atoms present in nature are commonly considered to have an atomic weight of 12 g and 1 g per mole and are respectively denoted as ^{12}C and ^{1}H. However, some naturally occurring stable C and H atoms have an additional neutron in their nucleus. The additional neutron increases the molecular weight to 13 g and 2 g per mole, respectively denoted as ^{13}C and ^{2}H. These ^{13}C and ^{2}H atoms are called heavy isotopes and represent a proportion of 1.07% and 0.0115% of each of the C and H atom pool, respectively (IUPAC 1991). Since PHCs are organic molecules uniquely composed of C and H atoms, it is expected to observe inclusion of a heavy atom in some molecules. Isotopically different benzene molecules, i.e., ^{light}molecules and ^{heavy}molecules, are depicted in Fig. 11.1. The ^{13}C and ^{2}H atom can be found at various positions within the ^{heavy}molecules. Due to the naturally low abundance of ^{13}C and ^{2}H atoms, it is generally assumed that PHCs such as BTEX compounds will include only one heavy isotope per molecule.

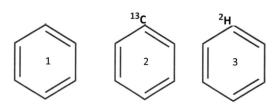

	^{light}molecule (molecule 1)	^{heavy}molecule (molecule 2)	^{heavy}molecule (molecule 3)
Chemical formula	C_6H_6	$^{13}C_1C_5H_6$	$C_6{}^2H_1H_5$
Molecular weight	78 g/mol	79 g/mol	79 g/mol
Degradation rate	normal	slower	slower

Fig. 11.1 Illustration of isotopically different benzene molecules with selected characteristics. The heavy isotope in molecules 2 and 3 can occupy other positions than those indicated. See text for additional specificities regarding degradation rate and heavy isotope positioning in the molecule

The presence of a heavy isotope confers to molecules slightly different physical properties. One key characteristic is that chemical bonds between two light isotopes are more rapidly broken compared to bonds between a light and a heavy isotope. More specifically, molecules with a heavy isotope present at the reactive position react more slowly than molecules containing only light isotopes (Elsner et al. 2005). This difference in isotopic composition hence causes slower degradation rates for heavymolecules compared to lightmolecules (Fig. 11.1). As a consequence, accumulation of heavymolecules in the remaining contaminant pool is expected over time, whether the reaction taking place is biotic (biodegradation) or abiotic (chemical oxidation or reduction) (Aelion et al. 2010; Elsner et al. 2005). For simplicity, degradation will further be used throughout the chapter as a general term referring to both biotic and abiotic reactions. This gradual accumulation of heavymolecules in the remaining fraction over time, reflected by isotope ratio changes with time, becomes unequivocal evidence of compound destruction. It is specifically the tracking of these isotope ratio changes that defines the essence of the CSIA method, which has proved to be a reliable means for assessing the fate of organic contaminants released in the environment.

The progressive change in isotope composition observed in the remaining contaminant pool is related to the change in contaminant concentration by the Rayleigh equation (Eq. 11.1) (Mariotti et al. 1981):

$$R_t = R_0 f^{\alpha-1} \qquad (11.1)$$

where R_t and R_0 are respectively the stable isotope ratio ($^{13}C/^{12}C$ or $^2H/^1H$) of the compound at time t or at initial time 0, f is the remaining contaminant fraction expressed by the ratio of concentrations measured at time t and at initial time (i.e., C_t/C_0), and α is the isotope fractionation factor. The isotope ratios (R_t and R_0) and VOC concentrations are obtained via analytical measurements carried out in the laboratory on groundwater samples collected on site (see Sect. 11.2.2). In contrast, α is a pre-determined parameter normally determined by experiments conducted in the laboratory. By definition, the parameter α quantifies the difference between the heavymolecules and lightmolecules degradation rates, and can be expressed by Eq. 11.2:

$$\alpha = \frac{^{heavy}k}{^{light}k} \qquad (11.2)$$

where $^{heavy}k$ and $^{light}k$ are the degradation rates of heavymolecules and lightmolecules, respectively (Fig. 11.1). As k is slower for heavymolecules compared to lightmolecules, the α coefficient is typically smaller than 1. The greater the difference between $^{heavy}k$ and $^{light}k$, the faster the isotope fraction occurs during the reaction and the greater the heavymolecule accumulates in the remaining contaminant pool. For convenience, α is often transformed into an enrichment factor (ε) using Eq. 11.3:

$$\varepsilon = (\alpha - 1) \qquad (11.3)$$

The ε is commonly multiplied by 1000 to avoid working with small fraction numbers, and consequently be expressed in unit of per mil (‰). The ε transformation and terminology will be used in this chapter.

11.2.2 Isotope Analysis and Delta Notation

Analytical measurements are conducted to determine the $^{13}C/^{12}C$ and $^{2}H/^{1}H$ isotope ratios for each compound of interest included in a mixture. The initial analytical procedure step consists of compound separation via a gas chromatography unit. The compounds of interest are then conveyed to a combustion oven (for $^{13}C/^{12}C$ isotope ratio) or to a pyrolysis oven (for $^{2}H/^{1}H$ isotope ratio) prior reaching the isotope ratio mass spectroscopy unit (IRMS). To insure accurate and precise measurement of the isotope ratio, a nominal mass of 1 nmol of C or 8 nmol of H for each compound of interest needs to reach the IRMS source detector (EPA 2008). To meet this mass requirement for environmental samples with low VOC concentrations, a common strategy is to combine a preconcentration device integrated in the analytical setup and processing a large sample volume.

By convention, measurements of isotope ratios are reported using the δ notation, and expressed relatively to the Vienna Pee Dee Belemnite (VPDB) or the Vienna Standard Mean Ocean Water (VSMOW) international reference standards for carbon and hydrogen, respectively (Eq. 11.4) (Clark and Fritz 1997):

$$\delta = \left(\frac{R}{R_{\text{std}}} - 1 \right) \tag{11.4}$$

where R and R_{std} are the isotope ratio of the sample and the international reference standard, respectively. The δ value obtained is then multiplied by 1000 to conveniently express the results in units of per mil (‰) or in milli-Urey (mUr) as used in some recent publications. The former unit of ‰ is used in this chapter.

11.2.3 Isotope Fractionation Processes and Quantification

As introduced earlier, bond-breaking reactions involving a heavy isotope located at the reactive position of the primary enzymatic attach will proceed at a slower rate, which leads to an observable isotope fractionation in the remaining substrate pool. However, bond-breaking reactions can be induced by a variety of biotic and abiotic degradation processes. As each degradation process can be carried out by different bond-breaking mechanisms, the extent of isotope fractionation will vary. In addition, the type of chemical bond uniting the atoms, dilution of the heavy atom inside increasing molecular structure, the commitment to catalysis (i.e., masking isotope effect due to several sequential reaction steps inside the cell) (Gandour and

Schowen 1978; Huskey 1991; Northrop 1981), and mass transfer limitation (Thullner et al. 2013) are also key factors influencing the extent of isotope fractionation. For these reasons, the ε values are compound-specific and process-specific (i.e., reaction mechanism-specific together with bacteria and enzyme-specific in the case of biodegradation). Isotope fractionation and corresponding ε values related to biological and chemical processes are briefly overviewed below. Detailed explanations on isotope fractionation variations are beyond the scope of this chapter and can be found in previous publications (Elsner et al. 2005; Aelion et al. 2010).

Isotope fractionation has also been assessed for non-bond-breaking processes (physical mass removal processes) such as volatilization from non-aqueous phase liquid (NAPL) or water (Julien et al. 2015; Kuder et al. 2009), sorption to organic matter (Imfeld et al. 2014), or aqueous and gaseous diffusion (Wanner and Hunkeler 2019; Bouchard et al. 2008a). However, when assessing the performance of in situ chemical oxidation (ISCO) or enhanced biodegradation treatment in saturated zones, physical mass removal processes are less likely to have a significant contribution and thus not considered in this chapter. For a field study assessing the performance of an air sparging system using CSIA (hence including physical mass removal processes), the reader is referred to Bouchard et al. (2018b). For further details on the fundamentals, design, and implementation of bioremediation and ISCO technologies, the reader is referred to Chaps. 14 and 15, respectively.

Laboratory-based determination of enrichment factors is commonly carried out using closed experimental setups, which ensure that solely the destructive process (biodegradation or chemical oxidation) is causing the concentration decrease. While single-isotope experiments used to be carried out, insight gained when combining measurements of two isotopes (for instance $\delta^{13}C$ and δ^2H) during the same experiment is substantial. Dual-isotope assessments allow coupling of $\delta^{13}C$ changes with δ^2H changes, hence revealing a specific isotope enrichment pattern associated to the destructive mechanism (Zwank et al. 2005; Kuder et al. 2005; Fischer et al. 2007). The combined $\delta^{13}C$ and δ^2H change can be quantified by the Λ value. The Λ value is equivalent to the slope of the linear regression obtained when plotting δ^2H changes as function of $\delta^{13}C$ changes, and approximated by Eq. 11.5 (Höhener and Imfeld 2021):

$$\Lambda = \frac{\ln\left[(\delta^2H/1000 + 1)/(\delta^2H_0/1000 + 1)\right]}{\ln\left[(\delta^{13}C/1000 + 1)/(\delta^{13}C_0/1000 + 1)\right]} \approx \frac{\Delta\delta^2H}{\Delta\delta^{13}C} \approx \frac{\varepsilon_H}{\varepsilon_C} \qquad (11.5)$$

where Δ (in ‰) is the difference between two δ measurements. The two approximate transformations diverge from the exact definition with increasing $\varepsilon_H/\varepsilon_C$, causing for instance an overestimated Λ by 5% when $\varepsilon_H/\varepsilon_C = 22$ (Höhener and Imfeld 2021).

11.2.3.1 Biological Process

As discussed in Chaps. 5, 9, and 14, biodegradation of PHCs and BTEX in particular may occur through a wide variety of bond-breaking mechanisms, which differ
whether the process is taking place under oxic or anoxic conditions, and also accordingly to which enzymatic pathway is being used. As a consequence, these specificities will generate different magnitudes of carbon and hydrogen fractionation,
and when both isotopes are combined, will exhibit specific carbon and hydrogen
enrichment patterns. Numerous laboratory studies were conducted to determine ε-
C and ε-H values for BTEX biodegrading under oxic or various anoxic conditions
and revealed these differences in enrichment patterns. A selection of literature ε-C
and ε-H values is provided in Table 11.1 for aerobic and anaerobic BTEX biodegradation. For each BTEX compound, the largest and the smallest ε-C values were
selected and are reported in Table 11.1 whether or not δ^2H changes were considered
during the experiment and regardless of the prevailing electron acceptor during the
anaerobic biodegradation. The latter ε sub-set will be used to interpret single-isotope
assessments further described in Sect. 11.4.2.1. In addition, experiments showing the
largest and the smallest Λ value (reported by the respective study or calculated using
Eq. 11.5 with ε-C and ε-H issued from the same experimental series) and regardless
of the electron acceptor for anoxic conditions are also reported in Table 11.1. This
second ε-C and ε-H value sub-set (or Λ) will be used to interpret dual-isotope assessments further described in Sect. 11.4.2.2. Note that among the published studies, the
isotopic evaluations were carried out either using a pure microbial strain, a microbial
consortium, or the native microbial population. The data selection presented in Table
11.1 was made regardless of these specificities to identify and use the widest possible
ranges. Finally, because only few Λ values are reported in the literature for xylene,
Table 11.1 reports the Λ values related to biodegradation regardless of the isomers
(*o*, *m*, or *p*). A recent compilation of ε-C and ε-H with corresponding Λ values can
be found in Vogt et al. (2016) reporting Λ values by enzymatic reaction.

11.2.3.2 Chemical Process

Chemical oxidation of VOCs initiated by oxidizing agents involves bond-breaking
reactions. An oxidation mechanism pathway is likely specific to each oxidizing agent,
which can furthermore depend on the activation method producing diverse radicals
(Zhang et al. 2016; Matzek and Carter 2016). In addition, the nature of radicals
formed can change due to contact with other species naturally present in groundwater
such as background organic matter, halide, phosphate, and carbonate (Lee et al.
2020; Li et al. 2017). Therefore, the chemical oxidation process will lead to ^{13}C and
^2H enrichment in the remaining VOC pool, and the extent of isotope fractionation
will depend on the selected oxidizing agent, activation method used, and prevailing
radicals. Since the ε value database is yet restrained for chemical oxidation reactions,
Table 11.1 lists the different ε-C and ε-H values available for BTEX when chemically
oxidized by persulfate, hydrogen peroxide, or permanganate under various activation

D. Bouchard et al.

Table 11.1 Carbon (ε-C) and hydrogen (ε-H) enrichment factors and related Λ values measured for BTEX compounds related to biodegradation (for aerobic and anaerobic) and chemical oxidation (for various oxidizing agents and activation method) processes

Compound	Type of degradation	Electron acceptor	Activation mode	ε-C ± uncertainty (‰)	ε-H ± uncertainty (‰)	Λ ± uncertainty[a] (unitless)	References
Benzene	Abiotic	Persulfate	None	-1.7 ± 0.1	n.s.	n.l.d.	Solano et al. (2018)
			None	-0.6 ± 0.1	-13.8 ± 5	23.0	Gusti (2018)
			None	-2	-10.9	5.5	Saeed (2011)
			Iron-activated (high citrate/Fe ratio)	-0.6 ± 0.2	-23 ± 5	38.3	Gusti (2018)
			Iron-activated (low citrate/Fe ratio)	-1.3 ± 0.7	-14 ± 3	10.8	Gusti (2018)
			Alkaline	-2	-10.6	5.3	Saeed (2011)
		H_2O_2	UV light	-0.7 ± 0.1	20 ± 2	-24 ± 4	Zhang et al. (2016)

(continued)

Table 11.1 (continued)

Compound	Type of degradation	Electron acceptor	Activation mode	ε-C ± uncertainty (‰)	ε-H ± uncertainty (‰)	Λ ± uncertainty[a] (unitless)	References
	Biotic	Aerobic (O_2)		-0.7 ± 0.1	n.s.	n.l.d.	Fischer et al. (2008)
				-1.46 ± 0.06	-12.8 ± 0.7	8.7	Hunkeler et al. (2001)
		Anaerobic (nitrate-reducing)		-2.2 ± 0.4	-35 ± 6	8 ± 2	Mancini et al. (2008)
		Anaerobic (methanogenesis)		-0.8 ± 0.2	-34 ± 8	39 ± 5	Mancini et al. (2008)
		Anaerobic (sulfate-reducing)		-3.6 ± 0.3	n.e.	n.e.	Mancini et al. (2003)
		Anaerobic (unspecified)		-0.6 ± 0.2	n.e.	n.e.	Fischer et al. (2009)
Toluene	Abiotic	H_2O_2	Fenton (Fe^{3+} and pH = 3.3)	n.s.	n.e.	–	Ahad and Slater (2008)
		OH radicals[b]	None	n.e.	29.8 ± 2.1	n.e.	Iannone et al. (2004)

(continued)

Table 11.1 (continued)

Compound	Type of degradation	Electron acceptor	Activation mode	ε-C ± uncertainty (‰)	ε-H ± uncertainty (‰)	Λ ± uncertainty[a] (unitless)	References
		Persulfate	None	-0.6 ± 0.1	-20 ± 3	35 ± 10	Solano et al. (2018)
			None	-1.6	-6.6	4.1	Saeed (2011)
			None	n.s.	n.e.	n.e.	Gusti (2018)
			Iron-activated (high citrate/Fe ratio)	-1.4 ± 0.1	n.e.	n.e.	Gusti (2018)
			Iron-activated (low citrate/Fe ratio)	-1.7 ± 0.1	n.e.	n.e.	Gusti (2018)
			Alkaline	-1.2	-6.4	5.3	Saeed (2011)
		Permanganate	None	-6 ± 0.4	-222 ± 9	34 ± 2.2	Wijker et al. (2013)
		H_2O_2	UV light	-0.36 ± 0.05	14 ± 2	-34 ± 6	Zhang et al. (2016)
	Biotic	Aerobic (O_2)		-1.8 ± 0.3	n.s.	n.l.d.	Vogt et al. (2008)
				-2.5 ± 0.3	-159 ± 11	63.6	Mancini et al. (2006)

(continued)

Table 11.1 (continued)

Compound	Type of degradation	Electron acceptor	Activation mode	ε-C ± uncertainty (‰)	ε-H ± uncertainty (‰)	Λ ± uncertainty[a] (unitless)	References
				-0.4 ± 0.3	n.e.	n.e.	Mancini et al. (2002)
				-3.3 ± 0.3	n.e.	n.e.	Mancini et al. (2002)
		Anaerobic (nitrate-reducing)		-5.7 ± 0.2	-78 ± 12	11 ± 3	Vogt et al. (2008)
		Anaerobic (sulfate-reducing)		-2.5 ± 0.5	-107 ± 23	41 ± 8	Herrmann et al. (2009)
		Anaerobic (methanogenic)		-0.5	n.e.	n.e.	Ahad et al. (2000)
Ethylbenzene	Abiotic	OH radicals[b]	None	n.e.	26.8 ± 3.5	n.e.	Iannone et al. (2004)
		H_2O_2	UV light	-0.31 ± 0.05	30 ± 3	-90 ± 24	Zhang et al. (2016)
		Persulfate	None	-1.6	-2.5	1.5	Saeed (2011)
			Alkaline	-1.1	-2.1	1.9	Saeed (2011)
	Biotic	Aerobic (O_2)		-0.5 ± 0.1	4 ± 3	-7 ± 3	Dorer et al. (2014)
				-0.5 ± 0.1	-28 ± 3	35 ± 9	Dorer et al. (2014)

(continued)

Table 11.1 (continued)

Compound	Type of degradation	Electron acceptor	Activation mode	ε-C ± uncertainty (‰)	ε-H ± uncertainty (‰)	Λ ± uncertainty[a] (unitless)	References
		Anaerobic (nitrate-reducing)		-4.1 ± 0.2	-111 ± 12	24 ± 5[d]	Dorer et al. (2014)
		Anaerobic (nitrate-reducing)		-4.1 ± 0.2	-156 ± 8	40 ± 3[e]	Dorer et al. (2014)
		Anaerobic (sulfate-reducing)		-0.6 ± 0.1	-76 ± 16	278 ± 123	Dorer et al. (2014)
m-xylene	Abiotic	H_2O_2	UV light	-0.30 ± 0.06	-2.8 ± 0.1	9 ± 2	Zhang et al. (2016)
o-xylene		Persulfate	None	-0.36 ± 0.04	-23 ± 2	55 ± 9	Solano et al. (2018)
			None	-1.4	-9.4	6.7	Saeed (2011)
			Alkaline	-1.3	-7.1	5.5	Saeed (2011)
		H_2O_2	UV light	-0.27 ± 0.02	-3.2 ± 0.8	11 ± 4	Zhang et al. (2016)
p-xylene		OH radicals[b]	None	n.e.	10.5	n.e.	Iannone et al. (2004)
		H_2O_2	UV light	-0.31 ± 0.05	-6.3 ± 0.7	19 ± 3	Zhang et al. (2016)
		Persulfate	None	-0.6	-5.6	9.3	Saeed (2011)
			Alkaline	-0.4	-7.4	18.5	Saeed (2011)

(continued)

Table 11.1 (continued)

Compound	Type of degradation	Electron acceptor	Activation mode	ε-C ± uncertainty (‰)	ε-H ± uncertainty (‰)	Λ ± uncertainty[a] (unitless)	References
Xylene[c]	Biotic	Aerobic (O₂)		-0.6 ± 0.1	n.e.	n.e.	Bouchard et al. (2008b)
				-2.3 ± 0.3	n.e.	n.e.	Morasch et al. (2002)
		Anaerobic (sulfate-reducing)		-2.3 ± 0.4	-41 ± 9	15 ± 4	Herrmann et al. (2009)
		Anaerobic (sulfate-reducing)		-0.7 ± 0.1	-25 ± 3	29 ± 5	Herrmann et al. (2009)
		Anaerobic (sulfate-reducing)		-3.2	n.e.		Wilkes et al. (2000)

For biodegradation, a selection was made among literature values to report only the largest and the smallest Λ values related to aerobic or anaerobic biodegradation, regardless of the microbial environment used (i.e., the use of pure strain, microbial consortium, or native soil microbial population). For anaerobic biodegradation, the selection was made regardless of the electron acceptor. The largest and the smallest ε-C determined by experiments conducted without hydrogen measurements are also provided when different than those determined in combination with hydrogen

[a] Normal font: value from the original study. Italic font: calculated value based on Λ = ε-H/ε-C

[b] Carried out with gas phase OH radicals and VOC inside an air-tight reaction chamber

[c] Regardless of the isomer (o, m or p). n.s. = not significant. n.e. = not evaluated. n.l.d. = no Λ value determined as either ε-H or ε-C is not significant

[d] when Δδ²H < 100‰

[e] when Δδ²H > 100‰

methods. When possible, Λ values (reported by the respective study or calculated using Eq. 11.5 with ε-C and ε-H values issued from the same experimental series) are also reported in Table 11.1. Nevertheless, given the limited knowledge regarding isotope fractionation generated by chemical oxidation, it is highly recommended to carry out laboratory experiments with aquifer material at the prevailing redox conditions to derive site-specific ε-C and ε-H values.

11.3 CSIA Implementation for Field Site Evaluation

With the understanding of the processes leading to isotope fractionation for BTEX compounds, the next step is to take advantage of this knowledge to evaluate the performance of a remediation treatment at the field-scale level. Field implementation and sampling strategy to evaluate the performance of a remediation treatment using CSIA should be built given the objective of the remediation treatment, site-specific conditions, but also to CSIA specificities. This section aims to provide general and specific considerations to bear in mind when establishing a CSIA-based remediation performance monitoring program, with the focus being given to groundwater assessments.

11.3.1 Approach and Sampling Strategy Considerations

The general application goal for CSIA is to document an isotopic shift for BTEX compounds caused by the treatment being applied. For this purpose, a baseline characterization is essential to document pre-treatment isotopic compositions, to which post-treatment results will be compared to. In addition, to ensure capturing the most valuable and timely information leading to a successful application, the following specific considerations should be addressed:

(i) Regardless of the remediation treatment to be deployed, a CSIA baseline characterization should be considered for each sampling location to be included in the post-treatment sampling program.

(ii) The number of sampling locations included in the sampling program should be sufficient to meet the objectives and be appropriate for the size of the treated area.

(iii) The use of injection wells as monitoring wells is discouraged by technical performance monitoring guidance documents as injection of large fluid volumes may create water displacement leading to momentarily VOC concentration decrease (Payne et al. 2008). Since dilution does not create isotope fractionation, CSIA is actually the right tool to assess such potential VOC displacement. Nevertheless, field practitioners should keep in mind that unrepresentative excessive and prolonged reactions occurring inside (or near) the injection well might not be representative of distant areas (Huling and Pivetz 2006).

(iv) Following the CSIA baseline characterization, at least two or three post-treatment sampling events should be considered. Since it can be challenging to predict exactly when maximum degradation of VOC mass would be observed given potential presence of VOC concentration rebounds, planning several post-treatment events is a judicious strategy.

(v) Post-sampling event periodicity can be estimated using the expected lifetime of the emplaced amendment and estimated groundwater flow velocity. Chemical oxidizing agents with short half-lives should have shorter intervals between sampling events (few weeks), whereas oxygen-releasing products designed to last longer could have extended intervals between sampling events (few months). The groundwater flow velocity is also to be considered for long lasting amendments, or when forced by an engineered system.

(vi) BTEX concentration should be considered when pre-selecting the sampling locations. Following the treatment, the remaining BTEX concentration should be above the isotope analytical detection limits (see Sect. 11.2.2). In contrast, if BTEX concentration is initially too high (strongly suggesting presence of NAPL), the BTEX mass decrease due to treatment might not be sufficiently strong to counteract BTEX mass input by NAPL dissolution, which would impede observation of an isotope enrichment.

11.3.2 CSIA Sampling Requirements and Procedures

Groundwater sampling for CSIA is normally performed using the same sampling procedure, vial sizes (typically 40 mL VOA glass vials), and preservative agents (to inhibit microbial activity) as for the common VOC concentration analysis. In cases where BTEX concentrations are expected to be low, larger glass containers can be considered to allow the laboratory reaching better analytical detection limits. However, when an ISCO treatment is performed, a quenching agent should be added to the samples to consume the remaining oxidant, hence avoiding post-sampling oxidizing reaction with BTEX (EPA 2012). This specific quenching procedure or vial size selection should be discussed with the selected CSIA laboratory prior to the sampling event. Upon sampling completion, the vials can be stored, handled, and shipped to the CSIA laboratory in a standard cooler using the same procedures for samples dedicated for VOC analysis. Finally, BTEX concentrations are required to perform the CSIA measurement in the laboratory, and thus samples for BTEX concentration analysis should always be collected and analyzed during the same sampling event.

11.4 CSIA Field Data

This section introduces assessment approaches and interpretation of CSIA field data. First, general and specific considerations for sound CSIA interpretation are discussed, either introduced by recalling the state of knowledge, factors affecting the isotope fractionation, or by identifying potential pitfalls that should be avoided. Then, interpretation schemes making use of only one element (single-element isotope assessment) or two elements (dual-element isotope assessment) are introduced.

11.4.1 Interpretation Considerations and Pitfalls

The Rayleigh equation (Eq. 11.1) is the key mathematical feature to interpret isotope fractionation. This equation relates the isotope change to the decreasing VOC concentration. It is worth mentioning that the equation was originally developed to describe the isotope fractionation occurring in a system where a single process controls the change in concentration. However, for complex systems like contaminated aquifers, one can foresee potential for VOC concentration decrease due to several co-occurring processes (for instance biodegradation, dispersion, dilution, diffusion, and sorption). While some processes such as dilution and dispersion do not cause fractionation, other processes may result in some isotope fractionation, hence calling for cautions when evaluating field site data. As biodegradation was more often shown to be the dominant controlling isotope fractionation process in natural systems, application of Rayleigh equation to assess contaminant fate in groundwater is generally accepted (EPA 2008). Nonetheless, some studies are underlying the need for specific application limitations for complex and heterogeneous aquifer systems, especially when performing quantitative assessments (Thullner et al. 2012). In the present work, degradation is assumed to be the controlling isotope fractionation process in groundwater, and that sorption and aqueous diffusion are processes not likely to cause significant isotope fractionation. In cases where remediation systems are emphasizing physical processes, such as air sparging or soil vapor extraction systems, or being conducted in an organic rich aquifer, considering isotope fractionation caused by physical processes may be indicated.

Notwithstanding the Raleigh equation application concept and restrictions, there are notable additional interpretation considerations and pitfalls that deserve to be considered, and should be kept in mind when interpreting CSIA field data set:

(i) Due to uncertainty errors related to field and analytical procedures, an isotope shift of at least 1‰ or 10‰, respectively for $\delta^{13}C$ and for δ^2H, respectively, should be observed to suggest presence of compound degradation, whereas an isotope shift of 2‰ or 20‰ can be considered as strong evidence of degradation activity (EPA 2008). It should nevertheless be kept in mind that, especially for aerobic biodegradation of BTEX, large extent of biodegradation is required in some cases before observing carbon isotope shift larger than

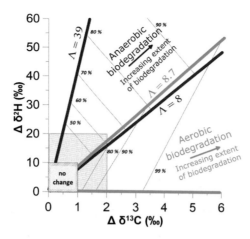

Fig. 11.2 Dual-carbon and hydrogen isotope plot illustrating expected isotope trends (delineated areas) for aerobic and anaerobic benzene biodegradation. Note that both areas are slightly overlapping. The Λ values were selected from Table 11.1. The grey dashed box represents no significant isotope change between two measurements whereas the grey dotted box is only suggestive of biodegradation effect. For each biodegradation delineated area, the magnitude of isotope fractionation expected for increasing extent of benzene biodegradation (from 50 to 99%) is indicated

2‰. Accordingly, observed field shifts smaller than 2‰ are not necessarily indicating an absence of biodegradation activity (see Sect. 11.4.2.2 and explanation of Fig. 11.2), but perhaps a dominant biodegradation mechanism generating insufficient isotope fractionation. This limitation hence underlines the importance of a multi-line of evidence approach.

(ii)   Interpretation of CSIA data should be carried out taking into account VOC concentration data and conventional field parameters (geochemical parameters and terminal electron acceptors) to validate the information supported by CSIA. As discussed in Chap. 10, additional specialized molecular biological tools (such as qPCR analyses or 16S rRNA gene amplicon sequencing to determine the structure of the microbial community) can also be included as an additional line of evidence complementary to CSIA.

(iii) The initial isotope composition (or source signature) of each contaminant before the occurrence of the spill is rarely known. To overcome this data gap, it is a common strategy to use the most negative $\delta^{13}C$ and δ^2H value measured at the site as the source value, for each targeted VOC.

(iv)   All suspected source zones on the site must be sampled to determine the likelihood of multiple sources (i.e., two distinct spills). Different spills would most likely have distinct BTEX isotope compositions. While in this case CSIA indicates the presence of multiple sources, the assessment of BTEX biodegradation will become challenging if both plume zones are overlapping.

(v)   A survey of carbon isotope composition of BTEX compounds collected from different gasoline sources showed variations ranging from −23.5 to −31.5‰ for benzene, from −22.9 to −30.4‰ for toluene, from −22.9 to −

31.1‰ for ethylbenzene, and from −22.4 to −30.6‰ for xylene (regardless of the isomer) (O'Sullivan and Kalin 2008). Accordingly, field measurements providing more positive $\delta^{13}C$ values than the higher-end value range would suggest that a degradation process has already affected the compounds.

(vi) Applying CSIA inside a source zone where NAPL is present may fail to demonstrate isotope fractionation although degradation may be occurring in the aquifer. Due to the limited biodegradation inside NAPL (because of toxicity) and negligible isotope fractionation related to dissolution process, the dissolution of VOCs from NAPL will consistently bring in groundwater VOCs with the original isotope signature. This process will thus influence the isotope ratio of dissolved VOC in groundwater. In some cases, the mass of freshly dissolved VOC can dominate the overall $\delta^{13}C$ (and $\delta^{2}H$) signature in groundwater, hence masking the isotope fractionation caused by the degradation process intended by the treatment. While this dissolution process impedes degradation demonstration, it however validates the presence of NAPL in the vicinity of the sampling location.

(vii) Given the limitation discussed in point *vi* above, application of the Rayleigh equation to quantify biodegradation extent or rate in the source zone should be done with precautions. While Rayleigh's law assumes a finite dissolved VOC mass, introduction of additional mass of VOC into groundwater (caused by NAPL dissolution and VOC desorption) will cause an underestimation of the degraded mass (see further discussion in Sect. 11.4.2.1).

(viii) Sampling a monitoring well with the screen positioned across the water table may result in groundwater samples mixing oxic water (upper part of the aquifer with BTEX undergoing aerobic biodegradation) with anoxic water (lower part of the aquifer with BTEX undergoing anaerobic biodegradation). Such mixing of reduction–oxidation conditions may affect evaluation of the dominant biodegradation mechanism controlling the isotope fractionation.

11.4.2   Assessment Approach

Two CSIA assessment approaches are discussed below, using either one or two isotopes. While a single-isotope assessment approach is meant to discriminate VOC concentration decrease due to degradation from dilution, the dual-isotope assessment approach has the advantage to provide more specific information by identifying the dominant VOC degradation mechanism. Both assessment approaches are described below by proposing CSIA data sorting and interpretation schemes to gain the most insight from the available information.

11.4.2.1 Single-Isotope Assessment Approach

The use of a single isotope (either $\delta^{13}C$ or δ^2H) will provide a direct line of evidence for BTEX mass destruction when dilution is potentially co-occurring during the treatment. Depending on site-specific assessment objectives, this can turn out to be sufficient to demonstrate treatment success. For such single-isotope assessment, comparison of isotopic shift between baseline and collected samples post-treatment for each BTEX compound is the basic interpretation scheme. The significance of the isotope shift should be interpreted based on threshold values as introduced above for $\delta^{13}C$ or δ^2H measurements. The larger the isotope shift, the more likely the treatment was effective for BTEX compounds. Nevertheless, the isotopic shift to be observed yet depends on the ε value, the degradation rate constant and, if the assessment compares the injection well to downgradient wells, groundwater velocity.

The interpretation can further be developed to evaluate the VOC mass proportion affected by the treatment and the degradation rate. The decrease in VOC concentration observed between two distal monitoring wells is usually due to a combination of destruction and dilution processes, which is expressed by Eq. 11.6 (Van Breukelen 2007b; Aelion et al. 2010):

$$f_{\text{overall}} = f_{\text{deg}} * f_{\text{dil}} \tag{11.6}$$

where the overall remaining mass fractionation (f_{overall}) is the product of remaining mass fractionation related to degradation (f_{deg}) and dilution (f_{dil}). The f_{dil} considers all physical processes, such as dispersion and sorption. It is worth recalling that f_{deg}, calculated based on the Rayleigh equation, only involves the mass of contaminant dissolved in groundwater. Since both f_{deg} and f_{dil} are reducing the VOC concentration, coefficients will be < 1.

When considering a monitoring well located in the source zone with presence of NAPL, the process of VOC dissolution will counteract both degradation and dilution processes. Accordingly, to evaluate the change in VOC concentration in groundwater at a monitoring well periodically sampled over time, Eq. 11.6 is amended to become Eq. 11.7:

$$f_{\text{overall}} = f_{\text{deg}} * f_{\text{dil}} * f_{\text{diss}} \tag{11.7}$$

where f_{diss} is the mass fraction input caused by VOC dissolution in groundwater. Since f_{diss} is a mass input process, the coefficient will be > 1. When groundwater flow velocity is negligible within the sampling periodicity, one can expect to have $f_{\text{diss}} \gg f_{\text{dil}}$, hence making f_{dil} negligible.

For both Eqs. 11.6 and 11.7, the f_{overall} can be calculated using concentration data measured either at the same well but at different sampling time (C_0/C_t) of for two distal wells aligned with respect to groundwater flow (C_0/C_x). As shown in Eq. 11.8, for f_{deg}, the latter can be determined using the isotope measurements and a rearrangement of Eq. 11.1:

$$f_{\text{deg}} = \exp^{(\delta^{13}C_t - \delta^{13}C_0)/\varepsilon} \qquad (11.8)$$

where $\delta^{13}C_t$ and $\delta^{13}C_0$ are isotopic field measurements obtained at time $= t$ and time $= 0$, respectively. Ideally, microcosm experiments are carried out with aquifer material at the prevailing redox conditions to determine site-specific ε values. The alternative option is to select an appropriate ε value from the literature. Since there are many different ε values related to BTEX biodegradation (as discussed in Sect. 11.2.3) the selection must be carefully made considering the specific conditions prevailing at the site. Conventional field parameters or specialized molecular biological tools (such as microbial genetics) carried out in parallel can assist the field practitioner in the selection. Nevertheless, it is good practice to calculate a range of estimated f_{deg} by using the largest and the smallest ε values available for each VOC (Table 11.1) to support interpretation or decision-making. The largest ε value will provide a more conservative estimate, hence reducing the risk of overestimating f_{deg}. If Eq. 11.7 applies to the field site conditions, the VOC mass input process (f_{diss}) will additionally bias the calculations, also leading to an underestimated degraded mass estimate.

Assuming a constant groundwater flow velocity and a constant first-order degradation rate in the aquifer, Eq. 11.1 can also be transformed to obtain an in situ degradation rate (k) for the targeted contaminant. Assuming a first-order degradation rate controlling the evolution of the remaining mass fraction as expressed by Eq. 11.9 (Aelion et al. 2010):

$$f_{\text{deg}} = \exp^{(-k*t)} \qquad (11.9)$$

Insertion of Eq. 11.9 into Eq. 11.8 provides Eq. 11.10:

$$k = -1 * \left(\frac{\delta^{13}C_t - \delta^{13}C_0}{\varepsilon * t} \right) \qquad (11.10)$$

where k is the in situ degradation rate (day^{-1}) and t is the time period (in days) between two sampling events within the same monitoring well. Equation 11.10 becomes applicable for evaluation using a single well over time as the injected product degrades VOCs faster than source input, which momentarily interrupts the steady state conditions. If Eq. 11.7 applies due to presence of NAPL in the vicinity of the monitoring well, the k value will be underestimated due to f_{diss}. To calculate k based on measurement from two distal monitoring wells (and measured during the same sampling event), the parameter t can be replaced by the associated site-specific distance (x) and groundwater velocity (v) to provide Eq. 11.11 (Aelion et al. 2010):

$$k = -1 * \left(\frac{\delta^{13}C_x - \delta^{13}C_0}{\varepsilon * (x/v)} \right) \qquad (11.11)$$

Equation 11.11 assumes the injected product has reached the distal monitoring well due to groundwater flow (i.e., not due to large radius of influence related to injection activity) and that steady state conditions are re-established. The calculated k value can then be used to estimate the length of the treatment area required to meet the groundwater criteria for BTEX concentrations. Since f_{dil} is not considered in Eq. 11.11, the estimated length of the treatment area will likely be conservative. To increase their reliability, these equations should only be applied when $\delta^{13}C_{t\,or\,x} - \delta^{13}C_0 > 2‰$ (or $20‰$ for δ^2H).

11.4.2.2 Dual-Isotope Approach

As described earlier, biological and chemical oxidation have specific $\delta^{13}C$ and δ^2H isotope fractionation processes which lead to different isotope enrichment patterns during the reaction. The use of dual isotopes ($\delta^{13}C$ and δ^2H) takes advantage of these specific isotope enrichment patterns associated with diverse degradation mechanisms. Therefore, a dual-isotope assessment not only demonstrates BTEX mass destruction, but also indicates the dominant degradation mechanism affecting BTEX mass while additional co-occurring process(es) may be present during the treatment. Compared to single-isotope assessments, dual-isotope assessments are more rigorous in demonstrating establishment of the intended mass removal process by the treatment.

For dual-isotope assessment, interpretation of CSIA field data is performed by comparing the trend observed with field measurements to the theoretical trends expected for biodegradation (aerobic and anaerobic) or chemical oxidation process (depending on the remediation treatment). The first step in the interpretation sequence is to establish a supportive $\delta^{13}C$ versus δ^2H reference plot illustrating theoretical enrichment trends for potentially occurring degradation processes. This is achieved by plotting calculated δ^2H data in function of $\delta^{13}C$, using respective Λ value for each degradation process listed in Table 11.1. The resulting timeless isotopic enrichment trend represents the expected $\delta^{13}C$ and δ^2H evolution as the degradation process is progressively occurring. To account for the isotope fractionation variability caused by diverse biodegradation mechanism pathways (due to site-specific microbial population composition), the minimum and the maximum Λ values reported in Table 11.1 for each selected process should be used to calculate extremity enrichment trends. Both extremity enrichment trends form a process-specific delineated area. An illustrative example is provided in Fig. 11.2, where process-specific delineated areas were calculated for aerobic and anaerobic biodegradation of benzene using the approximate transformation provided in Eq. 11.5. Note the simplification made for anaerobic biodegradation, where the selection of extremity Λ values was made regardless of the terminal electron acceptor. Figure 11.2 also indicates, for each biodegradation delineated area, the isotope shift expected for increasing extent of benzene biodegradation. For the selected Λ values, one can observe that at least 50% of anaerobic benzene biodegradation is needed to observe significant isotope shifts ($> 2‰$ for $\delta^{13}C$ or $> 20‰$ for δ^2H), whereas larger extents are needed for aerobic biodegradation. The proportion of biodegraded benzene (b) was calculated by a simple rearrangement of Eq. 11.8, leading to Eq. 11.12:

$$b(\%) = \left(1 - f_{\text{deg}}\right) * 100 \tag{11.12}$$

As a second step, field measurements are introduced in the reference isotopic plot. For each selected compound from a specific sampling location, the carbon and hydrogen isotope shift (or $\Delta\delta^{13}C$ and $\Delta\delta^2H$) between baseline and a post-injection event is calculated and the resulting data point (field measurement) is plotted in the reference isotopic plot. Additional points can be plotted for every post-treatment measurement conducted over time at the same monitoring well. The positioning of field measurements relative to process-specific delineated areas can then be evaluated. In addition, the significance of isotope shift should be interpreted based on threshold values (2‰ and 20‰) as introduced above for $\delta^{13}C$ and δ^2H measurements. The field measurements lying inside a process-specific delineated area suggest the related process as the dominant mass removing process. In some cases, field measurements may lie between two process-specific delineated areas, which would suggest a combination of the two processes. The evaluation can be deepened by calculating the contribution of each two different on-going processes using extended Rayleigh equations. These extended equations were previously developed to consider two concurrent degradation pathways with different ε values (van Breukelen 2007a).

11.5 Examples of Field Case Applications

Application of the CSIA method to evaluate the performance of different remediation approaches for BTEX-contaminated aquifers is presented below for two pilot scale tests. For each pilot test, CSIA was included to the remediation performance monitoring plan and used as a diagnostic tool to demonstrate that the BTEX mass degradation processes were established as initially intended.

11.5.1 In situ Chemical Oxidation Application

A pilot scale test was conducted at Site 1 to evaluate the potential of NaOH-activated sodium persulfate in reducing dissolved BTEX mass in a PHC source zone. CSIA was used as a performance metric to validate BTEX mass destruction although co-occurrence of dilution related to large volume fluid injections contributing to decrease BTEX concentration was expected. The discussion of the results will focus on $\delta^{13}C$ values measured for toluene, ethylbenzene, m,p-xylene, and o-xylene (TEXX) results. Benzene concentrations were below the detection limit for isotope analysis.

Table 11.2 Technical description of the pilot scale test conducted at Site 1

Item	Description
Treatment amendments	Sodium persulfate (injection solution 250 g/L), NaOH (as persulfate activator)
In situ injection	Recirculation system using two wells
Sampling event	• Baseline • Post 1 (26 days post-injection) • Post 2 (38 days post-injection)
Chemical parameters	• Geochemical parameters • TEXX concentration • TEXX $\delta^{13}C$

11.5.1.1   Field Approach

The injection of persulfate was performed through an injection–extraction system using two wells. The injection and extraction well were located 12 m apart and aligned with groundwater flow. The groundwater recirculation system was maintained in operation for the duration of persulfate injection and was then halted to leave the persulfate lingering in the aquifer. To demonstrate the destructive effect of ISCO treatment on dissolved TEXX, groundwater was collected before and during the treatment from monitoring well MW-S1 located midway between the two recirculation wells. The sampling event timing and the parameters analyzed are summarized in Table 11.2.

11.5.1.2   Results and Interpretation

For the four TEXX compounds in groundwater at MW-S1, a concentration decrease was observed during the post-injection event 1 (Post 1) compared to baseline level and was followed by an increase observed during the post-injection sampling event 2 (Post 2) (Fig. 11.3). In addition, baseline sulfate (SO_4) concentration increased from 32 mg/L to 1100 mg/L (Post 1) to then decrease to 810 mg/L (Post 2) (Fig. 11.3). The presence of SO_4 in groundwater is related to persulfate reaction as the former is a by-product of decaying persulfate. Similarly, the strong pH increase (from 7.2 to 12.5) also confirms the arrival of the NaOH-activated persulfate slug in the vicinity of the monitoring well. The sampling timing for Post 1 was established based on the persulfate decay rate previously determined during a bench test. The decay rate was used to ensure a sampling interval sufficiently long for persulfate to degrade a significant TEXX mass, but still with sufficient remaining persulfate restraining TEXX concentration to rebound. Based on TEXX and SO_4 concentration patterns observed (Fig. 11.3), the timing of Post 1 was adequate to capture persulfate effects on TEXX mass.

D. Bouchard et al.

Fig. 11.3 Changes in pH, SO$_4$, concentration and δ^{13}C for toluene, ethylbenzene, *m,p*-xylene, and *o*-xylene observed in groundwater at MW-S1 for baseline (day 0) and after 26 days (Post 1) and 38 days (Post 2) following NaOH-activated persulfate injection at Site 1. The dashed lines in panel C represent the original isotopic signature determined for respective compound at the site

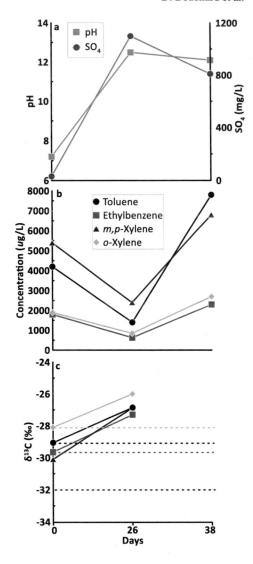

Single-isotope assessment: the δ^{13}C values measured for TEXX in groundwater at MW-S1 are presented in Fig. 11.3. The δ^{13}C values measured for TEXX at Post 1 were all more positive compared to their respective baseline value. The calculated Δ δ^{13}C for TEXX, respectively 2.2, 2.4, 3.2, and 2.1‰, were systematically larger than the prerequisite of 2‰ for strong evidence of degradation activity. These significant trends toward more positive values (i.e., enrichment of [heavy]molecules among the remaining TEXX pool), coinciding with the decline in TEXX concentrations, can be attributed to persulfate oxidation reactions. Accordingly, the δ^{13}C values measured strongly support that the temporal decrease in concentration was not solely caused by dilution, but by substantial degradation reactions affecting the TEXX mass. For

Table 11.3 Remaining mass fraction $f_{overall}$, f_{deg}, and f_{dil} calculated for toluene, ethylbenzene, and o-xylene using concentration or carbon isotope data

Calculations	Remaining mass fraction		
	Toluene	Ethylbenzene	o-xylene
$f_{overall}$ (Concentration data)	0.33	0.35	0.45
f_{deg} (Carbon isotope data)	0.16	0.11	0.20
$f_{dil} * f_{diss}$ (Eq. 11.7)	2.06	3.18	2.25

See text for related equations

Post 2, TEXX concentrations were observed to significantly increase. Although no $\delta^{13}C$ measurements were carried out, it is strongly expected that $\delta^{13}C$ values for TEXX would return to their original values (i.e., dashed lines in Fig. 11.3). In this case, freshly dissolved VOC mass coming from the NAPL (hence with the original isotope composition), coinciding with persulfate mass exhaustion, is expected to overwhelm the overall $\delta^{13}C$ signature in groundwater.

Remaining mass fraction: The isotopic evaluation can be further developed to evaluate the effect of persulfate treatment on TEXX mass. The remaining mass fraction $f_{overall}$ and f_{deg} were determined for toluene, ethylbenzene, and o-xylene using respective concentration and $\delta^{13}C$ (Eq. 11.8) values measured at baseline and Post 1 (Fig. 11.3). For f_{deg}, the respective ε-C value for toluene, ethylbenzene, and o-xylene related to NaOH-activated persulfate oxidation was used (Table 11.1). The resulting calculated f_{deg} values for toluene, ethylbenzene, and o-xylene are smaller than those respectively obtained for $f_{overall}$ (Table 11.3). We recall here that calculation of $f_{overall}$ indistinctively considers the processes of destruction (f_{deg}), dilution (f_{dil}), and dissolution (f_{diss}) (Eq. 11.7). However, in a source zone, VOC dissolution from NAPL can significantly contribute to increased dissolved concentrations, hence counteracting the former two processes. In addition, contribution of f_{dil} may also be reduced using temporal data measured from the same monitoring well. Accordingly, the $f_{dil} * f_{diss}$ values > 1 obtained for toluene, ethylbenzene, and o-xylene (Table 11.3) strongly suggest the occurrence of NAPL dissolution during the treatment period counteracting the apparent VOC destruction process. Accordingly, the $\delta^{13}C$ dataset reveal greater TEXX destruction by persulfate than concentration data alone would suggest. Due to the evidence of NAPL dissolution, calculation of degradation rate k (using Eq. 11.10) was not attempted as very likely to be underestimated.

11.5.1.3 Benefits of Using CSIA

The pilot scale test demonstrated that base-activated persulfate was effectively oxidized TEXX compounds. Although such remediation approach required injection of large amount of fluid, dilution was apparently not the only process controlling the VOC concentration variations observed. The significant changes in $\delta^{13}C$ to more positive values strongly support the occurrence of a TEXX mass destruction process. Furthermore, the remaining mass fraction calculations suggested that dissolution process (from residual NAPL) and oxidation process outweighed the dilution process as the calculated f_{deg} values were systematically smaller than $f_{overall}$ values. These findings contribute in supporting the decision to move forward with the full-scale application of the treatment inside this source zone area. Nevertheless, field practitioners should bear in mind that these isotopic calculations remain semi-quantitative, especially when treating a source zone. Since the process of VOC dissolution from NAPL represents a mass input to groundwater, the calculated mass proportion degraded is likely under predicted. In addition, the enrichment factor value used here was derived from a NaOH-activated laboratory experiment assuming BTEX degradation by SO_4 radicals. Under field conditions, degradation via unactivated persulfate and/or via OH radicals may also have occurred. Since different enrichment factors (Table 11.1) are observed due to different degradation mechanisms, the calculated f_{deg} values would then change accordingly. It is re-emphasized here that determination of site-specific ε value for NaOH-activated persulfate (through a laboratory bench test with site soil and groundwater) will allow more accurate isotope assessments, especially if a full-scale treatment is considered.

11.5.2 Bioremediation Application

A pilot scale test was conducted at Site 2 to evaluate the potential of stabilized hydrogen peroxide (H_2O_2) to enhance and sustain aerobic biodegradation of dissolved BTEX mass present in an anoxic plume zone. Hydrogen peroxide was used as source of dissolved oxygen (DO), and was combined with nutrients, a petroleum-specific facultative microbial inoculant (bioaugmentation), and a stabilizer reducing the decomposition rate of H_2O_2 reactions (hence maintaining an extended DO source). CSIA was used in combination with microbial gene assays as performance metrics to (i) assess the aquifer turnaround from well-established anoxic biodegradation conditions to oxic conditions, (ii) assess aerobic BTEX biodegradation, and (iii) assess the co-occurrence of dilution related to large volume fluid injections for the H_2O_2 delivery. The discussion of the results is focusing on toluene ($\delta^{13}C$), ethylbenzene ($\delta^{13}C$ and δ^2H), and m,p-xylene ($\delta^{13}C$ and δ^2H). Benzene concentrations were below the detection limit for isotope analysis.

11.5.2.1 Field Approach

The injection of stabilized H_2O_2 solution was conducted through an injection–extraction system as described above for Site 1. To assess toluene destruction and to assess aerobic biodegradation of ethylbenzene and m,p-xylene, groundwater was collected before and during the treatment from monitoring well MW-S2 located midway between the two recirculation wells. The sampling event timing and the sampling parameters analyzed are summarized in Table 11.4.

Table 11.4 Technical description of pilot scale tests carried out at Site 2

Item	Description
Treatment amendments	H_2O_2 (injection solution at 3%), H_2O_2 stabilizer, nutrients (N and P), and petroleum-specific facultative microbial inoculant
In situ injection	Recirculation system using two wells
Sampling event	• Baseline • Post 1 (15 days post-injection) • Post 2 (48 days post-injection)
Chemical parameters	• Geochemical parameters • TEX concentration • TEX-δ^{13}C • EX-δ^2H
Biological parameters (Genes)	• Toluene dioxygenase (TDO) • Benzylsuccinate synthase (BssA)

11.5.2.2 Results and Interpretation

Geochemical parameter results are shown in Fig. 11.4 for baseline, Post 1 and Post 2 sampling events at MW-S2. An aquifer under anoxic conditions was prevailing before the treatments, as suggested by a depleted DO level and strong negative ORP value (and by other geochemical parameter results not shown). Following the H_2O_2 solution injection, DO levels significantly increased up to 15 mg/L, and lasted above 2 mg/L for at least 40 days. Although high DO levels able to support oxic conditions were present, concentrations of Fe^{2+} and Mn^{2+} increased and concentration of SO_4 decreased following the injections, suggesting that anaerobic biodegradation was likely co-occurring in the aquifer (data not shown).

To assess the effect of oxygen addition on the microbial population dynamics, gene assays targeting toluene dioxygenase (TOD—indicative of aerobic BTE biodegradation) and benzylsuccinate synthase (BssA—indicative of anaerobic TEX biodegradation) were assessed (Fig. 11.4). The baseline conditions indicate low concentrations (gene copies/L) of both genes. After injection, a significant increase in TOD aerobic genes was observed only for the Post 2 sampling event, although high DO levels were already observed for the Post 1 sampling event. Also at Post 2 sampling event, the TOD gene concentration became larger than the BssA concentration by almost two orders of magnitude. These results suggest a slow transition to an aerobic population that has benefited from high DO concentration (Post 1 event) followed by depleted DO concentration (Post 2 event) in the aquifer. The BssA anaerobic population increased for the Post 1 event, benefiting first from the arrival of nutrients, but then decreased likely due to increasing oxic conditions.

Single-isotope assessment: Concentration and δ^{13}C values measured for toluene, ethylbenzene, and *m,p*-xylene in groundwater at MW-S2 before and after the injection are presented in Fig. 11.4. The ethylbenzene concentration for the Post 1 sampling

Fig. 11.4 Changes in
dissolved oxygen (DO) and
oxidation–reduction
potential (ORP) (panel
a), toluene, ethylbenzene and
m, p-xylene concentration
(panel b), gene copies of
toluene dioxygenase (TDO)
and Benzylsuccinate
synthase (BssA) (panel c),
and $\delta^{13}C$ values (panel d)
observed in groundwater at
MW-S2 during the treatment.
The dashed lines in panel
d represent the original
isotopic signature
determined for respective
compound at the site

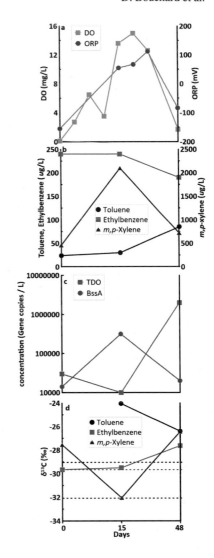

event was similar to the baseline condition, but then decreased at Post 2. The $\delta^{13}C$ value followed the same trend; remaining unchanged between baseline and Post 1 (equivalent to the original signature determined for ethylbenzene at this site), to then becoming enriched (more positive value) by 2.0‰ at Post 2. The enrichment observed at Post 2 is equal to the prerequisite of 2‰ for significant evidence of (bio)degradation process. In contrast, the m,p-xylene concentration at Post 1 significantly increased compared to baseline condition and coincided with $\delta^{13}C$ value changing to depleted value (more negative at Post 1). These changes for both parameters are indicative of m,p-xylene being released into solution (either due to desorption from the soil matrix and/or dissolution from some residual NAPL). The $\delta^{13}C$ value measured at

Post 1 was the most negative value measured, hence suggestive of the original carbon isotopic signature for *m,p*-xylene at this site. At Post 2, the *m,p*-xylene concentration decreased while the $\delta^{13}C$ value changed to significantly enriched value (by 5.6‰). Accordingly, the isotopic shifts measured for both ethylbenzene and *m,p*-xylene between Post 1 and Post 2 ($\geq$ 2‰) strongly support concentration decrease due to a destructive process. Lastly for toluene, while concentrations remained stable (and low) between baseline and Post 1, an increase was observed at Post 2. This increase in concentration coincided with a decreasing $\delta^{13}C$ value (between Post 1 and Post 2), suggesting that the toluene mass input into groundwater was momentarily greater than the mass biodegraded. Nevertheless, $\delta^{13}C$ values measured at Post 1 and Post 2 are more positive compared to the original signature (-29.1‰), hence indicating that toluene is nonetheless being biodegraded at the site. Note that toluene concentration at baseline was too low to conduct $\delta^{13}C$ analysis.

Dual-isotope assessment: In Fig. 11.5, the $\Delta\delta^2H$ calculated for ethylbenzene and *m,p*-xylene between Post 1 and Post 2 sampling events were plotted as a function of $\Delta\delta^{13}C$. The field measurements are compared to expected enrichment trends for aerobic and anaerobic biodegradation for the respective compounds. The delineated areas for aerobic and anaerobic biodegradation and the positioning of the field measurements were established as described in Sect. 11.4.2.2, using respective Λ values listed in Table 11.1. For ethylbenzene, the Λ value for $\Delta\delta^2H > 100$‰ was selected to represent the lower part of the anaerobic delineated area. The field measurements for both ethylbenzene and *m,p*-xylene positioned inside their respective delineated area expected for anaerobic biodegradation (Fig. 11.5), suggesting that these two compounds were mainly anaerobically biodegraded.

Biodegradation rates: To evaluate the first-order biodegradation rate (k) for ethylbenzene and *m,p*-xylene using carbon isotope data, Eq. 11.10 was used. For this purpose, the largest and the smallest ε-C values for anaerobic biodegradation of ethylbenzene and *m,p*-xylene (Table 11.1) were used to provide a range of estimated k values for each compound for the time period between Post 1 and Post 2 (33 days). The calculated k values for ethylbenzene and *m,p*-xylene are listed in Table 11.5. The latter in situ rates are further compared to reported in situ rates measured under natural attenuation conditions (unspecified reducing biodegradation conditions) and estimated by the conventional concentration-based approach (Aronson and Howard 1997). The in situ k values derived at Site 2 are in the high end, or larger, than the reported values for natural attenuation. These higher rates support the presence of enhanced biodegradation at the site due to the treatment, more likely due to injection of nutrients and microbial inoculants than dissolved oxygen.

11.5.2.3   Benefits of Using CSIA

Inclusion of the CSIA method (single- and dual-isotope assessments) in this pilot scale test evaluating the performance of biostimulation has provided valuable information that BTEX concentration data and geochemical parameters cannot reveal

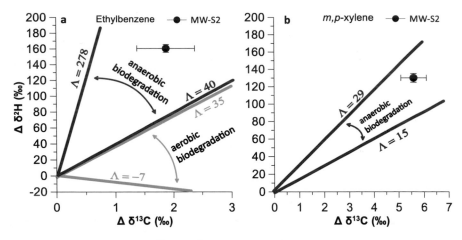

Fig. 11.5 Expected trends for $\delta^{13}C$ and δ^2H (reported as shift (Δ) relative to baseline value) during progressive aerobic and anaerobic biodegradation of ethylbenzene (**a**) and m,p-xylene (**b**), and the measured field value obtained for monitoring well MW-S2 during the post-injection sampling event 2 (reported as shift (Δ) relative to baseline value). The Λ values are taken from Table 11.1. Due to limited Λ values for m,p-xylene, the selection for anaerobic biodegradation was made regardless of the isomers (o, m, or p), whereas no Λ values are available for aerobic biodegradation

Table 11.5 In situ biodegradation rates (k) calculated for ethylbenzene and m,p-xylene using site-specific $\delta^{13}C$ data

Compound	In situ k rates (day^{-1})	
	Site[a]	Literature[b]
Ethylbenzene	0.014–0.096	0.0006–0.015
m,p-xylene	0.053–0.242	0.00103–0.0155[c]

See text for related equations
[a] Using Eq. 11.10. [b] From Aronson and Howard (1997). [c] Averaging values individually reported for m-xylene and p-xylene

or easily address. Certainly, the most valuable information is that CSIA provided evidence of ethylbenzene and m,p-xylene degradation during the treatment. The geochemical parameters and microbial genetic evaluation suggested evolving presence of both aerobic and anaerobic biodegradation. However, while the intended contaminant mass removal process by the treatment was aerobic biodegradation, CSIA suggested that ethylbenzene and m,p-xylene were still mainly anaerobically biodegraded during the course of the pilot test. The period with high DO concentration and evolving aerobic gene copies observed between Post 1 and Post 2 sampling events (for at least 33 days) suggest a transition from a prevailing BTEX-related anoxic microbial population to an oxic population. Nonetheless, the delivery of DO was not sufficiently long to sustain oxic condition and observing significant aerobic ethylbenzene and m,p-xylene biodegradation. Accordingly, numerous injection events (or a prolonged continuous injection event) will likely be required to sustain aerobic biodegradation, which will need to be considered in the treatment cost of the full-scale design. On the other hand, the in situ k rates derived using carbon isotope results for ethylbenzene and m,p-xylene were in the high end or larger than values reported in the literature for natural attenuation under reducing conditions. These higher degradation rates suggest the benefits of nutrients addition and microbial bioaugmentation. Finally, note that all the observations mentioned above are limited to groundwater samples repeatedly collected from a single monitoring well. Although this highlights the importance of sampling timing following the injection event to capture timely information, additional sampling points in space and time would have further supported the conclusions and strengthened the information gained from this pilot test.

11.6 Summary and Future Development

This chapter described a field procedure to support field practitioners when applying CSIA as a tool to assess the performance of in situ remediation treatments of BTEX-contaminated groundwater. The foremost benefit in applying this process-specific isotopic tool is the gain of direct destructive evidence for each BTEX compound assessed without relying on conventional VOC concentration analysis. When conducting dual-isotopes assessment, CSIA furthermore improves the understanding of the dominant mass degradation process(es) that are occurring during a treatment. Such valuable assessment tool confirming establishment of the intended mass removal process undeniably leads to cost-effective in situ treatment operations. Nevertheless, CSIA should be perceived as an assessment tool that needs to be used as part of a multi-line of evidence approach. Although the use of CSIA application is expected to increase in a near future, one current main application limitation is the limited availability of enrichment factor coefficients. As the enrichment factor is a key element to interpret the change in $\delta^{13}C$ and δ^2H observed during the treatment, the accuracy of the CSIA approach will undoubtedly benefit from enlarging the enrichment factor database (for both ε-C and ε-H) relatively to BTEX biodegradation, to chemical oxidation initiated by various oxidizing agents, and to physical mass removal processes enhanced by an engineered treatment (such as air sparging or soil vapor extraction). Finally, additional field scale demonstrations of CSIA application on treatments applied in saturated and unsaturated zones will also contribute to better disseminate CSIA use among the field practitioner community.

Acknowledgements The authors wish to acknowledge Global Remediation Technologies, Inc. (GRT) Project Team for designing and implementing the two pilot scale tests. We are thankful to GRT and Tersus Environmental, LLC, for fruitful result discussions, and to the Michigan Department of Environmental, Great Lakes and Energy - Remediation and Redevelopment Division (EGLE-RRD) for the permission to share the results.

References

Aelion CM, Höhener P, Hunkeler D, Aravena R (eds) (2010) Environmental isotopes in biodegradation and bioremediation. CRC Press, Taylor and Francis Group, Boca Raton, London, New York

Ahad JM, Slater GF (2008) Carbon isotope effects associated with Fenton-like degradation of toluene: potential for differentiation of abiotic and biotic degradation. Sci Total Environ 401:194–198

Ahad JME, Sherwood Lollar B, Edwards EA, Slater GF, Sleep BE (2000) Carbon isotope fractionation during anaerobic biodegradation of toluene: implications for intrinsic bioremediation. Environ Sci Technol 34:892–896

Aronson D, Howard PH (1997) Anaerobic biodegradation of organic chemicals in groundwater: a summary of field and laboratory studies. Draft final report. American Petroleum Institute, Washington, DC

Bouchard D, Hunkeler D, Gaganis P, Aravena R, Höhener P, Broholm MM, Kjeldsen P (2008a) Carbon isotope fractionation during diffusion and biodegradation of petroleum hydrocarbons in the unsaturated zone: field experiment at Værløse Airbase, Denmark, and modeling. Environ Sci Technol 42:596–601

Bouchard D, Hunkeler D, Hohener P (2008b) Carbon isotope fractionation during aerobic biodegradation of n-alkanes and aromatic compounds in unsaturated sand. Org Geochem 39:23–33

Bouchard D, Hunkeler D, Madsen EL, Buscheck T, Daniels EJ, Kolhatkar R, Derito CM, Aravena R, Thomson N (2018a) Application of diagnostic tools to evaluate remediation performance at petroleum hydrocarbon-impacted sites. Ground Water Monit Rem 38:88–98

Bouchard D, Marchesi M, Madsen EL, Derito CM, Thomson NT, Aravena R, Barker JF, Buscheck T, Kolhatkar R, Daniels EJ, Hunkeler D (2018b) Diagnostic tools to assess mass removal processes during pulsed air sparging of a petroleum hydrocarbon source zone. Ground Water Monit Rem 38:29–44

Clark ID, Fritz P (1997) Environmental isotopes in hydrogeology. Lewis Publishers, Boca Raton, FL

Dorer C, Vogt C, Kleinsteuber S, Stams AJ, Richnow HH (2014) Compound-specific isotope analysis as a tool to characterize biodegradation of ethylbenzene. Environ Sci Technol 48:9122–9132

Elsner M (2010) Stable isotope fractionation to investigate natural transformation mechanisms of organic contaminants: principles, prospects and limitations. J Environ Monit 12:2005–2031

Elsner M, Zwank L, Hunkeler D, Schwarzenbach RP (2005) A new concept linking observable stable isotope fractionation to transformation pathways of organic pollutants. Environ Sci Technol 39:6896–6916

EPA (2008) A guide for assessing biodegradation and source identification of organic ground water contaminants using compound specific isotope analysis (CSIA). Office of Research and Development, National Risk Management Research Laboratory. Unites States Environmental Protection Agency (U.S. EPA), Ada, Oklahoma

EPA (2012) Ground water sample preservation at in-situ chemical oxidation sites—recommended guidelines. Office of Research and Development, National Risk Management Research Laboratory. Unites States Environmental Protection Agency (U.S. EPA), Ada, Oklahoma

EPA (2017) In situ remediation: design considerations and performance monitoring. Technical guidance document. Version 1.0., New Jersey Department of Environmental Protection, United State Environmental Protection Agency (U.S. EPA)

Fischer A, Theuerkorn K, Stelzer N, Gehre M, Thullner M, Richnow HH (2007) Applicability of stable isotope fractionation analysis for the characterization of benzene biodegradation in a BTEX-contaminated aquifer. Environ Sci Technol 41:3689–3696

Fischer A, Herklotz I, Herrmann S, Thullner M, Weelink SAB, Stams AJM, Schlomann M, Richnow H-H, Vogt C (2008) Combined carbon and hydrogen isotope fractionation investigations for elucidating benzene biodegradation pathways. Environ Sci Technol 42:4356–4363

Fischer A, Gehre M, Breitfeld J, Richnow H-H, Vogt C (2009) Carbon and hydrogen isotope fractionation of benzene during biodegradation under sulfate-reducing conditions: a laboratory to field site approach. Rapid Commun Mass Spectrom 23:2439–2447

Gandour RD, Schowen RL (1978) Transition states of biochemical processes. Plenum Press, New-York, London, p 636

Gusti W (2018) The effectiveness of persulfate and hydrogen peroxyde on the oxidation of hydrocarbon contaminants at 30C: a study with focus on the performance of compound-specific isotope analysis. Master degree, University of Waterloo

Herrmann S, Vogt C, Fischer A, Kuppardt A, Richnow H-H (2009) Characterization of anaerobic xylene biodegradation by two-dimensional isotope fractionation analysis. Environ Microbiol Rep 1:535–544

Höhener P, Imfeld G (2021) Quantification of Lambda (Λ) in multi-elemental compound-specific isotope analysis. Chemosphere 267:129232

Huling SG, Pivetz BE (2006) In-situ chemical oxidation. EPA Office of Research and Development, EPA/600/R-06/072

Hunkeler D, Anderson N, Aravena R, Bernasconi SM, Butler BJ (2001) Hydrogen and carbon isotope fractionation during aerobic biodegradation of benzene. Environ Sci Technol 35:3462–3467

Huskey PW (1991) Enzyme mechanism from isotope effects. In: Cook PF (ed) Enzyme mechanism from isotope effect. CRC Press, Boca Raton, FL, USA, pp 37–72

Iannone R, Anderson RS, Vogel A, Rudolph J, Eby P, Whiticar MJ (2004) Laboratory studies of the hydrogen kinetic isotope effects (KIES) of the reaction of non-methane hydrocarbons with the OH radical in the gas phase. J Atmos Chem 47:192–208

Imfeld G, Kopinke FD, Fischer A, Richnow HH (2014) Carbon and hydrogen isotope fractionation of benzene and toluene during hydrophobic sorption in multistep batch experiments. Chemosphere 107:454–461

ITRC (2011) Compound specific isotope analysis. Interstate Technology & Regulatory Council. EMD team fact sheet

IUPAC (1991) Isotopic compositions of the elements. Pure Appl Chem 63:991–1002

Jochmann MA, Schmidt TC (2012) Compound-specific stable isotope analysis, 1st edn. Royal Society of Chemistry

Julien M, Nun P, Robins RJ, Remaud GS, Parinet J, Höhener P (2015) Insights into mechanistic models for evaporation of organic liquids in the environment obtained by position-specific carbon isotope analysis. Environ Sci Technol 49:12782–12788

Kuder T, Wilson JT, Kaiser P, Kolhatkar R, Philp P, Allen J (2005) Enrichment of stable carbon and hydrogen isotopes during anaerobic biodegradation of MTBE: microcosm and field evidence. Environ Sci Technol 39:213–220

Kuder T, Philp P, Allen J (2009) Effects of volatilization on carbon and hydrogen isotope ratios of MTBE. Environ Sci Technol 43:1763–1768

Kuntze K, Eisenmann H, Richnow HH, Fischer A (2019) Compound-specific stable isotope analysis (CSIA) for evaluating degradation of organic pollutants: an overview of field case studies. In: Boll M (ed) Anaerobic utilization of hydrocarbons, oils, and lipids. Springer International Publishing, Cham

Lee J, von Gunten U, Kim JH (2020) Persulfate-based advanced oxidation: critical assessment of opportunities and roadblocks. Environ Sci Technol 54:3064–3081

Li W, Orozco R, Camargos N, Liu H (2017) Mechanisms on the impacts of alkalinity, pH, and chloride on persulfate-based groundwater remediation. Environ Sci Technol 51:3948–3959

Mancini SA, Lacrampe-Couloume G, Jonker H, van Breukelen BM, Groen J, Volkering F, Lollar BS (2002) Hydrogen isotopic enrichment: an indicator of biodegradation at a petroleum hydrocarbon contaminated field site. Environ Sci Technol 36:2464–2470

Mancini SA, Ulrich AC, Lacrampe-Couloume G, Sleep B, Edwards EA, Lollar BS (2003) Carbon and hydrogen isotopic fractionation during anaerobic biodegradation of benzene. Appl Environ Microbiol 69:191–198

Mancini SA, Hirschorn SK, Elsner M, Lacrampe-Couloume G, Sleep BE, Edwards EA, Lollar BS (2006) Effects of trace element concentration on enzyme controlled stable isotope fractionation during aerobic biodegradation of toluene. Environ Sci Technol 40:7675–7681

Mancini SA, Devine CE, Elsner M, Nandi ME, Ulrich AC, Edwards EA, Lollar BS (2008) Isotopic evidence suggests different initial reaction mechanisms for anaerobic benzene biodegradation. Environ Sci Technol 42:8290–8296

Mariotti A, Germon JC, Hubert P, Kaiser P, Letolle R, Tardieux A, Tardieux P (1981) Experimental determination of nitrogen kinetic isotope fractionation: some principles and illustration for the denitrification and nitrification processes. Plant Soil 62:413–430

Matzek LW, Carter KE (2016) Activated persulfate for organic chemical degradation: a review. Chemosphere 151:178–188

Morasch B, Richnow HH, Schink B, Vieth A, Meckenstock RU (2002) Carbon and hydrogen stable isotope fractionation during aerobic bacterial degradation of aromatic hydrocarbons. Appl Environ Microbiol 68:5191–5194

NAVFAC (2013) Best practices for injection and distribution of amendments. TR-NAVFAC-EXWC-EV-1303

Northrop DB (1981) The expression of isotope effects on enzyme catalyzed reactions. Annu Rev Biochem 50:103–131

O'Sullivan G, Kalin RM (2008) Investigation of the range of carbon and hydrogen isotopes within a global set of gasolines. Environ Forensics 9:166–176

Payne FC, Quinnan JA, Potter ST (2008) Displacement concepts. Remediation hydraulics. CRC Press, Boca Raton

Saeed W (2011) The effectiveness of persulfate in the oxidation of petroleum contaminants in saline environment at elevated groundwater temperature. Ph.D. thesis, University of Waterloo

Solano FM, Marchesi M, Thomson NR, Bouchard D, Aravena R (2018) Carbon and hydrogen isotope fractionation of benzene, toluene, and o-xylene during chemical oxidation by persulfate. Ground Water Monit Rem 38:62–72

Thullner M, Centler F, Richnow H-H, Fischer A (2012) Quantification of organic pollutant degradation in contaminated aquifers using compound specific stable isotope analysis—review of recent developments. Org Geochem 42:1440–1460

Thullner M, Fischer A, Richnow H-H, Wick LY (2013) Influence of mass transfer on stable isotope fractionation. Appl Microbiol Biotechnol 97:441–452

USACE (2014) Design: in situ thermal remediation. Environmental Quality EM 200-1-21

Van Breukelen B (2007a) Extending the Rayleigh equation to allow competing isotope fractionating pathways to improvequantification of biodegradation. Environ Sci Technol 41:4004–4010

Van Breukelen B (2007b) Quantifying the degradation and dilution contribution to natural attenuation of contaminants by means of an open system Rayleigh equation. Environ Sci Technol 41:4980–4985

Vogt C, Cyrus E, Herklotz I, Schlosser D, Bahr A, Herrmann S, Richnow H-H, Fischer A (2008) Evaluation of toluene degradation pathways by two-dimensional stable isotope fractionation. Environ Sci Technol 42:7793–7800

Vogt C, Dorer C, Musat F, Richnow H-H (2016) Multi-element isotope fractionation concepts to characterize the biodegradation of hydrocarbons—from enzymes to the environment. Curr Opin Biotechnol 41:90–98

Wanner P, Hunkeler D (2019) Isotope fractionation due to aqueous phase diffusion—what do diffusion models and experiments tell? A review. Chemosphere 219:1032–1043

Wijker RS, Adamczyk P, Bolotin J, Paneth P, Hofstetter TB (2013) Isotopic analysis of oxidative pollutant degradation pathways exhibiting large H isotope fractionation. Environ Sci Technol 47:13459–13468

Wilkes H, Boreham C, Harms G, Zengler K, Rabus R (2000) Anaerobic degradation and carbon isotopic fractionation of alkylbenzenes in crude oil by sulphate-reducing bacteria. Org Geochem 31:101–115

Zhang N, Geronimo I, Paneth P, Schindelka J, Schaefer T, Herrmann H, Vogt C, Richnow HH (2016) Analyzing sites of OH radical attack (ring vs. side chain) in oxidation of substituted benzenes via dual stable isotope analysis ($\delta(13)C$ and $\delta(2)H$). Sci Total Environ 542:484–494

Zwank L, Berg M, Elsner M, Schmidt TC, Schwarzenbach RP, Haderlein SB (2005) New evaluation scheme for two-dimensional isotope analysis to decipher biodegradation processes: application to groundwater contamination by MTBE. Environ Sci Technol 39:1018–1029

Chapter 12
LNAPL Transmissivity, Mobility and Recoverability—Utility and Complications

G. D. Beckett

Abstract Light non-aqueous phase liquid (LNAPL) mobility, transmissivity (T_n), and recoverability are controlled by a suite of factors, namely release characteristics, the environment in which it resides, its age, physical/chemical properties, and others. LNAPL plumes are dynamic, but typically trend toward field stability after a release stops. Unlike groundwater mobility, finite LNAPL releases represent an ever-diminishing mobile mass due to degradation, residualization, and other factors that will be subsequently examined. For these and other reasons, mobility aspects vary both spatially and temporally. T_n is a useful, but complex, related parameter. It is commonly used to determine the applicability of LNAPL hydraulic recovery. But since it is not constant, a cutoff must be carefully considered, along with its manner of determination. Further, even if a plume is theoretically "recoverable" on a T_n rate basis that has very little relationship to whether that action would be expected to have any net environmental benefit. LNAPL hydraulic recovery is inherently self-limiting and has little net benefit unless a plume is mobile (and contained) or a significant mass recovery percentile is possible, which is rare.

Keywords LNAPL Mass Recovery · LNAPL Mobility · LNAPL Transmissivity · Relative Permeability · Residualization

12.1 Introduction

For plumes generated by fuel and related light non-aqueous phase liquid (LNAPL) releases, arguably the most important element is the genesis of that footprint over time. In the early stages of a release when LNAPL is migrating, it represents a direct potential risk to receptors controlled by the rate and ultimate locations of that transport. In the early stages of a release, LNAPL transmissivity (T_n) and mobility

G. D. Beckett (✉)
AQUI-VER, Inc, 9033 Cheyenne Way, Park City, UT 84098, USA
e-mail: g.d.beckett@aquiver.com

J. García-Rincón et al. (eds.), *Advances in the Characterisation and Remediation of Sites Contaminated with Petroleum Hydrocarbons*, Environmental Contamination Remediation and Management, https://doi.org/10.1007/978-3-031-34447-3_12

will generally be larger than later in its lifespan, often by orders of magnitude. The LNAPL release and its mobility is also affected by a variety of other natural processes including biodegradation, physical weathering, partitioning (collectively termed natural source zone depletion), residualization, and others aspects discussed elsewhere in this text.

Transmissivity, as a parameter in groundwater flow mechanics, has been around for much of the history of our field. In the mid-1800s, Henry Darcy framed aspects of water flow through sand as related to the pressure gradients and the properties of the sand (Simmons 2008). In the current era, framing T_n is an important element in determining a similar thing, the potential rates of LNAPL flow or recovery through geologic media. But unlike the perfect sand/water conditions of Monsieur Darcy, aspects of T_n are vastly more complex, and it is, at best, a distant relative to the transmissivity of groundwater mechanics. Its roots are in the underpinnings of Darcy's Law, petroleum and agricultural applications in multiphase mechanics and other related technical fields of interest.

Although controlled by the same physical processes, the LNAPL transport behavior in the vadose zone differs substantially from that around the water table or saturated zone where water contents are significantly greater and gradient interactions differ. Once a fuel release has stabilized, it represents a typically depleting source mass (often called the "source term") for dissolved and/or vapor-phase impacts, as well as being a source for potential methane generation (note, vapor-phase transport is also a multiphase process). LNAPL releases follow the gradients created by those release and related characteristics, like volume, rate of input, subsurface, and LNAPL properties, and so on, as discussed elsewhere in this text.

If an LNAPL release reaches the water table, its behavior differs substantially from the ambient groundwater conditions since the plume genesis is unrelated to the boundary conditions controlling groundwater and is vastly more sensitive to pore-scale lithologic variability. Many intuitive field observations may be at odds with the controlling realities. For instance, large thicknesses of floating LNAPL in the water table region are sometimes thought to represent the worst-case areas of a plume. But what if that large thickness is due to capillary and other resistive forces that do not allow the product to dissipate (i.e., the product is locally immobile)? What if the mobility in some sandy zones is vastly greater even though the observed thickness is quite small? What if the direction of LNAPL migration has little or sometimes nothing in common with the groundwater gradients? Yes, LNAPL generally follows the hydraulic gradients, but specifically the LNAPL gradients, which may differ substantially from groundwater gradients. Many apparently logical assumptions can lead to erroneous conclusions when it comes to LNAPL and multiphase mechanics, particularly in the real-world (applied, not theoretical).

Regardless of T_n or recoverability values, hydraulic LNAPL recovery can never "clean up" an aquifer, it is simply not possible. This is because of aspects related to LNAPL residualization (defined as non-mobile retention in the pore space in this chapter) and other factors. Residual LNAPL generally represents a persistent mass in the subsurface, the degree of persistence depending on setting, mass in-place, fuel characteristics, and others. Further, there are multiphase, formation, and operations-dependent limitations to hydraulic recovery such that even if there were no residualization, the method would still fail to recover all the LNAPL in-place. At best, it can only reduce or eliminate the spread of a mobile LNAPL plume and perhaps reduce its longevity in the environment if a sufficient percentage of the total mass is recovered (which almost never occurs). We will see that while sometimes useful, hydraulic recovery is generally ineffective at many sites to reduce risk, substantially reduce care and monitoring, or otherwise provide a net environmental benefit to those actions. That is certainly not to say that cleanups are not important in many instances, but rather that pump and treat is quite limited in its capacity to effect such cleanup goals, as will be clearly demonstrated here. Hydraulic recovery of LNAPL, in most cases, is like hitting a nail with a feather. One can do so, but to what beneficial effect?

The discussion will first provide a theoretical framework and overview identifying key parameters and other aspects of T_n that define the parameter. The second part of this chapter discusses the field and laboratory-based derivations of T_n and some of the complications in this task; other texts provide a description of how to perform these actions, which will not be repeated here. The last major section of this chapter deals with observed real-world complications and how nuances in porous media structure and makeup sometimes conflict with commonly applied multiphase theory and those effects with regard to T_n. Finally, the implications of T_n and its application to mobility and recoverability will be covered.

We hope to show the practical challenges to LNAPL characterization, the implementation, and effectiveness of LNAPL recovery in remediation applications so that practitioners can answer the question: "Is hydraulic recovery LNAPL useful in this situation, or should other methods of remediation be considered?" The author's suggestion? Use T_n and these related processes to explain important elements to the environmental protection equation, but test the key underlying assumptions against multiple lines of evidence, particularly field-based observations and data. It is usually an enlightening process that will point out deficiencies in the LNAPL conceptual site model (LCSM).

12.2   Quantitative Definition of T_n

As with groundwater flow in fully saturated aquifers, the LNAPL transmissivity is a bulk term that provides an indication as to the relative ease with which the liquid (LNAPL or oil in this case) may pass through a unit thickness of material. The USGS defines aquifer transmissivity as: "*The rate at which water of the prevailing kinematic viscosity is transmitted through a unit width of an aquifer under a unit hydraulic gradient. It equals the hydraulic conductivity multiplied by the aquifer thickness*" (USGS 2004).

The concepts and application of aquifer transmissivity was originally developed primarily for the evaluations of well hydraulics in confined aquifers (Freeze and Cherry 1979). As should be immediately clear to the reader, the scale and application of transmissivity to LNAPL migration and recovery is vastly smaller and coupled with nonlinear behaviors (as follows below). This chapter will delve into just a few of those complications following the development of idealized applications of T_n. This chapter will not redevelop the underlying fundamentals of transmissivity, storativity, specific yield, and other analogous aquifer characteristics. However, it is important to understand that each of these aspects play a role in LNAPL mobility and recovery and should be considered when appropriate to the evaluation at hand. Additional discussions of these basics are provided in several of the cited references.

To understand LNAPL transmissivity and its relationship to mobility, it is useful to define the parameter by its associated quantitative factors. Analogous to groundwater transmissivity, the LNAPL transmissivity can be defined similarly as shown by Eq. 12.1 below (Huntley 2000). Note "oil" is classically used as a general term for immiscible LNAPL, as in the "oil" phase of a 3-phase system consisting of the air/water, oil/water, oil/air phase couplets (e.g., see Parker 1989, an excellent overview of multiphase transport).

$$T_n = \int_{z_w}^{z_n} k_{rn} k \frac{\rho_n g}{\mu} \mathrm{d}z \tag{12.1}$$

where k is the porous media intrinsic permeability, k_{m} is the relative permeability of the LNAPL, ρ_n and μ are the LNAPL density and kinematic viscosity, respectively, and g is gravitational acceleration. Integration is between the base of the flowing LNAPL interval and the uppermost distribution. In an ideal equilibrium setting, that would be equal to the oil/water and oil/air interfaces represented by LNAPL thickness in a monitoring well.

Of the parameters shown in Eq. 12.1, only the fluid factors may be relatively constant in the local formation (i.e., could be moved out of the integral), assuming there is a single LNAPL source in the mobile interval. Even there, properties like viscosity and interfacial tension may change with weathering. Intrinsic permeability varies as a function of lithology, which tends to be naturally heterogeneous. The relative permeability, as discussed elsewhere in this text, varies as a function of saturation, which in turn varies as a function of capillary pressure, wettability, and other factors. The integration of these factors is across the vertical interval in which LNAPL flow occurs and not where immobile residual is present. These facets generally apply to the other fluid phases (water and gas) as developed elsewhere in this text, but this chapter is specifically focused on the mechanics of the LNAPL (aka, oil phase).

The value of T_n is usually most sensitive to the intrinsic and relative permeability ranges of a particular setting. Intrinsic permeability may range from 10^{-8} to 10^5 Darcy in natural earth materials (Freeze and Cherry 1979), a span of 13 orders of magnitude, although end-member values are relatively rare. The more significant complication with respect to T_n is the relative permeability scalar that varies nonlinearly from 0 to 1 as a function of the effective LNAPL saturation. LNAPL saturation, in turn, varies nonlinearly as a function of soil and fluid capillary characteristics, as discussed elsewhere in this text. A simplified form of the LNAPL relative permeability scalar is the square of the LNAPL effective saturation, shown in Eq. 12.2

below (Charbeneau et al. 2000). Later in this discussion, we will take a look at both theoretical and applied complexities of these and other controlling factors:

$$k_{rn} \approx \bar{S}_n^2$$

where k_{rn} is the LNAPL relative permeability scalar, and $\bar{S}_n$ is the effective LNAPL phase saturation, defined in Chaps. 2 and 3.

Finally, while less sensitive (linear effects), the density and viscosity of the LNAPL also contributes to the observed T_n, sometimes in unexpected ways. The kinematic viscosity of petroleum and its products varies over many orders of magnitude if one include crude oils, from less than 1.0 to more than 1 million centipoise (API 2006). The relative density of petroleum products is less variable, generally ranging from about 0.7–0.99 for common fuels; of course, any non-aqueous phase liquid (NAPL) with a density greater than water is defined as a dense NAPL (DNAPL). While the concepts discussed here often apply to DNAPL, it is less often observed as a free-phase lens that meets the testing and analysis requirements discussed for T_n below. Further, the concepts apply in related ways to air and water phase movement in a three-phase environment, again as discussed elsewhere in this text. Three-phase flow cannot be described without the coupling of capillarity, relative permeability, saturation, and underlying soil and fluid properties. T_n is merely one rather narrow facet of the broader family of related multiphase considerations.

Now that T_n has been defined by each of its integral parameters, we can see that if natural earth materials vary in permeability over roughly 10 orders of magnitude, then T_n should vary by several more orders (in theory). Later in this chapter we will discuss field observations and observe how the T_n field results align with these theoretical considerations and where they may not. First, however, let us review the implications of the mathematics of T_n from a theoretical and heuristic standpoint.

12.3 Theoretical Implications of T_n Factors

A suite of theoretical interrelationships is developed in this section that inspect the most sensitive and important factors affecting the values of T_n. This will help the reader understand the wide variability that might be expected and some of the practical implications of T_n and its application to site-specific LNAPL issues. As noted, the author views this discussion as heuristic, as these theoretical conditions may exist under ideal laboratory conditions, but as will be seen subsequently, rarely under field conditions. Nonetheless, the underlying theory is well-grounded in multiphase studies and applications.

The method of many of these T_n considerations is to invoke the assumption of vertical hydrostatic equilibrium (VEQ; i.e., no vertical head gradients in oil or water and that fluid levels in a well represent the formation heads and pressures, as described in Chap. 2). The VEQ assumptions provide an analytic basis for the comparisons that will be shown here (see a full VEQ discussion in Lenhard and Parker 1990; Farr et al. 1990). As the reader may suspect, whether VEQ is present or not is yet another facet to consider. Where T_n and relative permeabilities are small, it is mathematically and volumetrically unlikely that VEQ is ever attained. Subsequent discussion of field complications will touch on field challenges to some of these simplifying boundary assumptions.

12.4 Effect of Soil Type

The type of porous media and its capillary and permeability characteristics are synergistically related to both the effective LNAPL hydraulic conductivity (K_n) and T_n. Because capillary characteristics control the LNAPL saturation (S_n) as a function of the pore throat diameter distribution and other factors, and because capillarity is nonlinear, these facets exhibit strong interplay in the resulting K_n and T_n values. It is instructive to look at both parameters in parallel.

For five soils, ranging from predominantly fine-grained to clean coarse-grained materials,[1] the calculated theoretical VEQ LNAPL saturation distributions for 1 m of gasoline-free product in an observation well are shown in Fig. 12.1, along with the volume per unit area of free-phase LNAPL annotated. The differences in these profiles as a function of fuel type will be shown below. Table 12.1 provides the van Genuchten (1980) parameters used to generate each profile, based on API's LNAPL Parameters Database ranges (Beckett and Joy 2003), except for the irreducible residual water (S_{wr}) content was taken from Schaap et al. (2001)[1]. Note, the capillary fringe extends above 1 m but is not engaged in idealized flow (not part of T_n).

As noted above, the relative permeability with respect to LNAPL may be approximated as the square of its effective saturation (Charbeneau 2000). This particular function is selected for ease of explanation, and the reader should be aware that there are many relative permeability functions that might apply from agricultural and petroleum applications (e.g., Burdine 1953; Mualem 1976; Delshad and Pope 1989). As will be discussed in the real-world observations section, each of these functions is an empirical approximation and depends on the lithologic characteristics of the site-specific porous media. Due to the expense of relative permeability testing, these are not commonly performed for environmental applications. Also, to be discussed, there are likely laboratory and scaling issues that may need to be considered. Figure 12.2a and b show the effective permeability $(k_n = k_i * k_m)$ toward LNAPL on arithmetic and semi-log scales, respectively. The reason for plotting on

---

[1] The soil and fluid capillary parameters are derived primarily from the API LNAPL Parameters Database (API 2006). However, in the construction of that database, it was noted that residual saturation values were suspect as compared to other peer-reviewed data; that laboratory testing issue has yet to be resolved. Therefore, water residual saturation values were taken from the Rosetta model soil classifications (Schaap et al. 2001). Note the oil/air/water capillary fringe rises above 1 m.

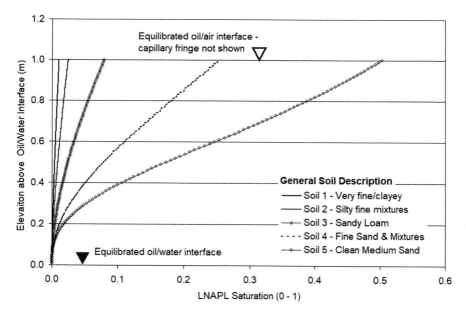

Fig. 12.1 VEQ gasoline saturation profiles for various soil types for 1 m free product

Table 12.1 Input parameters to VEQ estimates

α (m^{-1})	n	Percentile (%)	S_{wr}	Specific vol (L/m^3)	General soil type
0.33	1.40	10	0.30	2.74	Very fine/clayey
0.58	1.55	25	0.23	5.48	Silty fine mixtures
1.23	1.89	50	0.15	16.4	Sandy loam
2.49	2.21	75	0.12	42.5	Fine sand and mixtures
3.82	2.72	90	0.10	86.3	Clean medium to coarse sand

Source of input capillary parameter ranges: API LNAPL Parameters Database (Beckett and Joy 2003) and Schaap et al. 2001. The percentile applies to the database range of the van Genuchten parameters. α and n are van Genuchten (1980) capillary model parameters, as is S_{wr}—the irreducible residual water saturation

semi-log scale is visually clear, one cannot "see" the small values of effective permeability on an arithmetic scale when contrasting soil types. Obviously, the contrasts in soil capillary properties and the resultant LNAPL saturations are amplified by the relative permeability phenomenon.

The effective permeability tensor is highlighted because it provides a direct relationship to effective conductivity and its vertical integral, T_n. Several attributes of the effective permeability plots become immediately clear. First, at small S_n values (or correspondingly small capillary pressures), the k_n is very small for all soil/porous media types. It is of course smallest in fine-grained materials, but for any soil/porous

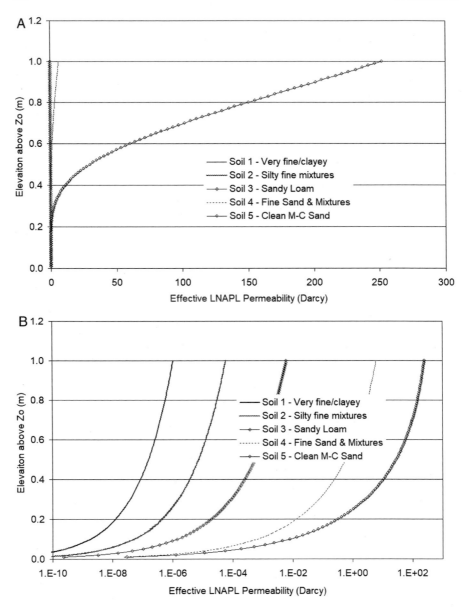

Fig. 12.2 a and **b**: Calculated effective permeability for a range of soils on arithmetic and semi-log scales. Intrinsic permeability ranges taken from Freeze and Cherry (1979)

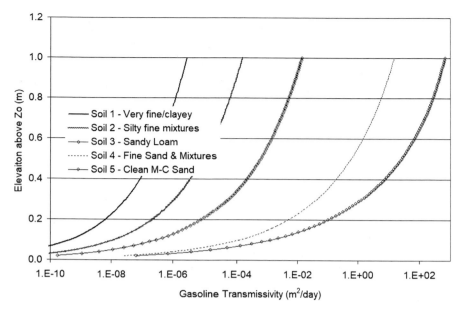

Fig. 12.3 Calculated gasoline transmissivity as a function of elevation and saturation above the oil/water interface

media type becomes negligible at some point on the curve. The effect is compounding with regard to T_n, because T_n is integrated over a mobile-zone thickness. As that LNAPL thickness and associated saturation decreases, so does T_n as a linear function of that thickness and a nonlinear function of saturation and relative permeability, as shown in Fig. 12.3, plotted again on semi-log scale.

As expected, the T_n values span many orders of magnitude. At low capillary pressures (or well thickness) and saturations, values become quite small regardless of soil type. Using the coarse soil Type 5 as an example, there would be about 525,000 times more flow potential at 1 m versus 0.1 m for the same porous media properties; "potential" because neither case may actually be mobile due to other impeding factors (see related discussions in this text). At the other end of the spectrum, because of differing capillary properties and saturation distributions, Soil Type 1 has a contrast of about 6000 times more flow for 1 m thickness versus 0.1 m. This exponential variability and other observations from Fig. 12.3 indicate that in real-world settings, T_n is a highly variable and complex parameter. It also implies that VEQ conditions may not always occur, as the time to fill a wellbore at low transmissivities, regardless of soil type, will be quite long and likely be overprinted by hydrologic variability. Unless the hydrogeologic system is relatively static (and most are not), VEQ assumptions and estimates themselves should be used with caution. More discussion on this and other complications will be discussed in the real-world observations section below.

Lastly, it is useful to consider how heterogeneous systems might behave (at least theoretically). If a system is under VEQ conditions, then the capillary pressure profile

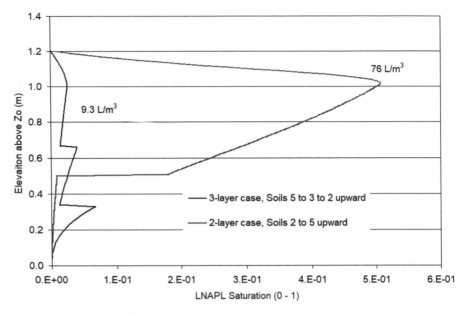

Fig. 12.4 VEQ LNAPL saturation estimates for two layered soil systems

will be linearly uniform across any sequence of soil/porous media types. For explanation purposes, we will look at two heterogeneous profiles. The first will be a three-layer system of coarse, mid-range, and fine-grained soils from the bottom-up (i.e., Soil Types 5, 3, and 2 above). The second will be a two-layer system with the finest-grade material comprising the bottom 0.5 m, and the coarse soil the upper 0.5 m. Each of these examples is again for 1 m of free-phase LNAPL under VEQ conditions.

The vertical position of soils upon one another has a significant impact on the estimated S_n distribution and the associated specific volume and T_n values. Because capillary pressures are smaller at the base of the LNAPL column, the coarse-grained soil contains much less LNAPL in Case 1 than in Case 2 even though the LNAPL thickness and soil characteristics are identical (Fig. 12.4). The resultant effect on the T_n values is shown in Fig. 12.5, which amplifies the noted saturation contrasts. In practical terms, the LNAPL in Case 2 is effectively perched on the lower permeability materials below that, even under ideal VEQ conditions, do not meaningfully contribute to the integrated T_n value. It is also interesting that, because of low saturations and intrinsic permeability, the upper two soil types for Case 1 add almost nothing to the T_n profile (too small to visualize). Similarly, one can see almost and 8-order increase in T_n above the 0.5 m mark for Case 2, supporting the observation above that this LNAPL condition is essentially perched on the finer-grained material.

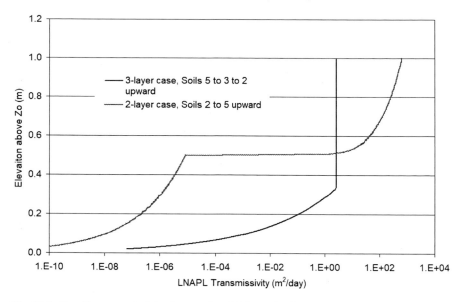

Fig. 12.5 Gasoline transmissivity for case 1 and 2 layered soil conditions under VEQ conditions

12.5   Effect of LNAPL Properties

As a final set of theoretical considerations, we will look at the role that LNAPL fluid characteristics play in T_n and its contrasts for different petroleum products and properties. From a T_n and mobility/recoverability standpoint, product viscosity is the most variable factor of petroleum products (linear effect). However, because the product density drives the capillary pressure between oil/water, that factor has a nonlinear influence in the expected T_n values. Another parameter, the interfacial fluid tension (IFT) between water and oil, often varies significantly in the field from ideal laboratory measurements; biodegradation and other aspects are likely causes, but more study is needed. Since the ratios of IFT are used to scale capillary properties for each phase couplet (oil/water, water/air, and oil/air), this too has a nonlinear impact to estimation of LNAPL saturation and T_n, as is explored here. Additional discussion of capillary processes and their basis and applications are provided elsewhere in this book.

Figure 12.6 provides contrasts in T_n values for the same 1 m of assumed LNAPL thickness, for five petroleum types, and using Soil #4. The fluid parameters for these products are provided in Table 12.2 (from API 2006). For the interested reader, Environment Canada has an oil properties database that has an extensive set of LNAPL properties (https://www.etc-cte.ec.gc.ca/databases/oilproperties/Def ault.aspx). As seen in this figure, the combination of density and viscosity differences combine to create order-of-magnitude differences in the expected T_n values as a function of product type. Per Eq. 12.1 above, the larger the viscosity, the lower the K_n and T_n values (inversely proportional on a linear basis). The density contrasts, although relatively small, roughly 0.7–0.9, affects the calculated capillary pressure between water and oil. That lower capillary pressure then nonlinearly decreases the calculated saturation with increases in density.

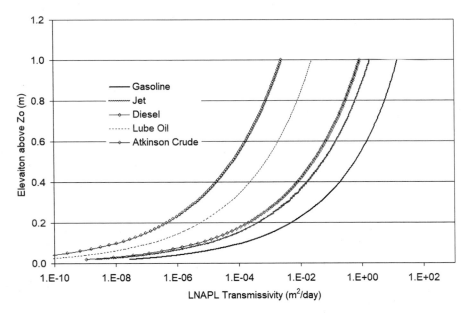

Fig. 12.6 LNAPL transmissivity under VEQ conditions for various petroleum products

Table 12.2 LNAPL physical properties

LNAPL type	Density (g/cc)	Viscosity (Centipoise)
Gasoline	0.73	0.62
Jet	0.84	1.0
Diesel	0.83	2.7
Lube oil	0.83	102
Atkinson crude	0.91	65

Note: For simplicity, only density and viscosity are shown and are assumed from general literature values. Other parameters, such as IFT, will be discussed subsequently and for this comparison were all assumed to be equal here (50 dynes/cm for oil/water and 25 dynes/cm for air/oil)

12.6   Transience of LNAPL Transmissivity

From a theoretical standpoint, T_n is transient, even absent complicating field aspects discussed subsequently and elsewhere in this book. As LNAPL releases develop in the subsurface, there is a sequential set of temporal and spatial changes that generally occur. In early time, the release will be located near its point of origin, as shown below in a multiphase model-derived schematic of LNAPL plume development for a finite release (Fig. 12.7, upper left panel). The associated LNAPL saturations and observed thicknesses are large early in time, as would be the resulting local area T_n. At this early time in the release progression, there is no LNAPL arrival yet at distal locations, where the T_n is temporarily zero. When there is a potential for significant recovery and migration control benefits, it is almost always in these early stages of a release.

As the LNAPL migrates, mass redistributes more widely, occupying a larger volume of porous materials. Along with that there are new LNAPL arrivals at further distances away from the release along with a corresponding decrease in mass and saturations in the release area. That causes a transiently increasing T_n with distance for some period, and decreasing T_n in the release zone. The transient sequencing of this scenario is shown in Fig. 12.8 for some example post-release times. As should be noted, different release and subsurface characteristics will generate analogous results, but with different migration characteristics and transient T_n conditions. Beyond the direct mass transfer effects discussed here, there are other field factors that transiently affect the T_n (natural mass losses, residualization, smear zone progression,

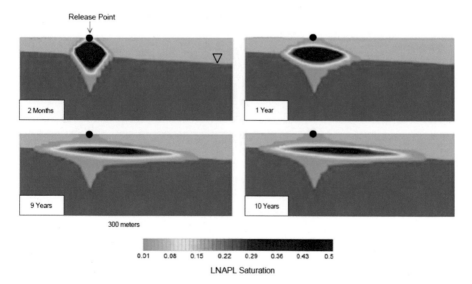

Fig. 12.7 Modeled LNAPL saturation distribution in a homogeneous scenario at example time stamps. Note that the plume is initially concentrated near its release point, migrating over time with saturations changing accordingly. The example does not consider processes, such as water table fluctuations

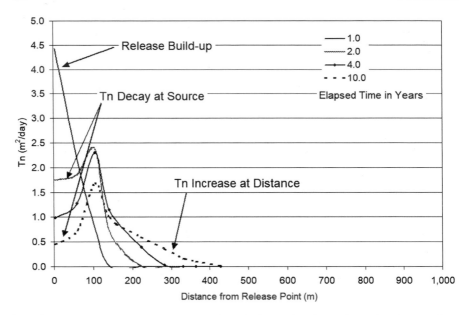

Fig. 12.8 Transient changes in T_n as a function of plume migration and genesis. Note that these types of changes will occur virtually everywhere, but vary significantly as a function of release setting and the characteristics of the fuels and porous materials involved

and others). Some of those are discussed further below in the field observations discussion and in other sections of this text. Remember, any process that influences LNAPL saturation, hydraulic conditions or other controlling properties discussed herein will also have a direct effect on the T_n and mobility conditions. For periodic releases at some active facilities, these facets become quite complex, but the principles discussed still apply.

12.7 Summary of Theoretical Observations

In summary, it is clear that T_n has linear and exponential sensitivities, both with respect to porous media and LNAPL properties. The K_n is similarly variable and that is important because it is a key component in discrete mobility potential within the more permeable and saturated zones in the subsurface, which often coincide due to related porous properties of permeability and capillarity. Darcy's Law can be used as a screening estimate by combining K_n (which is not a simple scaling of T_n), LNAPL gradients, and the effective porosity to result in an estimated average linear pore velocity. While heuristically useful, one must use such methods while remaining cognizant of other multiphase processes that often limit LNAPL plume migration, particularly for older plumes. In other words, plumes can reach field stability even when there is a conductivity and apparent gradient.

The above theoretical background leads to the following conclusions and implications, absent other complicating factors like water table fluctuation, residualization, natural source zone depletion (NSZD), and others. Additional (and sometimes non-intuitive) field- and data-based observations follow in the subsection below.

- Porous media that has low permeability and associated low LNAPL saturations (due to capillary properties) will not contribute meaningfully to T_n or K_n values and associated mobility or recoverability, particularly where higher permeability materials also reside in the LNAPL continuous-phase interval.
- Soil and oil types combine to produce order-of-magnitude variabilities in T_n/K_n values, even where initial observations of LNAPL thickness in monitoring wells are similar.
- Because of the nonlinear sensitivity to S_n, the T_n values will necessarily decrease over time with recovery actions, residualization, migration, and other phenomena (except under ongoing releases and/or transient hydraulic effects).
- While there is no thickness exaggeration implied by LNAPL observed in wells, only a portion of that total LNAPL saturation column will contribute meaningfully to T_n and mobility because lower saturation zones will have very low effective permeabilities. Therefore, the implications regarding mobility and recoverability are dependent primarily on the zones of high effective K_n.
- Because T_n is proportional to the time it will take to fill a given wellbore volume, caution needs to be applied when invoking VEQ assumptions. At small saturations and/or in fine-grained materials, VEQ likely does not occur under generalized field conditions. Even in high permeability materials, if there is significant water table variability, equilibrium may never occur.
- At some combination of saturation, permeability, and fluid properties, the effective mobility of LNAPL becomes *de minimis*. For instance, a value of practical impermeability in some clay liners is 1×10^{-6} cm/s, or a transmissivity across 1 m of 1×10^{-4} cm^2/s.
- Similar to the point above, at some lower limit of T_n, there is no practical way to field test that condition through general pumping or well-scale hydraulics. If a T_n estimate is needed in such conditions, it would be derived from its other fundamental parameters (which can be challenging, as discussed below). But, at such low T_n values both recoverability and likely mobility and will be negligible.
- As observed in the discussion of transience, the time where hydraulic recovery has the best opportunity to contain plume migration and recover the maximum mass possible is immediately following the release. The passing of time significantly diminishes the practicality of LNAPL recovery under most conditions (except active/ongoing releases). Many historic LNAPL plumes have been shown through field study and plume persistence to be practicably unrecoverable, regardless of the T_n values measured in the field. Other remediation methods may work, but hydraulic recovery will not, at least to any net environmental benefit.

12.8   Estimation of LNAPL Transmissivity

T_n may be estimated using analogous principles to aquifer testing, either as single-well tests or under pumping conditions. T_n may also be estimated through bench or laboratory petrophysical testing by determining the parameters of its integration defined in Eq. 12.1. While this is straightforward in concept, the complications discussed above, coupled with field observations in the next section, suggest that this determination is more complex than sometimes recognized. This portion of the chapter will focus on these complications rather than the nuts and bolts of performing T_n quantification testing. Other resources such as the API LNAPL Transmissivity Workbook (Charbeneau et al 2012), the ASTM Guide for Estimation of LNAPL Transmissivity (ASTM 2021), and Huntley (2000) all provide excellent descriptions of LNAPL T_n testing and parametric derivation.

One of the more common testing techniques for determining T_n is the LNAPL baildown test (aka, LNAPL slug test; ASTM 2021). It is relatively simple to execute. One quickly removes free product from a correctly screened observation well and then monitors its rate of return. The faster the LNAPL recovers, the greater the T_n (in general). The resources above detail the procedures and field considerations. Our discussion here is about dissecting the physical underpinning so that the complexities and limitations can be considered by the practitioner. As background, the author and the late Dr. David Huntley (Professor Emeritus, SDSU) have conducted many of these field tests since about the mid-1980s and have the benefit of observing both expected and unexpected results. The intent here is to provide some key observations and potential limitations to these methods of T_n estimation.

Foremost in our consideration is the fact that T_n is not a constant of the aquifer (Beckett and Huntley 2015), as it is for single-phase flow of groundwater across a large aquifer saturated thickness. For a finite LNAPL release, the transmissivity is expected to generally decrease over time due to a number of factors. One is progressive residualization and entrapment of the mobile LNAPL due to groundwater level variability, hysteresis, and occlusion. LNAPL plumes are also continuously losing mass and changing composition due to both partitioning and degradation mechanisms (e.g., Huntley and Beckett 2002; Garg et al. 2017). As the reader now recognizes, mass losses generally imply a reduction in LNAPL saturation and that causes associated declines in relative permeability, K_n, and the associated T_n value (see Eq. 12.1 above).

Further to T_n not being a constant are the conditions of testing. Baildown or pumping tests done under one set of free product and water level conditions will produce a different result than under others. One of those key factors is occlusion of LNAPL by rising water levels in unconfined settings. Figure 12.9 provides a general schematic of how that works. In Frame 1, T_n is at its greatest value before any residualization or occlusion. In Frame 2, T_n will be smaller based on the new residual mass held in the formation above. In Frame 3, T_n will be quite small because of occluded 2-phase mass below no longer in hydraulic communication with the well.

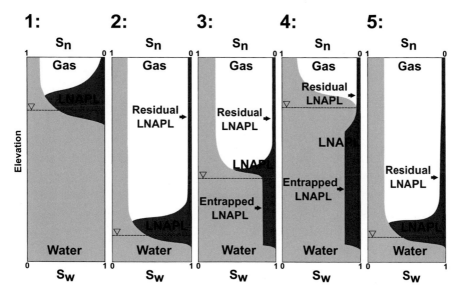

Fig. 12.9 Schematic of LNAPL residualization and occlusion with changing water table conditions. Both residualized and occluded LNAPL are described as "residual LNAPL." Figure courtesy of Richard Jackson

In Frame 4, the T_n is zero because there is no flowing free-phase product, and, in Frame 5, there is an approximate return to conditions in Frame 2.

To emphasize this important consideration, a baildown test for T_n in Frame 1 may imply an unacceptable mobility risk, whereas testing in Frame 4 is not necessary because no free-phase LNAPL means no mobility or recoverability ($T_n = 0$). But that would be inaccurate and highly misleading if the water table falls again as in Frame 5. Hydrologic conditions and the vertical distribution of LNAPL must be considered as part of the T_n characterization, along with plume mobility and stability aspects. The history of both release and hydrologic conditions are also important factors. So, unless one is in a demonstrably stable hydrologic environment, which is rare, one value of T_n will not confidently define conditions.

Figure 12.9 nicely captures the overall mechanics of unconfined water table variability on free product observations within a well. Under actual field conditions, these observations become even more complex, as expected. Figure 12.10 shows a long-period of fluid level gauging in an unconfined monitoring well where the estimated T_n is shown for a few different corrected water level elevations. When the LNAPL is fully submerged, the $T_n \sim 0$, and otherwise varies over several orders of magnitude depending on the observed free-phase thickness. While the LNAPL mobility/recoverability is negligible for about six years, from 1992 to 1998, it then returns to a temporary high mobility state when the water table falls to a low-level in early 2000.

The lesson is clear and often repeated. T_n is dependent on the overall timing and conditions of the aquifer acting on the LNAPL release, as shown in this example. The LNAPL plume mobility state may be negligible for many years, only to increase

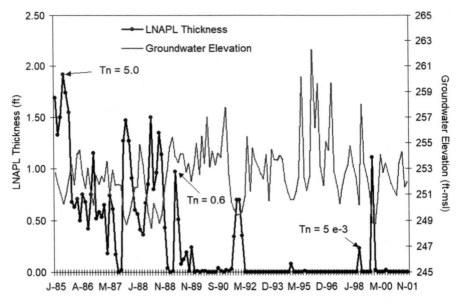

Fig. 12.10 Hydrograph of fluid levels in an unconfined monitoring well showing both the inverse relationship between LNAPL thickness and the corrected groundwater elevation. The estimated LNAPL transmissivity associated with different thicknesses is also shown, based on laboratory and field data for this location

dramatically on a change in conditions, particularly falling water tables. Clearly, a single set of T_n measurements at one point in time will not likely reflect the full range of potential mobility and recoverability conditions. If T_n testing is performed at periods of high entrapment, those will indicate a misleadingly small mobility and recoverability value. This is also why the quality of the LNAPL conceptual site model (LCSM, discussed elsewhere) is so important. If in-situ LNAPL conditions are not adequately understood, there is no context to evaluate whether and when T_n values will be at their greatest, commonly corresponding to the greatest mobility risk. Simply, T_n is transient and so is any risk associated with potential LNAPL plume mobility, even at late stages. One needs to look for transient events, not field steady-state conditions, if we wish to properly understand the mobility and risk context.

Other conditions of T_n testing include a number of variables and some fairly restrictive assumptions in the analytic methods of determining those values. Those include the linear and nonlinear parametric aspects discussed above, as well as practical aspects that may include those listed below (and others). In the final subsection of the chapter, we will review how some of those conditions affect interpretations regarding T_n estimation, LNAPL mobility, recoverability, and other aspects.

- Assumed well and LNAPL continuity with the formation

  - Potential for PVC swelling

- The effects of well filter packs and other construction materials
- Drilling rinds that may limit hydraulic connectivity
- Screen interval relative to the location of the mobile S_n profile in the formation

- Potentially confined LNAPL conditions
- Local area versus plume-wide characteristics
- Variations in permeability and saturation conditions that are not identified in well logs
- Variations in petrophysical and fluid properties at scales too small to visually identify
- Potential for multiple releases creating differing LNAPL conditions and physical properties
- Differential weathering of the LNAPL body and associated variations in liquid properties
- Hydrologic overprinting and development of a "smear zone" that contains mobile and immobile LNAPL and a resulting variable mobility state
- Release of occluded "immobile" LNAPL on changes in phase conditions, such as a falling water table.

Lastly, all aquifer testing analytic approaches have a number of constraining assumptions. For instance, the Bouwer–Rice slug test method is commonly used for T_n derivation from baildown tests (see references above). This slug test analytic method is based on the steady-state confined radial flow model, and the boundary requirements are: (a) aquifer/LNAPL zone has infinite areal extent; (b) aquifer and LNAPL zone is homogeneous and of uniform thickness; (c) aquifer/LNAPL potentiometric surface is initially horizontal; (d) the volume of LNAPL recovered to initiate the test is instantaneous; and (e) the return flow is steady state. Clearly, (a), (b), and (e) are never fully met, and (c) and (d) are often not met during LNAPL slug testing, introducing non-ideal conditions and error into most T_n derivations.

Blended numerical analytic approaches to T_n derivation have been developed (e.g., Zhu et al. 1993). These couple the capillary saturation analytic models (discussed elsewhere in this text) with a nonlinear parameter estimator to result in estimates of T_n and associated multiphase soil properties. One can also use multiphase models (e.g., MAGNAS3 1994; T2VOC 1995) to analyze test results by setting background saturation and physical property conditions that are then transiently overprinted by the initiation of the bail or pumping tests. Like an aquifer pump test, one then varies the controlling parameters to result in a model match to the observed LNAPL recovery. The author has found good utility in such numerical investigations, but those typically require a substantial work and computing effort to derive informative results. But that effort often results in better understanding via the struggle to adequately explain the data response. In any case, uncertainties and variability of in-situ field LNAPL conditions limit the resolution of even these more comprehensive solutions. The reader recognizes clearly by now that the number of controlling parameters is such that evaluations results are often non-unique (i.e., more than one parameter suite may explain the observed test results).

An alternate approach to estimating K_n and T_n is to measure the individual parameters comprising these porous media properties. Saturation testing is a direct physical measurement, as are fluid properties. Capillary testing can provide the relationship between pressure and saturation, usually on a 2-phase system that can then be scaled to the other fluid couplets through surface and interfacial tension ratios (see the capillary theory discussion elsewhere in this text). Relative permeability functions can be measured or assumed. Collectively then, one can estimate K_n directly for a given soil type, LNAPL type and saturation, with T_n being the integral of that over the usually variable LNAPL saturation profile in question. While that sounds simple enough, laboratory and field experience has shown there to be many issues with the scaling and application of petrophysical parameters to field conditions. Some of these issues will be discussed below in the field observations subsection of this chapter.

Taken collectively, it should be clear that T_n or K_n values and their derivation are probably best viewed as approximations of a highly complex set of variables and often nonlinear conditions. As will be shown in the next section, it is recommended that one inspect multiple lines of data when constructing an understanding of LNAPL plume conditions and the potential risks posed. Like most hydrogeologic endeavors, one will rarely find that inserting parameters into equations will result in representative estimates of T_n or other important attributes of LNAPL plumes. That will only result in decimal points on a derivation that is itself often invalid.

12.9   Field and Laboratory Testing Observations

This subchapter will provide some of the field and laboratory testing observations that the author and others have developed over the last few decades. As a researcher and primary author of the API LNAPL Parameters Database (Beckett and Joy 2003), the author has benefited by observing thousands of LNAPL plumes and overseen the LNAPL characterization and testing at many of those. The mix of the provided observations is meant to develop a deeper understanding of the real-world complexities that can impact multiphase mechanics and its application to important aquifer and environmental protection problems. Each aspect of the observations provided is important, and it is likely a site-specific matter as to when one area of observations become more important than others. It is also the case that while the observations themselves are often clear in their implication, the underlying causes may not.

12.10   Intrinsic Permeability and Fluid Type

It is a common applied assumption, as implied in the equations and discussions above, that intrinsic permeability is an invariant property of the porous media (as indicated by the term "intrinsic"). However, that assumption often conflicts with measurements. It has been recognized in the petroleum production industry that the oil permeability of different reservoirs can vary as a function of their mineralogical makeup. One supposes that should be similar in environmental and aquifer restoration applications since we encounter analogous geologic materials, albeit under differing conditions.

It is rare to find agreement between air-based and water-based intrinsic permeability measurements. Figure 12.11 shows some typical permeability laboratory data selected at random from the author's archives for a wide variety of soil types and permeability ranges. In all but one sample (of 29), the water-based permeability is less than the air-derived value, often by an order of magnitude or more. It is also clear that, in general, lower permeability/finer-grained materials exhibit larger contrasts in the measured permeability values between water and air. Indeed, it is in finer-grained materials, particularly those containing clay, where the polarity of water can interact with the porous media and produce the observed result. This working hypothesis explaining the observations would certainly benefit from additional scientific inquiry (as is true of much else in this text). But we see these effects in the field based on multiphase flow characteristics that are explainable only under non-ideal conditions.

While one can easily define an effective air transmissivity (e.g., Beckett and Huntley 1994), the purpose here are the implications to T_n and LNAPL mobility. From the observations above, one might suspect a similar outcome when one

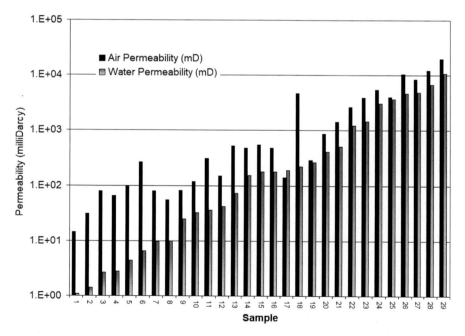

Fig. 12.11 Laboratory permeability measurements showing the common contrast between water- and air-derived intrinsic permeability values

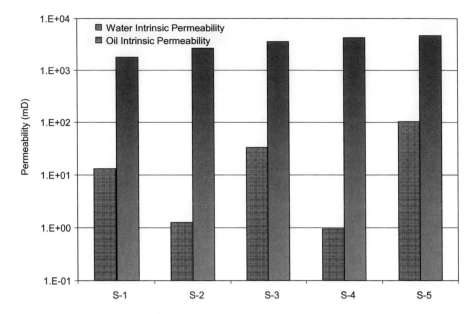

Fig. 12.12 Example contrast of intrinsic permeability measured on the same samples separately with water and kerosene

measures permeability to water relative to a petroleum product. Petrophysical laboratories often use kerosene for such tests. Figure 12.12 provides an example of permeability tests conducted on exactly the same core samples and under precisely the same test conditions. This site consists of mixed alluvial sediments with a significant clay/silt content in some horizons. As observed, the permeability contrast is quite large, up to several orders of magnitude for this example. The author has seen this outcome almost universally in samples tested in this way and suggests the reader try the same experiment. The primary implication is obvious; LNAPL is commonly much more mobile, particularly in fine-grained materials, than would otherwise be expected by applying standard ranges of intrinsic permeability. From a practical standpoint, it also means that field tests will tend to produce a smaller range of T_n variability than theory alone might suggest (more to follow on this issue). Coupled with this observation are difficulties in scaling laboratory-based measurements to the field (again, to be discussed).

12.11 Interfacial Tensions

Elsewhere in this text the reader will have been introduced to interfacial and surface tensions and how those are used to scale the capillary couplets between the phases of interest (air/water, water/oil, and oil/air). The IFT factors are often buried in other calculations or models and, in the author's experience, are viewed as relatively unimportant or otherwise being known through literature ranges. However, field conditions can substantially alter the IFT values, particularly for the oil/water couplet. The API LNAPL Parameters Database (API 2006) provides oil/water IFT values for 28 fluid pairs. Excepting one likely outlier at 65 dynes/cm, all IFT values are significantly smaller than a common literature value of about 50 dynes/cm; the measured range is from 8.0 to 39.4 dynes/cm. Because the IFT scaling parameter is used in the capillary functions, the ranges of IFT create another nonlinear effect. Figures 12.13 and 12.14 show this for four IFT values (low, average, and high values from the API LNAPL Database [API 2006]) relative to the commonly assumed value of 50 dynes/cm (S_n and T_n shown).

First, as the IFT value decreases, the LNAPL saturation and mass increase. As we know from above, that means that the relative permeability then increases exponentially, resulting in much higher LNAPL mobility than would be estimated using an assumed literature value of 50 dynes/cm. It is well known that petroleum hydrocarbons degrade in the environment, often producing polar compounds. While more research is needed, the author has observed generally smaller IFT values in plumes that have been weathered and exhibit a substantial footprint of polar compounds. One would generally expect that as the LNAPL interfaces become more polar through biodegradation, the IFT would decrease relative to non-polar LNAPL (i.e., fresh). Again, the effects of IFT are obvious and substantial, but more study is needed on

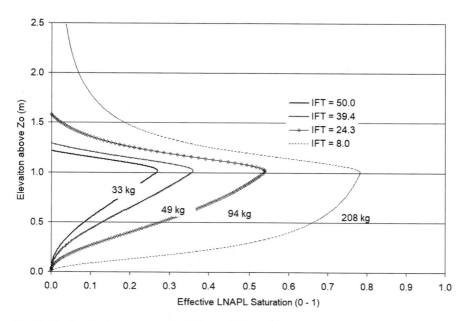

Fig. 12.13 Gasoline saturation under different IFT conditions for soil type 4. Note the zone above the free-oil surface (1 m) is shown because of the substantial mass present for low IFT values. The integrated LNAPL mass under the curves is given in kilograms

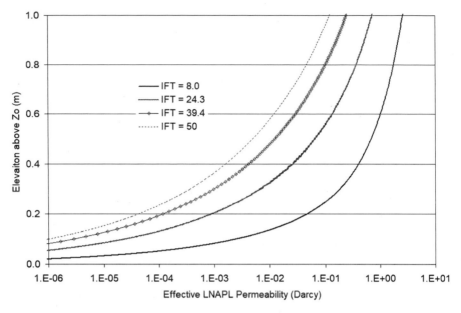

Fig. 12.14 Gasoline transmissivity as a function of elevation above the oil/water interface for varying IFT values, gasoline saturation under different IFT conditions for soil type 4

how it changes, rates, where (at phase interfaces or throughout the LNAPL body), its transient behavior and other aspects. As noted previously, the nonlinear aspects of multiphase theory make it important to understand the implications of parameter variability in the field and how that impacts mobility, recovery, and risk conditions.

12.12   Real-World Heterogeneity

Most people working in the field of groundwater contamination recognize the importance and ubiquitous presence of heterogeneity. This subsection will share data and observations from field sites that demonstrate, as suggested by theoretical considerations above, a complexity and scale that has direct implications to T_n, mobility, recoverability, and risk. In brief, the scale of complexity is such that most qualitative and quantitative evaluation approaches will not be reflective of this scale, but rather represent forms of averaging. As with other aspects discussed throughout this text, significant additional research is necessary if we are to better understand these processes to ultimately result in better groundwater protection and restoration for our public stakeholders.

Although this subsection is titled "real-world heterogeneity", it is useful to recall some interesting bench-scale observations. John L. Wilson and his colleagues at New Mexico Tech did some pioneering bench-scale modeling work in the late 1980s and early 1990s, including scanning electron microscopic (SEM) logging of some of those experiments (Wilson et al. 1990). One of the key observations of this wide-ranging investigation is that even in relatively homogeneous pore fields, heterogeneous LNAPL behavior is the norm. Only the pore pathways with continuous phase saturation and connectivity will contribute to mobility and the associated T_n values. For anyone wishing to "see" NAPL behavior, this study suite is exceptional and highly recommended.

12.13 T_n Field Observations

In this section, several sets of T_n tests will help to demonstrate the following observations:

- LNAPL thickness and T_n do not often correlate in the field, often due to heterogeneity;
- T_n from pumping test can be much larger than from baildown, and if the point of T_n is recoverability, then it likely needs to be measured under field pumping conditions;
- The quandary of stability; we often have clearly stable LNAPL plumes, yet T_n and gradients are both present that indicate mobility if one applies a simple Darcy approach. Balancing factors should be considered, such as pore-entry limitations, NSZD, and others when assessing stability.

In a uniform and relatively homogeneous system, T_n would be strongly correlated to the initial LNAPL thickness observed in a monitoring well. Figure 12.15 provides the results of T_n testing at a large release site as a function of the observed initial LNAPL thickness in each test well. This example site is in a dune sand system and is as homogeneous as one will ever find in the real world. For example, extensive permeability testing of these sands determined a range from about 2–13 Darcy, but within a factor of two for the 25th and 75th percentile values. Visually, one cannot distinguish any difference in permeability, and all boring logs simply identify a clean fine- to medium-grained sand. Further, the LNAPL released is also relatively uniform. In other words, there was just a single LNAPL type released. Yet review of the test results shows, contrary to theory, that there is no strong relationship between initial LNAPL thickness and the resulting T_n value. In fact, some of the smallest T_n values are associated with the largest thicknesses (no real surprise to our inner geologist).

What may explain the non-intuitive observations above? The multiphase mechanics discussed previously demonstrated that a wide variety of factors have nonlinear effects on the T_n results. For instance, although all the LNAPL released is of a very similar character, it was released over time allowing an opportunity for environmental changes in certain areas. For instance, the IFT values are variable and range from about 10–20 dynes/cm. The discussion above demonstrated the nonlinear effects of IFT on expected LNAPL saturation and T_n values. There is also the potential effects of fuel weathering and compositional changes that, in turn, affect the

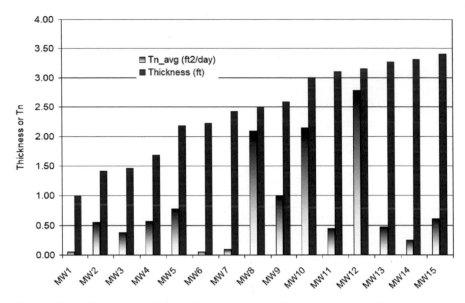

Fig. 12.15 LNAPL transmissivity results at a site with uniform dune sand aquifer and generally uniform LNAPL characteristics. Note, the wells names are pseudonyms to protect the privacy of the site owner

viscosity of the LNAPL in the subsurface. This site has a strong polar hydrocarbon footprint. Fluid measurements at this site suggest viscosity ranges from about 5–30 cP, or equivalently, influencing the T_n values by a factor up to six-fold (recall viscosity is a linear factor). The LNAPL density range, however, is quite uniform.

While there are many possible reasons for the observed T_n contrasts, one can easily observe that ideal theoretical outcomes are not always present in the field, even at the most uniform of sites. If direct theory application does not hold at this example "homogeneous" site, it is hard to imagine it being representative for almost any site.

Contrasting the above with a more typical site is also informative. This example site is in a large alluvial basin, and the soil types at the water table/LNAPL interval range from silty and clayey sands to coarser clean sand intervals. As shown in Fig. 12.16, more typical heterogeneous sites have wide range of T_n values as a function of the initial thickness (note the log scale because of that range). One can observe that the greatest T_n values correspond to two of the smallest initial thicknesses. Conversely, the smallest T_n values correspond to an observed well thicknesses of more than 3 m (or ~ 10-ft) of LNAPL. This is geology working in a multiphase way. From a mobility, risk, and recovery perspective, one may not have suspected that some of the smallest LNAPL thicknesses actually represent the largest mobility and risk potentials (i.e., thought is required).

A final observation regarding the example in Fig. 12.16 (and others like it) is as follows. Although the range of T_n values is relatively wide, covering a little more than two orders of magnitude, that range does not remotely begin to approach the

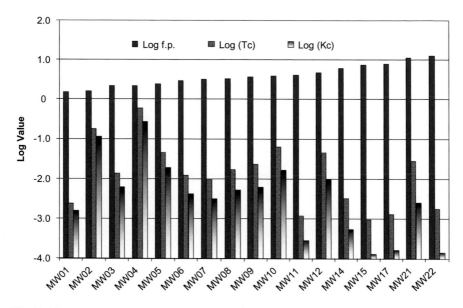

Fig. 12.16 LNAPL baildown test results plotted relative to increasing initial well LNAPL thickness (Log F.P., increasing to the right). Note, well names are pseudonyms for the privacy of the site responsible party

end-members that the theoretical discussion suggested. While this is a topic for more elaborate discussion elsewhere (and research), it has been the author's experience that T_n values from a wide range of sites and LNAPL types fall within a relatively narrow range that is inconsistent with theory, at least as commonly applied. Recall that earth materials are expected to have permeability ranges over many orders of magnitude, and those will be amplified by the capillary, relative permeability and fluid property aspects. We do not see that in field results. Why?

There are several possible explanations that combine to comprise the observed results. It was shown earlier how fine-grained materials have a much larger permeability to LNAPL than to water (Fig. 12.12). That would narrow the range of field T_n results in the direction observed above. Secondary permeability and other features in some sediments are present, but their effect on LNAPL mobility and T_n are commonly not recognized. The author worked on a site in a marine clay environment. Fuel impacts were not observed, directly beneath the tanks and pumps, from ground surface through about 30 m of vadose zone. Several feet of free-phase LNAPL was found within the water table zone. In this case, layering in the marine clay (fissility) allowed essentially a fracture-like transport regime to occur from the release source to groundwater. The only hint of impacts was very faint petroleum odors, but field laboratory sampling exhibited no detections in the vadose zone. If one had simply logged the subsurface without extending the investigation to the water table, it might have been reasonably (but wrongly) assumed that the thick zone of clay would impede any possible LNAPL transport downward. These and other field nuances control actual risks, and theory alone can be misleading.

In summary, the real-world exhibits heterogeneities in porous materials, in their fine-scale distributions, in secondary features, in LNAPL characteristics spatially and temporally, and others. Many of these aspects are not well studied, and some tend to be unsupportive of certain aspects of multiphase theory as commonly applied (per text above). The author's own take-away from all this is that LNAPL plumes will generally move farther and faster than the underlying multiphase mechanics suggest. Finite releases also typically stop migrating in relatively short periods, and old plumes are generally stable plumes (but, clearly with exceptions). In the pursuit of useful aquifer and environmental protections, we must all keep our eyes open to these possibilities. Those include the unexpected transport behaviors mentioned, as well as degree of heterogeneity that is probably best described as fractal in many settings (recall Wilson et al. (1990) groundbreaking experimentation and SEM work).

12.14   The (F)Utility of LNAPL Recovery

In this section, we will look at the utility and (often) futility of LNAPL hydraulic recovery as part of the T_n paradigm. This will be a combination of real-world observations coupled with the heuristic tendencies one might expect based on some of the theoretical underpinnings. A phrase that the author and late colleague Dr. David Huntley have long used is: "It is the LNAPL you leave behind and its characteristics that define the net benefit of any hydraulic recovery action." Dr. Huntley also used to say: "It's the LNAPL, stupid.", at times addressing this author. Meaning, do not overlook the importance of the LNAPL characteristics of multiphase, multicomponent conditions that are the key to understanding potential risk and potential benefit of any remedy action. This section considers only hydraulic recovery because of its prevalence and relationship to T_n.

T_n is often used as a metric for deciding on whether or not LNAPL hydraulic recovery might be practicable. For instance, the ITRC LNAPL guidance (ITRC 2009) on this matter suggests a cutoff range of 0.1–0.8 ft²/day, below which recovery is deemed likely impracticable. The API Interactive LNAPL Guide (API 2006) provides a number of LNAPL recovery nomographs as a function of equilibrated LNAPL well thicknesses (API, 2006) showing when recovery is theoretically possible (and where not). Figure 12.17 shows one example chart for an initial equilibrated (VEQ) LNAPL thickness of 2.5-ft for a variety of different fuel products. Given the discussion above on observed T_n variability, charts like this based on ideal theoretical conditions should be used with caution and assessed for whether site-specific T_n values align with the projections of recovery/mobility thresholds.

As the reader now knows, recovery conditions are proportional to the associated T_n value, which quickly diminishes as the saturations and observed thickness decreases. Said another way, LNAPL recovery always reduces T_n and is a self-diminishing process (absent new or ongoing releases). LNAPL recovery is also one

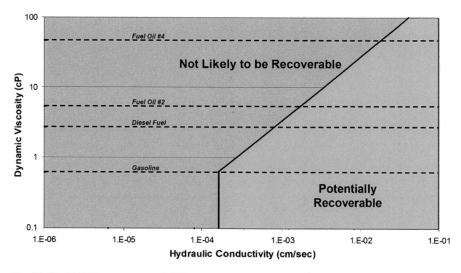

Fig. 12.17 LNAPL recovery probability chart showing the cutoff between potentially recoverable and non-recoverable product as a function of the LNAPL viscosity and the hydraulic conductivity of the soil materials. Source; API Interactive LNAPL Guide (API 2006)

of the most commonly applied remediation techniques. For instance, a review of the Los Angeles Basin refinery system found that hydraulic recovery was active (or had been historically) at almost all of the sites (Beckett et al. 2005). The same is true of many other refineries and large-scale terminals across the globe.

12.15   Recoverability Assessment of a Recent Release

This subsection will review a site-specific assessment of LNAPL mass and recoverability for a then recent release (relative to the time of the study; Beckett and Lyverse 2004). A synopsis of the study is provided followed by a few key findings as they relate to mobility and recoverability.

12.15.1   General Site Background and Findings

A field-based assessment of LNAPL and dissolved-phase mobility was undertaken after a pipeline valve rupture in early 2000 (light sweet crude) released approximately 13,000–16,000 barrels. The study is unique in that the time of release is precisely known, and it is a single release event with a uniform product and environmental sampling has provided an exceptional field-based tracking of the LNAPL plume migration. A comprehensive investigation using borehole geophysics (cone penetration testing [CPT], laser-induced fluorescence [LIF]) followed by continuous coring and petrophysical work produced a detailed understanding of site conditions. Consistent with the discussion above, based on observational data the LNAPL plume moved much farther and faster than would have been predicted based on soil and oil characteristics. The plume ceased migration after less than four years and was determined to be hydraulically non-recoverable at that stage of development.

Excavation and soil sampling work began within a short-time after the release was identified. That work was followed by boring and well installations, two time-separated CPT/LIF investigations, continuous coring with high resolution photography under white and UV light, and extensive petrophysical testing to define the soil and oil multiphase characteristics.

The site subsurface materials consist of marly sands and interbedded finer-grained materials. The laboratory-derived percentage of fines (silt and clay) ranged from about 40–85%. Oil saturations in the LNAPL interval ranged from non-detect to a maximum of about 17%. The LNAPL zones were targeted for sampling using the results of the LIF logs, coupled with high resolution core photography. Even within the LNAPL interval, it was observed that there were non-detects in some zones, which is at odds with the concept of a continuous-phase interval. Heterogeneity in material properties causes heterogeneity in saturations, and the real-world is often non-uniform in its behaviors.

The field-measured hydraulic conductivity of the water table zone ranged from about 0.15 to 40 ft/day, with a geometric mean of 4.5 ft/day, somewhat larger than might be expected based on the fine fractions present. The oil T_n ranged from about 0.02 to 0.4 ft²/day at the time of measurement about two years after the release and under declining mobility conditions, as developed further below.

The sweet crude has a density of about 0.84 g/cc, and a viscosity of about 4.5 cP, and an oil/water IFT of about 25 dynes/cm. It also has a large aromatic component that was used, in part, to track and understand plume development since the dissolved-phase plume is sourced by the LNAPL. Measured capillary properties indicated a relatively large capillary rise, or alternatively, a relatively high-water retention characteristic; an example characteristic curve is shown in Fig. 12.18. In general, given the observed and measured characteristics described, one would not expect high rates of LNAPL migration, but those were observed quite clearly in the data, as further discussed.

Because site investigation activities occurred coincident with the LNAPL plume migration, we were able to track that movement via a combination of data observations. One was new arrivals of free product at monitoring locations. Another was observed changes between two time-separated CPT/LIF investigation events (see below). Finally, increases in groundwater petroleum concentrations were also used to interpret the migration characteristics. Fig. 12.19 shows the general growth of the LNAPL plume over time, from February 2000 through the end of 2002, by which time, the majority of plume movement had ceased. As seen and expected, the growth in early time was the most significant. We estimate that the plume began its migration at rates of about 30–50 ft/day expansion. This is based primarily on field observations and limited sampling as the early movement did not yet have dedicated wells

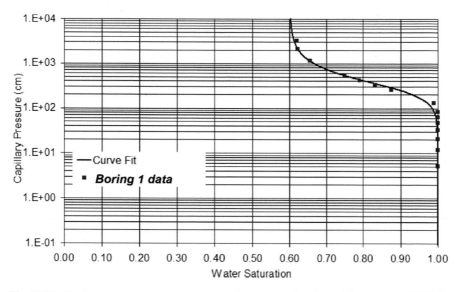

Fig. 12.18 Capillary characteristic curve from a boring sample collected in the zone of LNAPL migration. Note the relatively large capillary fringe equivalent of approximately 100 cm

and other investigative installations. By August 2001, when many more investigatory locations were available, the approximate rate of movement was about 1 ft/day, falling to about 0.1–0.2 ft/day by August with negligible migration after December 2002, or about three years following the release. The estimated incremental migration rates for the various available time stamps are shown in Fig. 12.20. Note the time gap from early 2000 to August 2001 corresponds to the time where investigation locations were being installed and data collected and assimilated.

The behaviors shown in Figs. 12.19 and 12.20 are heuristically consistent with the multiphase mechanics discussed previously. LNAPL mobility is greatest in early time and diminishes as the plume spreads into a larger volume of porous materials with an associated fall in average saturations and free-phase thickness. However, while the overall trends are consistent with theoretical expectations, the quantitative details are not (once again). For instance, the initial spreading rates on the order of tens of feet per day dwarf the range of groundwater flow. By the end of three years following the release, the long-axis of the plume had spread to about 1000-ft in length, or about 333-ft per year on average to give a general sense. The linear groundwater velocity is on the order of 20–80 ft/yr. This is a fine-grained setting and combined with the capillary properties and measured saturations that would otherwise suggest very limited LNAPL mobility, quite contrary to actual observations. For instance, if one used the analytic tools available in the API Interactive LNAPL Guide (2006), the estimated LNAPL velocity for this site's parameter ranges would be less than a tenth of a meter per day (accounting for the site parameter ranges). The key observation, once again, is that LNAPL generally migrates faster and farther than one might anticipate using solely theoretical relationships. This plume also ceased migration

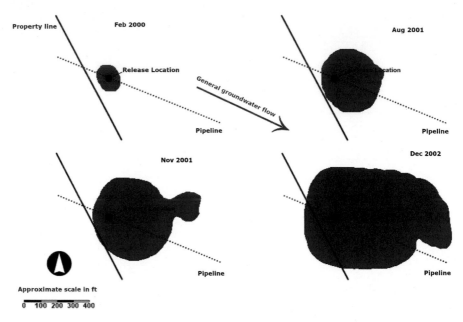

Fig. 12.19 LNAPL plume area outlines based on available data suites at sequential points in time

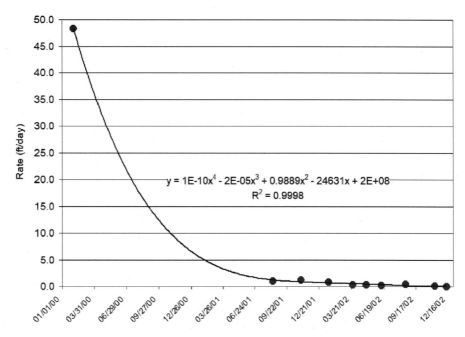

Fig. 12.20 Estimated incremental LNAPL migration rates based on the spatial distribution of LNAPL impacts. The progression is represented by a fourth-order polynomial fit

much faster than would be anticipated by theory alone and parameter application within the related equations or transient models.

Finally, there are several key site investigation observations that provide insights that could not be realized from standard site investigation techniques alone (such as borings, visual logging and related). Foremost among the critical data were the combination of CPT/LIF investigations, coupled with the laboratory core photography, to isolate core sampling target intervals that would be most relevant to the various technical questions regarding mobility, mass recovery, and potential environmental risks. First, the relative changes in the LNAPL plume morphology were clearly demonstrated by in-situ changes documented by the two separate CPT/LIF investigations. Figure 12.21 shows a plan view of the LNAPL plume distribution as indicated by LIF responses.

These same CPT/LIF data also provide a strong basis, when coupled with petrophysical photologs and test results, to evaluate the fine-scale distributions of lithologies and LNAPL within those. Figure 12.22 shows a geologic cross-section from prior site work by the well-qualified geologist who logged the borings. Figure 12.23 is approximately the same cross-section showing the fine-scale detail available from the combined CPT, LIF, and correlated petrophysical results. The contrast between the two could not be more pronounced. Figure 12.22, while accurate to its level of resolution, is not particularly useful in assessing the LNAPL properties or nuances that control this system. The tools and data suite in Fig. 12.23 is directly informative to the LNAPL mobility and risk questions under consideration.

In summary, it required the full suite of data collected, from geophysical, to continuous cores, to petrophysical and chemical analytics to adequately define the plume conditions and make a risk determination. Clearly there is an interpretive

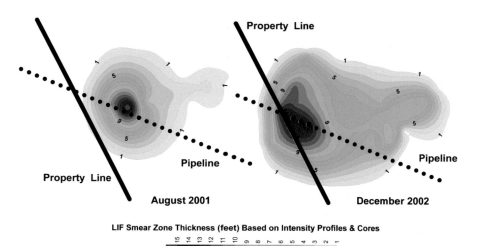

Fig. 12.21 Indicated smear zone thickness (ft) distribution in plan view. The smaller grey-shaded plume was defined in an August 2001 LIF investigation, the colored isopach by LIF results from December 2002

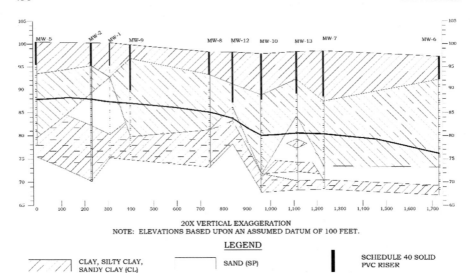

Fig. 12.22 Geologic cross-section from the site by a qualified geologist. This is the best that the "eye" can do

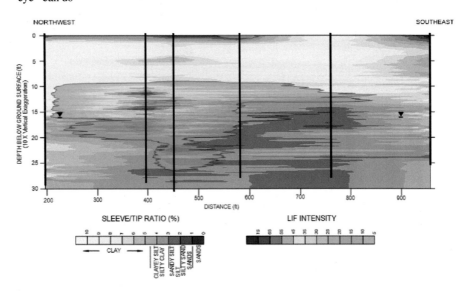

Fig. 12.23 Combined geologic, CPT, LIF, and petrophysical-related cross-section. This bears only slight resemblance to that above. LIF intensity is in percent of calibration, and tip/sleeve ratios (%) were correlated to continuous cores for the lithologic interpretation

component as well, which is part of the fun and challenge to multiphase mechanics. Where such data are unavailable (e.g., like in Fig. 12.22), there are many degrees of freedom around those interpretations, and the author highly recommends a holistic and thoughtful approach. One will not generally produce useful answers to mobility and risk questions by entering a static suite of parameters into a calculator or model. Clearly, LNAPL mobility/recoverability is vastly more complicated than that.

12.15.2   LNAPL Mobility and Recoverability

With the site background in mind, we can now consider the question of recoverability. As noted above, the T_n values in March 2002 were all below the upper limit of the ITRC recovery T_n cutoff value of 0.8 ft^2/day, and three of the five results were below the lower limit (see Fig. 12.24). As will be seen, the plume is indeed not recoverable by hydraulic methods to any practicable degree. However, it should also be recognized that this LNAPL plume was still mobile and laterally migrating at the time of these measured transmissivity values. Per the discussion above, this is why the conductivity, gradients and mobility potential are related, but represent a much different question than recoverability using T_n as its proxy. In other words, mobility can be present even at low T_n values if the discrete effective LNAPL conductivity and gradients are sufficient to allow migration to occur (other factors apply as well). The key question for this example site was whether there was any potential risk that might result from further migration. There was not because the plume was demonstrated to be stabilizing and could not reach any distal receptors. The shallow groundwater in this area is not potable. However, risk considerations are a discussion for a different chapter in this text.

From the site transmissivity values shown in Fig. 12.24, hydraulic recovery is unlikely to have any benefit. Recall, however, that the transmissivity is only relevant to the continuous phase LNAPL intervals, not the total of the plume body. From the data above, it can be determined that the phase-continuous LNAPL interval is about 10% of the total smear zone volume. Within that interval, about 17% of the LNAPL may be recoverable under ideal conditions. That assessment is based on two-phase and three-phase residual testing of capillary pressure-fluid saturation relationships and model evaluations of the recovery decay with distance from a pumping well (e.g., Fig. 12.25). The contrast in geometry between the phase-continuous interval and the total smear zone is shown in cross-section in Fig. 12.26.

Combined, the data and evaluations indicate that something on the order of only 2–4% of the LNAPL plume is recoverable as of the time plume migration ceased. This is consistent with site hydraulic recovery testing (trench and well pumping) that combined recovered less than 300 gallons of oil; recall, as much as 700,000 gallons were released. It does not require additional calculations to demonstrate that only a diminish fraction of the oil is recoverable and of too small a volume to be of any benefit. However, it is completely a different question as to whether hydraulic containment might have value. As noted, the plume was migrating over

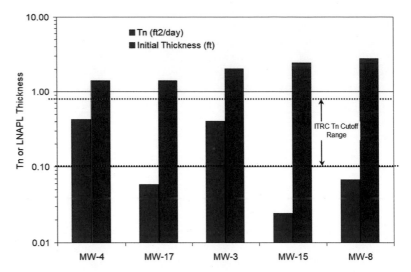

Fig. 12.24 LNAPL transmissivity values based on baildown tests at five monitoring locations. Once again, there is no relationship between T_n and initial LNAPL thickness, and all values are less than the upper ITRC cutoff value. Note the log scale

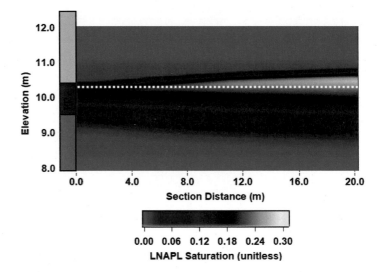

Fig. 12.25 Model simulation of LNAPL recovery at its asymptotic endpoint after 10 years of simulated LNAPL recovery. As seen, the radial effectiveness decays rapidly with distance and a substantial mass remains in the free-phase LNAPL zone. This particular simulation is analogous to, but is not site specific (i.e., heuristic)

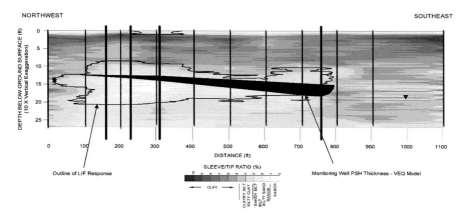

Fig. 12.26 Cross-section showing the outline of the LNAPL smear zone and within that, the zone of LNAPL phase continuity, which represents about 10% of the whole. At one time, there must have been phase continuity within the total smear zone, though not necessarily at the same point in time

about a three-year period. If there had been risks associated with that movement and the final plume distribution, then hydraulic containment with gradients sufficient to overcome the LNAPL migration gradients would be one protection measure to be considered. Cutoff trenches and others were also considered. In the end, the oversight agency concurred that the LNAPL plume had stabilized and there were no risks presented by this particular plume and its context. Obviously, this may not be the outcome at other release sites. As noted previously, even immobile LNAPL plumes can potentially present risks through fluxes in groundwater, vapor, or as a source for methane generation (to name just a few).

The conclusions and observations from this example site are compelling. A plume less than three years of age had already become impracticable to recover by hydraulic methods. The T_n implied under early migration conditions is many orders of magnitude greater than that observed at near-stable conditions. The residual smear zone is vastly larger than the free-phase interval and that will be the controlling mass relative to plume longevity. What might that suggest about other, older LNAPL plumes that have had vastly longer periods of residualization, entrapment, and weathering? In the next subsection, we will consider some hypothetical situations.

12.16 Laboratory-Derived versus Field LNAPL Transmissivity

A final case study is briefly presented here. At a site with an older suite of fuel releases, a detailed field investigation was conducted. Many of the approaches were similar to the case study above, and the approaches will not be developed in detail.

Continuous coring, petrophysical, CPT/LIF, and other investigation methods were applied to result a comprehensive LNAPL CSM. That CSM determined that the LNAPL was predominantly residualized, stable, and contained about one million gallons of free, occluded, and residual NAPL. Based on the data and multiphase evaluations, it was determined that less than 3% of the LNAPL in-place was recoverable, as shown in Fig. 12.27. The zone of potential recovery is also geographically limited, as heuristically anticipated by multiphase theory.

Lastly, a comparison was possible at this study site between the T_n values derived by baildown testing, versus the calculated values derived from the laboratory parameters described previously that define T_n (see Eq. 12.1). The laboratory-derived values were determined to be orders of magnitude smaller than the field measurements. The differentials ranged from a low of about 5 (field values 5 times greater than the laboratory value at that location) up to about 250 times greater. It is commonly recognized that laboratory scaling and other issues often produce results that are not field-comparable. The complications of multiphase mechanics make this laboratory testing issue even more prevalent in the author's study and testing experience. While petrophysical testing is valuable and provides insights, the parameter derivations generally cannot be put into a simple equation and result in realistic estimates of T_n (or mobility).

Fig. 12.27 Distribution of the recoverable LNAPL fraction based on detailed multiphase parameter collection, mass estimates, modeling, and related evaluations

12.17   Flux and Longevity Considerations on LNAPL Recovery

This last section will briefly explore physical and chemical considerations with respect to LNAPL mobility and recoverability. In other sections of this book, the physical and chemical properties and variability of LNAPLs were discussed. As is clear from this chapter, each of those has a direct or indirect effect on the mobility and recoverability of plumes in various settings. As shown by the example above and many others not discussed, the age of the LNAPL plume is often inversely related to its mobility and recoverability. Plumes that are stable, meaning not expanding geographically, are also not generally recoverable by hydraulic means. This is because the processes that cause stabilization are essentially the same that limit recoverability (residualization, weathering, pore entrapment, submergence, etc.). Before exploring these limitations, an editorial note is warranted. This discussion is about applying good science to set expectations on aquifer restoration measures. We have already observed that there are significant limitations to hydraulic recovery approaches and, where applicable, would often be part of a broader remedy solution. Recovery limitations do not imply that remediation is not viable, but rather that hydraulic recovery specifically may be incapable of achieving cleanup objectives and other options, or combined approaches, may be more appropriate. Where mitigation measures are necessary, recovery should not be the only method considered, particularly if there is no practical benefit. In short, remedy evaluations are about determining the most appropriate and effective measures and hydraulic recovery may often not be in that category.

For ease and consistency of comparison, we will explore the theoretical recovery aspects for a few different sets of conditions. The methods described in API #4715 (Hunley and Beckett 2002) will be used for the comparisons, using the toolkit in the API Interactive LNAPL Guide (API 2006). As should be clear by now, these comparisons are heuristic. Real-world plumes will be more complex than indicated by the simplified models applied. In general, these VEQ-based models will underpredict recovery in fine-grained materials and overpredict the same in coarser soils, per the earlier observations that field T_n/mobility conditions tend to coalesce around a much narrower range of values than indicated by multiphase theory. Further, these methods only consider the free-phase, vertical equilibrium LNAPL interval, not the full plume mass, which is critical to site-specific considerations as shown above. In other words, take the lessons learned and assume actual field conditions will be much less amenable to recovery attempts than what these models might suggest.

The evaluations here will use the three coarsest soil types defined in the theoretical discussion above; (i) Sandy loam; (ii) Sand mixtures; and (iii) clean coarse sands. Finer-grained materials will not be evaluated because they will not be predicted to have any recoverable LNAPL volumes. For each of these three soil types, LNAPL skimming and dual-phase (water and LNAPL) pumping recovery will be estimated on a mass basis, as well as inspecting for component chemical changes that might occur. Each recovery action is assumed to run for three years. For estimates, we will look at benzene and naphthalene as indicator compounds to give a sense for how recovery might influence the composition and longevity of these compounds.

Numeric values of mass are in kilograms for an LNAPL plume with dimensions of 30 × 30 m and an equilibrated initial thickness of 1 m. *The percentages in the bottom section are the estimate percent of the initial mass recovered.*

Table 12.3 presents the key mass and mass reduction results. On a bulk basis, one can quickly observe two key outcomes. Diesel is not as recoverable as gasoline, both because of its property differences, as well as differences in its distribution relative to the water table. Second, there is a vast difference in recovery potential, for this same initial LNAPL thickness (1 m), between these soils. The loamy sand presents a negligible recovery mass under these conditions. This observation is why there is no value or need to inspect finer-grained materials. Recovery in fine-grained materials will rarely have a net benefit for all the collective reasons discussed previously (keeping in mind that field conditions can and do differ from theory).

The estimated LNAPL recovery curves for each scenario are shown in Fig. 12.28 for the three years of recovery. One can observe the significant differences in both the cumulative totals between the different soil types and fuels, but also that the asymptotic break in each differs substantially. The smaller the initial and incremental T_n (and saturations), the less LNAPL that will be recovered and the longer it will take to reach an asymptote. This closely parallels general field observations of asymptotic recovery for hydraulic systems. The asymptote means that additional recovery is small and perhaps impracticable. It does not imply, however, that the recovery action has had any net benefit, it just means it is finished.

As noted previously, it is the LNAPL mass left behind that often defines the net benefit of hydraulic recovery as compared to other mitigation measures. This is particularly true if the LNAPL plume in question is stable and no longer mobile in the environment. As discussed in the earlier case example, although the LNAPL was not

Table 12.3 Mass and mass change percentages for each condition

	Initial conditions		Skimming—remains		Pumping—remains	
Soil type	gasoline	Diesel	gasoline	Diesel	Gasoline	Diesel
Sandy loam	33,147	15,168	32,677	15,168	32,443	15,168
Find sand and mixtures	82,155	48,374	32,200	34,932	31,655	28,139
Clean sands	124,062	89,993	27,055	29,428	26,966	25,031
Percent change						
Sandy loam (%)			1.42	0.00	2.12	0.00
Find sand and mixtures (%)			60.81	27.79	61.47	41.83
Clean sands (%)			78.19	67.30	78.26	72.19

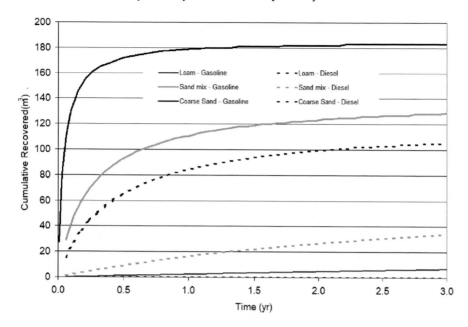

Fig. 12.28 Cumulative LNAPL volume recovery curves for each scenario (three soils and two fuels). Note that the recovery in the loamy sand is near zero for the diesel condition

practicably recoverable, there could have been a net benefit to hydraulic containment or other measures if the temporarily mobile plume had posed a risk.

For each of these example estimates, the benzene and naphthalene concentrations over time were tracked (methodologies per Huntley and Beckett 2002; API 2006). At the end of the three-year period, the estimated time reduction to reach 5 ug/l benzene and 20 ug/l naphthalene was calculated, as shown in Fig. 12.29. This chart shows that there is no net benefit to recovery in the loamy sand for either gasoline or diesel fuel relative to the longevity of chemical impacts. There is some potential benefit in the intermediate soil, more for gasoline than for diesel. The clean sands had the greatest time reduction by a factor of about 4–5 depending on the compound and fuel type.

The methodology of API (2002) accounts for multiphase, multicomponent partitioning, fluxes, and transport processes. It does not account for potential mass losses other than those partitioned fluxes or as mass removal from a recovery action (e.g., NSZD is not considered, nor are other complex processes). For those reasons, the absolute times of depletion are not as relevant as the comparative values noted. A blending of these types of evaluations with actual plume-specific chemical depletion observations is one way to assess some of the chemical longevity aspects. Qualitatively, however, one can still make some simple observations. Whatever the time to reach background concentrations may be, it could be scaled by the reduction factors above as a preliminary approximation.

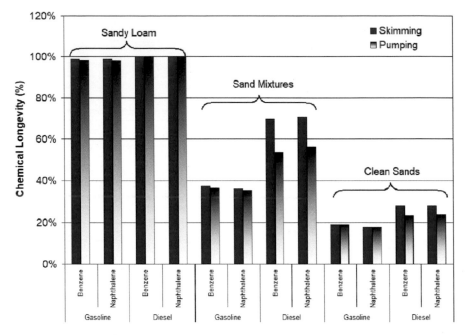

Fig. 12.29 Estimated chemical longevity as a percentage of the original for each fuel, soil type, and hydraulic recovery method. A value of 100% means the chemical longevity is unchanged relative to initial ambient conditions

While the heuristic evaluations above are quite useful, probably the most important consideration is not part of a simple VEQ-based estimate, but rather one that also incorporates estimates of the smear zone volume and residual mass. Taking the case example above, the free-phase zone was approximately 10% of the total volume of impacts at the time approaching plume stabilization (about three years). In such cases, even recovery of an idealized 80% of the mass in coarse-grained sands would be insignificant relative to the total mass in-place. Under a condition like that, perhaps 5–10% of the total mass might be recovered. Further, as shown in Fig. 12.25, more robust numerical modeling and field observations combine to indicate that simple uniform recovery around a well does not occur. The recovery effectiveness will decay with distance away from the recovery well, often quite strongly. That is why the methods described in API (2002) and API (2007) are idealized and not likely reflective of the actual field limitations to recovery (at least in most cases). The author performed numerical verification work for both these API toolkits and notes that in general, they are heuristic for the reasons described in prior sections (they are theory-based only with stringent assumptions). As noted above, one would expect such approaches to overestimate recoverability in coarse-grained materials and light-end fuels and the opposite for fine-grained materials and heavy-end products. The evaluation methods, as shown by demonstration, are useful in considering potential comparative outcomes and sensitivities to then lead to field testing and demonstration of actual conditions. Those actual conditions will, in general, be substantially more complex, nuanced and will look nothing like the theoretical estimates.

12.18 Conclusions

There are many conclusions and observations provided throughout this chapter. A few that come together as a result of the whole, coupled with decades of detailed field and laboratory investigations, are as follows:

- Hydraulic recovery has the highest potential benefits in plumes that were recently released and are still migrating in the environment. Little net benefit will be realized for plumes that are mostly residualized and stable in the environment because the processes that lead to that condition also severely limit LNAPL recovery.
- It is the mass left behind after recovery efforts (or others) that generally define the net benefit of the remedy action. Little benefit is expected when insignificant mass can be recovered by hydraulic methods, excepting certain containment objectives may have benefits if properly applied.
- T_n and mobility are highly transient and dependent on various in situ conditions. Fulsome consideration of those conditions and their implications are often warranted. Simplistic application of multiphase equations or models can provide heuristic value, but are unlikely to be representative of actual conditions. The value is dependent on the quality of the interpretations and LCSM.
- Old plumes are commonly stable, residualized, and non-recoverable plumes. If there are risks presented by such conditions, it is recommended that mitigation measures other than hydraulic recovery be considered, as it will do nothing to meaningfully alter those underlying conditions.
- Be aware of transient site conditions, like LNAPL entrapment, that may only temporarily immobilize plumes. Make assessments of risk and plume status holistic and thoughtful to ensure potential changes in conditions have been appropriately considered.
- Build LNAPL CSMs with a focus on processes, material properties, saturation distributions, and related observations. When there is apparent conflict between estimates and observations, clearly observations are favored and require plausible explanations.
- In places and jurisdictions where LNAPL recovery is expected because of perceived mass reduction benefits, challenge those assumptions if there is no net benefit to the action (in favor of actions with a net benefit). The purpose of mitigation measures should always lean toward resource restoration and risk protection (among others). LNAPL recovery usually achieves none of those goals.
- Tangential to the above, it is now widely recognized that LNAPL plumes degrade, sometimes at fairly high rates though natural source zone depletion mechanisms. Where LNAPL plumes are stable and NSZD rates are greater than initial LNAPL recovery rates, there is likely no benefit to hydraulic recovery. A plume in this

life-stage is typically mostly non-recoverable and that recovery, if applied, would be expected to diminish rapidly to less than the NSZD rates. If risks need to be managed, apply a different method.

References

API (2002) American Petroleum Institute Publication #4715. Evaluating Hydrocarbon Removal from Source Zones and its Effect on Dissolved Plume Longevity and Magnitude

API (2006) API Interactive LNAPL Guide Version 2.0.4

ASTM (2021) Standard guide for estimation of LNAPL transmissivity. Active Standard ASTM E2856-13

Beckett GD, Huntley D (1994) Characterization of Flow Parameters Controlling Soil Vapor Extraction: Ground Water, 32(2): 239–247

Beckett GD, Joy S (2003) Light non-aqueous phase liquid (LNAPL) Parameters Database-Version 2.0-User Guide. API #4731

Beckett, GD, Lyverse MA (2004) NAPL Immobilization; When & Why It Stops, & Other High-Level Observations. Presented at the 2004 Petroleum Hydrocarbons & Organic Chemicals in Ground Water, Baltimore, Maryland, sponsored by the National Ground Water Association & American Petroleum Institute

Beckett GD, Sale T, Huntley D, Johnson P (2005) Draft Report to the LARWQCB "Best Practices Study of Groundwater Remediation at Refineries in the Los Angeles Basin"

Beckett GD, Huntley D (2015) LNAPL transmissivity; a twisted parameter. J Groundwater Monit Remediation

Burdine NT (1953) Relative permeability calculations from pore size distribution data. Journal of Petroleum Technology, 5(3): 71–78

Charbeneau et al. (1999) Free Product Recovery of Liquid Hydrocarbons, API Publication #4682, Health and Environmental Sciences Department of the American Petroleum Institute.

Charbeneau RJ, Johns RT, Lake LW, McAdams III MJ (2000) Free-product recovery of petroleum hydrocarbon liquids. Ground Water Monit Remediat 20(3): 147–58

Charbeneau R, Beckett GD (2007) LNAPL distribution and recovery model (LDRM), API #4760, vol 2. User's Guide

Charbeneau R J, Kirkman A J, Muthu R (2012) API LNAPL Transmissivity Workbook: A Tool for Baildown Test Analysis.

Delshad M, Pope GA (1989) Comparison of the three-phase oil relative permeability models. Transp Porous Media 4(1): 59–83

Farr AM, Houghtalen RJ, McWhorter DB (1990) Volume estimates of light nonaqueous phase liquids in porous media. Ground Water 28(1): 48–56

Freeze, RA, Cherry JA, (1979) Groundwater. Prentice-Hall, Inc., Englewood Cliffs, New Jersey

Garg S, Newell C, Kulkarni P, King D, Adamson D, Irianni-Reno M, Sale T (2017) Oveview of natural source zone depletion. J Groundwater Monit Remediat 37(3)

Huntley D (2000) Analytic determination of hydrocarbon transmissivity from baildown tests. J Groundwater 38(1): 46–52

Huntley D, Beckett GD (2002) Persistence of LNAPL sources: Relationship between risk reduction and LNAPL recovery. J Contam Hydrol 59: 3–26

Lenhard RJ, Parker JC (1990) Estimation of free hydrocarbon volume from fluid levels in monitoring wells. Ground Water 28(1): 57–67

MAGNAS3, (1994) Multiphase Analysis of Groundwater, Non -aqueous Phase Liquid and Soluble Component in 3 Dimensions, Documentation and User's Guide, Hydrogeologic, Inc, Herndon, Virginia

Mualem Y (1976) A new model for predicting the hydraulic conductivity of unsaturated porous media. Water Resour Res 12: 513–522

Parker JC (1989) Multiphase flow and transport in porous media. Rev Geophy 27:311–328

Schaap M, Leij F, van Genuchten M (2001) Rosetta: a computer program for estimating soil hydraulic parameters with hierarchical pedotransfer functions

Simmons CT (2008) Henry Darcy (1803–1858): Immortalised by his scientific legacy. Hydrogeology Journal 16: 1023–1038

T2VOC User's Guide (1995) Ronald W. Falta, Karsten Pruess, Stefan Finsterle, Alfredo Battistelli. WIPP Project, Sandia National Laboratories, under Document No. 129847

USGS (2004) Basic Ground-Water Hydrology. U.S . Geological Survey, Water-Supply Paper 2220(86): 1983. Tenth printing 2004, revised.

van Genuchten MT (1980) A closed form equation for predicting the hydraulic conductivity of unsaturated soils. Soil Sci Soc Am J 44

Wilson J, Conrad S, Mason W, Peplinski W, Hagan E (1990) Laboratory investigation of residual liquid organics from spills, leaks, and the disposal of hazardous wastes. EPA report EPA/600/6–90/004

Zhu JL, Parker JC, Lundy DA, Zimmerman, LM (1993) Estimation of soil properties and free product volume from baildown tests. In Proc. Petroleum Hydrocarbons and Organic Chemical in Ground Water, National Water Well Association, Dublin Ohio, 99–111

Chapter 13
Incorporating Natural Source Zone Depletion (NSZD) into the Site Management Strategy

Tom Palaia and Sid Park

Abstract This chapter summarizes three petroleum remediation case study sites that incorporated natural source zone depletion (NSZD) into their site management strategy. The sites include: a light petroleum non-aqueous phase liquid (NAPL) pipeline release in an arid climate that transitioned to an NSZD-based monitored natural attenuation (MNA) remedy after several traditional remedial actions; a former refinery in a temperate climate that incorporated NSZD with phytoremediation; and a former industrial petrochemical facility in a tropical climate with waste dense petroleum hydrocarbon NAPL that transitioned to NSZD after several decades of groundwater extraction and treatment. The narratives provide examples of how NSZD was used as an effective and sustainable/resilient management option. Key points related to the use of NSZD on a broader range of site conditions are highlighted. The importance of acknowledging, accounting, and incorporating NSZD into remedies is emphasized because significant depletion rates are frequently observed to occur at petroleum NAPL release sites. The chapter demonstrates how NSZD can be incorporated into a conceptual site model and the implemented remedy. It provides details on the importance of site risk profile and regulatory framework as well as lines of evidence to support transition to a MNA remedy inclusive of NSZD.

Keywords Biodegradation · In-situ remediation · Natural attenuation · NSZD · Petroleum NAPL

T. Palaia (✉) · S. Park
Jacobs Solutions, Greenwood Village, CO 80111, USA
e-mail: Tom.palaia@jacobs.com

S. Park
e-mail: Sid.park@jacobs.com

© The Author(s) 2024
J. García-Rincón et al. (eds.), *Advances in the Characterisation and Remediation of Sites Contaminated with Petroleum Hydrocarbons*, Environmental Contamination Remediation and Management, https://doi.org/10.1007/978-3-031-34447-3_13

443

13.1 Introduction

The purpose of this chapter is to demonstrate natural source zone depletion (NSZD) as an effective and sustainable/resilient management remedy component for low-risk profile sites containing petroleum hydrocarbon non-aqueous phase liquid (NAPL). It stresses the importance of acknowledging, accounting, and incorporating NSZD into remedies because significant depletion rates are frequently observed to occur at petroleum NAPL release sites. This adds to the case study literature available on the subject to help practitioners see how NSZD can be used on their sites (Interstate Technology and Regulatory Council [ITRC] 2018; Cooperative Research Centre for Contamination Assessment and Remediation of the Environment [CRC CARE] 2020a, b). NSZD phenomena and measurement methods are discussed in further detail in Chap. 5.

This chapter presents a narrative of three diverse case study sites from the United States that incorporated NSZD into their site management strategy, both in the conceptual site model (CSM) and in the implemented remedy. Since NSZD occurs on most, if not all, petroleum NAPL sites, the NSZD evaluation results can be used to refine any CSM. The site-specific human health/ecological risk profile and regulatory framework are significant drivers that determine how NSZD is incorporated into the site management strategy, most logically in a monitored natural attenuation (MNA) framework. MNA has been traditionally thought of as a dissolved phase remedy, but, as described in Chaps. 1, 5, and 9, natural attenuation processes occur throughout the contaminated zone including the petroleum NAPL. It is fitting that NSZD be included in an MNA remedy because it is the term used to describe the collective, naturally occurring processes of dissolution, volatilization, and biodegradation of NAPL.

Using case studies, this chapter focuses on the relevance of the risk profile and regulatory framework to how NSZD is applied at a site. For those sites where NSZD monitoring may be included as part of the MNA remedy, this chapter also provides ideas on how NSZD rates can be assembled with other lines of evidence to support the use of or transition to MNA in various remedy configurations. Finally, it also shows how the NSZD remedy component can be implemented and monitored.

This chapter documents how NSZD was applied at diverse case study sites. At each site, NSZD was an accepted component of the site management strategy to control and degrade in-situ petroleum hydrocarbons. This chapter uses a combination of case study illustration and call out boxes of key points to relate important concepts to various other types of sites.

The case studies were chosen to show how NSZD can be applied in a wide range of regulatory, geographic, hydrogeologic, geochemical, and release conditions. While it is impossible to cover all situations, the three case studies cover typical scenarios. More importantly, they show the type of background information considered, provide a narrative on how the information was used to obtain a decision document with NSZD accepted, and highlight approaches that are likely typical of NSZD-oriented remedial programs. Extrapolation from these cases to readers' projects may help to facilitate more rapid integration of NSZD into remedial action plans and stakeholder acceptance. This chapter shows how NSZD can be used in a multitude of different ways. In doing so, it illustrates how adaptable and effective NSZD can be in the management and remediation of petroleum NAPL-contaminated sites.

The reader is advised to use discretion when interpreting the case studies and call out boxes, and to work closely with technical and regulatory teams when considering how to include NSZD in site management at their sites.

13.2 Overview of Case Study Site Setting

Three sites were selected to demonstrate a broad range in use of NSZD for site management.

Site A is a small petroleum light non-aqueous phase liquid (LNAPL) pipeline release site where NSZD-based MNA was applied as a treatment train transition remedy after excavation, soil vapor extraction (SVE), and LNAPL recovery.

Site B is a large former refinery site where a combination of NSZD with phytoremediation and hydraulic controls was used to manage LNAPL.

Site C is a medium-sized former industrial petrochemical facility with waste heavy petroleum hydrocarbon dense non-aqueous phase liquid (DNAPL) (no chlorinated compounds) where NSZD-based MNA was implemented as a transition remedy after several decades of groundwater extraction and treatment.

Table 13.1 further summarizes the petroleum releases that occurred and the general setting at each case study site. The CSMs are presented in Sect. 13.4 and timelines of events for each site are presented in Sect. 13.5.

Table 13.1 Summary of case study site conditions

	Site A: pipeline release, semi-arid climate	Site B: former refinery, temperate climate	Site C: former petrochemical plant, tropical climate
Release summary			
Petroleum Product(s)	Jet fuel (C9–C22)	Mixed petroleum releases and waste products (C5–35)	Petrochemical industrial process waste (primarily C11–C22 aromatics)
LNAPL or DNAPL (specific gravity)	LNAPL (0.80)	LNAPL (0.85)	DNAPL (1.08)
Release type (NAPL footprint size)	Leaking underground pipe (~0.8 hectare [2 acres])	Process and tank farm area spills (~61 hectares [150 acres])	Leaking surface impoundment/pond (~12 hectares [30 acres])

(continued)

Table 13.1 (continued)

	Site A: pipeline release, semi-arid climate	Site B: former refinery, temperate climate	Site C: former petrochemical plant, tropical climate
Approximate year(s) of release	Unknown: discovered in 1995	1917–1993	1959–1972
Contaminant of primary concern in groundwater[a]	Benzene up to 0.1 mg/L and up to 150 mg/L TPH-jet fuel (C9–22)	Benzene > 10 mg/L, MTBE up to 8.5 mg/L, and TPH-middle range hydrocarbons (C9–18) > 10 mg/L	BTEX and PAHs including benzene up to 21 mg/L, naphthalene up to 2,500 mg/L, and pyrene up to 200 mg/L

Physical site setting

Land use	Commercial, industrial	Former industrial, residential	Industrial
Current land disposition	Active commercial businesses, roadway, utility right-of-way	Decommissioned and vacated land, highly vegetated, some wetlands, adjacent residential	Decommissioned and vacated land, capped industrial landfill and former canal, and adjacent refinery and power plant
Geology	Simple: silt/clay underlain by fine to medium grained sand	Simple: silty clay loam alluvium underlain by weathered, fractured limestone	Complex: interlayered and interfingered gravel to clay over limestone
Hydrogeology	Unconfined	Unconfined in alluvium, overlying confined to semi-confined limestone, locally perched	Unconfined over semi-confined
	Water table rise from 22 mbg (71 ft bgs) (1997) to 17 mbg (57 ft bgs) (2022) (submerged LNAPL)	Water table depth with hydraulic containment system operational ranges from 1.5 to 7.6 mbg (5–25 ft bgs)	Tidally influenced, water table 1.8–3.0 mbg (6–10 ft bgs), saltwater wedge beneath site
	$Vs = 2.2E-5$ cm/s (0.06 ft/day) (fine to medium grained sand unit)	$Vs = 1.3E-2$ cm/s (0.36 ft/day) (weathered limestone unit)	$Vs =$ up to 1.1E-4 cm/s (0.3 ft/day) (unconfined overburden unit)
Groundwater redox state	Sulfate-reducing (100 mg/L SO_4 background), methanogenic (up to 34 mg/L dissolved CH_4)	Methanogenic (up to 12 mg/L dissolved CH_4)	Sulfate-reducing (4,000 mg/L SO_4 background), methanogenic (up to 19 mg/L dissolved CH_4)
Average groundwater temperature	23 °C (73 °F)	15 °C (59 °F)	30 °C (86 °F)
Historical average annual precipitation	30 cm (12 inches)	97 cm (38 inches)	102 cm (40 inches)

(continued)

Table 13.1 (continued)

	Site A: pipeline release, semi-arid climate	Site B: former refinery, temperate climate	Site C: former petrochemical plant, tropical climate
Risk and regulatory setting			
Exposure pathway driving remediation	*Human health*	*Ecological*	*Ecological*
	Groundwater, drinking water resource	Surface water	Surface water and sediment contact with marine mammals
		Human health	
		Groundwater consumption, vapor intrusion	
Regulatory program	Cleanup and abatement order	Consent order, site cleanup program	Hazardous waste management

[a] Results are NAPL-biased (i.e., laboratory analysis included entrained NAPL droplets) as most reported maximum concentrations exceed effective solubility
bgs = below ground surface
BTEX = benzene, toluene, ethylbenzene, and xylenes
mbg = meters below grade
MTBE = methyl tert-butyl ether
PAH = polycyclic aromatic hydrocarbons
TPH = total petroleum hydrocarbon
Vs = groundwater seepage velocity
LNAPL = light non-aqueous phase liquid
DNAPL = dense non-aqueous phase liquid

13.3 Importance of Site Risk Profile and Regulatory Framework

The site human health/ecological risk profile is perhaps the critical determinator of remedial approaches considered and, ultimately, selected for use. The regulatory framework is perhaps the most significant factor that drives how NSZD is selected and incorporated into a decision document.

While the application of this remedy component is maturing, Box #1 shows there are specific stakeholder concerns that pose a challenge to obtaining approval for NSZD in decision documents. Inherent to these concerns is the perception of risk. The following will outline how the regulatory framework, and various stakeholders' risk tolerances, will play important roles in the data requirements and how NSZD is incorporated into decision documents.

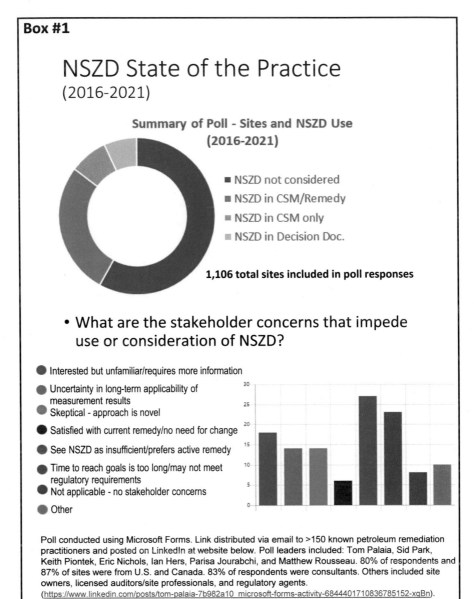

Box #2 highlights site risk profiles and regulatory frameworks that are more conducive to use of NSZD as a remedy.

Box #2

NSZD occurs on most sites, but is best suited as a remedial option at sites with:

- Low risk profile
- Stable NAPL and dissolved plumes
- Active facilities with risk management plans
- Exposure to impacts is restricted by engineering or administrative controls
- Remediation to the maximum extent practicable was achieved through recovery of mobile petroleum NAPL
- Regulatory agency amenable to risk-based remediation or risk management
- Stakeholder tolerance for longer remedial timeframes typically associated with residual petroleum hydrocarbon NAPL

13.3.1 Impacts of Regulatory Drivers and Risk Profile on NSZD Data Requirements

The regulatory environment and risk profile under which a site is managed can affect what data are needed to support the selection of NSZD as a component of the site remedy.

Five primary categories of regulatory drivers for the use of NSZD at petroleum NAPL sites, and NSZD's effectiveness at addressing each of these drivers, are shown in Table 13.2. Data collection activities should be focused on the site-specific regulatory driver(s).

For some regulatory programs, an estimate of the time needed to meet cleanup goals is required. In other programs, the rate of source reduction may be of primary concern to regulatory agencies. In the context of the risk profile, if exposure pathways are complete (current) or operable (potential future), then ensuring that risk is mitigated will drive data needs. If the risk profile and regulatory framework allow for an institutionally controlled restricted future land use and/or the typically long remedial timeframes for residual petroleum hydrocarbon remediation (i.e., a risk management plan), then the use of NSZD as part of a risk management plan may generate its own set of data requirements.

Table 13.2 Regulatory drivers and NSZD's effectiveness

Regulatory driver	NSZD effectiveness
Product removal	NSZD removes bulk mass of NAPL achieving prescribed in-well petroleum NAPL thickness and continues to remove mass after other recovery methods achieve their maximum extent practicable
Remedial timeframe	NSZD depletes petroleum hydrocarbon mass from the unsaturated and saturated zones at a rate exceeding other viable remedial approaches resulting in comparable remedial timeframes between alternatives
Risk management	NSZD protects human health and the environment through the reduction of bulk petroleum hydrocarbon mass and concentrations of contaminants to below site-specific risk tolerance limits at points of exposure
Media-specific cleanup standards	NSZD reduces concentrations of contaminants in groundwater, soil, vapor, or surface water to below regulatory established protection standards
Institutional controls	NSZD controls a stable extent of contamination or petroleum NAPL within a legally enforced boundary that restricts or mitigates receptor exposure to impacted media

The three case studies highlighted some of the data requirements that may need to be fulfilled when using NSZD. While most of the data required was typical of petroleum NAPL remediation projects, some was unique to NSZD. The reader is cautioned to carefully consider these requirements in relation to the project timing, logistics, budget, and planning.

13.3.2   Impacts of Project Phase on Incorporating NSZD into a Remedy

Project phase is a factor in how NSZD is applied in each regulatory environment. If the release is recent and the project is in the remedy-development phase, then measuring site-specific rates or using estimated NSZD rates based on the literature and institutional knowledge of previous projects may be useful. These rates can be used as a baseline mass removal rate to compare NSZD to other remedial approaches. Regulatory agency's receptivity to literature-based rates, or rates based on institutional knowledge, however, must be considered, as well as the uncertainties associated to these values instead of site-specific rates. The merits of using literature-based rates, or rates based on institutional knowledge, must be evaluated against the cost and time demands associated with developing a site-specific estimate of the rate. NSZD monitoring may also be a stand-alone component of an MNA remedy, in which case it would be considered when detailed remedial option analysis is being completed.

At historical release sites in a later project phase, such as remedy implementation and optimization, NSZD may be proposed as a final step in a treatment train. For many stakeholders, the use of NSZD may be challenging, as it may have been perceived as a "do-nothing" approach. This misperception may impede a switch to MNA at later stages in the project and delay approval of, or provision of, decision documents. Active and consistent education of stakeholders about the merits and effectiveness of NSZD can be key for projects at this phase.

Finally, NSZD can be incorporated in a multi-component (coupled) remedy at any phase. For example, it may be considered after other remedial approaches fail to meet remedial goals for the project. Again, stakeholders' perceptions and expectations would have to be carefully calibrated and addressed for a project to be successful.

13.3.3   Regulatory Pathways Taken at the Case Studies

The following is a summary of the regulatory drivers at each case study site, the time when NSZD was identified as a feasible component of the remedy, and the follow-up field activities performed to refine the CSM and support use of NSZD. They illustrate the effects of site risk profile and regulatory framework on decisions related to how NSZD was incorporated into site management. A more detailed description of the NSZD measurement methods is provided in Chap. 5.

Site A:

- Land use at time of release: commercial/industrial, groundwater available as an adjudicated resource.
- Description of original complete/operable exposure pathways: no complete (current) pathways, operable (potential future) pathways included drinking groundwater, vapor intrusion into unprotected buildings, or utility workers in a trench.
- 2015—governing authority issued a Cleanup and Abatement Order to cleanup groundwater to drinking water levels.
- Cleanup was implemented using excavation, LNAPL recovery, SVE, and semi-annual groundwater monitoring. After operating the LNAPL recovery system for 11 years and the SVE system for 5 years, system mass removal rates diminished.
- Monitoring with no further LNAPL recovery was proposed as a management strategy.
- The monitoring request was denied by the regulatory authority because it "... would not meet water quality objectives as LNAPL acts as a continuing source to groundwater."
- NSZD was identified as a viable component of the MNA remedy to address immobile LNAPL entrapped in saturated sediments in a way that balanced social, financial, and economic impacts.

- The sitewide NSZD rate was quantified by measuring surface carbon dioxide (CO_2) efflux and estimating ongoing source mass reduction. See Sects. 13.4.1 and 13.4.2 for additional details on the initial Site A NSZD monitoring program and results, and Sect. 13.6.1 for the routine NSZD monitoring performed to support the MNA remedy.

Site B:

- Land use at time of release: operating oil refinery.
- Description of original complete/operable exposure pathways: complete (current) pathway to ecological receptors in the river, operable (potential future) pathways included drinking groundwater, vapor intrusion into new unprotected buildings, or utility workers in a trench.
- 2003—original Consent Order was signed between the property owners and the regulatory agency.
- Required corrective action goals included:

  o Active product recovery of in-well LNAPL to the maximum extent practicable and to mitigate further LNAPL migration to the adjacent river.
  o Groundwater maximum contaminant level (MCL) or Tier 2 risk-based cleanup goal for constituents of concern without an MCL.

- As part of the MNA remedy already established for the dissolved phase, NSZD was identified as a feasible additional component to address sitewide remaining LNAPL impacts and to augment the portion of the remedy associated with the primary river receptor.
- LNAPL transmissivity was tested, and NSZD rates were measured. See Sects. 13.4.1 and 13.4.3 for additional details on the initial Site B NSZD monitoring program and results, and Sect. 13.6.2 for the routine NSZD monitoring performed to support the MNA remedy. The concept of LNAPL transmissivity is further discussed in Chaps. 2 and 12.

Site C:

- Land use at time of release: operating industrial petrochemical plant.
- Description of original complete/operable exposure pathways: complete (current) pathway to ecological receptors from sediment in, and groundwater seepage into, an adjacent canal; operable (potential future) pathways include drinking groundwater, vapor intrusion into new unprotected buildings.
- 1988—a Federal hazardous waste management program issued a Permit to operate the facility and established groundwater protection standards.
- The Permit required closure of the former surface impoundments, capping of canal sediments, and operation of a groundwater recovery system to recover contaminants and achieve site-specific risk-based groundwater protection standards.
- The impoundments were stabilized/closed, and the impacted canal sediment was capped to prevent ecological contact.

- The groundwater recovery system mass removal rates diminished over the ~28-year operation period. At about the same time, a major hurricane, a series of earthquakes, and a global pandemic significantly reduced system operability.
- MNA, inclusive of NSZD, was identified as a more resilient remedy to address residual petroleum DNAPL impacts trapped in saturated soil. Costs to operate and maintain the aging existing remediation system and dispose of the hazardous waste (including overseas shipping) had increased to over one million dollars per year.
- The CSM was refined using laser-induced fluorescence (LIF) delineation of the NAPL footprint, transmissivity tests modified for behavior of DNAPL on confining layers, and two rounds of NSZD measurements. See Sects. 13.4.1 and 13.4.4 for additional details on the initial Site C NSZD monitoring program and results, and Sect. 13.6.3 for the routine NSZD monitoring performed to support the MNA remedy. LIF technologies are further discussed in Chap. 8.

13.4 Integrating NSZD into the Conceptual Site Model

Because significant NSZD is frequently observed to occur at petroleum NAPL release sites, NSZD should be used to refine the CSMs. Box #3 calls out the multitude of ways NSZD measurements can be used in CSMs. When considering NSZD as a remedy component, it is important to understand the sensitivity of various elements of the CSM to the effects of NSZD and incorporate the NSZD-specific factors in all possible aspects of its development. Developing the CSM without considering NSZD can suggest that unmitigated risks are present (such as NAPL as a constant groundwater and vapor contaminant source) and timeframes for the completion of remediation can be much longer than estimates where NSZD rates are calculated and presented in the CSM. Including and estimating NSZD rates in the CSM may improve the assessment of risks and establishes a more defensible end date for the completion of remediation.

Box #3

NSZD - an important part of the petroleum NAPL CSM
NSZD measurements establish a remediation baseline and
support interpretation of contaminant delineation and
concentration trends.

> Refine the CSM with quantification of bulk petroleum NAPL and/or chemical constituent loss rates

> Determine whether NSZD is sufficient to address risk/concerns

> Delineate the NAPL footprint using vadose zone indicators of biodegradation

> Compare NSZD to historical or potential future remedial activities

> Support estimates of source zone remedial timeframes (in development[1])

> Assess NAPL stability through application of a mass balance of NSZD mass losses and measured mobile NAPL flux[2]

[1] There are various methods available to estimate petroleum NAPL source zone remedial timeframes, but no practical options exist that incorporate NSZD processes. The reader is advised to carefully select a method that appropriately represents the CSM and meets the data objectives.
[2] This is a less common use of the NSZD measurements due to its mathematical rigor; however, it is an option for project teams to explore. See bottom of Table 4-1 of ITRC (2018).

As described in detail in CRC CARE (2018, 2020a), petroleum NAPL type, site-specific lithology, moisture content, hydrogeology, site setting and climate, and extent of petroleum NAPL are all important factors necessary to support a CSM containing NSZD. Relevant information is provided for each case study site in Table 13.1 and described in more detail in this section.

Each of the case study sites are located mostly within non-residential areas where other petroleum infrastructure exists (e.g., pipeline corridor, refinery). The exception is Site B which is in a mixed residential area with vacant land to the north. Sites B and C have groundwater migration toward surface water bodies (a river and ocean, respectively).

As noted in Sect. 13.3.3, NSZD was identified as a feasible component of the remedies. This section describes how the NSZD measurements were performed and used to refine the CSM at each case study site. A simplified CSM with brief overview of salient site conditions is presented including how it was used as a basis to revise the remedy to be inclusive of NSZD.

13.4.1 Site-Specific Factors to Consider When Estimating NSZD Rates

Prior to performing the NSZD measurements, relevant components of the CSM were assessed to identify conditions that could affect NSZD rate estimates (CRC CARE 2018). These conditions, described in Table 13.3, were taken into consideration in the design of the NSZD measurement programs at each case study site.

NSZD rates were estimated at the case study sites using different CO_2 efflux methods.

In Site A, the sitewide NSZD rate was quantified using one round of 19 DCC measurements distributed across the LNAPL footprint with passive (sorbent) flux trap with ^{14}C measurements collected at 5 of the DCC locations. One round was deemed appropriate due to the monotonous climate and similar year-round subsurface temperature conditions. The results from the DCC and traps were analyzed/ geospatially integrated separately to derive a range of NSZD rates. It is important to understand the range of NSZD rates present at the site; there is inherent variability with NSZD measurements that should be quantified.

In Site B, NSZD rates were measured using DCC and passive (sorbent) flux trap with ^{14}C methods over spring and fall seasons. DCC measurements were collected from 106 locations over the LNAPL footprint and 18 background locations. Traps were deployed at 25 locations distributed across the LNAPL footprint. NSZD rates in the early fall in temperate climates are generally at the high end of the spectrum when subsurface temperatures are warmest and vice versa in the spring. The results from the DCC and traps were analyzed/geospatially integrated separately for each spring and fall event to derive a range of NSZD rates.

In Site C, two rounds of NSZD measurements using DCC and one round of soil gas sampling for ^{14}C analysis were completed at 35 locations over the petroleum DNAPL footprint and 4 background locations outside the DNAPL footprint. Two rounds of DCC measurements spanning a couple weeks apart was deemed appropriate due to the monotonous climate and similar year-round subsurface temperatures. ^{14}C analysis on samples from a barium carbonate ($BaCO_3$) field soil gas precipitation method was performed at all DCC locations because of the extreme variability in ground cover types and non-petroleum-related contributions to CO_2 efflux across the efflux survey area (i.e., areas within and proximal to a semipermeable landfill soil cap, arid and heavy vegetation, mangrove trees, and a surface water body). The ability to correct for background at all locations maximized the data quality of the survey.

Of note, while these NSZD measurement programs are similar, NSZD monitoring options vary significantly and should be based on site-specific conditions. The range of NSZD monitoring methods are discussed in more detail in Chap. 5. The fact that all of the case studies herein use CO_2-efflux-based programs should not be implied to mean that it suits all sites. If the site-specific indicator data do not include CO_2 efflux, then other methods (e.g., biogenic heat or chemical composition change) should be used instead.

Table 13.3 Site-specific factors considered to estimate NSZD rates at case study sites

	Site A: Pipeline release, semi-arid climate	Site B: Former refinery, temperate climate	Site C: Former petrochemical plant, tropical climate
Site-specific factors			
Factors requiring additional background CO_2 efflux measurements	Within the NSZD monitoring area, there is irrigated landscaping. Due to the relatively large coverage of asphalt and concrete, this also happens to be where the CO_2 efflux is focused and occurs adjacent to roadways, sidewalks, and buildings. Monitoring results in these thin, wet, highly vegetated areas was distinctly different than the drier right-of-way areas with sparse vegetation	Limestone beneath the site weathers to produce a background source of CO_2 in addition to natural soil organic respiration	The site is located downgradient of LNAPL impacts from the adjacent refinery and the LNAPL overlays the site petroleum DNAPL requiring additional correction considerations and measurements in areas where only the LNAPL exists, so it can be subtracted from the areas with both LNAPL and DNAPL
			Limestone beneath the site weathers to produce a background source of CO_2 in addition to natural soil organic respiration
			Shallow soil is silt and clay loam, naturally high in organic carbon requiring a higher density of carbon 14 (^{14}C) samples to improve background correction
Factors affecting measurement timing and annual rate calculations	Site is in a relatively monotonous climate zone. Therefore, there were no significant factors that affected measurement timing and extrapolating the annual NSZD rates	Areas include natural wetlands and a phytoremediation system that produce additional soil gas CO_2 during growing seasons	Intermittent flooding and rainfall events periodically restrict soil gas transfer requiring planning to ensure NSZD rates are estimated in drier periods outside of hurricane season to prevent low biased measurements
		Intermittent flooding and rainfall events periodically restrict soil gas transfer requiring planning to ensure NSZD rates are estimated in drier periods and adjusted to estimate annual rates	

(continued)

Table 13.3 (continued)

	Site A: Pipeline release, semi-arid climate	Site B: Former refinery, temperate climate	Site C: Former petrochemical plant, tropical climate
		Temperate climate with relatively shallow petroleum hydrocarbon impacts required two seasons of NSZD measurements because the rate is susceptible to seasonal changes due to warming and freezing of the soil	
Factors affecting field methods, investigation extent, and rate estimates	Ground cover over the LNAPL footprint is partially impermeable (asphalt/concrete) restricting CO_2 efflux to an area smaller than the LNAPL footprint (Fig. 13.1). This requires modification of the data integration approach to include only pervious areas to estimate sitewide NSZD [See Sect. 13.4.1 and Appendix E.1, Geospatial Integration in CRC CARE (2018)]	Locally perched groundwater and wetlands can restrict soil gas movement in some areas and may lead to false-negative measurements of ongoing underlying NSZD	Ground cover over the petroleum DNAPL footprint is partially impermeable (plastic landfill liner) restricting CO_2 efflux. Like Site A, this requires modification of the sitewide integration of dynamic closed chamber (DCC) efflux results to include only the pervious areas to estimate sitewide NSZD

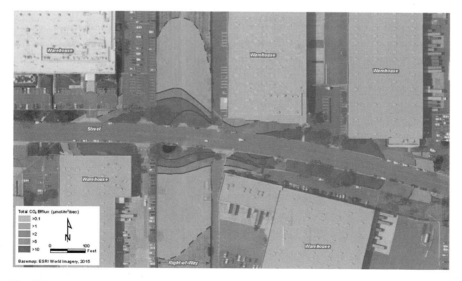

Fig. 13.1 Site A: illustration of CO_2 efflux at site with pervious and impervious ground cover (blue shading indicates areas where CO_2 efflux occurs over pervious ground cover areas and was included in the geospatial integration to estimate the sitewide NSZD rate)

13.4.2   Site A CSM: Pipeline Release, Semi-Arid Climate

The following summarizes the CSM for **Site A** (as shown on Fig. 13.2):

- The age of the jet fuel release is unknown (discovered in 1995)
- Primary constituents of concern are benzene and TPH-jet fuel (C9-22) in groundwater
- Elevated LIF responses limited to thin discontinuous, coarse-grained intervals below the water table near the release area (see Chap. 8 for more details on LIF technology)
- Significant regional water table rise occurred after the initial release and submerged the LNAPL body—pore fluid saturation test results from intact cores show that LNAPL is trapped in soil pores at residual saturation in the water-saturated media. LNAPL accumulates in wells by slow seepage of submerged LNAPL through screens.
- The LNAPL source is depleting and weathering as collectively indicated by:

  o The NSZD rate ranges between 2,700 and 4,900 L (L) (700–1,300 gallons [gal]) LNAPL degraded per year across the site (6,600–12,000 L per hectare per year [L/ha/year] [700–1,300 gal per acre per year (gal/ac/year)] equivalent). Note the areas of CO_2 expression varied between the data evaluation methods used to estimate the NSZD range (i.e., the DCC and trap results were integrated separately). The CO_2 efflux measured by DCC and traps at ground surface is derived from oxidized CH_4 generated by methanogenesis of the LNAPL present in both the capillary and saturated zones. The CH_4 by-product is directly outgassed through the saturated zone to the vadose zone through ebullition and is transported vertically upward to the ground surface.

  o After implementation of the past remedial activities (see Sect. 13.5.1), the recoverability of in-well LNAPL became impracticable[1] (transmissivity value of 0.005 square meters per day (m^2/day) (0.05 square feet per day [ft^2/day])

  o Dissolved phase benzene concentrations decreased >93% from 2007 to 2011

  o Drinking water quality objective exceedances in groundwater are limited to the LNAPL source area and immediate surrounding.

- Aqueous biodegradation is evident through observation of predictable differences in electron acceptors, biodegradation by-products, and redox conditions between upgradient and source area groundwater
- The perimeter monitoring wells surrounding the LNAPL body and dissolved phase plume halo remained absent of in-well LNAPL and below water quality objectives since 2013 when mechanical systems were shutdown
- The NSZD rate is ~10 times higher than the last recorded historical mechanical remediation rate (380 L [100 gal] LNAPL removed per year)

---

[1] Below the ITRC recoverability threshold range of 0.1–0.8 ft^2/day. https://lnapl-3.itrcweb.org/app endix-c-transmissivity-tn-appendix/.

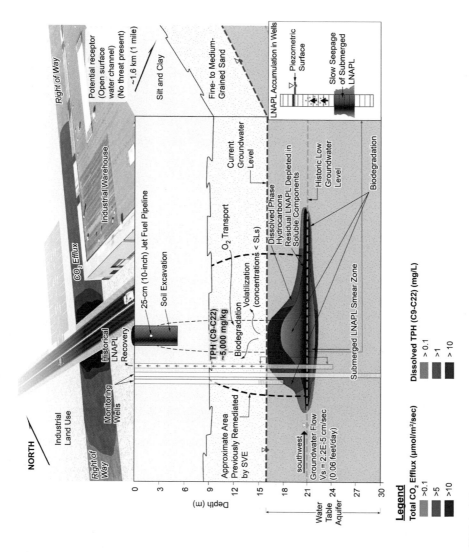

Fig. 13.2 Site A: CSM depiction

- Under the de facto NSZD remedy since 2013, there have been no complete (current) or operable (potential future) exposures due to:

    o Large depth of impacts (>5.5 mbg [18 ft bgs])
    o Soil gas survey results indicate concentrations do not pose a significant indoor or ambient air health risk
    o The portion of aquifer impacted by this release is in a groundwater use restriction area and a local water purveyor supplies potable water
    o The nearest municipal water supply well is 850 horizontal m (2,800 ft) hydraulically upgradient of the dissolved phase contaminant plume edge and screened within an aquifer that is >110 vertical m (370 ft) deeper than the LNAPL.

13.4.3   Site B CSM—Former Refinery, Temperate Climate

The following summarizes the CSM for Site B (as shown on Fig. 13.3):

- Petroleum hydrocarbon releases originated over 100 years ago and consist of crude oil and refined products
- Primary constituents of concern are benzene, MTBE, and TPH middle range hydrocarbons (C9–18) in groundwater
- Groundwater migrates toward an adjacent river
- Underlying weathered limestone is the aquifer matrix forming the primary lateral groundwater migration pathway; it is underlain by a competent limestone that retards vertical migration of groundwater
- Interceptor trenches were installed to prevent offsite migration of LNAPL and dissolved phase contaminants to neighboring properties and adjacent river

    o Treatment wetland is used to treat extracted groundwater before discharge to river

- Remediation is augmented with a phytoremediation plot to enhance rhyzodegradation of contaminants upland of the river
- LNAPL is depleted in zones of recovery, it has not been recovered from trenches since 2016
- Transmissivity values in wells and trenches are below the lower ITRC recoverability threshold metric of 0.01 m^2/day (0.1 ft^2/day) (see Chap. 12 for more details on transmissivity tests and recoverability endpoints)
- LNAPL seemed to be immobile only, trapped in soil pores as evidenced by negligible LNAPL recovery, lack of in-well LNAPL accumulation, and only intermittent observation of sheens and globules in water samples
- NSZD was associated to most of the NAPL depletion at the site; the estimated NSZD rate ranges between 220,000 and 265,000 L (58,000–70,000 gal) of LNAPL degraded per year (2,800–12,000 L/ha/year [300–1,300 gal/ac/year] equivalent) significantly exceeding historical mechanical remedial actions. Note the areas of

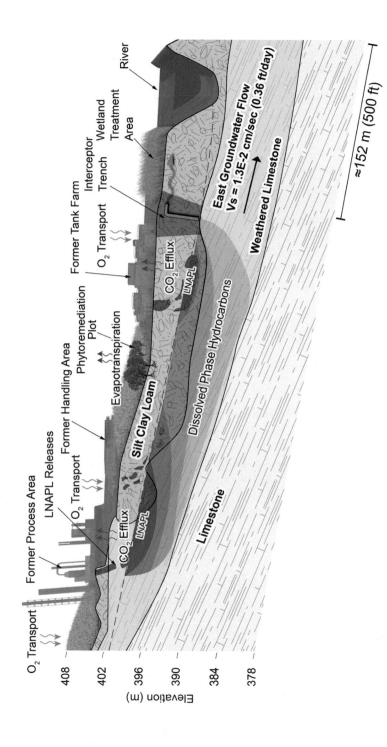

Fig. 13.3 Site B: CSM depiction

CO_2 expression varied between the data evaluation methods used to estimate the NSZD range (i.e., the DCC and trap results were integrated separately). The large magnitude of NSZD demonstrates that it has been the primary process historically reducing LNAPL saturations thereby limiting LNAPL migration. The range in the unitized NSZD rate is indicative of the relatively large difference of area of CO_2 expression on ground surface between the two monitoring events (i.e., the larger the integrated area of expression, the smaller the unitized NSZD rate).

- Under the interim measures including de facto NSZD, there are no complete (current) or operable (potential future) exposure pathways due to:

  o   Onsite residential land use is prohibited by institutional control
  o   Land is fenced and secured to avoid direct contact
  o   There are no occupied buildings
  o   The portion of aquifer impacted by this release does not qualify (due to low yield) for potable use and a local water purveyor supplies potable water
  o   Risk to river receptors is mitigated by historical interim measures, demolition of the facility, and hydraulic controls (trenches, wetlands treatment, and phytoremediation plot)

13.4.4   Site C CSM: Former Petrochemical Plant, Tropical Climate

The following summarizes the CSM for Site C (as shown on Fig. 13.4):

- Site is in an area vulnerable to extreme events (e.g., hurricanes, earthquakes)
- Primary constituents of concern are BTEX and PAHs in groundwater
- The petroleum hydrocarbon release is >40 years old and is waste material from a past petrochemical industrial process (viscosity >100 centipoise [cP])
- Shallow groundwater discharges to a canal that is tidally connected to the ocean and contains mammalian ecological receptors
- The petroleum DNAPL (1.08 specific gravity) migrated vertically downward from the waste impoundment via gravity- and density-driven forces
- The underlying limestone acts as an aquitard; very little DNAPL is observed at this interface. Most petroleum DNAPL is trapped in the overlying interbedded alluvium
- Petroleum DNAPL and the small, dissolved plume halo extend onto a neighboring industrial property
- Historical groundwater extraction focused on the downgradient edge of the DNAPL body
- Like Site A, the extent of groundwater protection standard exceedances is limited to the DNAPL source area and immediate surrounding
- Recoverability of DNAPL is low. Transmissivity tests (calculations modified for DNAPL behavior on a confining layer) at wells indicate a range between 0.001 and 0.2 m^2/day (0.01–2 ft^2/day). One well reported a transmissivity value greater

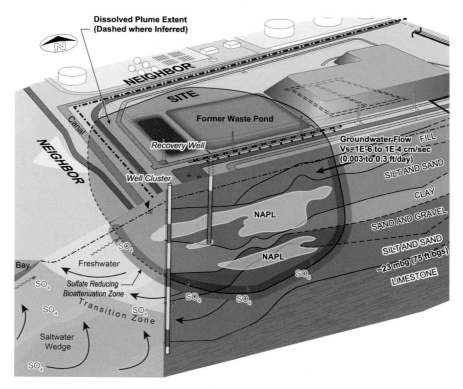

Fig. 13.4 Site C: CSM depiction

than the high end of the ITRC recoverability threshold metric (0.08 m^2/day [0.8 ft^2/day])

- A saltwater wedge transition zone facilitates control of plume migration through continuous supply of sulfate to support subsurface biodegradation at the hydraulically downgradient edge
- The NSZD rate ranges from 12,000 to 15,000 L (3,200–3,900 gal) petroleum DNAPL degraded per year (1,400–1,800 L/ha/year [150–190 gal/ac/year] equivalent). The warm 30 °C subsurface temperature facilitates NSZD of the weathered petroleum DNAPL. The tight range in the NSZD rate is indicative of the relatively small difference in the area and magnitude of CO_2 expression on ground surface between the two measurement rounds
- A 3-year plume stability evaluation demonstrated the effectiveness of NSZD and showed that the DNAPL and dissolved phase plume are not unacceptably unstable

in terms of DNAPL presence/absence at the perimeter and changes in plume center of mass migration distances and directions over time.

   o An exception was localized to a couple of wells within the pre-existing DNAPL footprint and only occurred because of an earthquake that compromised well-seal integrity

- There are no complete (current) or operable (potential future) exposure pathways due to:

   o The industrial land is fenced and secured to avoid direct contact and future land use over the DNAPL and dissolved plume footprint is prevented by deed restrictions limiting intrusive activities
   o There are no inhabited buildings that pose a risk of volatile organic vapor intrusion
   o The aquifer portion impacted by this release is not potable due to high salinity and low yield, and a local water purveyor supplies potable water
   o Risk to ecological receptors in the canal was mitigated by corrective measures (i.e., sediment cap).

13.5 Incorporating NSZD into the Site Management Strategy

As noted in Sects. 13.3.3 and 13.4, NSZD was identified as a feasible remedy component at each case study site and NSZD rates were quantified and woven into updated, better-defined CSMs. That was just the start of incorporating NSZD into site management. This section and the next discuss how these sites carried NSZD further into the remedy and decision document.

13.5.1 Comparison of NSZD Rates to Historical Remedial Actions

Box #4 shows results from various remediation projects comparing mass depletion rates from NSZD (13 sites) and mechanical remedial technologies (16 systems) (Palaia et al. 2021). It documents the significance of NSZD and its effectiveness in relation to mechanical systems such as skimming and aeration technology in their early- and later-stage operation. The results are supported by the literature. Garg et al. (2017) summarized published NSZD rates and reported the middle two quartiles between 6,500 and 26,000 L/ha/year (700–2,800 gal/ac/year) with a median of 16,000 L/ha/year (1,700 gal/ac/year). CRC CARE (2020a) demonstrated that NSZD significantly outperformed LNAPL recovery at four of the six studied sites and recorded average NSZD rates range of 1,900–8,400 L/ha/year (200–900 gal/ac/year) across six sites.

Box #4

Rates of NSZD vs. Mechanical Remediation

- Wide range of remediation rates

- NSZD generally falls within the middle two quartiles

- Scrutiny shows that mechanical mass removal rates decline over time and approach the median rate of NSZD

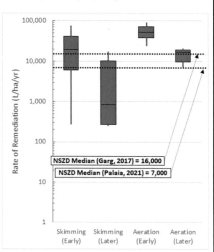

Ten "Skimming" systems included various types of manual and automatic LNAPL-only fluid recovery accounting for fluid mass recovery only. Six "Aeration" systems included multiphase extraction (MPE), combined air sparging and SVE, and SVE only accounting for both fluid and vapor-phase mass removal.

"Early" considers the approximate average mass removal rate during the first year of operation. "Later" considers the approximate average rate after the first year of operation. (Palaia, 2016).

To directly compare NSZD and mechanical system recovery for the case study sites, timelines and cumulative removal curves were created. Figures 13.5, 13.6 and 13.7 present an overview of the remedial activities conducted at each site and the petroleum NAPL removal achieved including a rough approximation of the volume removed by NSZD during each period. In each case, it shows NSZD removed more mass of petroleum hydrocarbon than the other remedial technologies. These data provide a strong line of evidence to support use of NSZD as a primary component of a site management strategy.

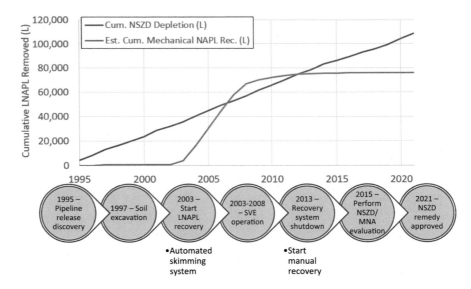

Fig. 13.5 Site A: remedial action and historical mass removal timeline (based on a constant NSZD rate simplifying assumption)

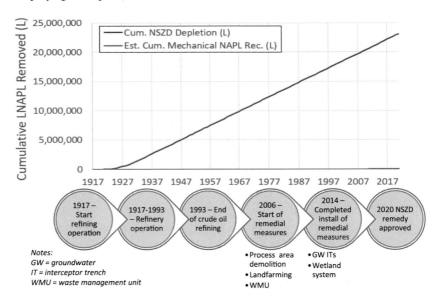

Fig. 13.6 Site B: remedial action and historical mass removal timeline (based on a constant NSZD rate simplifying assumption)

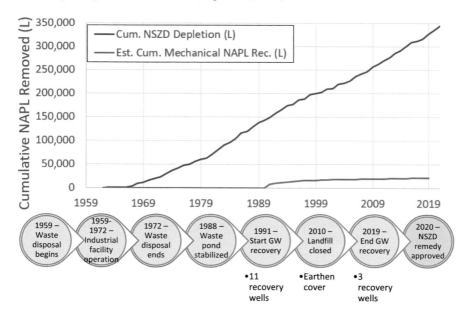

Fig. 13.7 Site C: remedial action and historical mass removal timeline (based on a constant NSZD rate simplifying assumption)

The NSZD removal values plotted in Figs. 13.5, 13.6 and 13.7 are based on a constant NSZD rate simplifying assumption in agreement with Garg et al. (2017) who suggested that NSZD rates typically are pseudo zero order over time. On the contrary, Davis et al. (2022) presented estimates of site-specific gasoline and diesel NSZD rates that exhibited nonlinear trends over a period of 21–26 years. Further research is encouraged to elucidate the long-term behavior of NSZD processes under various scenarios and provide practical solutions to predict NSZD rates over time. For the case studies presented in this chapter, the annual NSZD rates were estimated using randomized values derived from the measured ranges noted in Sects. 13.4.2 through 13.4.4 and extrapolated throughout the remediation life cycle. For Sites B and C associated with facilities and known startup dates, assumed NSZD rates were ramped up over a 12-year period starting from the inception of the operation noted on Table 13.1. The NSZD rates were assumed zero over years zero to two (facility was new, assumes no releases had yet occurred), then 10% of the measured rates over the next 5 years (some releases began), and 50% of the reported rates over the next 5 years (as the release volume started to accumulate).

Due to the large size of Site B (~61 hectares [150 acres]) and its 76-year historical use as a refinery that likely processed over 3.8 million L (1 million gal) of oil per day and had multiple large releases, the cumulative LNAPL removed by NSZD is very large.

13.5.2 Multiple Lines of Evidence to Support Use of NSZD in Remedy Transition

To assess whether the case study sites were ready for transition to an NSZD-based MNA remedy, multiple lines of evidence were used to assemble the proper rationale to obtain regulatory approval. This is typical to cover the various requirements of a remedy and address the inherent uncertainty associated with environmental data. It is analogous to presenting a case in a court of law; ample evidence must be presented to demonstrate the effectiveness of NSZD and the ability of the MNA remedy to protect human health and the environment and meet regulatory requirements. Box #2 can be used as a starting point to help craft the rationale.

As discussed above, the site-specific NSZD rates were estimated to demonstrate to each respective regulatory agency that mass reduction rates are both significant and greater than rates achieved from historical physical removal. However, the case to proceed with a management strategy that includes NSZD is not made solely with an NSZD measurement and comparison to past remedial activities.

The following elements provided a well-founded basis for the inclusion of NSZD into the remedy (the relevant site, as detailed in the CSM Sects. 13.4.2 through 13.4.4, is noted in parenthesis):

- CSM of the nature/extent and fate/transport of the petroleum hydrocarbon release was well defined (each site)
- Past remedial efforts operated to practical limit of recovery (Sites A and C).
- Petroleum NAPL and dissolved phase migration is mitigated either by NSZD (Sites A and C) or by hydraulic controls (Site B)
- No complete (current) or operable (potential future) exposure pathways exist (each site)
- The remedial timeframe for NSZD is comparable to other remedial approaches (Sites A and B)
- Other than NSZD rates, ample evidence of petroleum NAPL weathering and biodegradation (Sites A and C)

Perhaps the most important element that supports the implementation of NSZD is elimination or mitigation of human health and/or ecological exposure. It is crucial to note that this can be accomplished using various means including NSZD processes, administrative and engineering controls, other remedial actions, demonstration of poor resource quality (e.g., low transmissivity or high total dissolved solids in groundwater), and physical separation (e.g., adequate oxygenated soil exists between NAPL and receptor to alleviate concern of vapor intrusion). The case studies used each of these means to eliminate exposure pathways.

In addition, the sustainability and resiliency of continued mechanical removal versus NSZD should be considered. Box #5 summarizes five sustainability factors and their associated relative ranking for NSZD and four remedial technologies that were employed at one or more of the case study sites. NSZD is considered an inherently sustainable and resilient remedial process as it:

- Is naturally occurring and effectively remediates NAPL at many petroleum sites;
- Requires minimal energy and non-renewable resource use,

- Generates minimal waste,
- Consists of nominal vulnerable infrastructure,
- Is resilient to periodic climate or natural upsets like hurricanes and earthquakes,
- Uses readily available methods/materials that can be supported solely by local labor, and
- Can be monitored by remote telemetry if using the soil gas or biogenic heat methods obviating the need for routine travel.

Box #5

Sustainable Resilient Remediation

An optimized remedial solution that

- limits environmental impacts
- maximizes social and economic benefits
- creates resilience to extreme weather events, sea-level rise, and wildfires

Excerpt from Appendix A of the ITRC LNAPL - 3 Guidance (ITRC, 2018). The table below summarizes the ITRC's ranking of sustainability related evaluation factors of remedial options used at Sites A, B, and C.

	NSZD	Excavation	Multi-Phase Extraction	AS/SVE	Phyto-technology
Safety Concerns	Low	Moderate	Moderate	Low to Moderate	Low to Moderate
Community Concerns	Low to Moderate	Low to Moderate	Moderate	Low to Moderate	Low
Carbon Footprint/ Energy Needs	Low	High	Moderate	Moderate to High	Low
Waste Management	Low	Moderate to High	Moderate	Low to Moderate	Low
Cost	Low to Moderate	High	Moderate	Low to Moderate	Low

In particular, at Site C, due to multiple historical extreme natural events, the benefits of remedy resilience to prolonged power outages and inaccessibility were an important factor in accepting the transition to an NSZD-based MNA remedy.

Table 13.4 summarizes the primary lines of evidence (technical, sustainability, and risk reduction factors) used to support NSZD as a remedial component for each case study site. While a NSZD-based remedy can result in significant cost savings, an important consideration during remedy evaluation, cost was intentionally excluded herein due to its inconsistent consideration by stakeholders as a selection criterion. A NSZD-based remedy is inherently low cost, typically lower than other remedial alternatives, and that is a strong advantage. However, the decision to select it is also typically weighted heavily on technical, risk, and sustainability-related factors.

It is important to note that while NSZD was brought into later phases of the three case study projects and coupled or paired with other remedial technologies, it can also be implemented as the sole process in an MNA remedy. Box #6 calls out the baseline, stand-alone, coupled component, or treatment train transition options that exist for incorporation of NSZD in an MNA remedy.

Box #6

NSZD-inclusive Remedial Options[1]

Baseline

- NSZD rates are used to evaluate the relative benefit of other options (e.g., by comparison to typical LNAPL skimming system's removal rates)

Stand-alone

- A NSZD-based MNA remedy at sites where LNAPL-related impacts are stable, potential receptors are not at risk, and NSZD timeframes are consistent with the goals of the site owner and with regulatory requirements

Component of an engineered remedy

- NSZD is a primary remedy component for portions of the LNAPL zone, and engineered remediation systems are focused on those portions of the LNAPL footprint where more aggressive removal is appropriate to meet remedial goals

Transition remedy

- A final step in a treatment train, NSZD-based MNA is used as a long-term risk management approach after other remedial options have achieved their design objectives

[1] Modified from Appendix B Executive Summary in ITRC, 2018. Light Non-Aqueous Phase Liquid (LNAPL) Site Management: LCSM Evolution, Decision Process, and Remedial Technologies. LNAPL-3. Washington, D.C. https://lnapl-3.itrcweb.org/.

Table 13.4 Summary of supporting lines of evidence for NSZD in case studies

	Site A: Pipeline release, semi-arid climate	Site B: Former refinery, temperate climate	Site C: Former petrochemical plant, tropical climate
Technical factors			
Transmissivity (Tn)	Tn reduced from 0.03 m²/day (0.3 ft²/day) in 2008 to 0.005 m²/day (0.05 ft²/day) in 2013	Insufficient in-well LNAPL recharge to calculate Tn, no LNAPL in recovery trenches	Tn estimated[2] below 0.01 m²/day (0.1 ft²/day) in all but one well due to DNAPL viscosity >100 cP
Dissolve phase mass reduction[3]	~76 L (20 gal) LNAPL/year	~4 L (1 gal) TPH/year	~114 L (30 gal) DNAPL/year
NAPL mass reduction, sitewide NSZD rate	NSZD rates range from 2,600 to 4,900 L (700–1,300 gal) LNAPL degraded per year (6,500–12,000 L/ha/year [700–1,300 gal/ac/year] equivalent)	NSZD rates range from 220,000 to 265,000 L (58,000–70,000 gal) LNAPL degraded per year (2,800–12,000 L/ha/year [300–1,300 gal/ac/year] equivalent)	NSZD rates range from 12,000 to 15,000 L (3,200–3,900 gal) petroleum DNAPL degraded per year (1,400–1,800 L/ha/year [150–190 gal/ac/year] equivalent)
Sustainability–resiliency factors			
Sustainability and resiliency factors	Mechanical systems are energy intensive, blowers and pumps have high carbon footprint; aboveground infrastructure is vulnerable to extreme weather	Natural treatment systems (wetlands, phytoremediation) reduce carbon footprint; mechanical systems are energy-intensive; aboveground infrastructure is vulnerable to extreme weather	Hazardous waste requires overseas shipping for disposal; regular power loss from extreme weather events and earthquakes reduces mechanical system runtime and results in excessive replacement costs
Risk reduction factors			
Institutional controls	Groundwater use is legally restricted over the impacted zone in perpetuity	Non-residential land use governed by environmental use control to prohibit onsite residents and building construction	All future intrusive activity and facilities atop the DNAPL and plume footprint are prevented by deed restrictions
NAPL migration	No threat of LNAPL migration (LNAPL saturation reduced by historical corrective actions)	Interceptor trenches prevent potential offsite migration; no NAPL migration observed since 2006	Three-year plume stability evaluation showed DNAPL is not laterally or vertically migrating; salt water wedge supports stability with constant high concentration supply of sulfate at downgradient edge of plume
Dissolved phase plume stability	No dissolved plume downgradient of LNAPL footprint	Plume extent and constituent of concern concentrations are stable to decreasing	Stable center of mass confirmed using >20 rounds of monitoring during a 3-year geospatial analysis; decreasing total mass; dissolved plume is not present beyond DNAPL extent

[2] The transmissivity analysis was modified for DNAPL behavior.

[3] Follows the mass budgeting approach to estimate assimilative capacity as prescribed in National Research Council (2000).

13.6 Implementation of the NSZD-Based MNA Remedy

After assembling multiple lines of evidence to support the transition to NSZD for each case study site, regulatory agencies were engaged to discuss, finalize, and codify NSZD as a component of the MNA remedy in a decision document. The following sections narrate the evolution of approving NSZD as a remedy component at each site and the associated routine NSZD monitoring plan.

As part of the NSZD monitoring program, clear objectives and endpoints for the NSZD remedy are essential. Box #7 calls out example objectives, endpoints, and indicator data to determine the endpoints.

Box #7

NSZD Remedial Objective	Example Endpoint[1]	Example Indicator Data
Remove petroleum NAPL to the maximum extent practicable	Demonstrate sustained NSZD mass loss rates that exceed other remedial options	Periodic NSZD rate measurements for a period that is acceptable to stakeholders (e.g., 5-10 years)
Risk management (e.g., at an active facility)	Demonstration of no unacceptable risk to human health or the environment	Monitoring of constituent of concern concentrations proximal to receptor locations (e.g., in sub-slab soil vapor) to prove they are sustained below risk-based criteria
Attainment of water quality standards protective of a nearby receptor (e.g., risk-based corrective action)	Achievement of risk-based cleanup goals in groundwater or surface water	Monitoring of constituent of concern concentrations in groundwater, optionally monitoring the chemical content of petroleum NAPL (e.g., every 5-10 years) to document decreasing source concentrations
Demonstrate containment of petroleum NAPL within a controlled land area (e.g., a residual management zone)	Demonstrate that petroleum NAPL is not laterally migrating and NSZD is effectively contributing to a receding plume extent	In-well petroleum NAPL gauging and groundwater monitoring to prove absence of the NAPL outside the limits of the controlled area and a shrinking aqueous phase plume

[1] Modified from Table NSZD-4 in ITRC (2018).

13.6.1 Site A: Transition to NSZD-Based MNA Remedy with Groundwater Monitoring

Following an update of the CSM incorporating NSZD, time-series trends were evaluated to confirm that after mechanical remediation ceased, contaminant concentrations in source area groundwater were decreasing and the dissolved phase and LNAPL plumes were not migrating. The regulatory agency accepted the NSZD rate measurements and approved a follow-up report that provided the proposed cleanup goals (drinking water quality), remedial timeframe of < 75 years (comparable to other alternatives), a contingency plan for biosparging in case NSZD does not meet expectations, and technical and economic rationale on why attainment of background water quality was not feasible with other alternatives.

Fundamental to the success of gaining endorsement of NSZD was the demonstration that NSZD was depleting constituents of concern in the LNAPL, refuting the agency's initial statements that LNAPL would serve as a perpetual source to groundwater. LNAPL samples were analyzed using whole oil analysis to show the depletion of benzene and the soluble components of TPH-jet fuel compared to fresh fuel.

Additionally, passive diffusion bag samplers were used to evaluate the magnitude of petroleum biodegradates (aka., polar metabolites) in groundwater samples collected within the LNAPL footprint. This demonstrated that the fraction of total "normal" laboratory TPH results (i.e., low-flow groundwater samples analyzed without silica gel cleanup) consisting of original petroleum constituents was very small (i.e., < 10%). It also helped further support observations that the LNAPL was weathering and the remaining LNAPL footprint was largely weathering residuals rather than regulated constituents of concern. This supported classification of the site as a low-risk profile.

NSZD will be monitored annually using DCC and trap methods for rate measurements and groundwater sampling and analysis of constituents of concern for the first 5 years and every 5 years thereafter. The DCC survey and trap locations are consistent with the initial characterization event (with a small offset to avoid prior soil disturbances) to maximize comparability between monitoring events. Traps are installed at ~20% of the DCC survey locations. The NSZD performance monitoring also includes LNAPL sample analysis every 10 years to monitor continued depletion of benzene for as long as LNAPL remains in-well. The LNAPL samples are chemically analyzed using U. S. Environmental Protection Agency direct oil injection Method 8260 gas chromatography using mass spectrometry (GC–MS) with high-resolution chromatograms. In addition, the analysis of natural attenuation indicator parameters will be performed every 5 years on groundwater samples to assess electron acceptors and biodegradation by-products and affirm redox conditions conducive to biodegradation surrounding the submerged LNAPL.

NSZD was determined suitable for a future planned land use of commercial/industrial, with groundwater use restrictions for the duration of the MNA remedy by institutional controls. Institutional controls are the risk mitigation measure accompanying the NSZD remedy.

13.6.2 Site B: NSZD with Phytoremediation and Hydraulic Control

Due to the early date of historical releases (circa 1917), weathering (i.e., NSZD) had already depleted a large fraction of the petroleum LNAPL by the time remedial efforts were undertaken. LNAPL is no longer observed in site recovery trenches, and thin thicknesses of mobile LNAPL was only measurable in a few site wells.

Following the update of the CSM incorporating NSZD, further testing confirmed LNAPL transmissivity values below 0.01 m^2/day (0.1 ft^2/day) (the ITRC lower-end threshold metric for effective recovery[4]) and appropriate evidence of ongoing natural attenuation of dissolved phase constituents of concern. The timeframe to achieve cleanup goals in groundwater for each remedial alternative was estimated at generally < 50 years with estimates of > 100 years in isolated LNAPL areas. A corrective action plan was provided to the regulatory agency containing a proposed MNA remedy including NSZD paired with phytoremediation and limited additional short-term interceptor trench operation along the river. The regulatory agency approved the plan and issued a new Consent Order that included specific endpoints for interceptor trench operation (e.g., LNAPL transmissivity of <0.08 m^2/day [0.8 ft^2/days]) and contingency actions. The MNA remedy was implemented. Contingencies included an additional interceptor trench along the river and sub slab vapor barrier in case remedial expectations were not met.

NSZD rates are measured every 5 years using 25 passive flux traps co-located with the initial measurement event to evaluate performance and document long-term effectiveness. Groundwater sampling and analysis of constituents of concern is performed annually. The results of NSZD and groundwater monitoring are used to determine when interceptor trench operation is no longer necessary and/or sufficiently beneficial to continue and conversely, when contingency actions are needed.

NSZD was determined suitable for future planned non-residential land use. Surrounding land remains residential, and groundwater use is prohibited due to low yield. Interim groundwater containment and onsite Environmental Use Controls are the risk mitigation measures accompanying the portion of the site with the NSZD remedy.

---

[4] See Appendix C of ITRC (2018).

13.6.3 Site C: Transition to NSZD-Based MNA Remedy with Groundwater Monitoring

NSZD was identified as a viable component of an MNA remedy after a significant hurricane knocked out the power at the site necessitating high-cost measures to maintain permit compliance. A search for more resilient and sustainable remedies ensued because the expense to supply generator power was excessive. Due to regulatory agency skepticism of NSZD at a petroleum DNAPL site, the CSM was updated using LIF, NSZD rate estimates, and a multiyear plume stability evaluation.

Further data were required by the regulatory agency to document DNAPL chemical weathering, DNAPL recoverability using transmissivity tests, and the effectiveness of the saltwater wedge in mitigating dissolved phase plume expansion. Results of the direct inject GC–MS analysis indicated very low and decreasing contents of the contaminants of concern in the DNAPL. Transmissivity tests confirmed values below the ITRC threshold for recoverability. Study of the saltwater wedge verified that an upward gradient associated with high deeper groundwater salinity limited both lateral and vertical migration. The wedge's effect on plume attenuation (by acting as a source of sulfate) combined with natural attenuation indicator parameter data indicate biodegradation sustained by the adjacent ocean is occurring. These results combined with the evidence of NSZD in the source area provided evidence that lateral migration of the dissolved phase plume is not occurring.

Geospatial statistical analyses of groundwater sampling and petroleum DNAPL gauging results confirmed that the dissolved phase and petroleum DNAPL plumes, respectively, were stable without mechanical remediation. NSZD was determined to be more resilient to the impacts of extreme weather events (primarily high winds, flooding, and electricity outages) than other remedial alternatives.

Based on these results, the project team worked with the regulatory agency to prepare a permit modification containing NSZD and limited petroleum DNAPL recovery with an ITRC-based transmissivity endpoint metric of 0.01 m^2/day (0.1 ft^2/day). The NSZD rates will be monitored continuously and remotely with web-enabled instrumentation at five locations across the petroleum DNAPL footprint.

NSZD was determined suitable for a future planned land use of undeveloped open space or possibly as a solar farm. Surrounding land remains industrially zoned and groundwater remains unusable due to high salinity. Deed and site access restriction (i.e., fencing) accompany the NSZD-based MNA remedy to prevent contact with residual DNAPL and impacted groundwater.

13.7 Recommendations for Improved Practice

The use of NSZD as a site management remedy component is evolving. For example, NSZD has not been quantified at many heavy hydrocarbon and/or dense petroleum NAPL sites. NSZD will become more widely understood as it is observed to occur at more petroleum NAPL release sites. This will result in expanded use and acceptance in decision documents. As a result, the weight of evidence (i.e., the number of lines of evidence) needed to support the effectiveness of NSZD and justify NSZD as a remedy component may change. As would be done with any unfamiliar remedial option, engagement with stakeholders at the outset of NSZD consideration is prudent. This will help inform starting knowledge and primarily identify the amount and types of data that will be needed to justify the inclusion of NSZD as a sole component of the MNA remedy or part of a multi-component remedy.

To emphasize earlier statements, active and consistent education of stakeholders about the merits and effectiveness of NSZD is key. Stakeholders' perceptions and expectations must be carefully calibrated and addressed for a project to be successful.

13.8 Summary: Incorporating NSZD Into the Site Management Strategy

The narratives of three diverse case study sites presented in this chapter provide examples of how NSZD was incorporated into site management strategies. At each case study site, NSZD was incorporated both in the CSM and the implemented remedial option. Key points related to use of NSZD on a broader range of site conditions were highlighted in call out boxes. The case studies were used to demonstrate NSZD as an effective and sustainable/resilient management remedy component for low-risk sites containing petroleum NAPL. This chapter stresses the importance of acknowledging, accounting, and incorporating NSZD into remedies because NSZD is observed to occur at most, if not all, petroleum NAPL release sites.

The reader is advised to use discretion when interpreting the case studies and call out boxes, and to work closely with technical and regulatory teams when considering how to include NSZD in site management at their sites.

References

CRC CARE (2018) Technical measurement guidance for LNAPL natural source zone depletion. In: CRC CARE technical report no. 44, CRC for contamination assessment and remediation of the environment, Newcastle, Australia

CRC CARE (2020a) The role of natural source zone depletion in the management of light non-aqueous phase liquid (LNAPL) contaminated sites. In: CRC CARE technical report no. 46, CRC for contamination assessment and remediation of the environment, Newcastle, Australia

CRC CARE (2020b) Australian case studies of light non-aqueous phase liquid (LNAPL) natural source zone depletion rates compared with conventional active recovery efforts. In: CRC CARE technical report no. 47, CRC for contamination assessment and remediation of the environment, Newcastle, Australia

Davis GB, Rayner JL, Donn MJ, Johnston CD, Lukatelich R, King A, Bastow TP, Bekele E (2022) Tracking NSZD mass removal rates over decades: site-wide and local scale assessment of mass removal at a legacy petroleum site. J Contam Hydrol 248:104007

Garg S, Newell CJ, Kulkarni PR, King DC, Adamson DT, Irianni Renno M, Sale T (2017) Overview of natural source zone depletion: processes, controlling factors, and composition change. Groundwater Monit Remed 37(3):62–81. https://doi.org/10.1111/gwmr.12219

Interstate Technology and Regulatory Council (ITRC) (2018) Light non-aqueous phase liquid (LNAPL) site management: LCSM evolution, decision process, and remedial technologies. LNAPL-3, Washington, DC. https://lnapl-3.itrcweb.org/

National Research Council (2000) Natural attenuation for groundwater remediation. National Academy Press, Washington, DC

Palaia T, Hachkowski A, Nichols E (2021) NSZD: moving past the perceptions. In: Presented at the 2021 environmental services association of Alberta (ESAA) remediation technologies (RemTech) symposium, Banff Springs, Alberta, Canada

Chapter 14
Bioremediation of Petroleum Hydrocarbons in the Subsurface

Sarah M. Miles, Ron Gestler, and Sandra M. Dworatzek

Abstract Due to human activity and, to a lesser extent, natural processes, petroleum hydrocarbons continue to pollute the environment. These contaminants of concern can be found globally and their remediation is key to restoring affected sites to safe and functional status. Conventional treatment of sites contaminated with petroleum hydrocarbons relies heavily on remediation approaches that are often financially prohibitive or may be technically impractical and that sometimes produce undesirable by-products. Using microbes that occur in nature (if not always at the site), can be a viable treatment with distinct advantages. Understanding the environment, contaminants, and natural biological processes occurring are key aspects for effective application of remediation techniques that rely on biological processes. Whether by stimulating the native microbial community, or, secondarily, by augmenting the native community with known degrader populations to degrade the target compounds, bioremediation is a practical, effective, and sustainable natural solution to a wide array of contamination around the globe. This chapter explores approaches to bioremediation of both soil and groundwater contaminated by petroleum hydrocarbons, describing how the approaches work and the benefits and challenges associated with them. It focuses on the use of aerobic and anaerobic microbial bioremediation, phytoremediation, and mycoremediation to address petroleum hydrocarbons.

Keywords Bioremediation · Mycoremediation · Natural attenuation · Petroleum hydrocarbons · Phytoremediation

S. M. Miles · R. Gestler
Geosyntec Consultants International, Inc., Guelph, ON, Canada
e-mail: SMiles@geosyntec.com

R. Gestler
e-mail: RGestler@geosyntec.com

S. M. Dworatzek (✉)
SiREM, Guelph, ON, Canada
e-mail: SDworatzek@siremlab.com

© The Author(s) 2024
J. García-Rincón et al. (eds.), *Advances in the Characterisation and Remediation of Sites Contaminated with Petroleum Hydrocarbons*, Environmental Contamination Remediation and Management, https://doi.org/10.1007/978-3-031-34447-3_14

14.1 Introduction

Petroleum hydrocarbon (PHC) pollution is a major and continuing environmental issue caused by human activity and, to a lesser extent, by natural processes. Water and soil contaminated by PHCs can harm local ecosystems and contribute to widespread environmental deterioration. Conventional treatment techniques for these contaminated sites rely on methods that remove, reduce, or mitigate the toxic effects of PHCs. These include pump-and-treat systems, air sparging, multiphase extraction, soil vapour extraction, and excavation. Excavated materials—typically soil—can be disposed of off-site or treated using one or more remedial technologies, including bioremediation, incineration (e.g. open pit burning), thermal desorption, chemical treatment, and containment systems. All of these treatments have the potential to be financially prohibitive, technically impractical, and may produce undesired by-products (Speight and Arjoon 2012); some of them may result only in transferring contaminants from one media to another, rather than degrading the contaminants to non-toxic end products.

Bioremediation, which uses bacteria or other biological means to degrade contaminants, has emerged at the forefront of treatment technologies for petroleum-contaminated systems over the years (Morgan and Watkinson 1989; Butnariu and Butu 2020; Ławniczak et al. 2020). Because the natural environment contains a diverse consortium of microbial and plant life, harnessing the potential of living organisms to degrade and remediate contamination is a viable and effective treatment option (Fig. 14.1). Bioremediation functions either through biodegradation, which is the mineralization of the organic contaminant into carbon dioxide, water, inorganic compounds, and cell proteins, or it may transform the target organic compound into other simpler organic compounds that are generally less detrimental to the environment, such as methane, fatty acids, alcohols etc. (Atlas 1991; Speight and Arjoon 2012; Balseiro-Romero et al. 2018; Butnariu and Butu 2020).

At contaminated sites, PHCs commonly undergo biphasic biodegradation. In the initial phase of biodegradation, the rate of removal is high and primarily limited by microbial degradation kinetics of the native microbial community. In the second phase, the rate of PHC removal is slower and is primarily limited by bioavailability of the contaminant, with slow desorption of compounds from the soil mineral and organic matter fractions (Huesemann et al. 2004; Megharaj et al. 2011). Remediation timelines vary widely based on site specific conditions and the characteristics of the PHCs (Balseiro-Romero et al. 2018).

Whilst aerobic bioremediation approaches effectively remediate a wide range of PHCs, there are some instances where their application is technically impractical or excessively expensive due to oxygen demand or delivery. In these cases, previously unknown anaerobic approaches can also be considered. For example, whilst anaerobic benzene, toluene, ethyl benzene, and xylene (BTEX) bioremediation was once not expected to occur, it is now known to be possible under a range of anaerobic conditions (Ulrich and Edwards 2003). Whether aerobic or anaerobic, for bioremediation to be effective, microorganisms (naturally present or added through bioaugmentation)

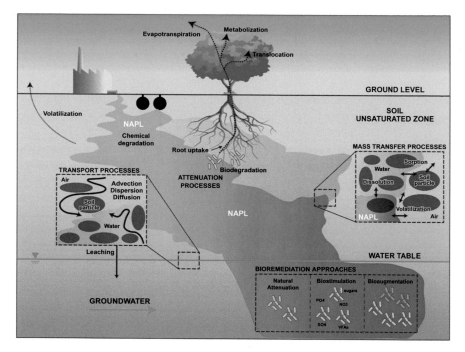

Fig. 14.1 Conceptual site model of PHCs and associated processes in the subsurface

must enzymatically break down contaminants and convert them into less hazardous products. This chapter gives an overview of bioremediation processes for PHCs and provides a summary of the current state of its use in the field. PHC contamination occurs in many matrices including soil, water, and vapour. This chapter covers the application of bioremediation processes in both soil and groundwater which are intrinsically linked to each other in the natural environment. The principles of bioremediation are applicable to both and thus are the focus of this chapter.

14.1.1   Background on Petroleum Hydrocarbons

Petroleum is produced through thermal maturation of buried organic material over millions of years. This process involves diagenesis (a short period of biological degradation after deposition), catagenesis (geothermal degradation and cracking), and metagenesis (further decomposition, mainly producing methane) (Varjani 2017; Ławniczak et al. 2020). Petroleum itself can be defined as any hydrocarbon mixture of natural gas, condensate, and crude oil. PHCs are a complex mix of compounds that contain varying proportions of hydrogen atoms on a carbon frame; they may also contain nitrogen, sulphur, and oxygen. After being extracted from the subsurface, naturally occurring raw unrefined oil, known as crude oil, is transported to refineries where it undergoes distillation to produce usable products, including gasoline, diesel, and other refined products (Atlas 1991; Varjani 2017; Truskewycz et al. 2019; Ławniczak et al. 2020). These distillation products are not distinct entities—they are composed of a complex mix of hundreds or thousands PHC compounds of different molecular weights and chemical properties. Hydrocarbon compounds can be sorted into four groups based on their structure: alkanes (or paraffins), cycloalkanes (or naphthenes), alkenes (or olefins), and arenes (or aromatics). Alkanes are saturated aliphatic hydrocarbons, where each carbon atom forms four single bonds with hydrogen and other carbons (a common example is hexane, or octane); cycloalkanes are saturated ring hydrocarbons (e.g. methyl-cyclopentane); alkenes are unsaturated aliphatic hydrocarbons with at least two carbon atoms joined by more than one bond (e.g. ethene); and arenes consist of one or more benzene rings containing hydrocarbons (xylene, e.g. BTEX). Compounds within this final group containing two or more benzene rings are termed as polycyclic aromatic hydrocarbons or PAHs (common examples are naphthalene, phenanthrene, and pyrene) (Colwell et al. 1977; Varjani 2017; Truskewycz et al. 2019; Kuppusamy et al. 2020a, b). Biodegradability trends of PHCs vary based on many factors, including size, weight, bonding, structures (ring and branching), solubility, and the affinity for sorption to organics (K_{oc}) or solids (K_d).

The composition of crude oil can vary based on location, age of the oil field, sources of organic matter, depth, and weathering (Atlas 1991; Ławniczak et al. 2020). Weathering is defined as processes that occur to the PHCs after their release into the soil, surface water, or groundwater. Source variability for environmental contamination can vary even more with potential organic or inorganic additives and with the age of contamination. Because sources and composition of contaminants of concern vary so greatly, environmental investigations often use the generalized term "total petroleum hydrocarbons" (TPH) to encompass all potential groups present at a location. The majority of PHC remedial efforts focuses on widely used fuels such as diesel or gasoline and specifically target BTEX and a limited number of PAHs due to their physical and chemical properties, toxicity, and adverse effect on human health (El-Naas et al. 2014). Other PHC compounds and associated products—including heavier PHCs, petroleum-derived metabolites, and arsenic (Cozzarelli et al. 2016)—may also pose negative impacts on the environment and exposed populations. Some of these impacts are not fully understood yet and their treatment may be particularly challenging.

To decide whether bioremediation is the best choice for remediation (and to decide when, where, and how to apply it), one must first understand which PHCs are present, how they may change over time, and their mobility. An accurate conceptual site model is critical to visualize the source, pathways, and receptors of the contaminated environment; such a model equips practitioners to choose an appropriate remedial plan to achieve clean-up targets. These targets are, in turn, usually developed based on potential endpoint uses, exposure risk to PHCs, and regulatory clean-up criteria for soil, surface water, and groundwater set by regulators and government agencies. Whilst a detailed discussion on the fate and transport of PHCs is not included herein, biodegradation of PHCs in soil and groundwater is key to controlling their fate and transport.

14.2 Microbial Degradation

14.2.1 Factors Controlling Biodegradation

Several factors affect the ability of microorganisms to successfully thrive and degrade target compounds, including the presence of such microbial populations, electron acceptors (e.g. oxygen, nitrate, sulphate, Fe^{3+}), water, and nutrients, as well as the temperature and pH of the media. In addition to considering the organisms necessary for remediation, one must also consider that the specific compounds to be remediated will also affect the ability of microorganisms to biodegrade them; these compound-specific factors include concentration, solubility, and chemical structure. Third, the local environment and particular media will also have an overarching effect on remediation; factors including porosity, permeability, lithology, and groundwater flow impact both the microbial population and compound fate and transport (Fig. 14.1). Consequently, biodegradation rates are typically variable across a plume (Davis, 2023), for instance between "hot spots" and the plume edge.

Microorganisms present must be metabolically capable of breaking down the organic compound of concern, as well as potential by-products of the process. Sufficient moisture in the subsurface is also critical; 30–90% moisture content is commonly required for optimal oil degradation (Vidali 2001). Microbial communities require water not only for cellular processes, but also as a medium for nutrient flux, since nutrients, electron acceptors, and carbon are present in the aqueous phase and since biochemical reactions typically occur in the aqueous phase.

Oxygen presence or absence is another critical aspect for living systems. For aerobic metabolism, oxygen is the terminal electron acceptor and must be present for degradation to occur. If oxygen is consumed faster than it is replenished, anaerobic conditions may occur, and availability of oxygen becomes rate limiting at dissolved oxygen (DO) concentrations below 1–2 mg/L (Shaler and Klecka 1986; Vidali 2001; Speight and Arjoon 2012). For anaerobic metabolism, alternative terminal electron acceptors (sulphate [SO_4^{2-}], nitrate [NO_3^-], and ferric iron [Fe^{3+}]) and carbon dioxide required for methanogenic reaction must be present in abundance for degradation to occur (Vidali 2001; Xiong et al. 2015). Because sulphate, nitrate, and some iron species may be more water soluble than oxygen, when oxygen cannot be supplied or sustained, anaerobic processes may have some advantage over aerobic processes. Regardless, bioavailable nutrients (including nitrogen, phosphorus, and sulphur) and trace minerals form the backbone of all necessary enzymes used to break down contaminants and are therefore required for all microbial life to survive (C:N:P = 100:10:10 optimal) (Vidali 2001; Khudur et al. 2015). Salinity is also key for microbial life—too much or too little can inhibit microbial degradation (Ulrich et al. 2009).

Microbial life can function in a wide range of temperatures. However, the optimal temperatures often fall within 0–40 °C (Atlas 1991; Vidali 2001). Temperature affects the biochemical reaction rates possible within an environment, with rates found to increase as temperatures rise from 7 to 35 °C, above which rates may no longer increase (El-Naas et al. 2014). Additionally, temperature impacts both microbial growth and contaminant bioavailability and solubility (Morgan and Watkinson 1989; Atlas 1991; Xiong et al. 2015). Alkalinity/acidity conditions represent another critical parameter, since many microorganisms can only survive in certain, often very narrow, pH ranges. Optimal pH is normally at neutral pH 7, but 6–8 is often acceptable. Additional factors may be affected by pH, like nutrient availability and heavy metal solubility; these factors can impact a community's capacity to degrade contaminants. The concentration of organic contaminants can also impact a microbial community's effectiveness in remediation, since high concentrations (approaching solubility in particular) of PHC may impart a toxic effect to key degrading microorganisms and limit their growth, whilst low concentration may be too low to sustain microbial growth (Truskewycz et al. 2019; Ławniczak et al. 2020). Please refer to Chaps. 1, 5, and 9 for a more complete discussion of the complex topic of fate and transport of the contaminant.

Because PHC bioavailability is essential to a microbial community's capacity for bioremediation, understanding bioavailability is crucial. Although there is much debate regarding the definition of the term, a good working definition for bioavailability is "the fraction of a chemical that is freely available to cross an organism's cellular membrane from the medium in which the organism inhabits at a given time" (Kuppusamy et al. 2020a, b). Of course, storage, assimilation, transformation, and degradation are only possible once the chemical has been taken within a cell. Two critical factors should be considered with respect to bioavailability of a compound: the mass transfer of the compound from the environment to the microorganism cell and the subsequent rate of uptake and metabolism of the chemical. On a cellular level, bioaccessibility may be defined as what chemicals are available from the outside environment to cross the membrane of a cell, assuming their transport between cell and environment can occur. Whilst a microorganism may be capable of promoting degradation of a contaminant, it can only do so if the contaminant is present and accessible.

Biodegradation of hydrocarbons from the aqueous phase is relatively efficient, in contrast to biodegradation of hydrocarbons that have sorbed to soil organic matter or that have been sequestered in low-permeability layers (Megharaj et al. 2011; Balseiro-Romero et al. 2018). Compound-specific characteristics that affect solubility not only reduce the concentration of PHC compounds in the aqueous phase but also cause preferential sorption to solid matrices and organic matter in the subsurface and subsequent slow release into the environment (Megharaj et al. 2011; Balseiro-Romero et al. 2018; Kuppusamy et al. 2020a, b). The extent to which hydrocarbons have partitioned to the solid phase is a factor that impacts how well biodegradation may occur at a site.

14.2.2 *Aerobic and Anaerobic Degradation*

Organic compounds—including PHCs—are used by microorganisms as the electron donors required for metabolic pathways under either aerobic or anaerobic conditions; in fact, PHCs are often the sole source of carbon for these organisms (Beauchamp et al. 1989; Liu et al. 2020). Transformation of PHCs is energetically more favourable under aerobic conditions as compared to anaerobic conditions and is often mediated by a wide range of naturally occurring bacteria and fungi. Oxidation of PHCs (and BTEX in particular) in groundwater and soil is well documented (Collins et al. 2002; El-Naas et al. 2014; Yu et al. 2022). PHCs are a relatively reduced chemical species and can serve as the energy source and the electron donor for aerobic microbial metabolism. Oxygen serves as the reactant to oxidize the substrate and as the electron acceptor for microbial metabolism. Aerobic bacteria use different types of oxygenases, including monooxygenase, cytochrome-dependent oxygenase, and dioxygenase, to insert one or two atoms of oxygen into their targets. Because anaerobic degradation of PHCs is less well known, it will be explored in depth below.

Anaerobic biodegradation uses other electron accepting processes that generally fall within two large categories. Microorganisms can either generate energy by coupling substrate oxidation to respiration via reduction of an alternate terminal electron acceptor (i.e. sulphate, nitrate, iron, manganese, or carbon dioxide) or they can generate energy through fermentation (Foght 2008). Alternatively, less common electron acceptors such as carbon dioxide, vanadium, cobalt, and uranium have been found to function for select species of microorganisms. For more information on microorganisms that use these less common electron acceptors and that are found in extreme sites with specific conditions, please refer to Morrill et al. (2014) or Greene et al. (2016).

Though PHCs are a source of carbon that can be used for anaerobic microbial growth, even under optimal conditions, their chemical inertness poses an energetic and mechanistic challenge for successful anaerobic microbial metabolism (Rabus et al. 2016). Microorganisms must overcome the high energy barriers of PHC in their initial activation and cleavage of non-polar carbon-hydrogen bonds. The lower energy activation of PHC under aerobic conditions favours O_2-dependent oxygenase-catalysed reactions for this initial C–H bond cleavage (Rabus et al. 2016). Whether aerobic or anaerobic, degradation of PHCs involves a multitude of possible reactions as described below.

The anaerobic transformation of toluene and xylenes is relatively well understood. Because the microbes that metabolize these compounds are widely distributed amongst nitrate- and sulphate-reducers and fermenting organisms, these compounds attenuate naturally in the environment given sufficient time (Toth et al. 2021). Conversely, the very narrow set of microbes that anaerobically degrade benzene have only recently been identified through genome sequencing (Vogt et al. 2011; Luo et al. 2014, 2016; Toth et al. 2021). *Deltaproteobacterium* ORM2 is one such

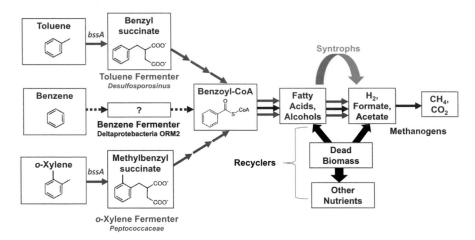

Fig. 14.2 Anaerobic biodegradation pathways for benzene, toluene, and xylene. While some of the degradation pathways are universal among anaerobic systems, the microbes shown are anaerobic and methanogenic bioaugmentation cultures unique to fermentative and methanogenic systems. Note that the mechanism for initial activation of the benzene ring is still unknown at this time. Figure courtesy of Dr. Courtney Toth, University of Toronto

organism. Toluene and xylene degradation typically occurs before benzene degradation because the responsible organisms tend to be more abundant in nature, however their presence may delay benzene degradation (Toth et al. 2021).

Anaerobic hydrocarbon-degrading microbial populations rely on a synergistic web of activities of diverse groups of microorganisms to achieve degradation of PHCs to less hazardous products, as illustrated by the pathways shown in Fig. 14.2. In the case of BTEX, these organisms include the following:

- Bacteria that ferment BTEX are the key components of the microbial community utilizing BTEX as a carbon and energy source.
- Archaea (primarily methanogenic archaea) are critical members of the community by virtue of their role converting acetate (a fermentation product of BTEX) into methane and carbon dioxide.
- Sulphate-reducing and/or fermentative bacteria that metabolize downstream fermentation products from BTEX (e.g. fatty acids and alcohols) are responsible for generating hydrogen (H_2), formate, and acetate, which are in turn metabolized by methanogenic archaea. We refer to these microorganisms as "syntrophs".
- Sulphate-reducing and/or fermentative bacteria that metabolize components of dead biomass (e.g. proteins, carbohydrates, and lipids) produce H_2, acetate, formate, and other methylated (C1) compounds, which are in turn metabolized by methanogenic archaea. We refer to these organisms collectively as "recyclers" (Lillington et al. 2020). Whilst not directly involved in anaerobic BTEX biodegradation, they generate micronutrients and co-factors required for BTEX-degrading enzymes and microorganisms.

- Low proportions of other organisms naturally present in many anaerobic ecosystems may contribute to the process in ways that are not yet known.

Anaerobic BTEX-degrading communities are often comprised of microorganisms unique to fermentative and methanogenic systems (e.g. *Deltaproteobacteria* ORM2, *Methanosaeta*, and *Methanoregula*) and may grow only under very specific conditions, including an oxidation reduction potential (ORP) below -100 millivolts (representing iron reducing conditions or lower). In fact, a significant amount of effort is typically required by a remediation practitioner to establish suitable conditions at a contaminated site in which these microorganisms will grow. For instance, the absence or depletion of benzene will cause concentrations of *Deltaproteobacteria* ORM2 to decrease to extinction, along with all other syntrophic microorganisms dependent on ORM2's growth and benzene fermentation products. This is because benzene is the only known substrate fermented by ORM2 (Luo et al. 2016; Toth et al. 2021) and benzene concentrations greater than 0.1 mg/L are required to stimulate ORM2 benzene degradation. Since naturally occurring or added electron donors and carbon sources (i.e. volatile fatty acids and other fermentable carbon substrates) will stimulate the proliferation of intrinsic and/or bioaugmented microorganisms, their presence may reduce or completely inhibit BTEX degradation by key organisms. Consequently, electron donors should typically not be added to BTEX-contaminated sites.

In addition, BTEX fermentation does not always require an exogenous or added electron acceptor. Known benzene-degrading microbes may metabolize benzene coupled syntrophically to sulphate reduction (if available) or via methanogenesis (Luo et al. 2016; Toth et al. 2021). Sulphate may also promote the growth of other (facultative) sulphate-reducing bacteria, including *Desulfovibrio* and *Geobacter* as it was seen in a recent study (Toth et al. 2021).

14.3   Bioremediation Technologies

The principles of bioremediation technologies are based upon key tenets of microbial life and degradation. This sort of treatment can be accomplished ex situ or in situ. In ex situ remediation, where contaminated soil or groundwater is excavated and removed for treatment; the soil or groundwater can be put to beneficial reuse or disposed of as a non-hazardous waste. In situ remediation, which is increasingly now a preferred approach, causes less disturbance of the area and normally comes with lower costs. Table 14.1 summarizes these technologies.

Creation of a conceptual site model is a first crucial step in understanding what microbial processes are or are not occurring at contaminated site and what can be done. In general, samples relating to aqueous geochemistry and microbial communities are gathered, geologic cores are taken, soil vapour and other parameters are measured, and historical data relating to the site and surrounding areas is reviewed.

Table 14.1 Summary of common bioremediation strategies

Technology	Examples	Benefits	Limitations	Considerations
In situ	Biosparging Monitored Natural Attenuation (MNA) Bioventing Bioaugmentation Biostimulation	Cost efficient A land preservation measure A natural process Can treat multiple media simultaneously (soil, groundwater, surface water) Causes limited disruption to site operation	Environmental constraints Longer duration Limited monitoring possible	Bioaccessibility of compound Metabolic capacity of community Field factors (permeability, depth, groundwater flow, etc.) Biodegradability of compound Compound-specific characteristics (solubility) Distribution of plume Whether bioaugmentation is beneficial
Ex situ soil treatment	Landfarming Composting Biopiles	Cost efficient Can usually be done on-site	Spatial constraints Longer duration Need to account for abiotic loss Bioavailability limitations	Bioaccessibility of compound Metabolic capacity of community Whether bioaugmentation is beneficial Field factors (climate, site size, access, and activity, etc.) Biodegradability of compound Compound-specific characteristics (solubility) Distribution of plume
Ex situ bioreactors	Slurry reactors Aqueous reactors	Involve a rapid metabolic process Environmental parameters that are able to be controlled and optimized Maximize mass transfer	Require excavation or pumping High upfront capital costs Relatively high capital and operation costs	Whether bioaugmentation is beneficial Toxicity of optimization amendments Toxicity of contaminants

Adapted from Vidali (2001)

This step establishes lines of evidence required for completing a feasibility assessment for a bioremediation approach. Additional assessments, such as treatability studies and advanced genomic studies, can also provide useful information.

14.3.1 Biostimulation and Bioaugmentation

For effective and efficient bioremediation, sufficient biomass to degrade the compound(s) of concern is required. This can be achieved through two approaches: biostimulation and bioaugmentation. Biostimulation amendments added to the subsurface stimulate the naturally occurring microbial community to degrade target contaminants. These amendments may include nutrients, such as phosphorus and nitrogen, or other trace minerals and electron acceptors that are absent or scarce, such as oxygen, nitrate, and sulphate (Anderson and Lovley 2000; Wolicka et al. 2009; Megharaj et al. 2011; Speight and Arjoon 2012; Brown et al. 2017; Müller et al. 2021; Primitz et al. 2021). Figure 14.3 shows a nutrient amendment infrastructure commonly required for injections at a treatment site. These amendments may help the microbial community to thrive and therefore increase the rate of biodegradation. For PHCs, this biodegradation typically occurs because these compounds serve as the electron donor necessary for microbial activity. This treatment option assumes the native community in the area was already metabolically capable of degrading the target compound, but was limited by lack of nutrients (Khudur et al. 2015).

Monitored natural attenuation (MNA) can be assessed where the natural populations of the microbial community and nutrients are sufficient for degradation of the

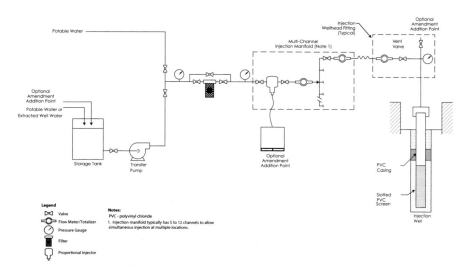

Fig. 14.3 In situ biostimulation—nutrient amendment infrastructure at ground surface required to inject amendment to contaminated groundwater via injection wells

target compound. This approach requires regular monitoring to track the remediation but requires no other amendments of the site (Chen et al. 2005; Kao et al. 2006). More information on monitoring natural attenuation processes and more specifically natural source zone depletion (NSZD) are provided in Chaps. 5 and 13.

If the naturally occurring microbial community does not already include microorganisms capable of promoting degradation of PHCs or if the population density is not great enough to support acceptable degradation, bioaugmentation may be an option (Megharaj et al. 2011; Varjani and Upasani 2021; Zang et al. 2021; Zuzolo et al. 2021a, b). Bioaugmentation is the process of amending an existing microbial community with microbial cultures or isolates known to degrade target compounds (Morgan and Watkinson 1989; Singer et al. 2005; Wolicka et al. 2009; Varjani and Upasani 2021). Figure 14.4 demonstrates a common in situ bioaugmentation treatment for groundwater. Once injections are complete, very little permanent equipment remains at the injection site other than sampling points. To treat the whole area and avoid localized effects, the introduced microorganisms or nutrients must be distributed throughout the contaminated matrix. In addition, the bioaugmented microorganisms must also be able to thrive alongside existing native microorganisms, or the treatment effect will be short lived, and the degradation of the compounds may not be sustained (Singer et al. 2005). This is because effective treatment also relies on bioaugmented microorganisms spreading as their population densities increase and moving with ambient groundwater, all of which occurs after injections are complete.

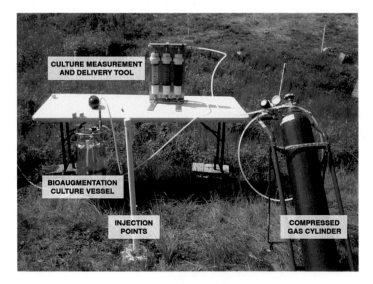

Fig. 14.4 In situ bioremediation—equipment required for bioaugmentation includes culture vessel, culture dispenser (on table), and compressed nitrogen cylinder allowing one-time culture injections into injection points which also can be combined with biostimulation amendments as required

14.3.2 In Situ Biological Technologies

In situ biological technologies influence remediation by changing the environmental conditions of the subsurface to enhance or alter the microbial community present. Common techniques, their benefits, and their limitations are summarized in Table 14.1. In situ treatment technologies include bioventing, biosparging, and low-energy thermal (further explored in Chap. 18) technologies.

A common treatment, bioventing uses wells to supply the contaminated subsurface with air and nutrients to stimulate the natural microbial community to degrade contaminants under aerobic conditions. This process provides enough air movement to stimulate microbial degradation whilst limiting the volatilization and release of hydrocarbons to the atmosphere (Frutos et al. 2010; Johnston et al. 2010; Tzovolou et al. 2015). In biosparging, air is pumped through wells under the groundwater table to increase the dissolved oxygen concentrations within the aqueous phase and promote microbial metabolism. Figure 14.5 demonstrates a common schematic of an in situ biosparging remediation system for PHC-contaminated groundwater. In addition to increasing oxygen in groundwater, this process can enhance mixing within the saturated zone and thereby increase the dissolution of sorbed hydrocarbon into the groundwater (Strzempka et al. 1997; Vidali 2001; Kao et al. 2008; Kabelitz et al. 2009). Low-energy heating is another way to support the microbial community; this technology aims to provide optimal temperature to maximize microbial kinetics whilst also enhancing LNAPL mobility and partitioning of hydrocarbons into other phases (Imhoff et al. 1997; Vermeulen and McGee 2000; Macbeth et al. 2012).

14.3.3 Ex Situ Biological Technologies

Ex situ treatment technologies bring contaminated soil and groundwater to the surface to enhance or alter microbial community and speed degradation. Such technologies include landfarming, composting, biopiles, and bioreactors. Landfarming is a very simple technique that has been widely applied to soil where contamination is generally shallow and accessible from the ground surface. Contaminated soil is excavated, spread over a large area, and periodically tilled to enhance oxygen penetration and keep the system aerobic (Vidali 2001; Bergsveinson et al. 2019). This method takes up a large, open land surface, which may not always be economical or feasible when space is at a premium. Composting is similar to landfarming but involves amending the excavated contaminated soil with organic materials like manure or agricultural waste to increase microbial community diversity and density (Hwang et al. 2001; Megharaj et al. 2011; Syawlia and Titah 2021). Biopiles are a hybrid of landfarming and composting, in which an engineered aerated composting cell is created from the excavated soil. Biopiles may require a smaller surface area footprint than conventional landfarming does (see Fig. 14.6). These large piles are designed to limit the leaching and volatilization loss of hydrocarbons experienced by landfarming and

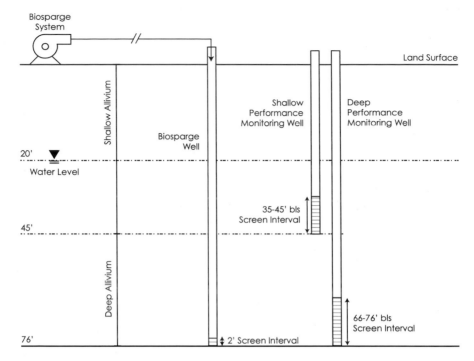

Fig. 14.5 Schematic of an in situ biosparge remediation system

composting whilst promoting microbial diversity, increasing temperature, and optimizing air penetration for both aerobic and anaerobic communities (Singh et al.
2017; Wang et al. 2021; Zhang et al. 2021). Bioreactors are another form of ex situ
technology but are often considered a separate category because they are a highly
controllable closed treatment system, separate from the natural environment. Both
slurry (mixed phase) and aqueous-phase reactors can be used to treat soil excavated
from the contaminated site or water pumped from it. Since bioreactors are an external
closed environment, parameters can be tightly controlled to foster biodegradation and
the process can be maximized for rapid microbial degradation kinetics (Chiavola et al.
2010; Das and Kumar 2018). Figure 14.7 shows potential infrastructure required for
treatment of PHC-contaminated groundwater.

14.3.4 Mycoremediation of Complex Organics

Interest in remediation using fungi as a sole degradation driver has grown in recent
years. Numerous studies indicate that microbial diversity, including fungal species,
is critical for remediation of PHCs (Obuekwe et al. 2005; Mohsenzadeh et al. 2009;
Zafra et al. 2014; Lee et al. 2015; Andreolli et al. 2016; Marchand et al. 2017).

Fig. 14.6 Ex situ bioremediation—enhanced bioremediation of PHC in biopiles. The most common application is to treat soils contaminated with PHCs by adding nutrients, moisture, and oxygen to the soil. Photos courtesy of Vertex Environmental Inc

Fungal species are a critical component to a healthy diverse rhizosphere, as discussed below and illustrated in Fig. 14.1. Though fungal degradation of PHCs can occur in isolation, diverse communities create a mutually beneficial system where fungi and other organisms work together to degrade organic contaminants more efficiently than they could in isolation. Investigations of microbial diversity within highly PHC-contaminated soil at a former petrochemical plant characterized degradation potential of 95 bacterial and 160 fungal identified isolates. *Fusarium oxysporum* and *Trichoderma tomentosum* significantly degraded all PAH compounds tested (anthracene, phenanthrene, fluorene, and pyrene). *Sordariomycetes* has often demonstrated high affiliation with hydrocarbon degradation, and fungal species studied belonging to *Sordariomycetes* class, *Trichoderma*, and *Fusarium* were found to be more efficient degraders than those of other classes studied (Hong et al. 2010; Wu et al. 2010; Argumedo-Delira et al. 2012). Bioaugmentation with six known potential hydrocarbon-degrading fungi was shown to significantly increase degradation over biostimulation of the native community alone in TPH-contaminated soil mesocosms (Medaura et al. 2021). Both bioaugmentation with fungal species and biostimulation with nutrients had TPH degradation rates above unamended contaminated soil (39.90

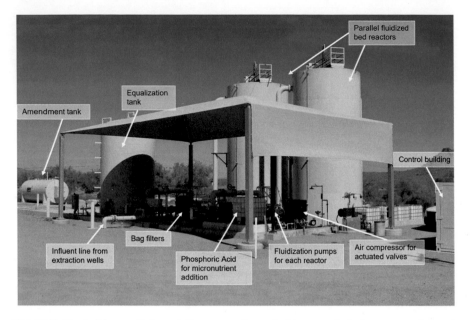

Fig. 14.7 Ex situ bioremediation—infrastructure for ex situ bioreactor includes pumps and amendment addition system (located under canopy) and two fluidized bed reactors for treatment of impacted water

± 1.99%, and 24.17 ± 1.31%, respectively). In addition to increased TPH degradation with bioaugmented fungal species, the bacteria within the native soil shifted to a more diverse community, enriched in known hydrocarbon-degrading bacteria orders Cytophagales, Bacteroidales, and Rhodocyclales. This fact indicates that a complex and synergistic relationship between the native community and bioaugmented species can occur, increasing overall capacity and rates of degradation of complex compounds (Medaura et al. 2021).

A study conducted by Argumedo-Delira et al. (2012) tested the tolerance of 11 strains of *Trichoderma* to naphthalene, phenanthrene, and benzo(α)pyrene. Several fungal strains of *Trichoderma* tested were capable of tolerating concentrations of phenanthrene and naphthalene above 250 mg/L, as well as benzo(α)pyrene concentrations of 100 mg/L. Although established potential for PAH remediation by fungi is present, enzymatic activity and pathways are not well understood. Andreolli et al. (2016) isolated *Trichoderma longibrachiatum* from uncontaminated forest soil through selective enrichment for hydrocarbon degraders. In a diesel-contaminated soil microcosm, soil inoculated with *Trichoderma longibrachiatum* demonstrated the fastest removal of $C_{12\text{-}40}$ hydrocarbon fraction, at 54.2 ± 1.6% in 30 days, compared to 7.3 ± 6.1% removal in controls (Andreolli et al. 2016). Additionally, Andreolli et al. (2016) characterized the potential for PAH removal, with 69–71% removal of phenanthrene, anthracene, pyrene, and fluoranthene, potentially indicating that *Trichoderma longibrachiatum* is a strong hydrocarbon degrader. This result is promising, not only because of the efficacy of the fungal strain, but also because it indicates fungal species not previously exposed to PHCs can be effective in their biodegradation.

Fungal species are found in all environments; a comprehensive study by Richardson et al. (2019) revealed a genetically diverse microbial community even in oil sands tailings water, which contains a multitude of bitumen-associated organics, including a toxic naphthenic acid fraction (Qin et al. 2019). Next-generation community sequencing revealed that, although limited in classification below the phylum level, two of the most abundant operational taxonomic units of the entire data set were fungi (Richardson et al. 2019). This presence of major fungal activity within the water fraction of the tailings pond indicates that fungi are able to resist such harsh and toxic environments and, potentially, are also able to metabolize organic compounds found in association with bitumen. Repas et al. (2017) isolated *Trichoderma harzianum* from plant roots growing in coarse tailings and found it had the capacity to remediate complex petrochemical residues present within the tailings. The fungal isolate *T. harzianum* was isolated in OSPW by Miles et al. (2019) and demonstrated the ability to withstand high salinity conditions ($\geq$ 60 g/L), a pH range of 2–9, and a naphthenic acid fraction compound-inhibitory concentration of 2400 mg/L. Further testing revealed this isolate, sourced from the environment, was able to grow on an agar plate using a single pure drop of naphthenic acids as its sole source of carbon; this indicates a strong potential for fungal remediation of toxic organic compounds like naphthenic acids (Miles et al. 2019, 2020). With vast fungal diversity and ubiquitous presence in the soil and groundwater environment, in situ mycoremediation is a compelling treatment method for toxic, otherwise recalcitrant organic contaminants in soil and groundwater.

14.3.5 Phytoremediation of Petroleum Hydrocarbons

Phytoremediation is a remediation technology in use since at least the 1980s to degrade, extract, contain, or immobilize contaminants from soil, groundwater, and other media using plants (Landmeyer 2012). Whilst a wide array of contaminants can be effectively treated using phytoremediation, PHCs are particularly amenable to this strategy, and the use of phytoremediation for PHCs has increased in recent decades. Phytoremediation sites commonly look like natural landscapes with rows of various species of trees and/or monitoring well stick-ups. This aspect is an additional benefit of this technique: it maintains the aesthetic of the natural environment and may even limit the amount of reclamation necessary after remediation.

Effective phytoremediation of PHCs typically occurs through several mechanisms, which may happen simultaneously, including rhizodegradation, phytodegradation, and to a lesser extent, phytovolatilization. Rhizodegradation refers to the microbial biodegradation of contaminants in the soil surrounding plant roots (the so-called rhizosphere effect), whilst phytodegradation describes contaminant degradation within plant tissue. Phytovolatilization, defined as root uptake and transfer of contaminants or their metabolites to atmosphere through plant transpiration, and other plant-related mechanisms may also contribute to PHC treatment in some circumstances as illustrated in Fig. 14.1 (McCutcheon and Schnoor 2003; Landmeyer 2012). Additionally, for sites with impacted groundwater, phytohydraulics—the interception of impacted groundwater and control of contaminant plume migration through transpiration—can be an important contributor to site remediation by reducing or preventing plume migration (Landmeyer 2012). Other specific mechanisms may also be involved, depending on contaminant types, the impacted medium (e.g. soil or groundwater), and the design of the phytoremediation system.

Of particular significance to PHC phytoremediation is rhizodegradation because it is often the single greatest contributor to phytoremediation effectiveness at PHC-impacted sites (Landmeyer 2012). Rhizodegradation may be thought of as plant-assisted bioremediation. A plant's roots release a variety of compounds known as root exudates, including simple sugars, polysaccharides, amino acids, and microbial growth factors. Exudates encourage the colonization and proliferation of bacteria and fungi nearby (Barac et al. 2004; Al-Zaban et al. 2021). As the plant thrives, the rhizosphere becomes a zone of enhanced microbial density and activity as compared to surrounding bulk soil (Zuzolo et al. 2021a, b; Zuzolo et al. 2021a, b). Notably, some root exudates may serve to induce the expression of microbial enzymes useful for cometabolic biodegradation of PHCs (Dagher et al. 2019; Davin et al. 2019).

The interaction between plant roots and their associated microbial communities and the mechanisms by which they interact is highly complex. Though its full significance to PHC phytoremediation is beyond the scope of this chapter, interested readers would be well served to further explore this topic using the references cited above.

Several factors must be considered in order to successfully design and apply phytoremediation at PHC-impacted sites (Susarla et al. 2002). These include the concentration, distribution, depth, and types of contaminants, as well as hydrogeological and soil conditions and remedial goals. These aspects and their interplay affect the choice of appropriate plants, the configuration of plantings, and locations chosen for implementation. Properties of the soil and groundwater also influence phytoremediation success; these include nutrient availability, particle size and soil classification, bulk density, salinity, redox potential, pH, cation exchange capacity, organic matter content, and the presence of microorganisms for degradation (Gerhardt et al. 2009). As with PHC remediation achieved through bacterial degradation, phytoremediation is largely dependent on environmental factors. However, plants often have a higher tolerance to rapid changes in the environment, including temperature, moisture, and salinity.

As a general rule, for phytoremediation be effective, plant roots must grow close to impacted media. As a result, impacted soil at depths below typical rooting depths may not be effectively treated with this approach. Though it depends on plant species and site conditions, roots generally cannot be expected to naturally penetrate deeper than 5–10 ft below ground surface (bgs) (McCutcheon and Schnoor 2003). PHC-impacted groundwater also must be accessible to plant roots for a remediation to occur through phytohydraulics. In some cases, an engineered phytoremediation approach has been demonstrated to effectively target impacted groundwater at depths greater than 5–10 ft bgs (Geosyntec Consultants 2022).

As with all technologies in this chapter, phytoremediation requires a proper understanding of this unique and evolving field, and less-experienced practitioners may want to consult with phytoremediation specialists to ensure appropriate system design, implementation, operation, and maintenance to aid in long-term effectiveness. Successful application of phytoremediation depends upon a complete conceptual site model that takes into account all available phytoremedial options. Phytoremediation alone, or in conjunction with other treatment technologies, has proven an effective remediation option available.

14.4 Conclusion

Given the continuing global demand for petroleum products, understanding the environmental implications of PHC contamination is critical. Remediation of PHCs released into the environment will continue for decades, if not centuries. Although PHCs are a complex mix of organic compounds, natural microbial communities have adapted to these carbon sources and can attenuate or remediate them, though they often need human intervention to achieve remedial goals. Harnessing the natural potential of microorganisms, fungi, and plants, under many different conditions, is key to the future of remediation.

With knowledge gained from the study of natural systems, practitioners can and should incorporate nature-based solutions as they develop technologies and treatment plans for PHC-impacted sites. As regulators increasingly mandate sustainable and green treatment technologies for such sites, practitioners must be well versed in these biologically mediated solutions and able to create a well-characterized conceptual site model based on a comprehensive understanding of the current environment and how it can be amended.

Treatment options for PHC-contaminated sites have grown drastically in the last 40 years. PHC compounds previously thought to be recalcitrant to anaerobic biological treatment have been proven amenable to it, given the correct conditions and organisms are present. Identification and analysis of PHCs has improved, and, in the near future, cheaper, faster, and more accurate analyses will likely enhance conceptual site models and improve remediation treatment plans. In addition, improvements and innovations regarding effective distribution of amendments to the subsurface will also increase effectiveness. In fact, as the leading edge of science advances, biological treatment of contaminated sites could become the principal treatment option for PHC contamination. For all this to occur, a bias towards traditional remedial technologies on the part of developers and landowners must be overcome, and a broader understanding and acceptance of biological treatments by regulators and by the public must be achieved.

By 2030, major growth within this field of study is expected to occur and technologies described in this chapter as innovative or exploratory may likely be commonplace. Once biological mechanisms for PHC remediation are better understood, these technologies can be applied with better confidence and efficacy. The broader application of molecular biological tools (discussed in Chap. 10), including microbial community analysis and advanced chemical analysis, (including compound-specific isotope analyses, as presented in Chap. 11) will contribute greatly to better harnessing biological processes for both in situ and ex situ remediation. Genetically modified organisms may well be designed and used to remediate PHCs in difficult environmental conditions (such as high salinity, low temperature, high clay content, or low oxygen) once they are accepted by the public and regulators and proven with successful field demonstrations.

Given the pace of growth in the field, in a few short years, biologically mediated remediation of PHC-contaminated sites will likely be possible on shorter timelines and at lower costs. Future innovation and research will lead to the coupling of biologically mediated remediation options with compatible and synergistic technologies to achieve even better remedial outcomes. Biotic and abiotic solutions can be applied in tandem to reduce costs, since abiotic remediation alone often occurs in a non-linear fashion, with the last phase of contaminant removal typically taking significantly more time, resources, and money than the first stages; adding a biotic component could vastly reduce cost and increase sustainability.

The work in the laboratory and the field to expand our understanding of bioremedial technologies, their effectiveness, and their applications is ever developing. Research into complex sites, difficult conditions, and complex contaminant mixtures will lead to better, cleaner, and greener solutions for practitioners to apply to contaminated environments.

Acknowledgements The authors would like to acknowledge and thank Dr. Courtney Toth of University of Toronto and Vertex Environmental Inc. for their contributions to the figures.

References

Al-Zaban MI, AlHarbi MA, Mahmoud MA (2021) Hydrocarbon biodegradation and transcriptome responses of cellulase, peroxidase, and laccase encoding genes inhabiting rhizospheric fungal isolates. Saudi J Biolog Sci 28(4):2083–2090

Anderson RT, Lovley DR (2000) Anaerobic bioremediation of benzene under sulfate-reducing conditions in a petroleum-contaminated aquifer. Environ Sci Technol 34(11):2261–2266

Andreolli M, Lampis S, Brignoli P et al (2016) Trichoderma longibrachiatum Evx1 is a fungal biocatalyst suitable for the remediation of soils contaminated with diesel fuel and polycyclic aromatic hydrocarbons. Environ Sci Pollut Res 23(9):9134–9143

Argumedo-Delira R, Alarcón A, Ferrera-Cerrato R et al (2012) Tolerance and growth of 11 Trichoderma strains to crude oil, naphthalene, phenanthrene and benzo[a]pyrene. J Environ Manag 95:S291–S299

Atlas RM (1991) Microbial hydrocarbon degradation—bioremediation of oil spills. J Chem Technol Biotechnol 52(2):149–156

Balseiro-Romero M, Monterroso C, Casares JJ (2018) Environmental fate of petroleum hydrocarbons in soil: review of multiphase transport, mass transfer, and natural attenuation processes. Pedosphere 28(6):833–847

Barac T, Taghavi S, Borremans B et al (2004) Engineered endophytic bacteria improve phytoremediation of water-soluble, volatile, organic pollutants. Nat Biotechnol 22(5):583–588

Beauchamp EG, Trevors JT, Paul JW (1989) Carbon sources for bacterial denitrification. Adv Soil Sci 10:113–142

Bergsveinson J, Perry BJ, Simpson GL et al (2019) Spatial analysis of a hydrocarbon waste-remediating landfarm demonstrates influence of management practices on bacterial and fungal community structure. Microb Biotechnol 12(6):1199–1209

Brown DM, Okoro S, van Gils J et al (2017) Comparison of landfarming amendments to improve bioremediation of petroleum hydrocarbons in Niger Delta soils. Sci Total Environ 596–597:284–292

Butnariu M, Butu M (2020) Bioremediation: a viable approach for degradation of petroleum hydrocarbon. Springer International Publishing, pp 195–223

Chen KF, Kao CM, Wang JY et al (2005) Natural attenuation of MTBE at two petroleum-hydrocarbon spill sites. J Hazard Mater 125(1–3):10–16

Chiavola A, Baciocchi R, Gavasci R (2010) Biological treatment of PAH-contaminated sediments in a sequencing batch reactor. J Hazard Mater 184(1–3):97–104

Collins C, Laturnus F, Nepovim A (2002) Remediation of BTEX and trichloroethene—current knowledge with special emphasis on phytoremediation. Environ Sci Pollut Res 9(1):86–94

Colwell RR, Walker JD, Cooney JJ (1977) Ecological aspects of microbial degradation of petroleum in the marine environment. CRC Crit Rev Microbiol 5(4):423–445

Cozzarelli IM, Schreiber ME, Erickson ML, Ziegler BA (2016) Arsenic cycling in hydrocarbon plumes: secondary effects of natural attenuation. Groundwater 54(1):35–45

Dagher DJ, de la Providencia IE, Pitre FE et al (2019) Plant identity shaped rhizospheric microbial communities more strongly than bacterial bioaugmentation in petroleum hydrocarbon-polluted sediments. Front Microbiol 10

Das AJ, Kumar R (2018) Bioslurry phase remediation of petroleum-contaminated soil using potato peels powder through biosurfactant producing Bacillus licheniformis J1. Int J Environ Sci Technol 15(3):525–532

Davin M, Starren A, Marit E et al (2019) Investigating the effect of *Medicago sativa* L. and *Trifolium pratense* L. root exudates on PAHs bioremediation in an aged-contaminated soil. Water Air Soil Pollut 230(12)

Davis GB (2023) Reviewing the Bioremediation of Contaminants in Groundwater: Investigations over 40 Years Provide Insights into What's Achievable. Frontiers in Bioscience-Elite 15(3):16

El-Naas MH, Acio JA, El Telib AE (2014) Aerobic biodegradation of BTEX: progresses and prospects. J Environ Chem Eng 2(2):1104–1122

Foght J (2008) Anaerobic biodegradation of aromatic hydrocarbons: pathways and prospects. J Mol Microbiol Biotechnol 15(2–3):93–120

Frutos FJG, Escolano O, Garcia S et al (2010) Bioventing remediation and ecotoxicity evaluation of phenanthrene-contaminated soil. J Hazard Mater 183(1–3):806–813

Geosyntec Consultants (2022) Phytoremediation using treeWell® technology, 27 Apr 2022

Gerhardt KE, Huang X-D, Glick BR et al (2009) Phytoremediation and rhizoremediation of organic soil contaminants: potential and challenges. Plant Sci 176(1):20–30

Greene A, Wright M, Aldosary H (2016) Bacterial diversity and metal reducing bacteria in Australian thermal environments. In: The spotlight: recent progress in the understanding of beneficial and harmful microorganisms. Microb Infect 32

Hong JW, Park JY, Gadd GM (2010) Pyrene degradation and copper and zinc uptake by *Fusarium solani* and *Hypocrea lixii* isolated from petrol station soil. J Appl Microbiol 108(6):2030–2040

Huesemann MH, Hausmann TS, Fortman TJ (2004) Does bioavailability limit biodegradation? A comparison of hydrocarbon biodegradation and desorption rates in aged soils. Biodegradation 15(4):261–274

Hwang EY, Namkoong W, Park JS (2001) Recycling of remediated soil for effective composting of diesel-contaminated soil. Compost Sci Utiliz 9(2):143–148

Imhoff PT, Frizzell A, Miller CT (1997) Evaluation of thermal effects on the dissolution of a nonaqueous phase liquid in porous media. Environ Sci Technol 31(6):1615–1622

Johnston CD, Woodbury R, Bastow TP et al (2010) Biosparging successfully limited fugitive VOCs while remediating residual weathered gasoline in a shallow sand aquifer. In: 7th international groundwater quality conference, vol 342, pp 225–228

Kabelitz N, Machackova J, Imfeld G et al (2009) Enhancement of the microbial community biomass and diversity during air sparging bioremediation of a soil highly contaminated with kerosene and BTEX. Appl Microbiol Biotechnol 82(3):565–577

Kao CM, Huang WY, Chang LJ et al (2006) Application of monitored natural attenuation to remediate a petroleum-hydrocarbon spill site. Water Sci Technol 53(2):321–328

Kao CM, Chen CY, Chen SC et al (2008) Application of in situ biosparging to remediate a petroleum-hydrocarbon spill site: field and microbial evaluation. Chemosphere 70(8):1492–1499

Khudur LS, Shahsavari E, Miranda AF et al (2015) Evaluating the efficacy of bioremediating a diesel-contaminated soil using ecotoxicological and bacterial community indices. Environ Sci Pollut Res 22(19):14809–14819

Kuppusamy S, Maddela NR, Megharaj M et al (2020a) Ecological impacts of total petroleum hydrocarbons. In: Total petroleum hydrocarbons. Springer, Cham

Kuppusamy S, Maddela NR, Megharaj M et al (2020b) Total petroleum hydrocarbons: environmental fate, toxicity, and remediation. Springer, Cham

Landmeyer JE (2012) Introduction to phytoremediation of contaminated groundwater. [Electronic resource]: historical foundation, hydrologic control, and contaminant remediation. Springer

Ławniczak Ł, Woźniak-Karczewska M, Loibner AP et al (2020) Microbial degradation of hydrocarbons—basic principles for bioremediation: a review. Molecules 25(4):856

Lee H, Yun SY, Jang S et al (2015) Bioremediation of polycyclic aromatic hydrocarbons in creosote-contaminated soil by *Peniophora incarnata* KUC8836. Bioremediat J 19(1):1–8

Lillington SP, Leggieri PA, Heom KA et al (2020) Nature's recyclers: anaerobic microbial communities drive crude biomass deconstruction. Curr Opin Biotechnol 62:38–47

Liu X, Li Z, Zhang C et al (2020) Enhancement of anaerobic degradation of petroleum hydrocarbons by electron intermediate: performance and mechanism. Bioresour Technol 295

Luo F, Gitiafroz R, Devine CE et al (2014) Metatranscriptome of an anaerobic benzene-degrading, nitrate-reducing enrichment culture reveals involvement of carboxylation in benzene ring activation. Appl Environ Microbiol 80(14):4095–4107

Luo F, Devine CE, Edwards EA (2016) Cultivating microbial dark matter in benzene-degrading methanogenic consortia. Environ Microbiol 18(9):2923–2936

Macbeth TW, Truex MJ, Powell T et al (2012) Combining low-energy electrical resistance heating with biotic and abiotic reactions for treatment of chlorinated solvent DNAPL source area

Marchand C, St-Arnaud M, Hogland W et al (2017) Petroleum biodegradation capacity of bacteria and fungi isolated from petroleum-contaminated soil. Int Biodeterior Biodegradation 116:48–57

McCutcheon SC, Schnoor JL (2003) Phytoremediation: transformation and control of contaminants. Wiley-Interscience

Medaura MC, Guivernau M, Moreno-Ventas X et al (2021) Bioaugmentation of Native Fungi, an efficient strategy for the bioremediation of an aged industrially polluted soil with heavy hydrocarbons. Front Microbiol 12

Megharaj M, Ramakrishnan B, Venkateswarlu K et al (2011) Bioremediation approaches for organic pollutants: a critical perspective. Environ Int 37(8):1362–1375

Miles SM, Hofstetter S, Edwards T et al (2019) Tolerance and cytotoxicity of naphthenic acids on microorganisms isolated from oil sands process-affected water. Sci Total Environ 695:133749

Miles SM, Asiedu E, Balaberda A-L et al (2020) Oil sands process affected water sourced Trichoderma harzianum demonstrates capacity for mycoremediation of naphthenic acid fraction compounds. Chemosphere 258:127231

Mohsenzadeh F, Naseri S, Mesdaghinia A et al (2009) Identification of petroleum resistant plants and rhizospheral fungi for phytoremediation of petroleum contaminated soils. J Jpn Petrol Inst 52(4):198–204

Morgan P, Watkinson RJ (1989) Hydrocarbon degradation in soils and methods for soil biotreatment. Crit Rev Biotechnol 8(4):305–333

Morrill PL, Brazelton WJ, Kohl L et al (2014) Investigations of potential microbial methanogenic and carbon monoxide utilization pathways in ultra-basic reducing springs associated with present-day continental serpentinization: the Tablelands, NL, CAN. Front Microbiol 5

Müller C, Knöller K, Lucas R et al (2021) Benzene degradation in contaminated aquifers: enhancing natural attenuation by injecting nitrate. J Contam Hydrol 238:103759

Obuekwe CO, Badrudeen AM, Al-Saleh E et al (2005) Growth and hydrocarbon degradation by three desert fungi under conditions of simultaneous temperature and salt stress. Int Biodeterior Biodegradation 56(4):197–205

Primitz JV, Vazquez S, Ruberto L et al (2021) Bioremediation of hydrocarbon-contaminated soil from Carlini Station, Antarctica: effectiveness of different nutrient sources as biostimulation agents. Polar Biol 44(2):289–303

Qin R, Lillico D, How ZT et al (2019) Separation of oil sands process water organics and inorganics and examination of their acute toxicity using standard in-vitro bioassays. Sci Total Environ 695

Rabus R, Boll M, Heider J et al (2016) Anaerobic microbial degradation of hydrocarbons: from enzymatic reactions to the environment. J Mol Microbiol Biotechnol 26(1–3):5–28

Repas TS, Gillis DM, Boubakir Z et al (2017) Growing plants on oily, nutrient-poor soil using a native symbiotic fungus. PLoS ONE 12(10)

Richardson E, Bass D, Smirnova A et al (2019) Phylogenetic estimation of community composition and novel eukaryotic lineages in Base Mine Lake: an oil sands tailings reclamation site in Northern Alberta. J Eukaryot Microbiol

Shaler TA, Klecka GM (1986) Effects of dissolved oxygen concentration on biodegradation of 2,4-dichlorophenoxyacetic acid. Appl Environ Microbiol 51(5):950–955

Singer AC, van der Gast CJ, Thompson IP (2005) Perspectives and vision for strain selection in bioaugmentation. Trends Biotechnol 23(2):74–77

Singh P, Jain R, Srivastava N et al (2017) Current and emerging trends in bioremediation of petrochemical waste: a review. Crit Rev Environ Sci Technol 47(3):155–201

Speight JG, Arjoon KK (2012) Bioremediation of petroleum and petroleum products. Wiley, Incorporated, Somerset, USA

Strzempka CP, Woodhull PM, Vassar TM et al (1997) In-situ biosparging and soil vapor extraction for JP-4 contaminated soils and groundwater: a case study. In: 4th International in situ and on-site bioremediation symposium, vol 4(1), pp 245–250

Susarla S, Medina VF, McCutcheon SC (2002) Phytoremediation: an ecological solution to organic chemical contamination. Ecol Eng 18(5):647–658

Syawlia RM, Titah HS (2021) Removal of hydrocarbons from contaminated soils using bioremediation by aerobic co-composting methods at ship dismantle locations. J Ecol Eng 22(6):181–190

Toth CRA, Luo F, Bawa N et al (2021) Anaerobic benzene biodegradation linked to the growth of highly specific bacterial clades. Environ Sci Technol 55(12):7970–7980

Truskewycz A, Gundry TD, Khudur LS et al (2019) Petroleum hydrocarbon contamination in terrestrial ecosystems-fate and microbial responses. Molecules 24(18)

Tzovolou DN, Theodoropoulou MA, Blanchet D et al (2015) In situ bioventing of the vadose zone of multi-scale heterogeneous soils. Environ Earth Sci 74(6):4907–4925

Ulrich AC, Edwards EA (2003) Physiological and molecular characterization of anaerobic benzene-degrading mixed cultures. Environ Microbiol 5(2):92–102

Ulrich AC, Guigard SE, Foght JM et al (2009) Effect of salt on aerobic biodegradation of petroleum hydrocarbons in contaminated groundwater. Biodegradation 20(1):27–38

Varjani SJ (2017) Microbial degradation of petroleum hydrocarbons. Biores Technol 223:277–286

Varjani S, Upasani VN (2021) Bioaugmentation of *Pseudomonas aeruginosa* NCIM 5514-A novel oily waste degrader for treatment of petroleum hydrocarbons. Bioresour Technol 319

Vermeulen F, McGee B (2000) In situ electromagnetic heating for hydrocarbon recovery and environmental remediation. J Can Pet Technol 39(8):24–28

Vidali M (2001) Bioremediation. An overview. Pure Appl Chem 73(7):1163–1172

Vogt C, Kleinsteuber S, Richnow HH (2011) Anaerobic benzene degradation by bacteria. Microb Biotechnol 4(6):710–724

Wang M, Garrido-Sanz D, Sansegundo-Lobato P et al (2021) Soil microbiome structure and function in ecopiles used to remediate petroleum-contaminated soil. Front Environ Sci 9

Wolicka D, Suszek A, Borkowski A et al (2009) Application of aerobic microorganisms in bioremediation in situ of soil contaminated by petroleum products. Biores Technol 100(13):3221–3227

Wu Y-R, Luo Z-H, Vrijmoed LLP (2010) Biodegradation of anthracene and benz[a]anthracene by two *Fusarium solani* strains isolated from mangrove sediments. Biores Technol 101(24):9666–9672

Xiong S, Li X, Chen J et al (2015) Crude oil degradation by bacterial consortia under four different redox and temperature conditions. Appl Microbiol Biotechnol 99(3):1451–1461

Yu B, Yuan Z, Yu Z et al (2022) BTEX in the environment: an update on sources, fate, distribution, pretreatment, analysis, and removal techniques. Chem Eng J 435:134825

Zafra G, Absalón ÁE, Cuevas MDC et al (2014) Isolation and selection of a highly tolerant microbial consortium with potential for PAH biodegradation from heavy crude oil-contaminated soils. Water Air Soil Pollut 225(2)

Zang T, Wu H, Zhang Y et al (2021) The response of polycyclic aromatic hydrocarbon degradation in coking wastewater treatment after bioaugmentation with biosurfactant-producing bacteria *Pseudomonas aeruginosa* S5. Water Sci Technol 83(5):1017–1027

Zhang K, Wang S, Guo P et al (2021) Characteristics of organic carbon metabolism and bioremediation of petroleum-contaminated soil by a mesophilic aerobic biopile system. Chemosphere 264

Zuzolo D, Guarino C, Tartaglia M et al (2021a) Plant-soil-microbiota combination for the removal of total petroleum hydrocarbons (TPH): an in-field experiment. Front Microbiol 11

Zuzolo D, Sciarrillo R, Postiglione A et al (2021b) The remediation potential for PAHs of *Verbascum sinuatum* L. combined with an enhanced rhizosphere landscape: a full-scale mesocosm experiment. Biotechnol Rep (Amst) 31:e00657

Chapter 15
In Situ Chemical Oxidation of Petroleum Hydrocarbons

Neil R. Thomson

Abstract In situ chemical oxidation (ISCO) is a mature treatment technology that involves the delivery of a chemical oxidant into a target treatment zone (TTZ) to destroy petroleum hydrocarbon (PHC) compounds, and thereby reduce risk to human health and the environment. Commonly used chemical oxidants include hydrogen peroxide, sodium percarbonate, ozone, sodium or potassium permanganate, and sodium or potassium persulfate. All these oxidants can degrade environmentally relevant PHCs except that permanganate is non-reactive toward benzene. Ozone is delivered into the TTZ as a gas while the other oxidants are typically delivered as a concentrated liquid. ISCO should be considered part of integrated remediation strategy and not used in isolation. This chapter provides a reader with an introduction to key aspects of ISCO that are relevant to applications at sites contaminated with PHCs. Following a discussion of the TTZ, it then examines the fundamentals of the common oxidants used. Next the interaction of chemical oxidants with aquifer materials is explored, and a description of relevant transport considerations is provided. This is followed with an overview of methods that can be used to deliver an oxidant to treat PHCs in a TTZ. This chapter closes with a summary of the important takeaway messages.

Keywords Amendment delivery · Aquifer material interactions · In situ chemical oxidation · Integrated treatment · Oxidants

N. R. Thomson (✉)
Department of Civil and Environmental Engineering, University of Waterloo, 200 University Ave., West Waterloo, ON N2L 3G1, Canada
e-mail: neil.thomson@uwaterloo.ca

© The Author(s) 2024   503
J. García-Rincón et al. (eds.), *Advances in the Characterisation and Remediation of Sites Contaminated with Petroleum Hydrocarbons*, Environmental Contamination Remediation and Management, https://doi.org/10.1007/978-3-031-34447-3_15

15.1 Introduction

An improved awareness of the health risk associated with petroleum hydrocarbons (PHCs), and an increased scientific understanding of the underlying processes related to their fate and transport over the past 30 years has fostered the growth and introduction of a number of remedial technologies. These technologies range from passive methods (e.g., monitored natural attenuation, and permeable reactive barriers) to more active methods [e.g., soil vapor extraction (SVE), in situ air sparging (IAS), and pump-and-treat (P&T)]. Technology selection depends on a number of factors including site characteristics (e.g., type and extent of PHC contamination, and hydrogeological conditions), and is influenced by robustness and economic viability (FRTR 2022).

In situ chemical oxidation (ISCO) is based on an extension of the already well-established water and wastewater treatment processes wherein chemical oxidants are added to water to destroy target contaminants and thereby improve water quality (Colthurst and Singer 1982; Cleasby et al. 1964; Houston 1918). Similarly, ISCO involves the delivery of a strong oxidizing reagent into the subsurface to transform PHC compounds into less harmful intermediates or end-products such as carbon dioxide (CO_2), and thereby reduce potential risk to public health and the ecosystem. While complete mineralization is a worthwhile objective, partial chemical oxidation where long-chain PHCs are converted into less-complex, more water-soluble, and more biodegradable compounds or intermediates is a beneficial objective (Karpenko et al. 2009).

ISCO is potentially able to operate over a wide range of conditions, and provides economic advantages over other conventional technologies due, in part, to its aggressive nature (ITRC 2005). There are a variety of chemical oxidants that are currently being used in ISCO applications, including hydrogen peroxide (H_2O_2), sodium percarbonate ($Na_2CO_3 \cdot 1.5H_2O_2$), ozone (O_3), sodium or potassium permanganate ($NaMnO_4$, $KMnO_4$), and sodium or potassium persulfate ($Na_2S_2O_8$, $K_2S_2O_8$).

The design of an ISCO system involves a comprehensive understanding of the underlying physical and chemical processes, and in general, requires an oxidant to have a high rate and extent of reactivity toward target PHCs, and a high stability and persistence in the subsurface. These criteria serve to establish the usefulness, effectiveness, and efficiency of an oxidant.

Since the initial application of ISCO in 1984 (Brown et al. 1986), there has been substantial research and development activities that have generated a tremendous amount of published literature dealing with oxidant chemistry, oxidant interactions with aquifer materials, transport processes, delivery methods, and combined or integrated treatment systems (e.g., see Siegrist et al. 2011; USEPA 2006; ITRC 2005). In addition to the published literature, the number of field applications has grown throughout the world to a point where ISCO is considered a mature technology and has been added to the remedial toolbox.

The aim of this chapter is to provide a reader with an introduction to key aspects of ISCO that are relevant to applications at sites contaminated with PHCs. It begins with a discussion of the target treatment zone, and then examines the fundamentals of the most commonly used oxidants. Next the interaction of chemical oxidants with aquifer materials is explored, and a description of relevant transport considerations is provided. This is followed with an overview of methods that can be used to deliver an oxidant to treat PHCs in the subsurface. Finally, the chapter closes with a summary of the important takeaway messages. There are numerous ISCO case studies available in the literature for an interested reader to consult (see USEPA 2022).

15.2   The Target Treatment Zone

Typically, ISCO should not be considered until all attempts to remove mobile light non-aqueous phase liquid (LNAPL) from the subsurface have been exhausted. The mass of oxidant required to oxidize mobile LNAPL, at most sites, will not be economically feasible. Since chemical oxidants are not able to mix with the LNAPL (i.e., they are not miscible), all the important chemical oxidation reactions occur in the aqueous phase. Thus, mass associated with the LNAPL must dissolve into the aqueous phase for it to be accessible to the oxidant. Mass transfer from the LNAPL to the aqueous phase is controlled by a mass transfer rate coefficient and the aqueous solubility of the PHC compound of interest. For some PHCs and LNAPL architectures, the mass transfer rate is limiting rather than the chemical oxidation reaction rate. Therefore, the removal of mobile LNAPL using recovery methods such as skimming, vacuum enhanced recovery, and multiphase extraction (ITRC 2018) is commonly recommended before any ISCO application. ISCO should not be viewed as a stand-alone treatment technology, but rather as integral part of an integrated treatment approach.

After the majority of the mobile LNAPL has been removed, ISCO can be used to treat the remaining LNAPL source zone and the associated dissolved phase plume. Figure 15.1 conceptually illustrates a residual LNAPL source zone and dissolved phase plume in profile and plan views. The objective of an effective ISCO application is to create an in situ reactive zone (IRZ) where target PHC compounds are oxidized to less toxic end products. The design of an IRZ requires that the delivered oxidants can react with the target PHCs of concern at a sufficient rate to overcome competing reactions, and that enough oxidant mass is distributed throughout the IRZ to support the demand (Sutherson and Payne 2005). ISCO treatment objectives need to be clearly stated so that expectations are non-ambiguous. The target treatment zone (TTZ) may be the residual LNAPL in the unsaturated zone, the entrapped LNAPL mass in the saturated capillary fringe and the saturated groundwater zone (smear zone), and/or the dissolved phase plume. While the IRZ design principles for each of these TTZs are similar, the way in which they are implemented is different. Site characterization data must be carefully reviewed to determine if, and where, LNAPL is likely present. Persistent dissolved PHCs plumes typically suggest the presence of LNAPL, and the TTZ should encompass those areas of the site where LNAPL sources are suspected to be present.

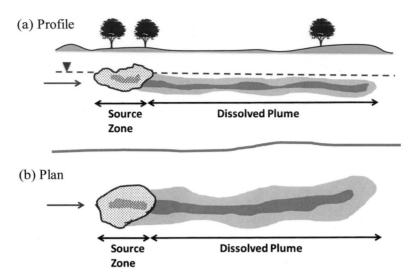

Fig. 15.1 A conceptual illustration of an LNAPL source zone and dissolved phase plume in **a** profile view, and **b** plan view. The source zone is presented as a smear zone in the capillary fringe and saturated groundwater zone with a core of higher LNAPL saturation. The dissolved phase plume is also shown with a higher concentration core

In addition to PHC mass distribution, consideration must also be given to the delivery of a liquid or gas oxidant into the TTZ. A TTZ in unconsolidated porous media (e.g., a mixture of sand, silt, and clay) is considered suitable for ISCO if the average saturated hydraulic conductivity (K) is > 10^{-6} m/s. The spatial variation or heterogeneity of K in the TTZ is also important with oxidant delivery into highly heterogeneous TTZs being problematic (e.g., variations in K of > 10^3 or 10^4 m/s).

15.3 Oxidant Fundamentals

The kernel of the any ISCO system is its capability to quickly react with the target PHC compounds of interest, and degrade them into innocuous end-products such as CO_2 and water. While complete mineralization of the target PHCs is the ultimate goal, often there are many reaction steps required to reach this goal and numerous intermediates or oxidation by-products will be produced. Usually these by-products are more water-soluble, less toxic, and susceptible to biodegradation. For example, a chemical oxidation pathway for benzene involves the production of phenol which is then transformed to catechol (a benzenediol) (Fu et al. 2017). While catechol can also be chemically oxidized, it is readily biodegraded since it is also formed during the aerobic degradation of benzene (e.g., Yu et al. 2001).

For the commonly used oxidants discussed in this chapter, Table 15.1 indicates their ability to degrade benzene, ethylbenzene, toluene, and xylenes (BTEX), mixtures of PHCs as represented by total petroleum hydrocarbons (TPH), and polycyclic aromatic hydrocarbons (PAHs) (e.g., naphthalene, fluorene, and anthracene). As shown, all oxidants are amenable to degrade the target PHCs listed except that permanganate is essentially non-reactive toward benzene. TPH is a bulk measure of the petroleum-based aromatic and aliphatic hydrocarbons found in gasoline, diesel, and oil range organics [e.g., see US EPA Method 8015D (SW-846)]. While Table 15.1 indicates that TPH can be degraded by the oxidants listed, this does not imply that complete mineralization of the total mass associated with TPH is possible, rather that TPH has been observed to decrease following exposure to one of the oxidants. Similarly, only some individual PAHs have been reported to be susceptible to oxidation and thus the total PAH sum, which is often used as a treatment metric, will decrease following exposure to an oxidant.

Often the standard reduction potential (E°) is used to compare the potential of various oxidant species. The standard reduction potential is the potential in volts (V) generated by a half-reaction relative to the standard hydrogen electrode (SHE) under standard conditions (25 °C, 1 atm, concentration of 1 M). The larger the potential the greater desire the species has to acquire electrons, or the more the species tends to be reduced (Snoeyink and Jenkins 1980). While E° values are informative, they do not provide information on reaction stoichiometry or kinetics for a specific PHC.

The rate of reaction is an important consideration for ISCO treatment effectiveness, and hence numerous reaction rate coefficients have been reported in the literature for a variety of PHCs, oxidant systems, and reaction conditions. Because many of these reactions occur in the aqueous phase, the chemistry of the oxidants and their interaction with PHC compounds have been the subject of intensive study. Typically, the second-order mass action law shown in Eq. (15.1) is used to represent the aqueous phase reaction kinetics for a PHC compound exposed to an oxidant (Forsey et al. 2010; Levenspiel 1999)

$$\frac{d[C_{PHC}]}{dt} = -k''[C_{PHC}][C_{ox}] \qquad (15.1)$$

Table 15.1 General ability of the most commonly used oxidants to degrade BTEX compounds, TPH, and PAHs

Target PHC	Peroxide	Percarbonate	Ozone	Permanganate	Persulfate
Benzene	✔	✔	✔	✗	✔
Ethylbenzene	✔	✔	✔	✔	✔
Toluene	✔	✔	✔	✔	✔
Xylenes	✔	✔	✔	✔	✔
TPH	✔	✔	✔	✔	✔
PAHs	✔	✔	✔	✔	✔

Adapted from USEPA (2006)

where $[C_{PHC}]$ is the molar concentration of the PHC compound (mol/L), $[C_{ox}]$ is the molar concentration of the oxidant (mol/L), and k'' is the second-order rate coefficient (L/mol s or M^{-1} s^{-1}). Tratnyek et al. (2019) complied a dataset of k'' values that were available as of 2005 which include data for a select number of PHC compounds. Treatability studies which are used to generate kinetic data are often limited since they require simplifying assumptions that may ignore changing reaction conditions (pH shifts and radical complexities), and generation and consumption of intermediates and by-products. In these cases, a pseudo first-order mass action law (Eq. 15.2) is often used to fit the kinetic data as given by

$$\frac{d[C_{PHC}]}{dt} = -k'_{obs}[C_{PHC}] \tag{15.2}$$

where k'_{obs} is the observed first-order rate coefficient (s^{-1}) associated with the degradation of the parent or target PHC compound. In addition to the concentration of the reactants (Eq. 15.1), the in situ degradation rate of a specific PHC compound is dependent on many variables that also play a role including temperature, pH, interaction with aquifer solids or materials, and the concentration of catalysts, reaction by-products, natural organic matter, and scavengers.

The following sub-sections provide a high-level overview of the commonly used oxidants (hydrogen peroxide, percarbonate, ozone, permanganate, and persulfate) since a basic understanding of the reactions involved is required. For some oxidants the suite of reactive species is too complicated, so a simplified perspective is discussed. Also provided is a short synopsis of the capability of each oxidant to degrade target PHCs based on laboratory studies. By no means is the description of each oxidant presented in this section complete, but it is considered to be sufficient to appreciate their unique qualities. For additional insight, an interested reader is encouraged to consult Siegrist et al. (2011, 2014), USEPA (2006) and ITRC (2005), and the vast list of references contained within.

15.3.1 Hydrogen Peroxide

Hydrogen peroxide (H_2O_2) is a strong liquid oxidant that is used in many industrial applications, and in water treatment processes. Hydrogen peroxide is often applied in combination with a catalyst or activator, and the term catalyzed hydrogen peroxide (CHP) has been used to refer to these hydrogen peroxide systems. When catalyzed or activated in water, hydrogen peroxide generates a wide variety of free radicals and other reactive species. A radical or free radical is a cluster of atoms that contains an unpaired electron. This configuration is extremely unstable, and hence radicals quickly react to achieve a stable configuration. Reactive species formed in a CHP system can include both oxidants and reductants.

As an alternative to the liquid form of H_2O_2, various forms of solid peroxides (e.g., percarbonate, and peroxide) have been used. Solid forms of H_2O_2 are more easily stored and transported than the liquid form of H_2O_2. At ambient temperature, sodium percarbonate ($Na_2CO_3 \cdot 1.5H_2O_2$) will yield H_2O_2, sodium (Na^+), and carbonate (CO_3^{2-}) when dissolved in water as given by Eq. (15.3):

$$2Na_2CO_3 \cdot 3H_2O_2 \rightarrow 3H_2O_2 + 4Na^+ + 2CO_3^{2-} \tag{15.3}$$

Although calcium peroxide (CaO_2, or magnesium peroxide, MgO_2) is usually considered an oxygen releasing compound used to promote aerobic biodegradation, it produces H_2O_2 when dissolved in water (Eq. 15.4)

$$CaO_2 + 2H_2O \rightarrow H_2O_2 + Ca^{2+} + 2OH^- \tag{15.4}$$

By itself, H_2O_2 can directly react with a PHC compound through a direct two-electron transfer (Eq. 15.5a). While this reaction has a high standard reduction potential ($E^\circ = 1.8$ V) the rate of these reactions is deemed slow and not relevant for ISCO applications (Watts and Teel 2005).

The chemistry of the CHP system grew from the well-established *Fenton's reagent* (Fenton 1894) in which ferrous iron (Fe^{2+}) salts were used to catalyze a dilute solution of H_2O_2 at a pH between 3 and 5 to yield hydroxyl radicals ($\cdot OH$), ferric iron (Fe^{3+}), and hydroxyl ions (Eq. 15.5b). The generated ferric iron then reacts with H_2O_2 or, with another produced radical, the hydroperoxyl radical ($HO_2^\cdot$, $E^\circ = 1.7$ V) to produce Fe^{2+} (Eqs. 15.5c and 15.5d). The reactions (Eqs. 15.5b–15.5d) will essentially deplete the available H_2O_2 in a self-destructing manner. The hydroxyl radical is a non-selective oxidant that has an unpaired electron making it highly reactive ($E^\circ = 2.8$ V) with a wide variety of PHC compounds.

$$H_2O_2 + 2H^+ + 2e^- \rightarrow 2H_2O \tag{15.5a}$$

$$H_2O_2 + Fe^{2+} \rightarrow OH^- + Fe^{3+} + \cdot OH \tag{15.5b}$$

$$Fe^{3+} + H_2O_2 \rightarrow Fe^{2+} + H^+ + HO_2^\cdot \tag{15.5c}$$

$$Fe^{3+} + HO_2^\cdot \rightarrow Fe^{2+} + H^+ + O_2 \tag{15.5d}$$

$$HO_2^\cdot \leftrightarrow O_2^{\cdot-} + H^+ \tag{15.5e}$$

$$Fe^{2+} + HO_2^\cdot \rightarrow Fe^{3+} + HO_2^- \tag{15.5f}$$

$$\cdot OH + Fe^{2+} \rightarrow Fe^{3+} + OH^- \tag{15.5g}$$

$$H_2O_2 + {}^{\cdot}OH \rightarrow H_2O + HO_2^{\cdot} \tag{15.5h}$$

As the concentration of H_2O_2 is increased, several other important radicals and ions are generated including the superoxide radical ($O_2^{\cdot-}$, Eq. (15.5e); $E^o = -2.4$ V) and the hydroperoxide anion [HO_2^-, Eq. (15.5f); $E^o = -0.9$ V]. In contrast to the hydroxyl and hydroperoxide radicals which are oxidants, the superoxide radical and hydroperoxide anion are reductants providing a reduction degradation pathway for some PHC compounds (Furman et al. 2009; Smith et al. 2004). The reactions given by Eqs. (15.5b)–(15.5h) occur nearly simultaneously and will continue until one of the reactants becomes limiting which is normally H_2O_2.

One of the important by-products of CHP reactions is the generation of large quantities of dissolved oxygen (O_2) that is available in the TTZ after H_2O_2 has been depleted. Since O_2 is an electron acceptor for the aerobic biodegradation of many PHC compounds, its presence may lead to further degradation of target PHCs and intermediates of chemical oxidation. However, O_2 evolution in some cases may result in the mobilization of volatile PHCs leading to unacceptable exposure pathways. CHP reactions are exothermic, and therefore they can result in significant temperature increases within the TTZ. Heat production is dependent on the concentration of H_2O_2 used and the rate of reaction. This excess heat and perhaps steam is a health and safety hazard which may need to be addressed.

To enable important CHP reactions to occur in the TTZ, ferrous sulfate ($FeSO_4$) or other Fe^{2+} salts are typically co-injected with H_2O_2 at concentrations significantly above background levels. In porous media, both Fe^{2+} and Fe^{3+} are susceptible to numerous reactions (e.g., precipitation, complexation) which eliminate them from the CHP reactions. More importantly perhaps, is that the solubility of Fe^{2+} at typical groundwater pH levels (7–9) is limiting. To overcome these issues, stabilizers are typically used (Checa-Fernandez et al. 2021). Phosphate was one of the initial stabilizing compounds used; however, since it is a strong complexing agent, it restricted the availability of Fe^{2+}. Watts et al. (2007) suggested that the most effective stabilizers are citrate, malonate, and phytic acids.

Numerous non-productive or scavenging reactions compete with the target PHCs for the radicals generated in a CHP system. The most common scavenger anions prevalent in groundwater systems are carbonate, bicarbonate, chloride, and sulfate. These anions react with ${}^{\cdot}OH$ to form the bicarbonate radical (Eq. 15.6a), the carbonate radical (Eq. 15.6b), the hypochlorite radical (Eq. 15.6c), and the sulfate radical (Eq. 15.6d). While these scavenging reactions are generally considered unproductive, evidence suggests that carbonate radicals are capable of oxidizing some aromatic compounds (e.g., benzene) (Umschlag and Herrmann 1999), and the sulfate radical has a high reactivity toward PHC compounds (see Sect. 15.3.4).

$$ {}^{\cdot}OH + HCO_3^- \rightarrow OH^- + HCO_3^{\cdot} \tag{15.6a}$$

$$ {}^{\cdot}OH + CO_3^{2-} \rightarrow OH^- + CO_3^{\cdot-} \tag{15.6b}$$

$$\cdot OH + Cl^- \rightarrow ClOH^- \tag{15.6c}$$

$$\cdot OH + SO_4^{2-} \rightarrow OH^- + SO_4^- \tag{15.6d}$$

In some instances, the presence of minerals associated with aquifer materials may act like a catalyst similar to the addition of Fe^{2+} discussed above. The minerals of relevance to ISCO are those associated with Fe or Mn since they are typically observed at the highest concentration in aquifer materials. Goethite (α-FeOOH), hematite (Fe_2O_3), and pyrolusite (β-MnO_2) have been shown to catalyze H_2O_2 (e.g., Kwan and Voelker 2003, Teel et al. 2001). In contrast to the homogeneous reaction between H_2O_2 and dissolved Fe^{2+}, the reaction between H_2O_2 and minerals is heterogeneous since it occurs at the mineral surface.

Prior to delivery, percarbonate is typically mixed with a combination of silicates and $FeSO_4$. When added to water, an alkaline pH version of the CHP system will be generated. In this base-catalyzed peroxide system, the predominant species that is generated is the hydroperoxyl radical ($HO_2^\cdot$). Na_2CO_3 released will form carbonate (CO_3^{2-}) and bicarbonate (HCO_3^-) which will scavenge radicals [see Eqs. (15.6a) and (15.6b)]. Ma et al. (2020) reported that the lifetime of percarbonate is relatively short which limits its availability in the subsurface.

CHP systems have been demonstrated in many laboratory studies to treat PHCs with success. For example, Watts and Dilly (1996) investigated the ability of phosphate stabilized H_2O_2 catalyzed by six different iron compounds to oxidize diesel sorbed to soils (1000 mg/kg). All iron catalysts promoted > 70% degradation in < 1 h. Xu et al. (2006) used H_2O_2 and a mixture of ferrous salts to oxidize diesel spiked soils, and reported that the mass removed (as high at 93%) was dependent on the concentration and volume of H_2O_2 as well as the soil composition. Usman et al. (2012) investigated the capability of magnetite (Fe_3O_4) catalyzed H_2O_2 to treat soil spiked with fresh crude oil and weathered crude oil. The results showed that 84% of the weathered crude oil and 92% of the crude oil was degraded in one week. Surprisingly, these authors also report that no by-products were observed and thus complete degradation of PHCs was achieved. Aromatic compounds (i.e., BTEX) have been found to be easily degraded by $\cdot OH$. Table 15.2 provides the second-order reaction rate coefficients for the reaction between $\cdot OH$ and BTEX compounds (Buxton et al. 1988).

Table 15.2 Second-order reaction rate coefficients for the oxidation of BTEX compounds by hydroxyl radicals (from Buxton et al. 1988)

Compound	k'' (M^{-1} s^{-1})
Benzene	7.8×10^9
Toluene	7.5×10^9
Ethylbenzene	3.0×10^9
m-, o-, p-xylene	6.7–7.5×10^9

Since PAHs are comprised of multiple aromatic benzene rings, oxidation of PAHs by CHP systems is possible. Lundstedt et al. (2006) reported that some PAHs (anthracene, benzo[a]pyrene, and perylene) were extensively degraded compared to other PAHs with fewer or equal number of benzene rings during CHP treatment of contaminated soil collected from a former gasworks site indicating selectively. Nevertheless, considering the complex structure of the higher ringed PAHs, it would be expected that numerous ˙OH reactions would be required to fully mineralize the parent PAH and, in this process, various intermediates would be generated (Lundstedt et al. 2006).

15.3.2 Ozone

Ozone (O_3) is a highly reactive and unstable gas, and must be generated at the point of use with an ozone generator. Onsite O_3 generators that use an air supply can produce an O_3 concentration of 1–2% by volume, so the bulk of the injected gas is atmospheric air. This gas can be delivered directly into the unsaturated zone or sparged into the saturated zone. In the latter case, the distribution of O_3 will be controlled by the same processes that control the air distribution during in situ air sparging (Tomlinson et al. 2003; Thomson and Johnson 2000). Figure 15.2 provides a schematic representation of the injection of ozone gas below a LNAPL source TTZ, and the observed spatial distribution of gas release locations at the water table. The nature of the discrete gas pathways within the TTZ suggests that treatment will be limited by mass transfer of O_3 to the aqueous phase.

Oxidation reactions can occur both in the gas and aqueous phases. Gas phase reactions will occur when PHC compounds have volatilized into the gas phase and react with O_3. Aqueous phase reactions require O_3 to partition into the groundwater which is driven by the O_3 concentration gradient between the gas and aqueous phases; however, the aqueous solubility of O_3 is low (~28 mg/L at 10 °C from a 5% O_3 gas concentration) which limits the mass of O_3 available in the aqueous phase. This limitation is compounded by the discrete gas channels that develop in the saturated zone as discussed above. Once in the aqueous phase, a number of decomposition processes involving organic and inorganic reactants exert a constant demand for O_3.

Tomiyasu et al. (1985) proposed the ozone decomposition pathway given by Eqs. (15.7a–15.7i). This pathway begins with a one-electron transfer from H_2O resulting in the formation of hydroxyl radicals (˙OH) and the ozonize anion (O_3^-) (Eq. 15.7a), or a two-electron transfer resulting in the formation of the hydroperoxide anion (HO_2^-) and oxygen (O_2) (Eq. 15.7b). Following this initial step, propagation reactions with ozone and the ozonize anion yield the hydroperoxyl radical (HO_2) and superoxide radical (O_2^-) (Eqs. 15.7c–15.7i). This suite of strong radicals is similar to those generated in the CHP system (Eqs. 15.5a–15.5h).

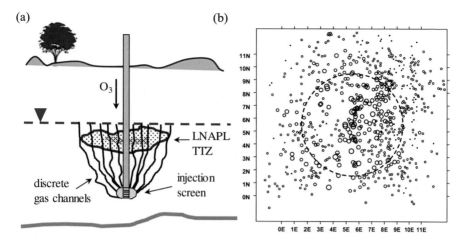

Fig. 15.2   a Schematic representation of the injection of ozone gas below a LNAPL TTZ. Near the injection screen there will be a region of elevated gas content; however, due to flow instabilities a network of isolated gas channels will form and the gas will migrate vertically upward to escape into the unsaturated zone. **b** Spatial distribution of non-uniform gas release locations at the water table during steady-state air sparging (from Tomlinson et al. 2003). Each observed air release location is indicated by a circle with a radius corresponding to the relative intensity of the air release at that location (smallest is the least intense). Grid spacing in northerly and easterly directions is 1 m

$$O_3 + OH^- \rightarrow O_3^- + \,^{\cdot}OH \tag{15.7a}$$

$$O_3 + OH^- \rightarrow O_2 + HO_2^- \tag{15.7b}$$

$$O_3 + HO_2^- \rightarrow O_3^- + HO_2 \tag{15.7c}$$

$$HO_2^- \leftrightarrow O_2^- + H^+ \tag{15.7d}$$

$$O_3 + O_2^- \rightarrow O_3^- + O_2 \tag{15.7e}$$

$$O_3^- + H_2O \rightarrow \,^{\cdot}OH + OH^- + O_2 \tag{15.7f}$$

$$O_3^- + \,^{\cdot}OH \rightarrow O_2^- + HO_2 \tag{15.7g}$$

$$O_3^- + \,^{\cdot}OH \rightarrow O_3 + OH^- \tag{15.7h}$$

$$^{\cdot}OH + O_3 \rightarrow HO_2^- + O_2 \tag{15.7i}$$

Ozone can react with target PHC compounds either through direction oxidation, or through the free radicals generated. Direct oxidation mechanisms include 1,3-dipolar cycloaddition of ozone to an alkene bond (a carbon–carbon double bond), and electrophilic attack on aromatic hydrocarbons (Smith and March 2007; Langlais et al. 1991).

Ozone has a long history of being used to treat PHCs associated with fuel releases. The use of IAS coupled with SVE has been used to remove mass from LNAPL source zones by extracting volatile compounds and stimulating aerobic biodegradation (Bouchard et al. 2018). As treatment progresses and most of the assessable volatile and semi-volatile compounds are depleted, the cumulative mass removal by an IAS/SVE system becomes asymptotic. At this point in the remedial effort, ozone can be used as an oxidizing gas to aggressively attack the remaining recalcitrant LNAPL mass. Investigations where ozone was injected into unsaturated columns packed with PHC impacted soils have yielded mass removal rates of 95% for crude oil (Dunn and Lunn 2002), and between 50 and 94% for diesel fuel (Yu et al. 2007; Jung et al. 2005). Since all these column systems had minimal water content the main degradation reactions were in the gas phase, presumably occurring at the soil/gas interfaces. The rates of reaction of the various PHC chains were noted to be similar; however, the C_{10} fraction took longer to degrade since higher carbon aliphatic compounds were oxidized to shorter-chain compounds before being oxidized to lower molecular weight organic compounds and CO_2. The majority of the lower weight organic compounds are suspected to be easily biodegraded (e.g., alcohols, aldehydes, ketones, and carboxylic acids).

Ozone has been shown to effectively degrade BTEX compounds both in the laboratory and in the field (Bhuyan and Latin 2012; Garoma et al. 2008; Black 2001). The direct reaction of ozone with BTEX compounds is given by Eqs. (15.8a)–(15.8d):

$$C_6H_6 + 11O_3 \rightarrow 6CO_2 + 3H_2O + 11O_2 \qquad (15.8a)$$

$$C_7H_8 + 18O_3 \rightarrow 7CO_2 + 4H_2O + 18O_2 \qquad (15.8b)$$

$$C_8H_{10} + 7O_3 \rightarrow 8CO_2 + 5H_2O \qquad (15.8c)$$

$$C_8H_{10} + 21O_3 \rightarrow 8CO_2 + 5H_2O + 21O_2 \qquad (15.8d)$$

Since ozone generates the similar suite of radicals as in the CHP system, ozone is expected to rapidly oxidize other PHCs in addition to BTEX. Tengfei et al. (2016) used O_3 as a pre-treatment method for weathered crude oil impacted soil, and observed 50% TPH reduction. In parallel with this reduction was a 20 times increase in Chemical Oxygen Demand (COD), and a 4 times increase in the 5-day biochemical oxygen demand (BOD) indicating that some PHCs were converted to partly oxidized products that easily biodegraded.

Ozone is very reactive with PAHs, and thus is very effective for the removal of PAHs from soils. PAHs with a few benzene rings (e.g., naphthalene, fluorine, phenanthrene) can be oxidized to a greater extent than PAHs with a larger number of rings (e.g., pyrene, chrysene, benzo(a)pyrene) (e.g., Nam and Kukor 2000; Masten and Davies 1997). Evidence suggests that a large number of intermediates may be formed depending on the parent PAH, and reaction conditions (gas or aqueous phase, pH, etc.). As a result, a series of oxidation steps are required to completely mineralize the parent PAH as illustrated by Yao et al. (1998) for the oxidation of pyrene.

15.3.3 Permanganate

Permanganate (MnO_4^-) has been used to treat both drinking water and wastewater (Eilbeck and Mattock 1987) for the removal of iron and manganese but has also been used for the treatment of trihalomethane precursors (Colthurst and Singer 1982) and phenolic wastes (Vella et al. 1990). MnO_4^- is a selective oxidant since it will degrade some organic compounds (Waldemer and Tratnyek 2006). There are two forms of permanganate that are widely used: potassium permanganate ($KMnO_4$ a crystal solid) and sodium permanganate ($NaMnO_4$ a liquid). Both forms have the same reactivity; however, the solubility of $KMnO_4$ is 65 g/L at 20 °C, and if higher concentrations are required $NaMnO_4$ is used (solubility of 900 g/L at 20 °C).

The MnO_4^- ion ($E^\circ = 1.7$ V) is a transition metal oxidant that reacts only through direct electron transfer rather than through any radical intermediates like in the CHP and ozone systems. As shown by the half-reactions in Eqs. (15.9a–15.9c), the number of electrons transferred depends on the system pH. In most natural groundwater systems, Eq. (15.9b) is dominant with MnO_4^- undergoing a three-electron transfer.

$$MnO_4^- + 8H^+ + 5e^- \rightarrow Mn^{2+} + 4H_2O \quad pH < 3.5 \tag{15.9a}$$

$$MnO_4^- + 2H_2O + 3e^- \rightarrow MnO_2(s) + 4OH^- \quad 3.5 < pH < 12 \tag{15.9b}$$

$$MnO_4^- + e^- \rightarrow MnO_4^{2-} \quad pH > 12 \tag{15.9c}$$

An unfortunate by-product of Eq. (15.9b) is the production of solid manganese dioxide (MnO_2) (Pisarczyk 1995). MnO_2 forms at the point of reaction as colloid size amorphous solids which coalesce into larger particles. The generated MnO_2 solids have the potential to agglomerate in pore bodies and throats which can result in a reduction of hydraulic conductivity (Schroth et al. 2001), and non-aqueous phase liquid (NAPL)/water mass transfer (Li and Schwartz 2004; MacKinnon and Thomson 2002). Observations from laboratory experiments suggest that the accumulation of MnO_2 occurs in regions of high NAPL saturations where there is a greater NAPL/water interfacial area for mass transfer to occur. For example, MacKinnon and Thomson (2002) reported that after exposure of a perchloroethylene (PCE) pool

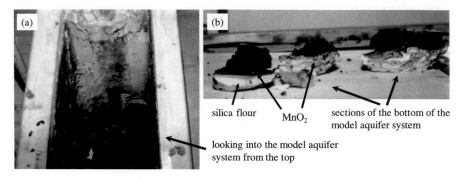

Fig. 15.3 MnO_2 crust formation at the interface of a NAPL pool in a 2-D model aquifer system used to study permanganate flushing to deplete a PCE DNAPL pool: **a** view into the 2-D model during the excavation process, and **b** several sections of the sand/silica flour interface removed from the model aquifer showing that the accumulation of MnO_2 had agglomerated, so that rock-like deposits of MnO_2 and sand were formed (from MacKinnon and Thomson 2002)

to permanganate, the MnO_2 accumulation at the NAPL/water interface was so thick and consolidated that the MnO_2 had effectively cemented sand into MnO_2 "rocks" (Fig. 15.3). An equivalent LNAPL architecture to a PCE pool would be at the top of the smear zone where higher LNAPL saturations are expected to be present (see Fig. 15.1).

As a result of mineralization, CO_2 will be generated and depending on the system pH, there is the potential for CO_2 gas to evolve from the dissolved phase (Lee et al. 2003; MacKinnon and Thomson 2002; Schroth et al. 2001). In some situations, the $CO_{2(g)}$ may fill pore bodies and decrease hydraulic conductivity and mass transfer, while in other situations volatile PHCs present in the system may partition into the $CO_{2(g)}$ causing spontaneous expansion and upward mobilization of the gas phase (Mumford et al. 2008). This latter situation may result in uncontrolled releases of PHC gases.

As described by Waldemer and Tratnyek (2006), the oxidation of an aromatic compound is through attack on a benzylic C–H bond (a hydrocarbon bond on an alkyl side chain of the aromatic ring). Toluene, ethylbenzene, and the xylene isomers, all contain benzylic C–H bonds and react with permanganate at slow but appreciable rates. Since benzene has no benzylic C–H bonds, it is almost unreactive with permanganate. The stoichiometric reaction between MnO_4^- and toluene, and between MnO_4^- and ethylbenzene is given by

$$C_7H_8 + 12MnO_4^- + 12H^+ \rightarrow 12MnO_2 + 7CO_2 + 10H_2O \qquad (15.10a)$$

$$C_8H_{10} + 14MnO_4^- + 14H^+ \rightarrow 14MnO_2 + 8CO_2 + 12H_2O \qquad (15.10b)$$

Table 15.3 Second-order
reaction rate coefficients for
the reaction between
permanganate and individual
BTEX compounds (from
Rudakov and Lobachev 2000)

Compound	k'' (M^{-1} s^{-1})
Benzene	NR
Toluene	2.3×10^{-4}
Ethylbenzene	1.1×10^{-2}
m-xylene	6.0×10^{-4}
o-xylene	8.9×10^{-4}
p-xylene	1.2×10^{-3}

NR not reactive

Table 15.3 provides the second-order reaction rate coefficients for the reaction between MnO_4^- and BTEX compounds (Rudakov and Lobachev 2000). These reaction rate coefficients are substantially smaller than those listed in Table 15.2 for the oxidation of BTEX compounds by hydroxyl radicals.

Only some PAHs are susceptible to oxidation by MnO_4^-. Forsey et al. (2010) performed a comprehensive investigation into the reactivity and kinetics of MnO_4^- toward PAHs. The oxidation of naphthalene, phenanthrene, chrysene, 1-methylnaphthalene, 2-methylnaphthalene, acenaphthene, fluorene, carbazole isopropylbenzene, and methylbenzene closely followed pseudo first-order reaction kinetics. The oxidation of pyrene was initially very rapid and then transitioned into a pseudo first-order rate at later time. Fluoranthene was only partially oxidized, and the oxidation of anthracene was too fast to be captured. Biphenyl, dibenzofuran, and tert-butylbenzene were non-reactive under the study conditions. The oxidation rate was shown to increase with the increasing number of polycyclic rings since less energy is required to overcome the aromatic character of a polycyclic ring than is required for benzene. Thus, the rate of oxidation increased in the series pyrene > phenanthrene > naphthalene.

15.3.4 Persulfate

Persulfate is used in many industrial processes and commercial products. For example, it is used as an initiator in the polymerization reaction during the manufacturing of latex and synthetic rubber, in the electronics industry to clean and microetch circuit boards, and in the pharmaceutical industry in the preparation of antibiotics. In general, the chemistry of persulfate is quite complex and not fully understood. The persulfate anion ($S_2O_8^{2-}$), also known as peroxydisulfate (PDS), is a strong oxidant ($E^o = 2.1$ V), and can react directly with some PHCs through a two-electron transfer process (Eq. 15.11a). Many important persulfate reactions occur when $S_2O_8^{2-}$ reacts with other species (i.e., activated) to form a suite of free radicals (Lee et al. 2018; Block et al. 2004). Reactive species formed can include both oxidants and reductants.

While activated persulfate leads to the potential degradation of a wider range of PHC compounds, additional $S_2O_8^{2-}$ is consumed and often not efficiently. Various forms of persulfate are available with the most common being solid potassium persulfate ($Na_2S_2O_8$). At room temperature, $S_2O_8^{2-}$ is quite stable at near-neutral pH (half-life of ~1800 days at pH of 7 and 20 °C), but under strong acidic conditions (pH < 0.3) $S_2O_8^{2-}$ hydrolyzes to generate Caros' acid (H_2SO_5) (Eq. 15.11b). Under acidic conditions, Caros' acid or peroxymonosulfuric acid oxidizes water into H_2O_2 (Eq. 15.11c) by transferring two-electrons, and the generated H_2O_2 is available to participate in the reactions given by Eqs. (15.5a)–(15.5h).

Persulfate can be activated by several different methods including the use of transition metals, heat, peroxide, or high alkaline conditions. Activation with dissolved transition metals (Fe^{2+}, Fe^{3+}, Ag^+, Cu^{2+}) is one of the most common methods and has parallels with Fenton's Reagent (Eq. 15.5b) (e.g., House 1962; Anipsitakis and Dionysiou 2004). Ferrous iron (Fe^{2+}) is the most frequently used transition metal and involves a one-electron transfer from Fe^{2+} to $S_2O_8^{2-}$ resulting in the generation of the sulfate radical (SO_4^-) (Eq. 15.11d) (Kolthoff et al. 1951). The sulfate radical is a powerful, non-selective oxidant ($E° = 2.4$ V). Heat activation of persulfate (Eq. 15.11e), which also yields SO_4^-, has been investigated at temperatures between 30 and 60 °C (e.g., Huang et al. 2005). Persulfate activation by peroxide in a dual oxidant system (Sra et al. 2013a; Crimi and Taylor 2007) is speculated to be initiated when ·OH reacts with $S_2O_8^{2-}$ (Eq. 15.11f). Finally, alkaline activation of persulfate involves that addition of a concentrated base (e.g., NaOH) to raise the pH of the persulfate solution to 11 or 12. At this elevated pH, $S_2O_8^{2-}$ undergoes base-catalyzed hydrolysis to generate SO_4^- and the superoxide radical (O_2^-) (Eq. 15.11g). The high concentration of OH^- will react with SO_4^- to yield ·OH (Eq. 15.11h). Finally, like H_2O_2, $S_2O_8^{2-}$ can be catalyzed by certain minerals (oxides of Fe or Mn) associated with the aquifer materials. Of all the persulfate activation methods, alkaline activation is considered the most aggressive.

$$S_2O_8^{2-} + 2e^- \rightarrow 2SO_4^{2-} \tag{15.11a}$$

$$S_2O_8^{2-} + H_2O \rightarrow H_2SO_5 + SO_4^{2-} \tag{15.11b}$$

$$H_2SO_5 + H_2O \rightarrow H_2SO_4 + H_2O_2 \tag{15.11c}$$

$$S_2O_8^{2-} + Fe^{2+} \rightarrow SO_4^- + SO_4^{2-} + Fe^{3+} \tag{15.11d}$$

$$S_2O_8^{2-} \xrightarrow{hv} 2SO_4^{\cdot -} \tag{15.11e}$$

$$S_2O_8^{2-} + \cdot OH \rightarrow SO_4^- + SO_4^{2-} + 1/2O_2 + H^+ \tag{15.11f}$$

$$2S_2O_8^{2-} + 2H_2O \xrightarrow{OH^-} SO_4^- + O_2^- + 3SO_4^{2-} + 4H^+ \tag{15.11g}$$

$$SO_4^- + OH^- \rightarrow SO_4^{2-} + {}^{\cdot}OH \tag{15.11h}$$

Similar to the CHP system, there are non-productive or scavenging reactions involving carbonate, bicarbonate, chloride or sulfate ions present in most groundwater systems that compete with the target PHCs for the radicals generated in an activated $S_2O_8^{2-}$ system [similar to Eqs. (15.6a–15.6d)].

The ability of $S_2O_8^{2-}$ to degrade PHCs depends on the activation approach used, and the PHC mixture (e.g., crude oil vs. gasoline). In general, $S_2O_8^{2-}$ can be used to treat BTEX and other aromatic PHCs. Huang et al. (2005) investigated the degradation of 59 compounds, including BTEX, using thermally activated $S_2O_8^{2-}$ and concluded that compounds with carbon–carbon double bonds or with reactive functional groups bonded to benzene rings were readily degraded. Sra et al. (2013a) conducted a comprehensive investigation that focused on the treatability of dissolved gasoline compounds by persulfate. Specifically, unactivated and activated $S_2O_8^{2-}$ (peroxide, chelated-ferrous, and alkaline) methods were used. The behavior of ten (10) gasoline compounds (benzene, toluene, ethylbenzene, o-xylene, m, p-xylene (meta- and para-xylene are reported together), 1,2,3-trimethylbenzene, 1,2,4-trimethylbenzene, 1,3,5-trimethylbenzene, and naphthalene) along with two carbon fractions (C_6–C_{10}, and C_{10}–C_{16}) and TPH were monitored. The use of unactivated persulfate resulted in almost complete oxidation of BTEX (> 99%), trimethylbenzenes (> 95%), and significant oxidation of naphthalene ($\sim 70\%$). Observed first-order rate coefficients (k_{obs}) were enhanced by 2–15 times using either peroxide or chelated-iron activation methods. Alkaline activation at pH 11 or 13 yielded a k_{obs} that was ~ 2 times higher than the unactivated case, except for BTE where the k_{obs} was reduced by 50% at pH of 13. The observed degradation trends for the lower carbon fraction (C_6–C_{10}) were consistent with BTEX behavior; however, the higher carbon fraction (C_{10}–C_{16}) did not show increased degradation by any of the persulfate activation methods explored. All trials demonstrated an initially fast destruction (<10 days) of the higher carbon fraction followed by a relatively slower oxidation rate or complete stalling suggesting the presence of recalcitrant PHCs. Overall, the oxidation of the higher carbon fraction was observed to be significant (60–85%) across most of the trials. The bulk gasoline stoichiometry for these experimental trials varied from 120 to 340 g-persulfate/g-TPH. The lower end of this stoichiometric range was for unactivated $S_2O_8^{2-}$ and clearly highlights the non-productive consumption of $S_2O_8^{2-}$ when activation methods are used.

Using a simulated mixture of gasoline PHCs (BTEX, nitrobenzene, and naphthalene) and a variety of porous media, Crimi and Taylor (2007) concluded that the degradation of the selected PHCs varied by $S_2O_8^{2-}$ activation method and porous medium used. The various activation methods resulted in different post-treatment oxidation gas chromatograph patterns suggesting that the organic by-products formed were dependent on the activation method used.

Sra et al. (2013b) conducted a pilot-scale experiment to investigate the fate and transport of an injected $S_2O_8^{2-}$ solution, and the concomitant treatment of a gasoline residual source zone. The high-resolution data collected indicated a 40–80% reduction in mass discharge in the monitored PHC compounds following treatment.

The majority of studies that have investigated the use of $S_2O_8^{2-}$ to treat diesel fuel have used fresh diesel fuel (e.g., Liang and Guo 2012; Do et al. 2010), and not weathered diesel fuel as expected to be encountered at most sites. Yen et al. (2011) used both soils spiked with fresh diesel fuel, and soil samples collected from a site impacted with diesel fuel to examine the ability of Fe activated $S_2O_8^{2-}$ to remove mass (TPH). Similar to the observations reported by Sra et al. (2013a), for the oxidation of dissolved gasoline compounds by $S_2O_8^{2-}$, they observed an initial rapid decrease in the TPH soil concentration over the first 40 days of ~ 40% which was followed by an additional ~20% decrease over the next 120 days of treatment. Yen et al. (2011) hypothesized that slow rate of mass removal was due to the consumption of Fe^{2+}, and the presence of recalcitrant PHCs.

Usman et al. (2012) investigated the capability of magnetite (Fe_3O_4) activated $S_2O_8^{2-}$ to treat soil spiked with fresh crude oil and weathered crude oil. They reported that 73% of the weathered crude oil and 83% of the fresh crude oil was degraded in one week. Apul et al. (2016) screened the ability of several oxidants, including thermal activated $S_2O_8^{2-}$ (45 °C), to degrade weathered crude oil present in two different soils. While the control (water only) removed ~40% of the TPH, the thermal activated $S_2O_8^{2-}$ treatment removed about 70% of the TPH.

The degradation of naphthalene, the simplest PAH, in water and soil has been reported by Huang et al. (2005), Crimi and Taylor (2007), and Sra et al. (2013a) using a variety of $S_2O_8^{2-}$ activation methods. Cuypers et al. (2000) used thermally activated $S_2O_8^{2-}$ (70 °C) to successfully oxidize sorbed PAHs in soils and sediments. Zhao et al. (2013) examined the ability of thermal, citrate chelated Fe^{2+}, alkaline, and H_2O_2 activated $S_2O_8^{2-}$ to treat PAHs in soil samples collected from a coking plant. Thermal activation (40–60 °C) resulted in the highest removal of PAHs (90–99%), followed by Fe^{2+} activated (80–90%), H_2O_2 activated (65–80%), unactivated (~70%), and alkaline activated (55–70%). In contrast to the batch experimental systems used in most studies, Lemaire et al. (2013) reported on a series of column experiments performed where a H_2O_2 activated $S_2O_8^{2-}$ solution was flushed into columns packed with homogenized soil collected from a former coking plant. The mass of PAHs removed (sum of the 16 US EPA regulated PAHs) was ~30%, presumably due to competition for the oxidants by the high carbonate and natural organic matter content soil.

15.4 Oxidant Interaction with Aquifer Materials

As discussed in Sects. 15.3.1 and 15.3.4, the presence of some minerals associated with aquifer materials will activate H_2O_2 and $S_2O_8^{2-}$, and generate radical species at the mineral surface that may degrade some PHCs if present. The radicals are unlikely to diffuse very far (perhaps a few nanometers) from the surface but will, nevertheless, result in productive or beneficial oxidant consumption. Despite this benefit, at most sites the oxidant will contact uncontaminated porous media. Thus, an important consideration in the design of an ISCO system is the interaction of an oxidant with naturally occurring reductants and catalysts that are associated with uncontaminated aquifer material. If these interactions are significant, they will influence oxidant persistence and treatment efficiency following delivery into the TTZ.

Typically, the impact of the dissolved groundwater species is overshadowed by the aquifer solids. For example, Barcelona and Holm (1991) concluded that the oxidation capacity of dissolved phase species (Mn^{4+}, Fe^{3+}, and dissolved organic carbon) was insignificant compared to the measured oxidation capacity of the aquifer solids of samples collected from two sites. Inorganic species (e.g., minerals) containing iron (Fe), manganese (Mn), sulphur (S), and the natural organic matter (NOM) associated with the aquifer solids are usually of concern (Mumford et al. 2005). NOM may be both refractory and labile toward oxidation, and therefore, the oxidation of NOM is highly dependent on the reactivity of the various functional organic groups that comprise the NOM. The possibility of multiple inorganic species, as well as a range of NOM, creates an extremely heterogeneous environment in which reactions may occur (see Fig. 15.4).

The result of the interaction between the selected oxidant and aquifer material leads to either a consumption of the oxidant by the aquifer solids, or an enhancement in the oxidant decomposition rate. When an oxidant is consumed, the reactive species associated with the aquifer solids are finite, and hence there exists a finite

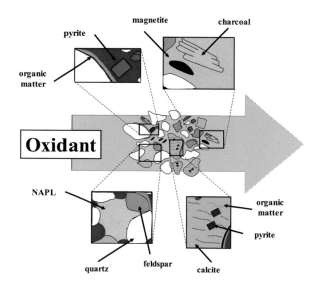

Fig. 15.4 Pore-scale conceptual model for natural oxidant interaction (NOI) showing the possibility of reaction with reduced aquifer solid species, reaction with dissolved NAPL species, and transport of un-reacted oxidant. Adapted from Mumford et al. (2005)

consumption or NOD (Natural Oxidant Demand). Once the maximum NOD is satisfied there is minimal additional reaction between the oxidant and the aquifer material. Conversely, an enhancement in the oxidant decomposition rate implies that there is infinite interaction capacity available.

To capture these behavioral differences and the associated underlying processes for all oxidant behavior, the term natural oxidant interaction (NOI) is used rather than natural oxidant demand (NOD). Regardless of the way in which NOI may manifest, it will decrease the mobility of the oxidant, the reaction rate with the target compound(s), and the mass of oxidant available thus creating an inefficient treatment system. Quantification of the NOI is a requirement for site-specific assessment and the design of cost-effective ISCO systems.

A general kinetic expression that can be used to capture the behavior of most oxidants in the presence of uncontaminated aquifer solids is given by Eq. (15.12):

$$\frac{d[C_{ox}]}{dt} = -k_o[C_{ox}] - \sum_{cat} k_{cat}[C_{ox}][C_{cat}] - \sum_{NOM} k_{NOM}[C_{ox}][C_{NOM}] \qquad (15.12)$$

where C_{ox}, C_{cat}, and C_{NOM} is the concentration of the oxidant, catalyst and NOM respectively, and k_o, k_{cat}, and k_{NOM} is the reaction rate coefficient for a decomposition reaction, catalytic reaction(s), and NOM reaction(s) respectively. The first term on the right-hand side of Eq. (15.12) captures oxidant depletion due to auto or thermal decomposition processes, the second term represents oxidant reactions with catalysts (by definition the catalysts involved in these reactions are not consumed), and the third term represents reactions with the various forms of NOM.

Table 15.4 describes the observed NOI behavior of four commonly used oxidants in the presence of uncontaminated aquifer materials. The decomposition of H_2O_2 in porous media follows a pseudo first-order mass action law. Petri et al. (2011) presented first-order rate constants from 18 studies where the persistence of H_2O_2 was tracked. While a range of CHP systems and porous medium types were used, the median first-order decomposition rate coefficient was 0.2/h (half-life of 3.5 h). Ozone is highly reactive with NOM (e.g., Hsu and Masten 2001) as well as with reduced minerals (metal iron and manganese oxides) (Jung et al. 2004; Lin and Gurol 1998). In general, when exposed to O_3, the oxidizable fraction of the NOM is depleted giving rise to a finite NOD, but the iron and manganese oxides act like catalysts and hence have an infinite interaction capacity.

In contrast to the behavior of H_2O_2 and O_3, MnO_4^- concentration data collected from batch experiments indicate that, as time progresses, the concentration of MnO_4^- decreases and asymptotically approaches a plateau (Xu and Thomson 2009). The decrease in MnO_4^- concentration and the mass of aquifer material in the test reactor are used to estimate the MnO_4^- NOD (mass of $KMnO_4$ consumed per mass of dried aquifer material in g/kg). NOD is analogous to the BOD which results from a test used to measure the amount of biodegradable organic content in water under aerobic conditions. Figure 15.5a shows a typical MnO_4^- NOD temporal profile and the characteristic fast and slow consumption rates. A high degree of correlation was observed

Table 15.4 Attributes of NOI behavior of the commonly used oxidants

Oxidant	NOI behavior
Hydrogen peroxide	Enhanced first-order degradation rate Insignificant change in NOM Repeated decomposition behavior Infinite interaction, infinite NOD
Ozone	Enhanced first-order degradation rate Significant change in NOM Infinite interaction with mineral catalysts, infinite NOD
Permanganate	Fast and slow consumption rates Significant change in NOM MnO_2 catalyzed decomposition Finite interaction Definitive NOD
Persulfate	Enhanced first-order degradation rate Slight to moderate decrease in NOM Repeated decomposition behavior Infinite interaction with mineral catalysts; infinite NOD

Modified from Siegrist et al. (2014)

by Xu and Thomson (2009) between the maximum NOD (NOD_{max}) observed and the NOM content implying that organic carbon is the major reduced species contributing to permanganate consumption for many aquifer materials. Once the NOD_{max} has been satisfied, no additional MnO_4^- will be consumed unless sufficient $MnO_{2(s)}$ has been deposited on the aquifer materials giving rise to $MnO_{2(s)}$ catalyzed decomposition (Xu and Thomson 2009; Stewart 1965). Figure 15.5b presents normalized breakthrough curves (BTCs) for a conservative tracer and MnO_4^- from a column packed with uncontaminated aquifer material. The delayed arrival of MnO_4^- is associated with the fast reaction rate, and the damped shape of the MnO_4^- BTC reflects the slow reaction rate.

The behavior of $S_2O_8^{2-}$ is similar to H_2O_2 and O_3 with the decomposition of $S_2O_8^{2-}$ following a first-order rate law (Sra et al. 2010) reflecting the interaction with catalysts and other reductants in the system. For example, Fig. 15.6 shows $S_2O_8^{2-}$ degradation profiles in the presence of three aquifer materials. A second dose of $S_2O_8^{2-}$ was added 125 days after the first dose. A first-order kinetic model was able to represent the $S_2O_8^{2-}$ temporal profiles reasonably well. For all aquifer materials, k_{obs} decreased following the second $S_2O_8^{2-}$ dose at Day 125 indicating that changes to the reactive capacity of the materials occurred during exposure to initial dose of $S_2O_8^{2-}$. The difference in k_{obs} values was attributed to the presence of a finite mass of oxidizable material that was initially consumed and gave rise to the accelerated $S_2O_8^{2-}$ degradation observed at early time (Oliveira et al. 2016).

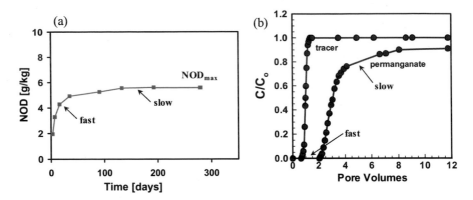

Fig. 15.5 **a** Typical permanganate NOD (g-KMnO$_4$/kg) profile collected from batch experiments (adapted from Xu and Thomson 2009). NOD$_{max}$ is the maximum NOD observed. **b** Conservative tracer and permanganate normalized breakthrough curves from a column packed with uncontaminated aquifer material (adapted from Xu and Thomson 2006). The role that the fast and slow reactions play is indicated on both panels

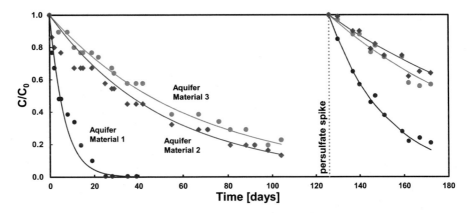

Fig. 15.6 Temporal persulfate concentration profiles following the exposure of three aquifer materials to a 1 g/L persulfate solution (adapted from Oliveira et al. 2016). A second persulfate dose (spike) was added at Day 125. The solid line is the best fit first-order kinetic model

To demonstrate the inherent differences between the persistence of H_2O_2, MnO_4^-, and $S_2O_8^{2-}$ in the presence of aquifer materials, consider the following injection scenario: (1) an uncontaminated aquifer is subject to the injection of an oxidant (H_2O_2, MnO_4^-, or $S_2O_8^{2-}$) (Fig. 15.7a) in two sequential episodes spaced 30 days apart; (2) following injection the oxidant solution remains immobile as it reacts with the aquifer material; (3) the controlling in situ kinetic parameters are taken from bench-scale efforts performed on the same aquifer material (for H_2O_2 and $S_2O_8^{2-}$ there was no finite mass of oxidizable material present); and (4) the injection concentration is adjusted so that the "oxidation strength" is identical for each oxidant. Figure 15.7b

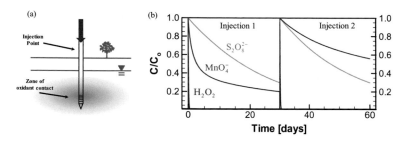

Fig. 15.7 a Injection of an oxidant into an uncontaminated aquifer. **b** Oxidant concentration profiles following two sequential pulse injection episodes into a synthetic aquifer for hydrogen peroxide (red), permanganate (purple), and persulfate (green). Adapted from Siegrist et al. (2014)

illustrates the oxidant concentration profiles over 60 days. The following observations are significant:

- H_2O_2 concentration is rapidly reduced.
- The H_2O_2 profiles following Injection 1 and Injection 2 are identical.
- The MnO_4^- concentration profile following Injection 1 decreases quickly and then slows down as it approaches an asymptote.
- The decrease in the MnO_4^- concentration profile following Injection 2 is much less than after Injection 1 reflecting the consumption of much of the fast-reacting NOM.
- The $S_2O_8^{2-}$ concentration profile is identical following Injection 1 and Injection 2.
- Since the $S_2O_8^{2-}$ reaction rate is substantially less than H_2O_2, an increase in $S_2O_8^{2-}$ persistence occurs.

15.5 Transport Considerations

For an ISCO system to be effective, contact is required between the delivered oxidant and the target PHC compounds. While a review of suitable oxidant delivery techniques is provided in Sect. 15.6.2, this section focuses on relevant oxidant transport considerations for solution-based oxidants (CHP, percarbonate, permanganate and persulfate). Ozone is delivered as a gas either into the unsaturated or saturated zone, and its transport behavior is controlled by multiphase flow theory (see Thomson and Johnson 2000).

Solution-based oxidants are frequently injected under elevated pressure and result in a rapid displacement of the pore water within the TTZ. During this injection phase, advective transport will dominate, and oxidant migration will be similar to a conservative tracer except it will react with both aquifer materials according to its natural oxidant interaction behavior (Table 15.4), and any oxidizable PHC compounds present. Groundwater velocities during this phase are expected to be substantially higher than under natural conditions. When catalysts or activators are co-injected with H_2O_2 or $S_2O_8^{2-}$ to produce powerful radicals, these reactions occur immediately, and the radicals are short-lived. For example, the quasi steady-state concentration of $\cdot OH$ is of the order of 10^{-12}–10^{-16} M, and the transport distance from the point of generation is only a few nanometers so they must be close to their target PHC to be effective. The injected oxidant solution will follow preferential pathways controlled by the presence of higher permeability or hydraulic conductivity (K) zones. Velocity variations within the TTZ will result in dispersion or apparent mixing of the oxidant solution. The generation of gases (e.g., O_2 in the CHP system) may alter flow pathways. At the interface of the dispersed oxidant solution and resident PHC laden groundwater, oxidation will occur based on the relative rates of reaction, and oxidant mass will be depleted.

After the injection phase is complete and elevated hydraulic pressures have dissipated to background levels, the injected oxidant solution will migrate with the natural groundwater flow. In order to maximize the rate of oxidation [see Eq. (15.1)] and provide sufficient oxidant mass to satisfy stoichiometric demands (e.g., see Sect. 15.6.1), and NOI requirements (Sect. 15.4), high oxidant concentrations are often used (50–100 g/L). These high concentrations result in a density difference between the oxidant solution and the surrounding groundwater, and give rise to a body force that will cause the oxidant solution to migrate downward. Schincariol and Schwartz (1990) show that significant instabilities in flow can occur even at density differences as low as 0.8 g/L. Density dependent transport of the oxidant solution plays an important role in its ultimate fate. As the oxidant solution migrates and mixes with groundwater by dispersion, the oxidant will react with target PHC compounds and be involved in NOI reactions. As a consequence, the oxidant will be consumed, and the spatial extent of migration affected. Persistent oxidants like permanganate and persulfate will perhaps be present for several weeks to months (if not consumed) while short-lived oxidants (H_2O_2) will be quickly depleted.

While the in situ destruction of target PHCs is one key attribute offered by ISCO, the second attribute is its capability to increase mass transfer from a LNAPL. Almost all of the reactions of interest that occur between an oxidant and a target PHC compound take place in the aqueous phase and not in the LNAPL. Therefore, LNAPL dissolution or mass transfer from the LNAPL to the aqueous phase is extremely important since it will, in part, control the degree of mass removal and hence treatment effectiveness. The rate at which PHC compounds dissolve from the LNAPL determines the dissolved phase plume concentrations and its longevity. The presence of a LNAPL results in dissolved phase PHC compounds that create a concentration gradient across a stagnant boundary layer between the LNAPL and the bulk aqueous solution (Fig. 15.8). The mass transfer of a PHC compound in the LNAPL into the aqueous phase is generally expressed by a macroscopic variation of the stagnate film model given by Schwarzenbach et al. 1993 (Eq. 15.13)

$$\frac{dC_{w,i}}{dt} = -k_d(C_{w,i}^{\text{eff}} - C_{w,i}) \tag{15.13}$$

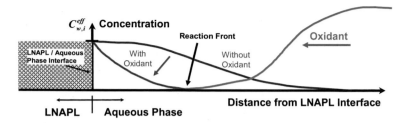

Fig. 15.8 Schematic of dissolved phase concentration gradients near a LNAPL/aqueous phase interface with and without an oxidant present in close proximity. As the oxidant and reaction front migrate toward the LNAPL/aqueous phase interface, the concentration gradient steepens (blue arrow) and mass transfer is enhanced

where $C_{w,i}$ is the aqueous phase concentration of the ith PHC compound in the LNAPL, $C_{w,i}^{eff}$ is the effective aqueous solubility estimated from modified Raoult's law (Lee et al. 1992), and K_d is the lumped mass transfer or dissolution rate coefficient. In a porous medium, k_d cannot be determined from first principles, and hence various methods have been developed to estimate it from system parameters such as pore size distribution metrics, LNAPL saturation, molecular diffusion coefficient, and groundwater velocity (e.g., Powers et al. 1994). As shown by Eq. (15.13), the rate at which LNAPL mass is depleted is a product of k_d, and the concentration gradient between $C_{w,i}^{eff}$ and $C_{w,i}$.

Oxidation reactions in the aqueous phase decrease the concentration of the PHC compounds in the bulk solution, and thus increase or steepen the concentration gradient (Fig. 15.8), which in turn will increase the overall rate of LNAPL mass removed from the system. Results from laboratory studies have shown that a 6–10 times increase in the mass transfer rate is possible to achieve when NAPL is exposed to an oxidant (Schnarr et al. 1998; MacKinnon and Thomson 2002). For enhanced mass transfer to be possible, the oxidant needs to be present near the LNAPL/aqueous phase interface.

In regions of the entrapped LNAPL source zone where preferential flow pathways exist, diffusive dominated transport is the principal mechanism for the migration of LNAPL compounds out of the non-advective zones and for the migration of oxidants into these zones. Persistent oxidants, those that are stable in the subsurface for an extended period of time, are particularly advantageous in these situations. When delivered into a TTZ by way of preferential flow pathways, these oxidants can diffuse from a preferential pathway toward more stagnant regions, driven by the oxidant concentration gradient as shown conceptually in Fig. 15.9. This is in the opposite direction as the gradient associated with the aqueous LNAPL compounds. This counter-diffusion or two-way diffusion process increases diffusive mass transfer by decreasing the distance between the LNAPL/aqueous phase interface and the area where the dissolved LNAPL compounds have been reduced as a result of oxidation.

Fig. 15.9 Illustration of a two-way diffusion process occurring between a preferential flow pathway and entrapped LNAPL mass. The oxidant must be available in the preferential flow pathway for sufficient time for diffusive transport to occur

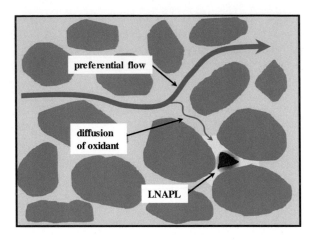

Similarly, for chemical oxidants to react with PHC mass stored (dissolved and sorbed) in lower K zones, the oxidant must diffuse from preferential flow zones. Since diffusion is a slow transport process compared to advection, a persistent oxidant is required. For example, Cavanagh et al. (2014) investigated the effectiveness of flushing unactivated $S_2O_8^{2-}$ in a higher K zone as a means to reduce BTEX mass flux from an adjacent lower K layer. After ~210 days, $S_2O_8^{2-}$ had diffused 60 cm into the lower K layer with a concomitant reduction in BTEX mass flux between 95 and 99%.

15.6 Delivery and Mixing Techniques

The success of an ISCO system depends on the ability to deliver the selected oxidant into the subsurface so that the oxidant will contact the PHC compounds in the TTZ. For this contact to result in a favorable outcome (i.e., oxidation to non-toxic end products, or by-products that are susceptible to biodegradation), this contact needs to be made with a sufficient concentration of oxidant present and for a sufficient duration that will allow the reaction to proceed to completion. Many sites will require several oxidant delivery episodes to achieve the desired treatment outcome. The delivery approach will vary depending on the defined TTZ and remedial objectives. As described in Sect. 15.2, the TTZ may be the residual LNAPL in the unsaturated zone, the LNAPL mass in the saturated capillary fringe and the saturated groundwater zone (smear zone), or the dissolved phase plume. Regardless of the TTZ, the mass of oxidant delivered should be sufficient to degrade the target PHCs as well as satisfy the NOI requirements.

15.6.1 Oxidant Dosing and Volume

To estimate the target PHC mass requires that adequate spatial (vertical and horizontal) site characterization data are available, and thus a conceptual site model (CSM) that is sufficiently robust must exist. Specifically, geologic and hydrogeologic as well as the PHC distribution (LNAPL, dissolved, and sorbed) including physical and chemical composition must be established. At some sites the CSM may be insufficiently developed and additional site characterization data specific to support the design of an ISCO system must be gathered. This is extremely important at sites with a high degree of physical heterogeneities. CSM uncertainties must be acknowledged and taken into consideration.

The oxidant mass required per mass of PHC depends on the functionality of the oxidant selected and the catalyst or activator used. Published stoichiometric data, if available, can provide an initial estimate, but bench-scale treatability testing using several samples collected from within the TTZ is highly recommended. NOI behavior can also be quantified using bench-scale testing on non-impacted aquifer material samples (e.g., Sra et al. 2010; Xu and Thomson 2008, 2009). Scale-up from bench-scale results to field-scale can be refined and optimized based on pilot test data and supported by experience. Often bench-scale tests are not conducted at representative in situ oxidant to solids mass ($M_{ox/s}$) ratios, and thus extracted parameters need to be scaled appropriately. In addition, consideration should be given to the various rates that affect oxidant mass following delivery into a TTZ; these include:

- dissolution rate from the multicomponent LNAPL (if present) to the aqueous phase,
- PHC desorption rate from aquifer materials to the aqueous phase,
- reaction kinetics between the oxidant and dissolved PHCs,
- reaction kinetics between the oxidant and the aquifer materials bearing in mind the different ways that NOI can manifest,
- the rate of oxidant decomposition through non-radical-producing pathways (applicable to H_2O_2 and $S_2O_8^{2-}$), and
- the PV exchange rate due to groundwater flow which will affect the oxidant residence time.

As an example of an estimate for permanganate dosing, consider a 1 m^3 block of porous medium with a mobile porosity of 30%, bulk density of 1800 kg/m^3, toluene saturation of 1%, and NOI of 5 g-KMnO$_4$/kg. Using the stoichiometry provided by Eq. (15.10a), 15.5 g of MnO_4^- or 20.6 g of $KMnO_4$ is required for complete mineralization of each g of toluene (C_7H_8). Therefore, to mineralize the 2.6 kg toluene (density of 0.87 g/mL) in this porous medium block requires 53.5 kg of $KMnO_4$, and to satisfy the NOI, and an additional 9 kg of $KMnO_4$ is required for a total of 62.5 kg. To meet this oxidant dosing requirement, ~7 pore volumes (PVs) of solution with a concentration of 30 g-KMnO$_4$/L needs to be delivered. This estimate assumes that all the oxidant solution remains in the porous medium block following delivery, and ideal conditions exist including no other oxidant demands and NAPL/

water mass transfer limitations. In addition, in this example it was assumed that the NOI value provided for $KMnO_4$ was NOD_{max} and representative of the finite consumption by the aquifer materials. In the case where the NOI is infinite, then an approximate oxidant residence time needs to be determined and used to estimate the mass associated with NOI over this time frame.

This dosing example clearly illustrates that several delivery episodes are likely required to supply enough oxidant mass to a TTZ. Following a delivery episode, rebound of PHC concentrations in the aqueous phase will indicate the presence of remaining PHC mass in the TTZ. After several delivery episodes when the rate of PHC mass removal diminishes and performance monitoring data (e.g., soil cores) indicate that the majority of PHC mass has been depleted, consideration should be given to a transition from ISCO to bioremediation. Bioremediation is often more cost-effective after ISCO has substantially decreased the mass in the TTZ.

Ideally, the volume of oxidant solution delivered should be equal or larger than the TTZ pore volume (PV_{TTZ}) to ensure maximum coverage. The PV_{TTZ} can be estimated by multiplying the TTZ area, the sum of the thicknesses of the higher K layers within the TTZ (if appropriate), and the mobile porosity. The higher K regions should be identified within the CSM. The mobile porosity is the portion of the total porosity that participates in advective flow and can be estimated from tracer test data (see Payne et al. 2008). As a result of preferential flow pathways, non-uniform oxidant distribution will occur at most sites. Therefore, the delivered oxidant solution volume needs to be > 1 PV_{TTZ} to attain the desired oxidant distribution across the TTZ. At sites with large variations in K within the TTZ, strategic delivery approaches may be required. Given the uncertainty inherent in a CSM and the capability of the selected delivery method to meet expectations, the oxidant distribution should be monitored during delivery. For example, Stevenson et al. (2020) used the elevated electrical conductivity (EC) signature of an activated $S_2O_8^{2-}$ solution to track its arrival and persistence in real time using a network of simple resistivity probes. This information can be used to refine and optimize the delivery strategy.

15.6.2 Delivery Methods

As shown schematically in Fig. 15.1, a TTZ for PHCs is typically located above the water table (unsaturated zone), and/or within the shallow saturated groundwater zone including the saturated capillary fringe. The delivery of solution-based oxidants in the unsaturated zone, either gravity-fed or injected under pressure, is associated with a number of challenges. First, solutions tend to migrate downward from their point of release due to gravity, which subsequently results in limited lateral spreading and a restricted radius of influence (ROI), and second, solutions will generally prefer to migrate along preferential paths. These difficulties lead to limited treatment and perhaps a loss of oxidant mass below a residual LNAPL TTZ in the unsaturated zone. As a result, the use of injection points (permanent or temporary) is typically restricted to the delivery of solution-based oxidants into a TTZ in the saturated zone.

As discussed in Sect. 15.3, H_2O_2 requires a catalyst, and $S_2O_8^{2-}$ may require an activator (if needed). Since H_2O_2 and the catalyst react very fast, and radicals are short lived, the mixing of these two reactants together should ideally occur in the TTZ. There are two general approaches used: (1) the catalyst and H_2O_2 solutions are co-injected, and mixing takes place at the injection point, and (2) the catalyst solution is injected first and is followed by the H_2O_2 solution and the catalytic reaction takes place at the interface of the two solutions. Since $S_2O_8^{2-}$ is more stable than H_2O_2, some of the species used to activate $S_2O_8^{2-}$ (e.g., NaOH for high pH) are mixed with the $S_2O_8^{2-}$ solution on ground surface just prior to injection.

15.6.2.1 Surface Application and Infiltration Trenches

For a residual LNAPL TTZ in the unsaturated zone, the use of ozone, the surface application of oxidants, or infiltration trenches are potential methods. Ozone is delivered as a gas either into the unsaturated or saturated zone using a distribution of sparge wells or points. A substantial volume of literature exists on the use of IAS to deliver air and other gases (e.g., see Johnson and Johnson 2012).

The surface application of oxidants may involve mixing a solid oxidant (MnO_4^- or $S_2O_8^{2-}$) into the top soil layer or irrigating an oxidant solution on the land surface above the TTZ. Following this initial application, infiltrating precipitation will dissolve the solids or mix with the applied oxidation solution and transport oxidant laden water downward through the unsaturated zone where it will contact the residual LNAPL. Clearly a surface application method requires the use of a persistent oxidant. A schematic of an infiltration trench or gallery to deliver an oxidant solution into a shallow LNAPL TTZ is shown in Fig. 15.10. In this approach, an oxidant solution is prepared in batch-mode in a surface facility, and then released into the system periodically. The solution is then distributed into a network of slotted pipes installed in high K material. Migration of the oxidant solution out of the infiltration trench is controlled initially by the pressure head in the slotted pipe network, but ultimately by gravity as it infiltrates through the unsaturated zone and into the TTZ.

15.6.2.2 Injection Points

The most common method to deliver a solution-based oxidant is through the use of temporary or permanent injection points. Temporary injection points typically involve Direct Push Technology (DPT) where a small-diameter rod equipped with an injection point is pushed to the desired depth interval and then a specified volume of oxidant solution is injected (see Fig. 15.11a). The screened or slotted open section of a temporary injection point is typically 30–150 cm long based on the tooling used and vertical target depth interval desired.

Depending on the mode of injection (top down or bottom up), the injection point is then advanced to the next depth interval and a specified volume of oxidant solution is again injected. This sequence is repeated for each target depth interval at that

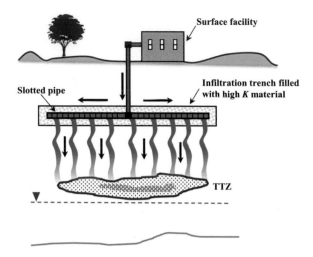

Fig. 15.10 Schematic of an infiltration trench used to deliver an oxidant solution into a shallow LNAPL TTZ in the unsaturated zone

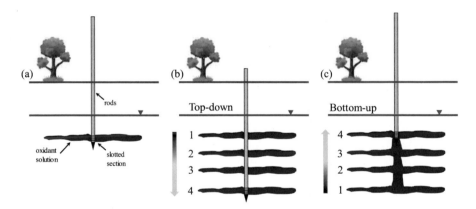

Fig. 15.11 Oxidant distribution **a** following an injection at a target depth interval using DPT, **b** using the top-down mode of injection at four target depth intervals, and **c** using the bottom-up mode of injection at four target depth intervals

location. Figure 15.11b illustrates the top-down injection mode where the process starts at the shallowest depth interval and progresses down. The bottom-up mode of injection is quite similar to the top-down mode; however, in this case the injection point is initially advanced to the deepest depth injection interval and then sequentially pulled-up to the next shallowest depth interval (Fig. 15.11c). In the bottom-up mode, the borehole created by the rods (if it remains open) may act as a pathway for the downward migration of oxidant solution injected at shallower target depth intervals producing an oxidant distribution as shown in Fig. 15.11c. Following injection at

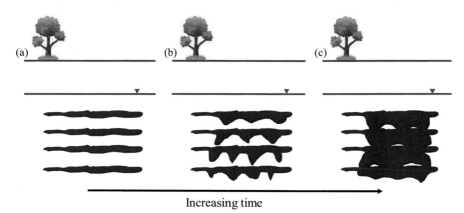

Increasing time

Fig. 15.12 Schematic illustrating the oxidant distribution following **a** initial top-down injection at four target depth intervals, and **b, c** the effects of density-driven transport resulting in the coalescence of the initial discrete depth intervals of oxidant

all the depth intervals at a given location, and regardless of the injection mode, the injected oxidant solution will be subjected to density-driven advection if it has sufficient density contrast (see Fig. 15.12).

In contrast, a permanent injection point is an injection well (e.g., continuous slotted screen, large open area, coarse sand pack) with the capability to accept large flow rates. The length of an injection well screen is usually 300–460 cm. Permanent injection points are used in situations where multiple injection episodes are expected at the same location and depth interval, and remobilization expenses are prohibitive. Once installed, permanent injection points provide less flexibility to respond to monitoring data compared to temporary injection points since the injection of additional episodes would be at the same location and depth interval. Temporary injection points provide the opportunity to strategically target "hot spots" identified in the CSM, partially overcome geological controls (i.e., preferential flow pathways), and allow for maximum flexibility for additional injection episodes to focus on areas of rebound. In addition, fouling of the screened interval by reaction by-products is less likely to occur in temporary injection points compared to permanent injection points which are used repeatedly. Typically, the injected oxidant volume delivered at a temporary injection point is much smaller than the volume delivered into a permanent injection well.

Under typical injection pressures, an oxidant solution is incompressible and when injected into an aquifer the pore water, which is also incompressible, is displaced. As a result, the injected solution will migrate radially away from the injection point and into the surrounding aquifer until injection has stopped and the hydraulic pressure increase has dissipated. The transitory pressure field that develops as a result of the injection will result in a groundwater mound since in a shallow groundwater system this is the path of minimal hydraulic resistance. The three-dimensional shape of the

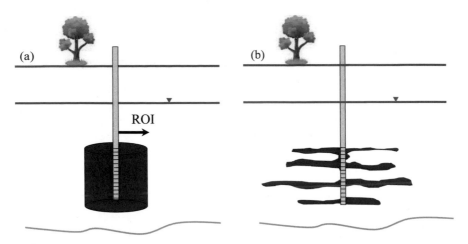

Fig. 15.13 **a** Oxidant distribution following injection into **a** homogeneous aquifer, and **b** layered heterogeneous aquifer from a permanent injection point. The theoretical radius of influence (ROI) is shown on panel (**a**) for the homogeneous aquifer

oxidant solution as it exits the injection point will be controlled by the near-well K-field with the solution entering the higher K zones. Figure 15.13 illustrates the oxidant distribution from a permanent injection well into a homogeneous aquifer, and layered heterogeneous aquifer. While the oxidant distribution around the injection well is uniform for a homogeneous aquifer, the oxidant distribution around the injection well for the layered heterogeneous aquifer shows preferential migration along the higher K layers. These higher K layers provide the path of least resistance from the wellbore into the porous medium.

For a TTZ within a source zone, the injection wells are typically located on an off-set grid pattern with overlapping ROI as shown in Fig. 15.14. An ideal ROI can be estimated from Eq. (15.14):

$$ROI = \sqrt{\frac{V_{inj}}{\pi h \theta_m}} \qquad (15.14)$$

where V_{inj} is the injected volume of oxidant solution, h is the injection zone thickness (normally length of the well screen), and θ_m is the mobile porosity. Again, the mobile porosity is the portion of the total porosity that participates in advective flow (Payne et al. 2008). The overlapping ROIs account for non-ideal conditions and to prevent dead zones between each injection point. A review of monitoring data collected during the injection phase is essential to ensure that the spatial distribution of the oxidant can be understood. Normally the sequence of injections starts at the perimeter locations and advances inward to avoid displacing PHC laden groundwater outside the TTZ.

Fig. 15.14 Schematic layout of 25 injection points across a source zone TTZ. The solid symbol (●) represents the location of an injection point, and the circle around each location represents the expected ROI for each injection point. The ROI at each injection point overlaps neighboring ROIs to ensure adequate distribution of the injected oxidant solution

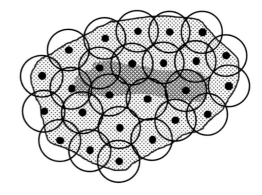

Delivery of an oxidant solution into a dissolved phase TTZ has additional challenges compared to a source zone TTZ. In a residual or entrapped source zone TTZ, the PHC mass is essentially immobile, while in a dissolved phase plume the PHC mass is a combination of mass in the aqueous and sorbed phases (depending on partitioning). In cases when the sorbed mass associated with the dissolved plume is minimal, the large volumes of oxidant solution injected will simply displace the target PHC laden groundwater since mixing is limited. Injection strategies that start at the edge of the plume and work inward will contain the PHC laden groundwater but will not enhance mixing. Following injection, and only if the oxidant is persistent, an oxidant slug will be transported by the ambient groundwater flow field and mixing along the edges will occur due to dispersion. Transverse dispersion (lateral or vertical) is weak compared to longitudinal dispersion (Gelhar et al. 1992) resulting in the majority of mixing occurring along the principal flow direction.

One potential approach to improve oxidant distribution in a dissolved phase plume is to use a cross-injection system (CIS) (Shayan et al. 2017; Gierczak et al. 2007; Devlin and Barker 1996). A CIS involves a series of injection-extraction wells used to deliver an oxidant solution perpendicular to the ambient flow direction in pulses (Fig. 15.15). Once the oxidant pulse between an injection well (IW) and extraction wells (EWs) has been established, the pumping system is stopped. The oxidant slug is then transported by the ambient flow field where dispersion will lead to spreading and mixing of the oxidant with PHC laden groundwater. In the example illustrated in Fig. 15.15, CIS would need to operate until such time at the source zone was treated.

To sustain a continual supply of oxidant solution into a TTZ, a recirculation system can be used as shown in Fig. 15.16 (see also Thomson et al. 2007; Lowe et al. 2002). This system is comprised of a number of injection and extraction wells, and a surface treatment system to amend the oxidant concentration, and remove problematic ions and by-products (e.g., Fe, MnO_2) that may foul injection wells. Some jurisdictions may have regulations that do not allow the injection of contaminated water, and hence any residual PHC compounds in the extracted groundwater would also need to be removed prior to re-injection. The continual supply of an oxidant solution ensures that reaction rates remain elevated, the rate of diffusion maximized, and adequate oxidant mass is being delivered to satisfy dosing requirements.

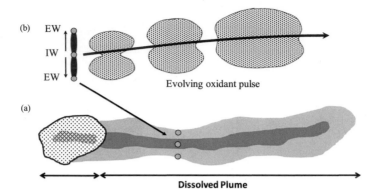

Fig. 15.15 Example of a CIS used to deliver an oxidant into a dissolved phase plume showing: **a** the location of CIS wells in the plume, and **b** the IW and EW, the initial oxidant pulse between the IW and EWs, and the spreading and mixing of the oxidant pulse

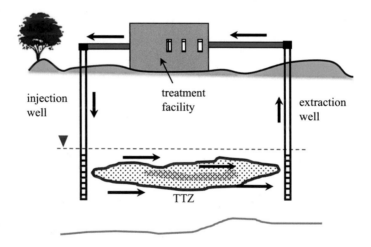

Fig. 15.16 Schematic cross-section of a recirculation system to maintain a continual supply of oxidant into a LNAPL TTZ. The treatment system amends the oxidant concentration and removes problematic compounds (PHCs and ions)

15.6.2.3   Other Methods

Electrokinetic Delivery

The inability to inject liquid oxidants into lower K TTZs significantly impacts the selection of ISCO as a preferred treatment technology. To overcome this limitation, electrokinetic (EK) enhanced delivery of chemical oxidants has been proposed (termed EK-ISCO) (Xu et al. 2020; Chowdhury et al. 2017a; Pham and Sillanpää 2015; Fan et al. 2014; Reddy 2013; Reynolds et al. 2008; Roach and Reddy 2006).

The EK delivery principle involves the application of a low-intensity DC voltage to an array of electrode wells installed in a low-K TTZ. Negatively charged ions (e.g., MnO_4^-, $S_2O_8^{2-}$) supplied at the cathodes (negative) will migrate to the anodes (positive) and be uniformly distributed across the porous medium between electrodes regardless of the K heterogeneity (Wu et al. 2012). EK transport is a combination of electromigration (movement of ions in response to the applied electric field) and electroosmosis (movement of pore fluid in response to the applied electrical field from the anode to the cathode) (Acar and Alshawabkeh 1993). The DC voltage gradient applied across the electrode field is the driving force for EK transport. A pH buffering system is required at each electrode well since the electrolysis of water will produce H^+ at the anode, and OH^- at the cathode. Cox et al. (2021) reported on a field demonstration designed to assess the ability of EK to delivery $S_2O_8^{2-}$ into a clay unit. Chowdhury et al. (2017b) described on a proof-of-concept laboratory experiment that extends EK-ISCO to include a thermal $S_2O_8^{2-}$ activation step. In this approach, $S_2O_8^{2-}$ is initially delivered by EK into a TTZ, and then the DC voltage is switched to AC, and the resistance of the aquifer material causes heating of the system. This activation step is similar to electric resistive heating (ERH, Beyke and Fleming 2005); however, the target temperature is lower than 90–100 °C.

Slow-Release Oxidant Materials

At some sites, it might be advantageous to control the rate of oxidant delivery. To meet this need a variety of slow-release oxidant materials have been developed to provide low-cost, passive, and long-term treatment. Since their initial introduction (Lee and Schwartz 2007; Ross et al. 2005; Kang et al. 2004) more than 20 laboratory studies have been published using waxes (e.g., paraffin), gels (e.g., chitosan), or a mixture of silica sand, cement, and bentonite as binding materials (O'Connor et al. 2018). Encapsulated in the binding material are MnO_4^- or $S_2O_8^{2-}$ solids. The rate of oxidant release (dissolution) depends on the binding material and the morphology (i.e., candles, pellets, or gels). Slow-release materials shaped like candles can be placed in a series of wells strategically located within a TTZ, or as the reactive media in a permeable reactive barrier (PRB) used to cut-off a dissolved phase plume. In either case, oxidant delivery is controlled primary by diffusion out of the well casing and into the surrounding groundwater. As the concentrated oxidant plume migrates downgradient it will slowly mix and react with the surrounding PHC laden groundwater. A major limitation of this type of slow-release delivery system is its narrow zone of influence associated with each release well (O'Connor et al. 2018). To expand the zone of influence, Christenson et al. (2012) used an airlift pump system to generate a continuous circulation pattern around each well thereby increasing the spreading of the oxidant as it dissolves from the slow-release material. Reece et al. (2020) present data from three field sites where this oxidant delivery system has been implemented. Slow-release gel materials can be injected directly into a TTZ where they can release low concentrations of oxidant for extended periods of time (e.g., Yang et al. 2016; Lee and Gupta 2014).

15.6.3 Mixing

Following the injection of an oxidant solution into a TTZ, the desired chemical reactions are initiated by transport processes that bring the oxidant and aqueous PHC compound together (Kitanidis and McCarty 2012). This contact between the oxidant and an aqueous PHC compound is the desired outcome of an effective delivery system; however, it can be one of the most difficult to achieve. While spreading increases the zone of contact between the oxidant and aqueous PHC compound, physical heterogeneities create preferential flow pathways at a range of spatial scales that lead to uneven spatial distribution of the injected oxidant.

Ineffective oxidant delivery can lead to partial removal of the PHC mass. The remaining PHC mass remains as immobilized, disconnected pockets that are trapped in diffusion-controlled zones (Conrad et al. 1992) or in the lower K zones where contact with the injected oxidant is limited. In these regions, molecular diffusion is the primary transport mechanism responsible for bringing the oxidant and PHC compound together (Gale et al. 2015; Jose and Cirpka 2004); however, diffusion is a slow process (see Fig. 15.9).

As noted in Sect. 15.6.2, in conventional recirculation systems, the extracted groundwater is amended before being re-injected. Conventional recirculation systems typically use steady and uniform flows; however, a time-dependent flow field is required to promote enhanced mixing (Ottino 1989; Aref 1984). Chaotic advection has the potential to improve oxidant distribution and promote enhanced mixing (Aref et al. 2017; Trefry et al. 2012). Chaotic advection can be defined as the creation of "small-scale structures" caused by the repeated stretching and folding of fluid elements in a laminar flow regime (Aref et al. 2017; Ottino 1989). This rapid stretching and folding creates highly complex particle trajectories which can improve the spatial distribution of oxidants in porous media and increase the zone of contact where reactions can occur (Mays and Neupauer 2012; Ottino 1989). Cho et al. (2019) employed a Rotated Potential Mixing (RPM) flow configuration in a proof-of-concept field investigation and used multiple lines of evidence to demonstrate that enhanced mixing using chaotic advection is possible. This is the first reported attempt to investigate aspects of chaotic advection at the field scale. The findings from this proof-of-concept study represent a critical step in the development of a novel method that can be used to improve the effectiveness of ISCO.

15.7 Options for Integrated Treatment

There is a myriad of in situ treatment technologies available for deployment by remediation scientists and engineers. The diversity of these technologies is a testament that no single technology is ideal for all sites as each has its own set of strengths and weaknesses making it suitable for a given hydrogeologic setting. To minimize clean-up costs and time and maximize remedial efficacy, an integration or combination of technologies at sites is desired. ISCO can be combined or integrated with other treatment technologies, either concurrently or sequentially. The concurrent use of a technologies to treat the same TTZ requires a clear understanding of technical compatibility and expected synergies, while the sequentially use of technologies (i.e., treatment train) requires a prescription of key transition points.

It is well established that oxidation of some PHC compounds generates bioavailable substrates that lead to enhanced rates of biodegradation suggesting that ISCO will enhance subsequent microbial activity (Tengfei et al. 2016; Sahl and Munakata-Marr 2006). In general, microbial communities are inhibited during exposure to oxidants, but both the microbial concentration and diversity increase following oxidant exposure (e.g., Kim et al. 2021; Richardson et al. 2011; Tsitonaki et al. 2008). Bartlett et al. (2019) reported that sulfate-reducing bacteria and the associated microbial community rebounded to pre-exposure levels following exposure to unactivated or alkaline activated $S_2O_8^{2-}$.

Considering these impacts on the microbial community, the sequential use of either monitoring natural attenuation (MNA) or enhanced in situ bioremediation (EISB) following ISCO is warranted. As shown in Eqs. (15.11a–15.11h), the decomposition of $S_2O_8^{2-}$ from reactions with PHC compounds or aquifer materials will generate excess amounts of sulfate (SO_4^{2-}). This excess SO_4^{2-} can serve as a terminal electron acceptor and enhance subsequent microbial sulfate reduction processes. Thus, following a $S_2O_8^{2-}$ treatment episode the generated SO_4^{2-} plume will migrate downgradient and enhance microbial sulfate reduction. Understanding the interplay between $S_2O_8^{2-}$ oxidation and subsequent enhanced microbial sulfate reduction is necessary to optimize the performance of this combined PHC treatment system (see Shayan et al. 2017). Similarly, peroxide, percarbonate, or ozone all generate a short-term footprint of O_2, a terminal electron acceptor for aerobic biodegradation. While the concentration of these oxidation end-products may be quickly depleted the delivery of additional electron acceptors may be required to sustain acceptable rates of biodegradation.

Surfactants and cosolvents can enhance the dissolution process, and will increase the concentration of PHC compounds in the aqueous phase where oxidation takes place (Wang et al. 2013; Li 2004). While a surfactant or cosolvent can be delivered prior to the oxidant, contact between the oxidant and the region of increased concentration of PHC compounds will suffer the same mixing limitations as discussed in Sect. 15.6.3. An alternate approach is to co-inject the surfactant or cosolvent along with the oxidant; however, both surfactants and cosolvents are organic compounds and are susceptible to oxidation. The compatibility of the surfactant or cosolvent with the selected oxidant is critical since this interaction will affect treatment effectiveness (see Dugan 2008, and Zhai et al. 2006).

Finally, sub-boiling thermal technologies can be used to thermally activate $S_2O_8^{2-}$ following delivery. Increasing the temperature in the TTZ will also enhance mass transfer (e.g., Imhoff et al. 1997), and increase the rate of reaction between the oxidant and the PHCs as described by the Arrhenius law (Levenspiel 1999). Sustainable methods (e.g., solar or geothermal heating) that can strategically increase subsurface temperatures to the sub-boiling range are currently available (Horst et al. 2018).

15.8 Closure

This chapter has presented a number of key aspects of ISCO that are relevant to application at sites contaminated with PHCs. The following are important takeaway messages:

- A robust CSM is required. This CSM must include geologic and hydrogeologic data, and an understanding of the PHC mass distribution in each phase (LNAPL, dissolved, and sorbed). If the CSM is insufficiently developed, then additional site characterization data specific to support the design of an ISCO system must be gathered. The degree of physical heterogeneity (spatial variation in K) is critical. CSM uncertainties must be acknowledged and taken into consideration.
- The TTZ and ISCO treatment objectives must be clearly defined. A distinction between residual LNAPL in the unsaturated zone, LNAPL mass in the saturated capillary fringe and the saturated groundwater zone (smear zone), and the dissolved phase plume is required.
- ISCO should be considered part of integrated remediation strategy and not in isolation. Attempts to remove mobile LNAPL from the TTZ should be exhausted before ISCO is implemented. Once mass of PHCs oxidized becomes limited, transition to biodegradation is recommended.
- The common oxidants used are CHP, percarbonate ($Na_2CO_3 \cdot 1.5H_2O_2$), ozone (O_3), permanganate (MnO_4^-), and persulfate ($S_2O_8^{2-}$). O_3 is delivered as a gas, and the other four are typically delivered as concentrated solutions.
- A basic understanding of the chemistry and main reactions for each oxidant is required. CHP, percarbonate, ozone, and persulfate can generate a suite of highly reactive radical species.
- Most of the important PHC degradation reactions occur in the aqueous phase.
- In general, the commonly used oxidants can degrade target PHC compounds in the aqueous phase at acceptable reaction rates; however, some longer chain PHC compounds are recalcitrant.
- Permanganate is unable to oxidize benzene, a common risk driver at most sites.
- Problematic reaction by-products from oxidants include gases and/or solid precipitates that may decrease aquifer K and mass transfer from the LNAPL to the aqueous phase.
- Volatile PHCs present in the system may partition into the gases causing spontaneous expansion and upward mobilization of the gas phase leading to uncontrolled releases of PHC gases.
- Reaction of radicals with common groundwater ions (carbonates, chloride, and sulfate) will affect treatment efficiency.

- Reactions with uncontaminated aquifer materials result in either a finite consumption of the oxidant or an enhancement in the oxidant decomposition rate. The interaction of an oxidant with aquifer materials will decrease the mobility of the oxidant, the reaction rate with the target PHCs, and the mass of oxidant available thus creating an inefficient treatment system.
- Stoichiometric mass requirements can be substantial and must be appreciated. Activation methods increase the stoichiometric mass requirements.
- Sufficient oxidant mass needs to be delivered into a TTZ to meet the productive and non-productive demands.
- Preferential flow pathways will control the oxidant distribution in a TTZ.
- A persistent oxidant is needed to allow the oxidant to either diffuse away from preferential pathways or diffuse into lower K regions.
- A variety of oxidant delivery methods are available with temporary injection points being the most flexible to surgically deliver an oxidant to hot spots identified in the CSM, and respond to monitoring data.
- Ineffective oxidant delivery will lead to incomplete treatment of the TTZ since mixing is limited. Novel mixing methods based on time-dependent flows offer a technique that potentially overcomes this limitation.
- Opportunities exist to combine or integrate ISCO with other treatment technologies.

References

Acar YB, Alshawabkeh AN (1993) Principles of electrokinetic remediation. Environ Sci Technol 272:638–2647
Anipsitakis GP, Dionysiou DD (2004) Radical generation by the interaction of transition metals with common oxidants. Environ Sci Technol 38(13):3705–3712
Apul OG, Dahlen P, Delgado AG, Sharif F, Westerhoff P (2016) Treatment of heavy, long-chain petroleum-hydrocarbon impacted soils using chemical oxidation. J Environ Eng 42(12):04016065
Aref H (1984) Stirring by chaotic advection. J Fluid Mech 143:1–21
Aref H, Blake JR, Budisic M et al (2017) Frontiers of chaotic advection. Rev Mod Phys 89(2):025007
Barcelona MJ, Holm RT (1991) Oxidation-reduction capacities of aquifer solids. Environ Sci Technol 25(9):1565–1572
Bartlett C, Slawson RM, Thomson NR (2019) Response of sulfate-reducing bacteria and supporting microbial community to persulfate exposure in a continuous flow system. Environ Sci Process Impacts 21:1193–1203
Beyke G, Fleming D (2005) In situ thermal remediation of DNAPL and LNAPL using electrical resistance heating. Rem J 15(3):5–22
Bhuyan SJ, Latin MR (2012) BTEX remediation under challenging site conditions using in-situ ozone injection and soil vapor extraction technologies: a case study. Soil Sediment Contam 21(4):545–556
Black H (2001) Ozone as a cleanup tool. Environ Sci Technol 35(13):283A–284A
Block PA, Brown RA, Robinson D (2004) Novel activation technologies for sodium persulfate in situ chemical oxidation. In: Proceedings of the 4th international conference on the remediation of chlorinated and recalcitrant compounds. Monterey, California, 24–27 May 2004

Bouchard D, Marchesi M, Madsen E, DeRito CM, Thomson NR, Aravena R, Barker JF, Buscheck T, Kolhatkar R, Daniels EJ, Hunkeler D (2018) Diagnostic tools to assess mass removal processes during pulsed air sparging of a petroleum hydrocarbon source zone. Groundwater Monit R 38(4):29–44

Brown RA, Norris RD, Westray M (1986) In situ treatment of groundwater. Paper presented at Haz Pro'86. Baltimore, Maryland, 1–3 Apr 1986

Buxton GV, Greenstock CL, Helman WP, Ross AB (1988) Critical review of rate constants for reactions of hydrated electrons, hydrogen atoms and hydroxyl radicals ($^{\cdot}$OH/$^{\cdot}$O^{-}) in aqueous solution. J Phys Chem Ref Data 17:513–886

Cavanagh BA, Johnson PC, Daniels EJ (2014) Reduction of diffusive contaminant emissions from a dissolved source in a lower permeability layer by sodium persulfate treatment. Environ Sci Technol 48(24):14582–14589

Checa-Fernandez A, Santos A, Romero A, Dominguez CM (2021) Application of chelating agents to enhance fenton process in soil remediation: a review. Catalysts 11(6):722

Cho MS, Solano F, Thomson NR, Trefry MG, Lester DR, Metcalfe G (2019) Field trials of chaotic advection to enhance reagent delivery. Groundwater Monit R 39(3):23–39

Chowdhury AIA, Gerhard JI, Reynolds D, Sleep BE, O'Carroll DM (2017a) Electrokinetic enhanced permanganate delivery and remediation of contaminated low permeability porous media. Water Res 113:215–222

Chowdhury AIA, Gerhard JI, Reynolds D, O'Carroll DM (2017b) Low permeability zone remediation via oxidant delivered by electrokinetics and activated by electrical resistance heating: proof of concept. Environ Sci Technol 51(22):13295–13303

Christenson M, Kambhu A, Comfort SD (2012) Using slow-release permanganate candles to remove TCE from a low permeable aquifer at a former landfill. Chemosphere 89:680–687

Cleasby JL, Baumann ER, Black CD (1964) Effectiveness of potassium permanganate for disinfection. J AWWA 56(4):466–474

Colthurst JM, Singer PC (1982) Removing trihalomethane precursors by permanganate oxidation and manganese dioxide adsorption. J AWWA 71(2):78–83

Conrad SH, Wilson JL, Mason WR, Peplinski WJ (1992) Visualization of residual organic liquid trapped in aquifers. Water Resour Res 28(2):467–478

Cox E, Watling M, Gent D, Singletary M, Wilson A (2021) Electrokinetically-delivered, thermally-activated persulfate oxidation (EK-TAP) for the remediation of chlorinated and recalcitrant compounds in heterogeneous and low permeability source zones. ESTCP ER-201626

Crimi ML, Taylor J (2007) Experimental evaluation of catalyzed hydrogen peroxide and sodium persulfate for destruction of BTEX contaminants. Soil Sediment Contam 16:29–45

Cuypers C, Grotenhuis T, Joziasse J, Rulkens W (2000) Rapid persulfate oxidation predicts PAH bioavailability in soils and sediments. Environ Sci Technol 34:2057–2063

Devlin JF, Barker JF (1996) Field investigation of nutrient pulse mixing in an in situ biostimulation experiment. Water Resour Res 32(9):2869–2877

Do S, Kwon Y, Kong S (2010) Effect of metal oxides on the reactivity of persulfate/Fe(II) in the remediation of diesel-contaminated soil and sand. J Hazard Mater 182(1–3):933–936

Dugan PJ (2008) Coupling surfactants/cosolvents with oxidants: effects on remediation and performance assessment. Dissertation, Colorado School of Mines

Dunn JA, Lunn SR (2002) Chemical oxidation of bioremediated soils containing crude oil. In: Proceedings of the 3rd international conference on the remediation of chlorinated and recalcitrant compounds. Monterey, California, 20–23 May 2002

Eilbeck WJ, Mattock G (1987) Chemical processes in wastewater treatment. Halsted Press, Toronto

Fan G, Long C, Fang G, Qin W, Ge L, Zhou D (2014) Electrokinetic delivery of persulfate to remediate PCBs polluted soils: Effect of injection spot. Chemosphere 117:410–418

Federal Remediation Technologies Roundtable (FRTR) (2022) Available at https://frtr.gov/default.cfm

Fenton HJH (1894) Oxidation of tartaric acid in presence of iron. J Chem Soc Trans 65:899–910

Forsey SP, Thomson NR, Barker JF (2010) Oxidation kinetics of polycyclic aromatic hydrocarbons by permanganate. Chemosphere 79(6):628–636

Fu X, Gu X, Lu S, Sharma VK, Brusseau ML, Xue Y, Danish M, Fu GY, Qiu Z, Sui Q (2017) Benzene oxidation by Fe(III)-activated percarbonate: matrix-constituent effects and degradation pathways. Chem Eng J 309:22–29

Furman O, Laine DF, Blumenfeld A, Teel AL, Shimizu K, Cheng IF, Watts RJ (2009) Enhanced reactivity of superoxide in water-solid matrices. Environ Sci Technol 43:1528–1533

Gale T, Thomson NR, Barker JF (2015) An investigation of the pressure pulsing reagent delivery approach. Groundwater Monit R 35(2):39–51

Garoma T, Gurol MD, Osibodu O, Thotakura L (2008) Treatment of groundwater contaminated with gasoline components by an ozone/UV process. Chemosphere 73:825–831

Gelhar LW, Welty C, Rehfeldt KR (1992) A critical review of data on field-scale dispersion in aquifers. Water Resour Res 28(7):1955–1974

Gierczak R, Devlin JF, Rudolph DL (2007) Field test of a cross-injection scheme for stimulating in situ denitrification near a municipal water supply well. J Contam Hydrol 89(1–2):48–70

Horst J, Flanders C, Klemmer M, Randhawa D, Rosso D (2018) Low-temperature thermal remediation: gaining traction as a green remedial alternative. Groundwater Monit R 38(3):18–27

House DA (1962) Kinetics and mechanisms of oxidations by peroxydisulfate. Chem Rev 62(3):185–203

Houston AC (1918) Rural water supplies and their purification. John Bale, Sons & Danielsson, Ltd., London

Hsu I-Y, Masten SJ (2001) Modeling transport of gaseous ozone in unsaturated soils. J Environ Eng 127:546–554

Huang KC, Zhao Z, Hoag GE, Dahmani A, Block PA (2005) Degradation of volatile organic compounds with thermally activated persulfate oxidation. Chemosphere 61:551–560

Imhoff PT, Frizzell A, Miller CT (1997) Evaluation of thermal effects on the dissolution of a nonaqueous phase liquid in porous media. Environ Sci Technol 31:1615–1622

ITRC (Interstate Technology and Regulatory Council) (2005) Technical and regulatory guidance for in situ chemical oxidation of contaminated soil and groundwater, 2nd edn. ISCO-2. Washington, DC. Available at http://www.itrcweb.org

ITRC (Interstate Technology and Regulatory Council) (2018) LNAPL site management: LCSM evolution, decision process, and remedial technologies. LNAPL-3. Washington, DC. Available at https://lnapl-3.itrcweb.org

Johnson RL, Johnson PC (2012) In situ sparging for delivery of gases in the subsurface. In: PK Kitanidis PK, McCarty PL (eds) Delivery and mixing in the subsurface: processes and design principles for in situ remediation. Springer, New York, pp 193–216

Jose SC, Cirpka OA (2004) Measurement of mixing-controlled reactive transport in homogeneous porous media and its prediction from conservative tracer test data. Environ Sci Technol 38(7):2089–2096

Jung H, Kim J, Choi H (2004) Reaction kinetics of ozone in variably saturated porous media. J Environ Eng 130:432–441

Jung H, Ahn Y, Choi H, Kim IS (2005) Effects of in-situ ozonation on indigenous microorganism in diesel contaminated soil: survival and regrowth. Chemosphere 61:923–932

Kang N, Hua I, Rao PSC (2004) Production and characterization of encapsulated potassium permanganate for sustained release as an in situ oxidant. Ind Eng Chem Res 43:5187–5193

Karpenko O, Lubenets V, Karpenko E, Novikov V (2009) Chemical oxidants for remediation of contaminated soil and water. A review. Chem Chem Technol 3(1):41–45

Kim E-J, Park S, Adil S, Lee S, Cho K (2021) Biogeochemical alteration of an aquifer soil during in situ chemical oxidation by hydrogen peroxide and peroxymonosulfate. Environ Sci Technol 55(8):5301–5311

Kitanidis PK, McCarty PL (2012) Delivery and mixing in the subsurface: processes and design principles for in situ remediation. Springer, New York

Kolthoff IM, Medalia AI, Raaen HP (1951) The reaction between ferrous iron and peroxides: IV. Reaction with potassium persulfate. J Am Chem Soc 73:1733–1739

Kwan WP, Voelker BM (2003) Rates of hydroxyl radical generation and organic compound oxidation in mineral-catalyzed Fenton-like systems. Environ Sci Technol 37(6):1150–1158

Langlais B, Reckhow DA, Brink DR (1991) Ozone in water treatment: application and engineering. Lewis Publishers, Boca Raton

Lee ES, Gupta N (2014) Development and characterization of colloidal silica-based slow-release permanganate gel (SRP-G): laboratory investigations. Chemosphere 109:195–201

Lee ES, Schwartz FW (2007) Characteristics and applications of controlled-release KMnO$_4$ for groundwater remediation. Chemosphere 66:2058–2066

Lee LS, Hagwall M, Delfino JJ, Rao PSC (1992) Partitioning of polycyclic aromatic hydrocarbons from diesel fuel into water. Environ Sci Technol 26(11):2104–2110

Lee ES, Seol Y, Fang YC, Schwartz FW (2003) Destruction efficiencies and dynamics of reaction fronts associated with the permanganate oxidation of trichloroethylene. Environ Sci Technol 37:2540–2546

Lee C, Kim H-H, Park N-B (2018) Chemistry of persulfates for the oxidation of organic contaminants in water. Membr Water Treat 9(6):410–419

Lemaire J, Laurent F, Leyval C, Schwartz C, Buès M, Simonnot M-O (2013) PAH oxidation in aged and spiked soils investigated by column experiments. Chemosphere 91(3):406–414

Levenspiel O (1999) Chemical reaction engineering, 3rd edn. Wiley, New York

Li Z (2004) Surfactant-enhanced oxidation of trichloroethylene by permanganate – proof of concept. Chemosphere 54:419–423

Li XD, Schwartz FW (2004) DNAPL mass transfer and permeability reduction during in situ chemical oxidation with permanganate. Geophys Res Lett 31:L06504

Liang C, Guo Y (2012) Remediation of diesel-contaminated soils using persulfate under alkaline condition. Water Air Soil Pollut 223(7):4605–4614

Lin S, Gurol MD (1998) Catalytic decomposition of hydrogen peroxide on iron oxide: kinetics, mechanism, and implications. Environ Sci Technol 32:1417–1423

Lowe KS, Gardner EG, Siegrist RL (2002) Field evaluation of in situ chemical oxidation through vertical well-to-well recirculation of NaMnO$_4$. Groundwater Monit R 22(1):106–115

Lundstedt S, Persson Y, Oberg L (2006) Transformation of PAHs during ethanol-Fenton treatment of an aged gasworks' soil. Chemosphere 65(8):1288–1294

Ma J, Yang X, Jiang X, Wen J, Li J, Zhong Y, Chi L, Wang Y (2020) Percarbonate persistence under different water chemistry conditions. Chem Eng J 389:123422

MacKinnon LK, Thomson NR (2002) Laboratory-scale in situ chemical oxidation of a perchloroethylene pool using permanganate. J Contam Hydrol 56:49–74

Masten SJ, Davies SHR (1997) Efficacy of in-situ ozonation for the remediation of PAH contaminated soils. J Contam Hydrol 28:327–335

Mays DC, Neupauer RM (2012) Plume spreading in groundwater by stretching and folding. Water Resour Res 48(7):W07501

Mumford KG, Thomson NR, Allen-King RM (2005) Bench-scale investigation of permanganate natural oxidant demand kinetics. Environ Sci Technol 39(8):2835–2849

Mumford KG, Smith JE, Dickson SE (2008) Mass flux from a non-aqueous phase liquid pool considering spontaneous expansion of a discontinuous gas phase. J Contam Hydrol 98:85–96

Nam K, Kukor JJ (2000) Combined ozonation and biodegradation for remediation of mixtures of polycyclic aromatic hydrocarbons in soil. Biodegradation 11:1–9

O'Connor D, Hou D, Ok YS, Song Y, Sarmah AK, Li X, Tack FMG (2018) Sustainable in situ remediation of recalcitrant organic pollutants in groundwater with controlled release materials: a review. J Control Release 283:200–213

Oliveira FC, Freitas JG, Furquim SAC, Rollo RM, Thomson NR, Alleoni LRF, Nascimento CAO (2016) Persulfate interaction with tropical soils. Water Air Soil Pollut 227:343

Ottino JM (1989) The kinematics of mixing: stretching, chaos and transport. Cambridge University Press, New York

Payne FC, Quinnan JA, Potter ST (2008) Remediation hydraulics. CRC Press Taylor & Francis Group, Boca Raton

Petri BG, Watts RJ, Teel AL, Huling S, Brown RA (2011) Fundamentals of ISCO using hydrogen peroxide. In: Siegrist RL et al (eds) In situ chemical oxidation for groundwater remediation. Springer, New York, pp 33–88

Pham TD, Sillanpää M (2015) Electrokinetic remediation of organic contamination. Environ Technol Rev 4:103–117

Pisarczyk K (1995) Manganese compounds. In: Encyclopedia of chemical technology. Wiley, New York, pp 1031–1032

Powers SE, Abriola LM, Weber WJ (1994) An experimental investigation of nonaqueous phase liquid dissolution in saturated subsurface systems: transient mass transfer rates. Water Resour Res 30(2):321–332

Reddy KR (2013) Electrokinetic remediation of soils at complex contaminated sites: technology status, challenges, and opportunities. In: Manassero M et al (eds) Coupled phenomena in environmental geotechnics. Taylor & Francis Group, London, pp 131–147

Reece J, Christenson M, Kambhu A, Li Y, Harris CE, Comfort S (2020) Remediating contaminated groundwater with an aerated, direct-push, oxidant delivery system. Water 12:3383

Reynolds DA, Jones EH, Gillen M, Yusoff I, Thomas DG (2008) Electrokinetic migration of permanganate through low-permeability media. Ground Water 46(4):629–637

Richardson SD, Lebron BL, Miller CT, Aitken MD (2011) Recovery of phenanthrene-degrading bacteria after simulated in situ persulfate oxidation in contaminated soil. Environ Sci Technol 45(2):719–725

Roach N, Reddy KR (2006) Electrokinetic delivery of permanganate into low-permeability soils. Int J Environ Waste Manag 1(1):4–19

Ross C, Murdoch LC, Freedman DL, Siegrist RL (2005) Characteristics of potassium permanganate encapsulated in polymer. J Environ Eng 131:1203–1211

Rudakov ES, Lobachev VL (2000) The first step of oxidation of alkylbenzenes by permanganates in acidic aqueous solutions. Russ Chem Bull 49:761–777

Sahl J, Munakata-Marr J (2006) The effects of in situ chemical oxidation on microbiological processes: a review. Remediation 16:57–70

Schincariol RA, Schwartz FW (1990) An experimental investigation of variable density flow and mixing in homogeneous and heterogeneous media. Water Resour Res 26:2317–2329

Schnarr MJ, Truax CL, Farquhar GJ, Hood ED, Gonullu T, Stickney B (1998) Laboratory and controlled field experiments using potassium permanganate to remediate trichloroethylene and perchloroethylene DNAPLs in porous media. J Contam Hydrol 29:205–224

Schroth MH, Oostrom M, Wietsma TW, Istok JD (2001) In-situ oxidation of trichloroethene by permanganate: effects on porous medium hydraulic properties. J Contam Hydrol 50:78–98

Schwarzenbach RP, Gschwend PM, Imboden DM (1993) Environmental organic chemistry. Wiley, New York

Shayan M, Thomson NR, Aravena R, Barker JF, Madsen EL, Buscheck T, Kolhatkar R, Daniels EJ (2017) Integrated plume treatment using persulfate coupled with microbial sulfate reduction. Groundwater Monit R 38(4):45–61

Siegrist RL, Crimi M, Simpkin TJ (eds) (2011) In situ chemical oxidation for groundwater remediation. Springer, New York

Siegrist RL, Crimi M, Thomson NR, Clayton W, Marley M (2014) In situ chemical oxidation. In: Kueper BH, Stroo HF, Ward CH (eds) Chlorinated solvent source zone remediation. Springer, New York, pp 253–306

Smith MB, March J (2007) Advanced organic chemistry: reactions, mechanisms, and structure, 6th edn. Wiley-Interscience, New York

Smith BA, Teel AL, Watts RJ (2004) Identification of the reactive oxygen species responsible for carbon tetrachloride degradation in modified Fenton's systems. Environ Sci Technol 38:5465–5469

Snoeyink VL, Jenkins D (1980) Water chemistry. Wiley, New York

Sra K, Thomson NR, Barker JF (2010) Persistence of persulfate in uncontaminated aquifer materials. Environ Sci Technol 44:3098–3104

Sra K, Thomson NR, Barker JF (2013a) Persulfate treatment of dissolved gasoline compounds. J Hazard Toxic Radioact Waste 17:9–15

Sra K, Thomson NR, Barker JF (2013b) Persulfate injection into a gasoline source zone. J Contam Hydrol 150:35–44

Stevenson D, Solano F, Wei Y, Thomson NR, Barker JF, Devlin JF (2020) Simple resistivity probe system for real-time monitoring of injected reagents. Groundwater Monit R 40(4):54–66

Stewart R (1965) Oxidation by permanganate. In: Wiberg KB (ed) Oxidation in organic chemistry, part A. Academy Press, New York

Suthersan SS, Payne FC (2005) In situ remediation engineering. CRC Press, Boca Raton

Teel AL, Warberg CR, Atkinson DA, Watts RJ (2001) Comparison of mineral and soluble iron Fenton's catalysts for the treatment of trichloroethylene. Water Res 35(4):977–984

Tengfei C, Delgado AG, Yavuz BM, Proctor AJ, Maldonado J, Zuo Y, Westerhoff P, Krajmalnik-Brown R, Rittmann BE (2016) Ozone enhances biodegradability of heavy hydrocarbons in soil. Environ Eng Sci 11(1):7–17

Thomson NR, Johnson RL (2000) Air distribution during in situ air sparging: An overview of mathematical modeling. J Hazard Mater 72:265–282

Thomson NR, Hood ED, Farquhar GJ (2007) Permanganate treatment of an emplaced DNAPL source. Groundwater Monit R 27(4):74–85

Tomiyasu H, Fukutomi H, Gordon G (1985) Kinetics and mechanism of ozone decomposition in basic aqueous solution. Inorg Chem 24:2962–2966

Tomlinson DW, Thomson NR, Johnson RL, Redman JD (2003) Air distribution in the Borden aquifer during in situ air sparging. J Contam Hydrol 67:113–132

Tratnyek PG, Waldemer RH, Powell JS (2019) IscoKin database of rate constants for reaction of organic contaminants with the major oxidants relevant to in situ chemical oxidation. Zenodo. https://doi.org/10.5281/zenodo.3596102Accessed08Mar2022

Trefry MG, Lester DR, Metcalfe G, Ord A, Regenauer-Lieb K (2012) Toward enhanced subsurface intervention methods using chaotic advection. J Contam Hydrol 127:15–29

Tsitonaki A, Smets BF, Bjerg PL (2008) Effects of heat activated persulfate oxidation on soil microorganisms. Water Res 42:1013–1022

Umschlag T, Herrmann H (1999) The carbonate radical ($HCO_3^{\cdot}/CO_3^{\cdot-}$) as a reactive intermediate in water chemistry: kinetics and modeling. Acta Hydrochim Hydrobiol 27:214–222

USEPA (U.S. Environmental Protection Agency) (2006) Engineering issue paper: in-situ chemical oxidation. EPA 600-R-6-072. Office of Research and Development. National Risk Management Research Laboratory, Cincinnati, OH, USA

USEPA (U.S. Environmental Protection Agency) (2022). In situ oxidation, application. https://clu-in.org/techfocus/default.focus/sec/In_Situ_Oxidation/cat/Application/. Accessed 08 Mar 2022

Usman M, Faure P, Ruby C, Hanna K (2012) Application of magnetite-activated persulfate oxidation for the degradation of PAHs in contaminated soils. Chemosphere 87(3):234–240

Vella PA, Deshinsky G, Boll JE, Munder J, Joyce WM (1990) Treatment of low level phenols with potassium permanganate. J Wat Pollut Control Fed 62(7):907–914

Waldemer RH, Tratnyek PG (2006) Kinetics of contaminant degradation by permanganate. Environ Sci Technol 40(3):1055–1061

Wang W, Hoag G, Collins J, Naidu R (2013) Evaluation of surfactant-enhanced in situ chemical oxidation (S- ISCO) in contaminated soil. Water Air Soil Pollut 224(12):1–9

Watts RJ, Dilly SE (1996) Evaluation of iron catalysts for the Fenton-like remediation of dieselcontaminated soils. J Hazard Mater 51:209–224

Watts RJ, Teel AL (2005) Chemistry of modified Fenton's reagent (catalyzed H_2O_2 propagations-CHP) for in situ soil and groundwater remediation. J Environ Eng 131:612–622

Watts RJ, Finn DD, Cutler LM, Schmidt JT, Teel AL (2007) Enhanced stability of hydrogen peroxide in the presence of subsurface solids. J Contam Hydrol 91:312–326

Wu MZ, Reynolds DA, Fourie A, Prommer H, Thomas DG (2012) Electrokinetic in situ oxidation remediation: assessment of parameter sensitivities and the influence of aquifer heterogeneity on remediation efficiency. J Contam Hydrol 137–135:72–85

Xu X, Thomson NR (2006) Oxidant fate in the subsurface environment: From batch to column systems. In: Proceedings of the 5th international conference on the remediation of chlorinated and recalcitrant compounds. Monterey, California, 22–25 May 2006

Xu X, Thomson NR (2008) Estimation of the maximum consumption of permanganate by aquifer solids using a modified chemical oxygen demand test. J Environ Eng 134(5):353–361

Xu X, Thomson NR (2009) A long-term bench-scale investigation of permanganate consumption by aquifer materials. J Contam Hydrol 110:73–86

Xu P, Achari G, Mahmoud M, Joshi RC (2006) Application of Fenton's reagent to remediate diesel contaminated soils. Pract Period Hazard Toxic Radioact Waste Manag 10:19–27

Xu H, Cang L, Song Y, Yang J (2020) Influence of electrode configuration on electrokinetic-enhanced persulfate oxidation remediation of PAH-contaminated soil. Environ Sci Pollut Res 27:44355–44367

Yao JJ, Huang Z, Masten SJ (1998) The ozonation of pyrene: pathway and product identification. Water Res 32:3001–3012

Yang S, Oostrom M, Truex MJ, Li G, Zhong L (2016) Injectable silica–permanganate gel as a slow-release MnO_4^- source for groundwater remediation: rheological properties and release dynamics. Environ Sci: Processes Impacts 18:256–264

Yen C, Chen K, Kao C, Liang S, Chen T (2011) Application of persulfate to remediate petroleum hydrocarbon-contaminated soil: feasibility and comparison with common oxidants. J Hazard Mater 186(2–3):2097–2102

Yu H, Kim BJ, Rittmann BE (2001) The roles of intermediates in biodegradation of benzene, toluene, and p-xylene by Pseudomonas putida F1. Biodegradation 12:455–463

Yu DY, Kang N, Bae W, Banks MK (2007) Characteristics in oxidative degradation by ozone for saturated hydrocarbons in soil contaminated with diesel fuel. Chemosphere 66:799–807

Zhai X, Hua I, Rao PSC, Lee LS (2006) Cosolvent-enhanced chemical oxidation of perchloroethylene by potassium permanganate. J Contam Hydrol 82:61–74

Zhao D, Liao X, Yan X, Huling SG, Chai T, Tao H (2013) Effect and mechanism of persulfate activated by different methods for PAHs removal in soil. J Hazard Mater 254–255(4):228–235

Chapter 16
Activated Carbon Injection for In-Situ Remediation of Petroleum Hydrocarbons

Scott Noland and Edward Winner

Abstract In-situ remediation of petroleum hydrocarbons (PHCs) using activated carbon (AC) is an emerging technology intended to enhance sorption and biodegradation mechanisms in soil and groundwater systems. The combination of pore types, source material, activation process, and grind of a particular AC influences its efficacy in subsurface remediation. When high-energy injection techniques are employed, installation of carbon-based injectate (CBI) slurries can be conducted in practically any geological setting, from sandy aquifers to low-permeability zones and weathered or fractured rock. Following an adequate CBI installation throughout the target treatment zone or as a permeable reactive barrier, dissolved PHC concentrations are typically observed to rapidly decrease. After a new equilibrium is formed, PHC concentrations typically decrease over time due to the biodegradation. PHC biodegradation, in association with the CBIs, is indicated by the presence of appropriate microbial communities found to grow on AC and is supported by multiple lines of evidence. Further research is encouraged to optimize the biodegradation and regeneration processes of CBI products for in-situ remediation of PHCs.

Keywords Activated carbon · Biodegradation · Carbon-based injectate · Direct-push injection · Permeable reactive barrier

16.1 Introduction

In-situ remediation of petroleum hydrocarbons (PHCs) using activated carbon (AC) is an emerging technology enhancing sorption and often treatment via biodegradation of soil and groundwater contaminants.

S. Noland (✉) · E. Winner
Remediation Products Inc., 6390 Joyce Drive, Golden, CO 80403, USA
e-mail: scott@remediationproducts.com

E. Winner
e-mail: ed@remediationproducts.com

© The Author(s) 2024
J. García-Rincón et al. (eds.), *Advances in the Characterisation and Remediation of Sites Contaminated with Petroleum Hydrocarbons*, Environmental Contamination Remediation and Management, https://doi.org/10.1007/978-3-031-34447-3_16

549

Installation of carbon-based injectates (CBI) can be done using direct-push technology (DPT) or other methods in practically any geological setting, from sandy aquifers to low-permeability zones and weathered or fractured rock when high-energy injection techniques are employed. Many variables must be determined before going to the field, including injection point grid spacing (both areal and vertical) across the target treatment zone, injection tips, pressure and injection flow rate, the mass of AC product to be injected at each depth and location, and top-down versus bottom-up injection technique.

It is well known that most adsorption of organic compounds occurs in the microporous structure of AC and that these pores are much smaller than bacteria capable of degrading these contaminants. Compelling evidence exists that compounds residing in micropores are assimilated by bacteria. AC has a number of characteristics that support microbial life and healthy growth including formation of microbial clusters and reduction of contaminant toxicity toward the microorganism. Specific microbes often found at PHC impacted sites are described along with lines of evidence that can be developed for demonstrating biodegradation is occurring. Use of compound-specific isotope analysis (CSIA) as a means of proving biodegradation of specific contaminants is discussed along with case study examples. See Chap. 11 for more information on the use of CSIA in petroleum-contaminated sites.

The concept of biological regeneration of AC is developed. This concept is of pivotal importance because effective regeneration of carbon creates a dynamic system that will continue to perform over long periods of time and not be limited by adsorption capacity. The chapter ends with a discussion of future research needs to address fundamental questions such as whether co-injection of bacterial consortia significantly improves performance and how indigenous microbes interact under bioaugmentation. In addition, we explore applications involving light non-aqueous phase liquid (LNAPL) from the perspective of adsorption capacity, toxicity, and biological response and limitations associated with low concentrations such as those often encountered with the fuel additive methyl tert-butyl ether (MTBE).

16.2   Characteristics of Activated Carbon

AC is composed of plates of hexagonal carbon rings. Many of the hexagonal rings are cleaved such that they are orientated out of the primary plane relative to most of the carbon rings. These ring cleavages also allow other rings structures such as pentagons and heptagons. These carbon rings, when viewed at a larger scale, form curved sheets, bowels, and ribbons (Harris et al. 2008). The interaction of these structures forms walls, resulting in cavities and passages commonly referred to as pores. Pores are both the passages into deeper sections of AC particles and the volume above surfaces upon which adsorbed compounds bind.

AC has a large hydrophobic surface that binds molecules through van der Waals forces and other physicochemical bonds related to the presence of different pore types within its structure. The pore structure is thus the primary physical characterization of AC. The pores are classified into three major groups based on pore diameter. The largest pores are macropores (pore diameter > 50 nm), followed by mesopores (pore diameter from 2 to 50 nm). These pores are important pathways for the diffusion of molecules into the carbon structure and to the micropores. The micropores (pore diameter < 2 nm) are where most of the adsorption occurs because in the micropores the gaps between the carbon plates are only a few molecular diameters. The binding forces, termed London forces, are weak and only influence a range of 1–5 molecular diameters from the carbon plates (Greenbank 1993). For perspective on the size of the AC pores, the smallest bacteria are about 200 nm, while the typical petroleum-degrading bacterium would be a rod around 600 by 400 nm (Weiss 1995).

The combination of pore types, source material, activation process, and grind of a particular AC influence its efficacy in subsurface remediation. AC derives from carbon source materials that have been chemically treated and/or heat-treated in a controlled atmosphere to drive off volatiles and increase porosity (Menendez Diaz 2006). AC source materials are carbon-rich substances such as bituminous coal, coconut shell, rice husks, and wood. The choice of material from which the AC is made significantly affects the adsorption character of the AC (Zango 2020). For example, despite being rich in binding pores (micropores), coconut-based AC exhibits slower kinetic adsorption performance due to having relatively fewer trans-port pores (McNamara 2018). The activation technique or process also affects the adsorptive characteristics of the resulting AC (Menendez Diaz 2006; Lillo-Ródenas 2005). Thus, oxygen-containing groups on the AC generally increase adsorption of compounds like toluene, and such groups are increased by acid modification but decreased with alkali modification (Ma 2019). On the other hand, alkali modifica-tion improves o-xylene adsorption to AC (Li 2011). The specific source material, activation technique, and post-activation processing combine to make many types of AC, each of which has unique adsorptive characteristics (Cheremisinoff 1978; Menendez Diaz 2006).

AC is available in a range of grinds or mesh sizes, but the products used for in-situ remediation are typically categorized as granular, powered, or colloidal (Fan 2017). As a broad statement, the smaller the grind, the faster the initial adsorption and subsequent desorption because the micropores increase as a percentage of all pores but are closer to the particle surface (Liang 2007; Piai 2019). For practical purposes in application to hydrocarbon remediation, the rate of adsorption across the various grinds is so fast that the grind is not selected based on adsorption rate.

Since the introduction of AC-based technologies for in-situ remediation of PHCs in early 2002, several products have been developed as shown in Table 16.1. Colloidal products are designed to be injected under low-energy conditions, while powdered AC (PAC) and granular AC (GAC) products are typically injected under pressure.

Table 16.1 Commercially available AC-based products for in-situ PHC remediation

Product	Property	PHC degradation pathway	Manufacturer
BOS 200®	PAC mixed with nutrients, electron acceptors, and facultative bacteria	Aerobic and anaerobic bioaugmentation	RPI
BOS 200+®	BOS 200® enhanced for application to LNAPL sites	Aerobic and anaerobic bioaugmentation	RPI
Plume Stop®	Colloidal AC suspension with an organic stabilizer	Anaerobic biodegradation via biostimulation	Regenesis
Petrofix®	Colloidal AC suspension with electron acceptors including sulfate and/or nitrate	Aerobic and anaerobic biodegradation	Regenesis
COGAC®	GAC or PAC mixed with calcium peroxide and sodium persulfate	Chemical oxidation. Aerobic and anaerobic biostimulation	Remington technologies
FluxSorb™	Colloidal carbon product	Aerobic and anaerobic biodegradation	Cascade

16.3   Application of Activated Carbon to In-Situ Petroleum Remediation

Compared to many conventional technologies for in-situ remediation, the performance of AC-based systems is relatively unaffected by certain subsurface conditions. For example, the adsorption characteristics of AC are the same in seawater and freshwater despite the different geochemical conditions. AC is inert, so it does not typically react with the geology to mobilize ions or metals. In addition, AC can be installed in a wide range of geological settings, from sandy aquifers to low-permeability zones and weathered or fractured rock when high-energy techniques are employed. Injecting AC into the subsurface may thus result in contaminant concentration reduction over a broad range of subsurface conditions, particularly when adequate contact between the AC and the contaminant is established and if a contaminant destruction mechanism such as biodegradation is simultaneously promoted. The objective of all injection designs should be to achieve contact between contaminant and CBI and match dosing of CBI to the distribution of contaminant mass.

16.3.1   Installation

AC amendments can be delivered to the subsurface via injection across the target treatment zone or as a permeable reactive barrier (PRB). A water-based carrier fluid is typically used to suspend AC of a wide particle size range prior to injection. Large

AC particle suspensions (5–50 μm mean diameter size) are typically referred to as slurries, while colloidal suspensions aided by dispersants are typically achieved using smaller AC particles (2–10 μm). Once injected, AC is expected to affix to the formation.

Colloidal carbon suspensions can be injected with low-pressure, low-flow techniques where the pump pressure is adjusted to ensure fracture pressures of the formation are not exceeded. Amendment distribution typically follows naturally existing fluid pathways when low-pressure injection is employed. The rate at which a colloidal carbon both agglomerates and affixes to the formation depends on the medium of suspension or dispersant employed (Haupt 2019). Slurries of AC are typically jetted into the subsurface, with the AC material being expected to settle out three to five meters from the injection points. High-pressure injection is commonly used to facilitate emplacement of CBIs in heterogeneous systems by creating secondary porosity within a region localized around the injection point.

Multiple techniques exist to deliver CBIs into the subsurface, including gravity feed, DPT, modified augers, soil mixing, and trenching. Straddle-packers are employed to allow pressurizing and isolating an injection interval in open hole bedrock installations or in other formations where DPT cannot penetrate the ground. Alternatively, boreholes can be grouted with bentonite clay (uniformly hydrated, distributed, and cured to avoid injection fluid to migrate up the well, i.e., daylighting).

Gravity feed relies on the hydraulic head of the injectate typically stored in a tank to create flow into a fixed well and through its screen (or through a hollow drill rod) into the subsurface. This approach is commended for its low cost and simplicity, yet it is not generally or effective. Gravity feed techniques are useful for large volume solution injections and require high-permeability formations. Injected fluids are preferably low in viscosity, and suspensions should be well dispersed and with small mean particle size that can flow past screens, well completion sands, and formation particles. This technique is typically considered inadequate of CBI particles settle out of suspension more quickly than they can be distributed. Preferential pathways observed during gravity feed techniques may also limit the capability of uniformly distributing CBI across a contaminated interval.

The most common CBI installation technique uses DPT with pump-supported injection with pump-supported injection to emplace the CBI in discrete intervals within the subsurface. Injection with DPT consists of driving hollow drill rods into the ground through hammering. The rod is configured with an injection tip. Once the tip is at a desired depth, an injection head is attached on surface and pumps force solutions, slurries, and gels through the rod and into the subsurface.

Injection pressures and flow rates primarily depend on nature of the geology and injection fluid. Low-pressure injection and high-pressure injection techniques can be employed to install CBIs. In high-permeability tills, either method can be effective (McGregor 2020). In low-permeability geologies (hydraulic conductivity below 10^{-5} m/s), it is difficult for low-pressure methods to achieve sufficient distribution and adequate contact with contaminants. High-pressure techniques are superior (Christiansen 2010).

Note that pressure gauges on most (if not all) injection pump systems are located near the pump discharge. At any given flow rate, the pressure at this location is mostly an indication of the pressure drop across the pipes, fittings, hoses, and tooling. Additional calculations or the use of a downhole pressure gauge are required to measure the pressure at the formation (i.e., at depth).

16.3.2 Injection Technique and Spacing

The relative difficulty of distributing AC in the subsurface results in implementation designs using low-volume injections across a high-density injection point grid within the target treatment zone or PRB. These points are laid out horizontally in a triangular grid along the direction of contaminant transport. The spacing between injection points should be adjusted based on geological setting, project goals, and costs. The distance between horizontally placed injection points is informed by experience. Efforts to control distribution out beyond three to six meters are rarely successful in overburden materials because moving closer to the surface correlates with a decreasing weight of the overburden and a corresponding decrease in resistence to flow.

Regarding the vertical spacing between discrete injection depths, a similar installation practice consists of alternating injection depths within the treatment zone on adjacent points. This injection depth "staggering" approach helps ensure that all targeted depth intervals receive CBI. Experience has shown that in low-permeability materials, injections can be completed every 0.5–0.6 m in depth. In high-permeability materials, the ideal spacing is around 1 m, and high-energy injection techniques are required to achieve good distribution of injectate in this case.

Regarding the injection technique, bottom-up injections should be avoided when fine-grained sediments are present. Imagine completing the first injection at depth, then as the injection rod is withdrawn to the next targeted zone, the borehole below the rod remains open. In this type of profile, it takes more pressure to open a fracture than it does to propagate the deeper one, and injectate will simply flow into the first injection location. Bottom-up injections are effective in heaving sand. In general, injections should be completed in a top–down manner so that as injections are completed, the rod is advanced, and it tends to seal off the previous location and prevent injectate cross-circuiting.

16.4 Patterns of AC Distribution

After injection, the CBI will ultimately follow the path of least resistance. As discussed in Chaps. 1, 4, 7, and 8 among others, subsurface properties can vary widely over small scales so it is challenging to predict where CBIs may travel when

forced into the formation. Through high-pressure injection (jetting and fracture generation), some control over placement of CBI can be exerted within a region localized
around the injection point. However, energy quickly dissipates, and further outward
movement of injectate is again relegated to low resistance pathways. In sedimentary environments, small gradations in particle size distributions can represent low
resistance pathways and channel flow of injectate.

CBIs is intensely black in color and can be recognized in most cores collected after
injection. As illustrated in Fig. 16.1, PAC injected under pressure is commonly found
in the form of thin horizontal seams and sometimes as angular or vertical features
following the path of least resistance. This is also depicted in Fig. 16.2, where CBI
follow subtle differences in sediment texture. CBI injection flow paths are often
found to preferentially follow strata interfaces or high-permeability pathways.

Injection of slurries into sands is more challenging than injecting solutions because
the suspended solids tend to be filtered from the injectate. Water readily moves
through saturated sands, leaving suspended solids localized in a small region around
the point of injection if low-energy injection is employed. High-energy injection

Fig. 16.1 Excavation post-CBI emplacement in the vadose zone demonstrates significant horizontal distribution. The CBI contained PAC. Note the horizontal carbon sheet above the bucket
mark traced by a yellow line (dashed). The carbon sheet was ~1.2 m across. A shallower horizontal
sheet is traced by an orange line (solid). Two angular carbon sheets having more vertical character
are marked by red lines (dotted)

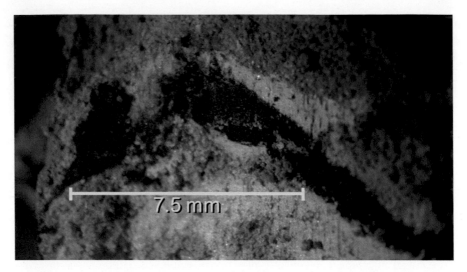

Fig. 16.2 PAC intercalated between materials exhibiting subtle texture variations

Fig. 16.3 Distribution of AC throughout aquifer sands after high-energy injection. The sleeve is a standard 3.6-cm diameter tube

techniques can overcome this difficulty as depicted in Fig. 16.3, where a slurry of granular AC was jetted into the aquifer sands at an exit velocity of over 60 m/s.

In summary, high pressure and flow can be employed to force slurries out from the point of injection and emplace it in both high- and low-permeability formations. On the contrary, low-flow and low-pressure (low energy) injection techniques will not distribute most CBI in clay lenses or other low-permeability zones. One strength of colloidal AC-based injectate is that vary uniform distributions in sands, and other high-permeability materials can be achieved under low-flow and low-pressure conditions.

16.5   Amount of Carbon Injected

The amount of carbon to be installed at a particular site will depend on the remedial goals established for that site. Figure 16.4 is an idealized model intended to help the reader understand the amount of carbon that needs to be injected to address a mass of contamination. Think of a mixing tank containing some volume (V) of water with of water with a dissolved contaminant at an initial concentration (Co). Adding and mixing a mass of AC (W) results in an equilibrium dissolved concentration (Ce) with a certain mass of contaminant adsorbed to the carbon. The concentration of contaminant sorbed in the AC is denoted Sc. In this system, a simple mass balance can be written as follows (Eq. 16.1):

$$CoV = CeV + WSc \qquad (16.1)$$

The left side of Eq. (16.1) is the mass of the contaminant in the vessel (CoV). The first term on the right is the mass of contaminant dissolved in water at equilibrium (CeV). The last term is the mass of contaminant adsorbed in carbon (WSc). Therefore, increasing the amount of carbon increases the contaminant adsorbed to the carbon and decreases the contaminant in the aqueous phase at equilibrium.

In order for the above mass balance to be useful, a relationship (isotherm) between dissolved phase equilibrium concentration and the concentration of contaminant adsorbed in the AC is needed. A variety of different isotherm equations have been proposed, some have a theoretical foundation, and some are empirical in nature. Commonly, the Langmuir or Freundlich isotherms are used to model adsorption behavior. The Freundlich isotherm is often applied to adsorption of organic compounds onto AC and is written as follows (Eq. 16.2):

$$Sc = kCe^{1/n} \qquad (16.2)$$

A plot of Ce versus Sc in log–log space yields a linear-plot whose slope is ($1/n$) and intercept is (k). Substituting into the mass balance for Sc yields the following equation (Eq. 16.3):

$$CoV = CeV + WkCe^{1/n}. \qquad (16.3)$$

Fig. 16.4 Idealized model to understand how much carbon may need to be injected consisting of a mixing tank containing some volume (V) of water with a dissolved contaminant at an initial concentration (Co) where some mass of AC (W) is added and mixed

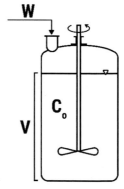

For example, if we have 10 ppm toluene in groundwater and our target is to reduce the concentration to 0.5 ppm, we can look up the Freundlich parameters (Dobbs 1980), $k = 26.1$ and $1/n = 0.44$. Alternatively, isotherms are often available from product vendors. Assuming an average porosity of 30%, there will be 300 L of groundwater per cubic meter of saturated formation. The only unknown is W, which can be calculated as in the equation below (Eq. 16.4):

$$W = 300(10 - 0.5)/26.1(0.5)^{0.44} = 148 \text{ g AC per cubic meter of saturated formation.}$$
$$(16.4)$$

This model is simplistic and does not reflect the complexities found in the subsurface. It is impossible to obtain complete mixing, and, usually, there are multiple contaminants competing for adsorption to AC. In the field, PHC impacted sites may be some of the most complex in that contamination consists of a multitude of adsorbates that interact very differently with AC that could require the development of isotherms specific to the PHC products found on a site. A simplification arises when concentration of absorbable species is low. In this case, each adsorbate equilibrium will be unaffected by the presence of other components. This presumption would likely apply at locations in the plume where only dissolved phase exists but not within or near an LNAPL source zone.

Typically, the practical experience of CBI vendors and installers suggests application of an engineering factor of safety (EFS) to estimated mass dosing from isotherm calculations. While efforts have been made to develop EFS systematically, the factors remain a matter of experience and commonly vary from two to tenfold the mass dose estimated based on isotherms. When very small amounts of AC are apparently needed, this is when the highest EFS are indicated. A more extensive AC loading increases the probability of contact between AC and a contaminant when the contaminant is present in low concentration.

The previous discussion regarding calculation of injection mass dosing applies to plume-wide treatment. With permeable reactive barriers (PRBs), one needs to consider the contaminant mass flux across and through the barrier, the design life of the barrier, and contaminant degradation during transition through the barrier. PRBs constructed with CBI can be thought of as a conventional carbon bed. In a conventional carbon bed, contaminant mass is adsorbed by the carbon as it enters the bed. Adsorption continues until the adsorption capacity of the carbon is approached, and the contaminant then travels a bit further into the bed. As this process continues, a narrow zone forms that slowly moves through the bed. This zone is called the mass transfer zone. The barrier design is based on the amount of carbon adsorption needed to meet the contaminant concentration goals on the downgradient side of the barrier for a defined time. Many references, e.g., (Thomas 1998; Ruthven 1984), describe how flow-through AC beds are designed. In a CBI PRB, biodegradation can contribute to increase the PRB operational lifespan. Biodegradation effects could also be accounted for during the design of the PRB to optimize the amount of carbon needed.

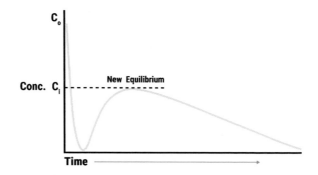

Fig. 16.5 Idealized general response in terms of contaminant concentration in groundwater after CBI installation

16.6 General Response to Installation of AC into a Contaminated Aquifer

After AC-based injectate for passive in-situ remediation of soil and groundwater was introduced to the industry (Noland 2002), initial expectations of some practitioners were that dissolved phase contaminants would simply disappear due to adsorption, and contaminated sites would clean up overnight. This expectation has been tempered over time as the complexities of subsurface environments and inefficiencies associated with CBI installation have come to be appreciated.

Figure 16.5 depicts an idealized general response in terms of contaminant concentration in groundwater after CBI installation. Dissolved contaminants are rapidly adsorbed, and the transient concentration reduction is often better than predicted based on design. This is because an equilibrium initially existed between the dissolved and sorbed mass of contaminant. The injection of AC disrupts this equilibrium as the rate of adsorption of contaminant from groundwater outpaces the rate of contaminant desorption from the formation. As contaminant desorbs from the formation into groundwater, adsorption continues, and the interplay between adsorption capacity of the AC and the contaminant mass flux from the formation results in rising concentrations in the aqueous phase until a new equilibrium is established (Ci on Fig. 16.5).

The magnitude of Ci and the time required for a new equilibrium to be established is a function of the contaminant mass initially present in the aquifer, the distribution of AC within the formation, and the stratigraphy. If the target treatment zone resides in a high-permeability environment, the time to achieve a new equilibrium will be short, often within a few weeks. If the aquifer sediments consist of silt or clay, the time required may span months. What is often incorrectly called "rebound" is a predictable partitioning of contaminant mass between the groundwater, the aquifer solids, and the AC as the system approaches a new equilibrium.

Presuming no significant contaminant mass is present that would cause overloading of the carbon and assuming robust biological activity is operating to degrade contaminants, a general decline from Ci should be observed. Given a general decline, a semilog plot of concentration over time should result in a straight line with a negative slope that will be related to the rate of contaminant degradation.

16.6.1 Effects Related to Contaminant Mass

The most important variable affecting the response to injection of AC into the aquifer is the amount of contaminant mass present. At locations far removed from a source area and where the contaminant mass sorbed to the formation is insignificant, observed post-injection concentration perturbances are slight if they occur.

In contrast, the other extreme is represented by significant non-aqueous phase liquid (NAPL) hydrocarbon in the formation. This condition is common in source areas, and it is unlikely that enough CBI can be injected to absorb the NAPL altogether. After injection of CBI, experience has shown that a short-term drop in dissolved phase concentration is often observed; however, recontamination of groundwater due to continued contact with NAPL may overcome the adsorption capacity of the CBI, and a return to original conditions is predictable.

The above-described static state should not be interpreted as a deterrent to using CBI to treat NAPL. Instead of employing a design based simply on adsorption capacity, the design should be based on the surface area of the injected AC. NAPL mass removal rates are a function of the total amount of AC injected due to the available surface area of the AC that supports degradation. Regardless of adsorption, more surface area correlates to increased biodegradation. In this manner, AC is like a catalyst. This allows one to predict the time required to eliminate NAPL and see the subsequent decline in groundwater concentration. This is very similar to measurement of carbon dioxide generation at NAPL sites to estimate the average mass removal rate and then use that rate to predict performance over time.

16.6.2 Competitive Desorption

There is a widely held belief within the environmental industry that organic compounds, once adsorbed to AC, are stabilized for, perhaps, centuries. An article published in 2015 stated: "organic compounds are sorbed to AC so strongly that it is almost certain that the contamination will be stable and unavailable for leaching for at least 50–100 years" (Fox 2015). The reality is that the adsorption of organic compounds is a reversible process controlled by the heat (enthalpy) of adsorption (Kolasinski 2019). This feature has significant consequences.

AC will efficiently remove contaminants from groundwater and soil until the adsorption capacity is reached. Refined petroleum, whether gasoline, diesel, or other fuels, contains many different organic compounds. AC interacts with all these compounds differently as heats of adsorption are specific to each compound, and there will be competition for available absorption sites on the AC. In addition, naturally occurring organic matter (NOM) also must be considered. Some dissolved organic compounds (NOM) are present in all groundwater, and they typically have high heats of adsorption and will also compete for adsorption sites.

When AC is first injected, adsorption sites within the carbon are unoccupied, and there is a vast space for compounds to bind. It does not matter if some compounds bind more energetically to AC than others; initially, there is room. As the space begins to fill up, those compounds with higher heats of adsorption tend to displace those having lower heats of adsorption. Now, those compounds with low heats of adsorption begin to desorb from the carbon into solution, while other compounds with high heats of adsorption continue to be adsorbed. This process is called "roll-over" within the AC industry, i.e., displacement of compounds having lower heats of adsorption.

Important examples of this are benzene and 1,2-dichloroethane. It is common for benzene concentration to be a site risk driver. Benzene has a relatively low heat of adsorption and can be displaced from AC by any number of common contaminants, including xylene, naphthalene, and NOM. To mitigate bleeding of contaminants from the AC, robust degradation mechanisms or pathways must be present to effectively remove and degrade adsorbed compounds from the carbon. Ideally, the rate at which compounds are degraded should effectively maintain unoccupied adsorption sites on the AC. This facilitates continuing adsorption of new contaminants without displacing those already bound to the AC. The most common means for accomplishing this with PHCs is biological degradation. The AC is biologically regenerated as bound contaminants are degraded. This mechanism will be discussed in the next section.

The proposed degradation mechanisms should be carefully evaluated whenever CBI technology is considered. Any application of CBI to contaminants where no viable degradation mechanism is known should be conscientiously weighted because the installed CBI may become a new source in the future. The clearest example of this is the use of AC at sites impacted by per and polyfluoroalkyl substances (PFAS), known as "forever chemicals" in the vernacular. A complete CBI remediation system must be supported by a treatment mechanism. Biodegradation is the preferred mechanism for most CBIs on the market intended for remediation of PHCs.

16.6.3 Adsorption by Activated Carbon

Literally every municipality in the USA uses beds of activated carbon to polish drinking water before it is released for consumption. Because of its widespread use and the number of large manufacturers competing to make effective carbons, there is a sense within the environmental industry that all carbons are the same (Fan 2017). As detailed in previous sections, activated carbons are manufactured from a wide variety of raw materials, and they have very different properties. The example in Sect. 16.5 used design parameters associated with F400 high grade carbon derived from bituminous coal manufactured by Calgon Carbon. One of the more important characteristics of AC is its adsorption capacity. This is a measure of the maximum mass of contaminant the carbon is capable of adsorbing at any given equilibrium concentration. The adsorption capacity is strongly dependent on contaminant concentration, the type of carbon (its source and activation method), and finally on the specific contaminant. There are organic compounds that are not adsorbed by AC at all. When dealing with multi-component products present at PHC sites, the adsorption dynamic becomes very complex, particularly in view of competitive desorption. Rules of thumb like AC can absorb roughly 15–20% by weight of organic contaminants are very misleading and should be discarded. PHC specific and AC product specific data should be relied upon.

One of the great advantages of AC is the ability to regenerate or reactivate spent carbons. Housekeeping on food-grade carbons is very good because when reactivated, the original customer will repurchase it. The same cannot be said for spent non-food-grade carbons. These carbons may originate from a multitude of sources ranging from wastewater to metal plating to manufacturing facilities. The reactivation process removes any organic contamination that may have been present, but inorganic contaminants will persist. Reactivated carbons that are not derived from food or drinking water applications should not be used for in-situ remediation because they may contain metal contaminants that could impact groundwater quality.

16.7 Microbial Growth on Carbon

16.7.1 Carbon Structure Providing Living Space

Natural organic matter, nutrient concentrations, and species richness decrease with depth, with notable increases in the transition zones such as the capillary fringe (Smith 2018). The typical decline in microbial abundance that accompanies increasing subsurface depth, at least in part, arises from decreasing transport pathways, the accompanying reduction in the influx of fresh organic carbon, and other required chemicals, such as electron donors or acceptors (Santamarina 2006). The size and shape of the cavities within the media and interstitial space between media grains restrain subsurface organisms' size, form, and mobility (Luckner 1991), and thus the population density. Population density is linked to the cavities and interstitial spaces, which are essential to the supply of nutrients, electron donors and acceptors, carbon sources, and the elimination of microbial metabolic wastes (Fredrickson 2001).

Essential requirements for subsurface microbial life include moisture, chemical factors such as electron donors, and media factors such as granularity. Since some of these elements are linked to the transport of water in the subsurface, microbial populations often reflect groundwater flow pathways (Maamar 2015; Graham 2017; Danczak 2018). The installation of AC into the subsurface by high-pressure, high-flow injection may establish new habitat by creating and expanding flow paths beyond that which existed in the native subsurface material (Lhotský 2021; CLU-IN U 2022).

The AC may accentuate water movement (Siegrist 1998; Sorenson 2019; ITRC 2020) due to the addition of three-dimensional networks of carbon seams and surfaces (Murdoch 1995; Bradner 2005; Or 2007). These seams can increase connectivity in the subsurface media as AC acts as a proppant holding the fractures open. The propped fractures can accommodate the introduction of additional microbes and microbial resources such as water and electron acceptors (Mangrich 2015; Zamulina 2020). Relatedly, AC's granular nature can also increase transport in the subsurface due to the formation of thick liquid films on the rough surfaces of AC (Or 2007; Massol-Deyá 1995). Together the AC seams and surfaces should support contaminant dispersion, movement of nutrients and terminal electron acceptors, the interchange of microbial metabolic products (Bures 2004), and microbial dispersal and colonization on newly accessible soil and aquifer compartments (Krüger 2019).

The surface of AC may provide recesses offering shelter for microbes, while its adsorptive properties enrich the environment of the AC with substrate (Weber et al. 1978). Microbial cells that inhabit the macropores and surface recesses are better sheltered against stressful environmental conditions, such as predation, than microbes in the native media (Kindzierski 1992; Żur 2016). Due to the surface recesses, more habitat exists for microbial growth (Liang et al. 2009). Concerning adsorptive capacity, by comparison to AC, organic resources in the subsurface are scattered within structurally heterogeneous media. AC consolidated resources result in higher substrate biodegradation and higher specific growth rates than non-adsorbing or weakly adsorbing media such as sand (Li 1983). Thus, AC consolidates organic resources (Weber et al. 1978) on a homogenous media, and microbes are drawn to the enriched environment of the AC (Liang 2007), which facilitates contact between degraders and contaminants (Chen 2012; Thies 2012; Bonaglia 2020).

It has long been observed that bacteria attach to and multiply upon the surface and in the macropores of AC (Board 1989; Morsen 1987; Liang 2008), and that biodegradation occurs on AC in the subsurface. (Peacock 2004) compared in-well coupons consisting of either glass wool or Bio-Sep® beads (75% PAC and 25% Nomex®) to investigate the in-situ bioreduction of technetium and uranium in a nitrate-contaminated aquifer. It was found that the beads supported microbes that reduced technetium and uranium, and the beads supported tenfold more microbial biomass than glass wool. It was stated that higher microbial biomass was most likely to exist due to the higher surface area of the beads and the concentration of nutrients on the beads as compared to the glass wool. Other studies have concluded that benzene, toluene, and phenol are effectively extracted from Bio-Sep® beads (Williams et al. 2013), metabolized by the microbes, and incorporated into fatty acids produced by microbes (Geyer 2005). Similar results were obtained for in-situ biodegradation of naphthalene at the McCormick and Baxter Superfund Site in Stockton, California (McCormick 2010). Nineteen well locations were tested, and all were positive for ^{13}C enrichment in the microbial lipid biomass.

PAC amended sediments in a microcosm study also demonstrated that bacteria attach to AC and multiply on the surface and within the macropores (Pagnozzi

2020). AC stimulated naphthalene biodegradation and mineralization under anaerobic conditions, as shown by the production of $^{14}CO_2$ from radiolabeled naphthalene (Pagnozzi 2021). The relative abundance of Geobacter, Thiobacillus, Sulfuricurvum, and methanogenic archaea, known to be associated with naphthalene degradation, increased after amendment with PAC. Under aerobic conditions, sediments containing 16 PAHs amended with PAC increased microbial community diversity, structure, and activity relative to both sand and clay controls. The alteration in the microbial community in response to PAC is consistent with that reported by Bonaglia (2020), who also examined and reported an increase in the biodegradation of high molecular weight PAHs for microbes associated with AC as did Acosta (Zapata Acosta 2019).

Furthermore, AC sustains microbial activity by decreasing toxic concentrations in the dissolved phase (Leglize 2008) and buffering the microbes against toxic shock (Sublette 1982). Specifically, the adsorption of PHCs onto AC significantly reduces their toxicity to microbes (Morsen 1990). The toxicity of petroleum compounds is reduced, as indicated by a sharp increase in the abundance of both native and inoculated petroleum degraders (Semenyuk 2014). Furthermore, AC added to sediments decreased toxicity as measured by ecological parameters (Kupryianchyk 2013). Likewise, adequately designed and deployed CBI should reduce contaminant toxicity to microbes.

16.7.2   AC Priming

Contaminant degradation is a function of the density of microbial microaggregates (Gonod 2006; Becker 2006). As discussed, AC installed in the subsurface is new habitat for microbial exploitation. Therefore, it may be advantageous to give hydrocarbon-degrading and degradation supporting microbes an opportunity to adhere to the AC before subsurface installation. This process is referred to as priming. Priming allows the AC to function as both a microbial inoculum delivery system and a readymade microbial habitat (Piai 2020). Similar approaches have proven successful in a variety of applications (Das 2017) and can result in more robust microbes (Meynet 2012) and microbial degradation that is sustained (Marchal 2013; Bonaglia 2020).

Priming stimulates formation of biofilms (term typically used to identify clusters of bacteria or archaea in a matrix of extracellular polymeric substances—EPS). Microorganisms embedded within biofilm achieve high resistance to toxic pollutants (Köhler 2006) and are durable, so biodegradation is persistent (Aktaş 2007; Ran 2018; Edwards 2013). In the subsurface, microbial-inoculated AC can establish a biofilm that promotes growth and diversity, degrades hydrocarbons and other contaminants, regenerates the AC, and persists over time (Lhotský 2021; Cho 2012). Figure 16.6 shows a case of microbial growth on AC in laboratory samples.

Fig. 16.6 Microbial growth on AC during a diesel bench test. A viscous material formed within days. The bubbles were gases produced by microbial activity

16.7.3 AC Adsorption as Barrier to Biodegradation

Studies have also shown that AC may have an inhibitory effect on microbial growth by decreasing the dissolved concentration of adsorbate available for microbial degradation. Marchal (2013) demonstrated for phenanthrene sorbed to soils that the stronger the adsorbent (AC, biochar, compost), the lower desorbed concentration, and the lower the microbial degradation (Marchal 2013). These results are not unique (Rhodes 2010), but they may be specific to adsorbates having high heats of adsorption like phenanthrene. Three-ring PAH, as is phenanthrene, degradation has been reported to be inhibited by AC, but 4 and 5-ring PAHs degradation is supported by AC (García-Delgado 2019). Aromatic compounds, having lower heats of adsorption, are not affected by AC to the degree reported for phenanthrene. Mangse (2020) found that AC carbon made toluene less bioavailable, but AC did not have a detrimental effect on toluene degradation (Mangse 2020). These observations are reasonable in that adsorbates that enter the AC are protected to some degree from biodegradation. The extent of that protection is influenced by the strength of the interaction with the AC. In the end, these observations follow from the adsorptive character of AC and they have not been demonstrated to be relevant to field application of AC, where multiple chemical adsorbates and NOM are present.

16.8   Evidence of Biodegradation at Sites with AC

16.8.1   Microbial Communities

Microbial attachment to AC is associated with an increase in the abundance of both native and inoculated petroleum degraders (Semenyuk 2014; Bonaglia 2020; Pagnozzi 2021). While the data to separate the function of AC from the CBI does not exist, an increase in microbial abundance is also associated with CBI emplacement. Figure 16.7 presents microbial data from groundwater samples collected from a CBI emplacement site 3 months post-emplacement. According to next-generation sequencing (NGS), the CBI emplaced sample has more total species, more individuals per species, and more rare species than either the control or the PHC impacted control. Figure 16.8 provides a stark visual confirmation of the abundance data presented in Fig. 16.7. More information on NGS can be found in Chap. 10.

A more definite question is whether the microbial populations in the CBI emplacement contain the individual microbes that are pertinent to PHC degradation. Figure 16.9 presents microbial data from groundwater collected from a CBI emplacement site 2 years post-emplacement. The microbial samples were examined

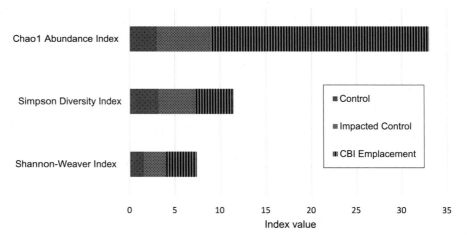

Fig. 16.7 Groundwater samples for both CBI treatment and controls analyzed by next-generation sequencing. Compared to controls, the CBI emplaced samples had more total species, more individuals per species, and more rare species. The Shannon–Weaver Index is the natural log of the total number of species in a sample. It reflects the total number of species present. The Simpson Diversity Index is the square of each species' abundance summed for the total sample population subtracted from 1; thus, proportional abundance is weighted more on the uniformity of individuals in the population, which is referred to as evenness. It indicates species dominance and reflects the probability of two individuals that belong to the same species being randomly selected. The scale is zero to 5, with 5 meaning that each species in a dataset is represented by the same number of individuals. The Chao1 abundance index is the number of individuals present and single or double individuals. It emphasizes individuals microbes present in small numbers

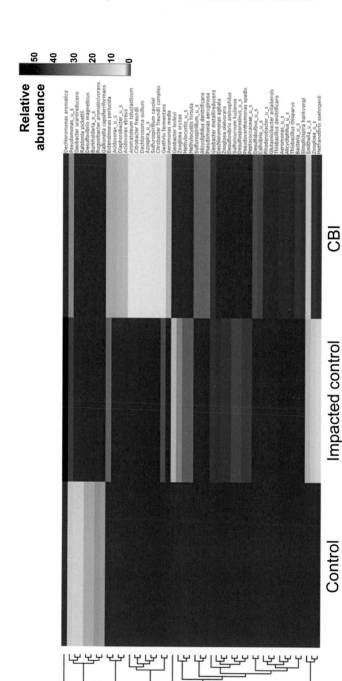

Fig. 16.8 Heat map presenting NGS microbial data at the species level by aligning taxa based on similarity. The three samples were from the same release site: the control sample was unimpacted by PHCs. The impacted control was impacted by PHCs but not treated with CBI. The CBI sample was drawn from an area in which CBI was emplaced. The colored scale indicated relative abundance taking into consideration genome size and number of reads

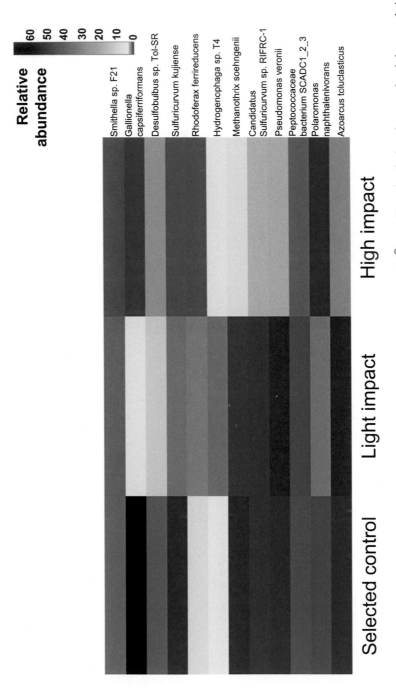

Fig. 16.9 NGS data displaying the most abundant twelve strains in samples within the CBI (BOS 200®) area. The colors in the columns and scale bar are relative intensity. The "control" sample was taken from the footprint of the injection but was intermittently impacted by petroleum. Due to site boundary constraints, an unimpacted control was not available. In the heavy impact sample, Smithella, Desulfobulbus, and Methanothrix are 48-, 36-, and 132-fold greater in absolute abundance than the low-impacted control

by NGS. Unambiguous control samples were not available. One side-gradient monitoring well, which had a history of intermittent PHC impacts, was selected as the best "control" available. Two consistently impacted monitoring wells were selected for comparison. All three monitoring wells were within the footprint of the in-situ injection of CBI. The samples were collected 2 years post-injection with a CBI containing nitrate and sulfate as electron acceptors. Added nitrates are short-lived being depleted in 1–3 months. The sulfate levels at the time of sampling ranged from 110 to 483 mg/L. The species that increased relative to the control monitoring well were catabolically and thermodynamically linked syntrophic communities. Figure 16.9 presents the most abundant species relative to the selected control. The colored scale indicated relative abundance taking into consideration genome size and number of reads. The most abundant microbes, Smithella, is a known alkane-degrading microbe (Ji 2020; Embree 2015). Under sulfate- and nitrate-reducing conditions, the primary condition at the site, Smithella degrades alkanes to acetate with the production of H_2 (Gieg 2014). *Desulfobulbus* sp. Tol-SR is distinguished by suites of genes (BSS, BBS, and BAM) associated with anaerobic toluene metabolism, which leads to acetate production (Laban 2015). Methanothrix utilizes acetate as a substrate for methanogenesis, while *Candidatus sulfuricurvum* derives energy from H_2 oxidation (Handley 2014). The connection between Smithella, Desulfobulbus, and H_2-utilizing methanogens is a typical hydrocarbon-degrading syntrophic relationship.

The difficulty with data presented in Fig. 16.9 is the lack of a control not impacted by PHCs. That does not prevent the observation that the microbes present were consistent with a petroleum degradation syntrophy. It does, however, make it difficult to determine how the CBI injection changed the population apart from the observation that the groundwater sample collected from the intermittently impacted monitoring well was less abundant than the samples drawn from the consistently impacted monitoring wells and that the microbial population present was capable of degrading PHCs.

The site data presented in Fig. 16.10 had appropriate control samples. The control samples were from areas that were not treated with CBI. Control 1 was drawn from an area that was unimpacted by PHC. Control 2 was drawn from an area impacted by PHC. Control 3 was drawn from an impacted area that bordered the CBI emplacement. It should be viewed as potentially influenced by the CBI. The three CBI emplacement groundwater samples were drawn from monitoring wells within the CBI treatment area. The samples were analyzed by NGS.

The microbial populations presented in Figs. 16.9 and 16.10 shared some characteristics. Smithella was the most consistently abundant microbe, but it was identified in all the impacted samples without regard to CBI emplacement status. Desulfobulbus was consistently present across the CBI emplaced samples. Methanothrix soehngenii was present but not confined to the CBI emplacement samples. *Candidatus sulfuricurvum* was not present. Two microbes absent from the Fig. 16.9 site were present at the Fig. 16.10 site. *Geobacter metallireducens* was present in the CBI samples and control 3, which was about 3 m from the CBI emplacement. *Geothrix fermentans* was present in the same pattern. *Geobacter metallireducens* uses electron donors such as acetate and uses Fe(III) as an electron acceptor (Coates 1999). *Geobacter*

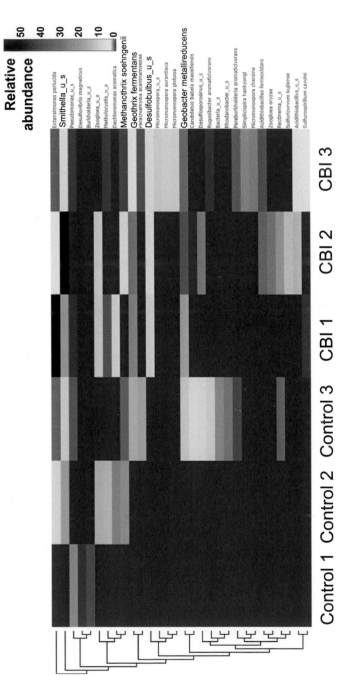

Fig. 16.10 NGS data from monitoring wells. The data is displayed at the species level. Multiple controls are present. Control 1 was drawn from an area that was unimpacted by PHC. Control 2 was drawn from an impacted area. Control 3 was drawn from an impacted area that bordered the CBI emplacement and should be viewed as potentially influenced by the CBI. The three CBI samples (CBI 1–3) were drawn from monitoring wells within the CBI emplacement. Smithella and Desulfobulbus were elevated relative to control. Methanothrix, however, did not have a consistent pattern. In this dataset, Geothrix and Geobacter were elevated

metallireducens couples the oxidation of aromatic compounds to the reduction of Fe(III)-oxides (Butler 2007), but acetate and ethanol are preferred over benzoate, although, benzoate is co-consumed with toluene and butyrate (Marozava 2014). These microbes are known to participate in direct interspecies electron transfer (DIET) using the AC. And the presence of these microbes may establish a second syntropic pathway to oxidize acetate and reduce CO_2 (Rotaru 2018). Both sites had a microbial population capable of degrading PHCs.

While describing the microbes present in the monitoring wells 2–3 years' post-injection and demonstrating that they were consistent with a hydrocarbon-degrading syntrophy is a good qualitative indicator of biodegradation in response to a CBI, the presence of genes encoding hydrocarbon enzymes in a microbial population can also be used to characterize biodegradation potential. Figure 16.11 presents qPCR data (QuantArray®-Petro from Microbial Insights, Inc.). The array quantifies a select suite of 25 functional genes involved in aerobic and anaerobic biodegradation of PHCs. Samples were collected from unimpacted monitoring wells and from impacted monitoring wells prior to injection with CBI (BOS 200®). Of the 25 functional genes, five of them were not present in any of the samples. Of the remaining 21, eight function genes were present in the unimpacted background, six additional ones were expressed in the impacted background post-injection, and post-inject seven additional function genes were observed. No identified function genes were lost by the first-month post-injection. Two previously identified functional genes were not identified at month five. At month five, 19 of the 21 quantified function genes were observed and on average the microbial count using the functional genes as markers increased 56-fold in cells/ml (range from 3.4 to 178.5-fold cells/ml). The average and range did not include three functional genes that increased from a low estimated value to 106-fold the estimated value. The gene array demonstrated that, for the period examined, the petroleum-degrading population expanded post-injection of CBI relative to both impacted and unimpacted controls. More information on qPCR can be found in Chap. 10.

Examining a subset of the data from Fig. 16.11, benzylsuccinate synthase (BSS in the QuantArray®) catalyzes the addition of fumarate to toluene (Biegert 1996) and potentially the monoaromatics xylene and ethylbenzene (Kharey 2020) to generate benzylsuccinate, the first step in anaerobic toluene metabolism (Biegert 1996). Benzylsuccinate becomes benzoyl-CoA and succinyl-CoA. Succinyl-CoA is recycled back to fumarate, while benzoyl-CoA proceeds through benzoyl-CoA reductase (BCR in the QuantArray®) to acetate, a small fatty acid easily used by many microbes. This is the toluene anaerobic pathway previously described above for site presented in Fig. 16.9. Thus, a selected subset of the gene expression from the Fig. 16.11 site provides a similar picture in Fig. 16.9 and has overlap to the site presented in Fig. 16.10. Other enzymes coding genes, associated with both aerobic and anaerobic pathways, are increased in number post-CBI emplacement, and the total microbial count in the impacted area also increases post-injection (Fig. 16.11) Notice that the total eubacteria, sulfate-reducing bacteria, iron-reducing bacteria, and denitrifying bacteria are increased more than tenfold. Thus, the release site presented in Fig. 16.11 confirms that CBI emplacement increases microbial abundance, increases microbes

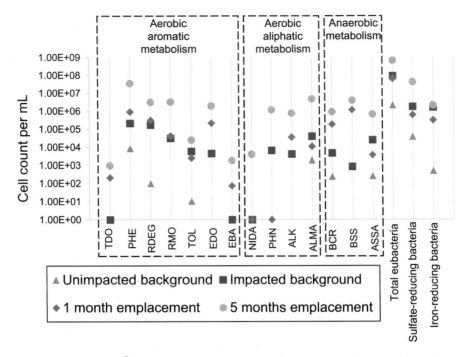

Fig. 16.11 QuantArray®-Petro data from a site where CBI was injected. Microbes having selected genes associated with both aerobic and anaerobic petroleum hydrocarbon metabolism increased in number relative to background. The increase was higher at 5 months post-emplacement than at 1 month

having genes known to be important in PHC degradation, and confirms the data conclusions drawn from Figs. 16.8, 16.9, and 16.10.

16.8.2 Additional Lines of Evidence of Microbial Activity

16.8.2.1 Decrease in Electron Acceptors and Increase in Gases

For CBI that provide electron acceptors, decreasing concentration in potential electron acceptors (such as oxygen, nitrate, and sulfate) is indicative of microbial activity. Figure 16.12 is from a petroleum release site. The decrease in electron acceptors followed an often-observed pattern. Contaminant concentration significantly decreased upon installation of CBI. Early post-emplacement, the system reaches equilibrium. When electron acceptors such as sulfate and nitrate were added with

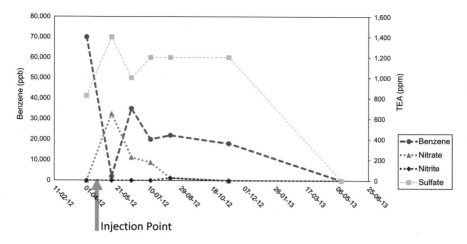

Fig. 16.12 Electron acceptors at a site where CBI was installed. Nitrate concentrations dropped first followed by a slight rise in nitrite. Sulfate concentrations persisted longer. Benzene concentrations were added to the graph for comparison

the AC, nitrate concentrations decreased quickly, leading to a slight, transient elevation of nitrite. Sulfate dropped slowly over time. The nitrate and sulfate consumption patterns were consistent with anaerobic metabolism using nitrate and sulfate as electron acceptors.

Increasing respiration products (such as CO_2, Fe(II), sulfide, and CH_4) is also consistent with microbial activity. Figure 16.13 demonstrates the rising concentration of methane (associated to anaerobic biodegradation processes as explained in Chap. 5) and other gases at a site where CBI was installed.

16.8.2.2 Decreasing Concentrations in Soil and Groundwater

The most often presented data to demonstrate decreases in contaminant concentrations are monitoring well data. Figure 16.5 depicted a typical groundwater concentration pattern after subsurface CBI emplacement. Figure 16.14 presents groundwater data for benzene and TVPH post in-situ emplacement of CBI. Contaminant concentrations in groundwater decrease due to PHCs adsorption onto the AC. The most water-soluble constituents can move into the AC quickly, but those same constituents may have lower heats of adsorption than constituents presenting a lower solubility, so there is a period over which a new equilibrium is established. The total amount of contamination relative to the AC determines the time it will take to reach equilibrium. From that point forward biodegradation controls the final phase of site remediation.

Core samples collected periodically at adjacent locations from sites treated with CBI typically show a decrease in contaminant concentration in soil, often exhibiting consistent patterns despite the high variability commonly associated to physical sampling of soil and aquifer materials. Figure 16.15a present ethylbenzene data

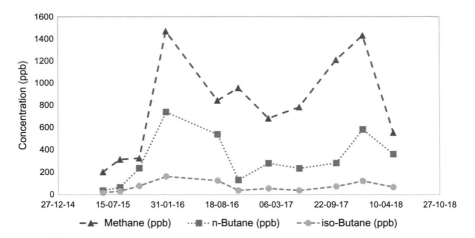

Fig. 16.13 Prior to injection, which was completed in June of 2015, gases, except for methane, were not detected above detection limits. Three months post-CBI emplacement, the concentration of gases linked to microbial metabolism like methane, *n*-Butane, and iso-Butane rose. Gases such as cis-2-Butene, propane, ethane, *t*-2-Butene, 1-Butene, and iso-Butylene (not displayed) exhibited similar patterns

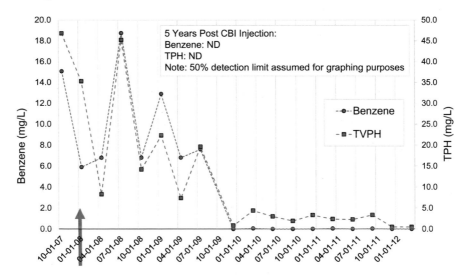

Fig. 16.14 The red arrow marks the injection of the CBI BOS 200®. The initial decrease in TVPH concentrations was followed by a re-equilibration phase. Biodegradation may have initiated early post-CBI emplacement but did not seem to dominate the PHC concentration pattern until the final stages of site remediation

a)

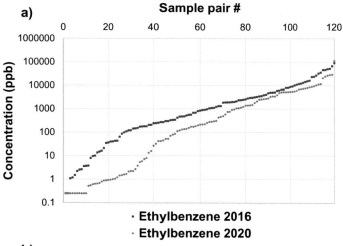

· **Ethylbenzene 2016**
· **Ethylbenzene 2020**

b)

Raw Statistics		
40 Months	2016	2020
Number of Valid Observations	120	120
Number of Distinct Observations	115	109
Minimum (Detection Limit)	0.25	0.25
Maximum	91100	112000
Mean	6285	3628
Median	866.5	215.5
Standard Deviation	14165	11431
Standard Error of Mean	1293	1044

Wilcoxon-Mann-Whitney (WMW) Test	
H0: Mean/Median of 2016 <= Mean/Median of 2020	
P-Value (Adjusted for Ties)	9.38E-04
Conclusion with Alpha = 0.05	
Reject H0, Conclude 2016 > 2020	

c)

Constituent	Median Difference	P-value
Benzene	3.6 fold lower	0.0009
Toluene	1.5 fold lower	0.0018
Ethylbenzene	4.0 fold lower	0.0009
m/p Xylene	2.2 fold lower	0.0091
o-Xylene	2.5 fold lower	0.019
1,2,4-TMB	1.2 fold lower	0.0289
Naphthalene	3.5 fold lower	0.0019

Fig. 16.15 Ethylbenzene concentration from soil samples was significantly lower in 2020 than in 2016. All other PHCs analyzed were also significantly lower in the 2020 soil samples than in the 2016 soil samples

from a BOS 200® in-situ injection into fat clay geology at an LNAPL site. The raw statistics for ethylbenzene shown in Fig. 16.15b are illustrative of the data for all other contaminants examined from the same site. All constituents examined were significantly decreased in concentration in the aquifer solids tested at 40 months (see Fig. 16.15c).

16.8.2.3   Compound-Specific Isotope Analysis (CSIA)

As discussed in Chap. 11, CSIA tracks the stable isotope composition of selected volatile organic compounds (VOCs) to obtain information on attenuation processes. In tracking the degradation of benzene due to biological degradation, benzene should become increasingly enriched for 13C isotopes as the benzene containing 12C isotopes preferentially has its carbon bond broken.

The international standard for the ratio of 13C–12C isotopes is the Pee Dee Belemite (PDB) standard and is 0.01123720 (IAEA 1993). So about 1% is 13C and 99% is 12C. To determine if biodegradation is indicated, one can measure the increase in 13C above the PDB standard. It has become convention to present the change (δ) in 13C, (δ13C) as the isotopic ratio of the sample (Rs) divided by the isotopic ratio of the standard (Rstd) (Meckenstock 1999; Hayes 2004). Next, because the fraction is small the value is multiplied by 1000 to shift the decimal place and present the data as part per thousand often reported as permill (‰). Thus, for isotopic data presented as the δ13C, as δ13C moves closer to zero the value becomes less negative because the molecules are enriched for δ13C. See Eq. 16.4 from (Mackenstock 2004).

$$\delta 13C(‰) = (Rs/Rstd - 1) \times 1000$$

$$(Rs/Rstd) = [(13C/12C)s - (13C/12C)std]/(13C/12C)std \qquad (16.5)$$

In Fig. 16.16, δ13C enrichment data are presented form a CBI injected site. To determine if benzene sorbed to the formation was being degraded, impacted but untreated core samples were collected in a benzene impacted area upgradient of the CBI emplacement and within the CBI emplacement area. These samples were examined for isotopic carbon enrichment under the assumption that the upgradient and downgradient samples were immediately connected by a shared flow path. The isotopic fractionation from the extracted mass indicated enrichment for δ13C in the downgradient, within the injection area, smear zone samples. The difference between three of the four points exceeds δ13C $\geq$ 2‰. The data were indicative of biodegradation.

To compare δ13C enrichment prior to interaction with the CBI emplacement versus the isotopic enrichment post interaction with the CBI emplacement, a stable isotope enrichment factor, epsilon (ε), may be calculated (see Eq. 16.7). The calculated epsilon for the data presented in Fig. 16.16 is (ε) = -3.5. This value is consistent with the stable isotope enrichment factors reported for a sulfate-reducing culture,

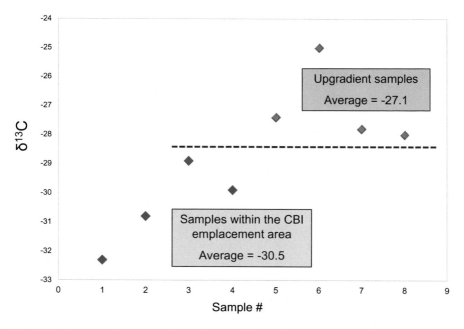

Fig. 16.16 Core samples were collected from a CBI emplacement on a petroleum release site. Points below the red line were upgradient of the CBI emplacement. Points above the red line were samples collected within the CBI emplacement. The data are indicative of biodegradation

which ranges from -2.7 to -3.6 (Mancini 2003). The CBI used, BOS 200®, contains significant sulfate; thus, the ε value is consistent with expectations and indicated biodegradation.

$$\varepsilon = \left\{ \left[\frac{(1000 + \delta13Cprior)}{1000 + \delta13Cpost} \right] - 1 \right\} \times 1000 \qquad (16.7)$$

To be thorough, neither abiotic nor non-degradative processes are known to result in significant stable isotopic fractionation of benzene relative to biological processes. Non-degradative processes such as sorption, volatilization, and dissolution are typically smaller than the analytical uncertainty associated with CSIA, which has been reported as ± 0.5‰ (Hunkeler 2008). Specifically, the sorption of benzene to activated carbon does not result in stable isotope fraction greater than this ± 0.5‰ accuracy (Slater 2000; Schüth 2003). Abiotic degradation processes associated with benzene can be similarly disregarded. Plotting the carbon stable isotope concentration δ13C against the natural logarithm of the concentration can be used to support that a single process is the main control behind the change in contaminant concentration (Hunkeler 2008).

Data from another site where CSIA was conducted are presented in Fig. 16.17, which depicts a sustained reduction of BTEX contamination downgradient of a CBI PRB despite BTEX continuing to enter the barrier from the upgradient side.

The CBI PBR was installed because an existing groundwater extraction and re-injection system was failing to perform. CSIA was conducted to determine whether the decrease in concentration was due only to adsorption to the AC or if the decrease was also affected by biodegradation. Groundwater samples were collected from monitoring wells located on a transect through the CBI PBR. The monitoring well upgradient of the CBI PBR was impacted by the same PHCs as the monitoring wells downgradient of the CBI PBR. The groundwater extraction and re-injection system remained active after the CBI PBR was installed. It withdrew groundwater from a series of wells located ~55 m downgradient from the control monitoring well and re-injects oxygenated water at ~18 m upgradient from the same control well.

Figure 16.18 shows $\delta 13C$ enrichment data. Of the five VOCs examined by CSIA, three (*n*-propylbenzene, ethylbenzene, m,p-xylene) were enriched, greater than analytical uncertainty, relative to the upgradient monitoring well. O-xylene did not demonstrate enrichment, and 1,2,4-trimethylbenzene enrichment was within laboratory error (data not shown). Determining the cause of the observed isotopic enrichment was complicated by the presence of the extraction and re-injection wells because re-injection may enhance degradation by oxygenating the re-injected groundwater. Nevertheless, enrichment is indicated.

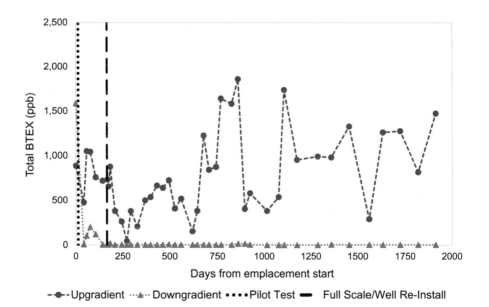

Fig. 16.17 Groundwater sampling data from upgradient and downgradient of emplaced PRB. BTEX on the downgradient side of the barrier remains suppressed despite BTEX continuing to enter the barrier from the upgradient side

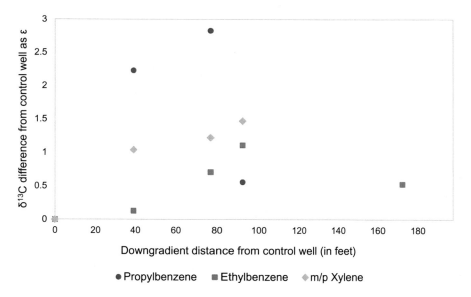

Fig. 16.18 Samples of groundwater collected upgradient of a PRB and at multiple distances down-gradient. *N*-propylbenzene, ethylbenzene, and *m, p*-xylene showed 13C enrichment. The results indicated the occurrence of biodegradation by having an enrichment of δ13C ≥ 1‰

16.9   Biological Regeneration

As a general process, biological regeneration (BR) of AC is well documented (Aktaş 2007; El Gamal 2018). Adsorption, biofilm formation, and biodegradation can occur simultaneously on AC (Piai 2020; Betsholtz 2021), with the combination of microorganisms and the adsorptive capacity of AC being synergistic in promoting BR (Nath 2011a, b). Upon emplacement in the subsurface, AC in CBIs adsorbs hydrophobic chemicals such as PHCs and natural organics. Microbes in the resource-rich regions near the AC bloom. As the biomass on the AC expands, desorption (Leglize 2008; Ying 2015) and regeneration rates increase (Betsholtz 2021) such that the adsorptive capacity of AC is extended (Chan 2018; Piai 2022). In time, some of the products of degradation as well as natural organics fill a percentage of the AC micropores, but they have not been demonstrated to either completely fill the micropores nor cause biodegradation to cease (Smolin 2020; Díaz de León 2021). This general scenario has been developed from in vitro experiments and the examination AC used in water treatment (FRTR 2021; El Gamal 2018). The development of microbial growth on AC is referred to as biologically activated carbon (BAC). The development of BAC is well documented and has been presumed to specifically occur on AC in CBIs emplaced in the subsurface (Regenesis 2022).

The specific source material (coal, coconut, etc.), activation method (Coelho 2006), and particle size (GAC, PAC, colloidal) (El Gamal 2018) all influence BR. For example, coconut carbon does not bioregenerate as well as wood-based (Piai 2019) or coal-based AC (Zhang 2013). The activation method, e.g., chemical versus physical, influences the AC's surface area, pore type distribution, and adsorptive capacity, which all affect BR. As concerning particle size, BR has been demonstrated for GAC and PAC (El Gamal 2018). Generally, increasing particle size lowers substrate diffusion within the biofilm while increasing biodegradation (Liang 2007; Rattier 2012; El Gamal 2018).

AC porosity is likely the most important characteristic influencing both the rate and extent of AC BR (El Gamal 2018; Lu 2020; Aktaş 2007). Larger pores generally support BR (Aktaş 2007; Díaz de León 2021) and may provide microbes, enzymes, and surfactants with more access to the micropores or better habitat for microbes. The microbial community that develops on AC is influenced by the pore structure and the compounds available for biological degradation (Yu 2021).

Individual chemicals vary in how they affect the degree and rate of BR (Nath 2011a, b; Chan 2018). The biodegradability of a specific contaminant varies due to molecular structure among other factors (Cirja 2008). In general, large molecular weight compounds having low heats of adsorption are biodegraded easier than small molecules having high heats of adsorption (García-Delgado 2019).

The influence of multiple factors makes it difficult to estimate the degree to which AC can be recycled. The percent of dissolved organic carbon (DOC) degraded by biological activity varies from 25 to 42% (Fundneider 2021). Bioregeneration ranges from 12 to 95% (Nath 2011a, b). The percentages of pore restoration reported in the literature are broad (Aktaş 2007; Nath 2011a, b). A model system of biofilm on AC degrading 2-nitrophenol was run for 38 months after which it was reported that 39% of the initial pore volume was still available (Smolin 2020). Thus, AC is not typically fully restored by microbial regeneration. This may be due to binding products of microbial metabolism and non-organic carbon (Smolin 2020; Díaz de León 2021). Nevertheless, the apparent adsorption capacity of microbial biofilm on AC exceeds that of the AC alone.

16.9.1 Process of Bioregeneration: Adsorption–Desorption, Exoenzymes, and Surfactants

It has been argued that biodegradation occurs in the dissolved phase external to the AC after which the PHCs or other chemicals on the AC re-equilibrate with the dissolved phase (de Jonge 1996; Abromaitis 2016; Díaz de León 2021). However, not all chemicals that bind to AC are released with equal ease (Nath 2011a, b) or released as rapidly as BR has been reported to occur (Piai 2021). Therefore, this process is insufficient to explain BR of AC.

Conceivably, multiple processes are involved in BR, including adsorption and desorption and the participation of exoenzymes (Perrotti 1974) and surfactants (Marchal 2013; Edwards 2013). Biofilm producing microbes have been demonstrated to degrade phenanthrene adsorbed to AC, while non-biofilm forming microbes failed to do so (Leglize 2008). Microbes create a scaffolding made of EPS, which adheres to surfaces and allows microbes to build a biofilm habitat that supports microbial cooperation. One aspect of the formation of the biofilm habitat is the ability for the microbes to release extracellular enzymes (exoenzymes) into the biofilm forming a kind of extracellular digestive system (Flemming 2010).

The exoenzyme hypothesis proposes that enzymes secreted by the bacteria diffuse into the micropores and catalyze a transformation that desorbs the adsorbate from the micropores so that it moves out of the AC (Perrotti 1974; Nath 2011a, b). The argument against the exoenzyme theory is that enzymes are too large to enter the micropores (Xiaojian 1991), which is generally correct since most enzymes range from 3 to 7 nm in diameter if viewed as a sphere (Erickson 2009). However, it may not be necessary for the exoenzymes to enter the micropores. Focusing on the aqueous phase versus the adsorbed phase may overlook biocenoses within the biofilm. If the concentration of PHCs is reduced in spaces within the biofilm, PHCs may be released from the micropores (Klimenko 2003a, b) and be degraded within the extracellular space. The release of PHCs from AC may be enhanced by the presence of biosurfactants, which are also common in biofilms and have been demonstrated to facilitate BR (Klimenko 2003a, b).

16.9.2 Diet

It has been proposed that PHCs in the micropores of AC are still available for biodegradation through DIET (Liu 2012; Summers 2010). Under methanogenic conditions, the biodegradation of PHCs requires the syntrophic cooperation between bacteria and archaea (Gieg 2014). Syntropic bacteria oxidize PHCs to CO_2, H_2, and acetate (Kim 2019). The methanogenic archaea remove the hydrogen, which helps the fermenters because the fermentation reaction is only thermodynamically favorable at low hydrogen ion concentrations (Barton 2005; Dolfing 2008). Microbial interactions whereby electrons are exchanged through intermediates such as hydrogen and acetate are well described (Kim 2019). A substantial body of evidence, however, supports that electron exchange can occur without chemical intermediates, also in the case of bacteria and methane-producing archaea (Summers 2010; Chen 2022), also in the case of bacteria and methane-producing archaea (Lovley 2017; Yee 2019). As a result of this process, bacteria may not need to be in direct contact with PHCs to be able to degrade them.

16.9.3 Conclusions and Future Research Needs

While the general efficacy of AC and the derived CBIs has been demonstrated in field applications and laboratory studies, further research is needed. Failure to meet remediation goals occasionally occurs. Most of those failures are due to insufficient site characterization and inadequate distribution of the CBI within the hydrocarbon mass. Therefore, technological improvements are needed to optimize remedial design characterization and CBI installation. Further research is also encouraged to better understand the adsorption–desorption mechanisms of different AC types for in-situ remediation of PHC fuel mixtures.

On the biological side, how the syntrophic relationships form, which individuals become most numerous in the syntrophy, and how the relationships evolve as degradation proceeds are appreciated, but more needs to be done to understand these processes in CBI treatment zones as well as questions like whether co-injection of bacterial consortia significantly improves performance and how indigenous microbes interact under bioaugmentation. Further metagenomic and CSIA studies across hydrocarbon types and geological settings could be enlightening. The cooccurrence of microorganisms has helped unwind some associations, but employing transcriptomics and proteomics will move us beyond descriptions or cooccurrences.

High contaminant concentrations and NAPLs are difficult to bioremediate quickly. The capacity of AC to adsorb contaminants thereby lowering dissolved phase contamination can decrease the negative impacts to the soil biota by hydrocarbon contamination. Adsorption onto AC can reduce the bioavailability and, thus, the toxicity of hydrocarbons to subsurface microbes (Meynet 2012). While the adsorptive capacity is helpful and biodegradation proceeds despite the presence of NAPL, the remediation market craves faster degradation rates. Therefore, optimization of both products and installation practices to increase the rate of NAPL degradation is a pertinent area of research.

On the other hand, many remediation technologies that are initially successful with fuel hydrocarbons suffer from mediocre performance when contaminant concentration becomes low. For example, MTBE is very soluble in water. Consequently, it moves readily with groundwater and often creates expansive low concentration plumes. Biodegradation will stall as the concentration of the MTBE drops too low to sustain a robust population of microbial degraders (USEPA 1999; Muller 2007; Li 2014). AC overcomes this limitation by concentrating the oxygenate in its pores. Optimizing CBI remedies to address very low-concentration plumes is an additional area of research.

AC is versatile and will function under aerobic or anaerobic conditions. Mixers used to prepare CBI slurries typically entrain air and oxygenate the water. AC readily adsorbs dissolved oxygen, so the state within the pore structure is aerobic at the time of injection. This feature suggests that injection of CBI at sites where sparging systems are installed may significantly enhance remedial performance. Typical sparging systems are inefficient as most of the oxygen added is lost in the vapor leaving the saturated formation. The addition of AC may enhance sparging efficiency because of its ability to adsorb oxygen. This illustrates the possibility of many different applications of AC in subsurface remediation that are yet to be considered or investigated.

References

Abromaitis VR (2016) Biodegradation of persistent organics can overcome adsorption–desorption hysteresis in biological activated carbon systems. Chemosphere 149:183–189. https://doi.org/10.1016/j.chemosphere.2016.01.085

Aktaş OA (2007) Adsorption, desorption and bioregeneration in the treatment of 2-chlorophenol with activated carbon. J Hazard Mater 59(4):769–777. https://doi.org/10.1016/j.ibiod.2007.01.003

Barton L (2005) Structural and functional relationships in prokaryotes. Springer, New York

Becker JM (2006) Bacterial activity, community structure, and centimeter-scale spatial heterogeneity in contaminated soil. Microb Ecol 51:220–231. https://doi.org/10.1007/s00248-005-0002-9

Betsholtz AK (2021) Tracking 14C-labeled organic micropollutants to differentiate between adsorption and degradation in GAC and biofilm processes. Environ Sci Technol 55(15):11318–11327. https://doi.org/10.1021/acs.est.1c02728

Biegert TFG (1996) Evidence that anaerobic oxidation of toluene in the denitrifying bacterium *Thauera aromatica* is initiated by formation of benzylsuccinate from toluene and fumarate. J Biochem 238(3):661–668. https://doi.org/10.1111/j.1432-1033.1996.0661w.x

Board TW (1989) Final report hazardous waste management approaches to protect water quality. Texas Water Developmnet Board, Austin. https://www.twdb.texas.gov/publications/reports/contracted_reports/doc/8483541.pdf

Bonaglia SB (2020) Activated carbon stimulates microbial diversity and PAH biodegradation under anaerobic conditions in oil-polluted sediments. Chemoshpere 248:126023. https://doi.org/10.1016/j.chemosphere.2020.126023

Bradner GM (2005) Effects of skin and hydraulic fractures on the performance of an SVE well. J Contam Hydrol 77:271–297. https://doi.org/10.1016/j.jconhyd.2005.02.001

Bures G (2004) Assessment of chitin distribution and fracture propagation during bio-fracingTM. Remediation of chlorinated and recalcitrant compounds—2004. In: Proceedings of the fourth international conference on remediation of chlorinated and recalcitrant compounds. Battelle, Monterey, CA. https://projects.battelle.org/chlorinated-conference/2004Chlor_Proceedings/Papers/3D-22.pdf

Butler JH (2007) Genomic and microarray analysis of aromatics degradation in *G. metallireducens* and comparison to a Geobacter isolate from a contaminated field site. BMC Genom 8:180. https://doi.org/10.1186/1471-2164-8-180

Chan PL (2018) Bioregeneration of granular activated carbon loaded with phenolic compounds: effects of biological and physico-chemical factors. Int J Environ Sci 15(8):1699–1712. https://doi.org/10.1007/s13762-017-1527-4

Chen BY (2012) Enhanced bioremediation of PAH-contaminated soil by immobilized bacteria with plant residue and biochar as carriers. J Soils Sedim 14:1350–1359. https://doi.org/10.1007/s11368-012-0554-5

Chen LF (2022) Improvement of direct interspecies electron transfer via adding conductive materials in anaerobic digestion: mechanisms, performances, and challenges. Front Microbiol 13:1051. https://doi.org/10.3389/fmicb.2022.860749

Cheremisinoff PN (1978) Carbon adsorption handbook. Ann Arbor Science Publishers, Ann Arbor

Cho Y-MW-J (2012) Long-term monitoring and modeling of the mass transfer of polychlorinated biphenyls in sediment following pilot-scale in-situ amendment with activated carbon. J Contam Hydrol 129:25–37. https://doi.org/10.1016/j.jconhyd.2011.09.009

Christiansen CM (2010) Comparison of delivery methods for enhanced in situ remediation in clay till. Groundwater Monit Remed 30(4):107–122. https://doi.org/10.1111/j1745-6592.2010.01314.x

Cirja MI (2008) Factors affecting the removal of organic micropollutants from wastewater in conventional treatment plants (CTP) and membrane bioreactors (MBR). Rev Environ Sci Biotechnol 7:61–78. https://doi.org/10.1007/s11157-007-9121-8

CLU-IN U (2022) clu-in.org. Retrieved from Contaminated site clean-up information. https://clu-in.org/techfocus/default.focus/sec/Environmental_Fracturing/cat/Overview/

Coates JE (1999) *Geothrix fermentans* gen. nov., sp. Nov., a novel Fe(III)-reducing bacterium from a hydrocarbon-contaminated aquifer. Int J Syst Bacteriol 49(4):1615–1622. https://doi.org/10.1099/00207713-49-4-1615

Coelho CO (2006) The influence of activated carbon surface properties on the adsorption of the herbicide molinate and the bio-regeneration of the adsorbent. J Hazard Mater B138:343–349. https://doi.org/10.1016/j.jhazmat.2006.05.062

Danczak RE (2018) Microbial community cohesion mediates community. Msystems 3(4):e00066-e118. https://doi.org/10.1128/mSystems.00066-18

Das NB (2017) Application of biofilms on remediation of pollutants: an overview. J Microbiol Biotechnol Res 14:783–790

de Jonge RB (1996) Bioregeneration of powdered activated carbon (PAC) loaded with aromatic compounds. Water Res 30:875–882. https://doi.org/10.1016/0043-1354(95)00247-2

Díaz de León G (2021) Impact of anaerobic biofilm formation on sorption characteristics of powdered activated carbon. University of Waterloo, Waterloo. https://uwspace.uwaterloo.ca/bitstream/handle/10012/17164/RochaDiazdeLeon_GriseldaRaquel.pdf?sequence=3

Dolfing JL (2008) Thermodynamic constraints on methanogenic crude oil biodegradation. ISME J 2:442–452

Edwards SJ (2013) Applications of biofilms in bioremediation and biotransformation of persistent organic pollutants, pharmaceuticals/personal care products, and heavy metals. Appl Microbiol Biotechnol 2:9909–9921. https://doi.org/10.1007/s00253-013-5216-z

El Gamal MM-N (2018) Bio-regeneration of activated carbon: a comprehensive review. Separ Purif Technol 197:345–359. https://doi.org/10.1016/j.seppur.2018.01.015

Embree EL-B (2015) Networks of energetic and metabolic interactions. PNAS 112(50):15450–15455. https://doi.org/10.1073/pnas.1506034112

Erickson H (2009) Size and shape of protein molecules at the nanometer level determined by sedimentation, gel filtration, and electron microscopy. Biol Proced Online 11(32):32–51. https://doi.org/10.1007/s12575-009-9008-x

Fan DG (2017) Current state of in situ subsurface remediation by activated carbon-based amendments. J Environ Manag 204:798–803. https://doi.org/10.1016/j.jenvman.2017.02.014

Flemming HW (2010) The biofilm matrix. Nat Rev Microbiol 8(9):623–633. https://doi.org/10.1038/nrmicro2415

Fredrickson JF (2001) Subsurface microbiology and biochemistry. Wiley-Liss Inc., New York

Fundneider TA (2021) Implications of biological activated carbon filters for micropollutant removal in wastewater treatment. Water Res 189:116588. https://doi.org/10.1016/j.watres.2020.116588

García-Delgado CF-S-J (2019) Co-application of activated carbon and compost to contaminated soils: toxic elements mobility and PAH degradation and availability. Int J Environ Sci Technol 16(2):1057–1068. https://doi.org/10.1007/s13762-018-1751-6

Geyer RP (2005) In situ assessment of biodegradation potential using biotraps amended with 13C-labeled benzene or toluene. Environ Sci Technol 39:4983–4989

Gieg LF-C (2014) Syntrophic biodegradation of hydrocarbon contaminants. Curr Opin Biotechnol 27:21–29. https://doi.org/10.1016/j.copbio.2013.09.002

Gonod LV (2006) Spatial distribution of microbial 2,4-dichlorophenoxy acetic acid mineralization fromfield to microhabitat scales. Soil Sci Soc Am J 70:64–71. https://doi.org/10.2136/sssaj2004.0034

Graham EC (2017) Deterministic influences exceed dispersal effects on hydrologically-connected microbiomes. Environ Microbiol 19(4):1552–1567. https://doi.org/10.1111/1462-2920.13720

Greenbank MA (1993) Effects of starting material on activated carbon characteristics and performance. In: Paper presented at WATERTECH Expo '93. Houston, Texas. https://p2infohouse.org/ref/33/32785.pdf

Handley KM (2014) The complete genome sequence for putative H_2- and S-oxidizer *C. sulfuricurvum* sp., assembled de novo from an aquifer-derived metagenome. Environ Microbiol 147:3443–34462

Harris PJF, Liu Z, Suenaga K (2008) Imaging the atomic structure of activated carbon. J Phys Condens Matter 20(36):362201. https://doi.org/10.1088/0953-8984/20/36/362201

Haupt SJ (2019) Migration of chlorinated solvent groundwater plumes with colloidal activated carbon. University of Rhode Island, Kingston. https://digitalcommons.uri.edu/cgi/viewcontent.cgi?article=2481&context=theses

Hayes J (2004) An introduction to isotopic calculations. https://www.whoi.edu/cms/files/jhayes/2005/9/IsoCalcs30Sept04_5183.pdf. Accessed 13 April 2022

Hunkeler DM (2008) A guide for assessing biodegradation and source identification of organic ground water contaminants using compound specific isotope analysis (CSIA). USEPA Office of Research and Development, Ada, Oklahoma. https://cfpub.epa.gov/si/si_public_record_report.cfm?Lab=NRMRL&dirEntryId=202171

IAEA (1993) Reference and intercomparison materials for stable isotopes of light elements. International Atomic Energy Agency, Vienna. https://www-pub.iaea.org/MTCD/publications/PDF/te_825_prn.pdf

ITRC (2020) Optimizing injection strategies and in situ remediation performance. https://ois-isrp-1.itrcweb.org. Accessed 12 Feb 2021

Ji JZ (2020) Methanogenic biodegradation of C9 to C12n-alkanes initiated by Smithella via fumarate addition mechanism. AMB Expr 10(23):956. https://doi.org/10.1186/s13568-020-0956-5

Kharey GSG (2020) Combined use of diagnostic fumarate addition metabolites and genes provides evidence for anaerobic hydrocarbon biodegradation in contaminated groundwater. Microorganisms 8(10):1532. https://doi.org/10.3390/microorganisms8101532

Kim BG (2019) Prokaryotic metabolism and physiology. Cambridge University Press, Cambridge

Kindzierski WG (1992) Activated carbon and synthetic resins as support material for methanogenic phenoldegrading consortia: comparison of surface characteristics and initial colonization. Water Environ Res 64(6):766–775

Kjellerup BA (2013) Application of biofilm covered activated carbon particles as a microbial inoculum delivery system for enhanced bioaugmentation of PCBs in contaminated sediment. In: SERDP Project ER-2135. https://clu-in.org/download/contaminantfocus/pcb/PCBs-Seds-ER-2135-FR-PI.pdf

Klimenko NG (2003a) Bioregeneration of activated carbons by bacterial degraders after adsorption of surfactants from aqueous solutions. Colloids Surf A Physicochem Eng Aspects 230(103):141–158. https://doi.org/10.1016/j.colsurfa.2003.09.021

Klimenko NS (2003b) Bioregeneration of activated carbons by bacterial degraders after adsorption of surfactants from aqueous solutions. Colloids Surf 230(1–3):141–158. https://doi.org/10.1016/j.colsurfa.2003.09.021

Köhler AH (2006) Organic pollutant removal versus toxicity reduction in industrial wastewater treatment: the example of wastewater from fluorescent whitening agent production. Environ Sci Technol 40:3395–3401. https://doi.org/10.1021/es060555f

Kolasinski KW (2019) Surface science: foundations of catalysis and nanoscience, 4th edn. Wiley, Hoboken. https://www.perlego.com/book/1284212/surface-science-pdf

Krüger US (2019) Bacterial dispersers along preferential flow paths of a clay till depth profile. Appl Environ Microbiol 85(6):e02658-e2718. https://doi.org/10.1128/AEM.02658-18

Kupryianchyk DR (2013) In situ treatment with activated carbon reduces bioaccumulation in aquatic food chains. Environ Sci Technol 11:4563–4571. https://doi.org/10.1021/es305265x

Laban NA-F (2015) Draft genome sequence of uncultivated toluene-degrading desulfobulbaceae bacterium Tol-SR, obtained by stable isotope probing using [13C6]toluene. Microbiol Resour Announc 3(1):1–2. https://doi.org/10.1128/genomeA.01423-14

Leglize PA (2008) Adsorption of phenanthrene on activated carbon increases mineralization rate by specific bacteria. J Hazard Mater 151(2–3):339–347. https://doi.org/10.1016/j.jhazmat.2007.05.089

Lhotský OK (2021) The effects of hydraulic/pneumatic fracturing-enhanced remediation (FRAC-IN) at a site contaminated by chlorinated ethenes: a case study. J Hazard Mater 417:125883. https://doi.org/10.1016/j.jhazmat.2021.125883

Li AY (1983) Availability of sorbed substrate for microbial degradation on granular activated carbon. J Water Pollut Control Feder 12:392–399

Li LL (2011) Surface modification of coconut shell based activated carbon for the improvement of hydrophobic VOC removal. J Hazard Mater 192(2):683–690. https://doi.org/10.1016/j.jhazmat.2011.05.069

Liang CC (2007) Modeling the behaviors of adsorption and biodegradation in biological activated carbon filters. Water Resour 41:3241–3250

Liang YZ (2008) Porous biocarrier-enhanced biodegradation of crude oil contaminated soil. Int Biodeter Biodegrad 63:80–87. https://doi.org/10.1016/j.ibiod.2008.07.005

Lillo-Ródenas MC-A-S (2005) Behaviour of activated carbons with different pore size distributions and surface oxygen groups for benzene and toluene adsorption at low concentrations. Carbon 43(8):1758–1776. https://doi.org/10.1016/j.carbon.2005.02.023

Liu FR-E (2012) Promoting direct interspecies electron transfer with activated carbon. Energy Environ Sci 47:8982. https://doi.org/10.1039/C2EE22459C

Lovley D (2017) Syntrophy goes electric: direct interspecies electron transfer. Annu Rev Microbiol 117:643–664. https://doi.org/10.1146/annurev-micro-030117-020420

Lu ZS (2020) Effect of granular activated carbon pore-size distribution on biological activated carbon filter performance. Water Res 177(15):115768. https://doi.org/10.1016/j.watres.2020.115768

Luckner LS (1991) Migration process in soil and groundwater zone. Bundesrepublik Deutschland, Leipzig

Ma XZ (2019) Adsorption of volatile organic compounds at medium-high temperature conditions by activated carbons. Energy Fuels 34(3):3679–3690. https://doi.org/10.1021/acs.energyfuels.9b03292

Maamar SA-C-A (2015) Groundwater isolation governs chemistry and microbial community structure along hydrologic flowpaths. Front Microbiol 6:13. https://doi.org/10.3389/fmicb.2015.01457

Mackenstock RM (2004) Stable isotope fractionation analysis as a tool to monitor biodegradation in contaminated acquifers. Contam Hydrol 75:215–255. https://doi.org/10.1016/j.jconhyd.2004.06.003

Mancini SA-C (2003) Carbon and hydrogen isotopic fractionation during anaerobic biodegradation of benzene. Appl Environ Microbiol 69(1):191–198. https://doi.org/10.1128/AEM.69.1.191-198.2003

Mangrich AS (2015) Improving the water holding capacity of soils of northeast Brazil by biochar augmentation. In: Satinder Ahuja JB (ed) Water challenges and solutions on a global scale (ACS symposium series), 1st edn. American Chemical Society, New York, pp 339–354

Mangse GW (2020) Carbon mass balance model to investigate biochar and activated carbon amendment effects on the biodegradation of stable-isotope labeled toluene in gravelly sand. Environ Adv 2:100016. https://doi.org/10.1016/j.envadv.2020.100016

Marchal GS (2013) Impact of activated carbon, biochar and compost on the desorption and mineralization of phenanthrene in soil. Environ Pollut 181:200–210. https://doi.org/10.1016/j.envpol.2013.06.026

Marozava SR (2014) Physiology of G. metallireducens under excess and limitation of electron donors. Part I. Batch cultivation with excess of carbon sources. Syst Appl Microbiol 37(4):277–286. https://doi.org/10.1016/j.syapm.2014.02.004

Massol-Deyá AA (1995) Channel structures in aerobic biofilms of fixed-film reactors treating contaminated groundwater. Appl Environ Microbiol 6(12):769–777. https://doi.org/10.1128/aem.61.2.769-777.1995

McCormick (2010) McCormick and baxter superfund site stockton: study report. Study Report Task Order No. 0018

McGregor R (2020) Distribution of colloidal and powdered activated carbon for the in situ treatment of groundwater. J Water Resour Protect 12(12):1001–1018. https://doi.org/10.4236/jwarp.2020. 1212060

McNamara JF (2018) Comparison of activated carbons for removal of perfluorinated compounds from drinking water. JAWWA 110:E2–E14. https://doi.org/10.5942/jawwa.2018.110.0003

Meckenstock RU (1999) 13C/12C isotope fractionation of aromatic hydrocarbons during microbial degradation. Environ Microbiol 1(5):409–414. https://doi.org/10.1046/j.1462-2920.1999.000 50.x

Menendez Diaz JA-G (2006) Types of carbon adsorbents and their production. Elsevier, Amsterdam

Meynet PH (2012) Effect of activated carbon amendment on bacterial community structure and functions in a PAH impacted urban soil. Environ Sci Technol 46:5057–5066. https://doi.org/10. 1021/es2043905

Morsen ARH (1987) Degradation of phenol by mixed culture of *Pseudomonas putida* and *Crytococcus elinovii* adsorbed on activated carbon. Appl Microbial Biotechnol 26:283–288. https:// doi.org/10.1007/BF00286325

Morsen AR (1990) Degradation of phenol by a defined mixed culture immobilized by adsorption on activated carbon and sintered glass. Appl Microbial Biotechnol 33:206–212. https://doi.org/ 10.1007/BF00176526

Murdoch L (1995) Forms of hydraulic fractures created during a field test in fine-grained glacial drift. Q J Eng Geol 28:23–35. https://doi.org/10.1144/GSL.QJEGH.1995.028.P1.03

Nath KB (2011a) Bioregeneration of spent activated carbon: effect of physico-chemcial parameters. J Sci Ind Res 70:487–492

Nath KB (2011b) Microbial regeneration of spent activated carbon dispersed with organic contaminants: mechanism, efficiency, and kinetic models. Environ Sci Pollut Res Int 18(4):534–546. https://doi.org/10.1007/s11356-010-0426-8

Noland S (2002) United States Patent No. 7,787,034 B2

Or DS (2007) Physical constraints affecting bacterial habitats and activity in unsaturated porous media: a review. Adv Water Resour 30:1505–1527. https://doi.org/10.1016/j.advwatres.2006. 05.025

Pagnozzi G (2020) Evaluating the influence of capping materials on composition and biodegradation activity of benthic microbial communities: implications for designing bioreactive sediment caps. Texas Tech University, Civil Engineering, Texas Tech, Lubbock

Pagnozzi GC (2021) Powdered activated carbon (PAC) amendment enhances naphthalene biodegradation under strictly sulfate-reducing conditions. Environ Pollut 268:115641. https://doi.org/10. 1016/j.envpol.2020.115641

Peacock AD-J (2004) Utilization of microbial biofilms as monitors of bioremediation. Microb Ecol 47:284–292

Perrotti AR (1974) Factors involved with biological regeneration of activated carbon. Am Inst Chem Eng Symp Ser 144:316–325

Piai LD (2019) Diffusion of hydrophilic organic micropollutants in granular activated. Water Res 162:518–527. https://doi.org/10.1016/j.watres.2019.06.012

Piai LB (2020) Biodegradation and adsorption of micropollutants by biological activated carbon from a drinking water production plant. J Hazard Mater 388:122028. https://doi.org/10.1016/j. jhazmat.2020.122028

Piai LV (2021) Melamine degradation to bioregenerate granular activated carbon. J Hazard Mater 414:125503. https://doi.org/10.1016/j.jhazmat.2021.125503

Piai LL (2022) Prolonged lifetime of biological activated carbon filters through enhanced biodegradation of melamine. J Hazard Mater 422:126840. https://doi.org/10.1016/j.jhazmat.2021. 126840

Ran JS (2018) Remediation of polychlorinated biphenyls (PCBs) in contaminated soils and sediment: state of knowledge and perspectives. Front Environ Sci 14:79. https://doi.org/10.3389/ fenvs.2018.00079

Rattier MR (2012) Organic micropollutant removal by biological activated carbon filtration: a review. Urban Water Security Research Alliance, Queensland

Regenesis (2022) regenesis.com: https://regenesis.com/wp-content/uploads/2019/02/brochure_4.1.pdf

Rhodes AH (2010) Impact of activated charcoal on the mineralisation of 14C-phenanthrene in soils. Chemosphere 79(4):463–469. https://doi.org/10.1016/j.chemosphere.2010.01.032

Rotaru AC-W (2018) Conductive particles enable syntrophic acetate oxidation between geobacter and methanosarcina from coastal sediments. Mbio 9(3):e00226-e318. https://doi.org/10.1128/mBio.00226-18

Ruthven D (1984) Principals of adsorption and adsorption processes. John Wiley & Sons, New York

Santamarina VR-L (2006) Mechanical limits to microbial activity in deep sediments. Geochem Geophys Geosyst 7:1525–2027. https://doi.org/10.1029/2006GC001355

Schüth CT (2003) Carbon and hydrogen isotope effects during sorption of organic contaminants on carbonaceous materials. J Contam Hydrol 64(3–4):269–281. https://doi.org/10.1016/S0169-7722(02)00216-4

Semenyuk NN (2014) Effect of activated charcoal on bioremediation of diesel fuel-contaminated soil. Microbiology 83:589–598. https://doi.org/10.1134/S0026261714050221

Siegrist RL (1998) X-231A demonstration of in situ remediation of DNAPL compounds in low permeability media by soil fracturing with thermally enhanced mass recovery or reactive barrier destruction. Oak Ridge National Laboratory and Collaborators, Oak Ridge. https://clu-in.org/download/techfocus/fracturing/X231A-3445604441067.pdf

Slater GA-K (2000) Carbon isotope effects resulting from equilibrium sorption of dissolved VOCs. Anal Chem 72(22):5669–5672. https://doi.org/10.1021/ac000691h

Smith HJ (2018) Impact of hydrologic boundaries on microbial planktonic and biofilm communities in shallow terrestrial subsurface environments. FEMS Microbiol Ecol 94(12):191. https://doi.org/10.1093/femsec/fiy191

Smolin SK (2020) New approach for the assessment of the contribution of adsorption, biodegradation and self-bioregeneration in the dynamic process of biologically active carbon functioning. Chemosphere 248:126022. https://doi.org/10.1016/j.chemosphere.2020.126022

Sorenson KN (2019) Use of permeability enhancement technology for enhanced in situ remediation of low-permeability media. Department of Defense Environmental Security Technology Certification Program (ESTCP), Alexandria, VA. file:///C:/Users/winne/Downloads/ER-201430%20Guidance%20Document%20(2).pdf

Sublette KS (1982) A review of the mechanism of powdered activated carbon enhancement of activated sludge treatment. Water Res 16(7):1075–1082. https://doi.org/10.1016/0043-1354(82)90122-1

Summers ZF (2010) Direct exchange of electrons within aggregates of an evolved syntrophic coculture of anaerobic bacteria. Science 330(6009), pp. 1413–1415. https://www.science.org/doi/10.1126/science.1196526

Thies JE (2012) In biochar for environmental management. Routledge, London. https://doi.org/10.4324/9781849770552

Thomas WJ (1998) Absorption technology and design. Reed Educational and Professional Publishing Ltd., Oxford

Weber W, Pirbazari M, Melson G (1978) Biological growth on activated carbon: an investigation by scanning electron microscopy. Environ Sci Technol 12(7):817–819. https://doi.org/10.1021/es60143a005

Weiss TH (1995) Effect of bacterial cell shape on transport of bacteria in porous media. Environ Sci Technol 29(7):1737–1740. https://doi.org/10.1021/es00007a007

Williams N, Hyland A, Mitchener R, Sublette K, Key K, Davis G et al (2013) Demonstrating the in situ biodegradation potential of phenol using Bio-Sep® Bio-Traps® and stable isotope probing. Remediation 23(1):7–22. https://doi.org/10.1002/rem.21335

Xiaojian ZZ (1991) Simple combination of biodegradation and carbon adsorption—the mechanism of the biological activated carbon process. Water Res 25(2):165–172. https://doi.org/10.1016/0043-1354(91)90025-L

Yee MO-W (2019) Extracellular electron uptake by two methanosarcina species. Front Energy Res 29:19. https://doi.org/10.3389/fenrg.2019.00029

Yu YL (2021) Influence of pore structure on biologically activated carbon performance and biofilm microbial characteristics. Front Environ Sci Eng 15(6):1–13. https://doi.org/10.1007/s11783-021-1419-1

Zamulina IP (2020) The effect of granular activated carbon on the physical properties of soils at copper contamination. In: E3S web of conferences. EDP Sciences, Rostovon-Don, pp 1–6. https://doi.org/10.1051/e3sconf/202017509003

Zango ZS (2020) An overview and evaluation of highly porous adsorbent materials for polycyclic aromatic hydrocarbons and phenols removal from wastewater. Water 12(10):2921. https://doi.org/10.3390/w12102921

Zapata Acosta KC-M (2019) Immobilization of *P. stutzeri* on activated carbons for degradation of hydrocarbons from oil-in-saltwater emulsions. Nanomaterials 9(4):500. https://doi.org/10.3390/nano9040500

Zhang WD (2013) Biological activated carbon treatment for removing BTEX from groundwater. J Environ Eng 139(10):1246–1254. https://doi.org/10.1061/(ASCE)EE.1943-7870.0000731

Żur JW (2016) Metabolic responses of bacterial cells to immobilization. Molecules 21:958. https://doi.org/10.3390/molecules21070958

Chapter 17
Application of Foams as a Remediation and Blocking Agent

Olivier Atteia, Henri Bertin, Nicolas Fatin-Rouge, Emily Fitzhenry, Richard Martel, Clément Portois, Thomas Robert, and Alexandre Vicard

Abstract Foam consists of a mixture of water loaded with surfactant and gas. Injected into the porous medium, foam has many useful properties for soil remediation. The properties of surfactants facilitate the mobilization of pollutants, and the presence of gas greatly reduces the consumption of reagents but also makes it possible to block the passage of water. The foam rheology also leads to specific effects such as the stabilization of the front. This chapter first describes the characteristics of the foam in air and then in the porous medium. Subsequently, a review of the literature on the experiments carried out in the laboratory makes it possible to highlight all the effects of the foam. The following section is devoted to rare foam injection experiments carried out in a real environment. Finally, a section is devoted to the modeling of foam displacement in a porous medium.

O. Atteia (✉) · A. Vicard
UMR EPOC, Bordeaux-INP, Talence, France
e-mail: olivier.atteia@ensegid.fr

A. Vicard
e-mail: alexandre.vicard@ensegid.fr

H. Bertin
I2M-CNRS, University of Bordeaux, Nouvelle-Aquitaine, France
e-mail: henri.bertin@u-bordeaux.fr

N. Fatin-Rouge
IC2MP, CNRS, Université de Poitiers, Poitiers, France
e-mail: nicolas.fatin-rouge@univ-poitiers.fr

E. Fitzhenry · R. Martel · T. Robert
INRS-ETE, Quebec, QC, Canada
e-mail: richard.martel@inrs.ca

T. Robert
e-mail: thomas.robert@inrs.ca

C. Portois
Colas Environnement, Bordeaux, France
e-mail: clement.portois@cer.colas.fr

J. García-Rincón et al. (eds.), *Advances in the Characterisation and Remediation of Sites Contaminated with Petroleum Hydrocarbons*, Environmental Contamination Remediation and Management, https://doi.org/10.1007/978-3-031-34447-3_17

Keywords Experiments · Foam · Mobilization · Modeling · Remediation

17.1 Introduction

Foam is defined as a dispersion of gas bubbles within a surfactant solution. In bulk, gas bubbles are separated by a continuous liquid phase composed of interconnected thin films called *lamellae*. Foam lamellae are stabilized by amphiphilic surfactant molecules characterized by hydrophobic tails and hydrophilic heads, a configuration which enables the molecules to rapidly adsorb onto the gas–liquid interface. Surfactants reduce the interfacial tension of biphasic systems, thereby promoting gas–liquid emulsification. As the active matter concentration of a surfactant solution is increased, the interfacial tension of the system will decrease, until the point at which surfactant molecules fully saturate the gas–liquid interface. At this concentration, referred to as the *critical micelle concentration* (CMC), the interfacial tension of the system is minimized and will remain relatively constant despite any further increases in active matter concentration. At concentrations above the CMC—the latter which typically ranges from 0.01 to 0.3 wt%—surfactant molecules will self-associate to form structures called *micelles*. Any given surfactant's CMC value will depend on the charge of its hydrophilic head (cationic, anionic, or non-ionic), as well as on the ionic strength of the solution and its temperature. The interfacial tension induced by surfactant stabilizes the lamellae; foam lamellae can be further stabilized through the addition of polymers and/or nanoparticles to the surfactant solution.

An important characteristic of injected foam is its gas fractional flow rate (equivalent to the volume of gas divided by total foam volume), commonly referred to as *foam quality* (f_q). Given that the gas phase is usually the major component of foam, f_q typically ranges from 30 to 99%. Foams have large surface areas (~200 m^2/L of solution) and can occupy large volumes (~0.1 m^3/L of solution), therefore making it useful in many applications (Cantat et al. 2013). Foams are often described in terms of their foamability defined as the capacity of the surfactants to form foam irrespective of the special foam properties. Foam is generated (i.e., lamellae are formed) by mechanical friction in presence of both phases. In bulk, foamability can be measured using normalized tests, such as the Ross-Miles or the Bikerman (Denkov et al. 2020; Longpré-Girard et al. 2019). Foamability depends on surfactant type and concentration and is optimal when the surfactant solution concentration is slightly greater than the CMC. However, foams are thermodynamically metastable systems that undergo constant cycles of bubble generation and destruction. Foam destruction can result from several phenomena: (i) gravitational liquid drainage in the lamellae, (ii) foam coarsening due to lamellae thinning (Ostwald ripening), and (iii) coalescence due to the pressure differences that exist between bubbles of different sizes.

17.2   Foam in Porous Media

In porous medium, the liquid phase is continuous and covers the solid walls, while the gas bubbles, separated by liquid films, are located within the pore space. The presence of a solid phase and a disordered porous network changes the properties of foam from the one measured outside the porous medium (liquid/gas ratio, viscosisty...).

17.2.1   Adsorption of the Surface-Active Agents

Adsorption of surfactants can be important because of the small molecule sizes and depends much on both the nature of the surfactant molecule and the solid material. When the liquid phase containing surfactant molecules is in contact with the solid surface, the rapid adsorption of the surfactant molecules may arise due to several interactions: (i) H-bond and acid–base interactions, (ii) electrostatic interactions between ionic groups, and (iii) chemical dispersion forces. Adsorption has been studied extensively and summary of the mechanisms as well as concatenation of results have been reviewed (Rosen and Et Kunjappu 2012; Schramm 2000). In practice, a surfactant slug should be injected before gas injection to ensure local availability of surfactant molecules at concentrations able to make lamellae.

17.2.2   Foam Generation

Following the pioneering work of Roof (1970), the topic of foam generation has been extensively studied in the literature. The following four mechanisms are responsible for bubble generation in porous media, the first of which, *snap-off*, has the most important influence on the resulting foam rheology of all the mechanisms.

This phenomenon is illustrated in Fig. 17.1a. When gas flows from a pore body toward a pore throat, the velocity increases locally, and the liquid film thickness increases until the two surfaces join and form a lamella that is stabilized by the presence of surfactant.

In the *leave behind* configuration (Fig. 17.1b), a lamella is generated parallel to the direction of flow due to the pore geometry.

In *Lamellae division a* moving gas bubble is separated into two parts when it encounters an obstacle (Fig. 17.1c).

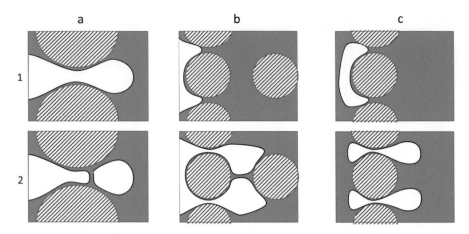

Fig. 17.1 Schematic of the snap-off (**a**), the leave behind (**b**) and lamellae division mechanism (**c**), 1 is before the mechanism action and 2 after the action

17.3 Pinch-Off

Liontas et al. (2013) described this mechanism by observing foam experiments performed in micromodels and showed that *Pinch-off* differs from *snap-off* in that it occurs in presence of several bubbles interacting (Fig. 17.2).

Fig. 17.2 Neighbor-wall pinch-off (**a**) and neighbor-neighbor pinch-off (**b**) (from Liontas et al. 2013)

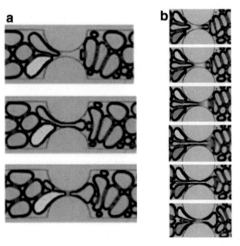

Fig. 17.3 Sketch of
diffusion process in a pore
(from Jones et al. 2017)

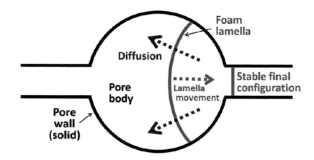

17.3.1   Foam Stability in Porous Media

When flowing in a porous medium, the overall stability of foam is directly dependent on the stability of its individual lamellae, which in turn depends on parameters such as surfactant concentration, gas diffusion, and petrophysical properties. Capillary pressure (P_c) plays a key role, as it causes destabilization of liquids film at high values. Khatib et al. (1988) found that foams collapse when capillary pressure exceeds a critical value, P_c^*, called *limiting capillary pressure*. Given that P_c is in general inversely proportional to soil permeability, foam destruction is more significant in fine-grained materials.

Foam structure changes with time due to gas diffusion between bubbles (Fig. 17.3). This mechanism, called *foam coarsening* or *Ostwald ripening*, leads to bubble coalescence. It depends on several parameters, including but not limited to surfactant concentration, gas type, pressure, and temperature (Marchalot et al. 2008; Jones et al. 2018).

17.3.2   Foam Rheology in Porous Media

When gas is injected into a porous medium, there are three possible flow behaviors, as illustrated in Fig. 17.4. First, in absence of surfactant in the liquid solution, foam cannot be generated, and gas and liquid flow separately (Fig. 17.4a), because of gravity, the gas flows with upward vertical fingering through the liquid phase. Second, foam may occur with a low lamellae density (density is the number of lamellae per unit volume, see Fig. 17.4b) because of lamellae instability or lack of generation efficiency. The low lamellae density only slightly diminishes the gas flow, and the given foam is called *weak foam*. Third, the lamellae density is high, resulting in a high resistance to foam flow with flat front displacement into the porous medium (Fig. 17.4c). In this case, the given fluid is called *strong foam*.

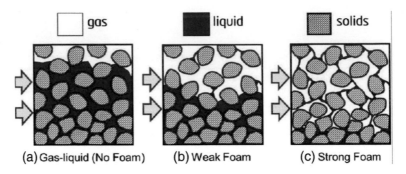

(a) Gas-liquid (No Foam)   (b) Weak Foam   (c) Strong Foam

Fig. 17.4 Different regimes of foam flow in porous media (from Sheng 2013)

Starting from the rheology of gas bubbles in capillary tubes where viscosity is due to porous medium resistance to lamellae movement, interface deformation, and surfactant gradient, Hirasaki and Lawson (1985) established the following relationship:

$$\mu_f = \mu_g + \frac{\alpha n_f}{v_g^{1/3}}$$

where μ_f and μ_g are, respectively, foam and gas viscosities, α is an empirical parameter that depends on the porous medium, surfactant, and flow conditions, n_f is the density of flowing lamellae, and v_g the local gas velocity.

In soils, gas mobility also depends on the size ratio of gas bubbles to pore throats, soil permeability, and gradient as well as the local saturations of the gas and liquid phases (Falls et al. 1988; Maire et al. 2018a, b). For strong foams at steady state, the trapped gas fraction typically ranges from 85 to 99%, and bubbles move irregularly according to the pore geometry, some bubbles being trapped other move more rapidly (Friedmann et al. 1991).

17.3.3 Resistance Factor

A foam's lamella density is directly associated with the viscosity it exhibits when flowing in porous media: High densities of lamella are associated with highly viscous foam. *Resistance factor* (RF) is commonly used to express the increase in pressure drop, ΔP, during foam flow by comparing it to the pressure drop induced by a single-phase water flood at the same volumetric flow rate, Q:

$$RF = \frac{\Delta P_{foam}}{\Delta P_{water}}\bigg|_Q$$

Similarly, the *mobility reduction factor* (MRF) can be used when the gas phase is used for the comparison:

$$\text{MRF} = \left.\frac{\Delta P_{\text{foam}}}{\Delta P_{\text{gas}}}\right|_{Q}$$

These parameters are useful for describing foam flows in porous media. Typically, strong and weak foams are characterized by MRF values above and below 100, respectively.

17.3.4   Foam Injection in Porous Media

The following three methods can be used to inject foam into a porous medium (Fig. 17.5). The first and third methods listed below result in the highest and lowest injection pressure values, respectively, while the second method is associated with injection pressures that are intermediate between the two (Maire et al. 2018a, b):

- Pre-generation: Foam is first generated outside the porous medium, then injected into the porous medium. This method is applicable in soils with permeabilities above 10^{-9} m^2 (>150 darcy), and/or in presence of agents that destabilize foams, such as hydrocarbons (Osei-Bonsu et al. 2015, 2017).
- Co-injection: Simultaneous injection of gas and surfactant solution directly into the porous medium, where foam is formed via the different mechanisms described in the foam generation section.
- Surfactant alternating gas (SAG): First a surfactant solution slug is injected, followed-up with the injection of a given amount of gas, and the whole operation is repeated many times over. This method of injection allows for better injectability, which is required in soils with permeabilities below 3×10^{-9} m^2 (<150 darcy). However, reaching steady state using this method takes a significant amount of time.

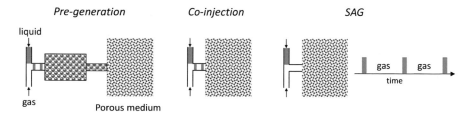

Fig. 17.5 Illustration of the three mechanisms of injection of foam in porous medium

17.4   Non-Aqueous Phase Liquid (NAPL) Recovery Mechanisms

At the microscopic scale, foam is a two-phase material that is subjected to many complex mechanisms in porous media, which taken together give foam its macroscopic properties. In an environmental remediation context, foam (i) has improved sweep efficiency in heterogeneous aquifers due to its non-Newtonian properties, (ii) eliminates displacement front instabilities caused by viscous forces, and (iii) lowers the capillary forces that otherwise maintain non-aqueous phase liquid (NAPL) droplets trapped, by decreasing the interfacial tension between the aqueous and organic phases in the presence of surfactants. The main characteristics of foam and a description of how each one contributes to contaminant mass removal are summarized in Fig. 17.6 (Fitzhenry 2021). Surfactant foams are able to recover NAPL present in an aquifer through three mechanisms: solubilization, mobilization, and volatilization.

17.4.1   Solubilization (Miscible Displacement)

The particular structure of amphipathic surfactant molecules allows them to position themselves at the interface between two immiscible liquids, and to lower the interfacial tension of the system (see Fig. 17.7) as also discussed in Chapter 16.

In a contaminated aquifer, micelles have the ability to both encapsulate NAPL droplets in their cores (emulsification) and dissolve the molecules of the NAPL

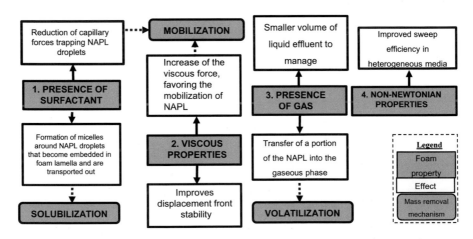

Fig. 17.6   Foam characteristics and NAPL recovery mechanisms (adapted from Fitzhenry 2021)

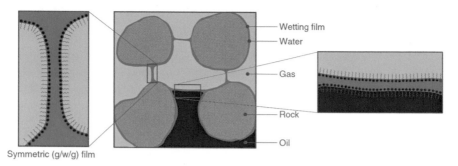

Fig. 17.7 Structure of foam bubble walls, with gas (green), surfactant solution (blue), and NAPL (red) adapted from Farajzadeh et al. (2012)

constituents directly into the surfactant solution (solubilization). Micellar solubilization thus increases the concentration of the contaminant in the liquid fraction of the foam, making the contaminant more available for removal.

According to the mechanism proposed by Schramm and Novosad (1990), when foam in a porous medium encounters a NAPL, the decrease in interfacial tension between the surfactant solution and the NAPL can lead to the formation of fine micelles around the NAPL droplets which become embedded in lamellae, as shown in Fig. 17.8. Although this allows the droplets to be transported for some distance, the incorporation of emulsified NAPL into lamellae can negatively affect foam stability, causing lamellae to eventually collapse and release the NAPL. Subsequently, an intact lamella may arrive, pick up the NAPL and carry it a little further, and so on. This mechanism of transport is very effective, leading to high ratio of mass of hydrocarbon recovered to mass of surfactant used (10:1; Maire et al. 2015). Moreover, the displacement of initially trapped NAPL droplets in the form of an emulsion results in a reconnected continuous phase of non-wetting fluid, which increases its mobility when subjected to pressure.

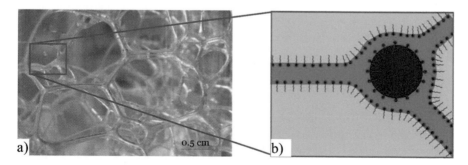

Fig. 17.8 Left, visualization of n-hexadecane droplets, stained with Oil Red-O, embedded in lamellae of a foam outside of porous media (Simjoo et al. 2013b). Right, theoretical diagram illustrating the configuration of surfactant monomers (gas in green, surfactant solution in blue, and organic phase in red), adapted from Denkov (2004)

17.4.2 Mobilization (Immiscible Displacement)

Entrapped NAPL saturation (S_{oe}) in an aquifer takes the form of discontinuous and near-immobile NAPL droplets that are trapped in pores by capillary forces. By definition, P_c is the difference between the pressures of the organic (typically the non-wetting phase) and aqueous (wetting) phases ($P_c = P_o - P_w$). In order for NAPL trapped at residual saturation to flow into a neighboring pore, the capillary pressure must exceed the entry capillary pressure of said pore P_{ce} (Joekar-Niasar et al. 2008):

$$P_{ce} = \frac{2\sigma_{ow}}{r}\cos\theta$$

where θ is the contact angle at the interface between the two liquids and the pore surface, and r is the pore radius (m). The efficiency of immiscible displacement of a NAPL by a washing solution is controlled by the capillary number (N_{Ca}), which is defined as the ratio of the viscous forces that promote displacement of NAPL droplets, to the capillary forces that maintain NAPL droplets trapped in the pores (Pennell et al. 1996):

$$N_{Ca} = \frac{q_w \mu_w}{\sigma_{ow}\cos\theta}$$

where q_w and μ_w correspond to the Darcy flux (m/s) and dynamic viscosity (Pa·s) of the displacing fluid (water), respectively, σ_{ow} to interfacial tension between the displaced (NAPL) and displacing fluid (N/m), and $\cos\theta$ to the wettability condition of the medium. In order to promote mobilization of NAPL as a free phase bank in the porous medium, it is necessary to increase N_{Ca} significantly by lowering interfacial tension by several orders of magnitude using surfactants.

NAPL recovery by mobilization can be represented by a capillary desaturation curve (CDC), a semi-logarithmic graph that shows the relationship between residual saturation in soils and capillary number (Lake 1989). A typical CDC (Fig. 17.10left) is characterized by a constant residual saturation (S_{or}) of the non-wetting phase at low capillary numbers, a critical capillary number ($N_{Ca}*$) required to initiate desaturation, and a full desaturation capillary number ($N_{Ca}**$) when S_{or} becomes zero (Robert et al. 2017).

The mobilization capacity of a NAPL depends on the properties (viscosity, density, wettability, and interfacial tension) of both the NAPL and the displacing fluid, as well as on the properties of the porous medium (permeability and pore connectivity). In fact, the wettability of the system and the particle size distribution of the soil both have an important influence on the shape and position of the CDC. For example, poorly sorted soils result in more inclined CDCs that cover a wide range of capillary numbers, due to large variations in pore size.

In addition, Dwarakanath et al. (2002) observed that the use of anionic surfactants can render an aquifer preferentially water-wet, resulting in a shift of the associated CDC to the left by 1–2 orders of magnitude, as shown in Fig. 17.9. A left-shifted

CDC in a water-wetting system means that complete desaturation will be more readily achievable than in a mixed wetting or oil-wetting system, since the same saturation value can be achieved with a lower capillary number. It should be noted that the effect that the adsorption of different types of surfactants onto the porous material has on system wettability is complex. Wettability depends on the way that the solid phase and the surfactant chemically interact, on the degree of saturation of the available adsorption sites, as well as on the way that the surfactant molecules interact with each other (monolayer *vs.* multilayers).

In the literature, alternatives to the N_{Ca} have been proposed, in an effort to improve the assessment of mobilization capacity by accounting for the buoyancy force. The Bond number (N_B) is the ratio of buoyancy to capillary forces, while the total trapping number account for both N_{Ca} and N_B contributions (Pennell et al. 1996).

Despite the destabilizing effect that light hydrocarbons have on foams, the latter still demonstrate better sweep efficiency in heterogeneous media when compared to conventional soil washing techniques that employ liquid solutions (Wang and Chen 2012). This is the result of the foam behavior presented in the first section of this chapter. This phenomenon provides a more uniform sweep of the porous medium. Nevertheless, foam flow in porous media is very complex and difficult to predict

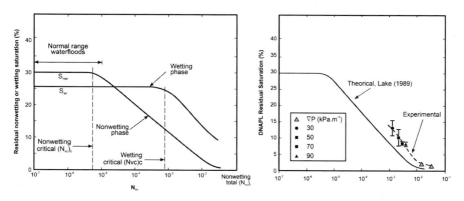

Fig. 17.9 Left: Schematic capillary desaturation curve (from Lake 1989), and right: Effect of foam on capillary desaturation of a dense NAPL (DNAPL) (adapted from Maire and Fatin-Rouge 2017)

Before Foam Injection   After Foam Injection

Fig. 17.10 Recovery of p-xylene (red) in 2-D sand tank within 2-layer silica sand (white): left p-xylene at a residual saturation of 0.17 before foam injection; right p-xylene at negligible saturation after foam injection

at the microscopic scale, which makes the total desaturation difficult to achieve in practice, due in part to bypassing of NAPL due to pore blocking by trapped bubbles.

17.4.3   Volatilization

A foam that is stable enough to flow through the porous medium can be composed of gas up to 98%, which represents the majority of the foam volume. Thus, the presence of gas bubbles in the saturated zone during foam flow makes it possible for some of the contaminant to be transferred into the gas phase by volatilization, depending on the vapor pressure of the contaminant. Mulligan and Eftekhari (2003) demonstrated that a large portion (66%) of pentachlorophenol in a fine sand was removed by volatilization during foam injection tests in columns. Similarly, Yan et al. (2019) conducted foam injection tests in columns of nitrobenzene-contaminated sand and found that 56% of the contaminant was removed by volatilization. Experiments studying the removal of hydrocarbons from creosote-contaminated soils involved mass balances calculations (C_{10}–C_{40} TPH index) that indicated 80% of the contaminant was removed in the gas phase (Fatin-Rouge 2020). A foam injection experiment conducted by Longpré-Girard et al. (2016) in a 2D sandbox contaminated with p-xylene at residual saturation indicated that 65% of the contaminant mass was removed by volatilization.

It should be noted that the degree of volatilization depends on the physico-chemical properties of both the contaminant and the mobilizing fluid (in this case, foam).

Regarding the contaminant, its vapor pressure and Henry's constant are of main importance. Both values tend to decrease the longer the hydrocarbon chain, and the higher the degree of H-bonds or stacking interactions. For mixtures of contaminants, the molar composition also plays an important role, as dictated by Raoult's law. Temperature also influences these values and is important to consider, especially when considering coupling foam injection with in situ thermal desorption. The octanol–water partition constant (K_{ow}), which quantifies a contaminant's affinity for the aqueous phase, is equally important.

Regarding the foam, both the foam quality as well as the extent of micellar solubilization are important. Higher foam quality is associated with increased transport in the gas phase, whereas high concentrations of micelles will stabilize the contaminants in the liquid phase. Some authors have reported that the degree of volatilization of a contaminant tends to decrease the higher the surfactant concentration (Bouzid et al. 2019; Roy et al. 1995), which is likely due to the isolation of the compound within the micelles, making it stabilized in the liquid phase.

17.5 Laboratory Experiments

17.5.1 Column Experiments Examining Foam–Oil Interactions

The various experimental approaches employed in the literature have examined different aspects of foam–oil interactions and oil mobilization mechanisms in porous media. 1D experiments carried out in well-controlled laboratory conditions can provide excellent insights into pressure drop versus injection velocity and imbibition-drainage versus capillary pressure relationships. These types of experiments also allow for easy exploration of different parameters, such as chemical formulation of foams. In the enhanced oil recovery (EOR) literature, the behavior of foam in porous media in the presence of hydrocarbons has been studied in detail. Most of these studies were performed in Bentheim sandstone core, a material whose permeability (on the order of $2-10^{-12}$ m^2) and mineralogy are considered representative of petroleum reservoirs (Peksa et al. 2015). In order to replicate the typical conditions of an EOR process, the common practice is to perform tests at elevated temperatures and pressures on columns initially saturated with brine.

Farajzadeh et al. (2010) performed foam tests by alternating surfactant (AOS) and gas (N$_2$ or CO$_2$) injection in sandstone cores (SAG method). The front half of the cores were saturated with brine, while the back half of the core were at residual oil saturation (Isopar H, C9–C11). This configuration was used in order to reproduce the water–oil transition encountered in oil reservoirs. They observed that at low pressures (1 bar), the presence of oil does not allow foaming in the core. At higher pressures (90 bar), the CO$_2$ foam reduced gas mobility in the region of low oil saturation; however, upon contact with the oil-containing zone, the gas and surfactant separated, and there was no oil recovery at the outlet. At even higher pressures (137 bar), the foam formed in the front part of the core persisted in contact with the oil, and the resulting increasing pressure increased the recovered mass.

Simjoo et al. (2013a) sought to improve the understanding of how the presence of an organic phase in a porous medium affects the rate of removal of that organic phase by foam. Previous studies often involved cores saturated with oil over only a portion of their length, in order to create a clean "antechamber" in which foam can form in situ prior to contact with oil. In this case, foam injection tests were carried out in sandstone cores containing a residual hexadecane phase along its entire length; this configuration is relevant in an ER context, where injection wells are placed within a source zone in order to displace NAPL and direct it to surrounding recovery wells. Thus, foam was formed by co-injecting gas (N$_2$) and surfactant solution (1 wt%) at a constant total flow rate ($Q = 1.1$ mL/min, $f_g = 91\%$). The surfactants employed included an AOS, as well as a mixture (1:1 by weight) of AOS and a polymeric fluorosurfactant (FC), and all solutions were prepared using brine (0.5 M NaCl). From their tests, Simjoo et al. (2013a) identified two regimes of oil displacement. The first involves an increase in the N_{Ca} and the formation of an oil bank. During this regime, oil saturation decreases until a value is reached at which oil cannot be displaced by the viscous force alone. This is when the second regime occurs, where the N_{Ca} remains nearly constant, and the oil is moved droplet by droplet as a dispersed phase.

Janssen et al. (2019) investigated a combined foam and alkaline solution injection process in sandstone cores for hexadecane recovery. They sought to understand the effects that different salt contents in a surfactant solution might have on the formation of an oil bank and its subsequent mobilization by foam. The sandstone cores used, initially at residual oil saturation, were previously treated by injection of a solution bank composed of brine (0–2.5 wt% NaCl), an alkali (Na_2CO_3), a surfactant (internal olefin sulfonate, or IOS), and an alcohol (sec-butanol). This first step led to an increase in capillary number, which allowed the mobilization of oil droplets that coalesced into an oil bank. Downstream of the oil bank, some oil was solubilized and formed a microemulsion. The next step was to simultaneously inject gas (N_2) and surfactant solution (IOS and Na_2CO_3 mixed with brine), which resulted in the formation of a strong foam and efficient oil displacement. The authors hypothesized that upon contact with the oil bank, a decrease in effective porosity caused a local increase in pressure drop, large enough to initiate foam generation. Thus, the results demonstrated that foam formation is possible in the presence of high oil saturation under these experimental conditions.

Tang et al. (2019) conducted sandstone core tests to compare how foams injected in different modes (i.e., pre-generation and co-injection) are affected by an organic phase present at residual saturation. The surfactant solution used was 0.5 wt% AOS with 3 wt% NaCl, and the gas used was N_2. A series of tests were performed for each of the two oils studied: (1) hexadecane, and (2) a mixture of hexadecane and oleic acid (80:20 by mass). Tests with hexadecane (known to be weakly destabilizing) revealed similar results for both pre-generation and co-injection. In both cases, formation of a foam bank was observed, with gradual refinement of the foam texture along the core. Tests with the oleic acid/hexadecane mixture (known to be highly destabilizing) revealed significant differences between the propagation of the foams formed by the two injection modes. Co-injection had barely formed foam, even in the presence of low oil saturation ($S_{or} \approx 0.1$). Weak foam began to form near the column outlet, and the foam began a very slow propagation in the opposite direction of injection. On the other hand, the pre-generated foam showed two phases of propagation, the first characterized by weak foam that gradually lowers the oil saturation, to allow for the eventual formation of a strong foam front in the second phase.

17.5.2 Adding Particles to Foam in Presence of Oil

The molar mass and molecular structure of surfactants may influence the stability of foam lamellae in the presence of organic phases (Vikingstad et al. 2006), although general trends remain unclear. Some studies (Osei-Bonsu et al. 2015; Yu et al. 2019) indicate that the rate of liquid drainage from lamellae was slowed for foams created from viscous surfactant solutions. This suggests that additives included in the surfactant solution to increase viscosity could help maintain foam stability in the presence of NAPL. In contrast, in contaminated sand column tests, Forey et al. (2019) observed that the higher the polymer concentration (and thus the viscosity of the surfactant solution), the lower the foam RF, meaning that the foam collapses easily upon contact with the organic phase.

The incorporation of colloidal particles into surfactant solutions to improve the efficiency of the foam injection process has been studied extensively in petroleum engineering (Nguyen et al. 2014; Singh and Mohanty 2017; Sun et al. 2014; Yang et al. 2017) and in environmental settings (Forey et al. 2019; Karthick and Chattopadhyay 2017). It is necessary to optimize the surfactant solution formulation, including particle type (composition, hydrophobic/hydrophilic behavior) and concentration, to avoid the formation of complexes (precipitates and aggregates) that can interfere with foam formation and clog the porous medium. Moreover, solid particles used to reinforce a foam in an environmental context must be non-toxic. Finally, the particles must be small enough not to reduce the permeability of the medium by deposition.

Sun et al. (2014) conducted micropattern and sand column foam injection tests to investigate the effect of adding hydrophobic silica nanoparticles (diameter $= 14$ nm) on the recoverability of a crude oil. The surfactant solutions used for the sand column tests consisted of sodium dodecyl sulfate (2.2 CMC), NaCl (0.5 wt%), and a variable SiO_2 concentration (between 0 and 2 wt%). Foam was generated by co-injection of the surfactant solution and gas (N_2) at a constant total flow rate ($Q_{foam} = 0.005$ mL/min, $f_q = 50\%$). An increasing recovery efficiency was observed for SiO_2 concentrations of 0–1 wt%, but for concentrations ≥ 1.5 wt% the improvement was negligible. As confirmed visually in micropatterning, the adsorption of solid particles onto the walls of foam bubbles allows the bubbles to maintain a spherical shape in the medium, indicating that they are more difficult to deform. This increased viscoelasticity allows the bubbles to exert a greater "micro-force" on the trapped droplets and dislodge them in greater numbers, even from dead-end pores.

Karthick and Chattopadhyay (2017) compared the effect of adding hydrophobic and hydrophilic silica nanoparticles for soil column diesel recovery by pre-generated foam injection. The surfactant solution used was Tween-20 (0.1 vol%; non-ionic) and 0.5 wt% SiO_2 nanoparticles (diameter $= 50–80$ nm). The hydrophobic particles had formed aggregates in water and had to be pretreated with tetrahydrofuran. The results revealed recovery rates of 78 and 57.5% with the solutions containing hydrophobic and hydrophilic particles, respectively. The authors attribute this difference to a better adsorption capacity at air–water interfaces in hydrophobic particles. For comparison, a soil washing test with the Tween-20 solution without nanoparticles in the same column recovered only 42% of the initial diesel; no nanoparticle-free foam test was performed.

17.5.3 Layered 2D Sandboxes

Studies in 2D cells are useful for improved understanding, controlling, and modeling of the mechanisms that drive the complex behavior of foam and the redistribution of contaminants. 2D studies are ideal for studying fluid behavior in permeability/ wettability contrasted media, which reflect the subsurface anisotropy of complex contaminated sites where usual treatments can be ineffective. The interest of foam in that context is its sweeping efficiency coupled to the processes presented above in the interaction oil/foam paragraphs.

17.5.3.1 Longpré-Girard et al. (2016)

In this study, the custom 2D sandbox used consisted of a stainless steel tank (1.5 cm thick walls) with an internal volume of 2.5 L (26 cm long, 14 cm high and 8 cm thick) and a tempered glass window (1.5 cm thick) to enable visualization of foam flooding experiments (Longpré-Girard et al. 2016). The ability of surfactant foam to enhance mobility control and light NAPL (LNAPL) recovery in a layered heterogeneous porous medium was investigated. The 2D sandbox was filled with two layers, respectively, of coarse and medium silica sand. The surfactant solution used was Ammonyx Lo at a concentration of 0.1% (w/w). The experimental protocol included tracer tests in uncontaminated and contaminated (p-xylene residual saturation of 0.17) conditions, foam injection under uncontaminated conditions, as well as surfactant solution and foam injection under contaminated conditions. The foam was pre-generated and injected into the tank using increasing pressures (from 14.7 to 51.5 kPa). Tracer tests indicated that the permeability contrast between sand layers was increased by LNAPL contamination. Foam injection under uncontaminated conditions presented an S-shaped front that indicated a better mobility control than the piston-shaped front obtained during tracer tests. During foam injection, complete sweep of the sand tank was achieved with 1.8 pore volume (PV) compared to 2.8 PV during tracer injection, thus indicating better mobility control with foam. Pre-flush of the contaminated sand tank with surfactant solution initiated p-xylene mobilization, but no free phase was recovered at the effluent. LNAPL recovery mechanisms involved during foam injection in the contaminated sand tank were 19% by mobilization, 16% by dissolution, and 65% by volatilization. The p-xylene concentration measured in the sand after the foam injection experiment was < 16.3 mg/kg (Fig. 17.10).

17.5.3.2 Effect of Nanoparticles Addition

Singh and Mohanty (2017) performed 2D sandbox foam injection tests containing two layers of sand (6:1 permeability contrast), to study the effect of adding hydrophilic silica nanoparticles (diameter $= 10$ nm) on the recovery of a crude oil. The foam was generated in a pre-column, using gas (N_2) and surfactant solution

(0.5 wt% AOS with 0.3 wt% SiO_2 nanoparticles), and then injected into the tank at a constant velocity ($v_{foam} = 1.4–10^{-5}$ m/s, $f_q = 80\%$). The particle-stabilized foam (generated $\Delta P = 34.5$ kPa after 22 VP injected) recovered 9% more oil than the particle-free foam (generated $\Delta P = 27.6$ kPa after 23 VP injected) injected under the same conditions.

The goal of the Forey et al. (2021) study was to investigate how the presence of oil impacts foam formation during a liquid–gas co-injection process. As organic phases, such as LNAPL, usually act as defoaming agents that thwart the use of foam in remediation treatments, the synergy between a biological surfactant (saponin 10 CMC 0.5% wt) and silica colloidal particles was explored to obtain a strong and stable foam. In the investigations, gas–liquid co-injection experiments were carried out in a large 2D porous medium in the absence and in the presence of oil (MACRON 1821 F-4 oil). When no oil was present, although foam visually swept the porous medium in a piston-like way, buoyancy segregation still acts, and a gas/liquid saturation gradient takes place across the flow direction. As the surfactant concentration increases, foam strengthens due to a higher lamellae stability that prevents bubbles to coalesce. If SiO_2 colloidal particles are incorporated in the formulation, the resistance factor is slightly enhanced and the water saturation (S_w) is substantially decreased. SiO_2 addition probably arises in thicker lamellae without significantly affecting the foam structure, and bubbles are in a shield consisting of solid SiO_2 colloidal particles. By repeating the experiment in the presence of a layer initially at residual oil saturation, formed foams are destroyed when they come in contact with oil, thus increasing the local gas mobility while it seems unaffected in the rest of the container. Results showed that foam resistance to oil-induced destruction was less damaging when the surfactant concentration was increased and was damped when SiO_2 particles were added to the formulation. With this formulation, the pressure gradient within the tank was similar in the regions with and without oil, indicating that the particles provided efficient mobility control.

17.5.4 Fractured Media

Fractured media are among the most difficult systems to manage, because of their strong heterogeneity and anisotropy, lower connectivity, the size, and the traveling of fractures that make flow very heterogeneous both in magnitude and direction (NASEM 2015). Despite all this, the improved sweep efficiency of foams and the ability to generate them in situ by co-injection has been demonstrated (Brattekås and Fernø 2016). Fatin-Rouge and Visitacion Carillo (2018) compared the recovery of different hydrocarbons with densities ranging from 0.9 to 1.8 g/cm³ in a water-saturated cell made of calcareous rectangular blocks with a network of fractures with size of ~3.5 mm. Pre-generated foams ($f_q = 95\%$) were injected an compared to aqueous solutions. Conclusions revealed that foam should be preferred for light and volatile contaminants given the low density of both foam and LNAPLs and the importance of the mobilized fraction in the gas phase of foam.

17.6   Field Applications

Foam has been applied to very few sites, and this was on chlorinated solvents contamination. The cases are presented below as the foam behavior and field processes could be very useful to understand the behavior of foam in presence of floating LNAPL.

17.6.1   Hirasaki et al. (1997a; b)

The first application of foams for in situ remediation of soils at the field-scale was performed on a military base located in Ogden, Utah (Hirasaki et al. 1997b). The contamination consisted of trichloethylene (TCE), a chlorinated solvent used for degreasing metal parts, released to the environment as a result of its improper storage in trenches without impermeable bases. The contaminant was present at a concentration of 668 mg/kg as a residual saturation (no NAPL was detected in the test cell). The water table was located at approximately 7.6 m above ground level. The aquifer consisted of a sand horizon (0–15.2 m, $K > 9.8–10^{-4}$ m/s), with an underlying clay layer (>15.2 m, $K < 3.9–10^{-4}$ m/s) that acted as a capillary barrier and prevented vertical migration of contaminant. An environmental characterization of the site revealed the absence of TCE in areas of high permeability. Thus, the approach of the pilot test was to use the foam to regulate the hydraulic conductivity profile of the aquifer by selectively blocking the high permeability layer to improve the recovery by the surfactant solution in the low-permeability layer.

The test cell had an area of approximately 85 m^2, on which 10 wells were constructed, including 3 injection wells, 3 recovery wells, 2 observation wells, and 2 hydraulic control wells. The injection and recovery wells were arranged in rows. Two observation wells were installed in the center of the cell and were used to identify the arrival time of the various fluids. A groundwater pumping and re-injection system was set up to prevent the loss of fluids outside the treatment area. The surfactant solution used was 4 wt% dihexyl sulfosuccinate (Aerosol MA-80I) (>4 CMC) and 10,250 ppm NaCl, a formulation that was optimized to maximize solubilization. The test lasted 3.2 days and consisted of simultaneous injection of surfactant solution at a constant flow rate ($Q = 113$ L/min) into two injection wells at a time, and air supplied by a constant pressure compressor (manually adjusted to be between 55 and 76 kPa) into the 3rd injection well. Air injection was alternated between the three wells over a two-hour period. The amount of TCE measured in the soils after treatment was 77 mg/kg, representing a removal of ~88% of the initial mass.

17.6.2   Portois et al. (2018a; b)

A foam injection test was conducted by Portois et al. (2018b) at an active plant site in Spain contaminated by a storage tank leak during the 1980s. The aquifer beneath the site had a very heterogeneous stratigraphy, consisting of an impermeable clay layer from 0 to 5.5 m, a Quaternary aquifer composed of fine to medium sand ($K \approx 10^{-6} - 4.10^{-5}$ m/s) from 5.5 to 7.5 m, and an impermeable clay horizon from 7.5 to 11 m. The contamination was present as a dissolved plume of TCE (max. concentration 300,000 μg/L). Given the presence of low-permeability clay layers, there were concerns about exceeding the maximum safe injection pressure during the pilot test. The approach was therefore to inject the foam into the permeable sand layers, to provide temporary containment around the source area and to limit plume flow downstream. This study was a first step in the development of a method that allows for the delivery of reagents into a source zone using foams.

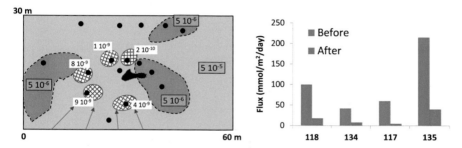

Fig. 17.11 Left: aerial view of foam injection system showing the position of the wells (black dots), the estimated position of the TCE NAPL source zone (red area), the hydraulic conductivity (K, m.s^{-1}) for the natural terrain and the estimated position and K of the foam areas (in squared black on white) from modeling. Right: value of the fluxes measured downstream the injection area, before and after the foam injection

The test cell had an area of ~1800 m^2 in which 6 injection wells (screened interval 5.7–7.7 m) were installed around the periphery of the source zone. The surfactant solution used was a mixture of sodium laureth sulfate, cocamidopropyl betaine, and lauryl glucoside, and air was supplied by a compressor. No surfactant injection was performed prior to foam injection to avoid mobilizing the contaminant. The test was conducted over a 96-h period, during which surfactant solution and air were co-injected into all 6 wells simultaneously. Compliance with the injection pressure limit of 300 kPa (to avoid fracturing) was ensured during the test by manually adjusting the total injection rate ($Q_{foam} = 12.1 - 35.4$ L/min), foam quality (95–99%), and surfactant solution concentration (0.3–0.9 wt%, i.e. 4–14 CMC). Blocking capacity was defined in terms of the ease of water flow through the medium: a decrease of more than 100 times the initial K_w was measured around the wells after foam injection. According to a numerical model, the radius of influence of the injected foam was 3.2 m (Fig. 17.11left). After foam injection, down gradient from the source zone, a three to fourfold reduction in the TCE mass discharged was measured (Fig. 17.11right).

17.6.3 Maire et al. (2017, 2018a; b)

Maire et al. (2018a, b) performed the first field-scale application of foam injection for NAPL displacement and recovery by pumping. The study site was an active chloralkali plant located in Tavaux, France. The simplified stratigraphy of the aquifer was as follows: a fine alluvial horizon (0–2 m, $K \approx 10^{-7}–10^{-9}$ m/s) at the surface, followed by coarse alluvium (2–9.5 m soil, $K \approx 10^{-2}–10^{-4}$ m/s), and an underlying impermeable marly bedrock ($K \approx 10^{-9}$ m/s). Over the course of a year, the water table varied between 2 and 4 m below ground surface. The contaminant at the site consisted of a complex mixture—which included hexachlorobutadiene (58%), hexachloroethane (16%), pentachlorobenzene (3.5%), and tetrachloroethylene (8%)—present as an 8.3 m^3 NAPL lens accumulated at the bottom of the aquifer (pool). In preparation for the pilot test, watertight bentonite cement walls were constructed along the perimeter of the cell (100 m^2 square) to prevent the loss of contaminants outside the treatment area. Nine wells were installed in the cell, including a central recovery well, an observation well in each of the four corners of the cell, and four foam injection wells. The foam injection wells were constructed around the central well, at a distance of 3 m from the central well, with injection sections of 9–9.5 m between packers. During the first phase of the test, which lasted ~1 month, groundwater was pumped from the central well and re-injected into the four corners of the cell, with the goal of recovering as much free phase as possible via a submersible pump installed in the central well: this configuration resulted in the recovery of 7.6 m^3 of NAPL. The second phase of the test consisted of foam injection in SAG mode, i.e., by alternating the injection of 1/6 VP of surfactant solution (sodium dihexylsulfosuccinate, 4 wt% or 2.7 CMC) followed by 1/3 VP of air. The injection was performed at constant pressure (value not specified).

Since the foam flows radially around an injection well, the NAPL was directed not only to the central well, but also to the four corners of the cell. In addition, due to technical limitations, the injection could only be performed in two wells at a time, instead of all four simultaneously. The configuration of the wells was found to not be optimal for the convergence of the NAPL to the central recovery well. No quantification of NAPL recovery was reported, but verification boreholes completed three weeks after injection revealed the presence of foam and the absence of NAPL droplets in the soils.

17.7 Modeling Foam Behavior

As the different mechanisms described above show, foam presents a complex behavior in porous media. Two main types of models have been proposed so far to simulate foam behavior. Note these approaches do not include multiphase simulation of NAPL transport. Firstly, the so-called bubble generation models describe in detail the generation and coalescence of bubbles according to parameters related to the dynamics of the different fluids present and the porous medium. The so-called local equilibrium (LE) models are an extension of the multiphase models in porous media which modify the properties of the gas phase when it is present in the form of foam. Globally, bubble models lead to a finer description of the processes and can be partly validated by micro-model experiments, even if this requires some averaging. On the contrary, LE models are rather intended for core, 2D pilot or reservoir scale simulations and do not include some detailed mechanisms.

Different articles highlight the convergence of the two types of approaches on generic cases, see in particular the excellent synthesis of Ma et al. (2015) and the Lotfollahi et al. (2016) comparative study. We will use here a LE formulation. This formulation is often referred to as the "STARS" approach because this model was one of the first to develop a generic foam model. However, this type of model is currently included in most reservoir models, and we further present an implementation in the OpenFoam free library.

This modeling part only deals with the specificity of foam flow. For the interaction of foam with contaminants or oil phase, other components of multiphase flow models shall be used. In the oil industry multiphase models like ECLIPSE or others include the interface tension between three phases and shall be able to model foam–oil interactions (e.g., Olabode et al. 2021). To our knowledge there are no models of foam injection with TOUGHREACT (Xu et al. 2012) which lacks of the foam properties part. Foam can be modeled with UTCHEM as a specific fluid, and the model may also simulate the foam–oil interactions. However, Janssen et al. (2020) showed that the parametric approach of UTCHEM could not reproduce the foam behavior observed in experiments.

17.7.1 Foam in Porous Media Theoretical Basis

In the framework of the generalized Darcy theory, the foam is represented as a gas with specific properties, mainly a higher viscosity. The gas permeability is modified according to the following equation:

$$q_g = -\frac{k k_{rg}^f}{\mu_g}(\nabla P - \rho g) \text{ with } k_{rg}^f = \frac{k_{rg}}{1 + fmmob \prod F_i}$$

where q_g is the Darcy flux of gas, k the permeability, k_{rg} the relative permeability for the gas phase, k_{rg}^f the relative permeability modified to account for foam, P is the pressure, g the gravity, ρ the density, $fmmob$ the maximum foam resistance, and F_i are foam factors.

In the STARS approach, each of the F_i function lies between 0 and 1 and the lower it is, the higher the k_{rg}^f will be, i.e., the lower the foam strength. Thus, this model allows, when one of the factors is zero, to return to a model of classical gas transport. In its formulation, the model also implicitly contains the fact that the presence of foam does not directly modify the capillary curves of water in the porous medium, which had been demonstrated by Bernard and Jacobs 1965 or Vassenden and Holt 1998 and more recently by Eftekhari and Farajzadeh (2017). Thus, the presence of foam mainly leads to a change in S_w which in turn modifies the gas relative permeability.

As shown in the previous sections, the major specificity of the foam is to present a maximum of viscosity for a given water content (often low), with a very fast decrease of apparent viscosity for low water contents (high quality) and much slower for the high water content part (low quality). This specific behavior is reproduced by a factor directly related to the water content, called F_{dry} for the dry-out effect of the foam at high gas contents. In the STARS model, F_{dry} is represented by:

$$F_{dry} = 0.5 + \frac{\arctan(epdry \cdot (S_w - fmdry))}{\pi}$$

where $fmdry$ is the critical water saturation under which foam collapses and $epdry$ controls the abruptness of the foam collapse (Vicard et al. 2022). This function creates a very steep front close to $fmdry$. Figure 17.12 represents on the left a modeled quality scan (QS) curve and on the middle the corresponding k_{rg}^f curve on a log scale. We can see that the maximum apparent viscosity point corresponds to a very steep drop in k_{rg}^f. In the k_r graph, the k_{rg} value for pure gas is presented, and one can see that the "gas as foam" relative permeability is several orders of magnitude lower than the pure gas, the difference depending mainly on $fmmob$ value. Although close to S_w^* the value of the water relative permeability is quite low, because the amount of water in the medium is low. This is illustrated by the right panel of the figure where the mobilities of the two phases are presented with the total mobility. It can be seen that close to S_w^* the total mobility is very small and is driven by the gas phase while when S_w increases, i.e., in the low-quality region (low f_g value in QS curve, high S_w) the water mobility becomes more important and finally drives the foam behavior.

The other functions have all the same formula in the STARS model:

$$F_i = \left(\frac{X}{f_i}\right)^{epi}$$

where X is a variable, and f_i and epi are constant. The functions studied here concern the effect of non-Newtonian behavior with the factor F_{cap}, the effect of the presence of oil with F_{oil}, and the surfactant concentration F_{surf}.

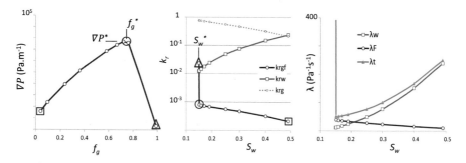

Fig. 17.12 Left: Quality scan curve with the maximum identified; middle: corresponding k_r (S_w) curves for the QS curve, large red symbols indicating the correspondence between the points, k_{rg} for pure gas, k_{rg}^f for gas as foam, k_{rw} for water; right: mobilities of the foam and water presented as function of S_w. Note that the high f_g region corresponds to the low S_w range

17.7.2 Fitting Foam Model Parameters

The STARS model includes in general two parameters for each factor, but the calibration is difficult because the observation is done on QS curves, whereas the real parameters of the model are the S_w and k_{rg}^f curves which are difficult to measure in this context. Moreover, the maximum of a QS curve is rarely obtained experimentally, and it does not correspond directly to the *fmdry* parameter. Therefore, there are different approaches that are all iterative. Ma et al. (2014) propose a method for fitting the *f*mdry parameters using optimization on a steady-state experiment supplemented with a transient experiment. Zeng et al. (2016) start with a quality scan experiment (fixed total velocity) to determine F_{dry} parameters and the transition quality point f_g^* followed by experiments at the given f_g^* value for several total velocities. However, their approach did not cover all the low-quality regime and lead to very low *epdry* values which are too low compared to the literature. Abbaszadeh et al. (2018) used a similar approach. Boeije and Rossen (2015b) used the high-quality part of the curve to determine graphically the parameters *fmdry* and *fmmob*, then using the curvature of the low-quality part chose a point to obtain a value of *epcap* and then *fmcap*. This procedure gave good results but has the disadvantage of not being automatic. Hosseini-Nasab et al. (2018) used a multi-parameter joint calibration, but it was performed on a single curve.

Vicard et al. (2022) proposed a more general approach which aims at calibrating a set of curves obtained for different total velocities. This involved a calibration of the non-Newtonian part using an Oswald's law type approach on the set of curves, followed by an iterative calibration of the parameters of the F_{dry} curve, differentiated according to the total velocities. This calibration approach gave satisfactory results on three different data sets.

Some authors did not use quality scan curves to retrieve foam model parameters. Ding et al. (2020) performed history matching showing a good fit to the model parameters, including the presence of oil.

The study by Valdez et al. (2021) detailed an estimate of the uncertainties on the major parameters of a quality scan curve. It turns out that with a few points and a single curve, the uncertainty is quite high, but it decreases sharply if the number of points on the curve is increased, justifying the need of several curves to calibrate the various model parameters.

17.7.3 Effects of Permeability

Luo et al (2019) gathered about ten studies with permeability variations highlighting, for most of them, a power law increase of apparent viscosity with permeability. However, as the authors pointed out, most of the studies cannot be used directly to determine the model parameters, as most of the experiments were conducted for a single foam quality. The more detailed study by Farajzadeh et al. (2015), confirmed by Gassara et al. (2017), similarly revealed a power relationship, implying small variations of *fmdry* as a function of k. In other papers a decrease of *fmdry* with permeability is also shown. If the apparent viscosity increases with the permeability, the global pressure gradient decreases slightly, as the increase in viscosity does not totally compensate the effect of the permeability. This viscous effect leads, however, to a major effect of the foam, which is the profile stabilization. This effect already reported in the theoretical part is explored using modeling by Wei et al. (2018) whom show blocking of the most permeable channels due to foam. Wang et al. (2019) present results showing an increase in the apparent viscosity of the foam with increasing permeability, up to a peak that occurs around 1 Darcy for a 3:5 mixture of LAO and OA-12 (lauramidopropylamine oxide and N,N-dimethyldodecylamine).

17.7.4 Effects of the Surfactant Type and Concentration

Most studies show major factors (*fmdry, fcap, epcap*) that differ widely with the type of surfactant and even with their concentrations. Farajzadeh et al. (2015) show, using moradi-araghi data that the model parameters change depending on the surfactant used for the same medium and injection conditions. Jones et al. (2016) show that the concentration of AOS significantly changes the shape of the quality scan curve during nitrogen foam injection. The curves also differ depending on the solution used (brine or fresh water). Zeng et al. (2016) compared an anionic surfactant (AOS1416) with to zwitterionic surfactants (LB and LS) using nitrogen and demonstrates a better foam strength with the anionic AOS. Presently, there is, to our knowledge, no way to extrapolate the parameters value from one surfactant to another.

17.7.5 Effects of Total Velocity

We performed a global study on the role of the total superficial velocity (u_t) on the variation of the quality scan curves and associated parameters gathering most of the studies providing such data (Vicard et al. 2022). This was done on three different media as no other studies report quality scan curves for different interstitial velocity. Figure 17.13 shows the major results as the variation of the transition point position between low and high quality with the total interstitial velocity. The value of $f_g{}^*$ is constant for the studies in quite high permeability media of Vicard et al. (2021) ($k = 5 \cdot 10^{-11}$ m^2) and Osterloh and Jante (1992) ($k = 6 \cdot 10^{-12}$ m^2) while significantly increases with u_t for lower permeability in the Alvarez et al. (2001) study ($k = 5 \cdot 10^{-13}$ m^2). As it is the first analysis of this type, we do not know if this effect is really due to the permeability and if the results in other media would be similar. These three studies illustrate that the value of $f_g{}^*$ largely varies according to surfactant (as already stated above), and quite low $f_g{}^*$ values were obtained with saponin (Vicard et al. 2022).

Concerning the pressure gradient (Fig. 17.12b) one thing clearly appears is that the pressure gradient is much higher in Alvarez et al. (2021) and Osterloh and Jante (1992) studies which operate in consolidated media and with non-environmental-friendly surfactant. It is known in the literature that the typical surfactants used in oil industry (AOS, SDS or similar) provide very strong foam and thus high pressure gradients.

In Alvarez et al. (2001), due to the small range of velocity tested, the shear thinning effect appears only at the highest velocity: the pressure gradient decreases (Fig. 17.12b). For the studies in higher permeability media, the relation between the pressure gradients shows a slight shear thinning effect for the Osterloh and Jante

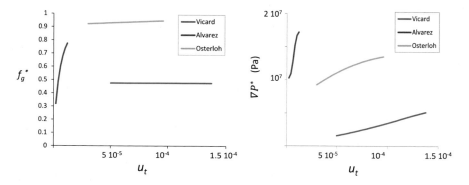

Fig. 17.13 Evolution of the breaking point of the quality scan curves as a function of velocity for three studies (Vicard et al. 2022, Alvarez et al. 2001, Osterloh and Jante 1992), left: the value of the gas fraction at the transition point ($f_g{}^*$), on the right value of the pressure gradient (U_t) at the same point (values of the Vicard's experiments were multiplied by 10 to see them better). U_t in m·s^{-1}

(1992) points but a simple linear relationship for Vicard et al. (2022)'s points. Therefore, as could be expected, the shear thinning effect varies with the medium permeability and certainly also with the type of surfactant associated with the pressure gradient value (Fig. 17.13).

17.7.6   Model Results to Understand Foam Behavior

With the limitations provided above, several papers used the STARS approach to detail the foam behavior in porous media.

17.7.6.1   Gravity Effect

Batot et al. (2016) illustrated the gravity effect on vertical 2D simulations (without radial effect) with a STARS-type foam model calibrated on 1D experiments. Using AOS and CO_2, the simulations showed a gravitational effect on the injection without foam, while the foam allowed eliminating almost totally this effect, with velocities of the order of 200 m/year (or about 0.6 m/day). However, in order to limit this gravity effect, the tendency is to increase the strength of the foam, which induces a low injectivity. Boeije and Rossen (2015a) showed that, by using SAG, a simplified model can determine an injection optimum that limits the gravity effect.

17.7.6.2   Heterogeneous Media

Tham (2015) compared simulation scenarios with and without foam, showing the profile stabilization effect with foam, but under the conditions presented, the amount of oil recovered in the presence of foam increases by less than 10% compared to the water alternating gas case, thus reducing heterogeneity is not always enough to remove all the oil from the porous media.

17.7.6.3   Foam Flow Around Injection Wells

Gong et al. (2018) pointed out through a simplified model that conventional reservoir models have difficulties in representing the process of foam injection in the vicinity of wells. Their study showed that these simulators, due to the large grid size close to the wellbore, tend to simulate too low injectivity values for both gas and water.

17.7.6.4 Unstability and Transient Effects

In the presence of oil or in coarse media, different authors show that several PVs are needed before obtaining a stable and efficient foam (e.g., Hosseini-Nasab et al. 2018). These processes were modeled by Lotfollahi et al. (2016), in a bubble model, the process being quite complex. To our knowledge this transient effect is not directly modeled in LE models or simplified by a foam formation approach. Instabilities within the foam in a SAG-type process are described by Farajzadeh et al. (2016), who states that they are due to spatially varying mobilities and can be limited by reducing the mobility at the injection point to a value close to the front.

17.8 Applying Foam Models for LNAPL-Contaminated Sites

The vast majority of published studies have been carried out under conditions similar to those of underground reservoirs for enhanced oil recovery (EOR). However, for an environmental remediation application, many differences exist:

- The permeability of the medium is often higher than deep sediments of EOR.
- The injection pressures are limited by the sediment resistance close to the soil surface (Maire et al. 2018a, b; Fitzhenry et al. 2022).
- Most experiments carried out for the EOR use CO_2, whereas this being more soluble than air or N_2 does not necessarily favor the desired behavior of the foam.
- The temperatures of oil reservoirs are much warmer, which a priori represents more difficult conditions than those for aquifers at the surface.

It appears that the conditions in ER are very different, which means that it will not be possible to use the same parameters as those used in the EOR conditions. However, different studies show that the model formulation can remain the same, and foam data are getting more abundant for these media.

17.9 Conclusion

This chapter presented the behavior of foam in porous media with the objective to use it as a remediation or blocking agent. The physics of foam is complex which justifies the introduction of several parameters to transform classical multiphase models into foam specific models. These parameters are quite empirical and require numerous experiments. Moreover, the foam parameters depend not only on the porous medium but also on the chemical properties of the foam and the injection velocity. Laboratory experiments helped to understand the behavior of foam and demonstrated that foam is able to remove NAPL content of the porous medium to various levels, often better than with conventional surfactants or other techniques. Other advantages of foam are its low cost and shear-thinning behavior favoring injection front stability and sweeping of the porous media.

Very few experiments have been conducted at field scale, although they have been relatively successful. Several explanations can be considered: (i) before injection, multiple experiments must be performed to select surfactant and foam types, (ii) the well injection itself is complex to be set, and (iii) in environmental application the surfactant choice is limited and those surfactants are less effective than in the case of EOR. Despite these difficulties, the recent increase of interest in foam applications and the increasing technical level reported in the literature suggest that foam injection is a very promising technique for in situ remediation at complex sites.

References

Abbaszadeh M, Varavei A, Solutions IP, Rodrı F, Garza D, Enrique A, Miller CA (2018) Methodology for the development of laboratory-based comprehensive foam model for use in the reservoir simulation of enhanced oil recovery. SPE Reser Evaluat Eng 21:26–28

Alvarez JM, Rivas HJ, Rossen WR (2001) Unified model for steady-state foam behavior at high and low foam qualities. SPE J 6(03):325–333

Batôt G, Delaplace P, Bourbiaux B, Nouvelles IFPE, Pedroni LG, Nabzar L (2016) WAG management with foams: influence of injected gas properties an surfactant adsorption

Bernard GG, Jacobs WL (1965) Effect of foam on trapped gas saturation and on permeability of porous media to water. SPE J 5:295–300

Boeije CS, Rossen WR (2015a) Gas-injection rate needed for SAG foam processes to overcome gravity override. SPE J 20:49–59

Boeije CS, Rossen W (2015b) Fitting foam-simulation-model parameters to data: I. Coinjection of gas and liquid. SPE Reser Evaluat Eng 18:264–272

Bouzid I, Maire J, Fatin-Rouge N (2019) Comparative assessment of a foam-based method for ISCO of coal tar contaminated unsaturated soils. J Environ Chem Eng 7(5):103346

Brattekås B, Fernø MA (2016) New insight from visualization of mobility control for enhanced oil recovery using polymer gels and foams. https://doi.org/10.5772/64586

Cantat I, Cohen-Addad S, Elias F, Graner F, Hohler R, Pitois O, Rouyer F, Saint-Jalmes A (2013) Foams: structure and dynamics. Belin. https://doi.org/10.1093/acprof:oso/9780199662890.001.0001

Denkov ND (2004) Mechanisms of foam destruction by oil-based antifoams. Langmuir ACS J Surf Colloids 20(22):9463–9505

Denkov N, Tcholakova S, Politova-Brinkova N (2020) Physicochemical control of foam properties. Curr Opin Colloid Interf Sci 50:81. https://doi.org/10.1016/j.cocis.2020.08.001

Ding L, Cui L, Jouenne S, Gharbi O, Pal M, Bertin H, Gue D (2020) Estimation of local equilibrium model parameters for simulation of the laboratory foam-enhanced oil recovery process using a commercial reservoir simulator. https://doi.org/10.1021/acsomega.0c03401

Dwarakanath V, Jackson RE, Pope GA (2002) Influence of wettability on the recovery of NAPLs from alluvium. Environ Sci Technol 36(2):227–231

Eftekhari AA, Farajzadeh R (2017) Effect of foam on liquid phase mobility in porous media. Nature Publishing Group, New York, pp 1–8. https://doi.org/10.1038/srep43870

Falls AH, Hirasaki GJ, Patzek TW, Gauglitz PA, Miller DD, Ratulowski T (1988) Development of a mechanistic foam simulator: the population balance and generation by snap-off. SPE Res Eng 3:884–892

Farajzadeh R, Andrianov A, Zitha PLJ (2010) Investigation of immiscible and miscible foam for enhancing oil recovery. Ind Eng Chem Res 49(4):1910–1919

Farajzadeh R, Andrianov A, Krastev R, Hirasaki GJ, Rossen WR (2012) Foam-oil interaction in porous media: implications for foam assisted enhanced oil recovery. Adv Colloid Interf Sci 183–184:1–13

Farajzadeh R, Lotfollahi M, Eftekhari AA, Rossen WR, Hirasaki GJH (2015) Effect of permeability on implicit-texture foam model parameters and the limiting capillary pressure. Energy Fuels 29:3011–3018. https://doi.org/10.1021/acs.energyfuels.5b00248

Farajzadeh R, Eftekhari AA, Hajibeygi H, Kahrobaei S, van der Meer JM, Vincent-Bonnieu S, Rossen WR (2016) Simulation of instabilities and fingering in surfactant alternating gas (SAG) foam enhanced oil recovery. J Nat Gas Sci Eng 34:1191–1204

Fatin-Rouge N, Visitacion Carillo S (2018) Récupération de phases liquides organiques en milieu karstique. Institut UTINAM (UMR 6213)

Fatin-Rouge N (2020) Contaminant mobilization from polluted soils: behavior and reuse of leaching solutions. In: van Hullebusch ED et al (eds) Environmental soil remediation and rehabilitation, applied environmental science and engineering for a sustainable future. Springer, New York. https://doi.org/10.1007/978-3-030-40348-5_1

Fitzhenry E, Martel R, Robert T (2022) Foam injection for enhanced recovery of diesel fuel in soils: sand column tests monitored by CT scan imagery. J Hazard Mater 3:128777. https://doi.org/10.1016/j.jhazmat.2022.128777

Fitzhenry E (2021) Application de la scanographie à l'étude de l'injection de mousse pour le traitement de sols contaminés au diesel, Mémoire de maîtrise, Institut national de la recherche scientifique. Centre Eau Terre et Environnement, Québec, p 174

Forey N, Atteia O, Omari A, Bertin H (2019) Saponin foam for soil remediation: on the use of polymer or solid particles to enhance foam resistance against oil. J Contam Hydrol 27:103560. https://doi.org/10.1016/j.jconhyd.2019.103560

Forey N, Atteia O, Omari A, Bertin H (2021) Use of saponin foam reinforced with colloidal particles as an application to soil remediation: experiments in a 2D tank. J Contam Hydrol 238:103761. https://doi.org/10.1016/j.jconhyd.2020.103761

Friedmann F, Chen WH, Gauglitz PA (1991) Experimental and simulation study of high-temperature foam displacement in porous media. Soc Petr Eng Res Eng 6:37–45

Gassara O, Douarche F, Braconnier B, Bourbiaux B (2017) Calibrating and interpreting implicit-texture models of foam flow through porous media of different permeabilities. J Petrol Sci Eng 159:588–602. https://doi.org/10.1016/j.petrol.2017.09.069

Gong J, Global S, International S (2018) SPE-190435-MS modelling of liquid injectivity in surfactant-alternating-gas foam enhanced oil recovery

Hirasaki GJ, Miller CA, Szafranski R, Lawson JB, Akiya N (1997a) Surfactant/foam process for aquifer remediation. In: Society of petroleum engineers, international symposium on oilfield chemistry. Houston, TX, p 10

Hirasaki GJ, Miller CA, Szafranski R, Tanzil D, Lawson JB, Meinardus H, Jin M, Londergan JT, Jackson RE, Pope GA, Wade WH (1997b) Field demonstration of the surfactant/foam process for aquifer remediation. In: Society of petroleum engineers, annual technical conference and exhibition. San Antonio, TX, p 16

Hirasaki GJ, Lawson JB (1985) Mechanisms of foam flow in porous media: apparent viscosity in smooth capillaries. SPE J 25(02):176–190. https://doi.org/10.2118/12129-PA

Hosseini-nasab SM, Douarche F, Simjoo M, Nabzar L, Bourbiaux B, Zitha PLJ, Roggero F (2018) Numerical simulation of foam flooding in porous media in the absence and presence of oleic phase. Fuel 225:655–662. https://doi.org/10.1016/j.fuel.2018.03.027

Janssen MTG, Pilus RM, Zitha PLJ (2019) A comparative study of gas flooding and foam-assisted chemical flooding in bentheimer sandstones. Transp Porous Media 131(1):101–134

Janssen MTG, Torres Mendez FA, Zitha P (2020) Mechanistic modeling of water-alternating-gas injection and foam-assisted chemical flooding for enhanced oil recovery. Ind Eng Chem Res 59:3606–3616. https://doi.org/10.1021/acs.iecr.9b06356Jones

Jones SA, Laskaris G, Vincent-bonnieu S, Farajzadeh R, Rossen WR (2016) Journal of industrial and engineering chemistry effect of surfactant concentration on foam: from coreflood experiments to implicit-texture foam-model parameters. J Ind Eng Chem 37:268–276. https://doi.org/10.1016/j.jiec.2016.03.041

Jones SA, Getrouw N, Vincent-Bonnieu S (2018) Foam flow in a model porous medium: I. The effect of foam coarsening. Soft Matter 14(18):3490–3496. https://doi.org/10.1039/C7SM01903C

Karthick RA, Chattopadhyay P (2017) Remediation of diesel contaminated soil by tween-20 foam stabilized by silica nanoparticles. Int J Chem Eng Appl 8(3):194–198

Lake LW (1989) Enhanced oil recovery. Prentice Hall, Englewood Cliffs, NJ, p 550

Liontas R, Ma K, Hirasaki GJ, Biswal SL (2013) Neighbor-induced bubble pinch-off: novel mechanisms of in situ foam generation in microfluidic channels. Soft Matter 9(46):10971

Longpré-Girard M, Martel R, Robert T, Lefebvre R, Lauzon JM (2016) 2D sandbox experiments of surfactant foams for mobility control and enhanced LNAPL recovery in layered soils. J Contam Hydrol 193:63–73

Longpré-Girard M, Martel R, Robert T, Lefebvre R, Lauzon J-M, Thomson N (2019) Surfactant foam selection for enhanced light non-aqueous phase liquids (LNAPL) recovery in contaminated aquifers. Transp Porous Media 131(1):65–84

Lotfollahi M, Farajzadeh R, Delshad M, Varavei A, Rossen WR (2016) Comparison of implicit-texture and population-balance foam models. J Nat Gas Sci Eng 31:184–197

Luo H, Ma K, Mateen K, Ren G, Neillo V, Blondeau C, Bourdarot G (2019) A mechanistic foam simulator incorporating systematic dependencies of various foam properties on permeability

Ma K, Farajzadeh R, Miller CA, Lisa S, George B (2014) Sensitivity in simulating steady-state and transient foam flow through porous media. Transp Porous Media 102:325–348. https://doi.org/10.1007/s11242-014-0276-9

Ma K, Ren G, Mateen K, Morel D, Cordelier P (2015) Modeling techniques for foam flow in porous media. SPE J 20:453–470. https://doi.org/10.2118/169104-PA

Maire J, Fatin-Rouge N (2017) Surfactant foam flushing for in situ removal of DNAPLs in shallow soils. J Hazard Mater 321:247–255

Maire J, Brunol E, Fatin-Rouge N (2018a) Shear-thinning fluids for gravity and anisotropy mitigation during soil remediation in the vadose zone. Chemosphere 197:661–669

Maire J, Joubert A, Kaifas D, Invernizzi T, Marduel J, Colombano S, Cazaux D, Marion C, Klein PY, Dumestre A, Fatin-Rouge N (2018b) Assessment of flushing methods for the removal of heavy chlorinated compounds DNAPL in an alluvial aquifer. Sci Total Environ 612:1149–1158

Marchalot J, Lambert J, Cantat I, Tabeling P, Jullien MC (2008) 2D foam coarsening in a microfluidic system. EPL 83(6):64006. https://doi.org/10.1209/0295-5075/83/64006

Mulligan CN, Eftekhari F (2003) Remediation with surfactant foam of PCP-contaminated soil. Eng Geol 70(3–4):269–279

NASEM, National Academies of Sciences, Engineering and Medecine. (2015) Characterization, modeling, monitoring, and remediation of fractured rock. The National Academies Press, Washington, DC

Nguyen P, Fadaei H, Sinton D (2014) Pore-scale assessment of nanoparticle-stabilized CO_2 foam for enhanced oil recovery. Energy Fuels 28(10):6221–6227

Olabode OA, Ogbebor VO, Onyeka EO (2021) The effect of chemically enhanced oil recovery on thin oil rim reservoirs. J Petrol Explor Prod Technol 11:1461–1474. https://doi.org/10.1007/s13202-021-01090-9

Osei-Bonsu K, Shokri N, Grassia P (2015) Foam stability in the presence and absence of hydrocarbons: from bubble- to bulk-scale. Colloids Surf A Physicochem Eng Aspects 481:514–526

Osei-Bonsu K, Grassia P, Shokri N (2017) Investigation of foam flow in a 3D printed porous medium in the presence of oil. J Colloid Interf Sci 490:850–858. https://doi.org/10.1016/j.jcis.2016.12.015

Osterloh WT, Jante MJ (1992) Effects of gas and liquid velocity on steady-state foam flow at high temperature. In: Society of petroleum engineers, enhanced oil recovery symposium. Tulsa, Oklahoma, p 12

Parlar M, Parris MD, Jasinski RJ, Robert JA (1995) SPE 29678 an experimental study of foam flow through berea sandstone with applications to foam diversion in matrix acidizing

Peksa AE, Wolf K-HAA, Zitha PLJ (2015) Bentheimer sandstone revisited for experimental purposes. Mar Petrol Geol 67:701–719

Pennell KD, Pope GA, Abriola LM (1996) Influence of viscous and buoyancy forces on the mobilization of residual tetrachloroethylene during surfactant flushing. Environ Sci Technol 30(4):1328–1335

Portois C, Boeije CS, Bertin HJ, Atteia O (2018a) Foam for environmental remediation: generation and blocking effect. Transp Porous Media 124(3):787–801

Portois C, Essouayed E, Annable MD, Guiserix N, Joubert A, Atteia O (2018b) Field demonstration of foam injection to confine a chlorinated solvent source zone. J Contam Hydrol 214:16–23

Ransohoff TC, Radke CJ (1998) Mechanisms of foam generation in glass bead pack. In: Society of petroleum engineers and research engineers, SPE Reservoir Engineering, pp 573–585

Robert T, Martel R, Lefebvre R, Lauzon JM, Morin A (2017) Impact of heterogeneous properties of soil and LNAPL on surfactant-enhanced capillary desaturation. J Contam Hydrol 204:57–65

Roof JG (1970) Snap-off of oil droplets in water-wet pores. SPE J 10(01):85–90

Rosen MJ, Et Kunjappu JT (2012) Surfactants and interfacial phenomena. John Wiley & Sons, Hoboken

Roy D, Kongara S, Valsaraj KT (1995) Application of surfactant solutions and colloidal gas aphron suspensions in flushing naphthalene from a contaminated soil matrix. J Hazard Mater 42(3):247–263

Schramm LL (2000) Surfactants, fundamentals and applications in the petroleum industry. Cambridge University Press, Cambridge

Schramm LL, Novosad JJ (1990) Micro-visualization of foam interactions with a crude oil. Colloids Surf 46(1):21–43

Sheng JJ (2013) Enhanced oil recovery. Field case studies. Gulf Professional Publishing, New York

Simjoo M, Dong Y, Andrianov A, Talanana M, Zitha PLJ (2013a) CT scan study of immiscible foam flow in porous media for enhancing oil recovery. Ind Eng Chem Res 52(18):6221–6233

Simjoo M, Rezaei T, Andrianov A, Zitha PLJ (2013b) Foam stability in the presence of oil: effect of surfactant concentration and oil type. Colloids Surf A Physicochem Eng Aspects 438:148–158

Singh R, Mohanty KK (2017) Foam flow in a layered, heterogeneous porous medium: a visualization study. Fuel 197:58–69

Sun Q, Li Z, Li S, Jiang L, Wang J, Wang P (2014) Utilization of surfactant-stabilized foam for enhanced oil recovery by adding nanoparticles. Energy Fuels 28(4):2384–2394

Tang J, Vincent-Bonnieu S, Rossen WR (2019) CT coreflood study of foam flow for enhanced oil recovery: the effect of oil type and saturation. Energy 188:1597

Tham SL (2015) A simulation study of enhanced oil recovery using carbon dioxide foam in heterogeneous reservoirs. In: SPE annual technical conference and exhibition. OnePetro

Valdez AR, Rocha BM, Maria J, Vilela A, Souza OD, Pérez-gramatges A, Weber R (2021) Foam assisted water–gas flow parameters: from core flood experiment to uncertainty quantification and sensitivity. Transp Porous Media 157:1–21. https://doi.org/10.1007/s11242-021-01550-0

Vassenden F, Holt T (1998) Experimental foundation for relative permeability modeling of foam. In: SPE Improved Oil Recovery Conference, Tulsa, April 1998.

Vicard A, Atteia O, Bertin H, Lachaud J. (2022) Estimation of local equilibrium foam model parameters as functions of the foam quality and the total superficial velocity. ACS omega 7(20):16866–76

Vikingstad AK, Aarra MG, Skauge A (2006) Effect of surfactant structure on foam–oil interactions: comparing fluorinated surfactant and alpha oelfin sulfonate in static foam tests. Colloids Surf A Physicochem Eng Aspects 279(1–3):105–112

Wang H, Chen J (2012) Enhanced flushing of polychlorinated biphenyls contaminated sands using surfactant foam: effect of partition coefficient and sweep efficiency. J Environ Sci 24:1270–1277. https://doi.org/10.1016/S1001-0742(11)60881-4

Wang Y, Yue X, Liu K, Zhang B, Ling Q (2019) Effect of permeability on foam mobility and flow resistance distribution: an experimental study. Colloids Surf A Physicochem Eng Aspects 582:123769. https://doi.org/10.1016/j.colsurfa.2019.123769

Wei P, Pu W, Sun L, Pu Y, Wang S, Fang Z (2018) AC SC. J Petrol Sci Eng 163:340–348. https://doi.org/10.1016/j.petrol.2018.01.011

Xu T, Sonnenthal E, Spycher N, Zhang G, Zheng L, Pruess K (2012) TOUGHREACT: a simulation program for subsurface reactive chemical transport under non-isothermal multiphase flow conditions, groundwater reactive transport models 1:74. https://doi.org/10.2174/978160805306 311201010074

Yan S, Wei-guo C, Ya-bin L, Xin W, Jian W (2019) Enhanced removal efficiency and influencing factors of nitrobenzene in soil by foam flushing. IOP Conf Ser Earth Environ Sci 295:12047

Yang W, Wang T, Fan Z, Miao Q, Deng Z, Zhu Y (2017) Foams stabilized by in situ-modified nanoparticles and anionic surfactants for enhanced oil recovery. Energy Fuels 31(5):4721–4730

Yu Y, Soukup ZA, Saraji S (2019) An experimental study of in-situ foam rheology: effect of stabilizing and destabilizing agents. Colloids Surf A Physicochem Eng Aspects 28:578

Chapter 18
Advances in Low-Temperature Thermal Remediation

Jonah Munholland, Derek Rosso, Davinder Randhawa, Craig Divine, and Andy Pennington

Abstract Remediation through traditional high-temperature thermal techniques (over 100 °C) are designed to remove contaminants like petroleum hydrocarbons via enhanced mobilization and volatilization. However, remedies of this nature can require significant infrastructure, capital, operational and maintenance costs, along with high energy demands and carbon footprints. Conversely, low-temperature thermal approaches (in the mesophilic range of ~15–40 °C) are an inexpensive and more sustainable method that can enhance the physical, biological, and chemical processes to remove contaminants. Heat transfer properties of subsurface sediments and other geological materials do not vary considerably and are relatively independent of grain size, unlike hydraulic properties that can vary several orders of magnitude within a site and often limit the pace of remediation of many in-situ technologies. Therefore, low-temperature thermal remediation is a promising alternative that can remediate contaminant mass present in both high- and low-permeability settings, including fractured rock. Emergence of risk-based non-aqueous phase liquid management approaches and sustainable best management practices further offer a platform for low-temperature thermal remedies to advance petroleum hydrocarbon remediation with lower capital and operational costs. Case studies demonstrating

J. Munholland (✉) · D. Rosso · D. Randhawa · C. Divine · A. Pennington
Arcadis U.S. Inc, 630 Plaza Drive (Suite 200), Highlands Ranch, CO 80129-2379, USA
e-mail: jonah.munholland@arcadis.com

D. Rosso
e-mail: derek.rosso@arcadis.com

D. Randhawa
e-mail: randhawa.40@osu.edu

C. Divine
e-mail: craig.divine@arcadis.com

A. Pennington
e-mail: andy.pennington@arcadis.com

D. Randhawa
Innovation and Economic Development, Ohio State University, 1524 N. High Street, Columbus, OH 43201, USA

© The Author(s) 2024
J. García-Rincón et al. (eds.), *Advances in the Characterisation and Remediation of Sites Contaminated with Petroleum Hydrocarbons*, Environmental Contamination Remediation and Management, https://doi.org/10.1007/978-3-031-34447-3_18

this approach along with preliminary sustainability comparisons of the associated reduced energy use and carbon footprint are described in this chapter.

Keyword Enhanced biodegradation · Heat transfer · Petroleum hydrocarbons · Solar energy · Sustainable site management

18.1 Emergence of Low-Temperature Thermal Remediation

In-situ thermal remediation (ISTR) technologies include various methods for applying energy to the subsurface to raise in-situ temperatures within a targeted treatment area. Traditional ISTR is generally considered an enhanced physical recovery technology for volatile organic compounds (VOCs) and select semi-volatile organic compounds because it is typically employed in combination with physical extraction methods (e.g., multiphase extraction, soil vapor extraction [SVE]) with ex-situ treatment of recovered vapors/fluids. The primary effect of introducing heat to the subsurface is enhanced VOC mass transfer to the vapor phase by increasing vapor pressures and volatilization rates, increasing the air permeability of the soil, and enhancing the gas-phase diffusion process, thereby significantly improving overall contaminant mass removal rates. Therefore, conventional "higher temperature" ISTR technologies (i.e., thermal conduction heating [TCH], electrical resistive heating [ERH], and steam-enhanced extraction) are designed to achieve subsurface temperatures (> 100 degrees Celsius [°C]) with the objective of rapid removal of high mass contamination including non-aqueous phase liquids (NAPLs) from source zones. However, it has been observed that elevated subsurface temperatures may persist up to two years after ISTR operations are terminated (Krauter et al. 1995) and this residual heat can enhance microbial activity, resulting in further degradation of contaminants and/or increased release of electron donors via hydrolysis (Suthersan et al. 2012; Macbeth 2019).

Over the past two decades, numerous field- and bench-scale studies have demonstrated accelerated rates of biotic and abiotic degradation mechanisms based on increased temperature (e.g., Dablow et al. 1995; Powell et al. 2007; Truex et al. 2007). Truex et al. (2007) referred to this as "fortuitous" remediation where enhanced contaminant degradation was seen as temperatures cooled but remained more than typical background temperatures, conditions that prevailed for months following the end of subsurface heating. Recently, low-temperature thermal remediation (LTTR) approaches have been developed that integrate all the above and (for situations appropriate) target subsurface temperatures far lower than temperatures achieved with conventional higher temperature thermal remedies. The infrastructure requirements to achieve lower temperature goals can be far simpler than a conventional thermal application, but the approach still relies on the fundamentals of heat transfer.

Until relatively recently, the focus of LTTR has mainly been on abiotic chemical reactions (such as hydrolysis and dehydrohalogenation) that can be greatly enhanced for some contaminants under comparatively low subsurface temperatures (generally less than 70 °C). However, the ability to stimulate biotic mechanisms for contaminant degradation has received less attention despite the available supporting fundamental research.

Studies have also indicated that for microbial consortia existing in the soil, a corresponding threefold increase in biodegradation rates of benzene, toluene, ethylbenzene, and total xylenes (BTEX) can occur by raising the temperature from 20 to 30 °C (Margesin and Schinner 2001; Dettmer 2002). Kulkarni et al. (2022) reported that increasing the temperature by 10 °C could double natural source zone depletion (NSZD) rates in NAPL-impacted sites. Another study on petroleum hydrocarbon biodegradation rates due to increase in temperature has shown peak degradation rates between 30 and 40 °C (Xu 1997). Temperatures above the operating range for a type of microorganism may result in a switch in the microbial consortia, and changes in microbial biochemistry may be observed. From a field remediation perspective, 15–40 °C seems to be the operational range for enhancing most biological treatment mechanisms that can be enhanced by increasing temperature.

Physical extraction systems (e.g., ex-situ groundwater treatment, air sparge [AS]/ SVE) remain important conventional solutions to address contaminant mass, particularly in zones of high concentration and fine lithological formations. Combining low-temperature thermal remediation with physical extraction systems has the potential to enhance mass recovery via liquid extraction, volatilization or gradually increasing NAPL dissolution. These enhanced mechanisms can provide a low-cost way to improve the effectiveness of an existing physical extraction systems while lowering cleanup timeframes.

In this chapter, we explore the basis for the approaches mentioned above through a discussion of the following:

- Effects of temperature on certain chemical properties and degradation potential of constituents of concern (COCs);
- The fundamentals of in-situ heat transfer important to LTTR, including thermal modeling;
- Types of contemporary and advanced LTTR approaches;
- Case studies that represent state-of-the-practice applications and performance of LTTR technologies to treat petroleum hydrocarbons; and
- Potential advantages and benefits of the approaches in treating sites with low-permeability geologic materials and in developing sustainable remedial strategies.

The last item includes opportunities in utilizing sustainable sources of heat, such as solar energy and waste heat, significantly enhancing the sustainable aspects of the approach. We explore both with a detailed review of how to implement such an approach and multiple example field applications to demonstrate the potential of the technique.

18.2   Temperature-Dependent Contaminant Properties

As the temperatures in the subsurface change, important aspects of the behavior of the COCs will also change. For the purposes of this discussion, we will focus on the enhancing effect on contaminant partitioning followed by contaminant degradation.

18.2.1   Contaminant Partitioning

In general, an increase in temperature can increase the solubility of organic contaminants in water, enhancing NAPL dissolution and contaminant recovery rates of groundwater-focused remedies. However, solubility dependence on temperature can be complex, particularly in the context of petroleum mixtures, where COCs for remediation are a small portion of the large number of compounds present. This needs to be carefully considered because the solubility enhancement of different contaminants may behave differently as temperatures change. Published solubility data for petroleum hydrocarbon compounds span a range of values, with authors reporting different results and noting that contradictory or conflicting values in literature are common. This variability may be due to differences in accounting for effects of phase change, biodegradation, and co-occurring compounds present in petroleum mixtures. However, published data can generally be helpful in assessing the potential effects of heating on dissolution. Figure 18.1 compares the approximate solubility of methyl tert-butyl ether (MTBE), benzene, and toluene. These concepts are discussed in detail by the US Environmental Protection Agency (USEPA) in the supporting documentation for their On-Site Assessment Tool (USEPA 2021).

The increase in temperature can also enhance dissolved phase/vapor phase partitioning for those same higher molecular weight chemicals as measured by changes in Henry's Law constants, as shown in Fig. 18.2. Henry's Law constants represent the proportionality between the gas phase and the dissolved phase for VOCs; therefore, as the Henry's Law constant increases so does the susceptibility of those contaminants to partitioning from the dissolved phase to the vapor phase.

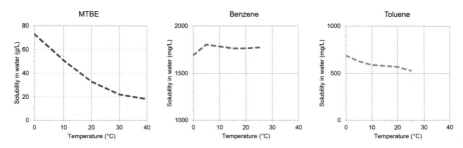

Fig. 18.1 Solubility of MTBE, benzene, and toluene relative to temperature (data from USEPA 2021, adapted from Shaw 1989 and Peters et al. 2002)

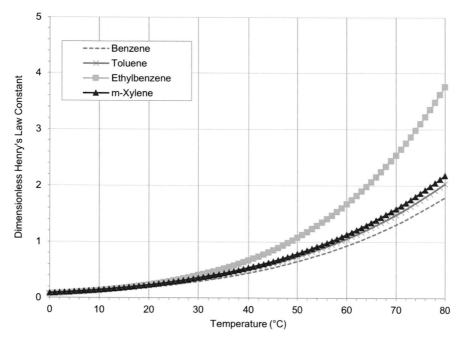

Fig. 18.2 Henry's Law constant relative to temperature (adapted from Sander 2015 and Staudinger and Roberts 2001)

When considering physical extraction systems (e.g., ex-situ groundwater treatment, AS/SVE) together with LTTR, the potential to increase volatilization can make the recovery of organic contaminants easier. The relationship between Henry's Law constants and temperature for COCs can be used to evaluate the potential for increased risk of vapor intrusion exposures during heating. This is a common concern raised during remedy evaluation and planning stages for LTTR approaches. It should be noted that temperature-dependent increases in volatilization or dissolution of constituents will likely be occurring simultaneously with temperature-dependent increases in biotic and abiotic degradation of volatilized or dissolved mass. Increases in mass transfer rates and increases in degradation rates can act to balance each other to some degree, making the net effect on exposure potentially lower than would be predicted through the evaluation of changes in mass transfer alone. As such, empirical evaluation of constituent concentrations in the medium or media of concern (e.g., monitoring of concentrations in soil gas or groundwater) is likely the most appropriate way to assess potential exposure concerns.

18.3   Temperature-Enhanced Biological Degradation

As remediation practitioners' understanding of biodegradation has grown, these natural processes are increasingly recognized as major contributors to positive remedial outcomes and observed successes at petroleum sites, even at sites where engineered mass transfer and mass removal (e.g., AS, SVE, and light non-aqueous phase liquid [LNAPL] pumping) were previously thought to provide the bulk of remedial benefit. Relatively modest increases in temperature can substantially accelerate biological reactions and offer important possibilities for engineering of in-situ bioremediation approaches.

Overall, the process of anaerobic, methanogenic biodegradation of organic material is well-understood through past work in engineered systems (e.g., study of landfills and anaerobic digesters in wastewater treatment) and the study of sites such as the U.S. Geological Survey's National Crude Oil Spill Fate and Natural Attenuation Research site located near Bemidji, Minnesota (Essaid et al. 2011; Garg et al. 2017). Microorganisms relevant to in-situ contaminant degradation are often classified based on the temperature range at which they are most active or prolific. Table 18.1 provides an example classification following this approach (Dettmer 2002). Mesophilic and thermophilic bacteria are promising for biodegradation of hydrocarbons with low water solubility, as solubility and subsequent bioavailability are enhanced at elevated temperatures (Margesin and Schinner 2001).

In general, a temperature range of 15–40 °C is conducive to favorable rates for typical biodegradation processes, with kinetics increasing toward the upper end of that range (Daniel et al. 2000). This is especially true for BTEX compounds as shown in Fig. 18.3. Maximum biodegradation rates in engineered anaerobic systems are often observed in the vicinity of 35–40 °C (Gerardi 2003; Xu 1997). Additionally, laboratory-observed biodegradation rates of LNAPL were found to be 0.002 mg L/h, 0.008 mg L/h, 0.012 mg L/h, and 0.015 mg L/h at increasing soil–water temperatures of 4 °C, 20 °C, 28 °C, and 36 °C, respectively (Yadav and Gupta 2022). Beyond 40 °C, biological degradation mechanisms may be limited by the decomposition or denaturing of enzymes involved in degradation pathways. Additionally, for in-situ systems where direct manipulation of the microbial population is difficult, the maximum attainable increase in biodegradation rates may come from modest increases in subsurface temperature (e.g., raising temperature to levels at the high end of historical natural ranges for a site, allowing the native microbial population

Table 18.1 Microorganisms classified based on temperature range (adapted from Dettmer 2002)

Classification	Low (°C)	High (°C)
Psychrophilic	0	20
Mesophilic	20	40
Thermophilic	40	80
Hyperthermophylic	80	> 100

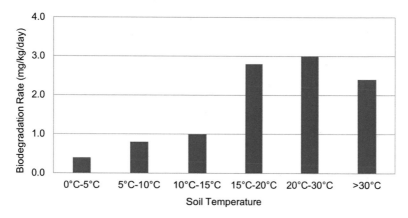

Fig. 18.3 BTEX biodegradation increase with soil temperature (adapted from Dettmer 2002)

to thrive rather than creating higher-temperature conditions suitable for microbes which may not be abundant) (Zeman 2013).

The natural degradation processes typically in place at petroleum hydrocarbon release sites are complex, and research into specific components and limiting factors is ongoing as part of work on natural source zone depletion. In general, however, a typical petroleum hydrocarbon site undergoing natural degradation will exhibit a strongly reducing zone of acetogenesis, fermentation, and methanogenesis in an anaerobic setting. Methane is often subsequently oxidized by subsurface bacteria as soil vapor reaches more aerobic conditions distant from the petroleum source, yielding carbon dioxide as the primary degradation product. The conversion of heavier petroleum compounds to methane and carbon dioxide as part of these processes can be tracked using established soil gas flux measurement methods (as described in American Petroleum Institute Publication 4784, API 2017) to evaluate approximate contaminant degradation rates.

Aerobic degradation processes can also be accelerated through temperature increases, leading to potential benefits from addition of low-temperature heating to traditional biosparging and bioventing systems designed to maintain oxygen-rich subsurface environments for petroleum degradation. The basic relationship between kinetics of aerobic hydrocarbon degradation and temperature, with increased oxygen utilization at temperatures of 20–30 °C, is well-documented in a variety of settings (e.g., Thamdrup et al. 1998). Furthermore, the exothermic nature of aerobic oxidation of hydrocarbons can create substantial heat, leading to a positive feedback loop and reducing the need for external energy inputs as biodegradation rates accelerate. A variety of performance indicators can be used to track progress of thermally enhanced aerobic biodegradation systems, including subsurface heat flux, oxygen utilization, and carbon dioxide generation.

18.4   In-Situ Heat Transfer Considerations

The application of LTTR generally involves the direct transfer of heat from an energy source to a delivery point (e.g., borehole heat exchanger [BHE]) installed within the subsurface. The fundamentals of heat transfer in porous media involve the following three primary mechanisms:

Conduction: This mechanism requires physical contact and is driven by temperature differences. The rate of heat flux is proportional to the magnitude of the temperature gradient, the surface area involved in the transfer, the properties of the matrices through which the heat must be transferred, and the length of the travel path.

Convection: This mechanism involves the transfer of heat associated with the macroscopic movement of fluids, which expand and become less dense when heated. This density change causes the hot fluid to rise, displacing cooler fluid that will sink. This can set up a circular current until the fluid temperature approaches the temperature of the heat source, which is generally a solid surface. Like conduction, convection is driven by the magnitude of the temperature gradient, the surface area of the heat source, and the properties of the fluid (including its bulk motion and permeability of the medium).

Radiation: This is the only transfer mechanism that does not require direct contact. All objects radiate thermal energy based on their temperature, the hotter the object the more it radiates.

   Conduction and convection are the main heat transfer mechanisms that govern the resulting temperature distributions during LTTR applications. More often, advection (the flow of fluids) can inhibit heating by continuously exchanging the fluid volume with fresh unheated fluid, a factor that must be identified and considered during the design stage. To help explain, we can consider a unit volume within the subsurface that contains solid material and void space (i.e., pores). The solid material can consist of different types of rocks, minerals, and organic materials of various shapes and sizes. The pore space can contain varying amounts of gases and liquids. The composition of the unit volume leads to variations in the thermal properties. Heat transfer by thermal conduction is related to its effective thermal conductivity, which is a measure of its ability to transmit heat by way of random molecular motion. This is represented in the energy conservation equation (Eq. 18.1) that governs subsurface temperature distribution (Hegele and McGee 2017):

$$\nabla \cdot [\lambda \nabla T] - \rho C_f q_f \cdot \nabla T + U = \frac{\partial}{\partial t}(\rho C T) \qquad (18.1)$$

where λ is the bulk thermal conductivity (watts per meter-kelvin [W/mK]), T is temperature (K), ρC_f is the volumetric heat capacity of the fluid, q_f is the Darcy velocity, U represents the heat source/sink term (watts per cubic meter [W/m^3]), and ρC is the bulk volumetric heat capacity of the unit volume. It should be noted that soil thermal conductivity is affected by water saturation. Figure 18.4 presents

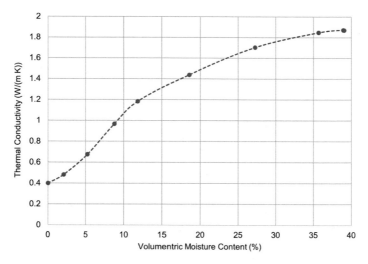

Fig. 18.4 Laboratory-measured relationship between thermal conductivity and saturation for a silty sand

a laboratory-measured relationship between thermal conductivity and the saturation of a silty sand. In this case, the relationship is nonlinear and the thermal conductivity of the soil under fully saturated conditions is more than six times higher than when it is completely dry. In some cases, it may be important to measure or use tools to predict this relationship (e.g., Likos 2014; Zhang and Wang 2017) and use this understanding to further parameterize predictive models.

Heat capacity is another important physical characteristic of any material relative to the ability to conduct heat and represents the ratio of heat added to (or removed from) an object to the resulting temperature change. This can be used to determine the thermal diffusivity, which is a measure of a material's ability to conduct heat relative to its ability to store heat (i.e., thermal conductivity divided by heat capacity). When heat is applied to a unit volume of soil, the heat flux through the unit volume will depend on the effective thermal conductivity and the corresponding rise in temperature will depend on the thermal diffusivity (in turn controlled by specific heat capacity) and amount of time involved in the heat transfer.

From a practical perspective, the heat transfer properties of a subsurface soil matrix do not vary considerably, unlike hydraulic properties that can vary many orders of magnitude. Consequently, where relevant, LTTR can reach contaminant mass that other technologies reliant on subsurface permeability characteristics cannot readily access. Compared to fluid injection processes, the conductive heating process is very uniform in its vertical and horizontal sweep (i.e., the entire treatment zone can be heated). This is governed by the thermal conductivity of a wide range of soil types only varying by a factor of about four over the complete range of soil types (sand, silt, clay, gravel, and even bedrock) and moisture contents. Thus, even in highly heterogeneous settings, heat transfer in both high- and low-permeability aquifer settings, including fractured rock, can be robust and highly predictable if designed,

Table 18.2 Comparison of thermal and hydraulic conductivity of varying lithologies (adapted from Marquez et al. 2016; Freeze and Cherry 1979; and Wong et al. 2009)

Media	Thermal conductivity range (W/mK)	Hydraulic conductivity range (m/s)
Basalt	1.3–2.3	5^{-7}–5^{-2}
Clay	0.9–2.3	9^{-12}–9^{-9}
Gabbro	1.7–2.5	2^{-9}–6^{-4}
Gneiss	1.9–4	2^{-9}–6^{-4}
Granite	2.1–4.1	2^{-9}–6^{-4}
Gravel	1.8–1.8	1^{-3}–1
Greenstone	2–2.9	2^{-9}–6^{-4}
Limestone	2.5–4	5^{-10}–2^{-6}
Loam	1.5–3.5	1^{-7}–1^{-3}
Marble	1.3–3.1	2^{-9}–6^{-4}
Mica schist	1.5–3.1	2^{-9}–6^{-4}
Peat	0.2–0.7	10^{-5}–10^{-8}
Peridotite	3.8–5.3	2^{-9}–6^{-4}
Quartzite	3.6–6.6	2^{-9}–6^{-4}
Salt	5.3–6.4	
Sand	1.7–5	6^{-6}–1^{-2}
Sandstone	1.3–5.1	1^{-10}–6^{-6}
Shale	1.5–2.1	1^{-13}–9^{-10}
Silt	0.9–2.3	2^{-10}–5^{-5}
Siltstone	1.1–3.5	

constructed, and implemented properly. Table 18.2 compares corresponding ranges to both thermal and hydraulic conductivities associated various subsurface lithologies.

In low-permeability formations, an entire targeted treatment zone can be heated to a desired treatment temperature regardless of the challenges in moving fluids through the same formation. While the above is true, careful consideration of the subsurface properties is important to ensure that a LTTR system design will ultimately raise the temperature of the subsurface to the desired target in the desired timeframe. Effective thermal conductivity does still vary (Brigaud and Vasseur 1989; Abu-Hamdeh 2003), which requires an understanding of subsurface composition and thermal modeling to simulate the subsurface temperatures expected to be achieved. The modeling can predict the magnitude of heating that can be achieved relative to remediation goals, by factoring in the heat source, the properties of the subsurface matrix, and the impact of, for example, heat loss due to radiation (although minimal) and advection. This is discussed in more detail below.

18.5 In-Situ Heat Transfer Modeling

Thermal modeling software is commonly used to simulate the subsurface temperatures expected to be achieved and support the overall design of a thermal remediation system. The model is used to determine the magnitude of impact that certain parameters have on heating, such as various sources of heat loss, which can lead to a more optimized design. To improve the accuracy of such predictions, there has been significant work to better predict thermal conductivity for such applications (Berdal et al. 2006; Rubin and Carlton 2017). Many different analytical and numerical models are available in the marketplace that can be utilized. For example, the Department of Energy TMVOC multiphase flow code (Pruess and Battistelli 2002) is comprehensive and robust, and considers the three-dimensional compositional flow of a gas, aqueous, and NAPL phases with full partitioning of contaminants between the phases. Each phase flows in response to gravitational and pressure forces with relative permeability and capillary pressure effects. Additionally, the heat transfer occurs by conduction, and by convection of both latent and sensible heat in each phase. The thermodynamic effects of evaporation and condensation are included in the energy balance, and the chemical properties (vapor pressure, Henry's Law constant, solubility) are temperature-dependent. Alternately, the simpler modeling software FlexPDE®, a two-dimensional simulation solving heat balance equations on a finite element grid has been used for Thermal In-situ Sustainable Remediation (TISR®) applications, with empirical temperature data used to refine the model and improve predictions. Figure 18.5 depicts the modeling output for an TISR® installation showing the ability to achieve temperatures between 40 and 60 °C (enhancing hydrolysis reactions) after nine months of operation at a site in Mexico where five BHE were placed 4 m apart (Horst et al. 2018).

Recently, previously published analytical solutions (e.g., Molina-Giraldo et al. 2011) were utilized to develop a full three-dimensional BHE model that is incorporated into the widely used Microsoft Excel program (Ornelles 2021). The Excel-based tool uses a Visual Basic code to utilize super positioning of an analytical model to predict changes in space and time associated with customizable BHE arrays (Ornelles 2021). In this tool, the propagation of heat applied to the subsurface through discrete BHEs can be simulated over time. The model can account for groundwater flow and considers vertical heat flow, both up to the ground surface and down below the heater. This program was used to perform a preliminary analysis of the degree of heating expected from a BHE array at a specific site located in California.

The Darcy velocity at the test location considered by Ornelles (2021) and Ornelles et al. (2023) is expected to be low, on the order of 0.03 m per day (0.1 foot per day), and the flow is in the westerly direction. For the purpose of this preliminary model, the porosity was assumed to be 0.3, the grain density was set to 2650 kg per cubic meter, the grain heat capacity was 1000 J per kilogram per °C, and the thermal conductivity was 2.5 W/m °C. Based on the experience from similar solar heating thermal projects, the average power to each borehole was assumed to be ranging from 53 to 66 watts per meter. The BHE array consists of eight heat exchangers that extend

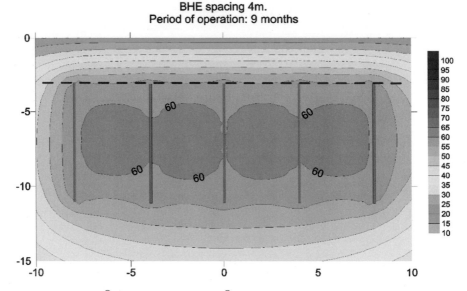

Fig. 18.5 FlexPDE® model output for a TISR® installation showing the ability to achieve temperatures between 40 and 60 °C in the treatment zone after nine months of operation at a site in Mexico (Horst et al. 2018)

from a depth of about 6 m to a depth of about 12 m (~20–40 feet). The boreholes are arranged in a triangular arrangement, and the spacing distance was optimized through iterative modeling. The final design was based on a spacing distance of 4.9 m (16 feet) between boreholes. The simulation was run for a total time of 1.5 years.

Figure 18.6 shows a plan view (elevation is 9 m below ground surface) of the predicted temperature increase (in °C) after 400 days of heating. Each contour line represents 1 °C of temperature increase over background. The effect of groundwater flow is evident, and the predicted average temperature increase inside the BHE array is about 10 °C (for 400 W average thermal power to each BHE), except immediately near the BHEs, where the temperature increase may be as high as 15–20 °C.

18.6 Low-Temperature Heating Methods

Currently, there are several conventional ISTR heating technologies that can be utilized for low-temperature thermal remediation and include ERH and TCH. Other traditional heating equipment that can be used include inline/instantaneous water heaters or geothermal heat pumps combined with open-loop (hot water flushing) and/or closed-loop (heat exchange) systems. Recently, sustainable sources of heat such as solar or capturing waste heat have come into the marketplace and are being utilized for reduced carbon footprint remediation. Selection of the appropriate heating

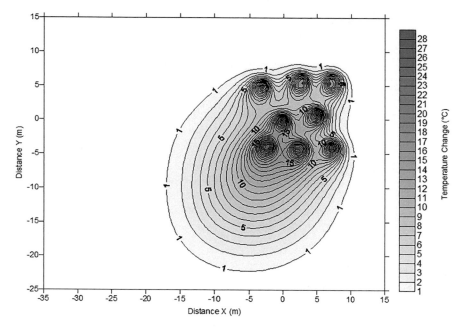

Fig. 18.6   Map view of simulated temperature increase (°C) after 400 days of heating with a thermal power of 400 watts per heater (figure courtesy of Justin Trainor and Ron Falta using the model tool described in Ornelles 2021)

technology is dependent on chemical and physical properties of the targeted COCs; concentration and mass distribution; the geologic/hydrogeologic setting in which heating will be employed; remedial objectives; source and magnitude of sustainable energy available; and the overall site location/infrastructure (i.e., space constraints, public perception/acceptance, power availability). Common heating methods used for low-temperature thermal remediation are discussed below.

18.6.1   Thermal Conduction Heating

TCH are soil remediation technologies in which a network of steel heater wells consisting of continuous, sealed steel casings are installed using conventional drilling techniques through the entire treatment interval. Heat is applied by installing electric- or gas-powered heaters inside the steel casings that heat the treatment area via thermal conduction. Traditionally, this approach heats to the boiling point of water (100 °C) or higher, a large amount of steam and vaporized contaminants are formed and need to be controlled, extracted, and treated. However, these traditional heaters as well as their associated power supply, distribution, control, and monitoring equipment are over-designed and over-engineered for low-temperature applications.

Recently, more cost-effective electrically powered heaters have been developed to be specifically used for LTTR. These heaters have a narrower profile compared to their higher temperature counterparts that can be installed in a smaller diameter pipe and borehole. This allows for the heaters to be installed to desired depths with less expensive and faster direct-push drilling technologies. While the low-voltage, low-power heaters require substantially less energy to operate, a power source is still required, usually supplied from the electrical grid. On smaller or remote systems, the power may be provided by solar photovoltaic panels. The low-voltage, low-power heaters can be easily customized in the field to the required heating lengths, eliminating costly fabrication and shipping charges. The heaters are also constructed of simple, easy to find, and inexpensive materials that can be re-used on future projects. This can significantly reduce the consumption of natural resources (steel, copper, concrete, equipment, fuel, chemicals, energy, etc.) and the carbon footprint associated with remediation activities.

18.6.2　Electrical Resistance Heating

Like TCH, traditional ERH will heat to 100 °C for the production of steam and volatilization of contaminants. However, it can be used to heat the subsurface to temperatures below 100 °C and utilized for LTTR applications. ERH directs low-frequency (60 Hz) three-phase electrical power to a network of subsurface electrodes installed in a repeating triangular pattern. Adjacent electrodes are out of phase such that gradients in electric potential are induced, which causes current conduction and resistive heat dissipation (i.e., Joule heating) throughout the treatment volume. ERH is typically best applicable in soil having electrical resistivity ranges from 5 to 500 Ω-meters (Ω·m) using standard power delivery equipment, which covers a wide range of naturally occurring fine- or coarse-grained media, and can be designed to account for heterogeneities in soil electrical properties. Both the saturated and vadose zones can be heated with ERH providing that sufficient soil moisture and electrical conductivity exist for efficient and safe current flow. In addition, ERH typically requires the addition of water and sometimes a conductive solute (e.g., potassium chloride) at the electrodes to maintain electrical continuity.

18.6.3　Geothermal Heat Pumps and Hot Water Flushing

Geothermal heat pump applications have been used for decades in the supplemental heating and cooling of commercial and residential buildings. This is done by embedding large heat exchangers into the subsurface and either circulating or injecting (closed loop or open loop) a transfer fluid from which heat is extracted (for heating) or to which waste heat is discharged (for cooling). Within a closed-loop system, the transfer fluid is re-circulated and heated/cooled via conduction through the heat exchanger and the subsurface. Conversely, open-loop systems typically extract groundwater that is then re-injected to the subsurface after above grade heating/cooling occurs. The advantages and potential challenges between closed-loop and open-loop systems should be considered when designing for remedial applications. Heat transfer via closed-loop systems will primarily be governed by thermal conductivity of the soil matrices, whereas open-loop systems will be heavily reliant on the hydraulic properties of the aquifer. However, when discharging heat, geothermal heat pumps have shown to increase the natural temperature of soils from 8 to 12 °C and as high as 40 °C in some regions of the world. Environmental engineers and policy makers have recognized the "positive side effects" of such extraction and injection-based heating–cooling systems for contaminant attenuation and enhanced biodegradation in urban groundwater legacy plumes (Slenders et al. 2010; Slenders and Verburg 2010).

One example is a former industrial facility in Europe that started its activities in 1915 and at its peak, employed more than 10,000 people. As a result of decades of industrial activity, soil and groundwater at the site were contaminated with chlorinated solvents to a depth of 60–70 m below ground surface. In 2009, an approach was implemented combining groundwater remediation with a geothermal energy system to support the facility, called "Sanergy". This was a first of its kind harnessing of a site mechanical system to benefit remediation. Since implementation, regulatory closure has been achieved for the remediation and the site continues to be part of an active technology headquarters for the owner.

18.6.4 *Thermal In-Situ Sustainable Remediation (TISR®)*

LTTR has a lower energy demand for heating compared to high temperature. Therefore, energy can be generated by capturing heat from a sustainable source (i.e., solar radiation or waste heat from commercial/industrial facilities) rather than more traditional electric or gas-fired energy sources. However, some electrically powered heating systems may use electricity generated by renewable sources. The conversion of solar radiation to electricity and/or heat continues to grow in popularity as the efficiency of the conversion equipment increases and costs decrease. Capitalizing on the improvement of solar collection and photovoltaic devices allows for LTTR to be implemented at a lower cost and solely dependent on renewable energy sources (i.e., self-sustainable). TISR® is a patented LTTR technology (US Pat. No. 10384246 and US Pat. No. 10,688,545) that transfers energy from solar radiation and/or waste heat to the subsurface by means of solar collectors (solar application), modified above grade heat exchangers (waste heat application), a closed-loop heat transfer fluid system, and BHEs designed to maximize the conductive heat transfer. The solar collectors or above grade heat exchangers heat a fluid generally consisting of propylene glycol and distilled water. The heated fluid is pumped by a small transfer pump through insulated manifolds and subsurface piping to BHEs as shown in Fig. 18.7. Because the system uses solar or waste heat energy, there are no utility costs incurred for the

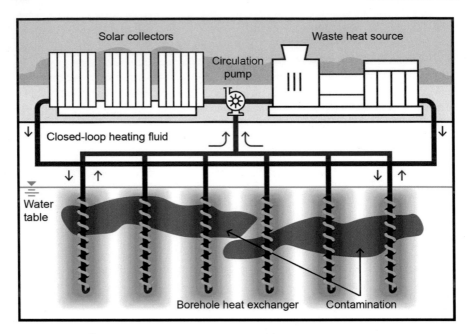

Fig. 18.7 TISR® process flow diagram using a solar energy and/or waste heat application

heating unit, and only minor power consumption to operate controls and the recirculation pump. In some instances, the minor electrical requirements for system controls and the recirculation pump can be powered by photovoltaic panels and invertors. This enables the system to be completely off the electrical grid and can be used for remote applications that are isolated from traditional power sources.

Solar-based TISR® systems use either evacuated tube or flat-plate collectors, which collect solar energy and focus it on a heat transfer fluid passing through the collector (Fig. 18.8). The use of solar energy is a sustainable application but is of course dependent on adequate solar radiation. The magnitude and duration of radiation varies based on such site characteristics as cloud cover, elevation, and latitude and changes seasonally due to sun angle and daylight length. Most of these factors can be overcome during the design phase but limitations of area to place collectors may hinder the ability to collect adequate solar energy to reach target heating temperatures. In some cases, supplemental heating via a traditional energy source may be necessary, particularly in the winter season in which day length is shorter and sun angle is decreased or at more extreme latitudes, but in all cases, utilization of a renewable energy source for at least part of the heating energy would provide a more sustainable approach and is likely to reduce overall energy and operational costs. Air temperature is less of a factor if the system is properly insulated to avoid heat loss. Table 18.3 shows the results from a case study in northern New York, United States, demonstrating the ability to capture significant heat energy away from the equator and when the sun is available, regardless of outside air temperature.

Fig. 18.8 TISR infrastructure for solar energy (solar collectors) and waste heating (blower heat exchanger) applications

Table 18.3 Field data demonstration of the ability to capture solar energy regardless of the ambient air temperature

Outside air temp (°F)	Weather conditions/ cloud over	Collector temp (°F)
38	Rain/overcast	39
37	Rain/overcast	43
36	Partly cloudy	115
57	Partly cloudy	114
15	Sunny	275

The availability of space for the installation of solar collectors, potential obstacles blocking the solar radiation for all or part of the day, and the possibility for the solar collectors to be periodically covered (leaves, snow, dust, etc.) all must be considered during the design phase. The direct use of solar energy for subsurface heating is a diurnal process which provides a heating period followed by a period of no heating during each 24-h period. This creates a pulsed approach to heating the subsurface. Extended periods of cloud cover and no or limited heating may have detrimental effects on sustaining temperatures, but the diurnal pulsing allows for temperature equilibration during the period of no heating. This increases temperature gradients at the borehole heat exchangers for the next day, which appears to allow for greater heat transfer compared to using a heat storage unit to support continuous heating.

The second sustainable energy source utilized for heating is the use of waste heat, where available. Waste heat sources can range from a large adjacent industrial or manufacturing facility to an existing remediation system which utilizes a heat exchanger for cooling (Fig. 18.8). The amount of waste heat and the impact on the process in which the waste heat is generated must be thoroughly understood and compared to the amount of heat needed for the remedial application. Integration of a waste heat system with an existing production operation could have effects on the production operation which can impact business and cause detrimental losses if not properly designed with adequate failsafe and alternate heat dissipating capabilities, should they be needed. Design and implementation are often straightforward depending on the location and characteristics of the process stream or equipment creating the waste heat.

Such systems generally require minimal operation and maintenance, typically involving inspections, failsafe testing, and performance monitoring. A system that uses waste heat provides a beneficial reuse for heating and will be dependent on the operation of the equipment or process that is generating the waste heat. Therefore, the potential for operation interruptions should be considered. Waste heat applications are generally operating as a continuous heating operation, unlike the solar application. A continuous steady source of heat is transferred to the subsurface. Waste heat offers relatively steady heat transfer to the subsurface compared to the diurnally or seasonally variable heat transfer provided by a solar collector based system.

An evaluation of available sustainable energy, energy requirements to reach target temperatures, as well as other factors previously outlined must be considered when determining the optimal sustainable energy source to use for low-temperature heating. Consideration of a hybrid approach utilizing both sustainable energy sources and supplementing a sustainable energy source with a traditional energy source can also provide benefit to achieve remedial goals within a shorter timeframe.

Custom BHEs are fabricated for each system based on the depth to water, extent of impacts, water and soil chemistry, and constructability. Typically, BHEs are installed in a vertical orientation within the saturated zone. Multiple BHEs are oriented in parallel to optimize the heat transfer by maintaining a maximum change in temperature between the subsurface and the BHE at all locations. During operation, the TISR® system is optimized through data collection from a combination of monitoring wells and thermocouples. Typically, thermocouple points will be installed at varying distances and depth intervals to gain a three-dimensional view of the temperature profile of the subsurface. In addition, groundwater samples for COCs, geochemical parameters, and microbial assessment tools can be used to evaluate contaminant degradation and treatment progress.

TISR can be applied in a complimentary function to existing remediation systems and is particularly well suited for accelerating treatment rates in systems designed to promote bioremediation. BHE heating coils can be installed inside or around the riser or screen of traditional remediation well including air sparge, biosparge, and injection wells. The combination of these technologies can greatly accelerate remediation due to the chemical property changes caused by increased temperatures as well as the increased desorption and dissolution which partitions the target compounds to a more accessible state for remediation. In addition, an increase in temperature can also increase the volatility of some compounds to allow them to be remediated via air sparging which may not be feasible at ambient conditions.

18.7 Sustainability and Resilience

As remediation practitioners continue to look for opportunities to reduce cleanup timeframes, reduce associated costs, and complete remediation in a sustainable and resilient manner, LTTR will continue to gain traction as a standalone option or as an option to enhance, replace, or augment other traditional remedial techniques. Offering innovative and resilient solutions to stakeholders will result in more beneficial ecosystem reuse and greater social impact. As we have reviewed, the value is derived from the fact that increases to subsurface temperatures can result in order-of-magnitude improvements in active biodegradation rates. The corresponding change in chemical properties can also be harnessed to enhance physical extraction (e.g., multiphase recovery of contaminant mass) by driving contaminant mass from immobile phases (sorbed to soils, trapped in silts and clays) to mobile phases (dissolved and vapor). This can allow extraction systems to work more efficiently and effectively, which reduces the time required for treatment and lifecycle costs. Similarly, traditional physical recovery methods that often reach asymptotic mass recovery may be either replaced or augmented to reach/accelerate the transition from active to passive remediation (e.g., natural source zone depletion, monitored natural attenuation).

Traditional ISTR as a standalone application is often not viewed as a sustainable remediation treatment technology due to material consumption for construction and the large energy demand required for operation (Horst et al. 2021). However, comparing the full lifecycle of the project entirety, combining or supplementing ISTR with sustainable technologies may minimize material input and significantly reduce energy consumption (Fig. 18.9).

Driven by a lower energy or sustainable heat source, incorporating LTTR into existing infrastructure reduces ecosystem restoration time, project lifecycle cost, and overall carbon footprint. Ecosystem restoration using sustainable technologies, such as TISR®, offers an innovative tool for environmental practitioners, facility owners, and our society. Utilizing solar and waste heat collection, TISR® has now been implemented at over a two-dozen sites worldwide including in the United States, Canada, Mexico, Brazil, and the Netherlands. The scale and magnitude of these systems has expanded, while best practices and guidance continue to be refined for greater efficiencies and optimization of heat transfer and energy use. Installation of a TISR® system utilizing an active manufacturing facility's waste steam is underway (at the time of publication) to integrate production and remediation in a symbiotic manner. Based on the preliminary results, the carbon footprint of TISR®, compared with other comparable in-situ remediation technologies, is significantly lower due to the low energy consumption. This is the most important parameter to indicate the difference between technologies. To date, there has been preliminary data indicating that TISR® has a carbon dioxide footprint that is 15–25 times lower compared with biosparging (comparable case). This has been roughly estimated comparing the

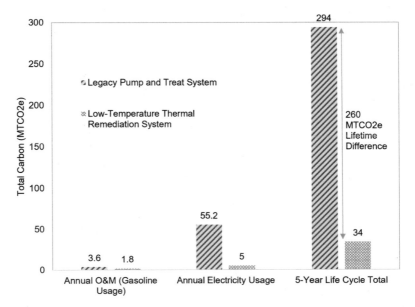

Fig. 18.9 Carbon footprint comparison of a legacy pump and treat system and a LTTR system (TISR)

amount of traditional energy consumption saved by using a renewable energy source during system operations. Another system has shown after five years of operation, the reduced carbon dioxide emissions have been tenfold lower when compared to a physical extraction system.

18.8   Case Studies

United States Army Corps of Engineers (USACE) Upstate, New York (Area 3805)

A site that has historical impacts from petroleum compounds located in upstate New York, United States. Main COCs were 1,3,5-trimethylbenzene, 1,2,4-trimethylbenzene, and BTEX. Lithology consists of medium to fine sand with depth and groundwater present at approximately 4.6 m below ground surface. A solar collector heating pilot test was implemented on a downgradient side of the plume with initial total petroleum concentrations more than 17 parts per million (ppm). The two-year pilot test was concluded in 2017 following two years of operation. The infrastructure layout is presented in Fig. 18.10 and consisted of three BHEs, six temperature monitoring points (TCs) and two 30-tube evacuated tube solar collectors circulating heated propylene glycol/water solution to the BHEs in a closed-loop system (Horst et al. 2018).

After approximately 18 months of heating, the temperature of the treatment area was increased from an average of approximately 12 °C to an average of approximately 20 °C as measured by the TCs. Baseline total BTEX groundwater concentrations near the BHEs ranged from 3 to 17 ppm (average 12 ppm). Concentrations during the last two sampling events of system operation (October 2017 and January 2018) were below detection limits—after approximately 18 months of heat application. The system was deactivated to evaluate post-treatment rebound, during which an average VOC concentration of 0.001 ppm was reported in the test area. As depicted in Fig. 18.11, the TISR solar application was able to reduce the VOC concentration by 99.9% in the two-year period (Horst et al. 2018).

The heating infrastructure (solar collectors, sensors, pump station, pump controller, and expansion tank) has since been trailer mounted and moved to address a different impacted area at the same site, which will continue to improve the cost effectiveness of the technology.

USACE Upstate, New York (Area 1795)

Located in another area at the site mentioned above, a full-scale application of waste heat coupled with AS/SVE was implemented in 2020. The main COCs were also 1,3,5-trimethylbenzene, 1,2,4-trimethylbenzene, and BTEX. Lithology also consisted of medium to fine sand with depth but groundwater was present at approximately 6 m below ground surface. This full-scale system was implemented following LNAPL recovery within the target zone and a comprehensive soil boring

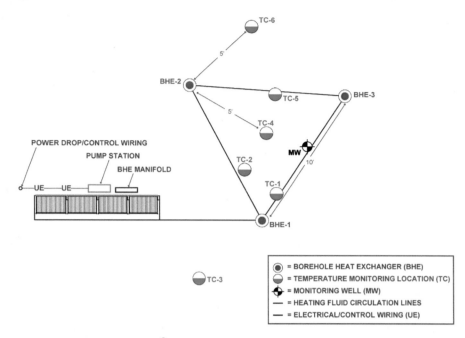

Fig. 18.10 USACE 3805 TISR® pilot test layout (adapted from Horst et al. 2018)

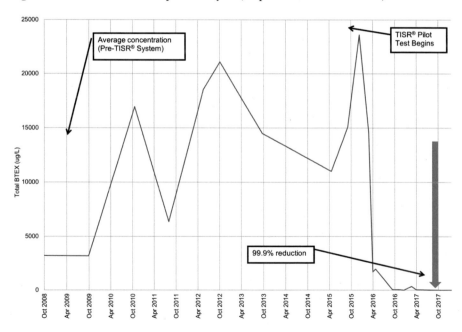

Fig. 18.11 USACE 3805 total BTEX concentrations over remedial timeline (Horst et al. 2018)

investigation. A pre-design soil boring investigation indicated that at least 60% of the target zone contained headspace readings from soil samples with greater than 1000 parts per million by volume (ppmv) based on field photoionization detector readings and LNAPL historically detected in wells. The system targeted impacts across an area approximately 5016 m^2 for AS operation of which approximately 2601 m^2 were heated utilizing the hybrid waste heat and inline heater application.

The AS system was designed and installed for the capture of waste heat from the AS blower, and heat was utilized to increase subsurface temperatures and increase contaminant desorption, dissolution, and volatility in an effort to reduce remedial system operation duration. The waste heat infrastructure included an alternate heat exchanger from the traditional air-to-air heat exchanger (which generally transfers heat to the atmosphere), to an air-to-liquid heat exchanger, a pump and control station, heating fluid conveyance tubing, and BHEs. Based on the volume within the heating target zone, a supplemental traditional electric inline heater was also utilized. When evaluating the costs of the waste heat operation, the AS/SVE system costs were not considered because that was the primary remedial approach; therefore, the additional costs associated with the addition of the heating were minimal relative to the costs associated with the full system installation because the same boreholes were utilized for the BHEs and AS wells and the air-to-air heat exchanger was simply replaced with an air-to-liquid heat exchanger.

Remedial operation of the AS and heating system began in September 2020. Operational data indicated that the waste heat alone was able to increase the heating fluid temperatures from 30 to 50 °C and with the supplemental inline heater increasing the temperature to 60 °C. The interpolated subsurface temperatures following 12 months of operation are presented in Fig. 18.12.

SVE operation recovered vapors at an average flow of 1000 normal meter cubed per hour (nm^3/hr) and a concentration of 2000 ppmv during the first weeks of SVE operation which occurred following approximately one month of heating operation. Following 3 months of SVE operation, influent concentrations declined to less than 100 ppmv and continued to drop until system deactivation. Several monitoring well headspace photoionization detector readings were initially greater than 10,000 ppmv and had dropped to less than 500 ppmv within 6 months of operation and continued to decline until system deactivation. Following one year of system operation, dissolved phase concentrations in performance monitoring wells declined well below target concentrations and close to drinking water standards following a temporary system deactivation to evaluate potential for concentration rebound. Two quarters of post-remediation groundwater sampling have been conducted and concentrations remain below target concentrations and near drinking water standards with no rebound observed.

Aerobic Hydrocarbon Treatment-Schenectady, New York (Information provided by TerraTherm Inc., 2022)

A site with historical impacts associated with an extensive LNAPL footprint was in New York. The primary COC was total xylenes. Lithology of the targeted impacted

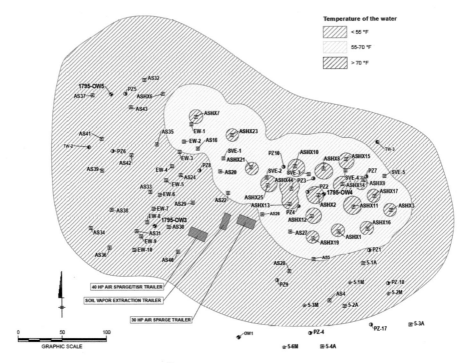

Fig. 18.12 USACE 1795C TISR® waste heat application wellfield temperature contours

zone consisted of low-permeability silts and clays with the water table located approximately 4.6 m below ground surface (bgs). A full-scale thermally enhanced bioremediation project was implemented using electrically powered TCH to address the xylenes.

Laboratory studies indicated that the native microbial population was capable of readily degrading the xylenes under aerobic conditions at 35 °C. However, the impacted interval had limited oxygen due to the relatively tight soils and over decades since the releases occurred, soil concentrations of total petroleum hydrocarbons and xylenes remained high, indicating little natural degradation under ambient conditions. The primary remedial objective for the site was to reduce soil concentrations of total xylenes to below 1000 ppm.

LTTR with electrically powered heaters was used to heat the subsurface to between 35 and 40 °C. A total of 143 heaters were installed at a 5-m spacing, to treat the 17,611 cubic yards of target volume (Fig. 18.13). To stimulate aerobic biological degradation, air/oxygen was delivered to the vadose zone via 143 passive air inlet wells and below the water table under pressure using 143 air injection wells. A network of SVE and dual-phase extraction (DPE) wells were used to keep pneumatic and hydraulic control of the vadose zone while recovering vapors and pushing air into the passive air inlet wells. A vapor treatment system was utilized to control vapor emissions while air was pulled into and from the targeted treatment zone. Waterloo

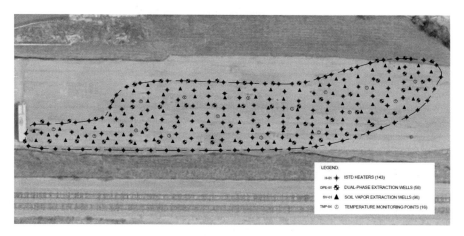

Fig. 18.13 Low-temperature heating via thermal conduction heating for thermally enhanced bioremediation of hydrocarbons-Schenectady, New York

profiling was used prior to construction to identify layers of high and low permeability within the treatment interval as the basis for selecting the screen intervals of the vapor recovery wells.

Heating progress was monitored through a network of temperature-monitoring points located both at the heaters and at the centroid locations between heaters (heaters were installed in a triangular pattern). Temperatures at centroid locations represent the furthest from any surrounding heaters and thus the slowest portions of the heated zone to increase in temperature. Additionally, vapor concentrations of COCs and biodegradation indicators (e.g., oxygen, carbon dioxide) were monitored within the well field at the SVE and DPE wells, along the section of the vapor collection manifold, and at the inlet to the treatment system. Interim soil sampling was used to track the progress.

Figure 18.14 shows the total average temperature at the centroid locations during operation. Heating was initiated in January 2017 with the heaters set initially at low power output (65–98 W/m) and then slowly ramped up over the next month to operational levels of between 262 and 328 W/m. Increases in temperature were observed at the centroid monitoring locations approximately one month after heating was initiated.

Over the course of 14 months of heating, the following observations were made with respect to concentrations of xylenes and biodegradation parameters sampled in the vapor stream: 1) maximum xylene concentrations were reported near the end of June 2017 which corresponded peak soil temperatures, 2) xylene concentrations decreased sharply one month later (July 2017) and then gradually over the next 4 months of heating, 3) oxygen concentrations in extracted vapors averaged 20.5% at the start of heating and then steadily decreased to < 19% (low of 17%) following 10 months of heating, 4) carbon dioxide concentrations started off low (<1%) and reached a peak after 7 months of heating (ranging between 2 and 24%). These data

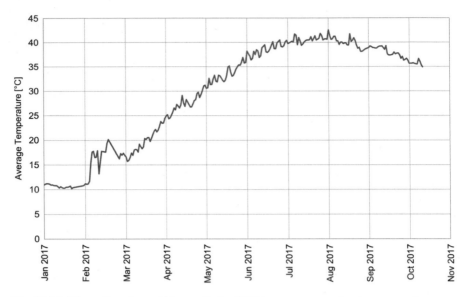

Fig. 18.14 Thermally enhanced bioremediation TCH low-temperature heating progression

are consistent with the thermally enhanced aerobic biodegradation of the xylene. Of the initial estimated starting COC mass of 21,000 pounds, only 3,800 pounds were removed by the vapor recovery system. The remaining COC mass was aerobically degraded in-situ using thermally enhanced biodegradation.

After 7 months of heating, the first interim soil sampling event revealed that the goals had been met in more than 75% of the treatment volume, with two hotspots remaining, where xylene concentrations remained above the treatment goals. The system was operated for an additional 6 months to improve the distribution of heat and degradation. Subsequent soil sampling events demonstrated that the goals were met, and New York State Department of Environmental Conservation granted site closure. It is estimated that over the 14 months of LTTR treatment, over 17,000 pounds of xylene were removed by thermally enhanced bioremediation alone. Importantly, the maximum temperature of the targeted treatment zone never exceeded 40 °C.

Former Gasoline Station-New Jersey, United States

The site located in New Jersey was a former gasoline station that operated between 1960 and 2007 before closing and subsequently being demolished. The station contained five gasoline underground storage tanks that were the source of soil and groundwater impacts. Concentrations of benzene were detected in soil more than 200 ppm, and LNAPL was measured in four monitoring wells during groundwater monitoring events. The site is currently vacant and maintained as a grassy field.

Lithology consisted of a low-permeability silt and sand mixture, and groundwater was present at approximately 2.5 m below ground surface. A TISR® system using solar collector heating was implemented on the downgradient side of the plume and

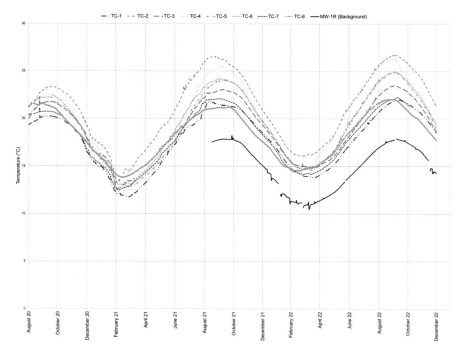

Fig. 18.15 Temperature monitoring data located 4.3 m below ground surface (approximately 2 m below the water able) during TISR® solar heating operations. Temperature monitoring points, TC-4, TC-5 and TC-6 were located approximately 0.8 m, 1.5 m and 4.6 m away from the closest BHE

within a portion of the LNAPL body. TISR was selected due to logistical challenges associated with other remedial options. Excavation was not feasible and was cost prohibitive, and the ineffectiveness of conventional mass removal technologies (e.g., AS/SVE and multiphase extraction) in low-permeability soils (silt and sand mixture) was demonstrated through pilot testing.

The TISR system comprised 12 solar collectors connected to seventeen BHEs (BHE-1 through BHE-17) installed downgradient of the former source area. Heating began in August 2020 and at the time of publication, operation was sustained through December 2022. Figure 18.15 displays temperature data from sensors located at 4.3 m below ground surface (approximately 2 m below the water table) in monitoring wells placed 0.8 m (TC-4), 1.5 m (TC-5), and 4.6 m (TC-6) away from the closest BHE. Groundwater temperature data were collected over two seasonal cycles where roughly a 10–13 °C fluctuation was observed between midwinter to midsummer (6–8 °C above ambient). Consequently, maximum heating of the subsurface and associated temperature increases occurred during the summer months but were not sustained throughout the winter months. This can be contributed to both the magnitude and the duration of solar radiation available to the solar collectors based on the geographical location of the site. Furthermore, heat energy is transferred to the subsurface during daylight hours and stopped overnight when there is no solar radiation available for the solar collectors to harvest and transfer. The seasonal variability in daily operating hours, which are generally higher in the summer, also matches the seasonal trend of solar radiation availability. These fluctuations associated with solar heating need to be assessed and integrated into the heating source design to ensure target temperatures can be obtained and maintained to increase contaminant degradation rates. Otherwise, a more sustainable heating source such as the sources introduced in the case studies above should be considered.

At the time of publication, soil re-sampling was not completed; however, operation of the TISR® system was able to achieve one of the primary project milestones, to reduce or eliminate measurable LNAPL at the site. Carbon dioxide and methane concentrations in soil gas have increased indicating enhanced microbial activity. Additionally, the results for dissolved-phase impacts have shown a decreasing trend in the core of the treatment area.

18.9   Conclusion

While LTTR continues to grow and evolve, certain key benefits have been identified and are as follows:

- Enhanced biodegradation: Biological degradation rates may double (some cases triple) for every 10 °C rise in temperature under mesophilic conditions.
- Application in tandem: Application in conjunction with source zone AS/SVE or multiphase extraction systems could significantly reduce treatment timeframes and complement natural attenuation.
- Effective in complex and heterogeneous geology: Heat transfer and corresponding treatment benefits not as limited by challenging hydrogeology (heterogeneous strata, tight clays, etc.) as other processes involved in the application of conventional remediation technologies.
- Cost savings: The potential to significantly supplement/enhance many existing remedial strategies can lead to reduced cleanup time and operational cost reductions.
- Reduced energy use/carbon footprint: As shown in one of the examples, five years of operation may reduce carbon dioxide emissions tenfold compared to a physical extraction system.
- Remote areas application: Remote areas present a challenge for traditional remedial approaches due to lack of availability of conventional energy sources. By utilizing solar energy to power subsurface heating, LTTR can help overcome that challenge.

Further research is encouraged to better characterize subsurface processes to optimize the implementation of LTTR systems and their integration with other site management strategies. With the combination of potential treatment enhancement and offering of a more sustainable solution, LTTR represents yet another area of significant opportunity for remediation practitioners to show their creativity—and one that is here right now.

References

Abu-Hamdeh N (2003) Thermal properties of soils as affected by density and water content. Biosys Eng 86:97–102

API (2017) Qualification of vapor phase-related natural source zone depletion processes, 1st edn. Publication 4784

Berdal T et al (2006) Experimental study of a ground-coupled heat pump combined with thermal solar collectors. Energy Build 38(12):1477–1484

Brigaud F, Vasseur G (1989) Mineralogy, porosity and fluid control on thermal conductivity of sedimentary rocks. Geophys J Int 98(3):525–542

Dablow J, Hicks R, Cacciatore D, van de Meene C (1995) Steam injection and enhanced bioremediation of heavy fuel oil contamination. Groundwater Technol 1183–1184

Daniel D, Loehr R, Webster M, Kasevich R (2000) Soil vapor extraction using radio frequency heating: resource manual and technology demonstration

Dettmer K (2002) A discussion of the effects of thermal remediation treatments on microbial degradation processes. National Network of Environmental Management Studies Fellow, Washington

Flanders C, Randhawa D, Visser P, LaChance J (2019) Thermal in-situ sustainable remediation system and method, US Pat. No. 10,384,246, issued August 20, 2019

Flanders C, Randhawa D, Visser P, LaChance J (2020) Thermal in-situ sustainable remediation system and method for groundwater and soil restoration, US Pat. No. 10,688,545, issued June 23, 2020

Freeze RA, Cherry JA (1979) Groundwater, vol 7632. Prentice-Hall Inc., Englewood Cliffs, p 604

Gerardi MH (2003) The microbiology of anaerobic digesters. Wiley

Hegele PR, McGee BCW (2017) Managing the negative impacts of groundwater flow on electrothermal remediation. Rem J 27(3):29–38

Horst J, Flanders C, Klemmer M, Randhawa DS, Rosso D (2018) Low-temperature thermal remediation: gaining traction as a green remedial alternative. Groundwater Monit Rem 18–27

Horst J, Munholland J, Hegele P, Klemmer M, Jessica G (2021) In situ thermal remediation for source areas: technology advances and a review of the market from 1988–2020. Groundwater Monit Rem 1–15

Krauter P, MacQueen D, Horn J, Bishop D (1995) Effect of subsurface electrical heating and steam injection on the indigenous microbial community. In: Spectrum 1996 conference. Lawrence Livermore National Laboratory, Seattle, pp 1–8

Kulkarni PR, Walker KL, Newell CJ, Askarani KK, Li Y, McHugh TE (2022) Natural source zone depletion (NSZD) insights from over 15 years of research and measurements: a multi-site study. Water Res 119170

Likos W (2014) Modeling thermal conductivity dryout curves from soil-water characteristic curves. J Geotech Geoenviron Eng 140

Macbeth T (2019) Heat-enhanced in situ degradation for treatment of energetic compounds impacting groundwater. Presentation at the fifth (5th) international symposium on bioremediation and sustainable environmental technologies, Baltimore, Maryland

Margesin R, Schinner F (2001) Biodegradation and bioremediation of hydrocarbons in extreme environments. Appl Microbiol Biotechnol 56:650–663

Marquez J, Bohórquez M, Sergio A, Sergio M (2016) Ground thermal diffusivity calculation by direct soil temperature measurement. Application to very low enthalpy geothermal energy systems. Sensors 16:306

Molina-Giraldo N, Blum P, Zhu K, Bayer P, Fang Z (2011) A moving finite line source model to simulate borehole heat exchangers with groundwater advection. Int J Therm Sci 50:2506–2513

Ornelles A (2021) Development and validation of an analytical modeling tool for solar borehole heat exchangers. Master's Thesis, Clemson University, p 121

Ornelles AD, Falta RW, Divine CE (2023) A design tool for solar thermal remediation using borehole heat exchangers. Groundwater 61(2):245–254

Peters U, Nierlich F, Sakuth M, Laugier M (2002) Methyl tert-butyl ether, physical and chemical properties, Ullmanns encyclopedia of industrial chemistry, 6th edn. Wiley-VCH Verlag GmbH & Co. KGaA

Powell T, Smith G, Sturza J, Lynch K, Truex M (2007) New advancements for in situ treatment using electrical resistance heating. Rem J 51–70

Pruess K, Battistelli A (2002) TMVOC, a numerical simulator for three-phase non-isothermal flows of multicomponent hydrocarbon mixtures in saturated-unsaturated heterogenous media. Lawrence Berkeley National Laboratory, Berkeley

Rubin A, Carlton H (2017) A review of two methods to model the thermal conductivity of sands, pp 800–808. https://doi.org/10.1061/9780784480472.085

Sander R (2015) Compilation of Henry's law constants (version 4.0) for water as solvent. Atmos Chem Phys 15:4399–4981

Shaw DG (1989) Solubility data series, hydrocarbon with water and seawater, part I: hydrocarbons C5 to C7, vol 37. International Union of Pure and Applied Chemistry, Pergamon Press, Oxford

Slenders H, Verburg R (2010) State of the environment. European Environment Agency, Copenhagen

Slenders et al. (2010) Sustainable synergies for the subsurface: combining groundwater energy with remediation. Rem J 143–153

Staudinger J, Roberts PV (2001) A critical compilation of Henry's law constant temperature dependence relations for organic compounds in dilute aqueous solutions. Chemosphere 44(4):561–576

Suthersan S, Horst J, Klemmer M, Malone D (2012) Temperature-activated auto-decomposition reactions: an under-utilized in situ remediation solution. Ground Water Monit Rem 34–40

Thamdrup B, Hansen J, Jørgensen B (1998) Temperature dependence of aerobic respiration in a coastal sediment. FEMS Microbiol Ecol 25(2):189–200

Truex M, Powell T, Lynch K (2007) In situ dechlorination of TCE during aquifer heating. Ground Water Monit Rem 96–105

USEPA (2021) EPA on-line tools for site assessment calculation. https://www3.epa.gov/ceampubl/learn2model/part-two/onsite/es-background.html. Accessed Apr 2022

Wong LS, Hashim R, Ali FH (2009) A review on hydraulic conductivity and compressibility of peat. J Appl Sci 9:3207–3218

Xu J (1997) Biodegradation of petroleum hydrocarbons in soil as affected by heating and forced aeration. J Environ Qual 26(6):1511–1516

Yadav BK, Gupta PK (2022) Thermally enhanced bioremediation of NAPL polluted soil-water resources. Pollut 2(1):32–41

Zeman N (2013) Thermally enhanced bioremediation of LNAPL. Master's thesis, Colorado State University, p 139

Zhang N, Wang Z (2017) Review of soil thermal conductivity and predictive models. Int J Therm Sci 117:172–183

Correction to: The Application of Sequence Stratigraphy to the Investigation and Remediation of LNAPL-Contaminated Sites

Junaid Sadeque and Ryan C. Samuels

Correction to:
Chapter 4 in: J. García-Rincón et al. (eds.), *Advances*
in the Characterisation and Remediation of Sites
Contaminated with Petroleum Hydrocarbons, **Environmental**
Contamination Remediation and Management,
https://doi.org/10.1007/978-3-031-34447-3_4

In the original version of the book, author provided belated corrections have been incorporated: Many text corrections and few figure corrections have been updated in this chapter 4. The correction chapter and the book have been updated with the changes.

The updated version of this chapter can be found at
https://doi.org/10.1007/978-3-031-34447-3_4

Glossary

Chapter 2: Historical Development of Constitutive Relations for Addressing Subsurface LNAPL Contamination

Capillary pressure (head) classically defined as the fluid pressure (units $F\,L^{-2}$) or head (units L) difference across the interfaces between immiscible fluids.

Constitutive relations critical properties needed to solve the governing equations. For fluid flow, it is the relations among fluid relative permeabilities, fluid saturations, and fluid pressures.

Effective saturation a fluid saturation scaled by a reduced pore volume (reduced total saturation). Typically, the reduced total saturation is 1 minus the irreducible (residual) water saturation (dimensionless)

Entrapped LNAPL LNAPL that is disconnected from continuous LNAPL in the pore spaces and occluded (surrounded) by water in either the vadose or water-saturated zones.

Fluid saturation the fluid volume per total pore volume (dimensionless)

Fluid transmissivity the rate of fluid movement per unit area under a unit hydraulic gradient through a vertical slice of the subsurface where the fluid is present (units $L^2\,T^{-1}$). For water, it typically refers to the water-saturated zone.

Free (mobile) LNAPL continuous LNAPL in the pore spaces that will move in response to a LNAPL pressure gradient.

Hydrologic property any physical property that affects fluid movement

Hydrostatic conditions the vertical distribution of fluid pressures is such that fluids are not moving upwards or downwards (vertically static).

Hysteresis different fluid saturations as a function of capillary pressure will result depending on whether fluids are draining or imbibing and the fluid saturation history.

Interfacial tension a measurement of unbalanced forces caused by the fluids' chemical properties across the fluid-fluid interfaces (immiscible fluids) (units $F\,L^{-1}$).

© The Editor(s) (if applicable) and The Author(s) 2024
J. García-Rincón et al. (eds.), *Advances in the Characterisation and Remediation of Sites Contaminated with Petroleum Hydrocarbons*, Environmental Contamination Remediation and Management, https://doi.org/10.1007/978-3-031-34447-3

Intrinsic permeability the conductance of fluids through the total pore volume expressed in L^2 units. It is a property of a porous medium.

Irreducible (residual) water saturation the water saturation at which there is minimal water drainage when the capillary pressure is significantly increased. It typically is the water saturation that no further drainage occurs when the air-water capillary pressure is increased to large values.

LNAPL-specific volume the volume of LNAPL per unit area in a vertical slice of the subsurface (units L^3 L^{-2}).

Multimodal pore-size distribution pore sizes that have distributions with two or more statistical modes.

Recoverable LNAPL LNAPL that can be extracted from the subsurface through wells by direct pumping, which will be LNAPL with positive pressures (above atmospheric) because only that LNAPL will enter boreholes. LNAPL with negative pressures (sub-atmospheric) will be retained in the pore spaces by capillary forces.

Relative permeability the fluid permeability at a given fluid saturation divided by the intrinsic permeability (dimensionless).

Residual LNAPL LNAPL that remains in the pore space following LNAPL drainage because its mobility is low. Typically, it is LNAPL in pore wedges, bypassed pores, and films that may be discontinuous or connected to continuous LNAPL by only thin films. It differs from entrapped LNAPL because it is not occluded by water.

Unimodal pore-size distribution pore sizes that have a single statistical mode of higher pore densities, typically following a normal or lognormal distribution.

Wettability the preference of a fluid to spread or adhere to a surface whether a solid or another fluid.

Wetting fluid the fluid phase that adheres to the porous medium surfaces in preference to any other fluid phases.

Chapter 4: The Application of Sequence Stratigraphy to the Investigation and Remediation of LNAPL-Contaminated Sites

Accommodation The space available for potential sediment accumulation. Also see Fig. 4.7.

Aeolian Arising from or related to the action of wind. Aeolian deposits refer to sediments driven/deposited by wind transport and erosion.

Aggradation The accumulation of sediment deposits that stacks beds atop each other, building upwards during periods of balance between sediment supply and accommodation. See also Fig. 4.14b-1.

Allostratigraphy A stratigraphic method that subdivides geological sequences based upon a hierarchical framework of bounding surfaces or discontinuities that serve to compartmentalize discrete packages of sediment or rock (North American Commission on Stratigraphic Nomenclature, 1983, Rev. 2005).

Allocyclic Geological processes external to sedimentary systems and include tectonic activity and global sea level change. Sedimentary responses to allocyclic processes (e.g., delta lobe switching) on geographic scales range from basinal to global. Also see Table 4.1.

Alluvial fan A fan-shaped accumulation of water-transported sediment; typically formed at the base of topographic features where there is a marked break in slope.

Autocyclic Geologic processes within a particular sedimentary system. Responses to autocyclic processes tend to be local and may range from millimeter-scale ripple migration to regional-scale events such as delta switching. Unlike allocyclic processes, autocyclic processes tend to be instantaneous geologic events that may be random in both time and space. Also see Table 4.1.

Avulsion The abandonment of a part or the whole of a channel belt in favor of a new course. An avulsion channel is the new channel that results from avulsion.

Back-stepping Landward movement of coastal sediments when sediment supply cannot keep pace with generated accommodation. Also known as retrogradation. See also Fig. 4.14b-3.

Barrier Island Coastal landforms and a type of dune system that forms by wave and tidal action parallel to the mainland coast.

Barrier spit A barrier spit is the landform (bar) resulting from the deposition of sediments in long ridges extending out from coasts. Barrier spits may partially block the mouths of bays.

Bay-head delta A sedimentary system of deposits that forms at the head of an estuary, or embayment, where sediment-laden freshwater enters brackish bay waters. Bayhead deltas form where the local rate of sediment input from rivers outpaces the rate of sea level rise. Also see Fig. 4.27.

Bed Laterally traceable, three-dimensional sedimentary unit of relatively uniform physical, chemical/mineralogical, and biological composition distinguishable from sedimentary units above and below. Bed size may range from very thin-bedded 1 cm to very thick-bedded, 1 m. See also Fig. 4.14a.

Bedset A relatively conformable succession of genetically related beds bounded by surfaces (called bedset surfaces) of erosion, non-deposition. See also Fig. 4.14a.

Bounding discontinuity Discontinuities in the sediment record (e.g., unconformities, ravinement surfaces, and flooding surfaces) that bound allostratigraphic units.

Braided River A network of river channels separated by small, often temporary islands called braid bars.

Chronostratigraphy The branch of stratigraphy that studies the ages of rock strata in relation to time.

Crevasse splay (crevasse mouth bar) A sedimentary fluvial deposit which forms when a stream breaks its natural or artificial levees and deposits sediment on a floodplain. Crevasse splays have a lobate or fan-shaped platform with distal thinning away from the levee. Also see Fig. 4.3.

Crevasse channel A channel that breaches and erodes the levee taking bedload materials from the primary channel and conveying them onto the floodplain at high flood stage. Also see Fig. 4.3.

Delta Deltas are discrete shoreline protuberances formed where rivers enter a still body of water (oceans, semi-enclosed seas, lakes, or lagoons) and supply sediments more rapidly than it can be redistributed by basinal processes. Generally, deltas are fed by distributary river channels that bring sediments from the continental realm.

Deltaic Of or relating to a delta environment where sediment is discharged from a river into a standing body of water in a lobate shape.

Debris flow Slurries of mud, rock debris, and just enough water to make the sediment into a viscous flow. Debris flows are typical of glacial and proximal alluvial fan settings.

Depositional environment Three-dimensional stratigraphic setting for typical geological processes resulting in district depositional entities. See Fig. 1.32 for illustration of common depositional environments encountered at contaminated sites.

Depositional facies An informal term for a group of facies (or facies associations) that are established as representative of certain depositional environments or subenvironments (e.g., channel bar depositional facies, delta mouth bar depositional facies).

Depositional system Three-dimensional assemblage of lithofacies, linked by active or inferred processes and environments. It encompasses depositional environments and the processes acting on them over a specific interval of time.

Diamicton A continental deposit (often of glacial origin) consisting of poorly sorted sand, gravel, and clay. A sandy diamicton has a high proportion of sand (e.g., more than 60% sand), whereas a clayey diamicton has a higher proportion of clay as matrix (e.g., more than 60% clay).

Distributary channel A stream channel that carries water away from the main river channel and distributes it to other area.

Estuarine Formed in an estuary. An estuary is a partially enclosed coastal body of brackish water with one or more rivers or streams flowing into it, and with a free connection to the open sea.

Facies analysis The widely used approach to the interpretation of sedimentary rocks that is based on the interpretation of the attributes of facies and their assemblages in terms of the processes responsible for their genesis. This is followed by the deduction of the most likely depositional environments in which the processes may have operated.

Facies architecture Hierarchy of facies from the smallest to the largest unit defined by a hierarchy of bounding surfaces. Also see Fig. 4.14a.

Facies association A group of depositional facies that are associated in interpreting a depositional environment.

Facies Lateral, mappable subdivision of a designated stratigraphic unit formed under common environmental conditions of deposition, distinguished from adjacent subdivisions on the basis of lithology and biology (fossils and trace fossils). This term is applicable to both consolidated and unconsolidated material. A term lithofacies is used when lithology is the only criterion for defining facies. Cf. Depositional Facies.

Facies models A general summary of a particular depositional system, based on many individual examples from recent sediments and ancient rocks.

Flooding surface A general term that refers to a surface that separates older rocks/sediments from younger rock/sediments and is marked by deeper-water strata resting on shallower-water strata.

Fluvial channel A conduit of water and sediments in a river of the continental environment.

Glacial Relating to, resulting from, or denoting the presence of ice, especially in the form of glaciers.

Global sea level change Sea level change that results in an alteration to the global volume of water in the world's oceans or net changes in the volume of the oceanic basins. Also known as eustatic change. It is measured as the height of the ocean surface above the center of the earth, without regard to whether adjacent land is rising or subsiding.

Graben An elongated block of the earth's crust lying between two faults and displaced downward relative to the blocks on either side.

Horst An elongated block of the earth's crust lying between two faults and displaced upward relative to the blocks on either side.

Hydrostratigraphic unit A body of sediment saturated with groundwater with limited connectivity to adjacent sediments. Sedimentary aquifers typically are composed of multiple hydrostratigraphic units due to heterogeneous geology.

Incised valley A valley incised by fluvial systems in response to a relative fall in sea level, with basinward movement of channels in a coastal setting that erode into underlying strata. Incised valleys are also a significant key to the identification of sequence-bounding unconformities. Also see Figs. 4.24a and 4.25.

Lacustrine Arising from or related to the action of a lake. Lacustrine deposits refer to sediments deposited in a lake.

Lithology The physical characteristics of rocks such as rock type, color, mineral composition, and grain size.

Lithostratigraphy Classification of bodies of rock based on the observable lithological properties of the strata and their relative stratigraphic positions.

Maximum flooding surface (MFS) A surface separating the transgressive systems tract from a highstand systems tract. It is commonly characterized by a condensed horizon reflecting very slow deposition. It corresponds to the time when the shoreline is at its most landward location (i.e., at the time of maximum transgression). Also see Fig. 4.8d.

Meandering river A type of river that has a single channel that winds sinuously through its valley.

Mouth bar Deltaic sediment deposit at the mouth of a distributary channel. Also see Fig. 4.27.

Overbank An alluvial geological deposit consisting of sediment that has been deposited on the floodplain of a river or stream by flood waters that have broken through or overtopped the banks.

Parasequence Relatively conformable succession of genetically related beds or bedsets bounded by marine flooding surfaces or their correlative surfaces. Also see Fig. 4.14b.

Point bar An alluvial deposit that forms by accretion on the inner side of an expanding loop of a meandering river. Point bars can be proximal, distal, and medial. Also see Figs. 4.2 and 4.3.

Progradation Seaward movement of sediments as sedimentation outpaces accommodation. Also see Fig. 4.14b-2.

Regression Seaward movement (progradation) of the shoreline either as a result of sea level fall ("forced" regression) or an increase in sediment supply ("normal" regression).

Relative permeability The fluid permeability at a given fluid saturation divided by the intrinsic permeability (dimensionless).

Relative sea level change Sea level change refers to how the height of the ocean rises or falls relative to the land at a particular location (considering local tectonic uplift or subsidence), as opposed to global sea level change. Cf. Global Sea Level Change.

Retrogradation See back-stepping.

Sand body geometry Shape of a stratigraphic unit as a result of sediment deposition by a transporting agent (e.g., a delta mouth bar has a lobate geometry).

Sequence A succession of strata deposited during a full cycle of change in accommodation or sediment supply between two sequence boundaries.

Sequence boundary (SB) The surface separating two sequences. It typically represents a subaerial erosion surface. Also see Figs. 4.8a and 4.25.

Sequence stratigraphy A correlation technique that subdivides the stratigraphic record into mappable rock bodies and examines the stacking patterns that result from changes in accommodation.

Sheet flood deposits Deposits of unconfined turbulent flows during episodic flash-floods, typical of an alluvial fan setting. Since the flows are turbulent, there is significant grain sorting and normally graded, fining-upward deposits are common. They produce broad deposits that are clast supported, with some imbrication of clasts.

Slicken slide A rock surface with a polished appearance and fine parallel scratches caused by abrasion during fault displacement.

Sorting The distribution of grain size of sediments, either in unconsolidated deposits or in sedimentary rocks. Sediments can be well sorted, moderately sorted or poorly sorted based on the variance seen in particle sizes. The degree of sorting may indicate the energy, rate, and/or duration of deposition, as well as the transport process (river, debris flow, wind, glacier, etc.) responsible for laying down the sediment.

Stratigraphic architecture Stratigraphic hierarchy and inter-relation of a depositional system marked by major stratigraphic surfaces. Cf. Facies Architecture. See also Fig. 4.14b.

Stratigraphy The study of sedimentary strata (layers) in order to understand the depositional history of a region.

Sub-environment A component of a depositional environment (e.g., levee and point bar are sub-environments of the fluvial depositional environment).

Systems tract A linking of contemporaneous depositional systems, forming the subdivision of a sequence. Systems tracts consist of conformable strata that were deposited during a particular segment of an accommodation cycle. The four different systems tracts established in sequence stratigraphy are, Falling Stage Systems tract (FSS) dominated by development of incised valleys and submarine canyons, Lowstand Systems Tract (LST) dominated by prograding fluvial systems and small deltas, Transgressive Systems Tract (TST) dominated by estuarine conditions and tidal deposits and Highstand Systems Tract (HST) dominated by large fluvial and deltaic systems through a full cycle of one stratigraphic sequence. See Fig. 4.8 for depositional characteristics of each systems tract.

Tidal bar (tidal mouth bar) Linear sand ridges that occur in the mouth of tide-dominated estuaries, as a result of tidal reworking of mouth bars. They represent coarsening upward sediments with mud-drapes. Also see Fig. 4.29.

Tidal channel An estuarine channel that is affected by ebb and flow of ocean tides. Deposits of tidal channels may resemble point bar deposits of fluvial channels, but typically exhibit mud-drapes related to the tidal environments. Also see Fig. 4.28.

Tidal flat Coastal wetlands that form in intertidal areas where sediments have been deposited by tides or rivers. Tidal flats can be muddy or sandy.

Till Unsorted material deposited directly or reworked by a glacier.

Time marker Penecontemporaneous stratigraphic surfaces that are typically used as markers for sequence stratigraphic correlations (e.g., flooding surfaces, sequence boundaries, etc.). Also see Fig. 4.5.

Transgression Drowning of depositional systems as a result of relative rise in sea level. Associated with retrogradation/back-stepping. Also see Figs. 4.8c and 4.14b-3.

Transgressive surface (TS) The first major flooding surface following the Lowstand Systems Tract and ensuing the Transgressive Systems Tract (TST). Also see Fig. 4.8c.

Walther's law The law that states that if two facies are found to be adjacent within the rock record, with no unconformities, then they must have been deposited laterally in succession. The most common example of Walther's law is shown in transgressive and regressive cycles seen in marine and coastal deposits. Also see Fig. 4.2.

Chapter 6: Petroleum Vapor Intrusion

Building pressure cycling (BPC) Technique that manipulates building air pressure and ventilation to promote or inhibit the intrusion of vapors into the building using either blower doors or by manipulation of existing HVAC systems.

Degassing Bubble formation because of methanogenesis in the saturated zone leading to super-saturated groundwater with gas (typically methane).

Ebullition Bubble transport of gas from groundwater to the vadose zone.

Exclusion distance Source to building separation screening distance beyond which the potential for vapor intrusion can be considered negligible.

Indoor attenuation factor Concentration of the compound in indoor air divided by the concentration of the compound in soil gas or groundwater.

Preferential pathways Sewer pipes and other utility conduits (e.g., fiber optics, cable television, and telephone cables) that can cause higher contaminant flux into a building compared to the average transport through the formation.

Reasonable Maximum Exposure (RME) Maximum exposure to a particular pollutant that is reasonably expected to occur at a particular site.

Chapter 10: Molecular Biological Tools Used in Assessment and Remediation of Petroleum Hydrocarbons in Soil and Groundwater

Biodegradation Breakdown of materials by microorganisms into products that are generally less detrimental to the environment.

Biostimulation Modification of the environment to stimulate existing bacteria capable of bioremediation.

Biotransformation Biochemical modification of one chemical compound or a mixture of chemical compounds.

Functional gene A portion of DNA coding for an enzyme or enzyme subunit that degrades a contaminant.

Molecular biological tools Tools targeting biomarkers (e.g., nucleic acid sequences, peptides, proteins, or lipids) to provide information about organisms and microbial processes relevant to the assessment or remediation of contaminants.

Primer Short strand of DNA that when bound to DNA or RNA serves as the starting point for DNA synthesis and amplification to be used in PCR, and some DNA sequencing, assays.

Printed in the United States
by Baker & Taylor Publisher Services